To Deborah, Gabriel and Jesse

A FIRST COURSE IN
ENGINEERING DYNAMICS

Second Edition

R. Valéry Roy

University of Delaware

A First Course in Engineering Dynamics
 ISBN-13: 978-0990696919 (R. Valery Roy)
Published by HYPERBOLIC PRESS

First printing, August 2016

Contents

PART I: KINEMATICS

PART II: KINETICS

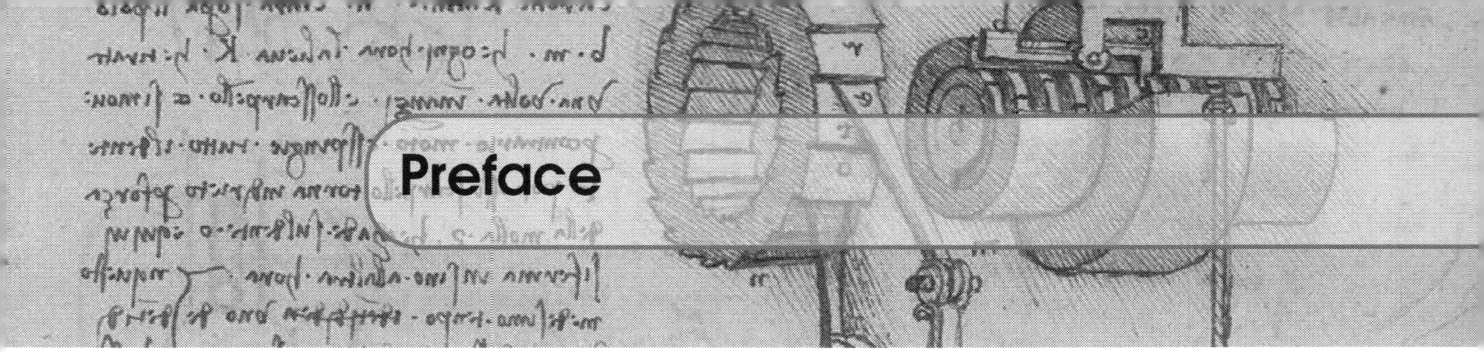

Preface

A First Course in Engineering Dynamics covers basic topics typically taught in a second mechanics course in a two-course sequence in most mechanical engineering curricula. Engineering Dynamics is a scientific branch of classical mechanics with a long and rich history. Its foundations were set long ago by Aristotle, Archimedes, Galileo, Kepler, Newton, Euler, Lagrange, Hamilton, Einstein and many other famous scientists. However, it is a continuously evolving discipline, as it is the driving force behind the research and development of many technological applications in robotics, biomechanics, aerospace, transportation, as well as in the medical and energy industries.

Arguably, modern science started in the 17th century with the work of Galileo, who may well be considered the first physical scientist. Indeed, with his seminal experiments on the laws governing the fall of bodies, Galileo emphasized quantitative measurements and mathematical modeling, and thus revolutionized the understanding of motion and gravity. As Galileo put it, "The book of Nature is written in mathematical language." Following Galileo's footsteps, scientists adopted his method of discovery of nature's physical laws, using the language and tools of mathematics. With Newton's work, mechanics truly became a physical theory based on a sound mathematical foundation. Newton laid out his ideas on mechanics, gravity, and other physical subjects in one of the greatest scientific books ever written, *Mathematical Principles of Natural Philosophy*, in 1687. Having developed the first ideas of differential calculus, his newly discovered laws of motion enabled him to derive many results, including Kepler's three laws of planetary motion. Up until the end of the 19th century, physicists confirmed the spectacular success of Newtonian mechanics. Investigations in many fields of science were patterned after those in mechanics, and in fact, other physical laws with the same universality as those of mechanics were discovered. Of course, 20th century physics showed that Newtonian mechanics has some deficiencies which were eventually addressed by the advent of the revolutionary ideas of relativity and quantum mechanics. However, classical mechanics provides a precise, well-defined, and powerful method of predicting and explaining many dynamical phenomena of greatly differing scales of time and space. It is truly extraordinary that the motions of the sun, the moon, the earth, rockets, footballs, cars and gyroscopes are all described with great accuracy by the same laws and governed by the same types of equations.

Why study Engineering Dynamics? First, its range of applicability in science and engineering is so enormous, the answers it has provided to long-standing and fundamental problems so numerous, that the study of mechanics is essential for all students wishing to understand the world around them.

Second, as a foundational course, it is pivotal to the intellectual development and to the education of scientists and engineers.

Third, it provides students with a great opportunity to learn how to describe a physical phenomenon in the language of mathematics. Learning mechanics is learning to develop a systematic modeling method to transform a physical system into a mathematical representation, with the help of physical laws and sound assumptions, so as to arrive at equations governing the system's evolution.

OVERVIEW OF BOOK STRUCTURE. The book is structured in two parts and six chapters:

Part I: Kinematics, Chapters 1 and 2,
Part II: Kinetics, Chapters 3 through 6.

The first two chapters are devoted to *Kinematics*, the branch of mechanics which deals with the study of motions without consideration of their causes. Kinematics is first applied to particle motion (Chapter 1), then to rigid body motion (Chapter 2), where more realistic engineering models will be studied. Kinematics is the foundation of this course.

Part II deals with *Kinetics*, that is, the branch of mechanics which studies the relationships between motion and the cause of motion. It is comprised of four chapters, each presenting a specific formulation of kinetics laws:

Chapter 3, Linear Momentum Formulation,
Chapter 4, Angular Momentum Formulation,
Chapter 5, Energy Formulation,
Chapter 6, Impulse/Momentum Formulation.

This structure offers many advantages, in that it allows presentations and proper justification of many fundamental topics usually not found in standard textbooks.

HOW TO USE THIS BOOK. The style of this book is resolutely concise and mathematical. For some students, the style will feel too dry and condensed. However, this is intentional. Unnecessarily long and convoluted derivations or explanations would be ill-suited to the elegance, simplicity and preciseness of Newtonian mechanics. The reader should also keep in mind that no matter how thick a textbook is, it will never take the place of the lecturer in live classrooms. Prior to each lecture, the student should study the planned material. Students would then be better prepared for the live lectures. Comprehension problems which would arise during prior reading of the upcoming lecture would be more efficiently addressed. After each lecture, students should return to the covered material in the textbook, as soon as possible, while their memory of the event is still fresh. Then, and only then, will the material become clear.

Throughout the book, critically important information will be organized into three categories.

1. DEFINITIONS OF FUNDAMENTAL QUANTITIES:

> **DEFINITION (MASS CENTER)**
>
> The *mass center* G of a system Σ of N particles is the point defined by
>
> $$\mathbf{r}_{AG} = \frac{1}{m}(m_1\mathbf{r}_{AP_1} + \cdots + m_N\mathbf{r}_{AP_N}) \tag{1}$$

2. FUNDAMENTAL RESULTS/THEOREMS:

Theorem of the Mass Center

The linear momentum of Σ is equal to the momentum of a fictitious particle of mass m located at the mass center G of Σ:

$$\mathbf{L}_\Sigma = m\mathbf{v}_G \tag{2}$$

3. Corollaries and Properties:

Properties

Conservation laws can be obtained from Euler's first principle under certain conditions:

①. If the resultant external force $\mathbf{F}_{ext\to\Sigma}$ vanishes at all time, then the linear momentum $m\mathbf{v}_G$ of the system is conserved, leading to

$$m_1\mathbf{v}_1 + m_2\mathbf{v}_2 + \cdots + m_N\mathbf{v}_N = \text{constant} \tag{3}$$

The strong emphasis of the book is on problem-solving. Throughout the book, at the end of most sections, the student will find practical guidelines for solution of the featured problems.

Before You Use This Book. Contrary to students' beliefs, Dynamics is a rational discipline, and thus requires a certain mathematical rigor both in the presentation of the subject matter and in the approach to problem-solving. Although Dynamics does not require advanced mathematical tools, it is necessary for students to be familiar with basic geometry and vector calculus, and with some elementary notions of ordinary differential equations. Vector calculus is generally familiar to students. However, it will be employed in a context and manner different from that taught in a mathematics course. It is also likely that some students will be confronted with unfamiliar mathematical notions. If this is the case, students will, in all likelihood, manage to learn them and become gradually at ease with them, as they will be continually exposed to them. It is likely that students will discover the importance and significance of these mathematical tools while solving engineering dynamics problems. The Appendix contains useful reviews of notions of vector calculus, ODE integration and conic sections.

Analytical Method of Problem-Solving. Students will quickly realize that the principles underlying engineering dynamics are relatively simple, involving very few formulas to memorize. The real challenges lie in problem-solving. Indeed, the application problems can be quite numerous and varied in their complexity, and their successful solution will have to rely, to a large degree, on a well-disciplined analytical method. We will see that often each problem can be analyzed by carefully breaking it down into smaller subproblems which are "easier" to solve. The solutions of these subproblems will then be combined to give the solution of the problem at hand. Solutions will never rely on simple memorization of formulas and substitution of data (familiarly known as "plug-and-chug").

A good solution always relies on a presentation of the results in a clear, logical, and orderly manner. These problem-solving skills will prove invaluable later in the student's engineering practice. A solution should ideally follow these steps:

①. Read and understand the problem carefully. Identify the stated assumptions (the knowns).

②. Identify the quantities requested (the unknowns).

③. Tabulate the information provided. Replace this information by a series of symbols, e.g. m for a mass, l for a length, h for a height, etc.

④. Formulate a solution method, a path leading to the answer. Identify the laws/principles to be used to obtain the result.

⑤. Draw the appropriate diagrams. Solutions of problems depend largely on graphical constructions.

⑥. Identify and apply the appropriate principles.

⑦. Report the answer first analytically by one or more formulas, and then numerically with the appropriate significant figures and units. Box the final answers.

⑧. Study the answer and determine if it is reasonable (analytically and numerically). Check for dimensional consistency.

A TYPICAL PROBLEM statement of particle kinetics might read: [1]

> A small cube P of mass 0.5 kg is set in motion on a plane inclined at angle 30° with the horizontal. It is sent from point O with initial velocity 2 m/s. Under the effect of gravity and friction forces, it reaches the highest position A and then slides back toward point O. The times to travel from O to A and from A to O are measured to be $T_1 = 5.2$ seconds and $T_2 = 9.8$ seconds, respectively.
>
> Show that the knowledge of T_1 and T_2 gives an expression for the kinetic coefficient of friction μ which characterizes the two surfaces in contact.

The solution to this problem is found by replacing all relevant parameters and variables by appropriate symbols. We first replace the mass 0.5 kg by the symbol m, the initial speed 2 m/s by the symbol v_0, the angle 30° by the symbol θ. Other symbols are already defined in the problem statement. A solution is then undertaken to find coefficient of friction μ analytically. We report the final answer in a box as follows

$$\boxed{\mu = \frac{k^2 - 1}{k^2 + 1}\tan\theta, \qquad k = \frac{T_2}{T_1}}$$

The advantages of this analytical approach versus the "numerical approach" are numerous:

①. We can check our answer for dimensional consistency: here a friction coefficient is expected to be dimensionless, and indeed our formula yields a dimensionless number. A numerical approach would make this task impossible, since all dependency on individual parameters would be lost. If the result is not consistent, then we may have made a mistake somewhere along our derivation. We can then return to it, and look for the mistake. [2] We will always check the dimensional consistency of our results.

②. We can use our formula to examine limiting cases. Here the limiting case $\mu = 0$ is interesting. Indeed, we expect that for $T_1 = T_2$ ($k = 1$), the contact would be frictionless,

---

[1] For a complete solution see Problem 3.1.

[2] Consult Appendix A for a review of the notion of dimension.

that is $\mu = 0$. You could not achieve this task with a single "numerical" solution.
③. Our formula is amenable to interesting graphical outputs: here we may want to plot μ versus $k = T_2/T_1$.
④. Our formula is amenable to computer programming: once a formula is found, it can be easily coded in `Python`, `Matlab`, `Fortran`, etc. Then not just one numerical value may be found, but as large a set as we may desire. The corresponding data files can then be conveniently plotted.
⑤. We may draw some interesting conclusions from our analytical answer: we note in our example that the friction coefficient is independent of the initial velocity or the mass of particle P. It does not depend on the individual values of T_1 and T_2 only on their ratio... Are such results to be expected? We could never raise such a question or draw such conclusions using a "numerical approach".

Every problem will be solved by following the approach outlined above.

A CLOSING WORD OF ADVICE. As with other subjects, learning Engineering Dynamics requires a great deal of practice. In order to master the subject matter, the student is encouraged to work and re-work all example problems. The solutions of these problems are not meant to be simply read, but to be worked through. It is only by attempting each problem, instead of passively reading the solution provided in the book, that the student will gain lasting benefits.

PART I: KINEMATICS

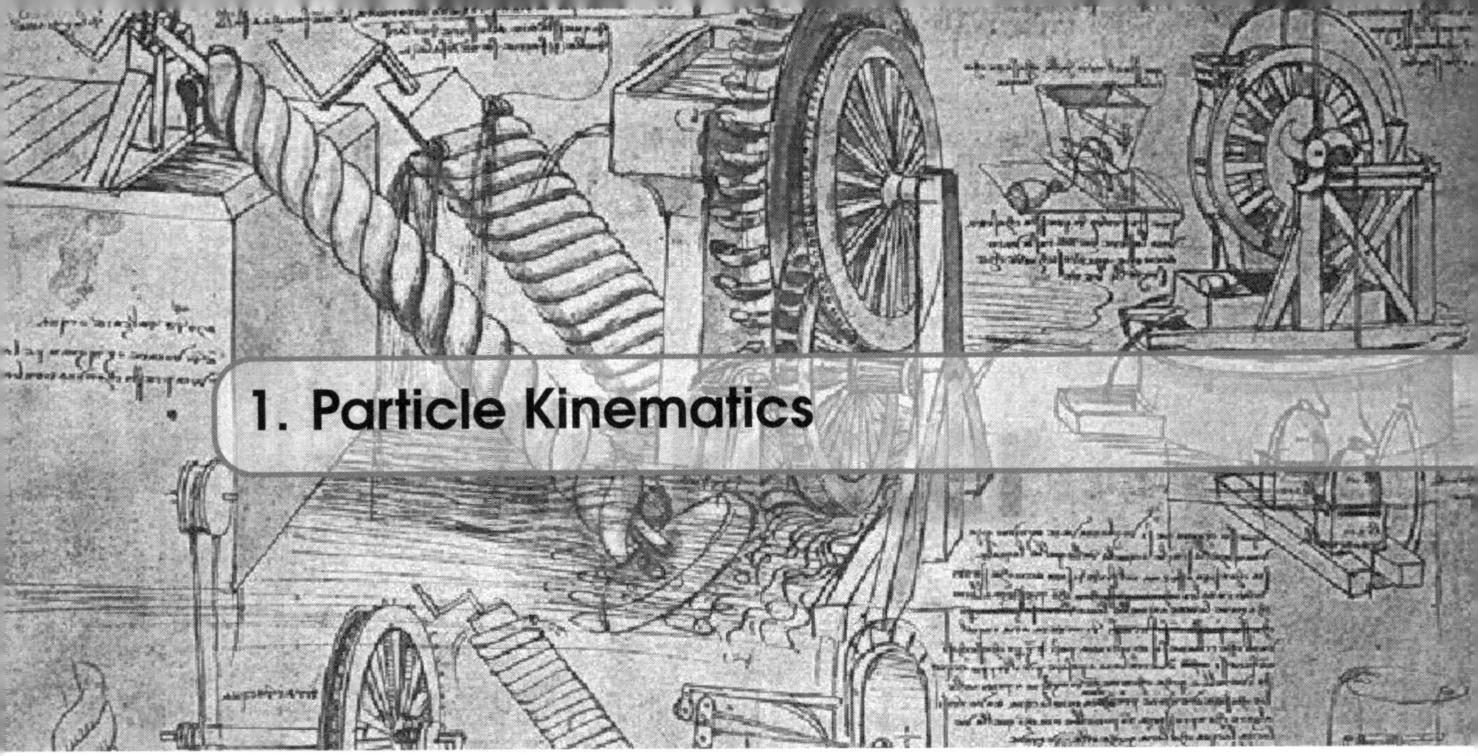

1. Particle Kinematics

KINEMATICS is the study of motions in themselves, that is, without consideration of their causes. In this chapter, the system is modeled as one or more *particles*, that is, as geometric points. The description of the motion of a particle is done relative to an observer attached to a *referential* (Section 1.1). Relative to a specific referential, a particle describes a *trajectory* along which its *velocity and acceleration vectors* characterize its kinematics. The simplest motion is *rectilinear*, corresponding to a straight line trajectory (Section 1.2). For more complex *curvilinear* motions (Section 1.3), such as circular motions, the determination of the kinematics of a particle is achieved by defining an adequate *coordinate system*, which may be Cartesian (Section 1.4), cylindrical (Section 1.5), or intrinsic to the trajectory itself (Sections 1.6 and 1.7). In many situations of engineering interest, particles are not moving independently of each other, but rather belong to a system: they become geometrically and kinematically dependent of each other (Section 1.8).

Kinematics, as the foundation of kinetics, is the backbone of this course. In Chapter 2, we will extend our study of kinematics to rigid body motions.

In this chapter, we will learn how

1. to characterize the kinematics of a particle,
2. to determine the kinematics of a particle in a variety of ways, given a particular description of its trajectory,
3. to find the trajectory of a particle, given a description of its kinematics.

1.1 Referentials

Consider a particle P (a point) in motion in space. Two observers located at locations A and B will make the same description of the motion of P if they are fixed relative to each other. If, however, they are in relative motion, their characterization of the motion of P will differ. It is thus necessary in kinematics to define a referential relative to which the observer is fixed.

Definition (Referential)

A *referential* $\mathcal{E}$ is a physical space defined as a set of points which remain at constant distance from each other.

The kinematics of a particle P depends on the choice of referential. Relative to a referential $\mathcal{F}$, itself in motion relative to $\mathcal{E}$, the trajectory of P (and the corresponding kinematics) will be different. Consider the example of Figure 1.1. A particular point P is fixed relative to the wheel of a bicycle $\mathcal{F}$, itself in motion relative to a referential $\mathcal{E}$ (the road). Point P has a circular motion relative to $\mathcal{F}$: its trajectory is a circle since the wheel is in pure rotation relative to $\mathcal{F}$. The motion relative to $\mathcal{E}$ is more complex, since it is seen to both rotate and translate in this referential: in fact it is known to describe a curve, aptly named *cycloid*, relative to $\mathcal{E}$. At any given time, the velocity (or acceleration) of P measured relative to referential $\mathcal{F}$ is drastically different to that measured relative to referential $\mathcal{E}$. This example demonstrates that it is fundamental to define the referential relative to which the kinematics of P is to be determined. Many of the (scalar or vector) quantities defined in mechanics are in fact dependent on the choice of referential.

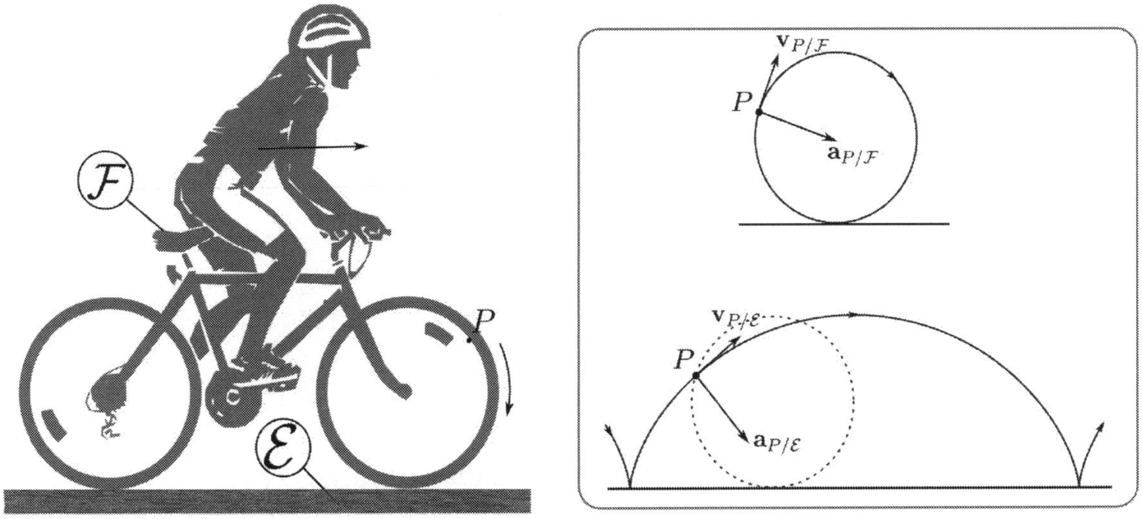

Figure 1.1 – *A point P in motion in referential $\mathcal{E}$ (road) and $\mathcal{F}$ (bicycle). The trajectory, velocity and acceleration of P will depend on the choice of referential.*

1.2 Rectilinear Motion

THERE is no simpler trajectory than a straight line. As such, motions on straight lines are a good introduction to particle kinematics.

DEFINITION (RECTILINEAR MOTION)

The motion of particle P relative to a given referential $\mathcal{E}$ is said to be *rectilinear* if its trajectory in referential $\mathcal{E}$ is a straight line.

This line can be defined by an origin O and a unit vector $\hat{\mathbf{e}}$. See Figure 1.2. Both O and $\hat{\mathbf{e}}$ are fixed in $\mathcal{E}$. Then we may define the *position vector* of P by its coordinate $x = x(t)$, that is,

$$\mathbf{r}_{OP} = x\hat{\mathbf{e}} \tag{1.1}$$

The *velocity* and *acceleration* of P relative to $\mathcal{E}$ are defined as the vectors

$$\mathbf{v}_P = \frac{d\mathbf{r}_{OP}}{dt} = \dot{x}\hat{\mathbf{e}}, \qquad \dot{x} = \frac{dx}{dt} \tag{1.2}$$

$$\mathbf{a}_P = \frac{d\mathbf{v}_P}{dt} = \ddot{x}\hat{\mathbf{e}}, \qquad \ddot{x} = \frac{d^2x}{dt^2} \tag{1.3}$$

The *speed* of P, denoted v_P, is the magnitude of the velocity vector $\mathbf{v}_P$: $v_P = |\dot{x}|$. It is easily seen that the velocity $\mathbf{v}_P$ always points in the direction of the motion.

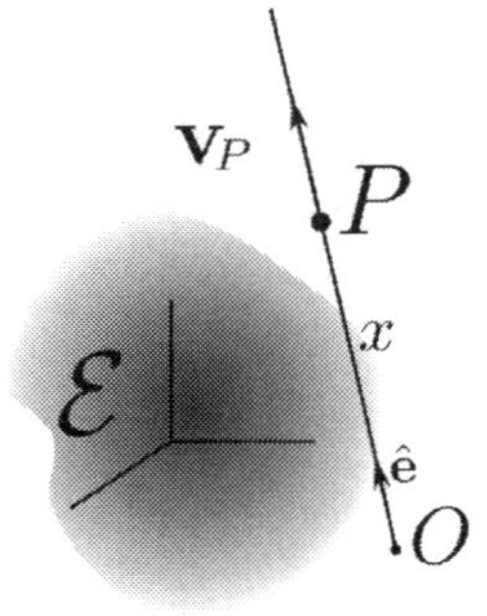

Figure 1.2 – *Rectilinear motion.*

Here are three particular rectilinear motions (see Figure 1.3):

1. A rectilinear motion is said to be *uniform* if it evolves at constant speed v_0, that is, $\mathbf{v}_P = v_0\hat{\mathbf{e}}$ at all time. Since v_0 is a constant, time-differentiation of $\mathbf{v}_P$ gives $\mathbf{a}_P = \mathbf{0}$ and time-integration gives $x(t) = x_0 + v_0 t$ with the initial value $x_0 = x(0)$.

2. A rectilinear motion is said to be *uniformly accelerated* if it evolves at constant acceleration $\mathbf{a}_P = \ddot{x}\hat{\mathbf{e}} = a_0\hat{\mathbf{e}}$. Two consecutive integrations of $\ddot{x} = a_0$ give the following quadratic x vs t relationship:

$$x(t) = a_0\frac{t^2}{2} + \dot{x}_0 t + x_0 \tag{1.4}$$

where $\dot{x}_0 = \dot{x}(0)$ and $x_0 = x(0)$ are constants of integration.

3. A rectilinear motion is said to be *harmonic* if its acceleration satisfies $\ddot{x} = -\omega x$, where ω is a constant. Integration of this ODE gives a sinusoidal x vs t relationship:

$$x(t) = A\sin(\omega t + \phi) \tag{1.5}$$

where the two constants A and ϕ are found by imposing the initial position $x(0) = x_0$ and the initial velocity $\dot{x}(0) = \dot{x}_0$:

$$A\sin\phi = x_0, \qquad A\omega\cos\phi = \dot{x}_0$$

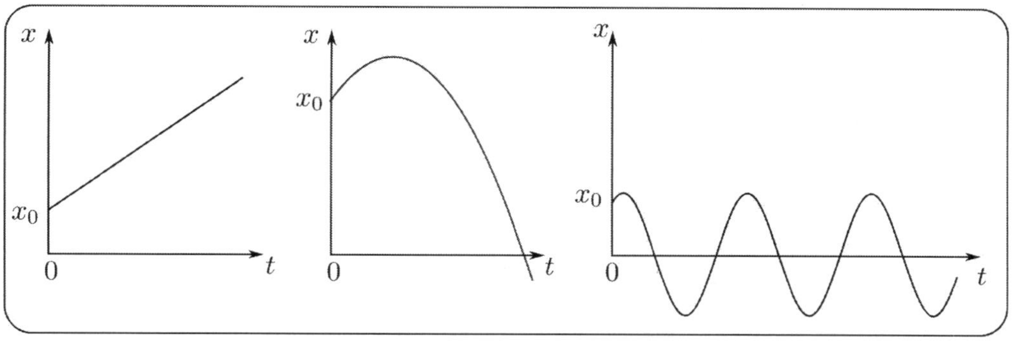

Figure 1.3 – *Rectilinear motions: uniform (left), uniformly accelerated (center), harmonic (right).*

Remark 1. In general, for a rectilinear motion, it is possible to relate x, $\dot{x}$ and $a = \ddot{x}$ by eliminating the time variable: starting with $a\,dx = \dot{x}\,d\dot{x}$ we integrate (the time variable disappears) to get

$$\int_{x_0}^{x} a\,dx = \tfrac{1}{2}(\dot{x}^2 - \dot{x}_0^2) \tag{1.6}$$

If a is a known function of x, then the integral $\int_{x_0}^{x} a(x)\,dx$ can be evaluated and equation (1.6) gives $\dot{x}$ versus x. For instance if $a = a_0$ is constant, then

$$x - x_0 = \tfrac{1}{2a_0}(\dot{x}^2 - \dot{x}_0^2) \tag{1.7}$$

Equation (1.7) can of course be obtained from (1.4) by eliminating the time variable between $x(t)$ and $\dot{x}(t)$.

Remark 2. Equation (1.6) is useful if $a(x)$ is given: what if a law $a(\dot{x})$ is given instead? To separate the variables x and $\dot{x}$ we now divide by $a(\dot{x})$ to find $dx = \frac{\dot{x}}{a(\dot{x})}\,d\dot{x}$ and obtain $x - x_0 = \int_{\dot{x}_0}^{\dot{x}} \frac{\dot{x}}{a(\dot{x})}\,d\dot{x}$. If the function $a(\dot{x})$ is specified, this would lead to an explicit relationship between x and $\dot{x}$.

Remark 3. Rectilinear motions are the simplest of motions: problems can be solved algebraically without the need of vectors whenever the motions of particles evolve along a common straight line.

Remark 4. These definitions will be generalized in section 1.3.

EXAMPLE PROBLEMS

General Procedure: Rectilinear Motions

To solve a rectilinear kinematics problem, follow these guidelines:

① Properly define the axis along which the motion takes place: this includes its orientation and origin. Define the initial conditions (position and velocity) in terms of the chosen coordinate.

② If the acceleration is given, integrate twice with respect to time to obtain position versus time by taking into account the initial conditions.

③ In problems for which a particle P moves along a vertical line under the effect of gravity, the particle's acceleration will be taken to be $a = g = 9.81$ m/s^2, directed downward.

Exercise 1.1 ▪A body P is dropped without initial speed from the location A atop a building. At the same instant, a body Q located at B, 50 m below A, is shot upward along a vertical trajectory.
a. Find the initial speed to give body Q so that the two bodies reach the same height at midpoint between A and B.
b. Find the initial speed to give body Q so that the two bodies reach the same height at the same speed. Find the corresponding height. ▪

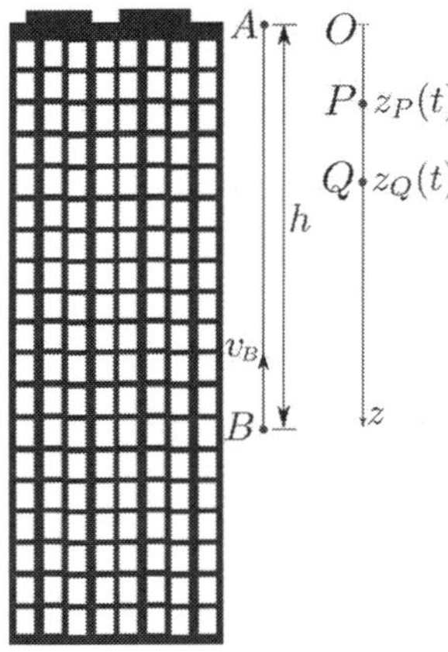

Solution. *We need to find the position $z_P(t)$ and $z_Q(t)$ of P and Q versus time by adopting a vertical axis Oz along which the motions take place. The condition that the bodies reach the same height is then established by setting $z_P(t) = z_Q(t)$ at a specific time.*

Let us adopt an axis Oz directed downward of origin O coinciding with A. The positions of particles P and Q are defined by their coordinate z_P and z_Q, respectively. The motions of P and Q are uniformly accelerated at acceleration $\ddot{z}_P = \ddot{z}_Q = g > 0$. Particle P is characterized by the initial conditions $z_P(0) = 0$ and $\dot{z}_P(0) = 0$ since at time $t = 0$ P coincides with A and is released without initial speed. Two integrations of the equation $\ddot{z}_P = g$ give

$$\dot{z}_P(t) = gt, \qquad z_P(t) = \tfrac{1}{2}gt^2$$

Similarly, the motion of Q is characterized by the initial conditions $z_Q(0) = h$ and $\dot{z}_Q(0) = -v_0 < 0$. This guarantees the initial position of Q at a distance h below P and an initial velocity directed upward. Two integrations of the equation $\ddot{z}_Q = g$ give

$$\dot{z}_Q(t) = gt - v_0, \qquad z_P(t) = \tfrac{1}{2}gt^2 - v_0 t + h$$

The unknown of the problem is speed v_0.

a. Here we set $z_P(t) = z_Q(t) = h/2$ at time $t = t_*$. We easily find

$$t_* = \sqrt{\frac{h}{g}}, \qquad \boxed{v_0 = h/t_* = \sqrt{gh}}$$

giving $v_0 = \mathbf{22.1}$ m/s.

b. Again we set $z_P(t) = z_Q(t)$ and $\dot{z}_P(t) = \pm\dot{z}_Q(t)$ at some time $t = t_*$. This last equation guarantees that the two particles meet at the same speed, either with the same direction or with opposite direction. This gives

$$v_0 t_* = h, \qquad g t_* = \pm(g t_* - v_0)$$

The first case (the particles meet with the same downward velocity) is impossible: it leads to $v_0 = 0$. The second case (the particles meet with the opposite velocity) leads to

$$\boxed{v_0 = \sqrt{2gh}}$$

giving $v_0 = \mathbf{31.3}$ m/s and $z_P = z_Q = \mathbf{12.5}$ m.

■

Exercise 1.2 ▪A balloon leaves the ground by ascending along a vertical line at a constant acceleration 0.5 m/s^2. Sixty seconds after the start, a ballast is dropped without initial velocity relative to the balloon. Find the time at which the ballast hits the ground. Then find the height reached by the balloon at this instant. ▪

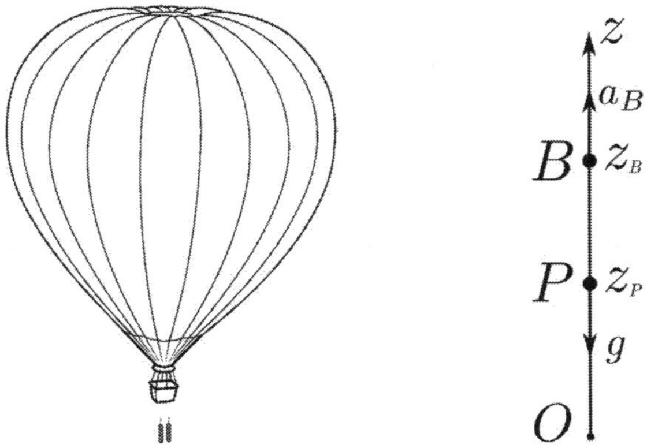

Figure 1.4 – *Problem 1.2*

SOLUTION. We refer to the balloon as particle B and the ballast as particle P. On a vertical axis Oz directed upward and with origin O at the ground level, we define the positions of B and P by the coordinates z_B and z_P. The time $t = 0$ corresponds to the start of the motion of B when $z_B = 0$ and $\dot{z}_B = 0$.

The motion of B is defined by $\ddot{z}_B = a_B$ with a_B = constant. Two consecutive time-integrations give $\dot{z}_B = a_B t$ (since $\dot{z}_B = 0$ at $t = 0$) and $z_B = a_B t^2/2$ (imposing $z_B = 0$ at $t = 0$).

The motion of P is defined by $\ddot{z}_P = -g$ with g the gravitational acceleration, in the time interval for $T_1 < t < T_2$. At the start of the motion, we have $z_P = z_B$ and $\dot{z}_P = \dot{z}_B$

at time $t = T_1$ (T_1 is given):

$$\dot{z}_P = a_B T_1, \qquad z_P = \frac{a_B}{2} T_1^2 \qquad [1-2]$$

We then integrate $\ddot{z}_P = -g$ twice to obtain $\dot{z}_P = -g(t - T_1) + a_B T_1$ (imposing $\dot{z}_P = a_B T + 1$ at $t = T_1$) and

$$z_P = -\frac{g}{2}(t - T_1)^2 + a_B T(t - T_1) + a_B T_1^2/2 \qquad [3]$$

imposing $z_P = a_B T_1^2/2$ at $t = T_1$. When the ballast hits the ground, we have $z_P = 0$: using [3], the corresponding time T_2 is found by solving equation

$$-g(T_2 - T_1)^2 + 2a_B T_1(T_2 - T_1) + a_B T_1^2 = 0 \qquad [4]$$

The solution of [4] is given by

$$\boxed{T_2 = T_1(1 + k + k(1 + 1/k)^{1/2})} \qquad \text{with } k = \frac{a_B}{g}$$

We have discarded the other solution since it would give $T_2 < T_1$. At $t = T_2$, the balloon reaches the height $h = a_B T_2^2/2$. With $a_B = 0.5\ \text{m/s}^2$ and $g = 9.81\ \text{m/s}^2$, we find $k = 0.051$, $T_2 =$ **77** s and $h =$ **1480** m. ■

Exercise 1.3 ■ A driver parks his jeep in an inclined street. He is located at a distance D from the jeep when the brake slips off. The jeep then rolls downward without initial speed at a constant acceleration 1.7 m/s². The driver can run at constant speed 2.5 m/s. If the driver is to catch up with his car, find the value that distance D must not exceed. ■

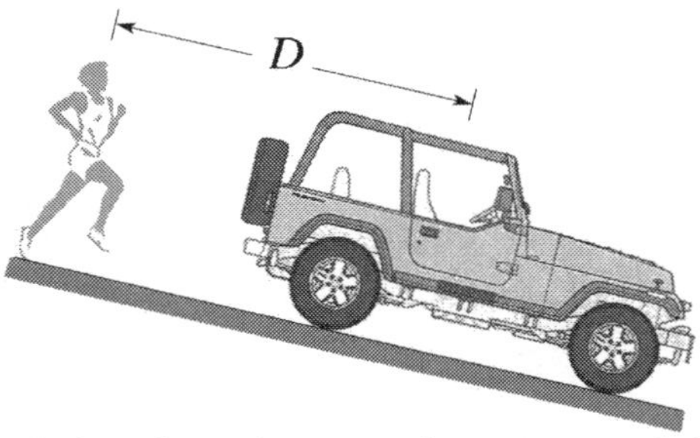

SOLUTION. Define a Cartesian axis Ox with O defined as the initial position of the driver P. On this axis, define the coordinates x_P of P and x_J of J (the jeep): at $t = 0$, we set $x_P(0) = 0$ and $x_J(0) = D$, assuming that both motions start simultaneously.

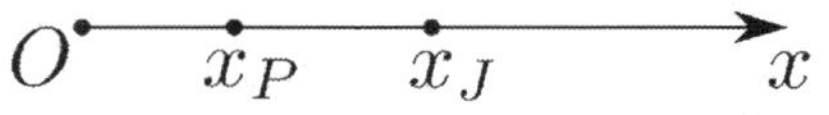

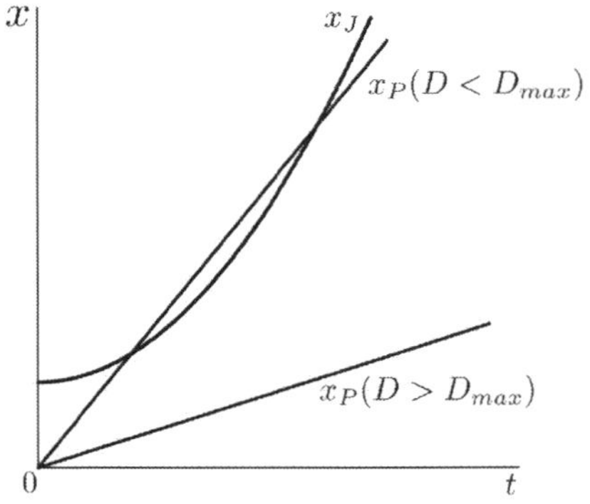

Figure 1.5 – *Problem 1.3 (Solution)*

The motion of P is defined by $\dot{x}_P = v_P$ (v_P is constant): integration gives $x_P = v_P t$ where the constant of integration has been found by imposing $x_P(0) = 0$.

The motion of J is defined by $\ddot{x}_J = a_J$: two integrations give $x_J = a_J t^2/2 + D$ where we have imposed $x_J(0) = D$ and $\dot{x}_J(0) = 0$.
P will catch up with J if there exists a solution $t = t_*$ to the equation $x_P(t_*) = x_J(t_*)$, that is,

$$v_P t_* = a_J t_*^2/2 + D \qquad [1]$$

This quadratic equation admits at least one solution if the condition $v_P^2 - 2Da_J \geq 0$ is satisfied. This gives the value that distance D must not exceed:

$$\boxed{D \leq D_{max} = \frac{v_P^2}{2a_J}}$$

Note that if $D < D_{max}$, [1] has two (positive) roots: P will pass J and is later caught up by J (see Figure 1.5). For $v_P = 2.5$ m/s and $a_J = 1.7$ m/s^2, we find $D_{max} =$ **1.8** m. ■

Exercise 1.4 ▪▪A hiker A is walking along a lake when he notices a swimmer B crying for help. He runs at speed u and swims at speed $v < u$ to reach B along straight paths. The paths make the angle α and β relative to the line normal to the shore. Assume u and v constant.

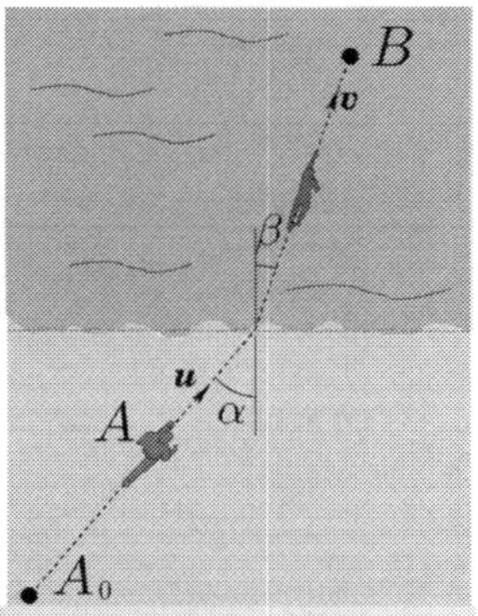

a. Show that the angles α and β must satisfy the equation

$$\frac{\sin\beta}{\sin\alpha} = \frac{v}{u}$$

for A to reach B in the shortest time.
b. Show that the straight path joining the initial position A_0 of A to B necessarily leads to longer time.
c. Find the optimal path of A when $v/u = 0.5$, line A_0B makes a 45° angle with the shoreline, and with A_0 and B equidistant from the shoreline. Determine the ratio T/T_{AB} of the optimal time to the time taken on the straight path A_0B. ■

Method: To find the time T to go from A_0 to B introduce the distance from A_0 and B to the shoreline, and the angle made by line A_0B with the normal to the shoreline. Then minimize T w.r.t. α by setting $dT/d\alpha = 0$ and by taking into account the dependence of β upon α.
SOLUTION. **a.** Call a and b the distances from A_0 and B to the shoreline, and θ the angle made by line A_0B (at the start of the motion). See Figure 1.6. Note that a, b and θ are constant parameters.

On the ground, A travels distance $a/\cos\alpha$ at constant speed u. On water, A travels distance $b/\cos\beta$ at constant speed v. The time to reach B is therefore

$$T(\alpha, \beta) = \frac{a}{u\cos\alpha} + \frac{b}{v\cos\beta}$$

We want to find the angles (α, β) that minimize time T. This is difficult to do mathematically if α and β are independent variables. But here β depends on α since the distance from A to C, $(a+b)\tan\theta$ (a constant) can be expressed as

$$a\tan\alpha + b\tan\beta = (a+b)\tan\theta \qquad [1]$$

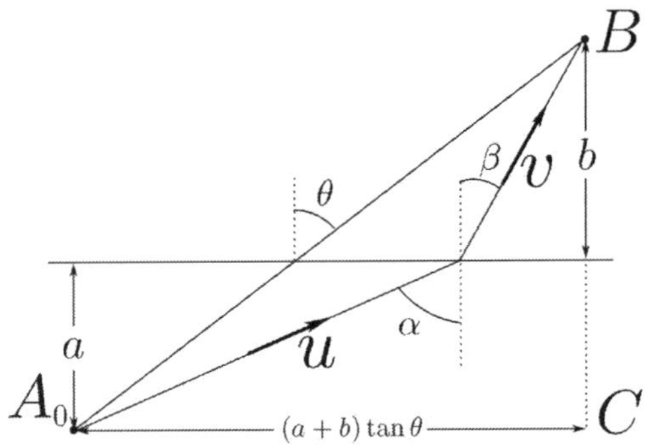

Figure 1.6 – *Problem 1.4 (Solution)*

In order to minimize $T(\alpha)$ w.r.t. α, we must find $dT/d\alpha$:

$$\frac{dT}{d\alpha} = \frac{a}{u}\frac{\sin\alpha}{\cos^2\alpha} + \frac{b}{v}\frac{\sin\beta}{\cos^2\beta}\frac{d\beta}{d\alpha}$$

taking into account the dependence of β with α. To find $d\beta/d\alpha$ we differentiate $a\tan\alpha + b\tan\beta$ =constant w.r.t. α to find:

$$\frac{d\beta}{d\alpha} = -\frac{a\cos^2\beta}{b\cos^2\alpha}$$

This then gives

$$\frac{dT}{d\alpha} = \frac{a}{\cos^2\alpha}\left(\frac{\sin\alpha}{u} - \frac{\sin\beta}{v}\right)$$

Hence we see that $dT/d\alpha$ is zero when α and β satisfy

$$\boxed{\frac{\sin\alpha}{\sin\beta} = \frac{u}{v}} \qquad [2]$$

This result is reminiscent of Snell-Descartes law for the paths of light wave across a boundary separating two regions of differing index of refraction. It can be verified that the second derivative $d^2T/d\alpha^2$ is positive if [2] is satisfied.

b. A necessarily takes a longer time to reach B if A travels on straight path A_0B: indeed this path corresponds to the case $\alpha = \beta$ and it cannot be a solution minimizing T since, if it were, this would imply $u = v$ according to [2].

c. For $v = 0.5u$, $a = b$ and $\theta = 45°$ we find that the angles α and β minimizing T are solutions of [1-2]:

$$\tan\alpha + \tan\beta = 2, \qquad \frac{\sin\alpha}{\sin\beta} = 2$$

We solve these equations numerically (using a numerical solver) and find $\alpha = \mathbf{56.9}°$ and $\beta = \mathbf{24.8}°$. The ratio T/T_{AB} is then found to be **0.95**. ■

Exercise 1.5 ■■■ A shell P is fired from a cannon located at the origin O of a plane Oxy and travels at constant speed $v = 465$ m/s along axis Ox. A fixed observer located in plane Oxy at point Q at distance r from O can hear the explosion and whistling sound of P along its trajectory. Denote by $c = 340$ m/s the speed of sound and by θ the angle of line OQ with line Ox ($0 < \theta \leq \pi$) and note that $v > c$.

a. Find the time $T(x)$ of perception of the sound corresponding with the passage of P at location x of its trajectory (in terms of parameters v, c, r and θ).
b. Assume that the observer at Q is in the cannon's line of sight ($\theta = 0$). Show that the explosion is not the first event heard by the observer.
c. Show that the phenomenon examined in b) occurs for all angles $\theta < \theta_{crit}$. Find the critical value θ_{crit}. ■

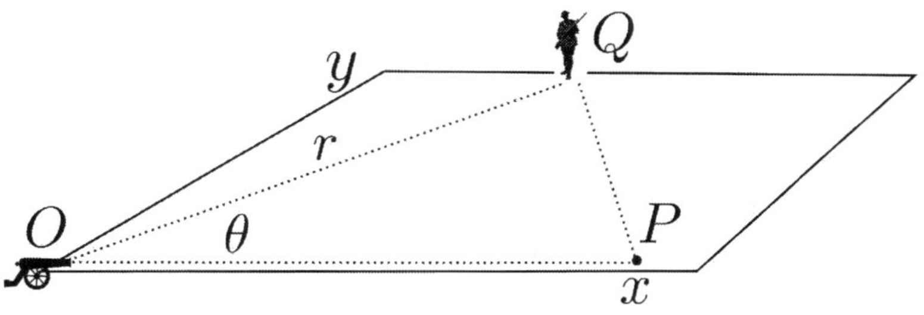

Figure 1.7 – *Problem 1.5*

SOLUTION. **a.** At a given time t, the position of P is defined by coordinate x. Since the speed of P is constant, we have $x = vt$, setting $t = 0$ at $x = 0$. At any given position of P, the sound travels at constant speed c along line PQ: it takes the time $|PQ|/c$ for the sound to travel the distance $|PQ|$. Hence the time taken for the sound to reach position Q corresponding to the passage of P at position x (that is, at time t) is given by (the passage of P at position x creates a sound wave which travels to Q (or any other point) at speed c)

$$T = t + \frac{|PQ|}{c}$$

We can express distance $|PQ|$ in terms of x, r, and θ

$$|PQ|^2 = x^2 + r^2 - 2rx\cos\theta$$

according to the law of cosines in triangle OPQ. Finally time T takes the expression

$$\boxed{T = \frac{x}{v} + \frac{1}{c}\sqrt{x^2 + r^2 - 2rx\cos\theta}} \qquad [1]$$

b. For $\theta = 0$, we find

$$T(x) = \frac{r}{c} + (\frac{1}{v} - \frac{1}{c})x, \quad 0 \le x \le r$$

We find that T is a decreasing function of x until $x = r$ is reached: $T(0) > T(r)$ since $v > c$. Hence the first event heard by Q is not the explosion at O, but rather the whistling sound of the shell!
c. The analysis of b) can be extended for arbitrary position of Q by examining the behavior of $T(x)$ near $x = 0$: a Taylor series gives the approximation

$$T \sim \frac{r}{c} + (\frac{1}{v} - \frac{1}{c}\cos\theta)x, \qquad x \to 0$$

We find again that T is not necessarily a monotonic function of x. We can distinguish two cases (see Figure 1.8):
CASE 1: If $\boxed{c > v\cos\theta}$, then T is an increasing function of x. The first event heard by

Q is the explosion of P at O, followed by the whistling of P along its trajectory.

Case 2: if $\boxed{c < v\cos\theta}$, then T is initially a decreasing function of x until the value T_* is reached, after which T increases monotonically. Again, the first event heard by Q is not the explosion at O, but rather the two whistling sound of the shell! At time $T(0) > T_*$, the explosion of P is then heard. The condition $c < v\cos\theta$ implies that Q is inside the region $\theta < \theta_{crit} = \cos^{-1}(c/v)$. *The critical angle θ_{crit} exists only for supersonic shells $(v > c)$.* For $c = 340$ m/s and $v = 465$ m/s, we find $\theta_{crit} = \cos^{-1}(c/v) = \mathbf{43}^\circ$.

Remark. In WW1, German high-velocity 77mm shells were known as "whizz-bangs" as soldiers would first hear the high-pitched noise of the shell before its explosion. Many soldiers in the cannon's line of sight would unfortunately only hear the "whizz".

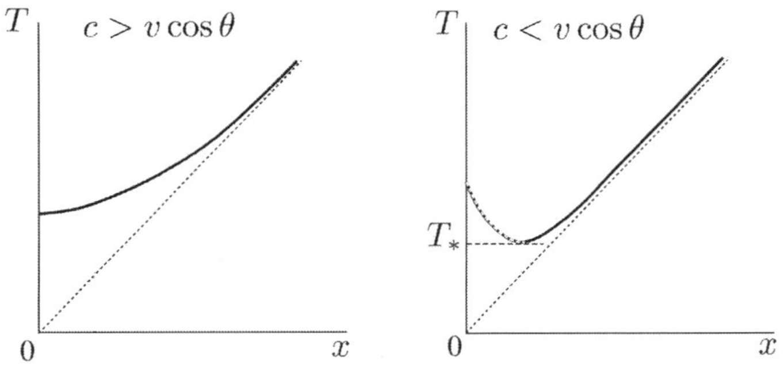

Figure 1.8 – *Problem 1.5 (Solution): for $c < v\cos\theta$, there exist two values of $x > 0$ for which $T(x)$ is smaller than $T(0)$.*

■

1.3 General Curvilinear Motion

We now generalize the kinematic definitions of Section 1.2. Let P be a particle in motion relative to a referential $\mathcal{E}$. See Figure 1.9.

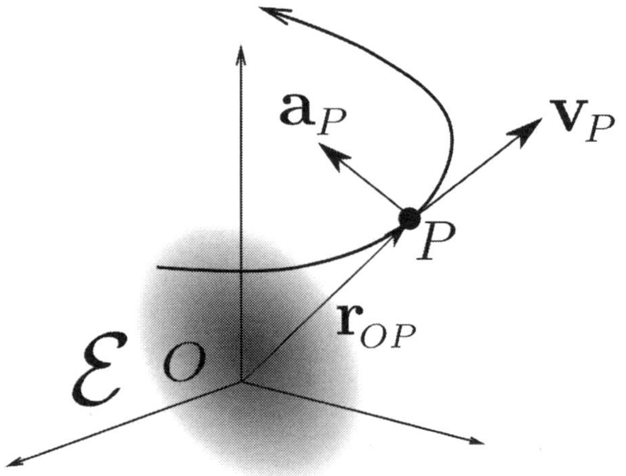

Figure 1.9 – *Curvilinear motion: The kinematics of P is characterized by its velocity $\mathbf{v}_P$ and acceleration $\mathbf{a}_P$.*

DEFINITION (VELOCITY/ACCELERATION)

Particle P is said to be in *curvilinear* motion relative to $\mathcal{E}$ if its trajectory is a curve of $\mathcal{E}$. Its position in $\mathcal{E}$ is defined by its *position vector* $\mathbf{r}_{OP}$ relative to an origin O of $\mathcal{E}$ (an arbitrary point fixed in $\mathcal{E}$). In general, $\mathbf{r}_{OP}$ must be viewed as a vector function of time. Then, the *velocity* and *acceleration* of P, denoted $\mathbf{v}_P$ and $\mathbf{a}_P$, respectively, are the vectors defined as

$$\mathbf{v}_P = \frac{d\mathbf{r}_{OP}}{dt}, \qquad \mathbf{a}_P = \frac{d\mathbf{v}_P}{dt} \tag{1.8}$$

Remark 5. The origin O is typically defined as the position of the observer in $\mathcal{E}$.

Remark 6. It may be necessary to use the notation $\mathbf{v}_{P/\mathcal{E}}$ and $\mathbf{a}_{P/\mathcal{E}}$ to specify the referential relative to which the motion is observed.

Remark 7. The velocity of P does not depend on the choice of origin. Consider another point O' fixed in $\mathcal{E}$: then $\frac{d}{dt}\mathbf{r}_{O'P} = \frac{d}{dt}\mathbf{r}_{O'O} + \frac{d}{dt}\mathbf{r}_{OP} = \mathbf{v}_P$ since vector $\mathbf{r}_{O'O}$ is constant. However, in practice, there is always a judicious choice of origin.

GEOMETRIC CHARACTERIZATION OF $\mathbf{v}_P$

At any given time, the velocity vector $\mathbf{v}_P$ is directed along the tangent direction to the trajectory of P. It always points in the sense of the motion.

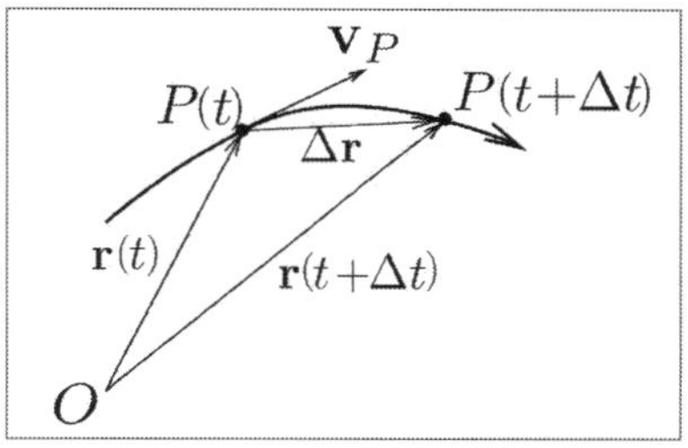

Figure 1.10 – *The velocity vector $\mathbf{v}_P$ is always tangent to the path of P (right).*

Proof. This can be seen graphically by interpreting $\mathbf{v}_P$ as the limit of the ratio

$$\frac{\mathbf{r}(t+\Delta t) - \mathbf{r}(t)}{\Delta t} = \frac{\Delta\mathbf{r}}{\Delta t}$$

as the time increment Δt tends to zero. We see in Figure 1.10 that, as $\Delta t \to 0$, $P(t+\Delta t)$ becomes closer to $P(t)$ and the vector $\Delta\mathbf{r}$ approaches the tangent direction at P while pointing in the sense of the motion. ■

DEFINITION (UNIFORM MOTION)

We call *speed* of particle P the magnitude of its velocity vector $\mathbf{v}_P$, denoted $v_P = |\mathbf{v}_P|$. The motion of P is said to be *uniform* if its speed v_P remains constant. In general, the acceleration of P in uniform motion is *not zero*. It is normal to $\mathbf{v}_P$ at all time.

Remark 8. The vector $\mathbf{v}_P$ may be of constant magnitude but its direction varies!

Proof. We can show that $\mathbf{a}_P$ is always normal to $\mathbf{v}_P$ in a uniform motion: indeed from $v_P^2 = \mathbf{v}_P \cdot \mathbf{v}_P = Cst$, we obtain by time-differentiation $\mathbf{v}_P \cdot \mathbf{a}_P = 0$. This property will be confirmed in Section 1.6. ■

The geometric angle ψ measured between the velocity and acceleration vectors indicates whether the speed increases or decreases: if this angle is acute (less than 90°), then the speed increases and, conversely, if this angle is obtuse, the speed decreases. This is easily shown by noting that $\frac{d}{dt}v_P^2 = 2\mathbf{v}_P \cdot \mathbf{a}_P = 2v_P|\mathbf{a}_P|\cos\psi$. Hence the sign of $\dot{v}$ is also the sign of $\cos\psi$.

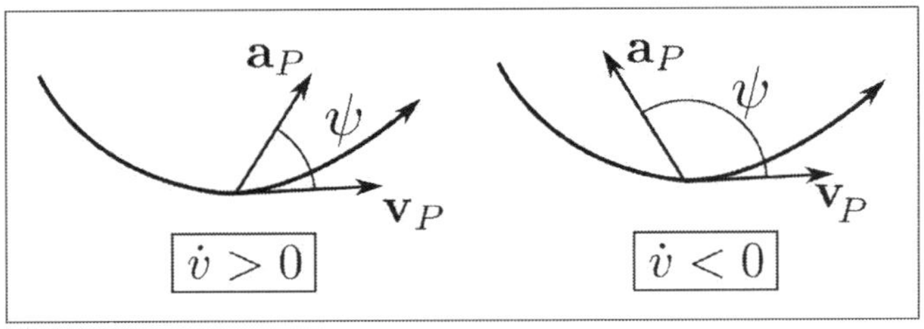

Figure 1.11 – *Accelerated ($\dot{v} > 0$) versus decelerated ($\dot{v} < 0$) motion.*

It is possible to obtain a geometric characterization of $\mathbf{a}_P$ similar to $\mathbf{v}_P$ as follows:

DEFINITION (HODOGRAPH)

Consider the fictitious particle Q of the velocity space $O'v_xv_yv_z$ whose position is defined by the velocity of P:

$$\mathbf{r}_{O'Q} = \mathbf{v}_P \tag{1.9}$$

As P moves in referential $\mathcal{E}$, Q describes a curve called *hodograph* of P. Since the velocity of Q is the acceleration of P, vector $\mathbf{a}_P = \mathbf{v}_Q$ is tangent to the hodograph at Q. See Figure 1.12.

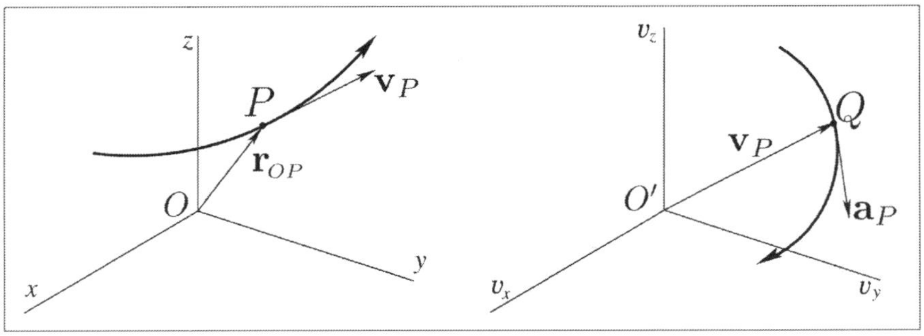

Figure 1.12 – *Trajectory (left) and hodograph of P (right): The acceleration of P is tangent at Q to the hodograph.*

Finally, we note that the definitions stated in equations (1.8) do not give a practical way to find the velocity and acceleration of a particle. We need to introduce a *coordinate system*, that is, a geometric construction which will provide three scalar quantities (q_1, q_2, q_3), called *coordinates* of P. The coordinates of P will allow a representation of the position vector $\mathbf{r}_{OP}$ of P. The motion of P is fully determined once the coordinates (q_1, q_2, q_3) are specified functions of time over an interval $t_1 \leq t \leq t_2$ of interest.

1.4 Cartesian Coordinate System

THE simplest way to define a coordinate system is to consider three mutually perpendicular axes intersecting at a fixed point O of referential $\mathcal{E}$. These axes are directed along three fixed unit vectors $\hat{\mathbf{e}}_x, \hat{\mathbf{e}}_y, \hat{\mathbf{e}}_z$. Then, we express the position vector $\mathbf{r}_{OP}$ of P on basis $(\hat{\mathbf{e}}_x, \hat{\mathbf{e}}_y, \hat{\mathbf{e}}_z)$ as follows:

$$\mathbf{r}_{OP} = x\hat{\mathbf{e}}_x + y\hat{\mathbf{e}}_y + z\hat{\mathbf{e}}_z \tag{1.10}$$

where (x, y, z) are called *Cartesian coordinates* of P. In general, $x = x(t)$, $y = y(t)$ and $z = z(t)$ are scalar functions of time since particle P is in motion.

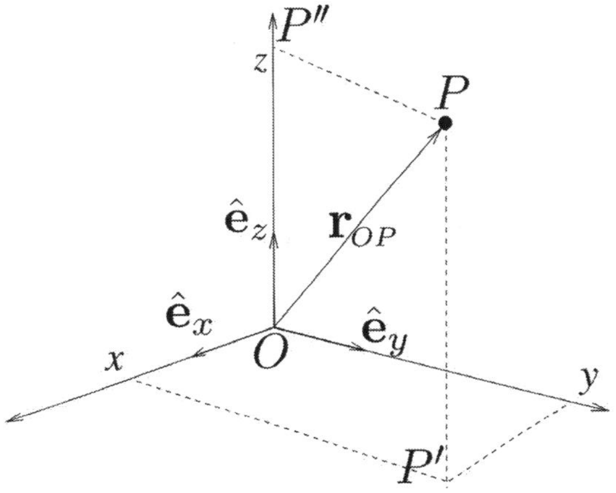

Figure 1.13 – *Cartesian coordinates*

The velocity and acceleration of P are then found to be

$$\mathbf{v}_P = \dot{x}\hat{\mathbf{e}}_x + \dot{y}\hat{\mathbf{e}}_y + \dot{z}\hat{\mathbf{e}}_z \tag{1.11}$$

$$\mathbf{a}_P = \ddot{x}\hat{\mathbf{e}}_x + \ddot{y}\hat{\mathbf{e}}_y + \ddot{z}\hat{\mathbf{e}}_z \tag{1.12}$$

with the notations $\dot{x} = dx/dt$, $\ddot{x} = d^2x/dt^2$, etc.

APPLICATION: PROJECTILE MOTION. We assume that the acceleration of particle P is given by $\mathbf{a}_P = -g\hat{\mathbf{e}}_y$ where $g = 9.81$ m/s^2 is the gravitational acceleration and $\hat{\mathbf{e}}_y$ is a unit vector pointing upward.[1] If we choose Cartesian axes Ox (horizontal) and Oy (vertical upward) with corresponding unit vectors $\hat{\mathbf{e}}_x$ and $\hat{\mathbf{e}}_y$ as shown in Figure 1.14, we can write

$$\mathbf{a}_P = \ddot{x}\hat{\mathbf{e}}_x + \ddot{y}\hat{\mathbf{e}}_y = -g\hat{\mathbf{e}}_y$$

in terms of the Cartesian coordinates (x, y) of P. This gives two equations

$$\ddot{x} = 0, \qquad \ddot{y} = -g$$

[1]The acceleration of P is a result of Newton's Second Law $m\mathbf{a}_P = m\mathbf{g} + \mathbf{F}$ after neglecting the friction force $\mathbf{F}$.

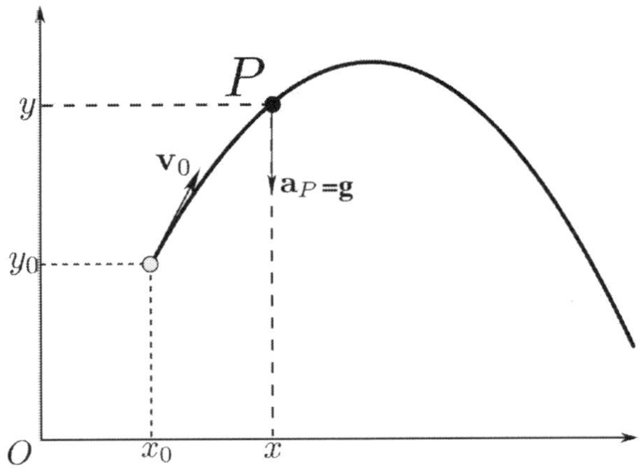

Figure 1.14 – *Projectile motion.*

which can be integrated to give

$$\dot{x} = v_{0x}, \qquad \dot{y} = -gt + v_{0y}$$

where (v_{0x}, v_{0y}) are the components of the velocity at $t = 0$. One more integration w.r.t. time gives

$$x = x_0 + v_{0x}t, \qquad y = y_0 + v_{0y}t - \tfrac{1}{2}gt^2$$

where (x_0, y_0) are the coordinates of P at $t = 0$. This gives a parabolic trajectory parametrized by time t. We can eliminate time between these two equations to obtain the Cartesian representation of the trajectory: $t = (x - x_0)/v_{0x}$ gives

$$y = y_0 + \frac{v_{0y}}{v_{0x}}(x - x_0) - \frac{g}{2v_{0x}^2}(x - x_0)^2$$

Remark 9. These equations should not be memorized but rather must be rederived for each particular problem.

Remark 10. More generally, it can be shown that the trajectory of a particle in motion at constant acceleration (in magnitude and direction) is a parabola. Assuming that $\mathbf{a}_P = \mathbf{a} =$ Constant, time-integration gives

$$\mathbf{v}_P = \mathbf{a}t + \mathbf{v}_0$$

and

$$\mathbf{r}_{P_0P} = \tfrac{1}{2}\mathbf{a}t^2 + \mathbf{v}_0 t$$

assume that $P = P_0$ and $\mathbf{v}_P = \mathbf{v}_0$ at time $t = 0$.

EXAMPLE PROBLEMS

General Procedure: Kinematics in Cartesian Coordinates

To solve a kinematics problem in Cartesian coordinates, follow these guidelines:

① Carefully define the Cartesian axes $Oxyz$. In a projectile motion, choosing Ox

horizontal and Oy vertical may not always be convenient. Identify the initial conditions in terms of the chosen coordinates, if needed.

② In a projectile motion, derive the trajectory by integration of $\mathbf{a}_P = \mathbf{g}$ on a set of convenient Cartesian axes. Do not memorize formulas.

③ Most problems are solved by expressing one or more geometric conditions imposed on the position, velocity or acceleration of P. This may involve the intersection of two curves, or conditions imposed on the direction of vectors $\mathbf{v}_P$ or $\mathbf{a}_P$ at a particular instant or at all time. If a condition has been found to be valid at all time, the resulting algebraic equation can be integrated or differentiated w.r.t. time.

Exercise 1.6 ▪A projectile P is launched with speed v_0 from point O to reach point A on an incline of angle α $(0 \leq \alpha \leq 90^o)$.
Find the range R and plot R versus launch angle θ.
Then determine the launch angle θ_* which maximizes R.
Verify that $\theta_* = \pi/4$ for $\alpha = 0$. ▪

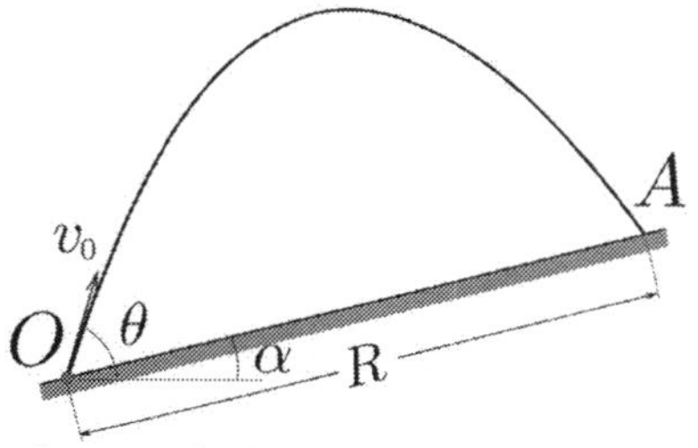

Solution. We choose a coordinate system Oxy with Ox directed along the incline. See Figure 1.15. We first need to write the gravitational acceleration $\mathbf{g}$ on the corresponding basis $(\hat{\mathbf{e}}_x, \hat{\mathbf{e}}_y)$:

$$\mathbf{g} = -g(\sin\alpha\hat{\mathbf{e}}_x + \cos\alpha\hat{\mathbf{e}}_y)$$

The trajectory of P of Cartesian coordinates (x, y) is then obtained by integrating the

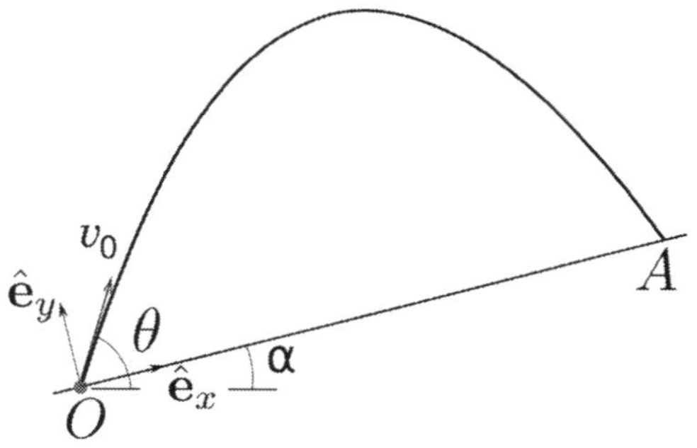

Figure 1.15 – *Problem 1.6 (Solution)*

equations:

$$\ddot{x} = -g\sin\alpha, \qquad \ddot{y} = -g\cos\alpha$$

with initial conditions $(\dot{x}, \dot{y}) = (v_0\cos(\theta-\alpha), v_0\sin(\theta-\alpha))$ at $t = 0$:

$$\dot{x} = -(g\sin\alpha)t + v_0\cos(\theta-\alpha), \qquad \dot{y} = -(g\cos\alpha)t + v_0\sin(\theta-\alpha)$$

One more integration gives

$$x = -(g\sin\alpha)\frac{t^2}{2} + v_0 t\cos(\theta-\alpha), \qquad y = -(g\cos\alpha)\frac{t^2}{2} + v_0 t\sin(\theta-\alpha)$$

with $x = y = 0$ at $t = 0$. The point of impact A is obtained by setting $y = 0$: this gives the time t_A taken to reach A

$$t_A = \frac{2v_0}{g\cos\alpha}\sin(\theta-\alpha) \qquad [1]$$

The range R is then obtained as $x(t_A)$:

$$R = -(g\sin\alpha)\frac{t_A^2}{2} + v_0 t_A\cos(\theta-\alpha)$$

Substitution of t_A gives, after simplifications

$$\boxed{R = \frac{2v_0^2}{g\cos^2\alpha}\cos\theta\sin(\theta-\alpha)} \qquad [2]$$

If we view R as a function of angle θ with $\alpha \le \theta \le \pi/2$, we see that $R(\alpha) = R(\pi/2) = 0$. Since $R \ge 0$, this function necessarily admits a maximum which is found by setting $dR/d\theta = 0$:

$$-\sin\theta\sin(\theta-\alpha) + \cos\theta\cos(\theta-\alpha) = \cos(2\theta-\alpha) = 0$$

We find that R is maximized for the value

$$\boxed{\theta_* = \frac{\pi}{4} + \frac{\alpha}{2}}$$

We recover the well-known result $\theta_* = \pi/4$ for $\alpha = 0$.

Remark. Attempt to solve this problem by choosing Ox horizontal and Oy vertical.

■

Exercise 1.7 ▪A weather balloon is launched from the ground ($z = 0$). Its velocity at any height z is the sum of two components: a vertical constant component $v_z = v_0$ and a horizontal component $v_x = kz$ due to air flow proportional to height z (where $1/k$ is a given time constant).

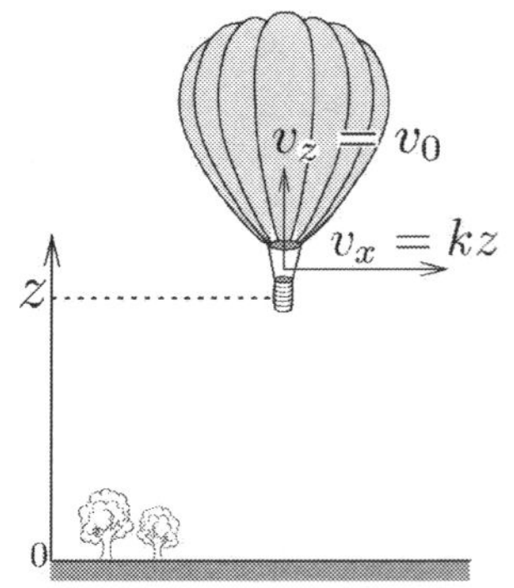

a. Find and then sketch the path taken by the balloon.
b. Find the acceleration of the balloon. ■

SOLUTION. **a.** We define the Cartesian coordinate system Oxz with Ox horizontal and Oz vertical (directed upward) and with the origin O chosen so that $x = z = 0$ at $t = 0$. Then the velocity vector $\mathbf{v}_P$ of P takes the expression

$$\mathbf{v}_P = kz\hat{\mathbf{e}}_x + v_0\hat{\mathbf{e}}_z = \dot{x}\hat{\mathbf{e}}_x + \dot{z}\hat{\mathbf{e}}_z$$

This gives two equations

$$\dot{x} = kz, \qquad \dot{z} = v_0 \qquad [1-2]$$

We first integrate [2] to find

$$z = v_0 t$$

We can now integrate [1]

$$x = \frac{k}{2}v_0 t^2$$

By substitution of $t = z/v_0$ we find the Cartesian expression of P's trajectory:

$$\boxed{x = \frac{k}{2v_0} z^2}$$

This is a parabola whose axis is the line Ox. It is easily sketched.
b. We differentiate $\mathbf{v}_P = kz\hat{\mathbf{e}}_x + v_0\hat{\mathbf{e}}_z$ w.r.t. time to obtain $\mathbf{a}_P = k\dot{z}\hat{\mathbf{e}}_x$ since v_0 is a constant. With $\dot{z} = v_0$ we obtain

$$\boxed{\mathbf{a}_P = kv_0\hat{\mathbf{e}}_x}$$

■

Exercise 1.8 ■■ A cannon fires successively two shells with initial speed $v_0 = 250$ m/s, the first at the angle $\theta_1 = 60°$ and the second at the angle $\theta_2 = 45°$ to the horizontal.

Find the time interval between the firings which would lead to the collision of the shells.
■

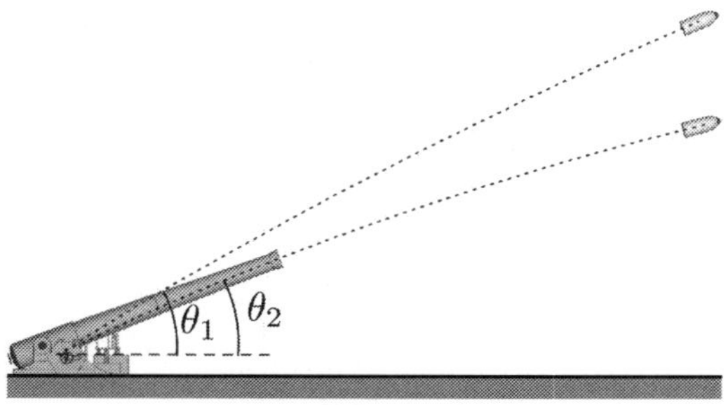

SOLUTION. With coordinate axes Ox (horizontal) and Oy (vertical upward), the path of the first shell P_1 is given by

$$x_1(t) = (v_0 \cos\theta_1)t, \qquad y_1(t) = (v_0 \sin\theta_1)t - \frac{gt^2}{2} \qquad [1-2]$$

where we imposed $x = y = 0$ and $(\dot{x}_1, \dot{y}_1) = (v_0\cos\theta_1, v_0\sin\theta_1)$ at $t = 0$. Similarly, the path of the second shell P_2 is given by

$$x_2(t) = (v_0 \cos\theta_2)(t-\tau), \qquad y_2(t) = (v_0 \sin\theta_2)(t-\tau) - \frac{g}{2}(t-\tau)^2 \qquad [3-4]$$

where we imposed $x = y = 0$ and $(\dot{x}_2, \dot{y}_2) = (v_0\cos\theta_2, v_0\sin\theta_2)$ at $t = \tau$. Here τ is the unknown time interval between the two firings. It is easy to see that equations [3-4] are obtained from equations [1-2] by the changes $t \to (t-\tau)$ (time shift) and $\theta_1 \to \theta_2$. For collision to occur we must impose the intersection of the two paths at time t_* (unknown):

$$(v_0\cos\theta_1)t_* = (v_0\cos\theta_2)(t_* - \tau) \qquad [5]$$

$$(v_0\sin\theta_1)t_* - \frac{gt_*^2}{2} = (v_0\sin\theta_2)(t_* - \tau) - \frac{g}{2}(t_* - \tau)^2 \qquad [6]$$

This gives two equations to solve for (t_*, τ): to find unknown τ, we use [5] to get $t_* - \tau = t_* \cos\theta_1 / \cos\theta_2$. We then substitute $t_* - \tau$ in [6] to find, after some algebraic manipulations,

$$\boxed{\tau = \frac{2v_0}{g} \frac{\sin(\theta_1 - \theta_2)}{\cos\theta_1 + \cos\theta_2}}$$

With $v_0 = 250$ m/s, $\theta_1 = 60°$ and $\theta_2 = 45°$, we find $\tau =$ **10.9** seconds.

Remark. To obtain an acceptable result, we must impose $\theta_1 > \theta_2$. For $\theta_1 < \theta_2$, the collision is not feasible.

■

Exercise 1.9 ■■A particle P moves in a referential $\mathcal{E}$ so that its Cartesian coordinates (x, y, z) satisfy

$$x = \sin(\phi + 1)$$
$$y = \sin(\phi + 2)$$
$$z = \sin(\phi + 3)$$

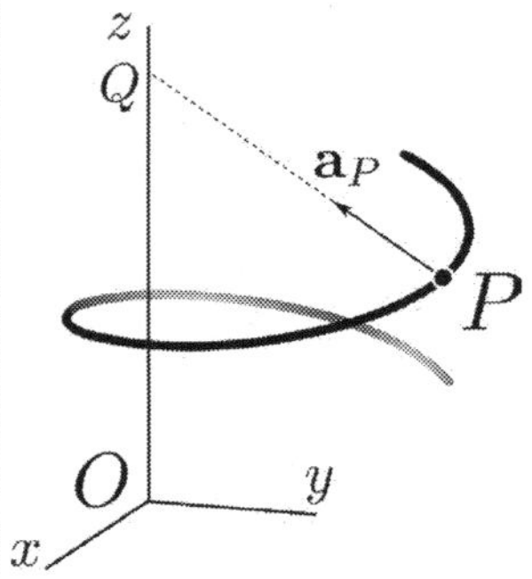

where ϕ is a function of time and x, y, z are measured in meters. At $t = 0$, P lies in the Oxy plane and its initial speed is $v_0 = 0.4$ m/s. Furthermore, the acceleration of P always points toward axis Oz.

a. Find the function $\phi(t)$. Conclude that the trajectory of P is periodic and find the period T.

b. Show that the path of P lies in a plane of $\mathcal{E}$ containing the origin.

c. Sketch the path of P. ■

SOLUTION. **a.** We need to impose that the acceleration of P always points toward axis Oz. This can be done by stating that there must exist a point Q lying on axis Oz such that $\mathbf{r}_{QP} = \lambda \mathbf{a}_P$ where λ is a non-zero scalar which depends on the position of P. This leads to 3 scalar equations

$$x = \lambda \ddot{x}$$
$$y = \lambda \ddot{y}$$
$$z - z_Q = \lambda \ddot{z}$$

where we have set $x_Q = y_Q = 0$ so that Q lies on axis Oz. Using the first two equations, we eliminate λ to find the condition

$$y\,\ddot{x} = x\,\ddot{y} \qquad [1]$$

Let us find the components $\ddot{x}$ and $\ddot{y}$: first we find (recall that ϕ is a function of t)

$$\dot{x} = \dot{\phi}\cos(\phi + 1), \qquad \dot{y} = \dot{\phi}\cos(\phi + 2)$$

leading to

$$\ddot{x} = \ddot{\phi}\cos(\phi + 1) - \dot{\phi}^2 \sin(\phi + 1), \qquad \ddot{y} = \ddot{\phi}\cos(\phi + 2) - \dot{\phi}^2 \sin(\phi + 2)$$

Now we can impose condition [1]:

$$\sin(\phi + 2)\left(\ddot{\phi}\cos(\phi + 1) - \dot{\phi}^2 \sin(\phi + 1)\right) = \sin(\phi + 1)\left(\ddot{\phi}\cos(\phi + 2) - \dot{\phi}^2 \sin(\phi + 2)\right)$$

After simplifications we obtain

$$(\cot(\phi + 1) - \cot(\phi + 2))\ddot{\phi} = 0$$

Since $\cot(\phi + 1) - \cot(\phi + 2)$ cannot possibly be zero, this implies $\ddot{\phi} = 0$ leading to

$$\boxed{\phi = \dot{\phi}_0 t + \phi_0}$$

where $(\phi_0, \dot{\phi}_0)$ are two constants of integration. To find these constants, we impose that

(i) P lies in the Oxy at $t = 0$: $z = \sin(\phi_0 + 3) = 0$ at $t = 0$ gives $\boxed{\phi_0 = -3}$.
(ii) its initial speed v_0 is known: $v_0 = (\dot{x}^2 + \dot{y}^2 + \dot{z}^2)^{1/2}$ gives

$$\boxed{v_0 = \dot{\phi}_0(\cos^2 2 + \cos^2 1 + 1)^{1/2}}$$

assuming $\dot{\phi}_0 > 0$. Numerically we find $\dot{\phi}_0 = 0.33$ rad/s and

$$\phi(t) = -3 + 0.33\,t$$

where t is measured in seconds. If we change ϕ to $\phi + 2\pi$, (x, y, z) is unchanged: P returns to its initial position. The path is periodic, of period $\boxed{T = 2\pi/\dot{\phi}_0} = \mathbf{19}$ seconds. Note that the path of P is necessarily a closed loop.

b. A plane passing through the origin takes the form $ax + by + cz = 0$ in Cartesian coordinates. This suggests to express $\cos\phi$ and $\sin\phi$ in terms of x and y and substitute these expressions in z. First we write

$$x = \sin\phi\cos 1 + \cos\phi\sin 1, \qquad y = \sin\phi\cos 2 + \cos\phi\sin 2$$

and solve for $(\sin\phi, \cos\phi)$:

$$\sin\phi = \frac{x\sin 2 - y\sin 1}{\sin 1}, \qquad \cos\phi = \frac{y\cos 1 - x\cos 2}{\sin 1}$$

Then we find z by substituting $\sin\phi$ and $\cos\phi$ in terms of x and y

$$z = \sin\phi\cos 3 + \cos\phi\sin 3 = \frac{x\sin 2 - y\sin 1}{\sin 1}\cos 3 + \frac{y\cos 1 - x\cos 2}{\sin 1}\sin 3$$

to find

$$\boxed{z\sin 1 = -x\sin 1 + y\sin 2} \qquad [2]$$

The trajectory of P is inscribed in the plane described by [2].

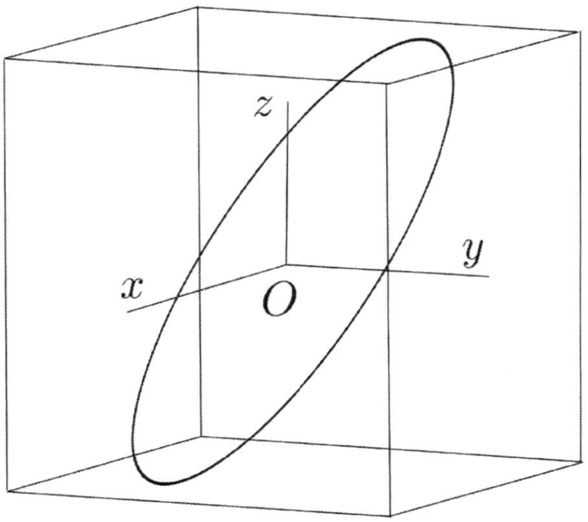

Figure 1.16 – *Problem 1.9 (Solution). The path is contained in the cube of center O and side length 2 m.*

c. The path of P is the intersection of the cylinder of axis Oz and elliptical cross-section described by

$$\sin^2\phi + \cos^2\phi = \frac{(x\sin 2 - y\sin 1)^2}{\sin^2 1} + \frac{(y\cos 1 - x\cos 2)^2}{\sin^2 1} = 1$$

and the plane described by [2]. The intersection is shown in Figure 1.16. It must be an ellipse! ■

Exercise 1.10 ■■■ A small "bouncy" ball is thrown from an inclined plane at initial speed v_0 and initial angle α_0. The ball then hits the surface with a perfect elastic bounce, that is, the ball's angle of incidence equals its angle of reflection at impact.

a. Show that the first bouncing trajectory corresponds to initial angle α_1 given by

$$\cot\alpha_1 = \cot\alpha_0 - 2\tan\theta$$

where θ is the angle of inclination of the plane with the horizontal.

b. Assuming that the ball exhibits k uphill bounces, find the initial angle α_k of the kth bounce. Then deduce the number N of uphill bounces of the ball before it bounces downhill. Find N for $\alpha_0 = \theta = 15°$. ■

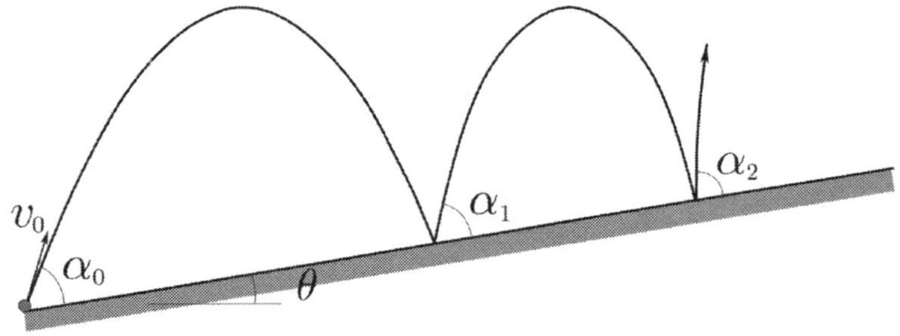

Figure 1.17 – *Problem 1.10*

Solution. **a.** We choose a Cartesian system Oxy with unit vector $\hat{\mathbf{e}}_x$ directed along the inclined plane. See Figure 1.18. It would be clumsy to choose axis Ox horizontal and axis Oy vertical since the angles $\alpha_0, \alpha_1, \ldots$ are defined relative to the incline. Then the acceleration of P can be written on $(\hat{\mathbf{e}}_x, \hat{\mathbf{e}}_y)$ as

$$\mathbf{a} = -g(\sin\theta\hat{\mathbf{e}}_x + \cos\theta\hat{\mathbf{e}}_y) = \ddot{x}\hat{\mathbf{e}}_x + \ddot{y}\hat{\mathbf{e}}_y$$

The ball's velocity can be obtained by integration of the equations $\ddot{x} = -g\sin\theta$ and $\ddot{y} = -g\cos\theta$:

$$\dot{x} = -gt\sin\theta + v_0\cos\alpha_0, \qquad \dot{y} = -gt\cos\theta + v_0\sin\alpha_0$$

using the initial conditions $\dot{x} = v_0\cos\alpha_0$ and $\dot{y} = v_0\sin\alpha_0$ at $t = 0$. One more integration leads to the ball's trajectory

$$x = -g\frac{t^2}{2}\sin\theta + v_0 t\cos\alpha_0, \qquad y = -g\frac{t^2}{2}\cos\theta + v_0 t\sin\alpha_0$$

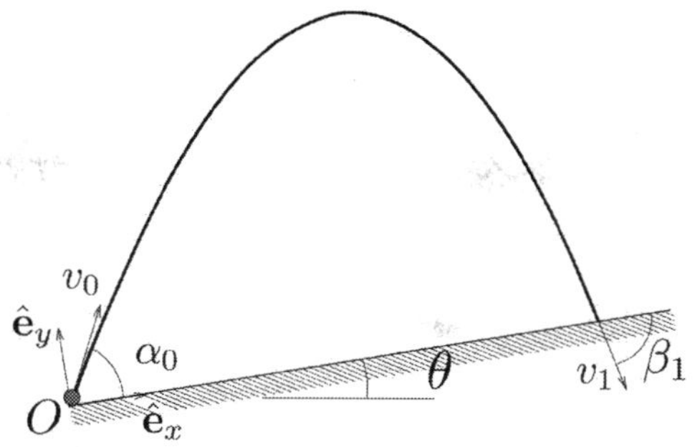

Figure 1.18 – *Problem 1.10 (Solution)*

The time of impact is obtained by setting $y = 0$:

$$t_1 = \frac{2v_0}{g} \frac{\sin\alpha_0}{\cos\theta}$$

The corresponding velocity is given by

$$\dot{x}_1 = -2v_0 \sin\alpha_0 \tan\theta + v_0 \cos\alpha_0, \qquad \dot{y}_1 = -v_0 \sin\alpha_0$$

from which we obtain the incident angle β_1 at impact (see Figure 1.18)

$$\cot\beta_1 = \frac{\dot{x}_1}{\dot{y}_1} = 2\tan\theta - \cot\alpha_0$$

The reflection angle $\alpha_1 = -\beta_1$ is thus given by

$$\boxed{\cot\alpha_1 = \cot\alpha_0 - 2\tan\theta} \qquad [1]$$

b. The initial angle α_2 of the second bounce can be obtained from equation [1] by replacing α_0 by α_1 since this formula is independent of the initial velocity v_0:

$$\cot\alpha_2 = \cot\alpha_1 - 2\tan\theta = \cot\alpha_0 - 4\tan\theta$$

For the kth bounce we easily get (by iteration)

$$\cot\alpha_k = \cot\alpha_0 - 2k\tan\theta \qquad [2]$$

As the ball further bounces uphill, the angle $\theta + \alpha_k$ approaches $\pi/2$ which corresponds to a vertical bounce. If $\theta + \alpha_k$ exceeds $\pi/2$, the ball bounces downhill. So to find the number N of uphill bounces, we impose $\alpha_k < \pi/2 - \theta$ or

$$\cot\alpha_k < \tan\theta$$

Using equation [2] we find the condition

$$\cot\alpha_0 < (2k+1)\tan\theta$$

The integer N is thus given by

$$\boxed{N = \left[\frac{1}{2}\left(\frac{\cot\alpha_0}{\tan\theta} - 1\right)\right]}$$

where $[x]$ denotes the integer part of real number x. For $\alpha_0 = \theta = 15°$ we find $N = \mathbf{6}$ bounces. ■

1.5 Cylindrical/Polar Coordinates

WE need to define a coordinate system better suited for simple or complex motions than the Cartesian coordinate system. The example of the circular path comes to mind: its Cartesian representation is not convenient to determine the particle's velocity and acceleration. Here we construct a new coordinate system as follows. Consider the Cartesian axes Ox, Oy and Oz of Figure 1.19. Given a point P, we define P' as its projection onto the Oxy plane and P'' its projection onto axis Oz. To define the position of P', we introduce the distance r from O to P' and the angle θ which line OP' makes with axis Ox. The quantities (r, θ) are called *polar* coordinates of P'. Finally the position of P'' is defined by (Cartesian) coordinate z. Point O is called *pole* of the coordinate system. Now we have

$$\mathbf{r}_{OP} = \mathbf{r}_{OP'} + \mathbf{r}_{OP''} = r\hat{\mathbf{e}}_r + z\hat{\mathbf{e}}_z$$

where $\hat{\mathbf{e}}_r$, the *radial unit vector*, defines the direction of vector $\mathbf{r}_{OP'}$. Note that angle θ is not explicit in this representation. However $\hat{\mathbf{e}}_r$ is a function of θ: $\hat{\mathbf{e}}_r = \cos\theta\hat{\mathbf{e}}_x + \sin\theta\hat{\mathbf{e}}_y$.

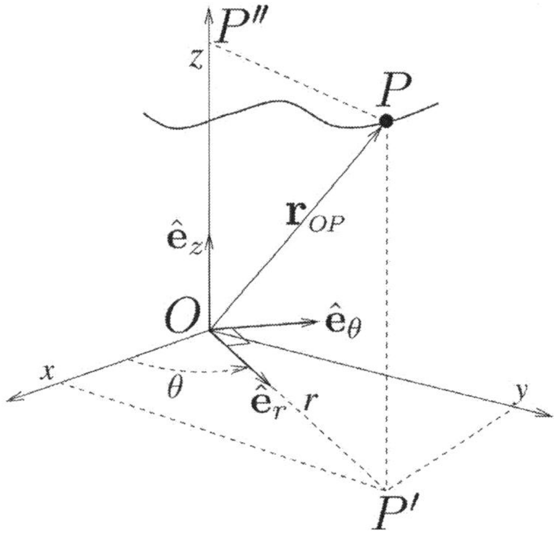

Figure 1.19 – *Cylindrical coordinates (r, θ, z) of P.*

DEFINITION (CYLINDRICAL COORDINATES)

The coordinates (r, θ, z) are called *cylindrical coordinates* of P. The position of P is then defined by

$$\mathbf{r}_{OP} = r\hat{\mathbf{e}}_r + z\hat{\mathbf{e}}_z \tag{1.13}$$

where the *radial unit vector* $\hat{\mathbf{e}}_r = \cos\theta\hat{\mathbf{e}}_x + \sin\theta\hat{\mathbf{e}}_y$ is defined by angle θ.

If P is in motion, then its cylindrical coordinates are functions of time and its velocity can be found by taking the time-derivative of $\mathbf{r}_{OP}$:

$$\mathbf{v}_P = \dot{r}\hat{\mathbf{e}}_r + r\frac{d\hat{\mathbf{e}}_r}{dt} + \dot{z}\hat{\mathbf{e}}_z$$

Keep in mind that the radial unit vector is a function of time. We need to find the expression of $d\hat{\mathbf{e}}_r/dt$:

$$\frac{d}{dt}\hat{\mathbf{e}}_r = \frac{d}{dt}(\cos\theta\hat{\mathbf{e}}_x + \sin\theta\hat{\mathbf{e}}_y) = \dot{\theta}(-\sin\theta\hat{\mathbf{e}}_x + \cos\theta\hat{\mathbf{e}}_y)$$

We recognize that the vector $(-\sin\theta\hat{\mathbf{e}}_x + \cos\theta\hat{\mathbf{e}}_y)$ is a unit vector, normal to $\hat{\mathbf{e}}_r$, and in the direction of increasing θ: we denote this unit vector $\hat{\mathbf{e}}_\theta$ and call it *transverse* unit vector. It is a function of time as $\hat{\mathbf{e}}_r$. This gives [2]

$$\frac{d}{dt}\hat{\mathbf{e}}_r = \dot{\theta}\hat{\mathbf{e}}_\theta \tag{1.14}$$

We can also show that

$$\frac{d}{dt}\hat{\mathbf{e}}_\theta = \frac{d\theta}{dt}\frac{d\hat{\mathbf{e}}_\theta}{d\theta} = \dot{\theta}(-\cos\theta\hat{\mathbf{e}}_x - \sin\theta\hat{\mathbf{e}}_y) = -\dot{\theta}\hat{\mathbf{e}}_r \tag{1.15}$$

The final expression of $\mathbf{v}_P$ in cylindrical coordinates is

$$\mathbf{v}_P = \dot{r}\hat{\mathbf{e}}_r + r\dot{\theta}\hat{\mathbf{e}}_\theta + \dot{z}\hat{\mathbf{e}}_z$$

Another time derivative gives the acceleration of P

$$\mathbf{a}_P = \ddot{r}\hat{\mathbf{e}}_r + \dot{r}\dot{\theta}\hat{\mathbf{e}}_\theta + (\dot{r}\dot{\theta} + r\ddot{\theta})\hat{\mathbf{e}}_\theta + r\dot{\theta}(-\dot{\theta}\hat{\mathbf{e}}_r) + \ddot{z}\hat{\mathbf{e}}_z$$

We collect the terms on each unit vector to find [3]

$$\mathbf{a}_P = (\ddot{r} - r\dot{\theta}^2)\hat{\mathbf{e}}_r + (2\dot{r}\dot{\theta} + r\ddot{\theta})\hat{\mathbf{e}}_\theta + \ddot{z}\hat{\mathbf{e}}_z$$

In conclusion,

Kinematics in Cylindrical Coordinates

The velocity and acceleration of a particle P of cylindrical coordinates (r, θ, z) are given by

$$\mathbf{v}_P = \dot{r}\hat{\mathbf{e}}_r + r\dot{\theta}\hat{\mathbf{e}}_\theta + \dot{z}\hat{\mathbf{e}}_z \tag{1.16}$$

$$\mathbf{a}_P = (\ddot{r} - r\dot{\theta}^2)\hat{\mathbf{e}}_r + (2\dot{r}\dot{\theta} + r\ddot{\theta})\hat{\mathbf{e}}_\theta + \ddot{z}\hat{\mathbf{e}}_z \tag{1.17}$$

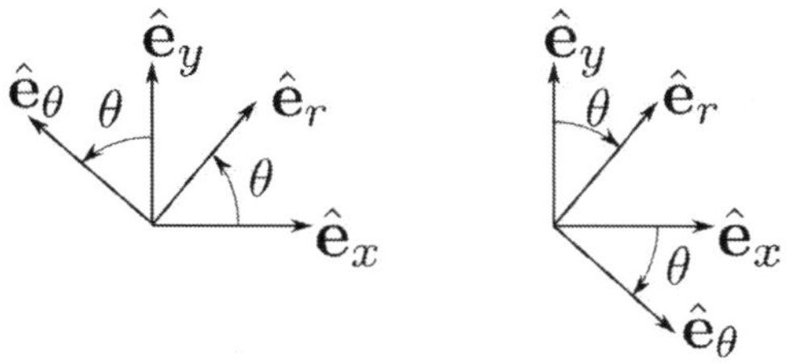

Figure 1.20 – *Two configurations of polar unit vectors.*

[2] A vector of constant magnitude is normal to its time-derivative: if $\mathbf{u}$ satisfies $\mathbf{u}\cdot\mathbf{u} = Cst$, then $\mathbf{u}\cdot\frac{d}{dt}\mathbf{u} = 0$.

[3] Note that the transverse component of $\mathbf{a}_P$ can be written in the form

$$a_\theta = \frac{1}{r}\frac{d}{dt}(r^2\dot{\theta})$$

This expression is key for the solution of central acceleration problems, that is, for which $\mathbf{a}_P$ points toward O at all time: then $a_\theta = 0$ implies $r^2\dot{\theta} =$ constant.

PROPERTIES

①. The key difference with the Cartesian coordinate system is that the position, velocity and acceleration vectors are expressed on time-dependent unit vectors (note though that $\hat{\mathbf{e}}_z$ is constant).

②. In practice, angle θ may be defined relative to axis Oy. See Figure 1.20. If the transverse unit vector $\hat{\mathbf{e}}_\theta$ is defined normal to $\hat{\mathbf{e}}_r$ in the direction of increasing θ, then formulas (1.16) and (1.17) remain unchanged.

③. If the path of P lies in plane Oxy, then set $z = \dot{z} = \ddot{z} = 0$ in (1.13,1.16,1.17). (r, θ) are the *polar coordinates* of P.

④. It is important to know how to derive (1.16) and (1.17): these two formulas hinge upon (1.14) and (1.15).

APPLICATION: CIRCULAR MOTION. The polar coordinate system is ideally suited for circular motions. Assume that a particle P stays at a constant distance $r = R$ from a fixed point O in a plane Oxy. Its path is a circle of center O and radius R. We can define its position vector as

$$\mathbf{r}_{OP} = R\hat{\mathbf{e}}_r$$

The quantities $(r = R, \theta)$ are the polar coordinates of P. We can simply differentiate $R\hat{\mathbf{e}}_r$ twice to obtain the velocity (using $d\hat{\mathbf{e}}_r/dt = \dot{\theta}\hat{\mathbf{e}}_\theta$)

$$\mathbf{v}_P = R\dot{\theta}\hat{\mathbf{e}}_\theta$$

(we verify that $\mathbf{v}_P$ is tangential to the path) and the acceleration of P (using $d\hat{\mathbf{e}}_\theta/dt = -\dot{\theta}\hat{\mathbf{e}}_r$)

$$\mathbf{a}_P = R\ddot{\theta}\hat{\mathbf{e}}_\theta - R\dot{\theta}^2\hat{\mathbf{e}}_r$$

We can also obtain these expressions from the general formulas (1.16) and (1.17) by imposing

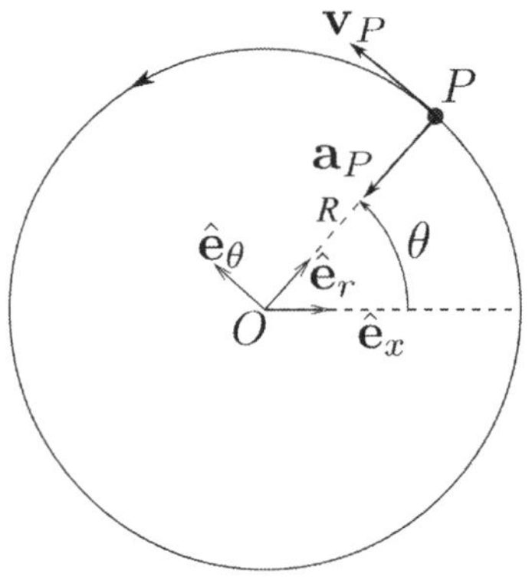

Figure 1.21 – *Circular motion: the velocity is tangent to the circle. Assuming angular velocity $\dot{\theta}$ constant, the acceleration is normal to the circle and points toward the center O.*

$\dot{r} = \ddot{r} = 0$. The speed of P is $v_P = R\dot{\theta} = R\omega$ (assuming $\dot{\theta} > 0$). The quantity $\dot{\theta} = \omega$ is called *angular velocity*. There is clearly no better way to represent the kinematics of P. Note that if P rotates at constant angular velocity ω, the speed of P stays constant (the motion is uniform) and the acceleration remains normal to the path of P: $\mathbf{a}_P = -\frac{v_P^2}{R}\hat{\mathbf{e}}_r$. See Figure 1.21. *This expression of the acceleration will prove to be valid for all uniform motion.*

EXAMPLE PROBLEMS

General Procedure: Kinematics In Polar/Cylindrical Coordinates

To solve a kinematics problem in polar/cylindrical coordinates, follow these guidelines:

① Define all fixed and moving unit vectors. The polar angle θ may be defined relative to axis Ox or Oy. In both cases, the transverse unit vector $\hat{\mathbf{e}}_\theta$ should always be defined in the direction of increasing θ so that formulas (1.14-1.17) remain valid.

② If the path of P is given, for instance in the form $\mathbf{r}_{OP} = r(\theta)\hat{\mathbf{e}}_r + z(\theta)\hat{\mathbf{e}}_z$, then the kinematics of P is conveniently obtained by direct time-differentiation and by using (1.14) and (1.15).

③ Problems are often solved by identifying all algebraic or geometric conditions imposed on the velocity or acceleration of P. These conditions must be expressed in terms of polar/cylindrical components of these vectors.

④ As an illustration of ③, the acceleration of P may be assumed *central*, that is, pointing toward a fixed point O: in this case, we must impose the transverse acceleration $a_\theta = 0$ (choosing O as the pole of the coordinate system).

⑤ The most challenging problems involve finding the trajectory in the parametric form $(r(t), \theta(t), z(t))$ by integration of kinematic equations. This is most often achieved by separation of the variables.

Exercise 1.11 ▪Ant A walks at constant speed v_0 along the second hand of length $R = 30$ cm of a clock. At $t = 0$, the ant is at the center of the clock which indicates 0 second. The ant arrives at the tip of the hand after 60 seconds.
a. Find and sketch the path of A.
b. Find and sketch the velocity of A at $t = 0, 15, 30, 45$ and 60 seconds.
c. Repeat b) for the acceleration of A. ▪

Solution. **a.** To parametrize the position of A we use the polar coordinates (r, θ) as shown in Figure 1.22. The radial unit vector is directed along the second hand. Note that angle θ defines the orientation of line OA relative to axis Oy in the clockwise direction in view of the motion of A.
The motion of A is then characterized by:
(i) a constant rate $\dot{r} = v_0$ in the radial direction, corresponding to the rectilinear motion of A relative to the second hand,
(ii) a constant angular rate $\dot{\theta} = \omega$ corresponding to one revolution per minute of the second hand.
We can integrate these two equations to obtain a (polar) parametric representation of the path of A with the initial conditions $r = 0$ and $\theta = 0$ at $t = 0$:

$$\boxed{r = v_0 t, \qquad \theta = \omega t} \qquad [1]$$

We then express r versus θ by eliminating the time variable

$$\boxed{r = \frac{v_0}{\omega}\theta} \qquad [2]$$

This is the equation of a *spiral* (see Figure 1.23).

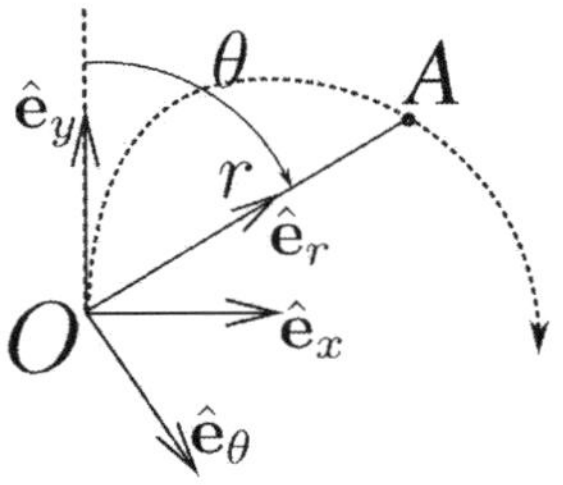

Figure 1.22 – *Problem 1.11 (Solution)*

b. The velocity is obtained by taking the time-derivative of the position vector $\mathbf{r}_{OA} = v_0 t\hat{\mathbf{e}}_r$ using

$$\frac{d\hat{\mathbf{e}}_r}{dt} = \dot{\theta}\hat{\mathbf{e}}_\theta = \omega\hat{\mathbf{e}}_\theta, \qquad \frac{d\hat{\mathbf{e}}_\theta}{dt} = -\dot{\theta}\hat{\mathbf{e}}_r = -\omega\hat{\mathbf{e}}_r$$

We first obtain the velocity

$$\boxed{\mathbf{v}_A = v_0(\hat{\mathbf{e}}_r + \omega t\hat{\mathbf{e}}_\theta)} \qquad [3]$$

The constant speed v_0 is found by imposing $r = R$ for $\theta = 2\pi$ (one revolution): this gives $v_0 = R\omega/2\pi$. The velocities of A at $t = 0, 15, 30, 45$ and 60 seconds are then given by

$$t = 0 \text{ s} \quad (r,\theta) = (0,0), \quad \mathbf{v}_A = v_0\hat{\mathbf{e}}_r = v_0\hat{\mathbf{e}}_y$$
$$t = 15 \text{ s} \quad (r,\theta) = (R/4, \pi/2), \quad \mathbf{v}_A = v_0(\hat{\mathbf{e}}_r + \frac{\pi}{2}\hat{\mathbf{e}}_\theta)$$
$$t = 30 \text{ s} \quad (r,\theta) = (R/2, \pi), \quad \mathbf{v}_A = v_0(\hat{\mathbf{e}}_r + \pi\hat{\mathbf{e}}_\theta)$$
$$t = 45 \text{ s} \quad (r,\theta) = (3R/4, 3\pi/2), \quad \mathbf{v}_A = v_0(\hat{\mathbf{e}}_r + \frac{3\pi}{2}\hat{\mathbf{e}}_\theta)$$
$$t = 60 \text{ s} \quad (r,\theta) = (R, 2\pi), \quad \mathbf{v}_A = v_0(\hat{\mathbf{e}}_r + 2\pi\hat{\mathbf{e}}_\theta)$$

with $R = 0.3$ m, $\omega = 0.105$ rad/s, $v_0 = \mathbf{5}$ mm/s. The velocity vectors are sketched in Figure 1.23.

c. The acceleration of A is obtained by differentiation of [3]:

$$\mathbf{a}_A = v_0(\dot{\theta}\hat{\mathbf{e}}_\theta + \omega\hat{\mathbf{e}}_\theta - \omega^2 t\hat{\mathbf{e}}_r) = \boxed{\omega v_0(2\hat{\mathbf{e}}_\theta - \theta\hat{\mathbf{e}}_r) = \mathbf{a}_A} \qquad [4]$$

The accelerations at time $t = 0, 15, 30, 45$ and 60 seconds are obtained by substitution of $\theta = 0, \pi/2, \pi, 3\pi/2$ and 2π in [4] with $\omega v_0 = 0.525$ mm/s^2. Note that the acceleration is always pointing toward the interior of the path.

■

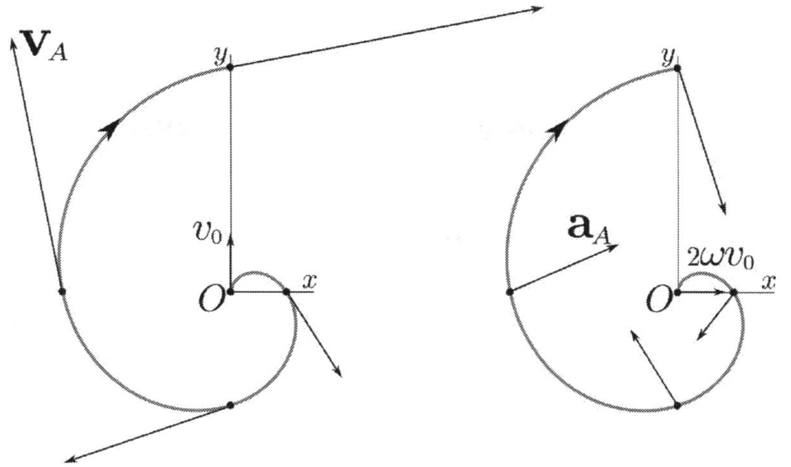

Figure 1.23 – *Problem 1.11 (Solution): trajectory of A with velocity and acceleration vectors at time $t = 0, 15, 30, 45$ and 60 seconds.*

Exercise 1.12 ▪A radar located at O tracks the flight of an airplane P flying at the constant height 300 m. At the instant shown, $\theta = 45°$ and the radar's rotation is characterized by $\dot{\theta} = 0.4$ rad/s and $\ddot{\theta} = -0.32$ rad/s^2.
Find the velocity and acceleration of P in polar coordinates.
a. Method 1: express the position vector in polar coordinates and take the time-derivative twice.
b. Method 2: use the fact that the motion of P is rectilinear and project $\mathbf{v}_P$ and $\mathbf{a}_P$ on the polar unit vectors. ■

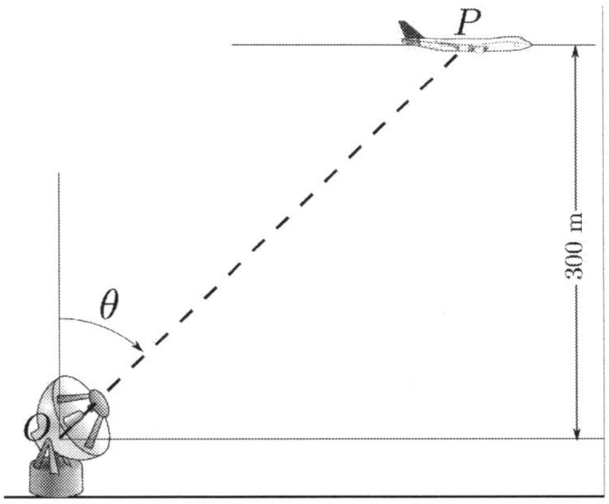

SOLUTION. First we sketch the polar unit vector associated with polar angle θ. See Figure 1.24. At any instant, the position of P is related to angle θ according to

$$r = \frac{h}{\cos\theta}, \qquad x = h\tan\theta$$

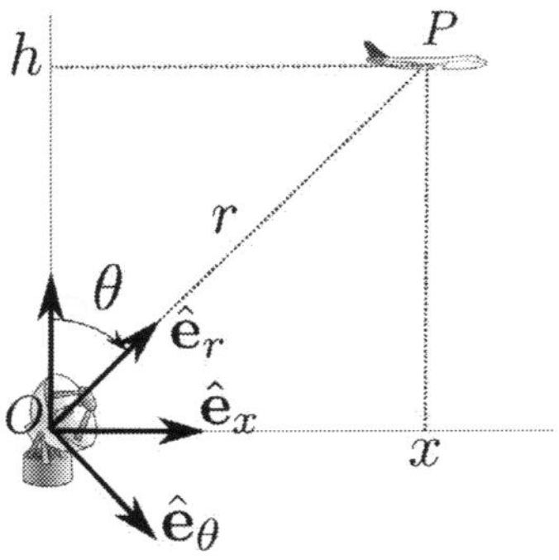

Figure 1.24 – *Problem 1.12 (Solution).*

a. Method 1: we start with the position vector $\mathbf{r}_{OP} = r\hat{\mathbf{e}}_r = \frac{h}{\cos\theta}\hat{\mathbf{e}}_r$ expressed in polar coordinates and differentiate w.r.t. time to obtain

$$\boxed{\mathbf{v}_P = \frac{h\dot\theta}{\cos^2\theta}(\sin\theta\hat{\mathbf{e}}_r + \cos\theta\hat{\mathbf{e}}_\theta)} = 169.7(\hat{\mathbf{e}}_r + \hat{\mathbf{e}}_\theta) \quad (\text{m/s})$$

where we used $\dot r = h\dot\theta\sin\theta/\cos^2\theta$ and $d\hat{\mathbf{e}}_r/dt = \dot\theta\hat{\mathbf{e}}_\theta$. The acceleration is obtained by taking the time-derivative of $\mathbf{v}_P$ (we use $d\hat{\mathbf{e}}_\theta/dt = -\dot\theta\hat{\mathbf{e}}_r$):

$$\mathbf{a}_P = \frac{h\ddot\theta}{\cos^2\theta}(\sin\theta\hat{\mathbf{e}}_r + \cos\theta\hat{\mathbf{e}}_\theta) + \frac{h\dot\theta}{\cos^2\theta}(\dot\theta\sin\theta\hat{\mathbf{e}}_\theta - \dot\theta\cos\theta\hat{\mathbf{e}}_r) + h\dot\theta^2(\frac{\cos^2\theta + 2\sin^2\theta}{\cos^3\theta}\hat{\mathbf{e}}_r + \frac{\sin\theta}{\cos^2\theta}\hat{\mathbf{e}}_r)$$

After simplifications, we obtain

$$\boxed{\mathbf{a}_P = (\frac{h\ddot\theta}{\cos^2\theta} + \frac{2h\dot\theta^2\sin\theta}{\cos^3\theta})(\sin\theta\hat{\mathbf{e}}_r + \cos\theta\hat{\mathbf{e}}_\theta)} = 0 \quad (\text{m/s}^2)$$

b. Method 2: we use the fact that the motion of P is rectilinear and express the position vector in Cartesian coordinates

$$\mathbf{r}_{OP} = x\hat{\mathbf{e}}_x + y\hat{\mathbf{e}}_y = h\tan\theta\hat{\mathbf{e}}_x + h\hat{\mathbf{e}}_y$$

Then we take the time-derivative twice to obtain

$$\mathbf{v}_P = \dot x\hat{\mathbf{e}}_x = \frac{h\dot\theta}{\cos^2\theta}\hat{\mathbf{e}}_x$$

and

$$\mathbf{a}_P = \ddot x\hat{\mathbf{e}}_x = (\frac{h\ddot\theta}{\cos^2\theta} + \frac{2h\dot\theta^2\sin\theta}{\cos^3\theta})\hat{\mathbf{e}}_x$$

To obtain $\mathbf{v}_P$ and $\mathbf{a}_P$ in polar coordinates, it remains to express $\hat{\mathbf{e}}_x$ on the polar basis $(\hat{\mathbf{e}}_r, \hat{\mathbf{e}}_\theta)$:

$$\hat{\mathbf{e}}_x = \sin\theta\hat{\mathbf{e}}_r + \cos\theta\hat{\mathbf{e}}_\theta$$

We the recover the same expressions of $\mathbf{v}_P$ and $\mathbf{a}_P$ of method 1. ■

Exercise 1.13 ▪A rod OA is in rotation at constant angular velocity $\dot\theta = 0.7$ rad/s about fixed point O. A second rod BP is in motion such that
(i) end point P is in contact with rod OA,
(ii) line BP is perpendicular to line OA,
(iii) rod BP passes through a fixed point Q located at a distance $h = 0.6$ m above O.
Find the velocity and acceleration of P when $\theta = 30°$. Sketch these vectors at this instant. ■

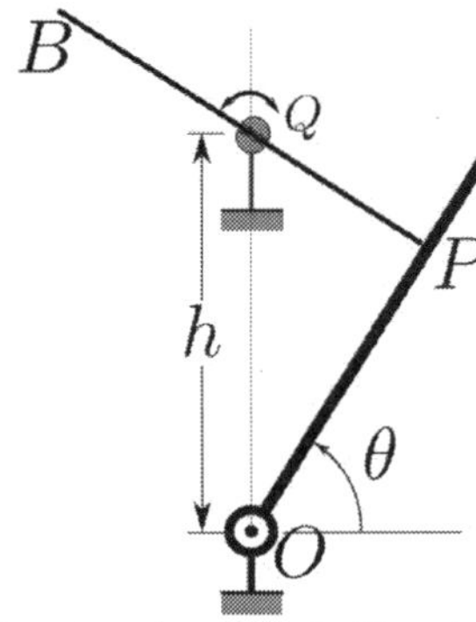

SOLUTION. We define the position of P by its polar coordinates (r, θ) relative to pole O. See Figure 1.25. At any given time, the quantities h, r and θ are related by $r = h\sin\theta$. This gives the position vector $\mathbf{r}_{OP} = r\hat{\mathbf{e}}_r = h\sin\theta\hat{\mathbf{e}}_r$ on the corresponding polar basis $(\hat{\mathbf{e}}_r, \hat{\mathbf{e}}_\theta)$. By time-differentiation, we immediately obtain the velocity of P:

$$\mathbf{v}_P = \frac{d}{dt}(h\sin\theta\hat{\mathbf{e}}_r) = \boxed{h\dot\theta(\cos\theta\hat{\mathbf{e}}_r + \sin\theta\hat{\mathbf{e}}_\theta) = \mathbf{v}_P}$$

Numerically, the velocity has magnitude $v_P = h\dot\theta = \mathbf{0.42}$ m/s and makes the angle

$\theta = 30^\circ$ relative to radial line $(O, \hat{\mathbf{e}}_r)$.

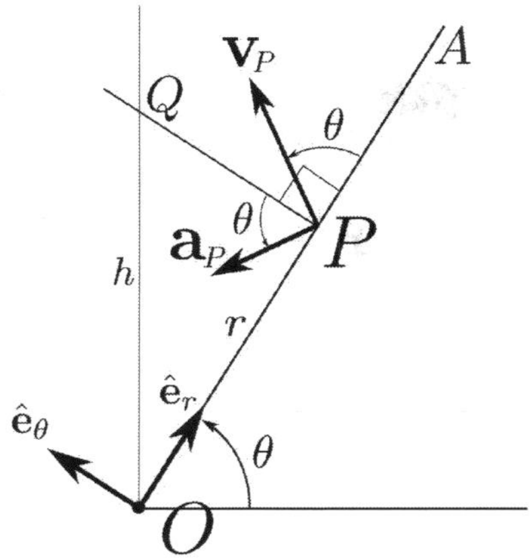

Figure 1.25 – *Problem 1.13 (Solution)*

Another time-differentiation gives the acceleration of P:

$$\mathbf{a}_P = \frac{d\mathbf{v}_P}{dt} = h\dot{\theta}(-\dot{\theta}\sin\theta\hat{\mathbf{e}}_r + \dot{\theta}\cos\theta\hat{\mathbf{e}}_\theta) + h\dot{\theta}(\dot{\theta}\cos\theta\hat{\mathbf{e}}_\theta - \dot{\theta}\sin\theta\hat{\mathbf{e}}_r)$$

which gives

$$\boxed{\mathbf{a}_P = 2h\dot{\theta}^2(-\sin\theta\hat{\mathbf{e}}_r + \cos\theta\hat{\mathbf{e}}_\theta)}$$

Numerically, the acceleration has magnitude $a_P = 2h\dot{\theta}^2 = \mathbf{0.588}$ m/s^2 and makes the angle $\theta = 30^\circ$ relative to transverse line $(O, \hat{\mathbf{e}}_\theta)$. ■

Exercise 1.14 ▪▪A satellite S is launched from Earth to reach an elliptic trajectory. In the polar coordinates (r, θ) of pole O (Earth's center), its trajectory takes the expression

$$r = \frac{p}{1 + e\cos\theta}$$

where e $(0 \leq e < 1)$ and p are constants. At the position of point D, the satellite's speed is 14,000 km/h.

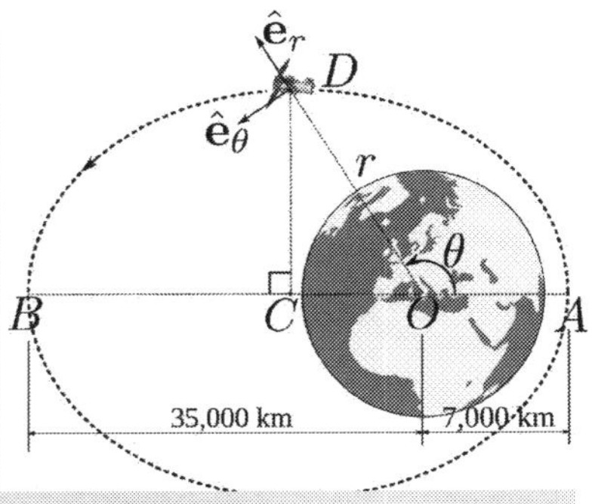

Furthermore, the acceleration of S is entirely radial, that is, $a_\theta = \mathbf{a}_S \cdot \hat{\mathbf{e}}_\theta = 0$ around the path (since the force of attraction due to Earth's gravity always points toward O).
a. Find the values of constant parameters e and p.
b. Show that $a_\theta = 0$ implies that the quantity $r^2\dot{\theta} = Cst$ where Cst is a constant to be found.
c. Conclude that the motion of S cannot be uniform. Find the speed of S at the positions $\theta = 0$ and $\theta = \pi$. ■

Solution. **a.** We find constants e and p by imposing the given values of r at A $(\theta = 0)$ and B $(\theta = \pi)$ (see Figure 1.26):

$$r_A = \frac{p}{1+e} \quad (\theta = 0), \quad r_B = \frac{p}{1-e} \quad (\theta = \pi)$$

Solving for e and p gives [4]

$$\boxed{e = \frac{r_B - r_A}{r_B + r_A}, \qquad p = \frac{2r_A r_B}{r_A + r_B}} \qquad [1-2]$$

With $r_B = 35,000$ km and $r_A = 7,000$ km, we find $e = \mathbf{0.667}$ and $p = \mathbf{11,667}$ km.

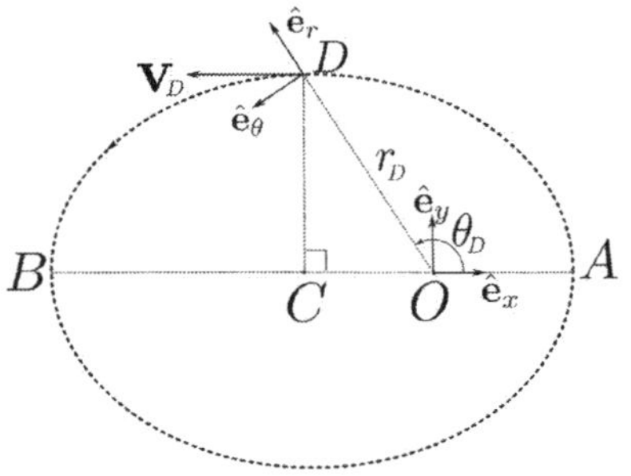

Figure 1.26 – *Problem 1.14 (Solution)*

b. If the acceleration points toward O at all time, then $a_\theta = 2\dot{r}\dot{\theta} + r\ddot{\theta} = 0$. However we know that $a_\theta = \frac{1}{r}\frac{d}{dt}(r^2\dot{\theta})$: hence $a_\theta = 0$ implies that the quantity $r^2\dot{\theta}$ remains constant along the path:

$$\boxed{r^2\dot{\theta} = Cst} \qquad [2]$$

To find constant Cst we must use the knowledge of velocity v_D (in both magnitude and direction) at the position D shown in Figure 1.26. But we are not given the coordinates (r, θ) at this location. We must use the fact that $\mathbf{v}_D = -v_D\hat{\mathbf{e}}_x$. First we derive the general expression of the velocity vector by taking into account that the path is given by the polar representation $r = p/(1 + e\cos\theta)$:

$$\mathbf{v} = \dot{r}\hat{\mathbf{e}}_r + r\dot{\theta}\hat{\mathbf{e}}_\theta = \frac{e}{p}r^2\dot{\theta}\sin\theta\hat{\mathbf{e}}_r + r\dot{\theta}\hat{\mathbf{e}}_\theta$$

To obtain this result, we differentiate $1/r = (1 + e\cos\theta)/p$ to find $\dot{r} = (e/p)r^2\dot{\theta}\sin\theta$. Then we consider the position D and write

$$\mathbf{v}_D = v_D(-\cos\theta_D\hat{\mathbf{e}}_r + \sin\theta_D\hat{\mathbf{e}}_\theta) = \frac{e}{p}r_D^2\dot{\theta}_D\sin\theta_D\hat{\mathbf{e}}_r + r_D\dot{\theta}_D\hat{\mathbf{e}}_\theta$$

This gives two equations

$$-v_D\cos\theta_D = \frac{e}{p}r_D^2\dot{\theta}_D\sin\theta_D, \qquad v_D\sin\theta_D = r_D\dot{\theta}_D \qquad [3-4]$$

We eliminate $\dot{\theta}_D$ by taking the ratio of [3] and [4] to find

$$-\cos\theta_D(1 + e\cos\theta_D) = e\sin^2\theta_D$$

after using $1/r_D = (1 + e\cos\theta_D)/p$. This gives, at the position D,

$$\boxed{\cos\theta_D = -e, \qquad r_D = \frac{p}{1 - e^2}} \qquad [5-6]$$

We also find, using the results of [1-2], that $r_D = (r_A + r_B)/2$. Now we can find

[4]The non-dimensional parameter e is known as the *eccentricity* of the ellipse: the larger e, the larger the distance from the ellipse center to the focus O.

constant Cst:

$$Cst = r_D^2 \dot\theta_D = r_D v_D \sin\theta_D = \frac{p}{\sqrt{1-e^2}} v_D \qquad [7]$$

With $v_D = 14,000$ km/h, we find $Cst = \mathbf{60,897}$ km^2/s.
c. The motion of S cannot possibly be uniform: for instance when $\dot r = 0$ (at position A or B) we have $v = r\dot\theta = Cst/r$ (using the constraint $r^2\dot\theta = Cst$) giving

$$v_A = \frac{Cst}{r_A} = \frac{p}{r_A\sqrt{1-e^2}} v_D, \qquad v_B = \frac{Cst}{r_B} = \frac{p}{r_B\sqrt{1-e^2}} v_D$$

With $v_D = 14,000$ km/h we find $v_A = \mathbf{2.23}\, v_D$ and $v_B = \mathbf{0.447}\, v_D$. ■

Exercise 1.15 ▪A particle P moves in the plane Oxy along a curve described in polar coordinates by

$$r = r_0 \exp(\theta), \qquad \theta = \omega t$$

where t represents time, and r_0 and ω are constants.
a. Find the velocity $\mathbf{v}_P$ in polar coordinates. Is the motion uniform? Show that the angle between the velocity of P and the radial line is constant. Sketch the trajectory of P.
b. Show that the polar coordinates of acceleration $\mathbf{a}_P$ are given by $a_r = 0$ and $a_\theta = 2\omega^2 r$. ■

SOLUTION. **a.** We start with $\mathbf{r}_{OP} = r\hat{\mathbf{e}}_r$ and use $\dot r = r_0 e^\theta \dot\theta = r\omega$ and $d\hat{\mathbf{e}}_r/dt = \omega\hat{\mathbf{e}}_\theta$ to obtain the velocity of P

$$\mathbf{v}_P = r\omega\hat{\mathbf{e}}_r + r(\omega\hat{\mathbf{e}}_\theta) = \boxed{\omega r_0 e^\theta(\hat{\mathbf{e}}_r + \hat{\mathbf{e}}_\theta) = \mathbf{v}_P} \qquad [1]$$

The motion is not uniform since the speed $v_P = \sqrt{2}\omega r_0 e^\theta$ varies along the path. We notice that $v_r = v_\theta$ at all time so that the angle ψ (see Figure 1.27, left) which $\mathbf{v}_P$ makes with the radial unit vector $\hat{\mathbf{e}}_r$ satisfies

$$\tan\psi = \frac{v_\theta}{v_r} = 1$$

leading to $\psi = 45°$ at all time. This is confirmed by plotting the trajectory of P in Figure 1.27.

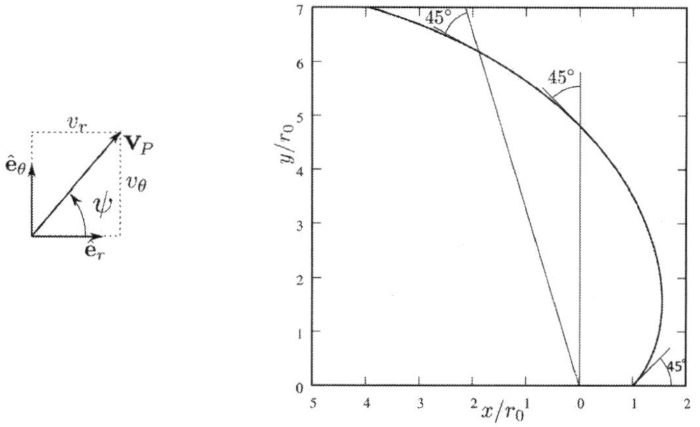

Figure 1.27 – *Problem 1.15 (Solution).*

b. We take the time derivative of [1] using $d\hat{\mathbf{e}}_\theta/dt = -\omega\hat{\mathbf{e}}_r$ to find

$$\mathbf{a}_P = \frac{d}{dt}(r\omega(\hat{\mathbf{e}}_r + \hat{\mathbf{e}}_\theta)) = \omega^2 r(\hat{\mathbf{e}}_r + \hat{\mathbf{e}}_\theta) + \omega r(\omega\hat{\mathbf{e}}_\theta - \omega\hat{\mathbf{e}}_r) = \boxed{2\omega^2 r\hat{\mathbf{e}}_\theta = \mathbf{a}_P}$$

■

Exercise 1.16 ■■ The motion of a particle P in the plane Oxy satisfies the following conditions:
(i) At time $t = 0$, P is at point A of axis Ox at a distance r_0 from O, and its initial velocity $\mathbf{v}_0$ makes the angle α ($0 < \alpha < \pi$) with axis Ox as shown.
(ii) At any time t, the speed of P is inversely proportional to the distance from O to P: $v_P = k/r$, (k is a constant).
(iii) The acceleration of P always points toward O.
Find the position of P at time $t = 2$ seconds for $r_0 = 0.8$ m, $v_0 = 0.55$ m/s, and $\alpha = \pi/4$. ■

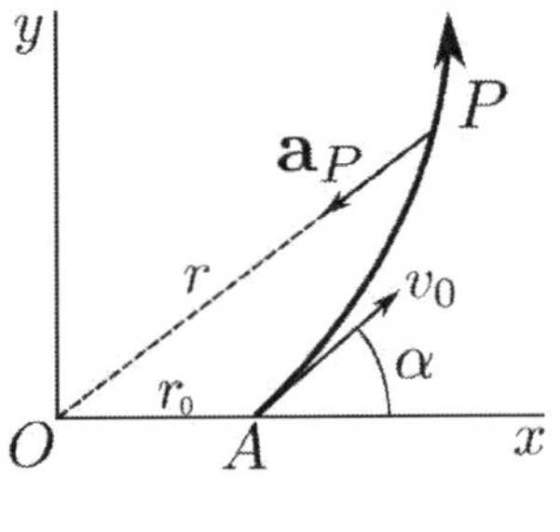

SOLUTION. We will find the trajectory of P in polar coordinates (r, θ). This coordinate system is ideally suited to impose the condition that the acceleration of P always points toward O: indeed this can be stated by writing $\mathbf{a}_P = a_r\hat{\mathbf{e}}_r$, that is, $a_\theta = \frac{1}{r}\frac{d}{dt}(r^2\dot{\theta}) = 0$. This shows that condition (iii) leads to

$$r^2\dot{\theta} = \text{constant} = r_0^2\dot{\theta}_0 \qquad [1]$$

where we have imposed $r = r_0$ (given) and $\dot{\theta} = \dot{\theta}_0$ at time $t = 0$. Since $\dot{\theta}_0$ is not given, we must find its value from the initial velocity: at $t = 0$ we have

$$\mathbf{v}_P = v_0(\cos\alpha\hat{\mathbf{e}}_x + \sin\alpha\hat{\mathbf{e}}_y) = \dot{r}_0\hat{\mathbf{e}}_x + r_0\dot{\theta}_0\hat{\mathbf{e}}_y$$

showing that

$$\dot{r}_0 = v_0\cos\alpha, \qquad r_0\dot{\theta}_0 = v_0\sin\alpha \qquad [2-3]$$

Next we impose condition (ii): $\mathbf{v}_P^2 = k^2/r^2 = \dot{r}^2 + r^2\dot{\theta}^2$ or using [1] and [3]

$$\dot{r}^2 + \frac{r_0^2 v_0^2\sin^2\alpha}{r^2} = \frac{k^2}{r^2}$$

After simplifications, we find that the quantity $r\dot{r}$ remains constant along the trajectory:

$$r\dot{r} = r_0\dot{r}_0 = r_0 v_0\cos\alpha$$

We can integrate this equation by separation of the variables $rdr = (r_0 v_0\cos\alpha)dt$ and imposing $r = r_0$ at $t = 0$:

$$\boxed{r^2 - r_0^2 = 2(r_0 v_0\cos\alpha)t} \qquad [4]$$

Now that $r(t)$ is found, we can return to [1] to find $\theta(t)$: first we find

$$\dot{\theta} = \frac{r_0 v_0\sin\alpha}{r^2} = \frac{r_0 v_0\sin\alpha}{r_0^2 + 2r_0 v_0 t\cos\alpha}$$

which is integrated by separation of the variables θ and t and by imposing $\theta = 0$ at $t = 0$:

$$\boxed{\theta = \frac{1}{2}\tan\alpha\ln\left(1 + \frac{2v_0\cos\alpha}{r_0}t\right)} \qquad [5]$$

Equations [4-5] define the position of P as a function of time. It is also possible to express r versus θ by eliminating the time variable

$$\boxed{r = r_0 e^{\theta \cot\alpha}}$$

This is the equation of a *logarithmic spiral*. For $r_0 = 0.8$ m, $v_0 = 0.55$ m/s, and $\alpha = \pi/4$, we find $(r, \theta) = (\mathbf{1.37}\ \mathrm{m}, \mathbf{31}^\circ)$ at $t = 2$ seconds. ■

Exercise 1.17 ■■A particle is motion in the interior of a cone of vertical axis Oz and half-angle α. Its position is defined in cylindrical coordinates (r, θ, z) relative to pole O. At $t = 0$, the position is given by $r = R$ and $\theta = 0$ and its velocity is $v_0 \hat{\mathbf{e}}_y$. The motion is characterized by a constant vertical component $a_z = -a_0 < 0$ of acceleration and a vanishing transverse acceleration $a_\theta = 0$.
a. Find the position of P as a function of time.
b. Find the terminal speed of P and the time taken to reach O for $a_0 = 0.65$ m/s², $\alpha = 30^\circ$, and $R = 1.2$ m. ■

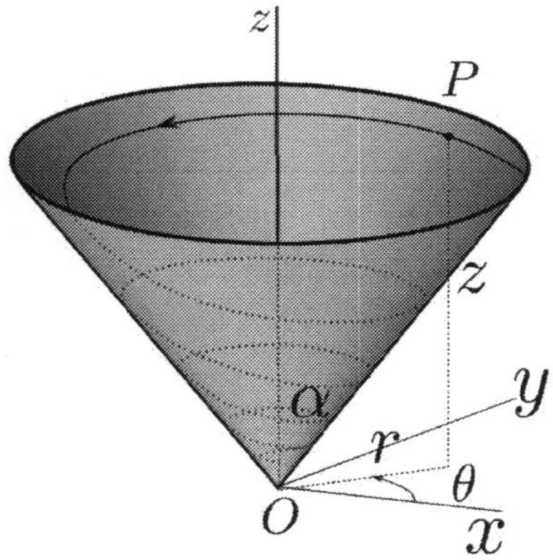

SOLUTION. **a.** First we note that the surface of the cone is characterized by $z = r\cot\alpha$. At $t = 0$, the value $r = R$ is imposed, giving $z = R\cot\alpha$ at $t = 0$. See Figure 1.28. In cylindrical coordinates, the velocity of P takes the expression

$$\mathbf{v}_P = \dot{r}\hat{\mathbf{e}}_r + r\dot{\theta}\hat{\mathbf{e}}_\theta + \dot{z}\hat{\mathbf{e}}_z$$

which becomes

$$\mathbf{v}_P = \dot{r}_0\hat{\mathbf{e}}_x + R\dot{\theta}_0\hat{\mathbf{e}}_y + \dot{z}_0\hat{\mathbf{e}}_z$$

at time $t = 0$. Since $\mathbf{v}_P = v_0\hat{\mathbf{e}}_y$, we find the initial conditions

$$\dot{r}_0 = \dot{z}_0 = 0, \qquad R\dot{\theta}_0 = v_0 \qquad [1-2]$$

Note that the conditions $\dot{r}_0 = \dot{z}_0 = 0$ are compatible with the geometric constraint $z = r\cot\alpha$ which necessarily implies $\dot{z} = \dot{r}\cot\alpha$ at all time.

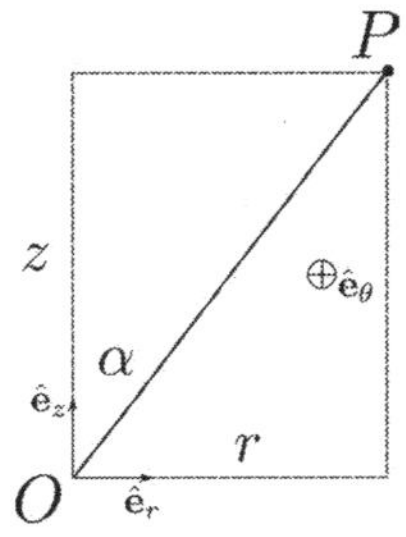

Figure 1.28 – *Problem 1.17.*

Next we impose the condition $a_z = \ddot{z} = -a_0$: with a_0 constant, we easily obtain z versus t by integrating twice

$$\boxed{z = R\cot\alpha - a_0\frac{t^2}{2}} \qquad [3]$$

where we have imposed $\dot{z} = 0$ and $z = R\cot\alpha$ at $t = 0$. This also gives r vs t:

$$\boxed{r = R - (a_0\tan\alpha)\frac{t^2}{2}} \qquad [4]$$

Finally we impose the condition $a_\theta = 0$: this leads to the condition $r^2\dot{\theta} =$ constant, that is,

$$r^2\dot{\theta} = Rv_0$$

We note that as $r \to 0$, $\dot{\theta} \to \infty$: P will spiral around O increasingly fast as it approaches O. Since r versus t is known, we can obtain angle θ versus t by integration:

$$\boxed{\theta = \frac{v_0}{R}\int_0^t \frac{dt}{\left(1 - \frac{a_0 \tan\alpha}{2R}t^2\right)^2} = \frac{v_0}{4kR}\left(\ln\frac{1+kt}{1-kt} + \frac{2kt}{1-k^2t^2}\right)} \qquad [5]$$

with $k^2 = \frac{a_0 \tan\alpha}{2R}$. Equations [3-5] determine the position of P vs time.
b. The time T taken by P to reach O (at infinite speed) is obtained by setting $z(T) = 0$ (or $r(T) = 0$): we find from [3]

$$\boxed{T = \sqrt{\frac{2R\cot\alpha}{a_0}}}$$

For $a_0 = 0.65\ \text{m/s}^2$, $\alpha = 30°$, and $R = 1.2$ m, $T = \mathbf{2.5}$ s is found.

■

1.6 Intrinsic Coordinates

Thus far, we have defined two coordinate systems to express the position vector $\mathbf{r}_{OP}$, velocity $\mathbf{v}_P$ and acceleration $\mathbf{a}_P$ of a particle P in motion. Other coordinate systems can be defined (the spherical coordinate system comes to mind). The drawback of these systems is that they are somewhat arbitrary [5] and provide little insight into the relationship between the path and the kinematics of the particle. The coordinate system we describe in this section remedies this limitation. It relates the kinematics to the intrinsic properties of the curve (that is, properties which depend solely upon the shape of the curve) along which the particle is traveling [6]. This coordinate system is said to be *intrinsic*. We limit this description to *planar* motions, and provide a generalization in Section 1.7. As will be seen, the components of $\mathbf{v}_P$ and $\mathbf{a}_P$ are determined on tangential and normal directions at P. They are referred to as the *NT components*.

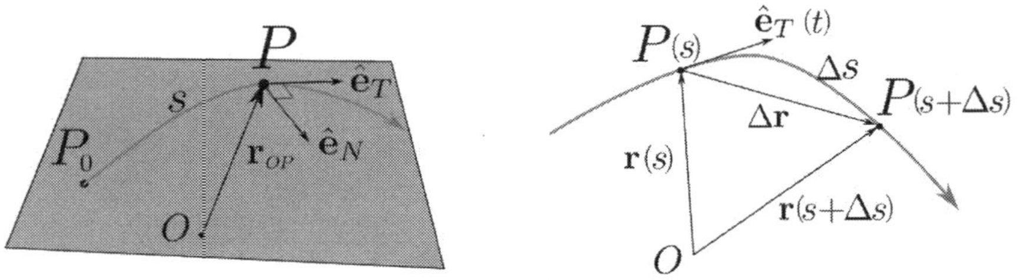

Figure 1.29 – *The position of particle P is defined by intrinsic coordinate s.*

[5]The exception is the polar coordinate system for circular motions.
[6]This description of kinematics is intimately connected to the *differential geometry* of curves.

Step 1. Intrinsic Expression of the Velocity

Assume that a particle P moves in a plane Oxy. See Figure 1.29. We orient the path of P in the sense of the motion and measure the *arc-length* s from a particular position P_0: the function $s(t)$ is an increasing function of time. We can now view the position vector of P as a function of s. We first examine the variation of $\mathbf{r}_{OP}$ with s.

Definition (Tangent Unit Vector)

The vector $d\mathbf{r}_{OP}/ds$ is a unit vector, tangent to the path, pointing in the sense of motion. This vector is denoted $\hat{\mathbf{e}}_T$ and is called *tangent unit vector* at P:

$$\hat{\mathbf{e}}_T = \frac{d\mathbf{r}_{OP}}{ds} \tag{1.18}$$

Proof. The properties of $\hat{\mathbf{e}}_T$ can be seen by using the expression

$$\hat{\mathbf{e}}_T = \lim_{\Delta s \to 0} \frac{\Delta \mathbf{r}}{\Delta s}$$

denoting $\Delta \mathbf{r} = \mathbf{r}_{OP}(s + \Delta s) - \mathbf{r}_{OP}(s)$. It is clear that, as $\Delta s \to 0$, the vector $\Delta \mathbf{r}$ becomes tangent to the path and points in the sense of motion. We also see that the magnitude of $\Delta \mathbf{r}/\Delta s$ approaches $\Delta s/\Delta s = 1$. See Figure 1.29. ∎

We can now express the velocity of P in the following form

$$\mathbf{v}_P = \frac{d\mathbf{r}_{OP}}{dt} = \frac{ds}{dt}\frac{d\mathbf{r}_{OP}}{ds}$$

or

$$\mathbf{v}_P = \dot{s}\hat{\mathbf{e}}_T$$

Note that $\dot{s} > 0$ is the speed v_P of P. As expected, we find that $\mathbf{v}_P$ is directed along the tangent direction.

Step 2. Intrinsic Expression of the Acceleration

To find the acceleration of P, we need the expression of $d\hat{\mathbf{e}}_T/dt$. As shown in Figure 1.30, we first introduce the angle φ which $\hat{\mathbf{e}}_T$ makes with the axis Ox. Angle φ is called the *turning angle* (or *tangential angle*). Let us assume that $d\varphi/ds > 0$, that is, angle φ increases as P moves along its path. Since $\hat{\mathbf{e}}_T = \cos\varphi\hat{\mathbf{e}}_x + \sin\varphi\hat{\mathbf{e}}_y$, then the vector $d\hat{\mathbf{e}}_T/d\varphi = -\sin\varphi\hat{\mathbf{e}}_x + \cos\varphi\hat{\mathbf{e}}_y$ is a unit vector, normal to $\hat{\mathbf{e}}_T$ and pointing toward the curve's interior. We call this vector *normal unit vector* $\hat{\mathbf{e}}_N$:

$$\frac{d\hat{\mathbf{e}}_T}{d\varphi} = \hat{\mathbf{e}}_N$$

Now we can write

$$\mathbf{a}_P = \ddot{s}\hat{\mathbf{e}}_T + \dot{s}\frac{d}{dt}\hat{\mathbf{e}}_T = \ddot{s}\hat{\mathbf{e}}_T + \dot{s}\frac{d\varphi}{dt}\frac{d\hat{\mathbf{e}}_T}{d\varphi} = \ddot{s}\hat{\mathbf{e}}_T + \dot{s}\frac{d\varphi}{dt}\hat{\mathbf{e}}_N = \ddot{s}\hat{\mathbf{e}}_T + \dot{s}^2\frac{d\varphi}{ds}\hat{\mathbf{e}}_N$$

The quantity $ds/d\varphi$, in analogy with the circle, is called *radius of curvature* ρ at P of the path (note that $\rho > 0$): the small arc of length ds is shaped like an arc of circle of radius ρ

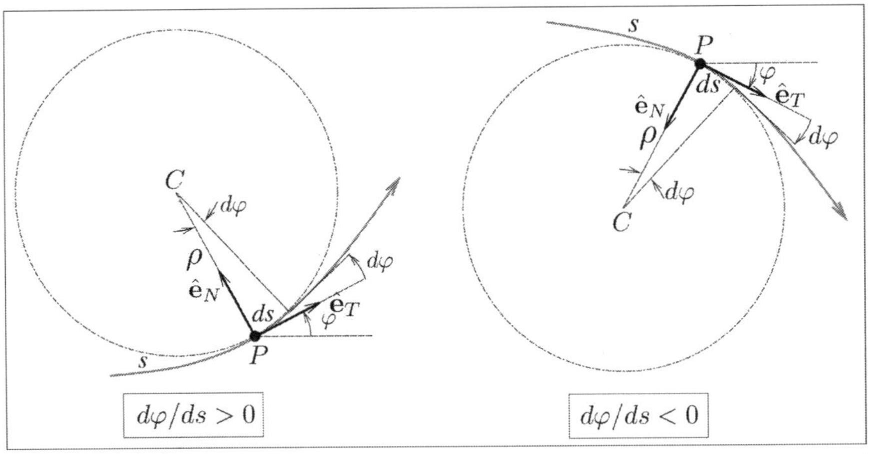

Figure 1.30 – *The path of particle P is defined by intrinsic coordinates s (arc-length) and φ (turning angle).*

and center C. The reciprocal of ρ is called *curvature*. See Figure 1.30 (left) for a graphical construction.

The radius of curvature is a local intrinsic property of the path. Finally, upon using $\dot{s} = v_P$ and $\ddot{s} = \dot{v}_P$ we obtain the expression

$$\mathbf{a}_P = \ddot{s}\hat{\mathbf{e}}_T + \frac{\dot{s}^2}{\rho}\hat{\mathbf{e}}_N = \dot{v}_P\hat{\mathbf{e}}_T + \frac{v_P^2}{\rho}\hat{\mathbf{e}}_N$$

Now consider the case $d\varphi/ds < 0$ shown in Figure 1.30 (right): the concavity at P is now downward. We now define the normal unit vector $\hat{\mathbf{e}}_N$ as

$$\frac{d\hat{\mathbf{e}}_T}{d\varphi} = -\hat{\mathbf{e}}_N$$

so that once again $\hat{\mathbf{e}}_N$ points toward the path's interior. We have now

$$\mathbf{a}_P = \ddot{s}\hat{\mathbf{e}}_T - \dot{s}^2\frac{d\varphi}{ds}\hat{\mathbf{e}}_N = \dot{v}_P\hat{\mathbf{e}}_T + \frac{v_P^2}{\rho}\hat{\mathbf{e}}_N$$

Remark 11. We can express the radius of curvature as $\rho = |ds/d\varphi|$ in both cases. Then $d\hat{\mathbf{e}}_T/ds = \hat{\mathbf{e}}_N/\rho$ whether $d\varphi/ds > 0$ or < 0.

In conclusion,

Kinematics in NT Components

At any given time, it is possible to define two unit vectors $(\hat{\mathbf{e}}_T, \hat{\mathbf{e}}_N)$, called *tangential and normal* unit vectors to the path at P. Then, the velocity and acceleration of P take the expressions

$$\mathbf{v}_P = v_P\hat{\mathbf{e}}_T, \qquad v_P = \dot{s} \tag{1.19}$$

$$\mathbf{a}_P = \dot{v}_P\hat{\mathbf{e}}_T + \frac{v_P^2}{\rho}\hat{\mathbf{e}}_N \tag{1.20}$$

where $s, \rho = |ds/d\varphi|, \varphi$ are the arc-length, the radius of curvature and turning angle of the path at P.

The formulas (1.19) and (1.20) are fundamental in kinematics. They are expressed in terms of the local, intrinsic characteristics s, ρ, $\hat{\mathbf{e}}_T$, and $\hat{\mathbf{e}}_N$ of the curve described by P.

Remark 12. The quantities s, ρ, φ, $\hat{\mathbf{e}}_T$ and $\hat{\mathbf{e}}_N$ vary along the path, and hence are functions of time.

PROPERTIES

①. The tangential acceleration $\mathbf{a}_{Pt} = \dot{v}_P \hat{\mathbf{e}}_T$ takes into account the change in the magnitude of $\mathbf{v}_P$: this vector vanishes when the motion is uniform.

②. The normal acceleration $\mathbf{a}_{Pn} = \frac{v_P^2}{\rho}\hat{\mathbf{e}}_N$ takes into account the change in the direction of $\mathbf{v}_P$: this vector vanishes when the motion is rectilinear (the direction of $\mathbf{v}_P$ is constant). The smaller the radius of curvature the larger the normal acceleration is. The normal acceleration always points toward the interior of the path.

③. In the neighborhood of P, the path of P behaves like a circle, called *osculating circle* of center C and radius ρ. C is referred to as the *center of curvature* of the path at P: its position relative to P is given by $\mathbf{r}_{PC} = \rho\hat{\mathbf{e}}_N$.

④. It is possible to relate the Cartesian and NT components of the velocity vector $\mathbf{v}_P = \dot{x}\hat{\mathbf{e}}_x + \dot{y}\hat{\mathbf{e}}_y = \dot{s}\hat{\mathbf{e}}_T$:

$$\dot{x} = \dot{s}\cos\varphi, \qquad \dot{y} = \dot{s}\sin\varphi, \text{ with } \varphi = (\hat{\mathbf{e}}_x, \hat{\mathbf{e}}_T)$$

or to relate its polar and NT components, $\mathbf{v}_P = \dot{r}\hat{\mathbf{e}}_r + r\dot{\theta}\hat{\mathbf{e}}_\theta = \dot{s}\hat{\mathbf{e}}_T$:

$$\dot{r} = \dot{s}\cos\psi, \qquad r\dot{\theta} = \dot{s}\sin\psi, \text{ with } \psi = (\hat{\mathbf{e}}_r, \hat{\mathbf{e}}_T)$$

See Figure 1.31.

⑤. **A geometric characterization of the acceleration:** We know that the velocity vector $\mathbf{v}_P$ is always tangent to the path of P. No such simple geometric relationship exists for the acceleration vector $\mathbf{a}_P$. However, we can sketch $\mathbf{a}_P$ relative to a planar path as a vector pointing in one of four quadrants at P depending on the sign of $\dot{v}_P$ and the concavity of the path at P. See Figure 1.32.

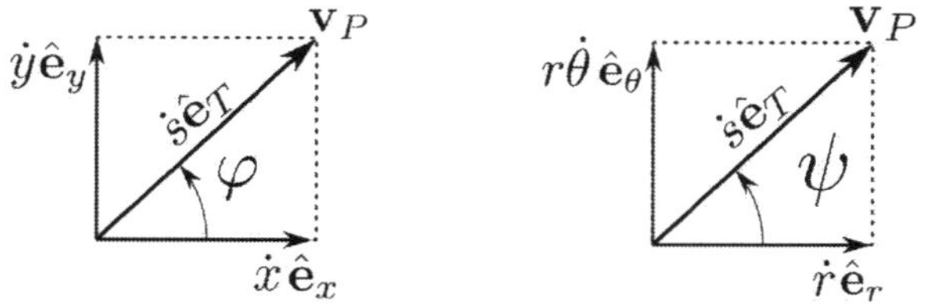

Figure 1.31 – *Relationship between NT, Cartesian and polar coordinates of velocity* $\mathbf{v}_P$.

STEP 3. DETERMINATION OF THE RADIUS OF CURVATURE

In practice, the radius of curvature ρ in equation (1.20) can often be found by using simple kinematics arguments. Since ρ is an intrinsic property of the curve, it is not a function of

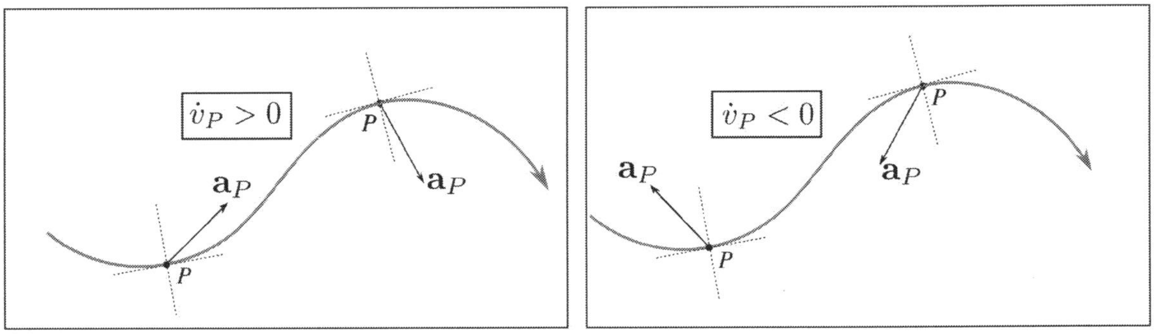

Figure 1.32 – *The acceleration of P must point in one of 4 quadrants defined by the tangential and normal directions. In the first case (left), the speed of P is assumed to increase along the path.*

the manner in which P travels along this curve: hence we can assume $v_P =$ constant, or $\dot{x} = \mathbf{v}_P \cdot \hat{\mathbf{e}}_x =$ constant or any other condition for the purpose of determining ρ. Then ρ is found from the normal acceleration $a_{Pn} = \mathbf{a}_P \cdot \hat{\mathbf{e}}_N = \frac{v_P^2}{\rho}$. We can outline a general procedure for the determination of ρ.

General Procedure: Determination of the Radius of Curvature (2D)

①. From the parametric representation of the curve (in Cartesian, polar or in any other representation) find the velocity $\mathbf{v}_P$ and the speed v_P of P.

②. From $\mathbf{v}_P$, find the tangent unit vector as $\hat{\mathbf{e}}_T = \frac{\mathbf{v}_P}{v_P}$.

③. Find the unit normal vector $\hat{\mathbf{e}}_N = \pm\hat{\mathbf{e}}_T \times \hat{\mathbf{e}}_z$, where $\hat{\mathbf{e}}_z$ is a unit vector normal to the plane of motion. If the concavity of the curve is known, choose the correct sign. If the expression found for ρ is negative, then the incorrect sign was chosen.

④. From the expression of $\mathbf{v}_P$, find the acceleration $\mathbf{a}_P$ of P: as this point it may be convenient to impose $v_P =$ constant or other condition on the motion of P.

⑤. From the expression of $\mathbf{a}_P$ and $\hat{\mathbf{e}}_N$ find the normal acceleration $a_{Pn} = \mathbf{a}_P \cdot \hat{\mathbf{e}}_N$.

⑥. Once a_N is found, the radius of curvature is obtained as $\rho = \frac{v_P^2}{a_{Pn}}$.

We can illustrate this simple procedure on two examples.

Example 1. Particle P moves on the parabola $2py = x^2$ $(p > 0)$. Find the radius of curvature ρ as a function of x.

First we find the velocity $\mathbf{v}_P = \dot{x}(\hat{\mathbf{e}}_x + \frac{x}{p}\hat{\mathbf{e}}_y)$ of P and its speed $v_P = \dot{x}(1 + \frac{x^2}{p^2})^{1/2}$. This leads to $(\hat{\mathbf{e}}_T, \hat{\mathbf{e}}_N)$

$$\hat{\mathbf{e}}_T = \frac{\hat{\mathbf{e}}_x + \frac{x}{p}\hat{\mathbf{e}}_y}{(1 + \frac{x^2}{p^2})^{1/2}}, \quad \hat{\mathbf{e}}_N = \frac{\hat{\mathbf{e}}_y - \frac{x}{p}\hat{\mathbf{e}}_x}{(1 + \frac{x^2}{p^2})^{1/2}}$$

To find $\mathbf{a}_P$, let us assume $\dot{x} =$ constant > 0: this leads to $\mathbf{a}_P = \frac{\dot{x}^2}{p}\hat{\mathbf{e}}_y$ and

$$a_{Pn} = \mathbf{a}_P \cdot \hat{\mathbf{e}}_N = \frac{\dot{x}^2}{p(1 + \frac{x^2}{p^2})^{1/2}}$$

and finally leading to $\boxed{\rho = \frac{v_P^2}{a_{Pn}} = p\left(1 + \frac{x^2}{p^2}\right)^{3/2}}$. ■

EXAMPLE 2. Particle P moves on the ellipse $(x/a)^2 + (y/b)^2 = 1$ which can be parametrized by θ according to

$$x = a\cos\theta, \qquad y = b\sin\theta \qquad (0 \le \theta \le 2\pi)$$

Assuming $\dot{\theta} > 0$, find the radius of curvature ρ as a function of θ.

First we find the velocity $\mathbf{v}_P = \dot{\theta}(-a\sin\theta\hat{\mathbf{e}}_x + b\cos\theta\hat{\mathbf{e}}_y)$ of P and its speed $v_P = \dot{\theta}(a^2\sin^2\theta + b^2\cos^2\theta)^{1/2}$. This leads to the unit vectors $(\hat{\mathbf{e}}_T, \hat{\mathbf{e}}_N)$

$$\hat{\mathbf{e}}_T = \frac{-a\sin\theta\hat{\mathbf{e}}_x + b\cos\theta\hat{\mathbf{e}}_y}{(a^2\sin^2\theta + b^2\cos^2\theta)^{1/2}}, \quad \hat{\mathbf{e}}_N = \frac{-a\sin\theta\hat{\mathbf{e}}_y - b\cos\theta\hat{\mathbf{e}}_x}{(a^2\sin^2\theta + b^2\cos^2\theta)^{1/2}}$$

To find acceleration $\mathbf{a}_P$, it is convenient to assume $\dot{\theta} =$ constant: this leads to

$$\mathbf{a}_P = \dot{\theta}^2(-a\cos\theta\hat{\mathbf{e}}_x - b\sin\theta\hat{\mathbf{e}}_y)$$

and

$$a_{Pn} = \mathbf{a}_P \cdot \hat{\mathbf{e}}_N = \dot{\theta}^2 \frac{ab}{(a^2\sin^2\theta + b^2\cos^2\theta)^{1/2}}$$

and finally to $\boxed{\rho = \dfrac{v_P^2}{a_{Pn}} = \dfrac{1}{ab}(a^2\sin^2\theta + b^2\cos^2\theta)^{3/2}}$. ∎

APPLICATION: THE DESIGN OF HIGHWAYS

In the design of roads, one is often faced with connecting straight segments to circular segments of roads. This is particularly true for highway interchanges. Imagine that you are driving your car on a straight road at constant speed v_0.

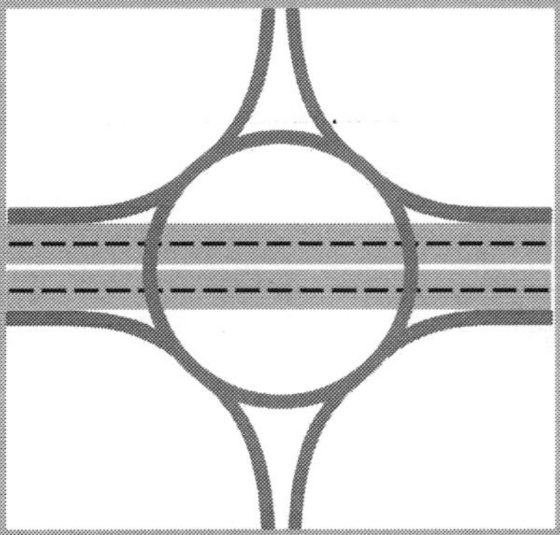

Figure 1.33 – *Highway interchange.*

Your body is not subjected to any acceleration. If you are then entering a curved road in the form an arc of circle of radius R, you would feel a sudden outward jolt due to the normal acceleration v_0^2/R. This can be the cause of discomfort and, over time, more wear and tear on your car! Furthermore, the steering angle must instantly pass from the value zero to a non-zero value. This implies that you would have to abruptly turn the steering wheel from its neutral position to a finite position to match the road's curvature. This is not a safe driving operation.

The solution to this problem is to use one or more transition roads which smoothly connect straight or circular segments. Consider the straight segment C_1 to be connected to the circular segment C_3 of radius R. As a potential transition segment C_2, one can choose a curve whose curvature κ varies continuously from the curvature $\kappa = 0$ of C_1 to the curvature $\kappa = 1/R$ of C_3. Curve C_2 thus ensures a smooth transition between C_1 and C_3.

With this choice, the driver will not feel a sudden centrifugal force while entering the curved road along which the steering wheel can be turned progressively, Such road design leads to a safer and more comfortable driving experience.
Such curves exist mathematically. In fact, the curve whose curvature κ varies linearly with its arc-length s from the zero value of $\mathcal{C}_1$ to the finite value of $\mathcal{C}_3$, has a long history dating back from the work of Bernoulli. It is known as *Euler spiral* or *Clothoid*. Euler spirals are widely used as transition curves in railroad and highway engineering. A study of a particular transition curve is the object of Project 1.

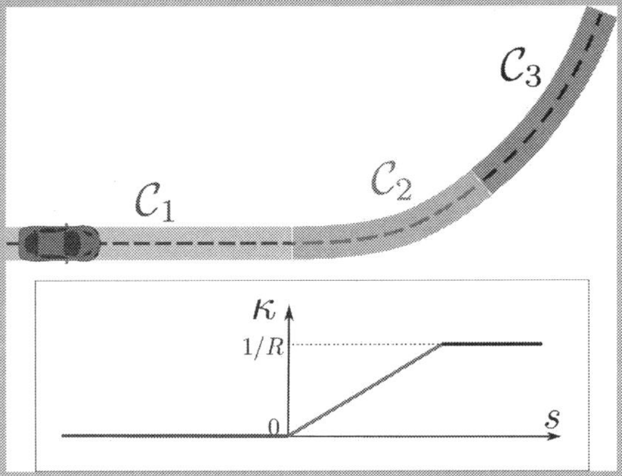

Figure 1.34 – *For a smooth driving experience, a section of clothoid $\mathcal{C}_2$ serves as a transition curve between a straight section $\mathcal{C}_1$ and a circular section $\mathcal{C}_3$.*

EXAMPLE PROBLEMS

General Procedure: Kinematics in NT Coordinates

① We now know that the velocity $\mathbf{v}_P$ and acceleration $\mathbf{a}_P$ of a particle P can take various expressions depending on the choice of coordinate system: for instance

$$\mathbf{a}_P = a_x\hat{\mathbf{e}}_x + a_y\hat{\mathbf{e}}_y = a_r\hat{\mathbf{e}}_r + a_\theta\hat{\mathbf{e}}_\theta = a_t\hat{\mathbf{e}}_T + a_n\hat{\mathbf{e}}_N$$

Any one of these components in a specific coordinate system can be obtained from the components in another coordinate system by simple projection. For instance, we can find normal component

$$a_n = a_x\hat{\mathbf{e}}_x\cdot\hat{\mathbf{e}}_N + a_y\hat{\mathbf{e}}_y\cdot\hat{\mathbf{e}}_N = a_r\hat{\mathbf{e}}_r\cdot\hat{\mathbf{e}}_N + a_\theta\hat{\mathbf{e}}_\theta\cdot\hat{\mathbf{e}}_N$$

if the orientation of $(\hat{\mathbf{e}}_x, \hat{\mathbf{e}}_y)$ (or $(\hat{\mathbf{e}}_r, \hat{\mathbf{e}}_\theta)$) is known relative to $(\hat{\mathbf{e}}_T, \hat{\mathbf{e}}_N)$.

② In Cartesian coordinates, two integrations of the acceleration components $(\ddot{x}, \ddot{y})$ lead to the trajectory of P. However the same procedure with the NT coordinates of $\mathbf{a}_P$ will not yield a complete description of the trajectory, unless the orientation of the unit vectors $(\hat{\mathbf{e}}_T, \hat{\mathbf{e}}_N)$ is specified (through the turning angle).

③ If the trajectory of a particle is known in some coordinate system, it is straightforward to find the tangential and normal unit vectors: start by finding $\mathbf{v}_P$ to find $\hat{\mathbf{e}}_T = \frac{\mathbf{v}_P}{v_P}$. Then $\hat{\mathbf{e}}_N$ is found as $\pm\hat{\mathbf{e}}_T \times \hat{\mathbf{e}}_z$ depending on the concavity of the curve at P ($\hat{\mathbf{e}}_z$ is a unit vector normal to the plane of motion). Once $\hat{\mathbf{e}}_T$ and $\hat{\mathbf{e}}_N$ are known, the tangential and normal components of the acceleration of P are found as $a_{Pt} = \mathbf{a}_P \cdot \hat{\mathbf{e}}_T$ and $a_{Pn} = \mathbf{a}_P \cdot \hat{\mathbf{e}}_N$.

④ If the radius of curvature of the trajectory of P is needed, use the fundamental result $a_{Pn} = \mathbf{a}_P \cdot \hat{\mathbf{e}}_N = \frac{v_P^2}{\rho}$ and recall that ρ is a geometric characteristic and that it does not depend on the manner in which P travels along its trajectory.

⑤ Conversely, if the tangential acceleration $a_{Pt} = \ddot{s}$ is known, it is possible to find the arc-length $s(t)$ by integration given initial conditions. Complete description of the trajectory of P requires knowledge of the turning angle $\varphi(t)$: if $\rho(s)$ is given, it can be obtained from the expression $\rho = \pm\frac{ds}{d\varphi}$.

Exercise 1.18 ▪A particle P moves in a plane so that at any time its tangential and normal accelerations are given as $a_{Pt} = 2\ \text{m/s}^2$ and $a_{Pn} = 4t^4\ \text{m/s}^2$, respectively. Time t is measured in seconds. At $t = 0$, the point is at rest. Find the radius of curvature ρ of P's path after traveling a distance of 6 m. Sketch the path of P. Can this path be completely found? ■

SOLUTION. Let us denote $a_{Pt} = a$ and $a_{Pn} = bt^4$ (a and b positive constants). From the knowledge of a_{Pt} we can get the arc-length s versus t: from $\ddot{s} = a$ with $\dot{s} = 0$ (P is at rest at $t = 0$) and $s = 0$ at $t = 0$ we obtain by integration

$$\dot{s} = at, \qquad s = \frac{a}{2}t^2$$

Then from the knowledge of $a_{Pn} = \dot{s}^2/\rho$ we can obtain the radius of curvature versus s:

$$\boxed{\rho = \frac{a^2}{bt^2} = \frac{a^3}{2bs}}$$

We notice that as s increases, the radius of curvature tends to 0: one possible way to sketch the path described by P is as a curve which converges toward a fixed point as $t \to \infty$ (see Figure 1.35).

There is no way to have a complete description of the path since we are not given any knowledge of the orientation of $(\hat{\mathbf{e}}_T, \hat{\mathbf{e}}_N)$ along the path. We would need the knowledge of both arc-length $s(t)$ and turning angle $\varphi(t)$ versus time to find the trajectory of P.

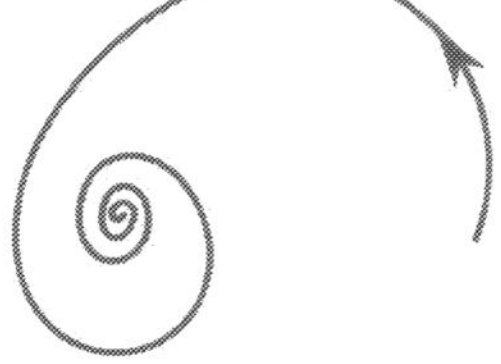

Figure 1.35 – *Problem 1.18 (Solution). Possible sketch of the path of P.*

When $s = 6$ m, we find $\rho = \mathbf{0.167}$ m. ■

Exercise 1.19 ▪A particle P departs from origin O of a plane Oxy and describes a trajectory characterized by the arc-length $s(t)$ and turning angle $\varphi(t)$ (relative to fixed axis Ox) given as

$$s(t) = s_0(1 - e^{-kt}), \qquad \varphi(t) = \frac{\pi}{2} - \omega t$$

where s_0, k and ω are positive constants.
a. Find the tangential and normal components of the acceleration of P.
b. By using the identity $\dot{x}\hat{\mathbf{e}}_x + \dot{y}\hat{\mathbf{e}}_y = \dot{s}\hat{\mathbf{e}}_T$, find the Cartesian coordinates $(x(t), y(t))$ of the trajectory of P. Show that P approaches a fixed point A as $t \to \infty$. Plot the trajectory of P for $k/\omega = 0.5$ in non-dimensional variables $X = x/s_0$, $Y = y/s_0$ and $\tau = \omega t$. ∎

SOLUTION. **a.** Since the arc-length s is given as a function of time, we can find the speed and the tangential component of the acceleration of P:

$$v_P = \dot{s} = ks_0 e^{-kt}, \quad \boxed{a_{Pt} = \dot{v}_P = -k^2 s_0 e^{-kt}}$$

We cannot yet find the normal acceleration from $a_{Pn} = v_P^2/\rho$ since the radius of curvature ρ is not known. To find ρ we use the formula $\rho = |ds/d\varphi|$ with $\dot{s} = ks_0 e^{-kt}$. Using $ds/d\varphi = \dot{s}/\dot{\varphi}$ and $\dot{\varphi} = -\omega$ we find:

$$\boxed{\rho = \frac{k}{\omega} s_0 e^{-kt}}$$

We note that $\rho \to 0$ as $t \to \infty$. Now we find

$$a_{Pn} = \frac{v_P^2}{\rho} = \frac{k^2 s_0^2 e^{-2kt}}{\frac{k}{\omega} s_0 e^{-kt}}$$

leading to

$$\boxed{a_{Pn} = \omega k s_0 e^{-kt}}$$

b. To obtain the trajectory in Cartesian coordinates, we express the velocity vector on $(\hat{\mathbf{e}}_x, \hat{\mathbf{e}}_y)$ and $(\hat{\mathbf{e}}_T, \hat{\mathbf{e}}_N)$:

$$\dot{x}\hat{\mathbf{e}}_x + \dot{\mathbf{y}}\hat{\mathbf{e}}_y = v_P \hat{\mathbf{e}}_T$$

Since $\hat{\mathbf{e}}_T = \cos\varphi \hat{\mathbf{e}}_x + \sin\varphi \hat{\mathbf{e}}_y$, we obtain $(\dot{x}, \dot{y})$ as functions of time:

$$\dot{x} = v_P \cos\varphi = ks_0 e^{-kt} \sin(\omega t)$$
$$\dot{y} = v_P \sin\varphi = ks_0 e^{-kt} \cos(\omega t)$$

Then coordinates (x, y) are obtained by simple integration and by imposing $x = y = 0$ at $t = 0$:

$$x(t) = s_0 \frac{k\omega}{\omega^2 + k^2} - s_0 \frac{ke^{-kt}}{\omega^2 + k^2} (\omega \cos(\omega t) + k \sin(\omega t))$$
$$y(t) = s_0 \frac{k^2}{\omega^2 + k^2} + s_0 \frac{ke^{-kt}}{\omega^2 + k^2} (\omega \sin(\omega t) - k \cos(\omega t))$$

where we used the integrals

$$\int e^{-kt} \cos(\omega t) dt = e^{-kt} \frac{\omega \sin \omega t - k \cos \omega t}{\kappa^2}, \int e^{-kt} \sin(\omega t) dt = -e^{-kt} \frac{\omega \cos \omega t + k \sin \omega t}{\kappa^2}$$

(with $\kappa^2 = \omega^2 + k^2$). The trajectory of P in terms of the non-dimensional variables

X, Y, τ is given by

$$X(\tau) = \frac{K}{1+K^2} - \frac{Ke^{-K\tau}}{1+K^2}(\cos\tau + K\sin\tau), Y(\tau) = \frac{K^2}{1+K^2} + \frac{Ke^{-K\tau}}{1+K^2}(\sin\tau - K\cos\tau)$$

with $K = k/\omega$. We see that as $t \to \infty$ point P approaches the point A of coordinates

$$X_A = \frac{K}{1+K^2}, \qquad Y_A = \frac{K^2}{1+K^2}$$

The trajectory of P is shown in Figure 1.36 for $K = 0.5$.

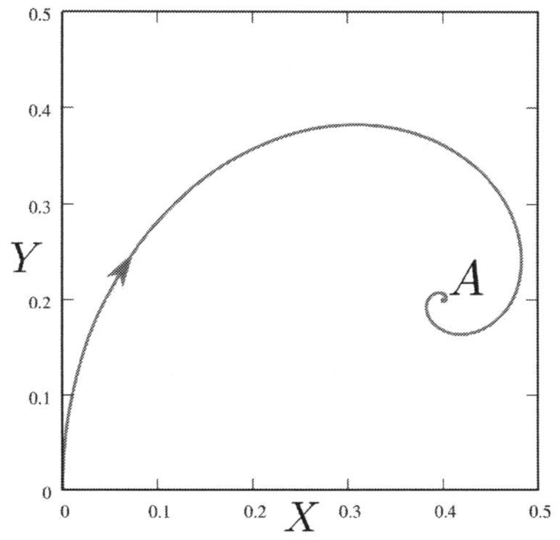

Figure 1.36 – *Problem 1.19 (Solution). The trajectory of P for* $\tau = 0 \ldots 15$. *Note that the path is tangent to axis OY at O since* $\varphi = \pi/2$ *at* $t = 0$.

■

Exercise 1.20 ■A weather balloon is launched from the ground ($z = 0$). Its velocity at any height z is the sum of two components: a vertical constant component $v_z = v_0$ and a horizontal component $v_x = kz$ due to air flow proportional to height z (where $1/k$ is a given time constant).

Find the normal and tangential components of its acceleration at any time t. Then find the radius of curvature of the trajectory as a function of time. ■

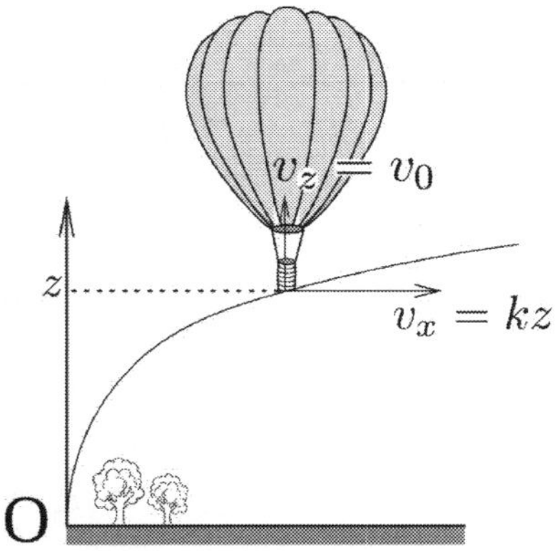

SOLUTION. We found the acceleration $\mathbf{a}_P = kv_0\hat{\mathbf{e}}_x$ of P in Problem 1.7. We can then determine the tangential and normal components of $\mathbf{a}_P$ as follows:

$$a_{Pt} = \mathbf{a}_P \cdot \hat{\mathbf{e}}_T, \qquad a_{Pn} = \mathbf{a}_P \cdot \hat{\mathbf{e}}_N$$

Hence we need to find the unit vectors $\hat{\mathbf{e}}_T$ and $\hat{\mathbf{e}}_N$ at any given time along the path.

We start with $\hat{\mathbf{e}}_T$ which is defined as $\mathbf{v}_P/v_P$:

$$\hat{\mathbf{e}}_T = \frac{kt\hat{\mathbf{e}}_x + \hat{\mathbf{e}}_z}{(1+k^2t^2)^{1/2}}$$

using the expression $\mathbf{v}_P = kv_0t\hat{\mathbf{e}}_x + v_0\hat{\mathbf{e}}_z$ found in problem 1.7. Now we have

$$\boxed{a_{Pt} = \mathbf{a}_P \cdot \hat{\mathbf{e}}_T = v_0\frac{k^2t}{(1+k^2t^2)^{1/2}}}$$

We can verify that $a_{Pt} = \dot{v}_P$! Then to find unit vector $\hat{\mathbf{e}}_N$, we use the fact that vectors normal to $\hat{\mathbf{e}}_T = (a\hat{\mathbf{e}}_x + b\hat{\mathbf{e}}_z)$ can be obtained as $\pm\hat{\mathbf{e}}_T \times \hat{\mathbf{e}}_y = \pm(a\hat{\mathbf{e}}_z - b\hat{\mathbf{e}}_x)$:

$$\hat{\mathbf{e}}_N = \pm\frac{-\hat{\mathbf{e}}_x + kt\hat{\mathbf{e}}_z}{(1+k^2t^2)^{1/2}}$$

To pick the right choice for $\hat{\mathbf{e}}_N$ we make sure that $\hat{\mathbf{e}}_N$ points toward the center of curvature: this is easily done here since we know the shape of the trajectory of P (a parabola). We pick the $-$ sign. Now we can find a_N:

$$\boxed{a_{Pn} = \mathbf{a}_P \cdot \hat{\mathbf{e}}_N = v_0\frac{k}{(1+k^2t^2)^{1/2}}}$$

The expression of the radius of curvature is then obtained from the identity $a_{Pn} = v_P^2/\rho$:

$$\boxed{\rho = \frac{v_P^2}{a_{Pn}} = \frac{v_0}{k}(1+k^2t^2)^{3/2}}$$

Remark. If the wrong sign is chosen for $\hat{\mathbf{e}}_N$, the radius ρ will be negative. The error is then easily corrected!

■

Exercise 1.21 ■■A small mass P is attached to an inextensible taut string whose other extremity is attached to the rim of a fixed cylinder. The particle lies on a horizontal plane. P is then set in motion while the string is wound around the cylinder in such a way as to keep $\dot{\theta} = 10$ rev/min at all time.
a. Find the velocity and acceleration of P at the instant shown ($\theta = \pi/2$).
b. Find the normal and tangential accelerations of P at the instant shown. Deduce the radius of curvature at P and the corresponding center of curvature.
c. Find a representation of the path of P parametrized by angle θ. Sketch this curve. ■

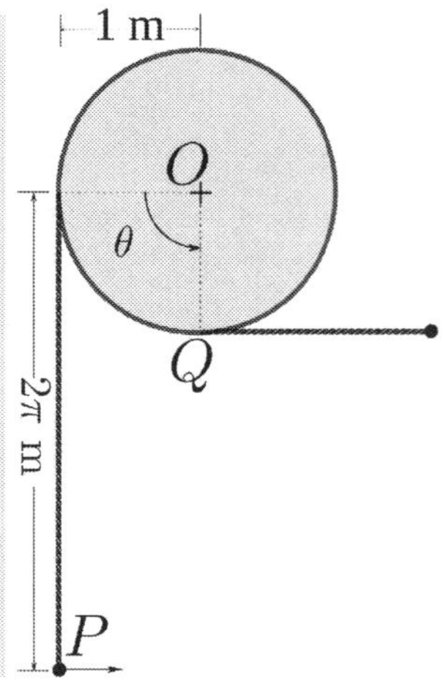

SOLUTION. **a.** To define the position of P at any given time, we introduce the two fixed unit vectors $\hat{\mathbf{e}}_x$ and $\hat{\mathbf{e}}_y$ as shown in Figure 1.37 (left) and the two moving unit vectors $\hat{\mathbf{e}}_r$ and $\hat{\mathbf{e}}_\theta$ associated with the circular motion of point Q defined as the point of contact of the string with the cylinder. We also introduce the constants R, l_0 and $\omega = \dot{\theta}$. Then we can express the position of P as

$$\mathbf{r}_{OP} = \mathbf{r}_{OQ} + \mathbf{r}_{QP} = R\hat{\mathbf{e}}_r + l\hat{\mathbf{e}}_\theta \qquad [1]$$

where l is the length of the unwound part of the string. This length can be found as a function of angle θ and total length $l_0 = s + l = R\theta + l$:

$$l = l_0 - R\theta \qquad [2]$$

To find the expression of velocity $\mathbf{v}_P$ on $(\hat{\mathbf{e}}_r, \hat{\mathbf{e}}_\theta)$ we differentiate the position vector $\mathbf{r}_{OP} = R\hat{\mathbf{e}}_r + l\hat{\mathbf{e}}_\theta$ w.r.t. time, using $d\hat{\mathbf{e}}_r/dt = \dot{\theta}\hat{\mathbf{e}}_\theta$ and $d\hat{\mathbf{e}}_\theta/dt = -\dot{\theta}\hat{\mathbf{e}}_r$:

$$\mathbf{v}_P = R\dot{\theta}\hat{\mathbf{e}}_\theta + \dot{l}\hat{\mathbf{e}}_\theta - l\dot{\theta}\hat{\mathbf{e}}_r = R\dot{\theta}\hat{\mathbf{e}}_\theta - R\dot{\theta}\hat{\mathbf{e}}_\theta - (l_0 - R\theta)\dot{\theta}\hat{\mathbf{e}}_r$$

Note that the velocity of P cannot be found by differentiating vector $\mathbf{r}_{QP}$ since point Q is in motion. With $\dot{\theta} = \omega$, we find

$$\boxed{\mathbf{v}_P = -(l_0 - R\theta)\omega\hat{\mathbf{e}}_r} \qquad [3]$$

Likewise, to find $\mathbf{a}_P$ we differentiate $\mathbf{v}_P = -l\omega\hat{\mathbf{e}}_r$:

$$\mathbf{a}_P = -\dot{l}\omega\hat{\mathbf{e}}_r - l\omega(\omega\hat{\mathbf{e}}_\theta)$$

to finally obtain

$$\boxed{\mathbf{a}_P = R\omega^2\hat{\mathbf{e}}_r - (l_0 - R\theta)\omega^2\hat{\mathbf{e}}_\theta} \qquad [4]$$

Note that when $l = l_0 - R\theta = 0$ points P and Q coincide, and we find $\mathbf{v}_P = \mathbf{0}$ and $\mathbf{a}_P = R\omega^2\hat{\mathbf{e}}_r$.
With $R = 1$ m, $l_0 = 2\pi$ m, $\theta = \pi/2$ and $\omega = 1.05$ rad/s, we have $\hat{\mathbf{e}}_r = \hat{\mathbf{e}}_y$ and $\hat{\mathbf{e}}_\theta = -\hat{\mathbf{e}}_x$. We find $\mathbf{v}_P = -\mathbf{4.9}\hat{\mathbf{e}}_y$ m/s and $\mathbf{a}_P = -\mathbf{5.2}\hat{\mathbf{e}}_x + \mathbf{1.1}\hat{\mathbf{e}}_y$ m/s^2.

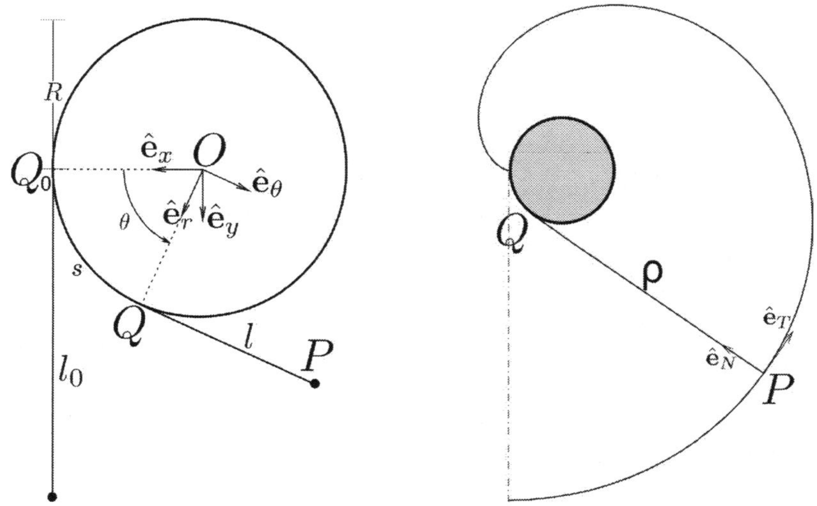

Figure 1.37 – *Problem 1.21 (Solution). The trajectory of P for $l_0/R = 2\pi$.*

b. From the expression of $\mathbf{v}_P = -l\omega\hat{\mathbf{e}}_r = v\hat{\mathbf{e}}_T$ we find that the tangential unit vector and the speed of P are given by

$$\hat{\mathbf{e}}_T = -\hat{\mathbf{e}}_r, \qquad v = l\omega = (l_0 - R\theta)\omega$$

We can then compare the expression [4] of $\mathbf{a}_P$ to $\dot{v}\hat{\mathbf{e}}_T + (v^2/\rho)\hat{\mathbf{e}}_N$ to find that

$$\hat{\mathbf{e}}_N = -\hat{\mathbf{e}}_\theta, \qquad \frac{v^2}{\rho} = (l_0 - R\theta)\omega^2$$

which shows that the tangential and normal components of the acceleration are given by

$$\boxed{a_{Pt} = -a_r = -R\omega^2, \qquad a_{Pn} = -a_\theta = (l_0 - R\theta)\omega^2}$$

radius of curvature is

$$\boxed{\rho = l}$$

This shows that *the center of curvature is located at point Q at all time*. When $\theta = \pi/2$, $a_{Pt} = -\mathbf{1.1}\ \mathrm{m/s^2}$, $a_{Pn} = \mathbf{5.2}\ \mathrm{m/s^2}$ and $\rho = \mathbf{4.7}$ m.
c. To find the trajectory of P, we express $\mathbf{r}_{OP}$ on the Cartesian basis $(\hat{\mathbf{e}}_x, \hat{\mathbf{e}}_y)$:

$$\mathbf{r}_{OP} = R(\cos\theta\hat{\mathbf{e}}_x + \sin\theta\hat{\mathbf{e}}_y) + (l_0 - R\theta)(\cos\theta\hat{\mathbf{e}}_y - \sin\theta\hat{\mathbf{e}}_x)$$

The Cartesian representation of the curve is then:

$$\boxed{x = R\cos\theta - (l_0 - R\theta)\sin\theta, \qquad y = R\sin\theta + (l_0 - R\theta)\cos\theta}$$

This trajectory is plotted in Figure 1.37 (right) for $l_0/R = 2\pi$. ■

1.7 The Frenet Basis

In three dimensions, the analysis of the intrinsic coordinate system remains nearly identical. The unit tangent and normal vectors at P are still defined by

$$\hat{\mathbf{e}}_T = \frac{d\mathbf{r}_{OP}}{ds}, \quad \hat{\mathbf{e}}_N = \rho\frac{d\hat{\mathbf{e}}_T}{ds} \tag{1.21}$$

where ρ is a *positive* scalar still referred to as *the radius of curvature* of the curve $\mathcal{C}$ described by P. However, since more than one normal direction to a 3D curve can be defined, $\hat{\mathbf{e}}_N$ is referred to as the *principal unit normal* at P. A second normal direction can be defined as $\hat{\mathbf{e}}_B = \hat{\mathbf{e}}_T \times \hat{\mathbf{e}}_N$ and is known as the *binormal unit vector*. The basis of unit vectors $(\hat{\mathbf{e}}_T, \hat{\mathbf{e}}_N, \hat{\mathbf{e}}_B)$ defines the so-called *Frenet basis* at location P of the curve $\mathcal{C}$ described by P. See Figure 1.38. Consequently, both NT representations (1.19) and (1.20) of the kinematics of P remain valid.
It is of interest to find the variation of the unit vectors $(\hat{\mathbf{e}}_T, \hat{\mathbf{e}}_N, \hat{\mathbf{e}}_B)$ along the curve $\mathcal{C}$. We already know how the tangent unit vector varies along $\mathcal{C}$ according to formula $d\hat{\mathbf{e}}_T/ds = \hat{\mathbf{e}}_N/\rho$. The answer is provided by the *Serret-Frenet* formulas.

Serret-Frenet Formulas

The variations of the unit vectors $(\hat{\mathbf{e}}_T, \hat{\mathbf{e}}_N, \hat{\mathbf{e}}_B)$ along the path $\mathcal{C}$ of P are given by the formulas

$$\frac{d\hat{\mathbf{e}}_T}{ds} = \frac{1}{\rho}\hat{\mathbf{e}}_N, \qquad \frac{d\hat{\mathbf{e}}_N}{ds} = -\frac{1}{\rho}\hat{\mathbf{e}}_T + \frac{1}{\tau}\hat{\mathbf{e}}_B, \qquad \frac{d\hat{\mathbf{e}}_B}{ds} = -\frac{1}{\tau}\hat{\mathbf{e}}_N \tag{1.22}$$

where τ is known as the *radius of torsion* of the curve at P. The reciprocal to τ is called *torsion* of curve $\mathcal{C}$ at P.

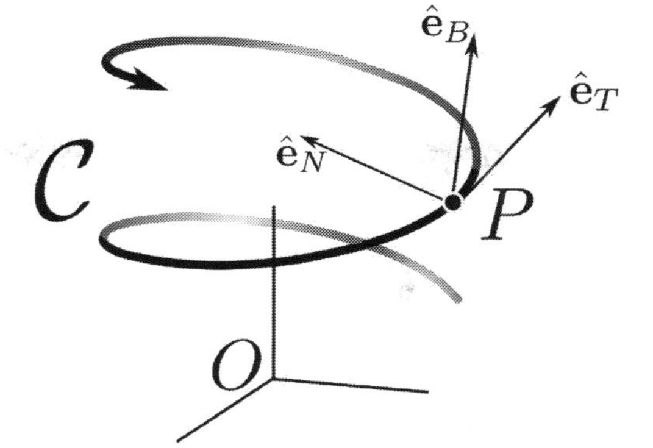

Figure 1.38 – *The Frenet basis at P.*

Proof. To prove the last two Serret-Frenet formulas, we first use the fact that $\hat{\mathbf{e}}_N$ and $\hat{\mathbf{e}}_B$, being unit vectors, are orthogonal to their derivatives. Hence there exist four scalars a, b, c and d such that

$$\frac{d\hat{\mathbf{e}}_N}{ds} = a\hat{\mathbf{e}}_T + b\hat{\mathbf{e}}_B, \qquad \frac{d\hat{\mathbf{e}}_B}{ds} = c\hat{\mathbf{e}}_T + d\hat{\mathbf{e}}_N$$

To find these scalars, we differentiate the relations $\hat{\mathbf{e}}_T \cdot \hat{\mathbf{e}}_N = 0$, $\hat{\mathbf{e}}_N \cdot \hat{\mathbf{e}}_B = 0$, and $\hat{\mathbf{e}}_T \cdot \hat{\mathbf{e}}_B = 0$ with respect to arc-length s and upon replacing $d\hat{\mathbf{e}}_T/ds$ by $\hat{\mathbf{e}}_N/\rho$, $d\hat{\mathbf{e}}_N/ds$ by $a\hat{\mathbf{e}}_T + b\hat{\mathbf{e}}_B$, and $d\hat{\mathbf{e}}_B/ds$ by $c\hat{\mathbf{e}}_T + d\hat{\mathbf{e}}_N$, we obtain

$$a + 1/\rho = 0, \quad b + d = 0, \quad c = 0$$

Hence, we obtain (1.22) after writing $b = 1/\tau = -d$. ■

Remark 13. Note that the radius of torsion, as opposed to the radius of curvature, is not necessarily positive. It measures the deviation of $\mathcal{C}$ from a planar curve. When the path is planar, $d\hat{\mathbf{e}}_B/ds = \mathbf{0}$ leading to vanishing torsion $(1/\tau = 0)$.

The following problem illustrates kinematics along a 3D curve.

Exercise 1.22 A particle P travels downward in a referential $Oxyz$ inside a tube which has the shape of a regular circular helix. In Cartesian coordinates its path is described by

$$x = R\cos\theta$$
$$y = R\sin\theta$$
$$z = \frac{h}{2\pi}\theta$$

where R, h are positive constants. In addition particle P is known to move at constant speed v_0.

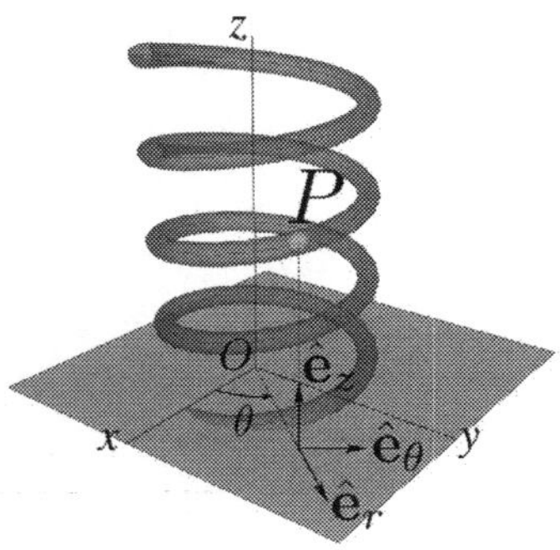

a. Write the position vector $\mathbf{r}_{OP}$ of P in cylindrical form, that is, along unit vectors $(\hat{\mathbf{e}}_r, \hat{\mathbf{e}}_\theta, \hat{\mathbf{e}}_z)$.
b. Determine the velocity $\mathbf{v}_P$ of P in cylindrical coordinates (by differentiating the expression of $\mathbf{r}_{OP}$). Deduce the relationship between $\dot{\theta}$ and the constants R, h, v_0. Find the distance traveled by P after one revolution.
c. Find the acceleration $\mathbf{a}_P$ of P in cylindrical coordinates (by differentiating the expression of $\mathbf{v}_P$). From this result, deduce the expression of the normal unit vector $\hat{\mathbf{e}}_N$, the radius of curvature ρ at any location of the trajectory of P.
d. Complete the solution by finding and sketching the Frenet basis at P. Find the radius of torsion τ. ■

SOLUTION. **a.** Since the radial unit vector is given by $\hat{\mathbf{e}}_r = \cos\theta\hat{\mathbf{e}}_x + \sin\theta\hat{\mathbf{e}}_y$, we can express the position vector $\mathbf{r}_{OP}$ as

$$\mathbf{r}_{OP} = R(\cos\theta\hat{\mathbf{e}}_x + \sin\theta\hat{\mathbf{e}}_y) + \frac{h}{2\pi}\theta\hat{\mathbf{e}}_z = R\hat{\mathbf{e}}_r + H\theta\hat{\mathbf{e}}_z \qquad [1]$$

where, to simplify notations, we have defined $H = h/2\pi$.
b. To find velocity $\mathbf{v}_P$ we take the time-derivative of equation [1] using $d\hat{\mathbf{e}}_r/dt = \dot{\theta}\hat{\mathbf{e}}_\theta$:

$$\mathbf{v}_P = R\dot{\theta}\hat{\mathbf{e}}_\theta + H\dot{\theta}\hat{\mathbf{e}}_z = \dot{\theta}(R\hat{\mathbf{e}}_\theta + H\hat{\mathbf{e}}_z) \qquad [2]$$

From this expression we can find the expression of the speed of P:

$$v_P = |\dot{\theta}|\sqrt{R^2 + H^2}$$

Since P travels downward we have $|\dot{\theta}| = -\dot{\theta}$ ($\dot{\theta} < 0$): then upon using $v_P = v_0$ we find

$$\boxed{\dot{\theta} = -\frac{v_0}{\sqrt{R^2 + H^2}}} \qquad [3]$$

Note that, since v_0 is constant, $\dot{\theta}$ is also a constant: $\ddot{\theta} = 0$. We can also obtain the tangential unit vector from the expression of $\mathbf{v}_P$:

$$\hat{\mathbf{e}}_T = \frac{\mathbf{v}_P}{v_P} = -\frac{R\hat{\mathbf{e}}_\theta + H\hat{\mathbf{e}}_z}{\sqrt{R^2 + H^2}}$$

The l distance traveled by P after one revolution is found from the expression of $ds = (dx^2 + dy^2 + dz^2)^{1/2}$:

$$l = \int_0^{2\pi} \left(R^2\sin^2\theta + R^2\cos^2\theta + H^2\right)^{1/2} d\theta = 2\pi\sqrt{R^2 + H^2}$$

c. To find acceleration $\mathbf{a}_P$ we take the time-derivative of equation [2] keeping in mind that $\dot{\theta} = Cst$ and $d\hat{\mathbf{e}}_\theta/dt = \dot{\theta}\hat{\mathbf{e}}_r$

$$\boxed{\mathbf{a}_P = -R\dot{\theta}^2\hat{\mathbf{e}}_r = -\frac{Rv_0^2}{(R^2 + H^2)}\hat{\mathbf{e}}_r} \qquad [4]$$

Since P moves at constant speed, its acceleration is given by $\mathbf{a}_P = \frac{v_0^2}{\rho}\hat{\mathbf{e}}_N$: comparison with [4] gives both $\hat{\mathbf{e}}_N$ and ρ:

$$\boxed{\hat{\mathbf{e}}_N = -\hat{\mathbf{e}}_r, \qquad \rho = R + \frac{H^2}{R}}$$

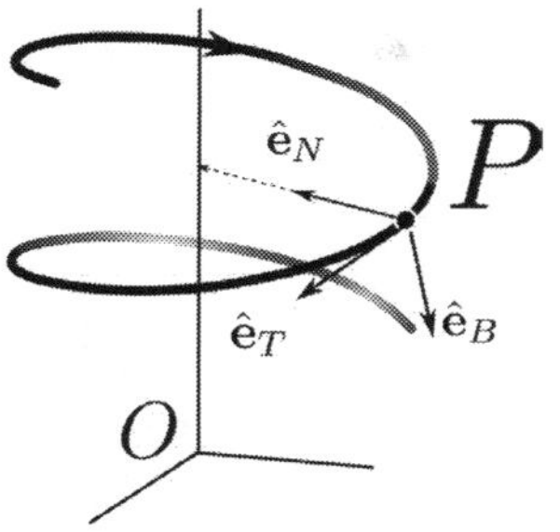

Figure 1.39 – *Problem 1.22 (Solution)*

d. To complete the solution, we need to find the binormal unit vector $\hat{\mathbf{e}}_B = \hat{\mathbf{e}}_T \times \hat{\mathbf{e}}_N$:

$$\hat{\mathbf{e}}_B = -\frac{R\hat{\mathbf{e}}_\theta + H\hat{\mathbf{e}}_z}{\sqrt{R^2 + H^2}} \times (-\hat{\mathbf{e}}_r) = \frac{-R\hat{\mathbf{e}}_z + H\hat{\mathbf{e}}_\theta}{\sqrt{R^2 + H^2}}$$

The Frenet basis is sketched in Figure 1.39. To find the torsion, we use the last of the Serret-Frenet formulas (1.22): hence we need to find

$$\frac{d\hat{\mathbf{e}}_B}{ds} = \frac{1}{v_0}\frac{d\hat{\mathbf{e}}_B}{dt} = \frac{1}{v_0}\frac{H(-\dot{\theta}\hat{\mathbf{e}}_r)}{\sqrt{R^2 + H^2}} = -\frac{H\hat{\mathbf{e}}_N}{R^2 + H^2}$$

We do indeed get $d\hat{\mathbf{e}}_B/ds$ colinear to $\hat{\mathbf{e}}_N$, and we can now identify the radius of torsion

$$\boxed{\tau = H + \frac{R^2}{H}}$$

■

1.8 Kinematics of Particles in Constrained Motion

In many situations of interest, the motion of a particle P is not free in space. Rather, it is constrained in some manner: for instance (see Figure 1.40),

1. P may be constrained to follow a given curve $(x(u), y(u), z(u))$ (parametrized by u in Cartesian coordinates): then its velocity $\mathbf{v}_P$ is given by

$$\mathbf{v}_P = \dot{u}\left(\frac{dx}{du}\hat{\mathbf{e}}_x + \frac{dy}{du}\hat{\mathbf{e}}_y + \frac{dz}{du}\hat{\mathbf{e}}_z\right)$$

2. P may be constrained to lie on a surface $f(x, y, z) = 0$ (in Cartesian coordinates): then its velocity components $(\dot{x}, \dot{y}, \dot{z})$ must satisfy the equation

$$\frac{\partial f}{\partial x}\dot{x} + \frac{\partial f}{\partial y}\dot{y} + \frac{\partial f}{\partial z}\dot{z} = 0$$

3. P may stay at a constant distance from a point A: $|AP| = l =$ constant. It can be shown[7] that this condition implies the following constraint between $\mathbf{v}_A$ and $\mathbf{v}_P$:

$$\mathbf{v}_A \cdot \hat{\mathbf{u}} = \mathbf{v}_P \cdot \hat{\mathbf{u}}$$

where $\hat{\mathbf{u}}$ is a unit vector directed along AP.

These examples show that simple kinematic conditions follow from geometric constraint equations.

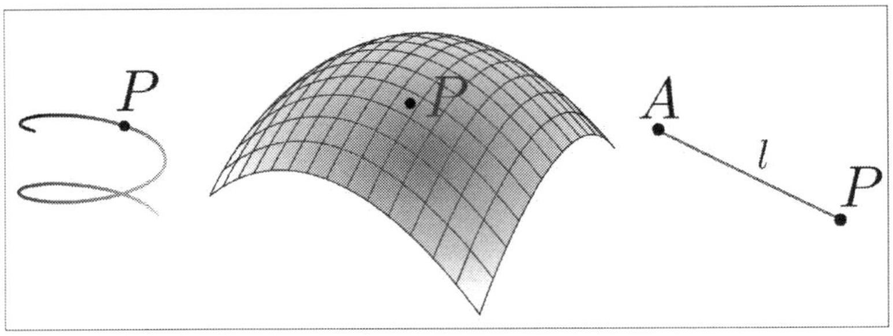

Figure 1.40 – *Possible geometric constraints.*

More generally, whenever a mechanical system can be modeled as a system of particles, the motions of its constituent particles are inevitably constrained relative to one another. Only certain (relative) motions are allowed and other motions are prevented. See Figure 1.41. A simple way to describe the kinematics of such systems consists of determining all geometric relationships which describe the constraints between the particles' coordinates. Assume that the position of a system of n particles P_1, P_2, ..., P_n is described by N coordinates $q_1, q_2, \ldots, q_N$ with $q_i = q_i(t)$ (these coordinates may be Cartesian, polar, etc).

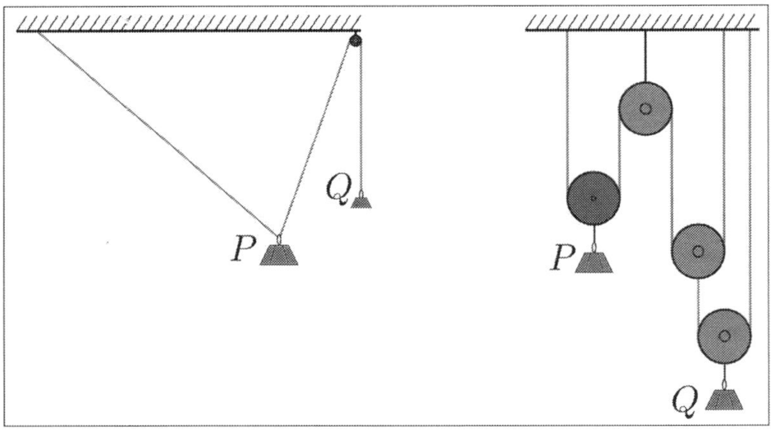

Figure 1.41 – *Systems of interconnected particles.*

[7]Time-differentiation of $\mathbf{r}_{AP} \cdot \mathbf{r}_{AP} = Cst$ gives $2\mathbf{r}_{AB} \cdot (\mathbf{v}_P - \mathbf{v}_A) = 0$.

Definition (Degrees of freedom)

If there exist M independent equations of the type

$$f(q_1, q_2, ..., q_N, t) = 0 \tag{1.23}$$

then the number $(N - M)$ is called the *number of degrees of freedom* of the system. It is the number of independent coordinates of the system.

Equations (1.23) reflect the geometric constraints satisfied by the system's constituents. Then a simple way of getting kinematic relationships consists of taking the time-derivative of equation (1.23). If this equation holds at all time, then we obtain (by using the chain rule)

$$\frac{\partial f}{\partial q_1}\dot{q}_1 + \cdots + \frac{\partial f}{\partial q_N}\dot{q}_N + \frac{\partial f}{\partial t} = 0 \tag{1.24}$$

Equation (1.24) represents a kinematic constraint equation. It expresses the fact that the particle velocities are not independent of each other. Similarly, acceleration constraint equations can be obtained by taking the time-derivative of (1.24).

EXAMPLE PROBLEMS

General Procedure: System of Particles

To solve a kinematics problem involving one or more particles in constrained motion, follow these steps.

① Define the system and the referential relative to which the motions are observed: identify the relevant particles and define a convenient origin O. Identify all known quantities and the desired (kinematic) unknowns.

② Parametrize the position of each particle of the system. This may involve a combination of coordinate systems depending on the motions involved. Do not "over"-parametrize: redundant coordinates will make the analysis more difficult.

③ Determine all possible geometric constraint equations of the type (1.23) governing the system's coordinates. This may involve any of the situations mentioned above. Other possible situations may include:

(i) a cable or rope connecting two or more particles of the system: a geometric constraint is found by guaranteeing that the cable's length remains constant during the motion,

(ii) two (or more) interacting particles P_1 and P_2: their positions can be related by a vector loop equation

$$\mathbf{r}_{OP_1} + \mathbf{r}_{P_1P_2} + \mathbf{r}_{P_2O} = \mathbf{0}$$

which, if valid at all time, can be time-differentiated. If the particles' positions have been parametrized, this equation will provide 3 scalar equations. It is important to derive geometric equations *valid at all time*: this requires that the configuration of the system be defined at an arbitrary time, rather than at a specific time.

④ If the geometric equations found in step 3 are valid are all time, then taking the time-derivative (once or twice) of these equations will yield kinematic equations. Solve for the kinematic unknowns.

Exercise 1.23 A rod is connected to two sliders at pins A and B ($|AB| = 70$ cm) which are constrained to move along perpendicular directions. At the instant shown, $v_A = 3$ m/s and $\theta = 50°$.
a. Find the velocity of B at the instant shown. Assuming a constant speed v_A of A, is the motion of B uniform?
b. Show that the velocity of midpoint I of rod AB is given by $\mathbf{v}_I = (\mathbf{v}_A + \mathbf{v}_B)/2$. Find the trajectory described by I.
c. Find the velocity of C at the instant shown (with $|CB| = |AB|$). Show that $\mathbf{v}_C$ is normal to line JC. Find the trajectory described by C. ■

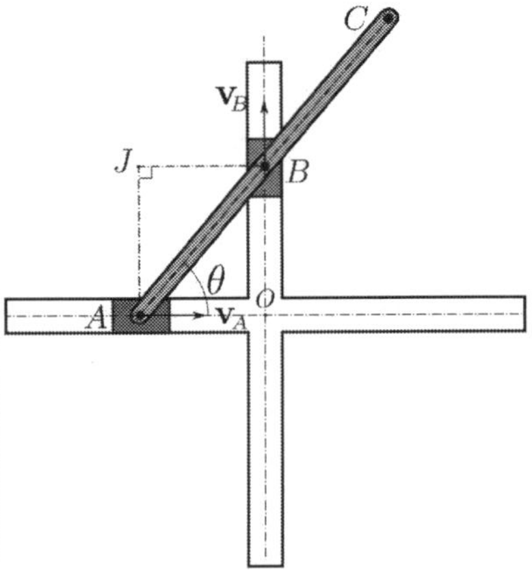

Solution. We introduce fixed unit vectors $(\hat{\mathbf{e}}_x, \hat{\mathbf{e}}_y)$ as shown in Figure 1.42 and denote $(x_A, 0)$, $(0, y_B)$, (x_C, y_C) the Cartesian coordinates of A, B and C respectively. The solution method consists of finding simple relationships between the positions of A, B, C, I and J.

a. We relate points point A to point B by writing the equation $\mathbf{r}_{AB} = L(\cos\theta\hat{\mathbf{e}}_x + \sin\theta\hat{\mathbf{e}}_y) = \mathbf{r}_{AO} + \mathbf{r}_{OB}$. This gives

$$x_A = -L\cos\theta, \qquad y_B = L\sin\theta$$

We can take the time-derivative of these geometric equations to obtain

$$\dot{x}_A = L\dot{\theta}\sin\theta, \qquad \dot{y}_B = L\dot{\theta}\cos\theta$$

We can then relate $\dot{y}_B$ to $\dot{x}_A$:

$$\dot{y}_B = \dot{x}_A \cot\theta$$

to obtain the velocity of B

$$\boxed{\mathbf{v}_B = v_A \cot\theta\ \hat{\mathbf{e}}_y} \qquad [1]$$

This result shows that $v_A =$ constant does not imply $v_B =$ constant since angle θ varies. With $v_A = 3$ m/s and $\theta = 50°$, we find $\mathbf{v}_B = \mathbf{2.52}\,\hat{\mathbf{e}}_y$ m/s.

b. From the equation $\mathbf{r}_{AI} = \mathbf{r}_{AB}/2$ we find

$$\mathbf{r}_{OI} = \frac{1}{2}(\mathbf{r}_{OA} + \mathbf{r}_{OB})$$

which can be time-differentiated to lead to

$$\boxed{\mathbf{v}_I = \frac{1}{2}(\mathbf{v}_A + \mathbf{v}_B)} \qquad [2]$$

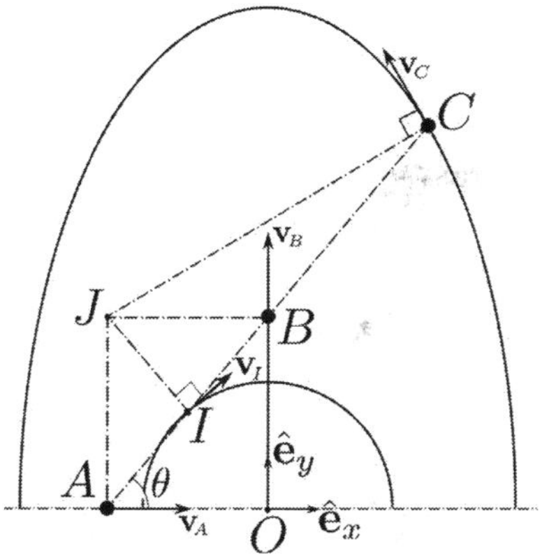

Figure 1.42 – *Problem 1.23 (Solution)*

We can also find the Cartesian coordinates of center I:

$$x_I = \frac{x_A}{2} = -\frac{L}{2}\cos\theta, \qquad y_I = \frac{y_B}{2} = \frac{L}{2}\sin\theta$$

Upon eliminating angle θ, we obtain the equation of a circle of center O and radius $L/2$:

$$\boxed{x_I^2 + y_I^2 = \frac{L^2}{4}} \qquad [3]$$

c. From the equation $\mathbf{r}_{AC} = 2\mathbf{r}_{AB}$ we obtain $\mathbf{r}_{OC} = -\mathbf{r}_{OA} + 2\mathbf{r}_{OB}$ from which we get the Cartesian coordinates of center C:

$$x_C = -x_A = L\cos\theta, \qquad y_C = 2y_B = 2L\sin\theta$$

Upon eliminating angle θ, we obtain the equation of an ellipse centered at O:

$$\boxed{x_C^2 + \frac{1}{4}y_C^2 = L^2} \qquad [3]$$

The velocity of C is obtained by differentiating $\mathbf{r}_{OC} = -\mathbf{r}_{OA} + 2\mathbf{r}_{OB}$:

$$\boxed{\mathbf{v}_C = -\mathbf{v}_A + 2\mathbf{v}_B = L\dot{\theta}(-\sin\theta\hat{\mathbf{e}}_x + 2\cos\theta\hat{\mathbf{e}}_y)}$$

To show that this vector is normal to line JC we first find vector $\mathbf{r}_{JC} = \mathbf{r}_{OC} - \mathbf{r}_{OJ} = \mathbf{r}_{OC} - \mathbf{r}_{OA} - \mathbf{r}_{OB}$:

$$\mathbf{r}_{JC} = -2x_A\hat{\mathbf{e}}_x + y_B\hat{\mathbf{e}}_y = L(2\cos\theta\hat{\mathbf{e}}_x + \sin\theta\hat{\mathbf{e}}_y)$$

We can check that $\mathbf{v}_C \cdot \mathbf{r}_{JC} = 0$ at all time.
At the instant shown, we find $\mathbf{v}_C = \mathbf{-3\,\hat{e}_x + 5\,\hat{e}_y}$ m/s. ■

Exercise 1.24 A mechanical system is assembled by connecting sliding collar A, block B, and rod AB by spherical joints. The translation of B along line $y = d = 40$ cm of plane Oxy causes A to translate along axis Oz. At the instant shown, $h = 50$ cm, $\mathbf{v}_B = v_B\,\hat{\mathbf{e}}_x$ with $v_B = 1.4$ m/s and $\mathbf{a}_B = a_B\hat{\mathbf{e}}_x$ with $a_B = -0.5$ m/s^2.
Find the velocity and acceleration of A. ■

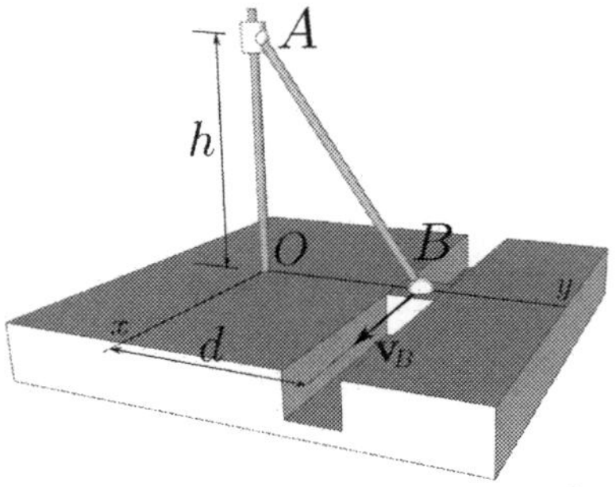

SOLUTION. First we note that we cannot parametrize the system in the given configuration since it is specific to a particular time. Figure 1.43 shows the same system in an arbitrary configuration. The parametrization is done by defining the Cartesian coordinates x_B and z_A on axes $Oxyz$ since both A and B have rectilinear motions:

$$\mathbf{r}_{OB} = x_B\hat{\mathbf{e}}_x + d\hat{\mathbf{e}}_y, \qquad \mathbf{r}_{OA} = z_A\hat{\mathbf{e}}_z$$

We can find an algebraic equation between x_B and z_A by writing the condition $|AB| = L = Cst$:

$$z_A^2 + x_B^2 + d^2 = L^2 \qquad [1]$$

Equation [1] is valid at all time and can be differentiated w.r.t. time to give

$$2z_A\dot{z}_A + 2x_B\dot{x}_B = 0 \qquad [2]$$

At the configuration shown, we have $x_B = 0$: this gives $\boxed{\dot{z}_A = 0}$. The result $\dot{z}_A = 0$ is

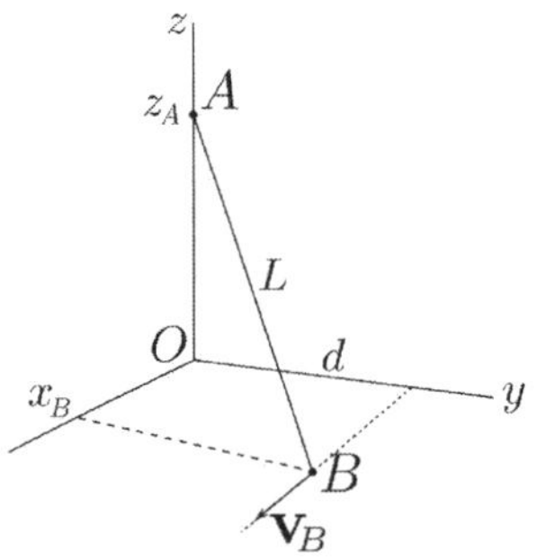

Figure 1.43 – *Problem 1.24 (Solution)*

expected since z_A reaches the maximum value $z_A = h$ at the instant shown. Equation [2] can be differentiated w.r.t. time to give

$$z_A\ddot{z}_A + \dot{z}_A^2 + x_B\ddot{x}_B + \dot{x}_B^2 = 0$$

At the configuration shown, we have $z_A = h$, $\dot{z}_A = 0$, $x_B = 0$, $\dot{x}_B = v_B$ and $\ddot{x}_B = a_B$: this gives $\boxed{\ddot{z}_A = -\dfrac{v_B^2}{h}}$. We find $\ddot{z}_A = -\mathbf{3.92}$ m/s^2. ■

Exercise 1.25 A pin P can move relative to the slotted square which slides along a horizontal support at constant speed 0.5 m/s. A thin, inextensible cable is connected between point A of the square and a fixed point Q located at a distance $h = 1$ m from the horizontal support. The cable is constrained to slide around the pin whose weight keeps it taut at all time. The cable's length is equal to the distance 3 m measured from A to the support.

Find the speed of P when $x = 1.5$ m. What is the trajectory of P? ■

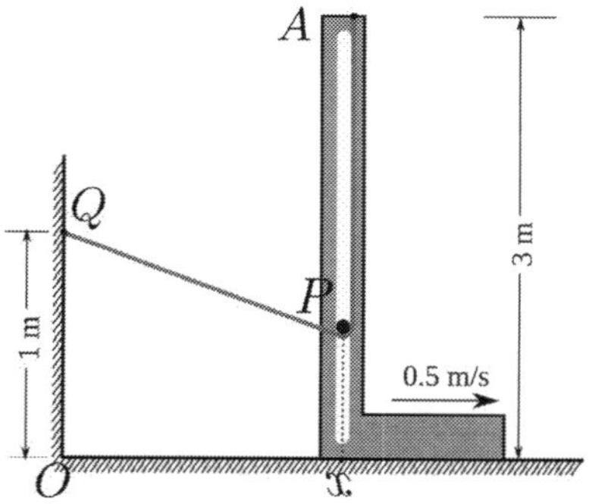

SOLUTION. On Cartesian axes Oxy of origin O we define as (x, y) the coordinates of pin P. P is constrained to move relative to the square due to the slot and relative to the support due to the cable. We can express these constraints by the geometric relationship $l = |AP| + |PQ| = Cst$ with $|AP| = l - y$ and $|PQ| = \sqrt{x^2 + (h - y)^2}$.

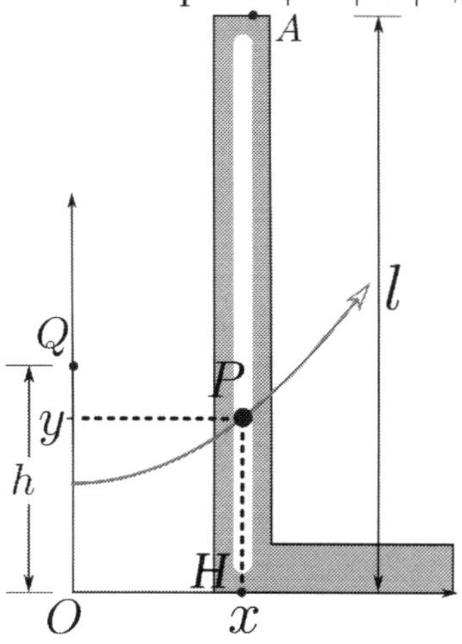

This expresses the fact that the cable's length l can be found as the sum of segments AP and PQ. This gives the equation

$$l = l - y + \sqrt{x^2 + (h - y)^2}$$

and, after simplifications,

$$\boxed{y = \frac{x^2 + h^2}{2h}} \qquad [1]$$

This shows that the trajectory of P is a parabola. We can differentiate [1] (valid at all time) to obtain $\dot{y} = x\dot{x}/h$. We can then find the speed of P as $v = (\dot{x}^2 + \dot{y}^2)^{1/2}$ at any location x:

$$\boxed{v = \dot{x}\left(1 + \frac{x^2}{h^2}\right)^{1/2}}$$

With $\dot{x} = 0.5$ m/s, $h = 1$ m, $x = 1.5$ m, we find $v =$ **0.90** m/s.
Equation [1] can also be found as the equality $|PQ| = |PH|$ where H is the projection of P onto the x-axis: the set of points P is known to be a parabola of focus Q and directrix Ox. ■

Exercise 1.26 A block P is suspended by a cable to a fixed horizontal support at O. A second cable connects P to block Q via a small pulley fixed at point A. At the instant shown, Q has velocity 1.2 m/s and acceleration 0.5 m/s^2 both directed upward.
Find the magnitude of the acceleration of P at this instant. Neglect the size of the pulley. ■

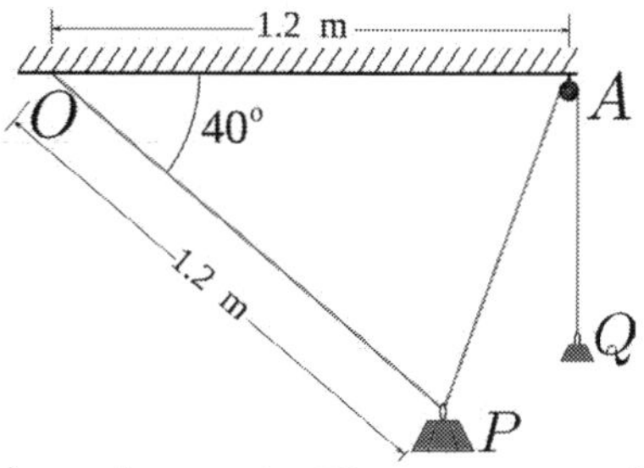

SOLUTION. We choose a fixed basis of unit vectors $(\hat{\mathbf{e}}_x, \hat{\mathbf{e}}_y)$ as shown in Figure 1.44 and origin O of the corresponding Cartesian coordinate system. We note that the path of P is the circle of center O and radius R. We then introduce the polar unit vectors $(\hat{\mathbf{e}}_r, \hat{\mathbf{e}}_\theta)$ associated with the motion of P. P has polar coordinates $(r = R, \theta)$ and its velocity and acceleration are given by

$$\mathbf{v}_P = R\dot{\theta}\hat{\mathbf{e}}_r, \qquad \mathbf{a}_P = R(\ddot{\theta}\hat{\mathbf{e}}_\theta - \dot{\theta}^2\hat{\mathbf{e}}_r)$$

Block Q is in rectilinear motion along line $(A, \hat{\mathbf{e}}_y)$. With the Cartesian coordinates (R, y_Q), its acceleration is given by $\mathbf{a}_Q = \ddot{y}_Q\hat{\mathbf{e}}_y$.

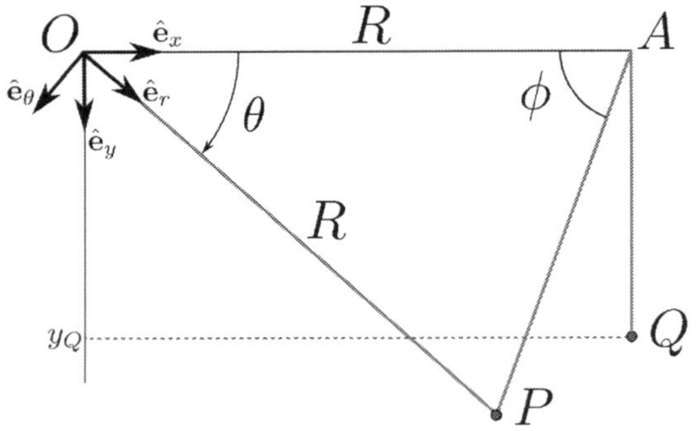

Figure 1.44 – *Problem 1.26*

We can relate coordinate y_Q to angle θ by guaranteeing that the length of cable PAQ remains constant along the motion. Since triangle OPA is isosceles, we find angle $\phi = \pi/2 - \theta/2$ and $|AP| = 2R\sin(\theta/2)$. We can now find the length of cable PAQ:

$$y_Q + 2R\sin\frac{\theta}{2} = Cst \qquad [1]$$

Since [1] is valid at all time, we can differentiate it w.r.t. time twice to obtain

$$\dot{y}_Q = -R\dot{\theta}\cos\frac{\theta}{2}, \qquad \ddot{y}_Q = \frac{R}{2}\dot{\theta}^2\sin\frac{\theta}{2} - R\ddot{\theta}\cos\frac{\theta}{2} \qquad [2]$$

This gives

$$R\dot{\theta} = -\frac{\dot{y}_Q}{\cos\frac{\theta}{2}}, \qquad R\ddot{\theta} = -\frac{\ddot{y}_Q}{\cos\frac{\theta}{2}} + \dot{y}_Q^2\frac{\sin\frac{\theta}{2}}{2R\cos^2\frac{\theta}{2}}$$

We now express the magnitude of $\mathbf{a}_P$ in terms of $\dot{y}_Q$ and $\ddot{y}_Q$:

$$|\mathbf{a}_P|^2 = R^2\dot{\theta}^4 + R^2\ddot{\theta}^2 = \frac{\dot{y}_Q^4}{R^2\cos^4\frac{\theta}{2}} + \left(-\frac{\ddot{y}_Q}{\cos\frac{\theta}{2}} + \dot{y}_Q^2\frac{\sin\frac{\theta}{2}}{2R\cos^2\frac{\theta}{2}}\right)^2$$

or

$$\boxed{|\mathbf{a}_P|^2 = \frac{\dot{y}_Q^4}{4R^2\cos^4\frac{\theta}{2}}(4 + \sin^2\frac{\theta}{2}) + \frac{\ddot{y}_Q^2}{\cos^2\frac{\theta}{2}} - \dot{y}_Q^2\ddot{y}_Q\frac{\sin\frac{\theta}{2}}{R\cos^3\frac{\theta}{2}}}$$

With $R = 1.2$ m and $\theta = 40°$, we find $|\mathbf{a}_P| = \mathbf{1.56}$ m/s^2. ■

Exercise 1.27 ▪In the block and tackle system shown below, the velocity and acceleration of Q are 0.55 m/s (downward) and 0.38 m/s^2 (upward), respectively.
Find the velocity and acceleration of block P at the instant shown. ▪

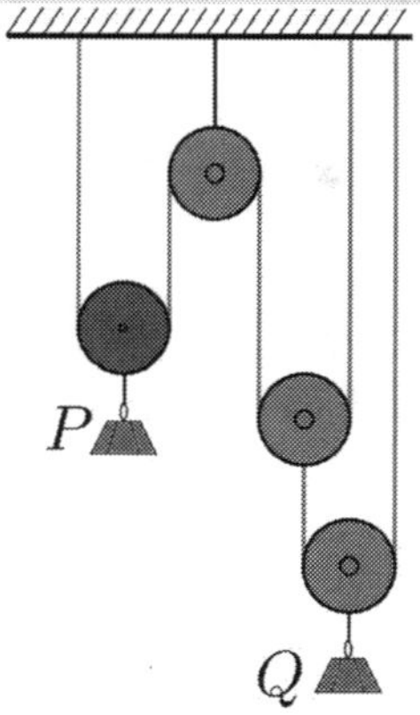

SOLUTION. The motion of both P and Q are rectilinear. To parametrize the position of the system, we need to define the position of the centers of the three translating pulleys. We introduce the coordinates z_P, z_Q and z_R of points P, Q and R on a vertical axis Oz, as shown in Figure 1.45.

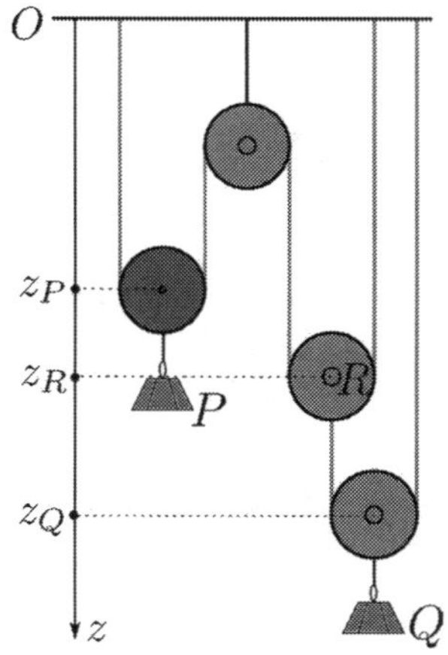

Figure 1.45 – *Problem 1.27*

These three points do not move independently. They are constrained by the two cables which interconnect the four pulleys. To express the constraints, we find the cable lengths which must remain constant during the motion. We start with the cable which connects P and R:

$$2z_P + 2z_R = Cst_1$$

Similarly, the cable connecting R and Q gives

$$(z_Q - z_R) + z_Q = Cst_2$$

We can eliminate z_P between these two equations to find

$$z_P + 2z_Q = Cst \qquad ([1])$$

This is a 1 DOF system. Since equation [1] is valid at all time, it can be time-differentiated to obtain $\boxed{\dot{z}_P = -2\dot{z}_Q, \quad \ddot{z}_P = -2\ddot{z}_Q}$. Numerically, we find $\dot{z}_P = \mathbf{1.1}$ m/s and $\ddot{z}_P = \mathbf{-0.76}$ m/s^2.

▪

PROJECT 1: Kinematic Design of Roads

In the design of roads, one is faced with the problem of connecting straight paths with appropriate curved paths. Referring to the figure displayed below, consider a vehicle P traveling at constant speed v_0 on a straight road Δ_1. The goal of this project is to design a transition road segment $\mathcal{C}$ to connect Δ_1 to straight road Δ_2 perpendicular to Δ_1. Cars are assumed to travel at maximum speed $v_{max} = 20.1$ m/s and must not exceed the acceleration $a_{max} = 8$ m/s^2.

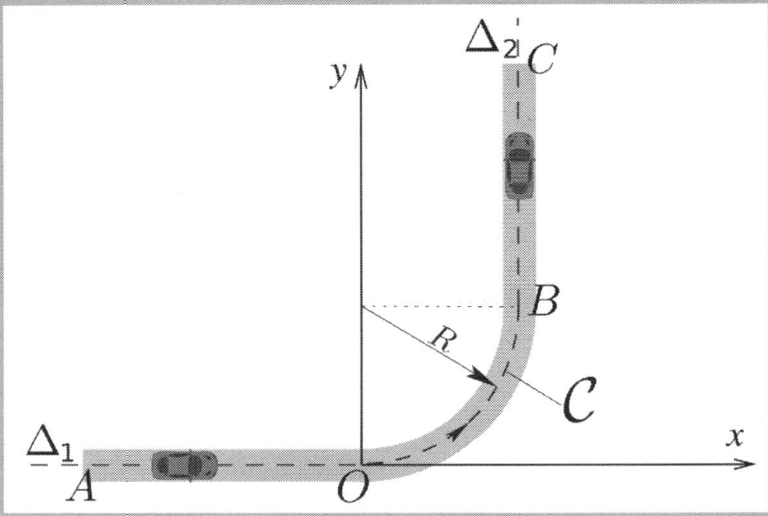

1. As a first design, consider the arc of circle of radius R shown in the figure which smoothly meets Δ_1 at O and Δ_2 at B. Show that this connecting path is not acceptable.
2. Propose a better design for $\mathcal{C}$ which connects O to point B, but which alleviates the problems of the circular arc. Give $\mathcal{C}$ in the form $(x(s), y(s))$ in terms of the arc-length $s = v_0 t$. Plot $\mathcal{C}$ along with the circular arc.
Hints: Assume that the curvature of $\mathcal{C}$ satisfies $\kappa = cs$, where the arc-length s is measured from origin O and c is a constant. Then deduce that $(x(s), y(s))$ are given by

$$x(s) = \sqrt{\frac{\pi}{c}}\,\mathrm{C}\left(s\sqrt{\frac{c}{\pi}}\right), \qquad y(s) = \sqrt{\frac{\pi}{c}}\,\mathrm{S}\left(s\sqrt{\frac{c}{\pi}}\right)$$

where the functions C and S are the Fresnel cosine and sine integrals

$$\mathrm{C}(z) = \int_0^z \cos\left(\frac{\pi}{2}u^2\right) du, \qquad \mathrm{S}(z) = \int_0^z \sin\left(\frac{\pi}{2}u^2\right) du$$

Then find constant c. Note that the two non-elementary functions C and S are available in most computer algebra systems.
3. Identify other engineering applications for which this design problem is relevant.

PROJECT 2: Kinematic Analysis of the Cycloid

Consider a point P of a bicycle wheel of radius R and center C. At $t = 0$, point P is in contact with the road. Center C moves at constant speed v_0. We assume that the wheel does not slip: hence the distance x traveled by C is equal to the arc-length $\widehat{IP} = R\theta$, where θ is the angle between line CI and line CP.

1. Find the Cartesian coordinates $(x(t), y(t))$ versus time t in terms of the parameters R and v_0. Sketch the trajectory of P (this curve is known as a *cycloid*).
2. Find the Cartesian components of the velocity and acceleration of P. Show that the angle between the acceleration and the velocity is $\theta/2$. In particular, find the velocity of P at the instants P comes in contact with the road. Sketch the velocity and acceleration vectors of P in the graph found in 1) at $\theta = \pi/4, \pi/2, 3\pi/4, 2\pi$.
3. Show that the radius of curvature of the path of P is given by

$$\rho = 4R \left| \sin \frac{v_0 t}{2R} \right|$$

Then show that the center of curvature K is given by $\mathbf{r}_{PK} = 2\mathbf{r}_{PI}$. Sketch the trajectories described by both P and K. Conclude that the path of K is also a cycloid.

1.9 Problems

Problem 1.1 A metro train can accelerate at the value $a_1 = 4$ km/(h.s) to reach a maximum speed of 90 km/h and brake at the deceleration of $a_2 = -3$ km/(h.s). Find the minimal time to reach two stations separated by the distance of 5.5 km.

Problem 1.2 A skydiver has reached the terminal velocity of $v_0 = 200$ km/h when he releases his parachute. This induces a constant deceleration of 15 m/s^2. For a safe landing the impact velocity of the skydiver with the ground must be below the value of 4 m/s. What the minimum safe height above which the skydiver can release his parachute?

Answer: 102.3 m

Problem 1.3 Two airplanes A and B are in rectilinear motion in the same plane at the constant speed $v_A = 550$ mph and $v_B = 1,200$ mph, respectively. Their relative direction is given by angle φ, with their initial relative position as shown in the figure.

a. Assume $\varphi = 60°$. Find the minimal distance reached between A and B.

b. With the same initial conditions, find the angle φ for B to intercept A.

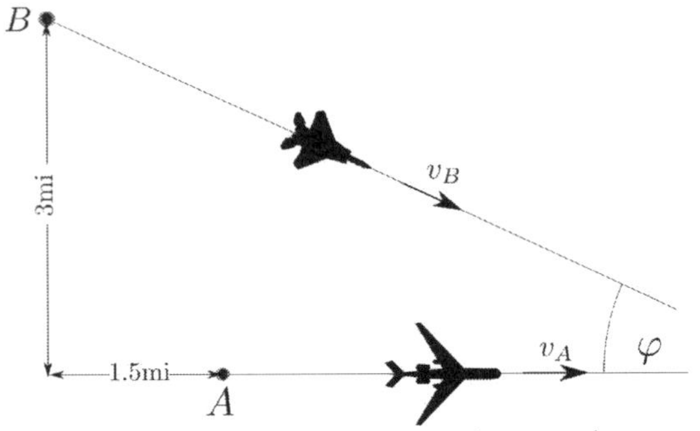

Answer: a). 1.35 mi.

Problem 1.4 The motion of a ship is observed from location O of the shore at three instants $t_0 =$ 12:20 pm, $t_1 =$ 12:40 pm, and $t_2 =$ 1:00 pm. The corresponding positions P_0, P_1 and P_2 of the ship are given by angles $\theta_0 = 79°.47$, $\theta_1 = 62°.18$, and $\theta_2 = 47°.14$. The ship is assumed to travel at constant speed $v_0 = 16$ km/h.

Find the distance OA from O to the landing point A and the corresponding arrival time.

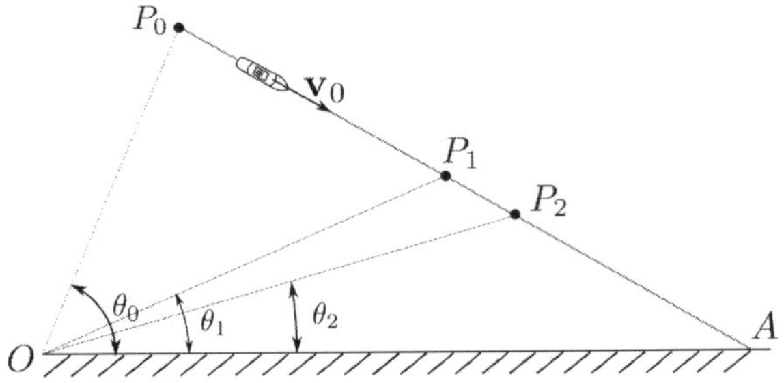

Problem 1.5 The *Doppler effect* is the change in frequency of a periodic event for an observer moving relative to its source. Assume that a sequence of acoustic beeps are produced at frequency f from a source A moving at constant speed v_A along axis Ox.

Find the frequency f_O of these beeps received by a detector at O. Assume that the acoustic waves travel at speed c.

Problem 1.6 A fighter aircraft travels at constant supersonic speed v and constant altitude h. An observer located at origin O at ground level receives sound waves traveling at speed c. Denote by $\tau(t)$ the time of reception of a sound signal emitted by P at time t. Assume that at time $t = 0$, P is directly above O. Use $v = 495$ m.s^{-1}, $c = 330$ m.s^{-1}, and $h = 1000$ m.

a. Determine $\tau(t)$. Plot this function for $-5 \leq t \leq 5$ (second).

b. Find the time $t*$ for which $d\tau/dt = 0$. Find the corresponding time τ_*. Find the position P_* of P at t_* and the position Q_* of P when the sound emitted at t_* is perceived at τ_*. Find the values of τ at $t_* - 0.1$ s and $t_* + 0.1$ s: conclude that the event occurring at t_* corresponds to a sonic boom.

c. Is the aircraft heard at O prior τ_*?

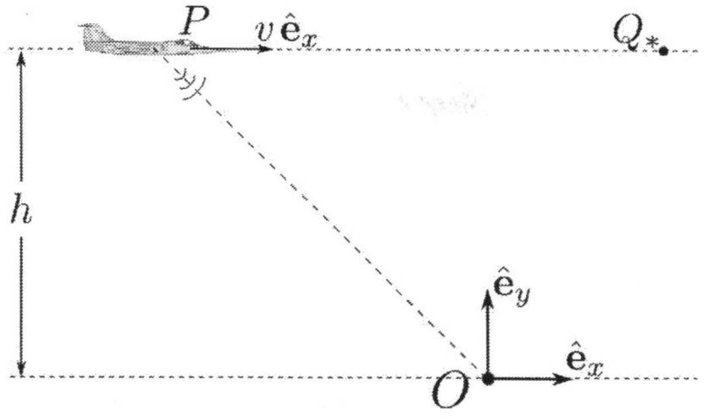

Problem 1.7 In a plane Oxy the acceleration of a particle A satisfies

$$\ddot{x} = x, \qquad \ddot{y} = y - 3x$$

with the initial conditions $x(0) = 1$, $y(0) = 0$, $\dot{x}(0) = -1$ and $\dot{y}(0) = \frac{3}{2}$. The coordinates x and y are measured in meters, and time t in seconds.
a. Find and plot the trajectory $(x(t), y(t))$. *Hint: first solve for $x(t)$, then look for a particular solution of the type $y = te^{-t}$.*
b. Find the time taken for A to travel from any position A_1 to position A_2 where the velocity is colinear to the position vector $\mathbf{r}_{OA_1}$.

Answer: b). $t_2 - t_1 = 1$ s.

Problem 1.8 An athlete can run at speed $U = 6$ mph and swim at speed $V = 3$ mph. He wishes to cross a river of width $D = 0.1$ mile to reach a position C a distance $L = 0.2$ mile downstream. He takes straight path AB along the river's edge and straight path BC across the river. Neglect the current speed. Determine the path which minimizes the time taken to reach C and find the corresponding time.

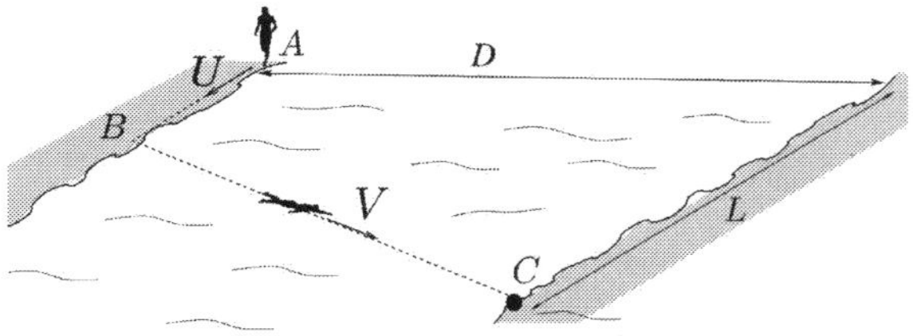

Answer: 3 min 44 s.

Problem 1.9 An aircraft A travels in a uniform rectilinear motion with velocity $\mathbf{v}_A$. A missile B is aimed from O toward A with constant velocity $\mathbf{v}_B$. At the time of release of B, the velocity $\mathbf{v}_A$ makes an angle α with line OA: $\alpha = (\mathbf{v}_A, \mathbf{r}_{OA})$.

a. At which angle $\beta = (\mathbf{v}_B, \mathbf{r}_{OA})$ should B be aimed so as to intercept A. Can B always intercept A?
b. Which values of (α, β) should be chosen so that B intercept A in minimum time. Characterize the solution geometrically.

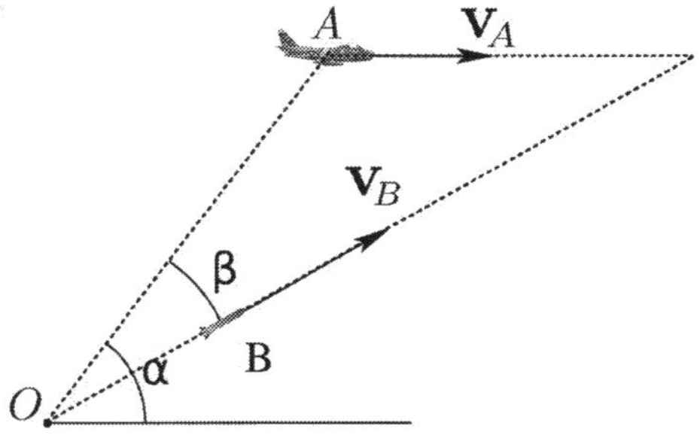

Problem 1.10 A rocket Q is launched from location A from rest, at constant acceleration γ. At a same instant, a projectile P departs from O at speed v_0 to intercept Q. Show that v_0 must satisfy the condition $v_0^2 > D(g + \gamma)$. Find the values of launch angle α which allow P to intercept Q. Under what condition P reaches Q while climbing?

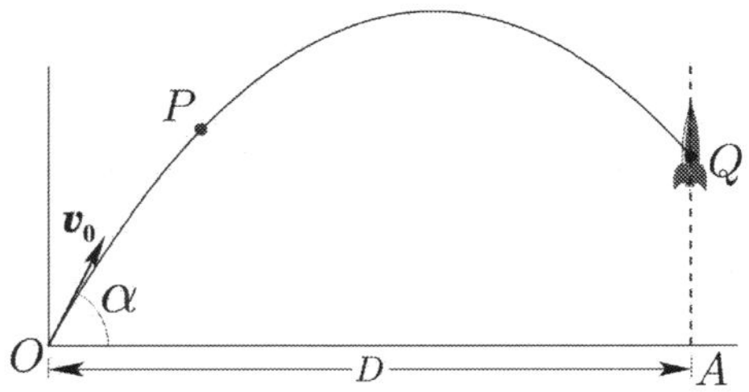

Problem 1.11 A bomb P is dropped from a height of 18 m from a airplane traveling at the constant horizontal speed 380 km/h. P is aimed at a dam of height 10 m measured above the water line. Denote by D the horizontal distance measured from the point of release of P to the dam.

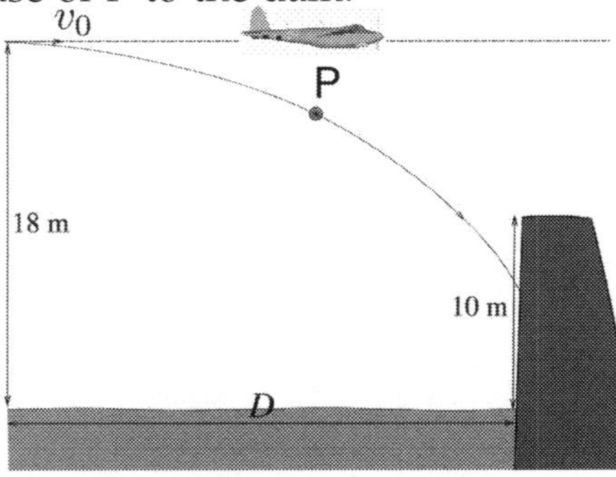

Find the allowable range of values taken by D for the bomb to hit the dam above the water line.

Answer: $135\ m \leq D \leq 202\ m.$

Problem 1.12 A particle P is in motion along the trajectory $\mathcal{C}$ described in Cartesian coordinates by

$$x = 3t, \qquad y = 3t^2, \qquad z = 2t^3$$

where $t \geq 0$ is measured in seconds and (x, y, z) in meters.
a. Find the velocity vector of P.
b. The line tangent to $\mathcal{C}$ at P intersects plane Oxz at A and plane Oxy at B. Find the trajectories of A and B and show that the ratio $|AP|/|BP|$ stays constant.

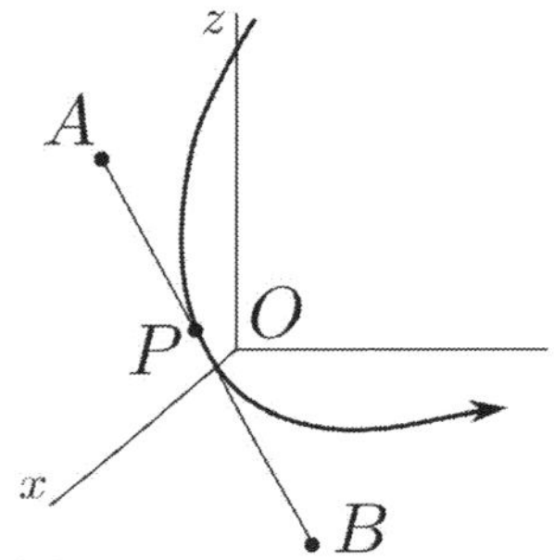

Answer: $|AP|/|BP| = 3/2$.

Problem 1.13 On May 24, 1941, the battleship Bismarck may have sunk the battle cruiser H.M.S. Hood by firing a shell at a distance of 15 km. Assume that the H.M.S. Hood travels at constant speed $v_B = 50$ km/h relative to the referential attached to the Bismarck. The shell initial velocity is $v_A = 800$ m/s.
Find the angle β relative to the line of sight (at the time of the firing) and the launch angle α relative to the horizontal which must be given to the gun so that the projectile fired from the Bismark hits the H.M.S. Hood.

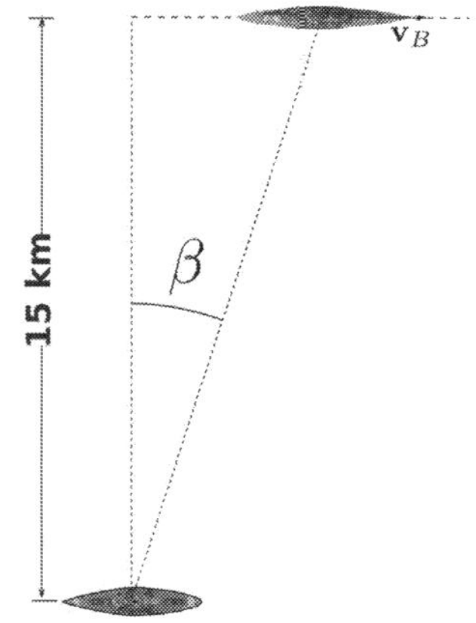

Problem 1.14 A projectile P is shot from a point O with an initial velocity of magnitude v_0 with launch angle α relative to the ground Ox. P travels at constant acceleration $\mathbf{a}_P = -g\hat{\mathbf{e}}_y$.

Show that the set of points of plane Oxy which can be reached by P by varying launch angle α (in the interval $0 < \alpha < \pi$) at constant initial speed v_0 is contained within the so-called *safety parabola* of equation

$$y = -\frac{g}{2v_0^2}(x^2 - \frac{v_0^4}{g^2})$$

Sketch this parabola along with the trajectories corresponding to $\alpha = 80°, 60°, 45°, 30°, 10°$ on the "normalized" axes $X = gx/v_0^2$ and $Y = gy/v_0^2$.

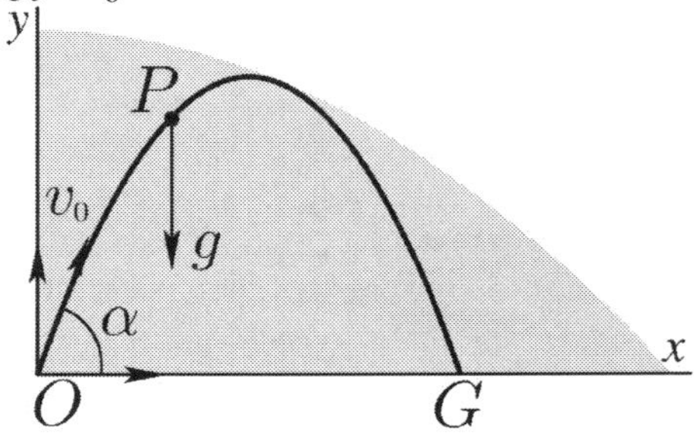

Problem 1.15 A dart is thrown at initial speed v_0 toward the bulls eye B of a target located at the distance D and height H from the launch point A.

a. Under what condition can the dart reach B? If this condition is satisfied, show that there exist two possible values of the launch angle α to reach B from A. Examine the case $v_0 = 8$ m/s, $D = 8$ m and $H = 2$ m.

b. Show that, if points A and B are at the same level ($H = 0$), the two trajectories satisfy $\alpha_1 + \alpha_2 = \pi/2$.

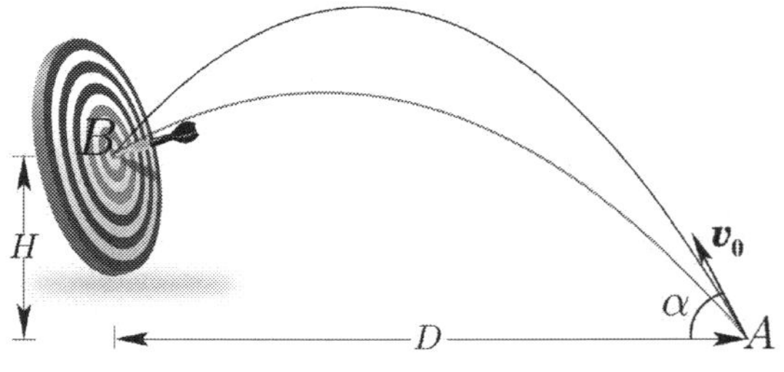

Problem 1.16 The trajectory $\mathcal{C}$ of a particle P

in plane Oxy is described by the *strophoid*

$$x(t) = \frac{t}{1+t^3}, \quad y(t) = \frac{t^2}{1+t^3}$$

where x, y are measured in meters, t in seconds with $-1 < t < \infty$.
a. Find the velocity of P at time t.
b. Plot $\mathcal{C}$: in particular study the shape of P in the limits $t \to -1$ and $t \to \infty$.
c. Show that three positions P_1 at t_1, P_2 at t_2 and P_3 at t_3 of P on $\mathcal{C}$ are aligned if the product $t_1 t_2 t_3$ is a constant k (to find).

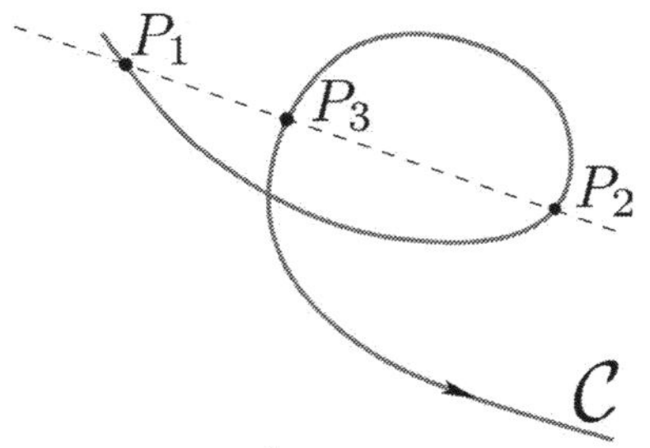

Answer: c) $t_1 t_2 t_3 = -1$ s^3.

Problem 1.17 The trajectory $\mathcal{C}$ of a particle P in motion in a referential $\mathcal{E}$ of Cartesian axes $Oxyz$ is described by

$$x(t) = \frac{1+t}{1+t^2}, y(t) = \frac{t(1+t)}{1+t^2}, z(t) = \frac{1-t^2}{1+t^2}$$

where x, y, z are measured in meters, t in seconds.
a. Show that curve $\mathcal{C}$ is contained in a plane OXY.
b. Show that $\mathcal{C}$ is necessarily closed.
c. Find the velocity and acceleration of P at time $t = 1$ second.
d. Find and then plot the trajectory of P in OXY.

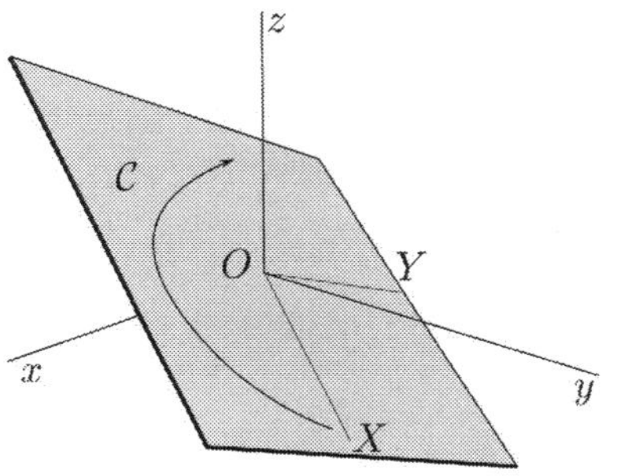

Problem 1.18 A dog B is chasing a robber A which runs at constant speed v_A along straight line $y = D$ of plane Oxy. At time $t = 0$, B is at origin O at a distance D from A. B chases A at constant speed $v_B > v_A$ so that its velocity vector $\mathbf{v}_B$ points toward A at all time. Denote by ϕ the angle between line Ox and line BA and by l the distance from A and B.
a. Show that distance l satisfies

$$\frac{dl}{dt} = (v_A \cos\phi - v_B) \qquad [1]$$

b. Show that the time T taken for B to reach A satisfies

$$v_A T = \int_0^T v_B \cos\phi \, dt \qquad [2]$$

c. Use [1-2] to find the expression of T in terms of D, v_A and v_B. Use $D = 100$ m, $v_A = 17$ km/h, $v_B = 45$ km/h.

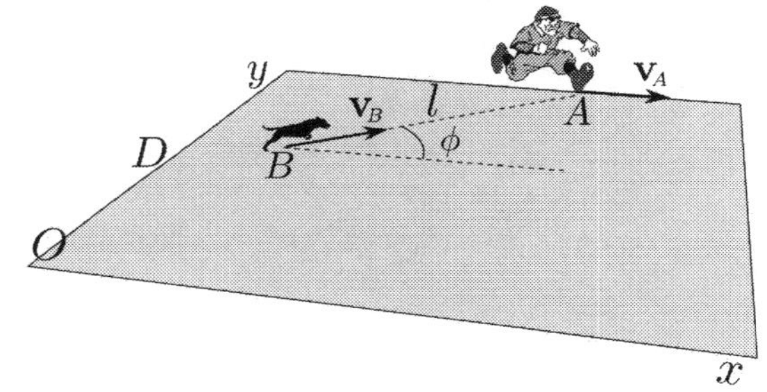

Answer: c) $T = 9.3$ s.

Problem 1.19 A particle P moves in a plane Oxy along a curve given by

$$x(t) = r_0 e^{-kt} \sin(\omega t), \quad y(t) = r_0 e^{-kt} \cos(\omega t)$$

where r_0, k and ω are positive constants. Consider the point Q whose position is defined by $\mathbf{r}_{OQ} = \frac{1}{\omega}\mathbf{v}_P$.
a. Find the polar coordinates (r, θ) and (ρ, ϕ) of P and Q.
b. Find the velocities $\mathbf{v}_P$ and $\mathbf{v}_Q$ in polar coordinates.
c. Find the acceleration $\mathbf{a}_P$ in polar coordinates.

Problem 1.20 A particle P is in motion along the curve $r = r_0(1 + \cos\theta)$. At $t = 0$, P departs from $(r, \theta) = (2r_0, 0)$ with the speed v_0. P moves toward O in such a way that its acceleration is central, that is, it always points toward O.
a. Show that the speed of P satisfies $v = v_0\sqrt{\frac{8r_0^3}{r^3}}$.
b. Find the time taken by P to reach O.

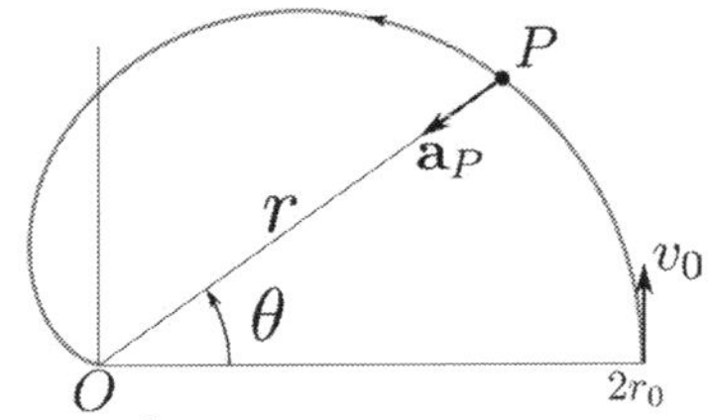

Answer: b). $T = \frac{3\pi}{4}\frac{r_0}{v_0}$

Problem 1.21 A particle P is in motion along the parabola $y^2 = 2px$ ($p = 0.5$ km) at the constant speed $v_0 = 25$ km/h. Find the acceleration of P at the positions $y = 4.5$ km, $y = 2.25$ km and $y = 0$.

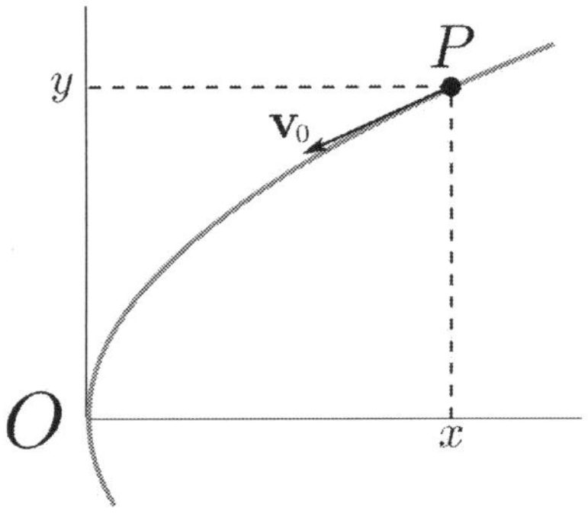

Problem 1.22 A particle P of polar coordinates (r, θ) is in motion in a plane Oxy. Its motion is characterized by the following properties:
(i) its acceleration always points toward O, that is, $a_\theta = 0$,
(ii) its hodograph is the circle centered at point $(0, v_0)$.
Show that the radial acceleration of P satisfies

$$a_r = \frac{k}{r^2}$$

where k is a constant. *Recall that the hodograph is the trajectory of point Q whose position is defined by $\mathbf{r}_{OQ} = \mathbf{v}_P$.*

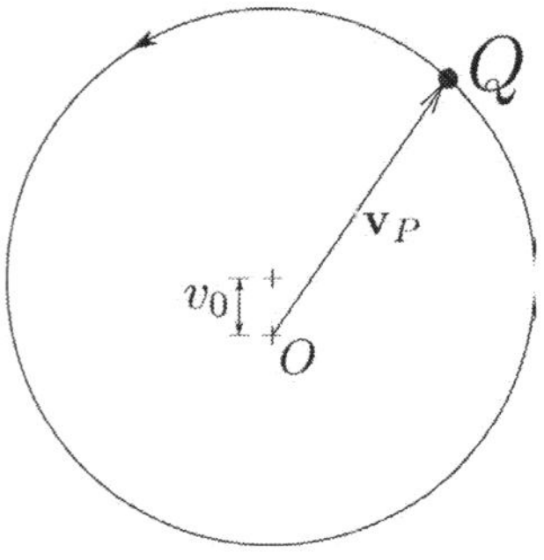

Problem 1.23 A satellite P of polar coordinates (r, θ) is in motion around the Earth O along the ellipse of equation:

$$r = \frac{p}{1 + e\cos\theta}$$

where p and e are constant parameters ($0 < e < 1$). At time $t = 0$, P has speed v_0 at the position $\theta = 0$.

a. Show that the velocity $\mathbf{v}_P$ of P can be written as the sum of two vectors $\mathbf{A}$ and $\mathbf{B}$ of constant magnitude, with $\mathbf{A}$ directed along the y-axis and $\mathbf{B}$ directed along the transverse direction. Find the expressions of $\mathbf{A}$ and $\mathbf{B}$.

b. Using the expression of $\mathbf{v}_P$ found in a), find the acceleration $\mathbf{a}_P$ of P as a function of r. Then find the hodograph of P.

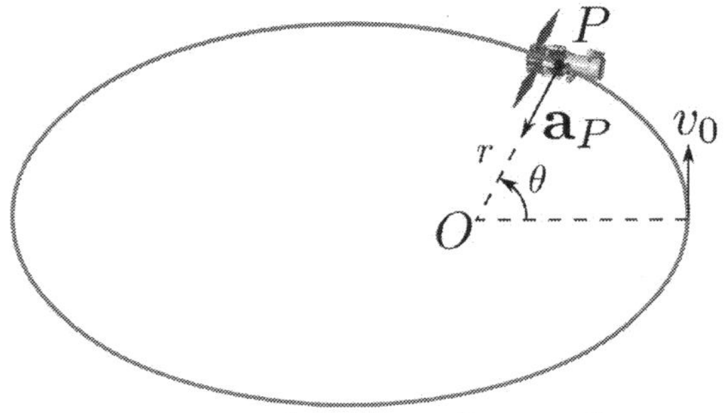

Answer: b) $\mathbf{a}_P = (C/r^2)\hat{\mathbf{e}}_r$, $C = pv_0^2/(1+e)^2$.

Problem 1.24 A slider A moves along a circular guide of radius $R = 0.5$ m by the rotation of crank arm OA at constant rate of rotation $\dot{\theta} = 0.40$ rad/s in the clockwise direction.

a. Find the trajectory of A in polar coordinates (r, θ).

b. Find the velocity and acceleration of A on the corresponding basis $(\hat{\mathbf{e}}_r, \hat{\mathbf{e}}_\theta)$ at any angle θ. Find the magnitude of these vectors for $\theta = 60°$. Is A in uniform rotation?

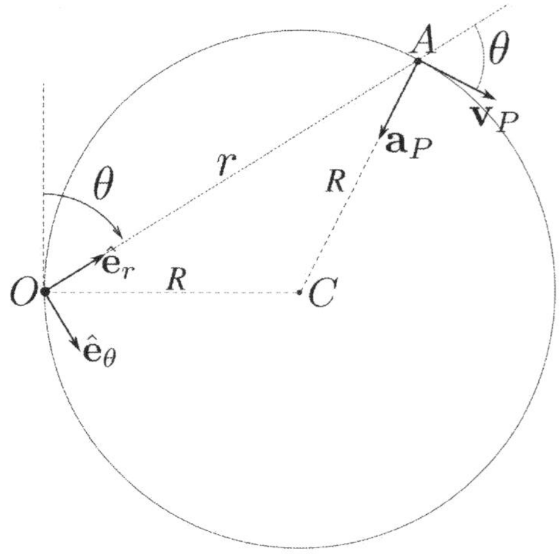

Problem 1.25 A point A moves along a circle $\mathcal{C}$ of radius R. Given a fixed point O on $\mathcal{C}$, consider the point P on line OA at a distance R from A as shown. Define θ as the angle between a fixed line (axis Ox) and line OA. Assume that $\dot{\theta} = \omega =$ is a positive constant.
a. Plot the trajectory of P along with that of A. Show that P passes through O twice when A travels once around $\mathcal{C}$ (find the corresponding values of θ, $-\pi \leq \theta \leq \pi$).
b. Find the velocity and acceleration of P when P passes through O.

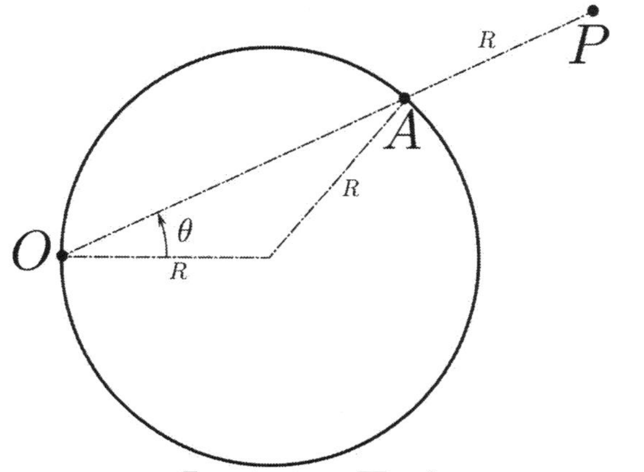

Answer: $v_P = \sqrt{3}R\omega$, $|\mathbf{a}_P| = \sqrt{13}R\omega^2$.

Problem 1.26 A car P climbs a road traced out on a mountain of height H and base radius R. The mountain is defined as the axisymmetric surface $z = H(1 - \frac{r^2}{R^2})$. P moves at constant speed v_0, and climbs the same constant height $\Delta z = h$ every turn around the mountain.
a. Express the velocity of P as a function of r, v_0, and the geometric constants h, H and R.
b. Find the time T taken to reach the top of the mountain (write T as a definite integral). Evaluate T numerically for $H = 1500$ m, $R = 1000$ m, $h = 100$ m and $v_0 = 25$ km/h.

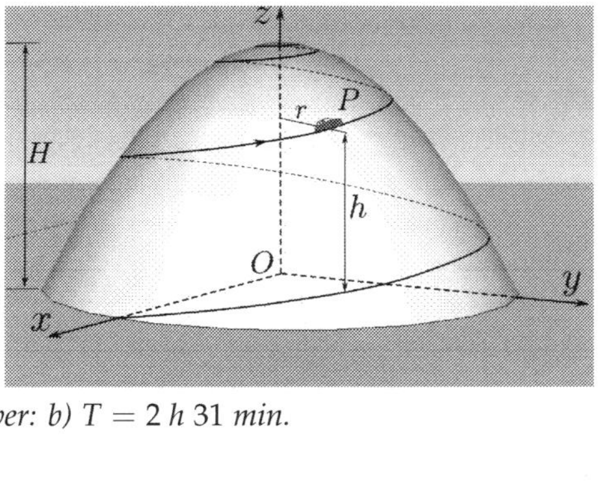

Answer: b) $T = 2$ *h 31 min.*

Problem 1.27 A particle P is in motion in a referential $\mathcal{E}$ of Cartesian axes $Oxyz$ in the following manner: P moves along a circular support $\mathcal{C}$ of center O and radius $R = 1$ m at constant angular velocity $\omega = 1$ rad/s, while $\mathcal{C}$ rotates about its diameter Oz at constant angular velocity $\omega = 1$ rad/s. At time $t = 0$, $\mathcal{C}$ lies in the plane Oxz and P is at point A on axis Ox.
a. Show that the Cartesian coordinates of P satisfy

$$x(t) = \cos^2 t,\ y(t) = \sin t \cos t,\ z(t) = \sin t$$

b. Find the velocity and acceleration of P in cylindrical coordinates.
c. Find the trajectory of Q projection of P onto plane Oxy. Deduce that the trajectory of P is the intersection of a sphere and a cylinder. Then sketch the trajectory of P.

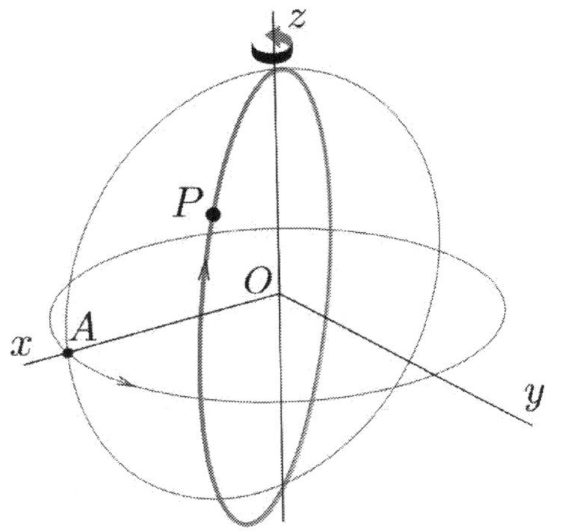

Problem 1.28 Consider the surface $2pz = x^2 + y^2$ of a referential $\mathcal{E}$ of Cartesian axes $Oxyz$ (p is positive constant). A particle P is in motion on this surface so that its acceleration $\mathbf{a}_P$ is always directed along line PQ, where Q is the projection of P onto axis Oz.

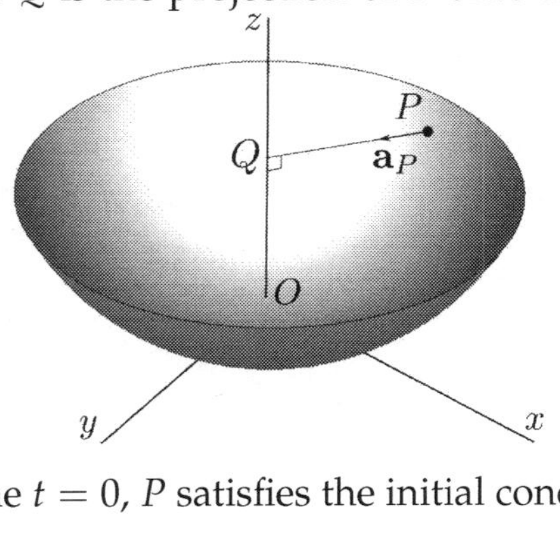

At time $t = 0$, P satisfies the initial conditions

(i) $z = z_0$ and $\dot{z} = \dot{z}_0$ ($\dot{z}_0 < 0$),
(ii) $y = 0$, $\dot{x} = 0$.
a. Find the time T taken by P to reach O.
b. Find the position of P as a function of time.

Problem 1.29 A particle P moves in a planar curve whose equation is described in polar coordinates (r, θ) by $r = 2R\cos\theta$. In addition, P moves at constant speed v_0 and $\dot{\theta} > 0$ at all time.
a. Find the velocity of P in polar coordinates and determine $\dot{\theta}$ in terms of θ and constants R and v_0.
b. Determine the tangent and normal unit vectors $(\hat{\mathbf{e}}_T, \hat{\mathbf{e}}_N)$ at any given angle θ.
c. Find the acceleration $\mathbf{a}_P$ on $(\hat{\mathbf{e}}_T, \hat{\mathbf{e}}_N)$.
Answer: c). $\mathbf{a}_P = (v_0^2/R)\hat{\mathbf{e}}_N$.

Problem 1.30 A point P of plane Oxy describes the trajectory

$$x(t) = t, \qquad y(t) = \cosh(t)$$

with t is measured in seconds, and (x, y) in meters. A point Q moves along P as follows

$$\mathbf{r}_{OQ} = \mathbf{r}_{OP} + l(t)\hat{\mathbf{e}}_N$$

where $\hat{\mathbf{e}}_N$ is the unit normal vector to the path of P. The distance $l(t)$ must be found so that $\mathbf{v}_Q$ is colinear to $\hat{\mathbf{e}}_N$ at all time.
a. Show that Q must take the position of the center of curvature of the path of P. Determine the Cartesian coordinates $(x_Q(t), y_Q(t))$ of Q.
b. Find the expression of the velocity of Q on $(\hat{\mathbf{e}}_T, \hat{\mathbf{e}}_N)$. Deduce the unit normal/tangential $(\hat{\mathbf{e}}_{QN}, \hat{\mathbf{e}}_{QT})$ vectors corresponding to the path of Q.
c. Find the expression of the acceleration of Q on $(\hat{\mathbf{e}}_{QN}, \hat{\mathbf{e}}_{QT})$.

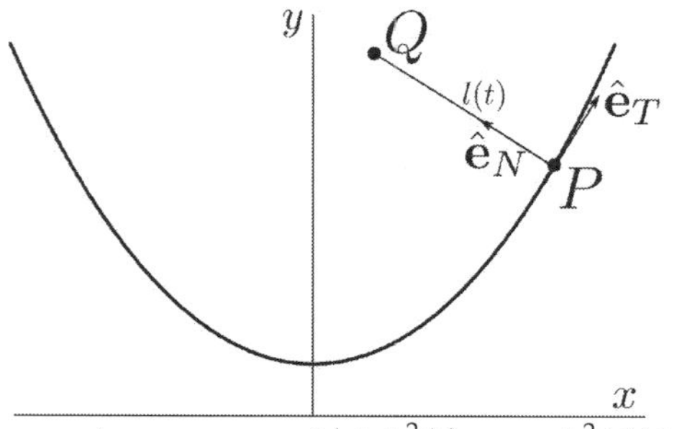

Answer: c). $\mathbf{a}_Q = 2(\sinh^2(t) + \cosh^2(t))\hat{\mathbf{e}}_{QT} + 2\sinh(t)\hat{\mathbf{e}}_{QN}$

Problem 1.31 A particle P moves in a plane Oxy along a curve whose polar form takes the expression

$$r = 4r_0\cos^3\frac{\theta}{3}, \qquad 0 \leq \theta \leq \frac{3\pi}{2}$$

at constant angular speed $\dot{\theta} = \omega > 0$ (r_0 is a positive constant).
a. Sketch the trajectory of P, then find the velocity of P. Determine speed v_P as a function of θ.
b. Find the total length L traveled by P and the duration T of the motion.
c. Find the acceleration of P. Deduce the radius of curvature of the path as a function of θ.
Answer: a). $v_P = 4r_0\omega\cos^2\frac{\theta}{3}$. *b).* $L = 3\pi r_0$. *c).* $\rho = 3r_0\cos^2\frac{\theta}{3}$.

Problem 1.32 The motion of a particle P in plane Oxy is characterized by the conditions
(i) P is at O and $\mathbf{v}_P = v_0\hat{\mathbf{e}}_x$ at time $t = 0$,
(ii) the tangential acceleration a_T takes the constant value a_0 at all time,
(iii) the normal acceleration takes the expression $a_N = \omega^2\rho$ at all time, where ω is a positive constant and ρ is the radius of curvature.
a. Find the velocity $\mathbf{v}_P$ of P as a function of time t.
b. Find the trajectory $(x(t), y(t))$ of P as a function of time t. Plot the XY-curve for $\frac{a_0}{\omega v_0} = 0.5$ in the variables $X = \omega x/v_0$ and $Y = \omega y/v_0$.
c. Choosing $a_0 = 0.4$ m/s^2, $v_0 = 1.15$ m/s and $\omega = 0.2$ rad/s, find speed v_P and radius of curvature ρ when P has traveled the distance of 10 m.

Answer: c). $v = 3.05$ *m/s*, $\rho = 15.3$ *m.*

Problem 1.33 A particle P is in motion in a plane Oxy in such a way that
(i) its acceleration always points toward O,
(ii) its velocity keeps a constant angle $\alpha = 45°$ with the radial line.
At $t = 0$, the position of P is defined by $r = 0.5$ m and $\theta = 0$.
a. Show that the radial acceleration of P follows a $1/r^3$ law.
b. Find and sketch the trajectory of P.

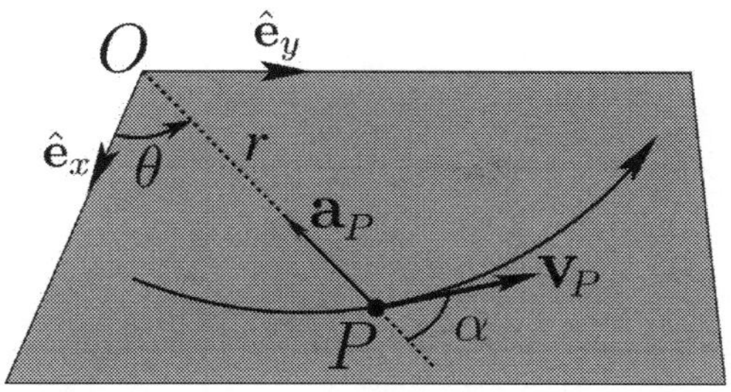

Problem 1.34 A particle P in planar motion is attracted to a fixed point O in such a way that its velocity $\mathbf{v}_P$ makes the angle $\theta/2$ with respect to transverse unit vector $\hat{\mathbf{e}}_\theta$ when the radial unit vector $\hat{\mathbf{e}}_r$ makes the angle θ with respect to axis Ox. Furthermore P moves at constant speed v_0 and is initially at point A on axis Ox at distance $r = r_0$ from O.

a. Show that the path of P is given in polar coordinates by the curve

$$r = \frac{r_0}{2}(1 + \cos\theta)$$

and find θ versus t. Sketch this path. Find the time taken to reach O (*use the integral* $\int \tan x\, dx = -\ln|\cos x| + Cst$).

b. Determine the acceleration $\mathbf{a}_P$ of P in polar coordinates, then find the normal and tangential components of $\mathbf{a}_P$.

c. Deduce the radius of curvature at any location of the path.

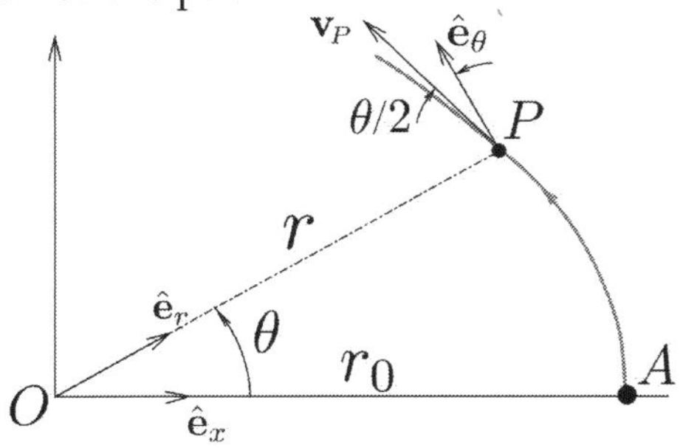

Problem 1.35 Given two perpendicular axes Ox and Oy, consider the point $A(a, 0)$ on axis Ox and point $B(0, b)$ on axis Oy, with a and b two positive constants.

a. Find the parabola tangent to axis Ox at A and tangent to axis Oy at point B.

b. A dog D and his master M are in motion in plane Oxy in the following way: (i) M travels along axis Oy at constant speed v leaving O at $t = 0$ toward B, (ii) D, starting at A at $t = 0$, travels along the parabola found in a) by following M at all time, meeting M at B. Find the motion of D as a function of time. Find the minimum speed of D.

Problem 1.36 Four boys A, B, C and D play a running game which consists of chasing each other as follows: starting at the corner points of a square of side length 50 m, A chases B, B chases C, C chases D and D chases A. Assume that the four boys run at the same constant speed $v = 6$ m/s. Denote by l the distance between any two adjacent boys.

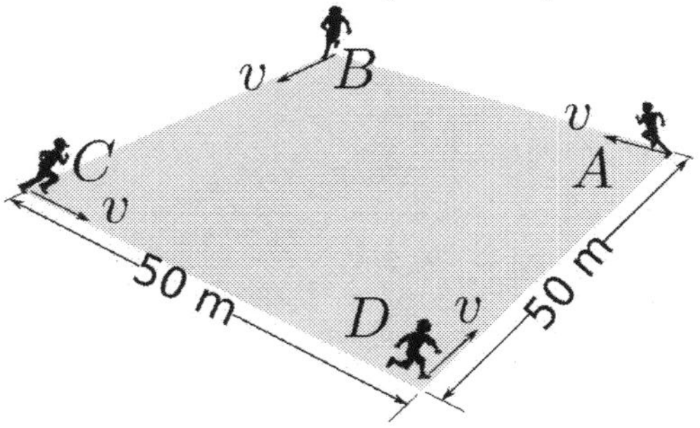

a. By writing $l^2 = \mathbf{r}_{AB} \cdot \mathbf{r}_{AB}$ show that

$$l\frac{dl}{dt} = \mathbf{r}_{AB} \cdot (\mathbf{v}_B - \mathbf{v}_A)$$

Then express dl/dt in terms of speed v.

b. Use the result of a) to find l versus t. Deduce the time T taken by the boys to meet. *Hint: use the fact that at any given time A, B, C and D form a square of side length l.*

Answer: b). $T = 8.3$ s.

Problem 1.37 In the crankshaft OAB, point A moves at constant speed v_A about O in the clockwise direction. The endpoint B of the connecting rod AB is in rectilinear motion along axis Ox. The geometry of the crankshaft is characterized by length $L = |OA| = |AB|$ and varying angle ϕ.

a. By finding geometric relationships between the positions of A and B, determine the velocity of B in terms of v_A, L and angle ϕ.

b. Consider the midpoint C of AB: from $\mathbf{r}_{AB} = 2\mathbf{r}_{AC}$ find the velocity of C in terms of v_A, L and angle ϕ. Show that $\mathbf{v}_C$ is always perpendicular to line IC where I is the intersection of line OA and the normal at B of line OB.

c. Find the acceleration of B in terms of v_A, L and angle ϕ.

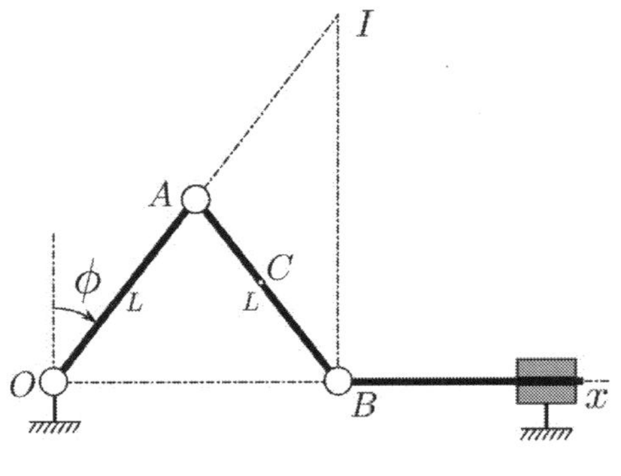

Answer: c). $\mathbf{a}_B = -2\frac{v_A^2}{L}\sin\phi\,\hat{\mathbf{e}}_x$

Problem 1.38 A rigid rod of length $2L$ supports two masses A and B at its endpoints. It is placed on top of a cylindrical support of radius R and exhibits oscillations about its horizontal position without slipping at contact point I. The position of the rod is defined by angle θ between the vertical and line OI. When $\theta = 0$, the contact point I coincides with the midpoint of AB.

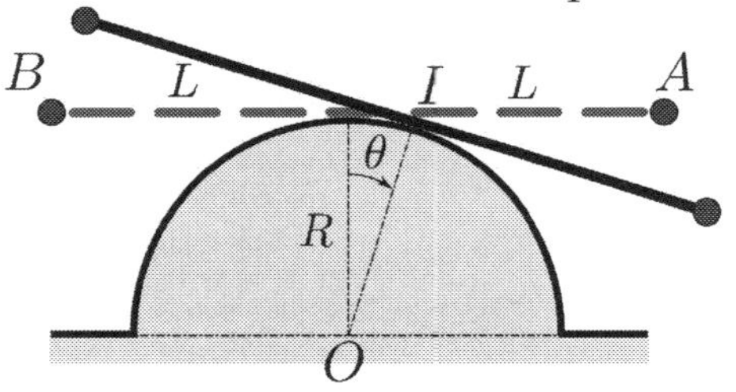

a. Find the Cartesian coordinates of A and B as a function of θ, R and L.

b. Determine the velocities of A and B as a function of θ, $\dot{\theta}$, R and l. Verify that both $\mathbf{v}_A$ and $\mathbf{v}_B$ are colinear to line OI.

Problem 1.39 A slider B moves in plane Oxy along a circular path of diameter $d = 50$ cm at uniform speed $v_B = 1.6$ m/s. It is connected to a collar A which can translate along axis Oz. At the instant shown, the position of A is defined by $h = 50$ cm and B lies on axis Oy. Neglect the size of the slider.

a. Find the velocity and acceleration of A at the instant shown.

b. At a later time, line OB makes the angle $\theta = 30°$ with axis Oy. Find the velocity and acceleration of A at this position.

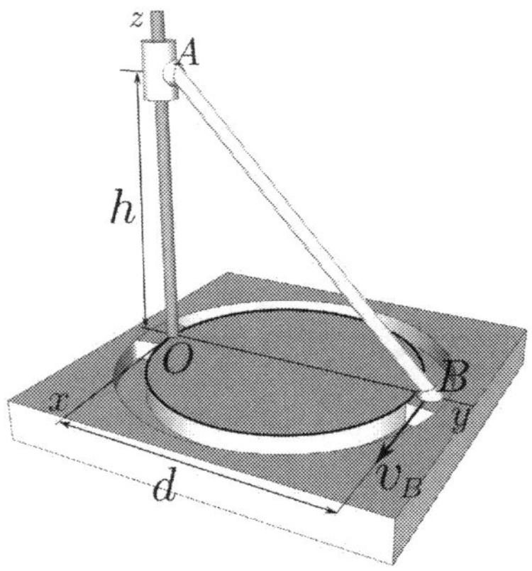

Problem 1.40 The collar B slides along a vertical guide. At the instant shown, B reaches the velocity $\mathbf{v}_B = 0.8$ m/s and acceleration $a_B = 0.4$ m/s^2, both directed downward. Thanks to the connecting rod AB of length $l = 1.0$ m, the end point A of arm OA travels along a circular path of center O and radius $R = 0.65$ m.

a. Find the velocity of A at the instant shown.

b. Find the acceleration of A at the instant shown.

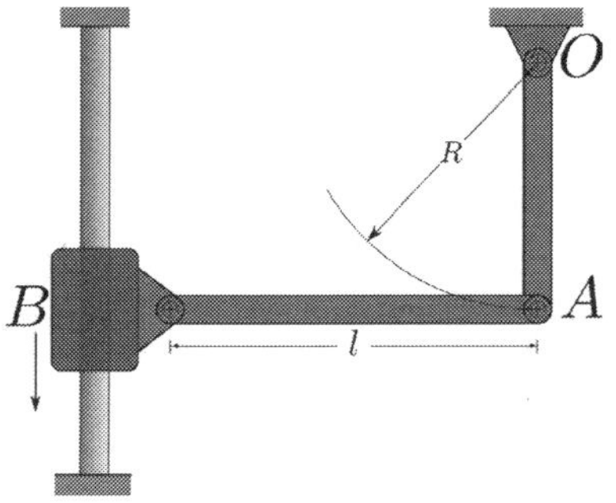

Answer: a). $v_A = 0$, *b).* $a_A = 0.64$ m/s^2 $\leftarrow$

Problem 1.41 Consider the mechanical system of problem 1.40. Now assume that B moves uniformly at speed $v_B = 0.8$ m/s downward in the same configuration. Find the velocity and acceleration of A 0.6 second after the instant shown. Use $l = 1.0$ m and $R = 0.65$ m.

Problem 1.42 Two particles A and B connected by a rod of length l are in motion in a horizontal plane Oxy. A moves along axis Ox at constant speed v_0. The orientation of the rod defined by angle θ must be such that the speed of B must be equal to v_0 at all time. At time $t = 0$, $\theta = \pi/2$ so that B lies on axis Oy.

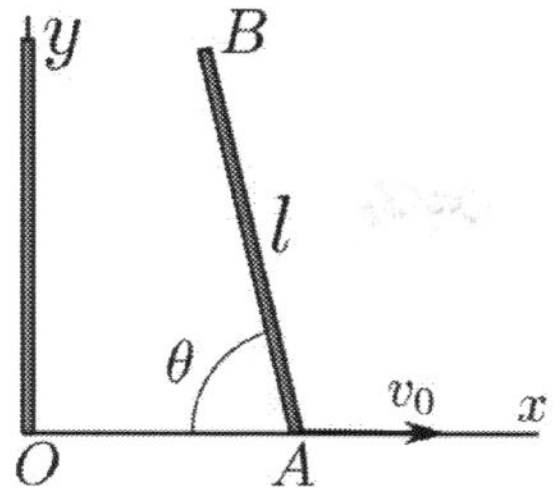

a. Find the equation which gives $\dot{\theta}$ versus θ. Show that two possible trajectories are possible. Find the trajectories of B in the form $(x_B(t), y_B(t))$. Sketch these trajectories. (use $\int \frac{d\theta}{\sin\theta} = \ln(\tan(\theta/2))$).

b. Show that $\mathbf{v}_A$ and $\mathbf{v}_B$ have equal projections on line AB.

Problem 1.43 In this slider-crank mechanism, the rotation of the crank OA is transformed into a reciprocating motion of the slider. The connecting rod AB is pinned at A with the crank and at B with the slider. At the instant shown, line OA is perpendicular to line OB and A has angular velocity $\dot{\theta} = 0.75$ rad/s and angular acceleration $\ddot{\theta} = -0.03$ rad/s^2. With $L = 0.8$ m, find the velocity and acceleration of B at the instant shown.

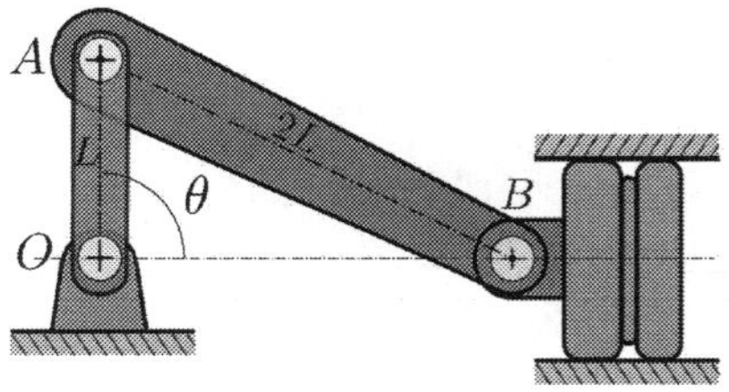

Problem 1.44 A block and tackle is a system of two or more pulleys connected by ropes or cables, usually used to lift or pull heavy loads. It provides mechanical advantage that amplifies the force applied to the rope. In this block and tackle a single rope is pulled downward at the speed 0.5 m/s and acceleration 0.1 m/s^2 both directed downward at a particular instant.

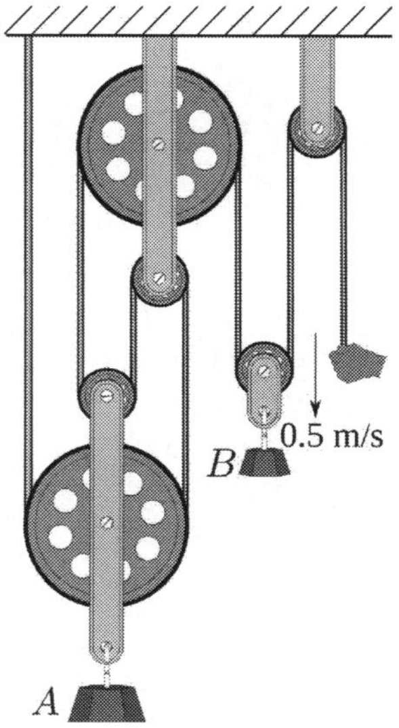

Find the velocity and acceleration of the masses A and B.

Problem 1.45 A nitrogen molecule P is in motion at constant velocity in the interior of a container in the shape of a rectangular cuboid of (dimensionless) size $10 \times 20 \times 30$ as shown in the figure. It departs from the center O of the container at the initial (dimensionless) velocity $\mathbf{v}_0 = (\hat{\imath} + \sqrt{2}\hat{\jmath} + \sqrt{3}\hat{\mathbf{k}})$. It moves along straight paths by bouncing elastically on the walls of the container. For instance, if P bounces off the face $x = 5$ with incoming velocity $\mathbf{v}^- = (a\hat{\imath} + b\hat{\jmath} + c\hat{\mathbf{k}})$, then the rebound velocity is $\mathbf{v}^+ = (-a\hat{\imath} + b\hat{\jmath} + c\hat{\mathbf{k}})$.

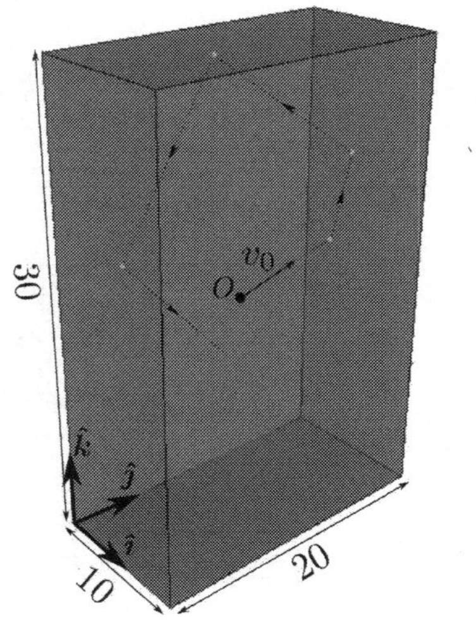

a. Which face of the container is hit first, and at what time?
b. Find the position of P at (dimensionless) time $t = 10$.

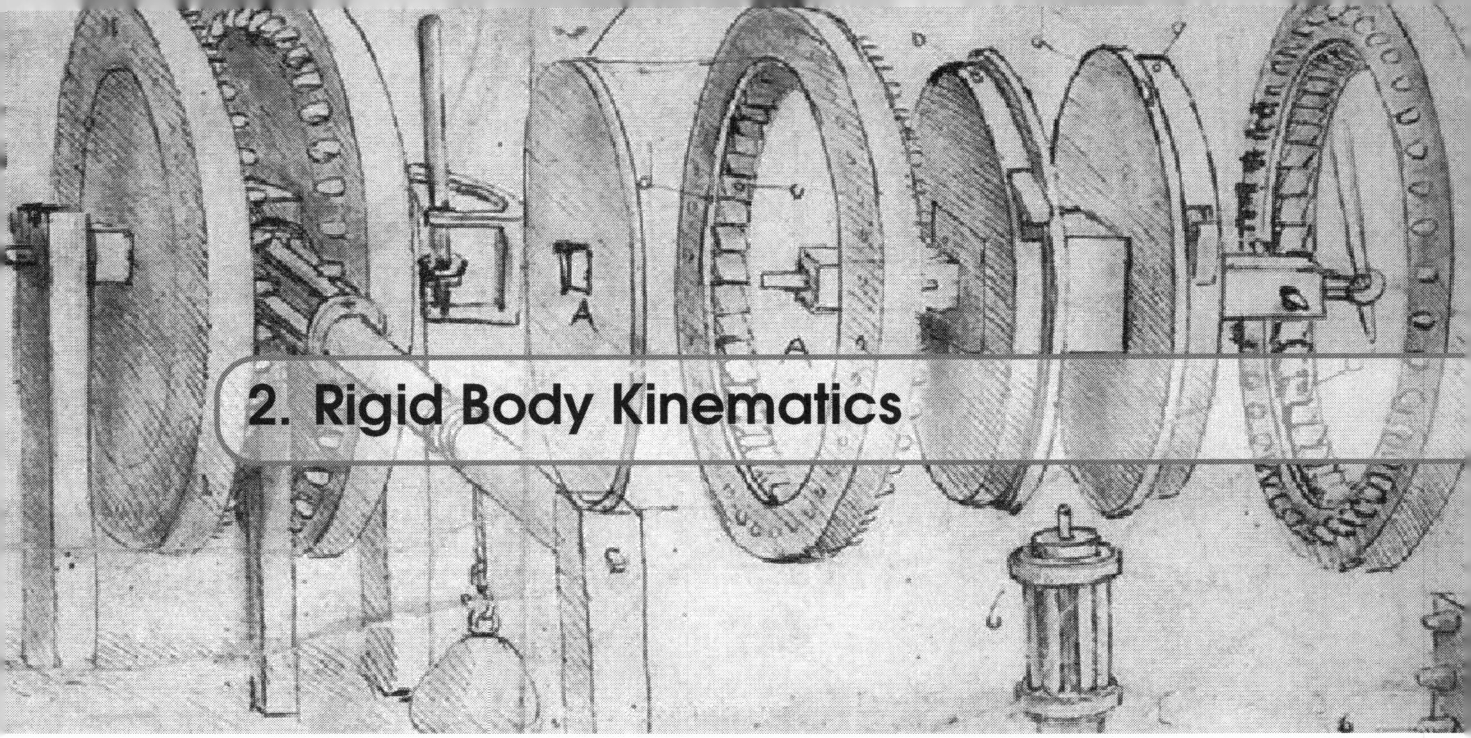

2. Rigid Body Kinematics

THE GOAL of Chapter 2 is to study the kinematics of rigid bodies in planar motion. Our task is to find the velocity and acceleration at all points of a body. At first sight, the task seems daunting. However, we will find that the velocity and acceleration fields of a rigid body are both governed by simple relationships. We will learn that the velocity field satisfies the identity $\mathbf{v}_B = \mathbf{v}_A + \boldsymbol{\omega} \times \mathbf{r}_{AB}$ between any two points A and B, where the *angular velocity* $\boldsymbol{\omega}$ characterizes the body's change of orientation. Hence the velocity field is entirely specified at any instant by the knowledge of: (i) the velocity of a particular point of the body and (ii) the angular velocity, a vector independent of position.

We start our study with two simple rigid body motions: *translation* (Section 2.2) and *rotation* (Section 2.3). In Section 2.3, we will learn how to determine the angular velocity of a rigid body. General planar motions are treated in Section 2.4. This section is devoted to the properties of the velocity field of a rigid body. We will also learn in Section 2.7 that a rigid body in general planar motion can always be characterized as being in *instantaneous rotation* about a particular point, the *instantaneous center of rotation*.

After a brief description of *joints* (Section 2.5) and *rolling motion* (Section 2.6), we will learn how to analyze planar *mechanisms*, that is, systems of rigid bodies assembled in such a way as to produce desired motion transformations and specific input/output kinematic laws. To accomplish this task, we will devise three methods of analysis: (i) a graphical method based on the notion of instantaneous center of rotation (Section 2.7), (ii) a geometric method (Section 2.8), and (iii) a general vectorial method based on *kinematic loop equations* (Section 2.10). Finally, in Section 2.9, we seek to resolve a problem of great practical importance: how the kinematics of a particle can be obtained from one referential to another. This problem has interesting connections to rigid body kinematics.

In this chapter, we will learn

1. to determine the velocity/acceleration fields of a rigid body at a particular instant,
2. to analyze mechanisms by a variety of methods.

2.1 Basic Definitions

First we start with a few key definitions.

> **Definition (Rigid Body in Planar Motion)**
>
> 1. A body $\mathcal{B}$ is said to be *rigid* if for any two points P and Q attached to $\mathcal{B}$ the distance $|PQ|$ remains constant. It is important to view $\mathcal{B}$ as not just a bounded set of points but also as the *rigid extension* of this set, that is, the set of *all* points of the space moving rigidly with $\mathcal{B}$. In kinematics, we make no distinction between rigid body $\mathcal{B}$ and its rigid extension.
> 2. A rigid body $\mathcal{B}$ is in *planar motion* relative to a referential $\mathcal{E}$ if points of $\mathcal{B}$ move in parallel planes. Hence it is sufficient to study the motion of $\mathcal{B}$ in a given (arbitrary) plane, called *reference plane*.

Remark 1. A interesting kinematic consequence of the rigidity assumption is that, for any two points P and Q attached to rigid body $\mathcal{B}$, the time differentiation of the constant scalar $|\mathbf{r}_{PQ}|^2 = \mathbf{r}_{PQ} \cdot \mathbf{r}_{PQ}$ leads to

$$\mathbf{r}_{PQ} \cdot \mathbf{v}_P = \mathbf{r}_{PQ} \cdot \mathbf{v}_Q$$

This is the fundamental property of *equiprojectivity* which we will revisit in Section 2.4.

Remark 2. In rigid body kinematics, we write $\mathbf{v}_P$ for $\mathbf{v}_{P/\mathcal{E}}$. However in some circumstances it is critical not to omit the dependence of velocity upon the chosen referential.

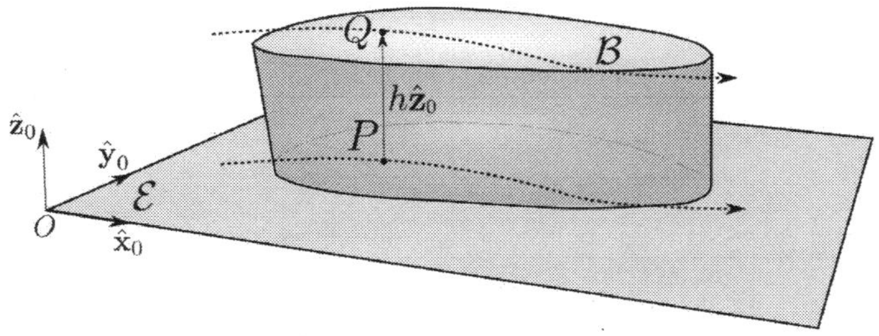

Figure 2.1 – *General planar motion: the trajectories of points of $\mathcal{B}$ lie in planes parallel to the reference plane $(O, \hat{\mathbf{x}}_0, \hat{\mathbf{y}}_0)$ fixed in $\mathcal{E}$. If P and Q satisfy $\mathbf{r}_{PQ}$ = constant, then $\mathbf{v}_P = \mathbf{v}_Q$ at all time: their trajectories are parallel.*

Proof. If $\hat{\mathbf{z}}_0$ is a unit vector normal to the reference plane, consider two points P and Q of $\mathcal{B}$ such that $\mathbf{r}_{PQ} = h\hat{\mathbf{z}}_0$. See Figure 2.1. Since h is a constant, we have

$$\frac{d}{dt}\mathbf{r}_{PQ} = \mathbf{0} = \frac{d}{dt}(\mathbf{r}_{OQ} - \mathbf{r}_{OP}) = \mathbf{v}_Q - \mathbf{v}_P$$

Hence $\mathbf{v}_P = \mathbf{v}_Q$ and $\mathbf{a}_P = \mathbf{a}_Q$: *points P and Q move along parallel paths* and it is sufficient to study the kinematics of $\mathcal{B}$ for points lying in the reference plane. ■

Remark 3. In general if P and Q both lie in the reference plane, the paths of P and Q are not, in general, parallel!

In most situations, it is sufficient to represent body $\mathcal{B}$ as a two-dimensional geometric figure. Three types of motion will be studied: (i) translational motion (section 2.2), (ii) rotational motion (section 2.3), and (iii) general planar motion (section 2.4).

2.2 Translational Motion

Definition (Translation)

A rigid body $\mathcal{B}$ is in translational motion relative to $\mathcal{E}$ if, for any two points A and B attached to $\mathcal{B}$, the position vector $\mathbf{r}_{AB}$ remains constant in $\mathcal{E}$.

This definition has important consequences on the kinematics of $\mathcal{B}$: The velocity and acceleration fields are invariant fields, that is, they are independent of position.

Translational Kinematics

All points of a body $\mathcal{B}$ in translational motion have the same velocity $\mathbf{v}(t)$ and acceleration $\mathbf{a}(t)$ at any given time:

$$\mathbf{v}_A = \mathbf{v}_B = \mathbf{v}(t), \qquad \mathbf{a}_A = \mathbf{a}_B = \mathbf{a}(t) \tag{2.1}$$

for any two points A and B of $\mathcal{B}$.

Proof. (2.1) is a direct consequence of $\mathbf{r}_{AB} = \mathbf{r}_{OB} - \mathbf{r}_{OA} = Cst$: assuming O an origin of $\mathcal{E}$, then, $\frac{d}{dt}\mathbf{r}_{AB} = \mathbf{0}$ implies $\mathbf{v}_B - \mathbf{v}_A = \mathbf{0}$ at all times, and for all points attached to $\mathcal{B}$. Then $\frac{d}{dt}(\mathbf{v}_B - \mathbf{v}_A) = \mathbf{0}$ leads to $\mathbf{a}_A = \mathbf{a}_B$. ■

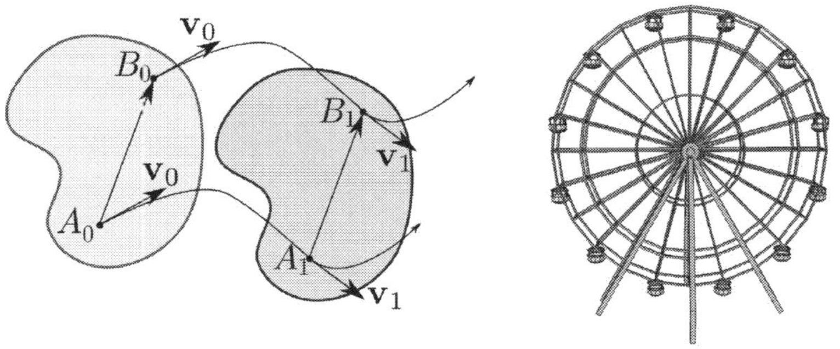

Figure 2.2 – *For any two points A and B of a body in translation, $\mathbf{r}_{AB}$ remains constant. At time $t = t_0$, all points have the same velocity $\mathbf{v}_0$. At time $t = t_1$, all points have velocity $\mathbf{v}_1 \neq \mathbf{v}_0$. The points move along parallel paths (left). The gondolas of this Ferris wheel are in translation: all points of a gondola have the same velocity and acceleration at a given time and their trajectories are circular (right).*

Remark 4. *The orientation of $\mathcal{B}$ remains constant.* If $(\hat{\mathbf{x}}_0, \hat{\mathbf{y}}_0, \hat{\mathbf{z}}_0)$ is a fixed basis of unit vectors, then $(\hat{\mathbf{x}}_0, \hat{\mathbf{y}}_0, \hat{\mathbf{z}}_0)$ is also fixed relative to $\mathcal{B}$.

Remark 5. The vectors $\mathbf{v}(t)$ and $\mathbf{a}(t)$ can vary in magnitude and/or direction. If $\mathbf{v}(t)$ (and thus $\mathbf{a}(t)$) remains constant in direction, the translation is said to be *rectilinear* and the points of $\mathcal{B}$ describe parallel lines. Otherwise, $\mathbf{v}(t)$ varies in direction, the translation is said to be *curvilinear* and the points of $\mathcal{B}$ move along parallel curves (See Figure 2.2).

2.3 Rotational Motion

Here we consider the motion of a rigid body about a fixed axis. Such motions are ubiquitous in mechanical systems. The description of the kinematics of rotation is more involved than for translation.

DEFINITION (ROTATION)

A rigid body $\mathcal{B}$ is said to be in *rotational* motion about a fixed axis Δ of a referential $\mathcal{E}$ if any point P attached to $\mathcal{B}$ remains at a constant distance from Δ. Hence, the points of $\mathcal{B}$ move along circular paths centered about Δ in parallel planes. Δ is the *axis of rotation* of $\mathcal{B}$.

Remark 6. Note that the existence of a particular point of $\mathcal{B}$ with a circular trajectory does not necessarily imply that $\mathcal{B}$ is in rotational motion.

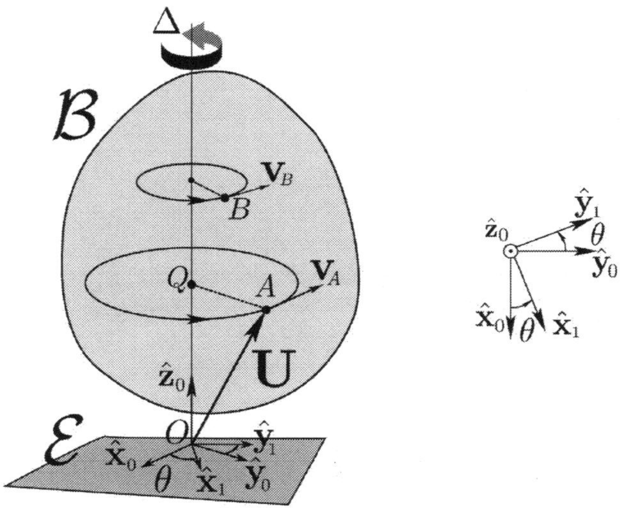

Figure 2.3 – *$\mathcal{B}$ in rotational motion about axis Δ of referential $\mathcal{E}$. Points A and B of $\mathcal{B}$ moves along parallel circular paths of axis Δ.*

ANGULAR VELOCITY OF $\mathcal{B}$. In order to characterize the orientation of $\mathcal{B}$ relative to $\mathcal{E}$, we introduce two right-handed bases of unit vectors: $(\hat{\mathbf{x}}_0, \hat{\mathbf{y}}_0, \hat{\mathbf{z}}_0)$ fixed in $\mathcal{E}$ and $(\hat{\mathbf{x}}_1, \hat{\mathbf{y}}_1, \hat{\mathbf{z}}_0)$ fixed in $\mathcal{B}$ with $\hat{\mathbf{z}}_0$ directed along the axis of rotation Δ. See Figure 2.3. We then define the *oriented* angle $\theta = (\hat{\mathbf{x}}_0, \hat{\mathbf{x}}_1) = (\hat{\mathbf{y}}_0, \hat{\mathbf{y}}_1)$ as shown. The change of orientation of $\mathcal{B}$ is then characterized by the following theorem.

ANGULAR VELOCITY

There exists a unique vector, called *angular velocity* of body $\mathcal{B}$ and denoted $\boldsymbol{\omega}_{\mathcal{B}}$, such that

$$\frac{d\mathbf{U}}{dt} = \boldsymbol{\omega}_{\mathcal{B}} \times \mathbf{U} \tag{2.2}$$

for *any* vector $\mathbf{U}$ attached to $\mathcal{B}$. It takes the expression

$$\boldsymbol{\omega}_{\mathcal{B}} = \dot{\theta}\hat{\mathbf{z}}_0 \tag{2.3}$$

Proof. We can write $\mathbf{U}$ on the basis attached to $\mathcal{B}$: $\mathbf{U}(t) = u_x\hat{\mathbf{x}}_1 + u_y\hat{\mathbf{y}}_1 + u_z\hat{\mathbf{z}}_0$. Note that the components (u_x, u_y, u_z) are time independent since $\mathbf{U}$ is fixed relative to $\mathcal{B}$. We then find $d\mathbf{U}/dt$:

$$\frac{d\mathbf{U}}{dt} = u_x\frac{d\hat{\mathbf{x}}_1}{dt} + u_y\frac{d\hat{\mathbf{y}}_1}{dt} = u_x\dot{\theta}\hat{\mathbf{y}}_1 - u_y\dot{\theta}\hat{\mathbf{x}}_1$$

where we have used $d\hat{\mathbf{x}}_1/dt = \dot{\theta}\hat{\mathbf{y}}_1$ and $d\hat{\mathbf{y}}_1/dt = -\dot{\theta}\hat{\mathbf{x}}_1$. We then express $\hat{\mathbf{y}}_1$ as the cross-product $\hat{\mathbf{z}}_0 \times \hat{\mathbf{x}}_1$ and $\hat{\mathbf{x}}_1$ as $\hat{\mathbf{z}}_0 \times (-\hat{\mathbf{y}}_1)$. Then we have

$$\frac{d\mathbf{U}}{dt} = u_x\dot{\theta}\hat{\mathbf{z}}_0 \times \hat{\mathbf{x}}_1 + u_y\dot{\theta}\hat{\mathbf{z}}_0 \times \hat{\mathbf{y}}_1 = \dot{\theta}\hat{\mathbf{z}}_0 \times (u_x\hat{\mathbf{x}}_1 + u_y\hat{\mathbf{y}}_1 + u_z\hat{\mathbf{z}}_0)$$

Then, recognizing that $\dot{\theta}\hat{\mathbf{z}}_0 = \boldsymbol{\omega}_{\mathcal{B}}$, we obtain (2.2). ■

Remark 7. Equation (2.2) shows that *the angular velocity characterizes the change of orientation of body $\mathcal{B}$* since it can be used for the determination of the time-derivative of vectors attached to $\mathcal{B}$.

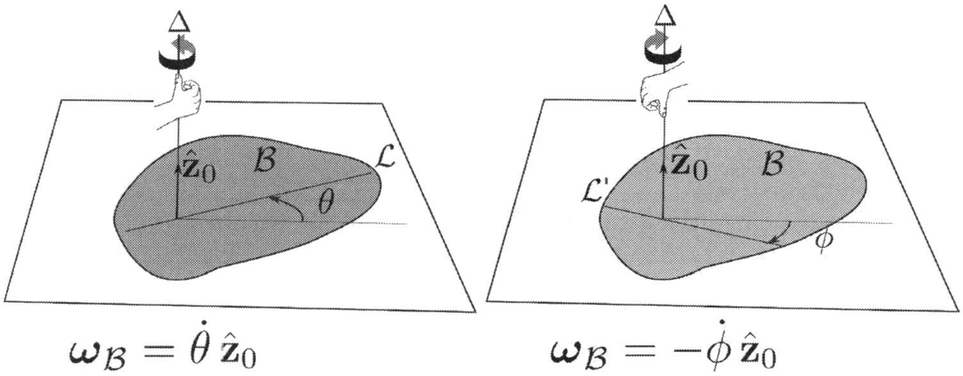

Figure 2.4 – *Angular velocity of body $\mathcal{B}$: the direction of $\boldsymbol{\omega}$ is obtained by the right-hand rule.*

Properties

①. The angular velocity is a unique vector of constant direction (it is always normal to the plane of motion): it is not a function of the choice of unit vector $\hat{\mathbf{z}}_0$ or of angle θ.

②. The angular velocity can be defined by choosing any oriented angle from a fixed line to a line $\mathcal{L}$ attached to $\mathcal{B}$. Figure 2.4 shows two angles θ and ϕ to define the orientation of two lines $\mathcal{L}$ and $\mathcal{L}'$ attached to $\mathcal{B}$. Since the sum $\theta + \phi$ is necessarily a constant, we have $\dot{\phi} = -\dot{\theta}$. Then the angular velocity takes the expressions $\boldsymbol{\omega}_{\mathcal{B}} = \dot{\theta}\hat{\mathbf{z}}_0 = -\dot{\phi}\hat{\mathbf{z}}_0$.

③. In general $\boldsymbol{\omega}$ is obtained by the right-hand rule: use your right hand, and rotate it in the sense of increasing θ (or ϕ). Then the direction of $\boldsymbol{\omega}_{\mathcal{B}}$ is given by the direction of your thumb. Using the same rule, $\boldsymbol{\omega}_{\mathcal{B}}$ gives the sense of rotation of the body: if $\dot{\theta} > 0$, angle θ increases and $\mathcal{B}$ rotates in the counterclockwise direction. In this case, $\boldsymbol{\omega}_{\mathcal{B}}$

points in the direction of $+\hat{\mathbf{z}}_0$ (since $\dot{\theta} > 0$) and the right-hand rule gives a counterclockwise rotation. Conversely if $\dot{\theta} < 0$, the right-hand rule gives a clockwise rotation.

④. In practice, $\boldsymbol{\omega}_\mathcal{B}$ need not be defined in terms of the rate of change of an angle. Once a unit vector $\hat{\mathbf{z}}_0$ is chosen along the axis of rotation, it is sufficient to simply write $\boldsymbol{\omega}_\mathcal{B} = \omega_\mathcal{B}\hat{\mathbf{z}}_0$ with $\omega_\mathcal{B} > 0$ or $\omega_\mathcal{B} < 0$ depending on the actual sense of rotation of $\mathcal{B}$.

A consequence of formula (2.2) is the derivation of the velocity and acceleration fields of $\mathcal{B}$.

Velocity/Acceleration Fields

If rigid body $\mathcal{B}$ is in rotation about axis $\Delta = (O, \hat{\mathbf{z}}_0)$, then the velocity and acceleration vector of any point A of $\mathcal{B}$ can be obtained from the formulas

$$\mathbf{v}_A = \boldsymbol{\omega}_\mathcal{B} \times \mathbf{r}_{OA} \tag{2.4}$$

and

$$\mathbf{a}_A = \boldsymbol{\alpha}_\mathcal{B} \times \mathbf{r}_{OA} + \boldsymbol{\omega}_\mathcal{B} \times (\boldsymbol{\omega}_\mathcal{B} \times \mathbf{r}_{OA}) \tag{2.5}$$

where

$$\boldsymbol{\alpha}_\mathcal{B} = \frac{d\boldsymbol{\omega}_\mathcal{B}}{dt} = \ddot{\theta}\hat{\mathbf{z}}_0 \tag{2.6}$$

is the *angular acceleration* of $\mathcal{B}$.

Proof. Considering an arbitrary point A of $\mathcal{B}$, we can use formula (2.2) by choosing $\mathbf{U} = \mathbf{r}_{OA}$ where point O is an arbitrary point of axis Δ. Then $d\mathbf{r}_{OA}/dt = \mathbf{v}_A$. Hence we have $\mathbf{v}_A = \boldsymbol{\omega}_\mathcal{B} \times \mathbf{r}_{OA}$. We then take the time-derivative of (2.4) to obtain $\mathbf{a}_A = \boldsymbol{\alpha}_\mathcal{B} \times \mathbf{r}_{OA} + \boldsymbol{\omega}_\mathcal{B} \times \mathbf{v}_A = \boldsymbol{\alpha}_\mathcal{B} \times \mathbf{r}_{OA} + \boldsymbol{\omega}_\mathcal{B} \times (\boldsymbol{\omega}_\mathcal{B} \times \mathbf{r}_{OA})$. ■

Remark 8. We can also use $\mathbf{U} = \mathbf{r}_{QA}$ where Q is the projection of A onto Δ: then $\mathbf{v}_A = \boldsymbol{\omega}_\mathcal{B} \times \mathbf{r}_{QA}$.

Properties

①. The advantage of formulas (2.4-2.5) is that they are applicable to any point attached to $\mathcal{B}$, independently of a coordinate system.

②. These formulas give the standard expressions of the velocity and acceleration of a particle in circular motion (see Figure 2.5):

$$\mathbf{v}_A = \boldsymbol{\omega}_\mathcal{B} \times \mathbf{r}_{OA} = \dot{\theta}\hat{\mathbf{z}}_0 \times r_A\hat{\mathbf{e}}_r = r_A\dot{\theta}\hat{\mathbf{e}}_\theta$$

and

$$\mathbf{a}_A = \ddot{\theta}\hat{\mathbf{z}}_0 \times r_A\hat{\mathbf{e}}_r - \dot{\theta}^2 r_A\hat{\mathbf{e}}_r = r_A\ddot{\theta}\hat{\mathbf{e}}_\theta - r_A\dot{\theta}^2\hat{\mathbf{e}}_r$$

③. If O and A lie in the same plane, then the second term of (2.5) takes the expression

$\boldsymbol{\omega}_\mathcal{B} \times (\boldsymbol{\omega}_\mathcal{B} \times \mathbf{r}_{OA}) = -\omega_\mathcal{B}^2 \mathbf{r}_{OA}$:

$$\mathbf{a}_A = \boldsymbol{\alpha}_\mathcal{B} \times \mathbf{r}_{OA} - \omega_\mathcal{B}^2 \mathbf{r}_{OA} \tag{2.7}$$

④. If A and B are two points of the same rigid body $\mathcal{B}$, then the two equations $\mathbf{v}_A = \boldsymbol{\omega}_\mathcal{B} \times \mathbf{r}_{OA}$ and $\mathbf{v}_B = \boldsymbol{\omega}_\mathcal{B} \times \mathbf{r}_{OB}$ lead to

$$\mathbf{v}_B = \mathbf{v}_A + \boldsymbol{\omega}_\mathcal{B} \times \mathbf{r}_{AB} \tag{2.8}$$

We will show that (2.8) holds true for arbitrary motion.

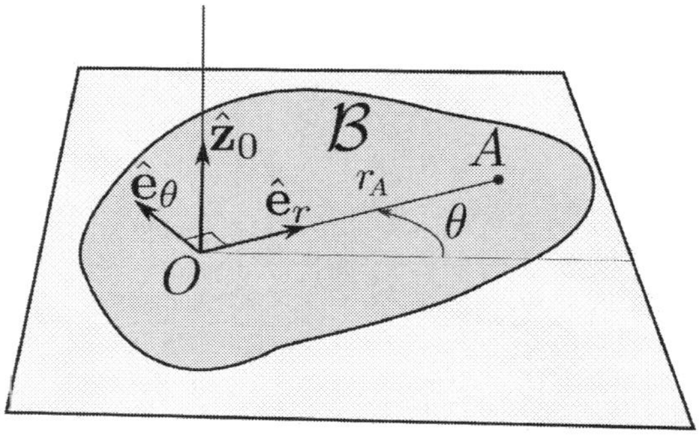

Figure 2.5 – *Point A of body $\mathcal{B}$ is in circular motion about O.*

Finally, formulas (2.4-2.5) have important geometric consequences for the velocity and acceleration vectors of points of $\mathcal{B}$ at any given time. Figure 2.6 (left) shows the velocity distribution of the body: along a given line OA the velocities are normal to this line and form a triangle as consequence of formula (2.4). Along another line OB, the velocity triangle is directly similar to that associated to line OA.

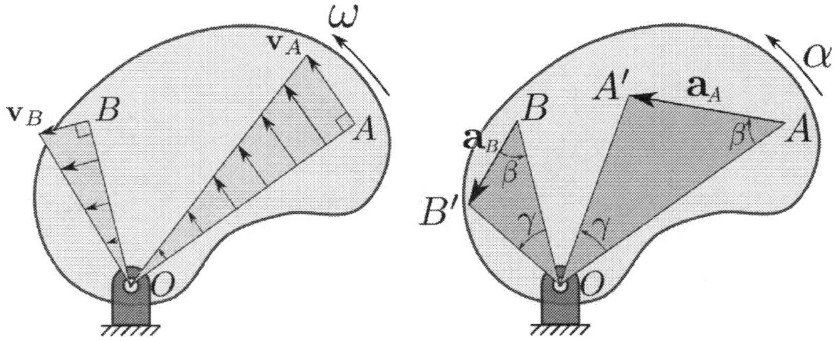

Figure 2.6 – *Graphical construction of velocities (left) and accelerations (right).*

A similar property exists for accelerations: Figure 2.6 (right) shows the relationship between the accelerations of any two points A and B of the body. The triangles OAA' and OBB' appear to be *directly similar*: all corresponding angles in triangles OAA' and OBB' are equal and described in the same rotational sense. This property is a consequence of (2.5) and is shown as follows: to find the angle β at A in triangle OAA' we determine $\cos\beta$ and $\sin\beta$ by finding the scalar product $\mathbf{r}_{AO} \cdot \mathbf{a}_A$ and the cross product $\mathbf{r}_{AO} \times \mathbf{a}_A$:

$$\cos\beta = \frac{\mathbf{r}_{AO} \cdot \mathbf{a}_A}{|OA||\mathbf{a}_A|} = \frac{\omega^2}{\sqrt{\alpha^2 + \omega^4}}$$

where we have used $|\mathbf{a}_A| = |OA|\sqrt{\alpha^2 + \omega^4}$,

$$\mathbf{r}_{AO} \times \mathbf{a}_A = \mathbf{r}_{AO} \times (\boldsymbol{\alpha} \times \mathbf{r}_{OA}) = -\alpha|OA|^2\hat{\mathbf{z}}_0 = -|OA||\mathbf{a}_A|\sin\beta\hat{\mathbf{z}}_0$$

leading to

$$\sin\beta = \frac{\alpha}{\sqrt{\alpha^2 + \omega^4}}$$

The expressions of $\cos\alpha$ and $\sin\alpha$ show that the angle β is not a function of the choice of A. It is easy to show that $\frac{|AA'|}{|BB'|} = \frac{|\mathbf{a}_A|}{|\mathbf{a}_B|} = \frac{|OA|}{|OB|}$. This shows that *the triangles OAA' and OBB' are directly similar.*

EXAMPLE PROBLEMS

General Procedure: Rigid Body Translation/Rotation

To solve a rigid body kinematics problem, follow these guidelines:

① Since a vectorial solution is sought, a set of fixed unit vectors $(\hat{\mathbf{x}}_0, \hat{\mathbf{y}}_0, \hat{\mathbf{z}}_0)$ with $\hat{\mathbf{z}}_0$ normal to the plane of motion will need to be defined.

② Identify whether a rigid body is in translation or rotation. Recall that the orientation of a body in translation is invariant during the motion.

③ The angular velocity of a body in rotation will be always be written on as $\omega_{\mathcal{B}} = \omega_{\mathcal{B}}\hat{\mathbf{z}}_0$, with the sign of $\omega_{\mathcal{B}}$ depending on the sense of rotation of $\mathcal{B}$.

④ Application of (2.4-2.5) requires point A to be viewed as a point attached to body $\mathcal{B}$. It is convenient and at times necessary to denote the velocity of A as $\mathbf{v}_{A\in\mathcal{B}}$, and its acceleration $\mathbf{a}_{A\in\mathcal{B}}$ to emphasize that A is to be viewed as attached to $\mathcal{B}$, although this may be so only in an instantaneous way (see example 2.3). We will come back to this point on many occasions.

Exercise 2.1 ▪A lifting mechanism is composed of three rigid bodies 1, 2 and 3 as shown in Figure 2.7 with $|AB| = |CD| = 3$ m and $|AC| = |BD| = 3.6$ m. It is actuated by a pneumatic cylinder 4 whose piston extends at the constant rate of 0.08 m/s ($|CE| = |CD|$ and $|CF| = |CD|/2$).
Find the magnitude of the velocity and acceleration of midpoint I of BD when the system reaches the vertical position ($\theta = \pi/2$). Describe the trajectory of I. ▪

Solution. First we note that the orientation of body 3 is invariant during the motion as a consequence of the geometry of bodies 1 and 2 ($|AB| = |CD|$). This implies that

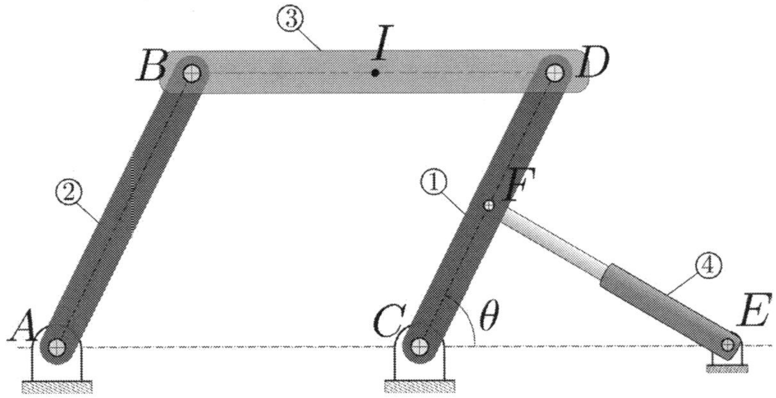

Figure 2.7 – *Problem 2.1*

3 is in translational motion: at any given time, the velocities of points of body 3 are all equal to $\mathbf{v}_D$. The motion of body 3 is coupled to that of body 1 through point D which satisfies $\mathbf{v}_{D\in 3} = \mathbf{v}_{D\in 1}$. Hence to find the velocity of D, we view D as a point of body 1 which is in rotation about C: with the choice of $(\hat{\mathbf{x}}_0, \hat{\mathbf{y}}_0, \hat{\mathbf{z}}_0)$ shown in Figure 2.8 we find by using formula (2.4)

$$\mathbf{v}_D = \boldsymbol{\omega}_1 \times \mathbf{r}_{CD} = \dot{\theta}\hat{\mathbf{z}}_0 \times 2L(\cos\theta\hat{\mathbf{x}}_0 + \sin\theta\hat{\mathbf{y}}_0) = 2L\dot{\theta}(\cos\theta\hat{\mathbf{y}}_0 - \sin\theta\hat{\mathbf{x}}_0) \qquad [1]$$

where we have denoted $2L = |CD|$ and $\boldsymbol{\omega}_1 = \dot{\theta}\hat{\mathbf{z}}_0$ the angular velocity of body 1. We are not given the value of angular speed $\dot{\theta}$: it must be found from the value of extensional velocity $\dot{s}$ of the piston: with $s(t) = |EF|$, $2L = |CE|$ and $L = |CF|$ we have in triangle CEF (see Figure 2.8):

$$s^2 = 4L^2 + L^2 - 4L^2\cos\theta = 5L^2 - 4L^2\cos\theta \qquad [2]$$

Since [2] is valid at all time, we can differentiate it w.r.t. time to obtain:

$$s\dot{s} = 2L^2\dot{\theta}\sin\theta \qquad [3]$$

giving the relationship between $\dot{\theta}$ and $\dot{s}$

$$2L\dot{\theta} = \frac{s\dot{s}}{L\sin\theta} = \frac{\dot{s}}{\sin\theta}(5 - 4\cos\theta)^{1/2}$$

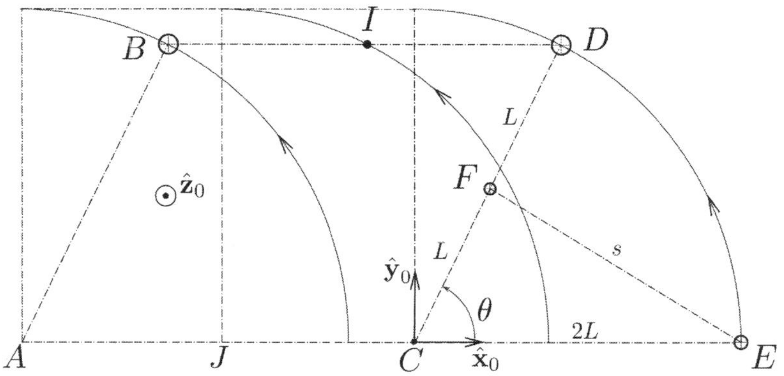

Figure 2.8 – *Problem 2.1 (Solution)*

We can now find the velocity of I:

$$\boxed{\mathbf{v}_I = \mathbf{v}_D = \frac{\dot{s}}{\sin\theta}(5 - 4\cos\theta)^{1/2}(\cos\theta\hat{\mathbf{y}}_0 - \sin\theta\hat{\mathbf{x}}_0)}$$

At $\theta = \pi/2$, we have $\mathbf{v}_I = -\sqrt{5}\dot{s}\,\hat{\mathbf{x}}_0 = -0.18\,\hat{\mathbf{x}}_0$ (m/s) giving $|\mathbf{v}_I| = \mathbf{0.18}$ m/s. Since body 3 is in translational motion, the trajectory of I is obtained by a translation of the trajectory of D of vector $\mathbf{r}_{DI} = -d\hat{\mathbf{x}}_0$ denoting $d = |DI| = |CJ|$. The trajectory of D is (arc of) the circle of center C and radius $2L$: this implies that the trajectory of I is the (arc of) circle of center J (midpoint of segment AC) and radius $2L$ (see Figure 2.8).
To find the acceleration of I, we must find the acceleration of D (as a point of body 1) by using formula (2.5)

$$\begin{aligned}\mathbf{a}_D &= \ddot{\theta}\hat{\mathbf{z}}_0 \times \mathbf{r}_{CD} - \dot{\theta}^2\mathbf{r}_{CD} = \ddot{\theta}\hat{\mathbf{z}}_0 \times 2L(\cos\theta\hat{\mathbf{x}}_0 + \sin\theta\hat{\mathbf{y}}_0) - 2L\dot{\theta}^2(\cos\theta\hat{\mathbf{x}}_0 + \sin\theta\hat{\mathbf{y}}_0) \\ &= 2L\ddot{\theta}(\cos\theta\hat{\mathbf{y}}_0 - \sin\theta\hat{\mathbf{x}}_0) - 2L\dot{\theta}^2(\cos\theta\hat{\mathbf{x}}_0 + \sin\theta\hat{\mathbf{y}}_0)\end{aligned}$$

To find angular acceleration $\ddot{\theta}$ we differentiate [3] w.r.t. time taking into account $\dot{s} =$ constant:

$$\dot{s}^2 = 2L^2(\ddot{\theta}\sin\theta + \dot{\theta}^2\cos\theta)$$

At $\theta = \pi/2$, using $2L\dot{\theta} = \dot{s}\sqrt{5}$ and $2L^2\ddot{\theta} = \dot{s}^2$ we find

$$\boxed{\mathbf{a}_D = -\frac{\dot{s}^2}{L}(\hat{\mathbf{x}}_0 + \frac{5}{2}\hat{\mathbf{y}}_0)}$$

Numerically, $|\mathbf{a}_D| = \mathbf{0.011}$ (m/s^2). ■

Exercise 2.2 ■In the belt drive of Figure 2.9, the belt is wrapped around the pulleys $\mathcal{A}$ and $\mathcal{B}$. The smaller pulley $\mathcal{A}$ is attached to a shaft rotating about A at the angular velocity 50 rev/s in the clockwise direction. The motion of the belt in turn drives the rotation of larger pulley $\mathcal{B}$ about B. The geometry of the system is given by the radius $R_A = 58$ cm of pulley $\mathcal{A}$, the angle of wrap $\gamma = 160°$ of $\mathcal{A}$ and the distance $e = |AB| = 225$ cm between the shafts' axes.
Find the velocity of points I and J. Deduce the angular velocity of the driven pulley. ■

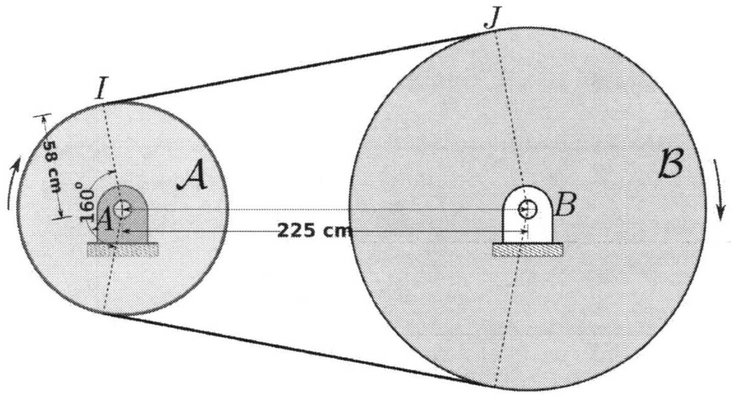

Figure 2.9 – *Problem 2.2*

Solution. First we note that the radius R_B of pulley $\mathcal{B}$ is not given. It can be found from the geometry of the system: if O denotes the intersection of line IJ and line AB, and $d = |OA|$ then,

$$\cos\frac{\gamma}{2} = \frac{R_A}{d} = \frac{R_B}{d+e}$$

This gives

$$\boxed{R_B = R_A + e\cos\frac{\gamma}{2}}$$

To solve this problem, we first choose a basis of unit vectors $(\hat{\imath}, \hat{\jmath}, \hat{\mathbf{k}})$ as shown in Figure 2.10. We write the angular velocities of $\mathcal{A}$ and $\mathcal{B}$ as $\boldsymbol{\omega}_{\mathcal{A}} = \omega_A\hat{\mathbf{k}}$ and $\boldsymbol{\omega}_{\mathcal{B}} = \omega_B\hat{\mathbf{k}}$. Note that $\omega_A < 0$ since the rotation of pulley $\mathcal{A}$ is in the clockwise direction. Then, by using formula (2.4), we find the velocity of I as a point of body $\mathcal{A}$ which is in rotation about A

$$\mathbf{v}_{I\in\mathcal{A}} = \omega_A\hat{\mathbf{k}} \times \mathbf{r}_{AI} = \omega_A\hat{\mathbf{k}} \times R_A(-\cos\frac{\gamma}{2}\hat{\imath} + \sin\frac{\gamma}{2}\hat{\jmath}) = R_A\omega_A(-\cos\frac{\gamma}{2}\hat{\jmath} - \sin\frac{\gamma}{2}\hat{\imath})$$

Similarly, for point J as a point of body $\mathcal{B}$:

$$\mathbf{v}_{J\in\mathcal{B}} = \omega_B\hat{\mathbf{k}} \times \mathbf{r}_{BJ} = \omega_B\hat{\mathbf{k}} \times R_B(-\cos\frac{\gamma}{2}\hat{\imath} + \sin\frac{\gamma}{2}\hat{\jmath}) = R_B\omega_B(-\cos\frac{\gamma}{2}\hat{\jmath} - \sin\frac{\gamma}{2}\hat{\imath})$$

Since the sections of the belt in contact the pulleys move rigidly with $\mathcal{A}$ or $\mathcal{B}$, the velocities $\mathbf{v}_{I\in\mathcal{A}}$ and $\mathbf{v}_{J\in\mathcal{B}}$ can be considered as velocities of points attached to the belt. Then we note that at any given time the section IJ of the belt is in translational motion since its direction is invariant: we can then state that $\mathbf{v}_{I\in\mathcal{A}} = \mathbf{v}_{J\in\mathcal{B}}$:

$$R_A\omega_A(-\cos\frac{\gamma}{2}\hat{\jmath} - \sin\frac{\gamma}{2}\hat{\imath}) = R_B\omega_B(-\cos\frac{\gamma}{2}\hat{\jmath} - \sin\frac{\gamma}{2}\hat{\imath})$$

giving

$$\boxed{\omega_B = \omega_A\frac{R_A}{R_A + e\cos\frac{\gamma}{2}}}$$

Numerically, with $\omega_A = -50$ (rev/s), we find $\omega_B = \mathbf{-29.9}$ **(rev/s)** ↻. ∎

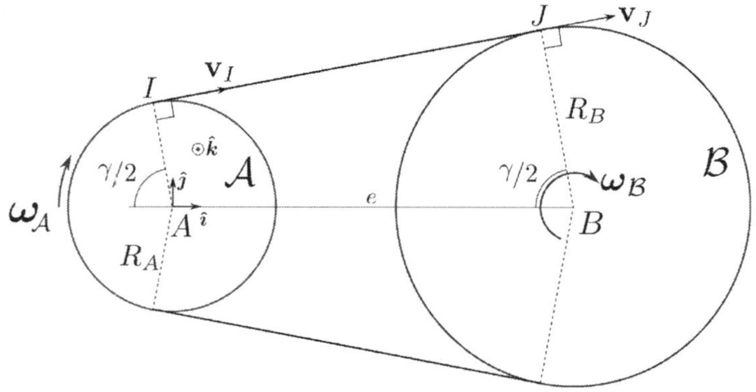

Figure 2.10 – *Problem 2.2 (Solution)*

Exercise 2.3 ■■Two disks $\mathcal{A}$ of radius R_A and $\mathcal{B}$ of radius R_B rotate in the clockwise direction at the constant angular velocity ω_A and ω_B about A and B, respectively. They are driven in rotation while in contact at a point I of their rims.

a. Find the velocities $\mathbf{v}_{I\in\mathcal{A}}$ and $\mathbf{v}_{I\in\mathcal{B}}$. Deduce the expression of the difference $\mathbf{v}_{I\in\mathcal{A}} - \mathbf{v}_{I\in\mathcal{B}}$, known as the slip velocity. Find the condition which guarantees no-slip between the disks.

b. Find the expression of $\mathbf{a}_{I\in\mathcal{A}} - \mathbf{a}_{I\in\mathcal{B}}$ and compare with $\frac{d}{dt}(\mathbf{v}_{I\in\mathcal{A}} - \mathbf{v}_{I\in\mathcal{B}})$. Comment. ■

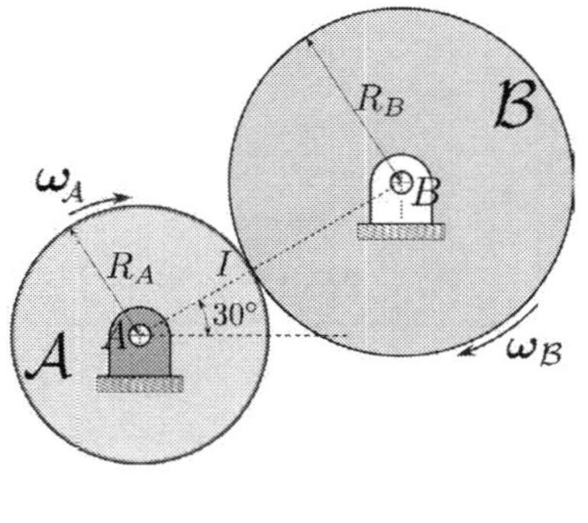

SOLUTION. Define a fixed basis $(\hat{\mathbf{x}}_0, \hat{\mathbf{y}}_0, \hat{\mathbf{z}}_0)$ as shown in Figure 2.11. The position of B relative to A is then given by $\mathbf{r}_{AB} = (R_A + R_B)(\cos\beta\hat{\mathbf{x}}_0 + \sin\beta\hat{\mathbf{y}}_0)$. Since body $\mathcal{A}$ is in rotation about A, the velocity of I as a point of $\mathcal{A}$ is then given by (with $\omega_A < 0$)

$$\mathbf{v}_{I\in\mathcal{A}} = \omega_A\hat{\mathbf{z}}_0 \times \mathbf{r}_{AI} = \omega_A\hat{\mathbf{z}}_0 \times R_A(\cos\beta\hat{\mathbf{x}}_0 + \sin\beta\hat{\mathbf{y}}_0) = R_A\omega_A(\cos\beta\hat{\mathbf{y}}_0 - \sin\beta\hat{\mathbf{x}}_0)$$

Similarly we denote $\boldsymbol{\omega}_B = \omega_B\hat{\mathbf{z}}_0$ the angular velocity of body $\mathcal{B}$ ($\omega_B < 0$): the velocity of I as a point of $\mathcal{B}$ is then given by

$$\mathbf{v}_{I\in\mathcal{B}} = \omega_B\hat{\mathbf{z}}_0 \times \mathbf{r}_{BI} = \omega_B\hat{\mathbf{z}}_0 \times R_B(-\cos\beta\hat{\mathbf{x}}_0 - \sin\beta\hat{\mathbf{y}}_0) = -R_B\omega_B(\cos\beta\hat{\mathbf{y}}_0 - \sin\beta\hat{\mathbf{x}}_0)$$

Then the slip velocity $\mathbf{v}_{I\in\mathcal{A}} - \mathbf{v}_{I\in\mathcal{B}}$ is given by

$$\boxed{\mathbf{v}_{I\in\mathcal{A}} - \mathbf{v}_{I\in\mathcal{B}} = (R_A\omega_A + R_B\omega_B)(\cos\beta\hat{\mathbf{y}}_0 - \sin\beta\hat{\mathbf{x}}_0)} \quad [1]$$

Remark. We note here that the notation $\mathbf{v}_{I\in\mathcal{A}}$ and $\mathbf{v}_{I\in\mathcal{B}}$ is critical for a geometric point such as I.

To guarantee no slip between the disks, we set $\mathbf{v}_{I\in\mathcal{A}} = \mathbf{v}_{I\in\mathcal{B}}$: this gives the condition $\boxed{R_A\omega_A + R_B\omega_B = 0}$.

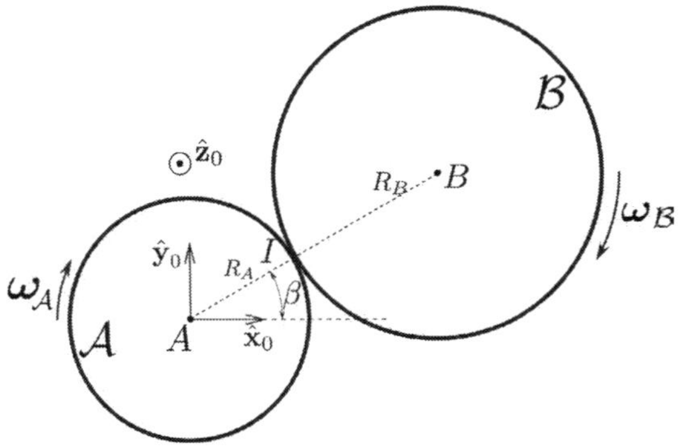

Figure 2.11 – *Problem 2.3*

b. We repeat the analysis done in a) for the acceleration $\mathbf{a}_{I\in\mathcal{A}}$, noting that $\boldsymbol{\alpha}_A = \mathbf{0}$ since the rotation of $\mathcal{A}$ is uniform:

$$\mathbf{a}_{I\in\mathcal{A}} = \alpha_A\hat{\mathbf{z}}_0 \times \mathbf{r}_{AI} - \omega_A^2\mathbf{r}_{AI} = -R_A\omega_A^2(\cos\beta\hat{\mathbf{x}}_0 + \sin\beta\hat{\mathbf{y}}_0)$$

by using formula (2.5). Similarly for $\mathbf{a}_{I\in\mathcal{B}}$:

$$\mathbf{a}_{I\in\mathcal{B}} = \alpha_B \hat{\mathbf{z}}_0 \times \mathbf{r}_{BI} - \omega_B^2 \mathbf{r}_{BI} = R_B \omega_B^2 (\cos\beta \hat{\mathbf{x}}_0 + \sin\beta \hat{\mathbf{y}}_0)$$

This leads to

$$\boxed{\mathbf{a}_{I\in\mathcal{A}} - \mathbf{a}_{I\in\mathcal{B}} = -(R_B\omega_B^2 + R_A\omega_A^2)(\cos\beta\hat{\mathbf{x}}_0 + \sin\beta\hat{\mathbf{y}}_0)}$$

We note that the expression [1] holds at all time and can be differentiated to give

$$\frac{d}{dt}(\mathbf{v}_{I\in\mathcal{A}} - \mathbf{v}_{I\in\mathcal{B}}) = \mathbf{0}$$

since ω_A, ω_B and β are constant. This shows that we cannot obtain $\mathbf{a}_{I\in\mathcal{A}}$ from $\frac{d}{dt}\mathbf{v}_{I\in\mathcal{A}}$. In fact we cannot obtain $\mathbf{v}_{I\in\mathcal{A}}$ by taking the time-derivative of $\mathbf{r}_{AI}$ (which is constant): $\mathbf{v}_{I\in\mathcal{A}} \neq \frac{d}{dt}\mathbf{r}_{AI}(=\mathbf{0})$ and $\mathbf{a}_{I\in\mathcal{A}} \neq \frac{d}{dt}\mathbf{v}_{I\in\mathcal{A}}(=\mathbf{0})$. This can be explained by the nature of I. Geometric point I is uniquely defined as the contact point at all time. However the contact point I is not fixed relative to an observer attached to $\mathcal{A}$ (in other words, $\mathbf{v}_{I/\mathcal{A}} \neq 0$). We can obtain the velocity of a point P as $\frac{d}{dt}\mathbf{r}_{AP}$ and its acceleration as $\frac{d}{dt}\mathbf{v}_P$ only if P is attached to body $\mathcal{A}$ *at all times*: this is not the case for I. ■

2.4 General Planar Motion

CONSIDER a rigid body $\mathcal{B}$ in general planar motion. We know that it is sufficient to study the restriction of $\mathcal{B}$ to a reference plane $(O, \hat{\mathbf{x}}_0, \hat{\mathbf{y}}_0)$ of $\mathcal{E}$. The unit vector $\hat{\mathbf{z}}_0 = \hat{\mathbf{x}}_0 \times \hat{\mathbf{y}}_0$ is normal to the plane of motion. To define the position of $\mathcal{B}$ relative to $\mathcal{E}$, we define (see Figure 2.12)
(i) the Cartesian coordinates (x_A, y_A) of a particular point A of $\mathcal{B}$,
(ii) the oriented angle θ between the fixed line $(O, \hat{\mathbf{x}}_0)$ and a particular line AB attached to $\mathcal{B}$.

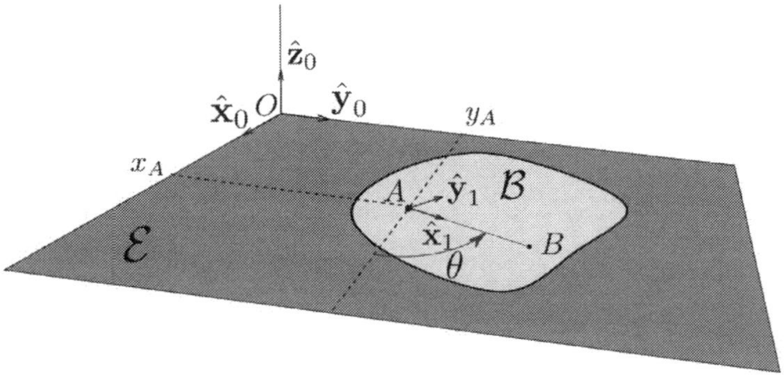

Figure 2.12 – *General planar motion.*

If the planar motion of $\mathcal{B}$ relative to $\mathcal{E}$ is *unconstrained*, then the coordinates (x_A, y_A, θ) are independent: *a rigid body in general planar motion has at most 3 degrees of freedom.*

Remark 9. If the motion of $\mathcal{B}$ is constrained, $\mathcal{B}$ will have fewer than 3 degrees of freedom.

We seek to relate the velocities and accelerations of two arbitrary points A and B of $\mathcal{B}$ and generalize the results obtained in Sections 2.2 and 2.3. As in the case of rotational motion, we define the angular velocity $\boldsymbol{\omega}_{\mathcal{B}}$ of $\mathcal{B}$ as the unique vector such that $d\mathbf{U}/dt = \boldsymbol{\omega}_{\mathcal{B}} \times \mathbf{U}$ for all vectors $\mathbf{U}$ fixed in $\mathcal{B}$. With the notations of Figure 2.12, the angular velocity takes the expression $\boldsymbol{\omega}_{\mathcal{B}} = \dot{\theta}\hat{\mathbf{z}}_0$.

Given points O, A and B, we can write the vector relationship $\mathbf{r}_{OB} = \mathbf{r}_{OA} + \mathbf{r}_{AB}$. The time-derivative of this equation gives

$$\mathbf{v}_B = \mathbf{v}_A + \frac{d\mathbf{r}_{AB}}{dt}$$

Since $\mathbf{r}_{AB}$ is a vector fixed in $\mathcal{B}$, we have $d\mathbf{r}_{AB}/dt = \boldsymbol{\omega}_{\mathcal{B}} \times \mathbf{r}_{AB}$. We can now state the main formulas of rigid body kinematics:

Velocity/Acceleration Field of a Rigid Body

The velocity and acceleration fields of a rigid body $\mathcal{B}$ of angular velocity $\boldsymbol{\omega}_{\mathcal{B}}$ satisfy the following property

$$\mathbf{v}_B = \mathbf{v}_A + \boldsymbol{\omega}_{\mathcal{B}} \times \mathbf{r}_{AB} \tag{2.9}$$

$$\mathbf{a}_B = \mathbf{a}_A + \boldsymbol{\alpha}_{\mathcal{B}} \times \mathbf{r}_{AB} - \omega_{\mathcal{B}}^2 \mathbf{r}_{AB} \tag{2.10}$$

for any two points A and B attached to $\mathcal{B}$. In (2.10), $\boldsymbol{\alpha}_{\mathcal{B}} = d\boldsymbol{\omega}_{\mathcal{B}}/dt$ is the angular acceleration of body $\mathcal{B}$.

Proof. Formula (2.10) is easily obtained by taking the time-derivative of (2.9): with $\frac{d}{dt}\boldsymbol{\omega}_{\mathcal{B}} = \boldsymbol{\alpha}_{\mathcal{B}}$ and $\frac{d}{dt}\mathbf{r}_{AB} = \boldsymbol{\omega}_{\mathcal{B}} \times \mathbf{r}_{AB}$ we find

$$\mathbf{a}_B = \mathbf{a}_A + \boldsymbol{\alpha}_{\mathcal{B}} \times \mathbf{r}_{AB} + \boldsymbol{\omega}_{\mathcal{B}} \times (\boldsymbol{\omega}_{\mathcal{B}} \times \mathbf{r}_{AB})$$

If both points A and B lie in the same (reference) plane, we find $\boldsymbol{\omega}_{\mathcal{B}} \times (\boldsymbol{\omega}_{\mathcal{B}} \times \mathbf{r}_{AB}) = -\omega_{\mathcal{B}}^2 \mathbf{r}_{AB}$. ■

Properties

1. Formulas (2.9-2.10) show that the knowledge of the velocity and acceleration of a particular point and the knowledge of the angular velocity and acceleration of $\mathcal{B}$ allow the determination of the velocity and acceleration of any other point of $\mathcal{B}$.

2. A consequence of (2.9) is the property of *equiprojectivity* of the velocity field: since $\mathbf{r}_{AB} \cdot (\boldsymbol{\omega}_{\mathcal{B}} \times \mathbf{r}_{AB}) = 0$ we can write

$$\mathbf{v}_B \cdot \mathbf{r}_{AB} = \mathbf{v}_A \cdot \mathbf{r}_{AB} \tag{2.11}$$

This shows that velocities $\mathbf{v}_A$ and $\mathbf{v}_B$ have the same projection on line AB: $\mathbf{r}_{AA_1} = \mathbf{r}_{BB_1}$. See Figure 2.13 (left). This property is useful in practice. It also shows that *the velocity of any point of $\mathcal{B}$ can be found from the knowledge of the velocity of one particular point and the direction of another*. See the construction of Figure 2.13 (right).

3. Equiprojectivity is fundamental to rigid body motion: it can be shown that if (2.11) is satisfied for all points of body $\mathcal{B}$, then its velocity field necessarily satisfies (2.9).

4. We recover the two cases examined in Sections 2.2 and 2.3:

(i) if $\mathcal{B}$ is in *translational motion*, $\boldsymbol{\omega}_{\mathcal{B}} = \boldsymbol{\alpha}_{\mathcal{B}} = \mathbf{0}$ and formulas (2.9-2.10) give $\mathbf{v}_B = \mathbf{v}_A$ and $\mathbf{a}_B = \mathbf{a}_A$.
(ii) if $\mathcal{B}$ is in *rotational motion* about O, then (2.9-2.10) give $\mathbf{v}_A = \boldsymbol{\omega}_{\mathcal{B}} \times \mathbf{r}_{OA}$ and $\mathbf{a}_A = \boldsymbol{\alpha}_{\mathcal{B}} \times \mathbf{r}_{OA} - \omega_{\mathcal{B}}^2 \mathbf{r}_{OA}$.
5. In practice, it can be useful and sometimes necessary to denote the velocity and acceleration of a point A of rigid body $\mathcal{B}$ as $\mathbf{v}_{A\in\mathcal{B}}$ and $\mathbf{a}_{A\in\mathcal{B}}$: this notation emphasizes that A must be considered a point of $\mathcal{B}$ even if this is meant only in an instantaneous way. In the latter case, $\mathbf{v}_{A\in\mathcal{B}}$ and $\mathbf{v}_A$ may be two different quantities as will be learned in Section 2.9.

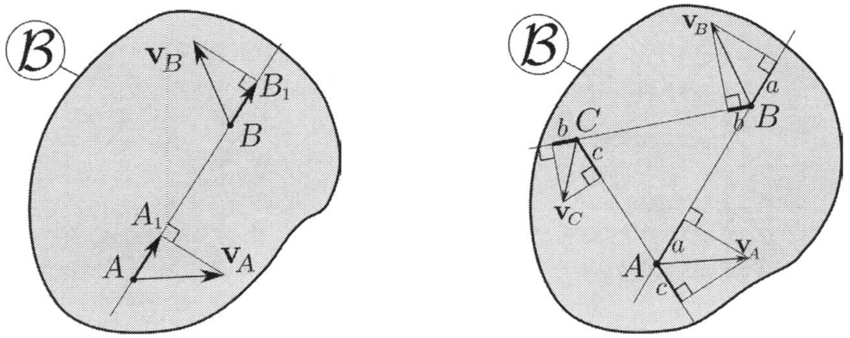

Figure 2.13 – *Equiprojectivity of the velocity field:* $\mathbf{r}_{AA_1} = \mathbf{r}_{BB_1}$ *(left). Velocity* $\mathbf{v}_C$ *can be determined from the knowledge of* $\mathbf{v}_A$ *and of the direction of* $\mathbf{v}_B$ *(right).*

HERE is an example to illustrate the use of (2.9-2.10).

Exercise 2.4 ▪A speed boat cruising at the constant speed of 45 km/h pulls a water-skier by the rope AB which remains taut. The extremity B of the rope describes a sine curve, as the water-skier slaloms between buoys.
Find the speed of the water-skier at the position shown. Then find its speed when extremity B coincides with the position C (one of the extrema of the sine trajectory). ▪

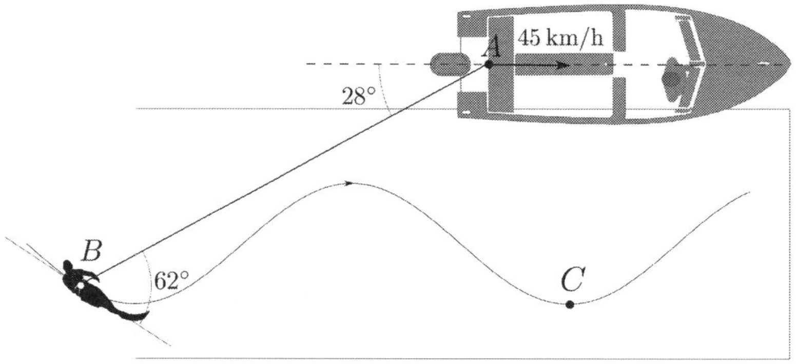

Figure 2.14 – *Problem 2.4.*

SOLUTION. This problem is solved by a simple application of formula (2.9). We choose unit vectors $(\hat{\imath}, \hat{\jmath}, \hat{\mathbf{k}})$ as shown in Figure 2.15 and introduce the angles α and β. Since the cable is taut, it behaves like a rigid body. The velocities of the two extremities A and B can then be related by the equation

$$\mathbf{v}_B = \mathbf{v}_A + \boldsymbol{\omega} \times \mathbf{r}_{AB} = v_A\hat{\imath} + \omega\hat{\mathbf{k}} \times l(-\cos\alpha\hat{\imath} + \sin\alpha\hat{\jmath}) = v_A\hat{\imath} + l\omega(-\cos\alpha\hat{\jmath} - \sin\alpha\hat{\imath})$$

Since the direction of $\mathbf{v}_B$ is given by angle $\beta - \alpha$, we can write

$$\mathbf{v}_B = v_B\big(\cos(\beta-\alpha)\hat{\imath} + \sin(\beta-\alpha)\hat{\jmath}\big)$$

This leads to two equations:

$$v_B\cos(\beta-\alpha) = v_A - l\omega\sin\alpha, \qquad v_B\sin(\beta-\alpha) = -l\omega\cos\alpha$$

We solve for v_B and find:

$$v_B\cos(\beta-\alpha) = v_A + v_B\frac{\sin(\beta-\alpha)\sin\alpha}{\cos\alpha}$$

or

$$\boxed{v_B = v_A\frac{\cos\alpha}{\cos\beta}} = \mathbf{84.6}\ \text{km/h}$$

We recognize here the property of equiprojectivity: $\mathbf{v}_A \cdot \mathbf{r}_{AB} = \mathbf{v}_B \cdot \mathbf{r}_{AB}$.

When B is at the position B, we have $\mathbf{v}_B = v_B\hat{\imath}$. Now we have the equation

$$v_B\hat{\imath} = v_A\hat{\imath} + \omega\hat{\mathbf{k}} \times l(-\cos\gamma\hat{\imath} + \sin\gamma\hat{\jmath})$$

which gives

$$v_B = v_A - l\omega\cos\gamma, \qquad 0 = l\omega\sin\gamma$$

We find $\omega = 0$ and $v_B = v_A = \mathbf{45}$ km/h which is also a consequence of equiprojectivity. ■

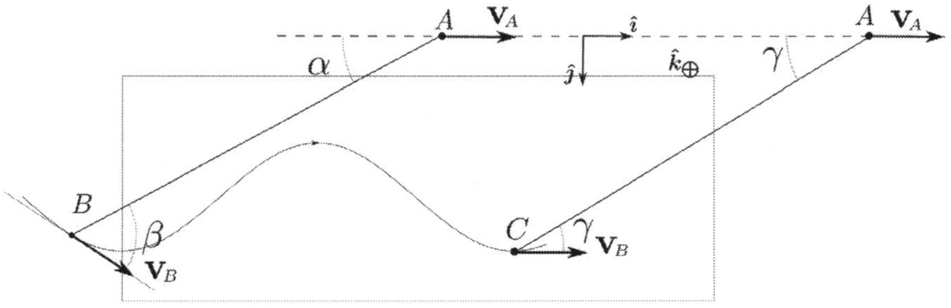

Figure 2.15 – *Problem 2.4.*

Exercise 2.5 ▪A 5-meter long ladder is in contact with a fixed support at points A and B. At the instant shown, the velocity and acceleration of A are 0.5 m/s and 4.5 m/s^2 both directed downward.
Find the velocity and acceleration of B, the angular velocity and acceleration by using formulas (2.9-2.10).
Verify the property of equiprojectivity between $\mathbf{v}_A$ and $\mathbf{v}_B$. ■

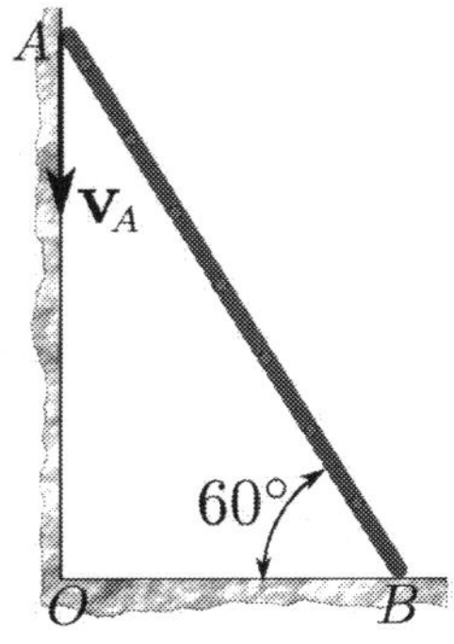

Solution. First we choose unit vectors $(\hat{\imath}, \hat{\jmath}, \hat{\mathbf{k}})$ as shown in Figure 2.16 (left). The vectorial method consists of using equation (2.9) to relate $\mathbf{v}_A$ and $\mathbf{v}_B$ (the velocities of two points of the *same* rigid body). The body is constrained to stay in contact with its support: $\mathbf{v}_A$ is directed on $\hat{\jmath}$ and $\mathbf{v}_B$ is directed on $\hat{\imath}$:

$$\mathbf{v}_A = -v_A\hat{\jmath}, \quad \mathbf{v}_B = v_B\hat{\imath}$$

After introducing the oriented angle θ as shown, we write the angular velocity of the body as $\boldsymbol{\omega} = -\dot{\theta}\hat{\mathbf{k}}$. With $L = |AB|$, formula (2.9) gives

$$\begin{aligned} v_B\hat{\imath} &= -v_A\hat{\jmath} - \dot{\theta}\hat{\mathbf{k}} \times \mathbf{r}_{AB} = -v_A\hat{\jmath} - \dot{\theta}\hat{\mathbf{k}} \times L(\cos\theta\hat{\imath} - \sin\theta\hat{\jmath}) \\ &= (-v_A - L\dot{\theta}\cos\theta)\hat{\jmath} - L\dot{\theta}\sin\theta\hat{\imath} \end{aligned}$$

This gives two equations

$$v_B = -L\dot{\theta}\sin\theta, \qquad -v_A - L\dot{\theta}\cos\theta = 0$$

which can be solved to give v_B and $\dot{\theta}$:

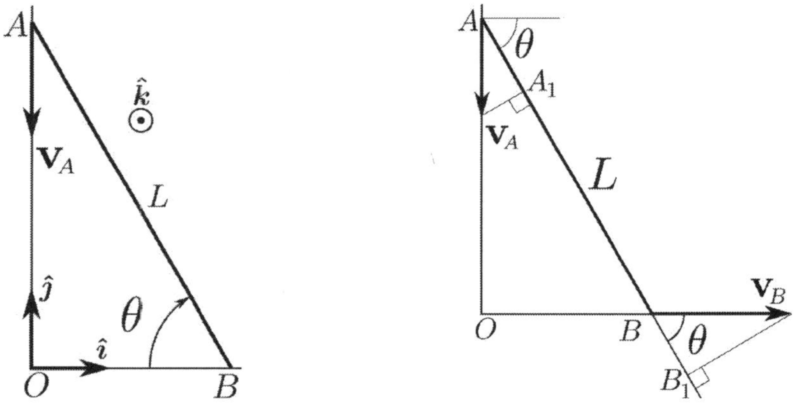

Figure 2.16 – *Problem 2.5*

$$\boxed{v_B = v_A\tan\theta, \qquad \dot{\theta} = -\frac{v_A}{L\cos\theta}} \qquad [1-2]$$

Note that $\dot{\theta} < 0$ as expected. With $L = 5$ m, $\theta = 60°$ and $v_A = 0.5$ m/s, we find $v_B = \mathbf{0.87}$ m/s and $\dot{\theta} = -\mathbf{0.2}$ rad/s.

To find the acceleration of B we proceed the same way: we set $\mathbf{a}_A = -a_A\hat{\jmath}$, $\mathbf{a}_B = a_B\hat{\imath}$ and $\boldsymbol{\alpha} = -\ddot{\theta}\hat{\mathbf{k}}$ and apply formula (2.10) between points A and B

$$\begin{aligned} a_B\hat{\imath} &= -a_A\hat{\jmath} - \ddot{\theta}\hat{\mathbf{k}} \times \mathbf{r}_{AB} - \dot{\theta}^2\mathbf{r}_{AB} \\ &= (-a_A - L\ddot{\theta}\cos\theta)\hat{\jmath} - L\ddot{\theta}\sin\theta\hat{\imath} - L\dot{\theta}^2(\cos\theta\hat{\imath} - \sin\theta\hat{\jmath}) \end{aligned}$$

This gives two equations

$$a_B = -L\ddot{\theta}\sin\theta - L\dot{\theta}^2\cos\theta, \qquad -a_A - L\ddot{\theta}\cos\theta + L\dot{\theta}^2\sin\theta = 0$$

which can be solved to give a_B and $\ddot{\theta}$:

$$\boxed{a_B = a_A \tan\theta - \frac{v_A^2}{L\cos^3\theta}, \qquad \ddot{\theta} = -\frac{a_A}{L\cos\theta} + \frac{v_A^2 \sin\theta}{L^2\cos^3\theta}} \qquad [3-4]$$

With $L = 5$ m, $\theta = 60°$, $v_A = 0.5$ m/s, $a_A = 4.5$ m/s^2, we find $a_B = \mathbf{7.4}$ m/s^2 and $\ddot{\theta} = -\mathbf{1.73}$ rad/s^2.

To verify the equiprojectivity of $\mathbf{v}_A$ and $\mathbf{v}_B$, we need to set the projections of $\mathbf{v}_A$ and $\mathbf{v}_B$ on line AB equal: $\overline{AA_1} = \overline{BB_1}$. See Figure 2.16(right). Since the angles made by $\mathbf{v}_A$ and $\mathbf{v}_B$ relative to line AB are known, we find

$$v_A \sin\theta = v_B \cos\theta$$

We recover equation [1]. ∎

Before considering more complex examples, we need to describe the manner in which rigid bodies are interconnected.

2.5 Joints between Pairs of Rigid Bodies

Mechanisms can be considered as systems of interconnected rigid bodies. It is convenient to denote the rigid bodies comprising a mechanism as 1,2,3,... in motion relative to a referential 0. The interconnection between any two rigid bodies is assured by a *joint* which guarantees a particular relative motion between two bodies. The analysis of mechanisms hinges on the study of joints which can be characterized by their geometry and the particular relative motion that they permit. The most general joint in planar kinematics is the 3-degree-of-freedom *planar joint* shown in Figure 2.1 whereby a planar surface of a rigid body is constrained to slide along another surface. However, three other types of joints with fewer degrees of freedom are encountered in planar mechanisms:

1. Revolute Joints or more simply *pivots* or *pins*: The relative motion between rigid bodies 1 and 2 is a rotation about a common point C which satisfies

$$\mathbf{v}_{C\in 1/0} = \mathbf{v}_{C\in 2/0}, \qquad \mathbf{a}_{C\in 1/0} = \mathbf{a}_{C\in 2/0}$$

at all time. See Figure 2.17. This is a one degree-of-freedom joint. Then the velocities of points of body 2 relative to body 1 are found according to

$$\mathbf{v}_{P\in 2/1} = \boldsymbol{\omega}_{2/1} \times \mathbf{r}_{CP}$$

where $\boldsymbol{\omega}_{2/1}$ is the angular velocity of body 2 relative to body 1.

2. Prismatic Joints or simply *sliders*: The relative motion between body 2 and body 1 is a rectilinear translation along a common direction $\hat{\mathbf{u}}$ (necessarily fixed relative to both 1 and 2). See Figure 2.18. The velocities of points of body 2 relative to body 1 satisfy

$$\mathbf{v}_{P\in 2/1} = v(t)\hat{\mathbf{u}}, \qquad \mathbf{a}_{P\in 2/1} = \dot{v}(t)\hat{\mathbf{u}}$$

for all points P. This is a one degree-of-freedom joint.

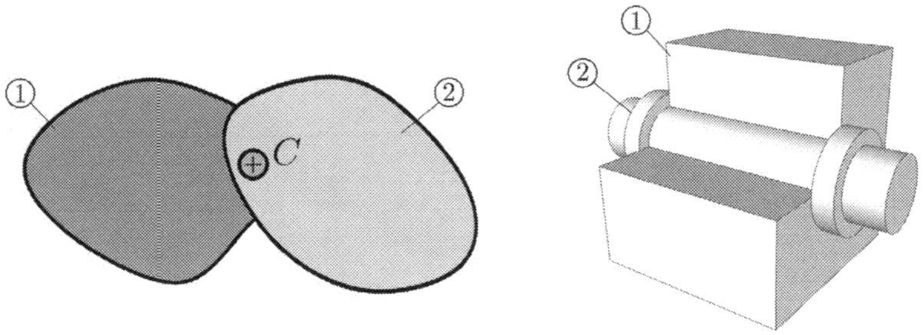

Figure 2.17 – *Pivot connection between rigid bodies 1 and 2 about point C (left). 3D-model of two revolute surfaces in relative rotation (right).*

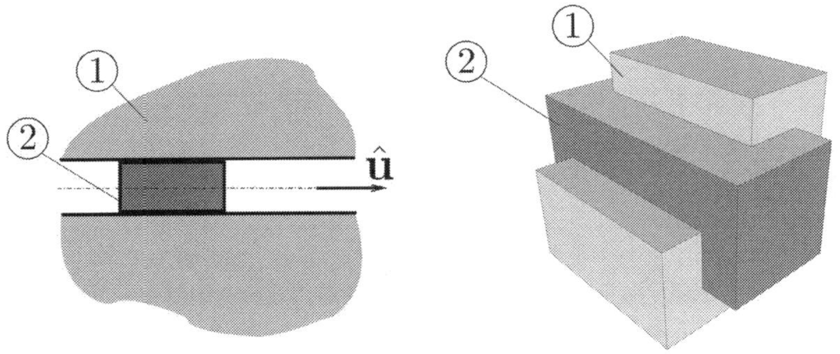

Figure 2.18 – *Slider connection between bodies 1 and 2 along vector* **û** *(left). 3D-model of two prismatic surfaces in contact (right).*

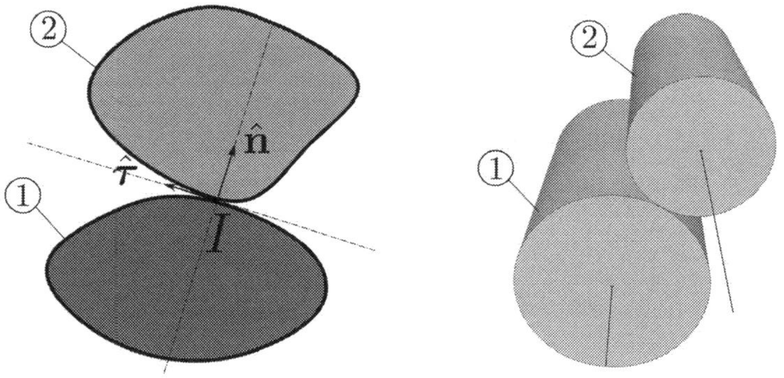

Figure 2.19 – *Point contact at I between two rigid bodies (left). 3D-model of two cylindrical surfaces in line contact (right).*

3. Point-Contact Joint: Two rigid bodies 1 and 2 are in relative motion in a referential 0 in such a way that, at any time, a point (or possibly a line) of the boundary of body 1 coincides with a point (or possibly a line) of the boundary of body 2. See Figure 2.19. Denoting by I the contact point, we assume that a tangent line at I to both boundaries exists. Call $\hat{\mathbf{n}}$ the common normal unit vector at I. Intuitively, the contact is maintained as long as the points of bodies 1 and 2 which coincide at I have the same velocity along the direction of $\hat{\mathbf{n}}$

$$\mathbf{v}_{I\in 1/0} \cdot \hat{\mathbf{n}} = \mathbf{v}_{I\in 2/0} \cdot \hat{\mathbf{n}} \tag{2.12}$$

This condition reduces the number of degrees of freedom of body 2 relative to body 1 to two. We will show in Sections 2.6 and 2.9 that the difference $\mathbf{v}_{I\in 2/0} - \mathbf{v}_{I\in 2/0}$ defines the *slip velocity* $\mathbf{v}_{I\in 2/1}$ which is necessarily directed along the tangential unit vector $\hat{\boldsymbol{\tau}}$. Condition (2.12) is thus specific to the motion of 2 relative to 1. However in practice, since the motions of 1 and 2 are described relative to a fixed referential 0, the no-loss of contact is imposed in the form of equation (2.12). We will return to point-contact joints in Section 2.6.

Note that in the case of point contact, the notation $\mathbf{v}_{I\in i/j}$ is vital. While there is a unique geometric point I at any given time, there are two ways of writing the velocity at I: $\mathbf{v}_{I\in 1/0}$ and $\mathbf{v}_{I\in 2/0}$. These two velocities may not be equal and should not be confused with $\mathbf{v}_{I/0}$.

Remark 10. There exist other joints such as ball-socket joints, helical joints, etc. But they do not lead to planar kinematics.

Next we illustrate how to analyze simple mechanisms with a few examples.

EXAMPLE PROBLEMS

General Procedure: General Planar Motions

To analyze the kinematics of a system of rigid body bodies in planar motion, follow these guidelines:

① At the start of the analysis, define the fixed unit vectors $(\hat{\mathbf{x}}_0, \hat{\mathbf{y}}_0, \hat{\mathbf{z}}_0)$ with $(\hat{\mathbf{x}}_0, \hat{\mathbf{y}}_0)$ in the reference plane and $\hat{\mathbf{z}}_0 = \hat{\mathbf{x}}_0 \times \hat{\mathbf{y}}_0$ normal to this plane. Then express all given and unknown kinematic quantities on $(\hat{\mathbf{x}}_0, \hat{\mathbf{y}}_0)$ (for velocities and acceleration) and $\hat{\mathbf{z}}_0$ (for angular velocities and accelerations). Choose an origin O if needed. Identify the nature of all joints which interconnect pairs of rigid bodies.

② The analysis typically consists in finding one or more input/output kinematic relationships: given the kinematics of an "input" body (often known as "link") of the system, find the kinematics of an "output" body. This is done by relating the velocities or accelerations of key points within each pair of interconnected bodies of the mechanism and by applying the two fundamental formulas (2.9-2.10).

③ An alternative method of analysis is based on the property of equiprojectivity: given two points of the *same* rigid body, the projections of their velocities along the line joining them must be equal. This method relies on a graphical construction and can be very efficient for the determination of kinematic input/output relationships.

④ Care must be taken to find out if the system's configuration is general or specific to a particular time, and hence, to a particular configuration. In the latter case, the time-derivative of geometric or kinematic relationships (in scalar or vector form) cannot be time-differentiated, since they are not valid at all time.

Remark 11. We will devise in Section 2.10 a general method for the kinematic analysis of mechanisms.

Exercise 2.6 ■The wheel 1 rotates at constant angular velocity ω_1 in the clockwise direction. It drives the motion of links 2 and 3. All joints are pivots centered at O, A B, and C. The geometry is given by the parameters R, l_2, l_3 and angle θ_2.

Determine the angular velocity and acceleration of links 2 and 3 as functions of ω_2 at the instant shown. ■

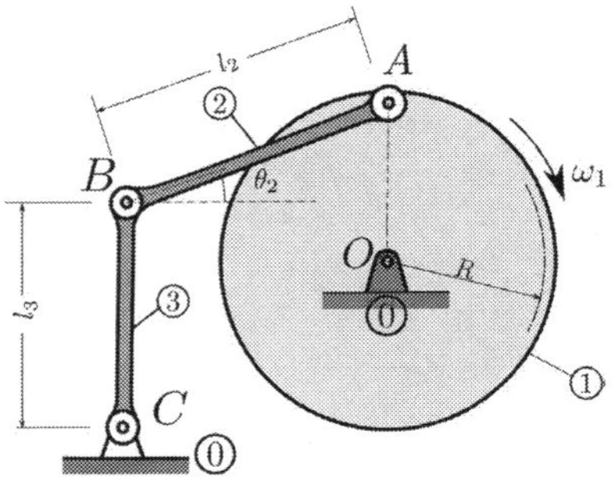

SOLUTION. The solution consists of relating the velocities of O, A, B and C by using formulas (2.9-2.10). At any pin P connecting bodies i and j we can write $\mathbf{v}_{P\in i} = \mathbf{v}_{P\in j}$ and $\mathbf{a}_{P\in i} = \mathbf{a}_{P\in j}$. This allows to connect the kinematics of body i to that of body j. We write the angular velocities/angular accelerations as $\boldsymbol{\omega}_1 = \omega_1\hat{\mathbf{k}}$, $\boldsymbol{\omega}_2 = \omega_2\hat{\mathbf{k}}$, $\boldsymbol{\omega}_3 = \omega_3\hat{\mathbf{k}}$, $\boldsymbol{\alpha}_1 = \mathbf{0}$, $\boldsymbol{\alpha}_2 = \alpha_2\hat{\mathbf{k}}$, $\boldsymbol{\alpha}_3 = \alpha_3\hat{\mathbf{k}}$. We seek the expressions of ω_2, ω_3, α_2 and α_3 as functions of ω_1.

STEP 1. We start with a velocity analysis. We first choose a basis $(\hat{\imath}, \hat{\jmath}, \hat{\mathbf{k}})$ as shown in Figure 2.20 with $\hat{\mathbf{k}} = \hat{\imath} \times \hat{\jmath}$. The analysis starts with pin A: $\mathbf{v}_{A\in 1} = \mathbf{v}_{A\in 2}$ with

$$\mathbf{v}_{A\in 1} = \omega_1\hat{\mathbf{k}} \times (-R\hat{\jmath}) = R\omega_1\hat{\imath} = \mathbf{v}_{A\in 2}$$

We then move to pin B whose velocity is related to that of pin A according to:

$$\mathbf{v}_{B\in 2} = \mathbf{v}_{A\in 2} + \omega_2\hat{\mathbf{k}} \times l_2(-\cos\theta_2\hat{\imath} + \sin\theta_2\hat{\jmath}) = R\omega_1\hat{\imath} + l_2\omega_2(-\cos\theta_2\hat{\jmath} - \sin\theta_2\hat{\jmath}) \quad [1]$$

Finally, we relate $\mathbf{v}_{B\in 3}$ to pin C on body 3

$$\mathbf{v}_{B\in 3} = \omega_3\hat{\mathbf{k}} \times (-l_3\hat{\jmath}) = l_3\omega_3\hat{\imath} \quad [2]$$

We conclusion the analysis by stating $\mathbf{v}_{B\in 2} = \mathbf{v}_{B\in 3}$ using the expressions of [1] and [2]:

$$R\omega_1\hat{\imath} = l_3\omega_3\hat{\imath} - l_2\omega_2(\cos\theta_2\hat{\jmath} + \sin\theta_2\hat{\imath})$$

This is easily solved to give:

$$\boxed{\omega_2 = 0, \omega_3 = \frac{R}{l_3}\omega_1}$$

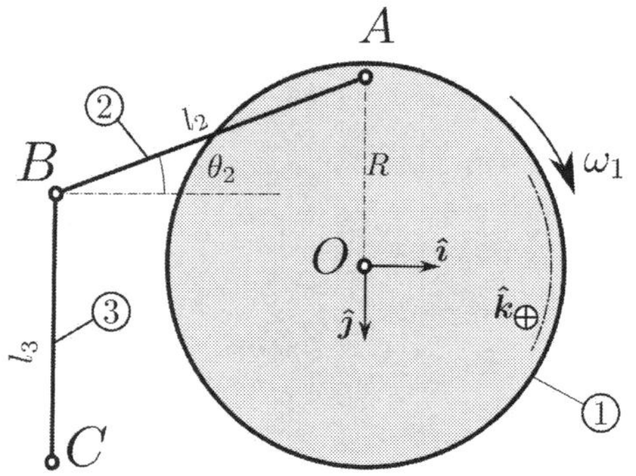

Figure 2.20 – *Problem 2.6*

STEP 2. We now repeat this analysis to find the angular accelerations of bodies 2 and 3.

We start by relating the accelerations of pin A and pin B:

$$\mathbf{a}_{A\in 1} = R\omega_1^2\hat{\jmath} = \mathbf{a}_{A\in 2} = \mathbf{a}_{B\in 2} + \alpha_2\hat{\mathbf{k}} \times l_2(\cos\theta_2\hat{\imath} - \sin\theta_2\hat{\jmath}) = \mathbf{a}_{B\in 2} + \alpha_2 l_2(\cos\theta_2\hat{\jmath} + \sin\theta_2\hat{\imath})$$

with (using $\omega_3 = 0$)

$$\mathbf{a}_{B\in 2} = \mathbf{a}_{B\in 3} = \alpha_3\hat{\mathbf{k}} \times (-l_3\hat{\jmath}) + l_3\omega_3^2\hat{\jmath} = l_3\alpha_3\hat{\imath} + l_3\omega_3^2\hat{\jmath}$$

This gives the equation

$$R\omega_1^2\hat{\jmath} = l_3\alpha_3\hat{\imath} + l_3\omega_3^2\hat{\jmath} + l_2\alpha_2(\cos\theta_2\hat{\jmath} + \sin\theta_2\hat{\imath})$$

which is solved to give:

$$\boxed{\alpha_2 = \frac{R(l_3 - R)}{l_2 l_3 \cos\theta_2}\omega_1^2, \qquad \alpha_3 = -\frac{R(l_3 - R)}{l_3^2}\omega_1^2 \tan\theta_2}$$

■

Exercise 2.7 ■The mechanism shown in Figure 2.21 is driven by the slider 3 at constant velocity $v_C = 1.75$ m/s in the upward direction. The geometry is given by $|AB| = 1.7$ m, $|BC| = 0.8$ m and the angle of 30° between line AB and BC.

Find the angular velocities and angular accelerations of links 1 and 2 at the instant shown. ■

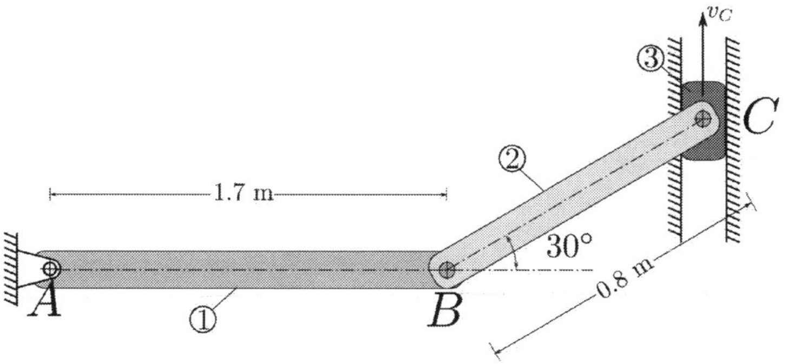

Figure 2.21 – *Problem 2.7*

SOLUTION. First we define the fixed basis $(\hat{\mathbf{x}}_0, \hat{\mathbf{y}}_0, \hat{\mathbf{z}}_0)$ as shown in Figure 2.22 with $\hat{\mathbf{z}}_0 = \hat{\mathbf{x}}_0 \times \hat{\mathbf{y}}_0$ normal to the reference plane. We first determine the angular velocities $\boldsymbol{\omega}_1 = \omega_1\hat{\mathbf{z}}_0$ and $\boldsymbol{\omega}_2 = \omega_2\hat{\mathbf{z}}_0$ of links 1 and 2. We start with point C whose velocity $\mathbf{v}_C = \mathbf{v}_{C\in 3} = \mathbf{v}_{C\in 2}$ is known and relate $\mathbf{v}_{C\in 2}$ to $\mathbf{v}_{B\in 2}$ as the velocities of two points of the same body:

$$\mathbf{v}_C = v_C\hat{\mathbf{y}}_0 = \mathbf{v}_{C\in 2} = \mathbf{v}_{B\in 2} + \omega_2\hat{\mathbf{z}}_0 \times \mathbf{r}_{BC} = \mathbf{v}_{B\in 2} + \omega_2\hat{\mathbf{z}}_0 \times l_2(\cos\theta\hat{\mathbf{x}}_0 + \sin\theta\hat{\mathbf{y}}_0)$$

giving

$$v_C\hat{\mathbf{y}}_0 = \mathbf{v}_{B\in 2} + l_2\omega_2(\cos\theta\hat{\mathbf{y}}_0 - \sin\theta\hat{\mathbf{x}}_0) \qquad [1]$$

Then we express the velocity $\mathbf{v}_{B\in 2} = \mathbf{v}_{B\in 1}$ as that of a point of body 1 in rotation about A:

$$\mathbf{v}_{B\in 1} = \boldsymbol{\omega}_1 \times \mathbf{r}_{AB} = \omega_1\hat{\mathbf{z}}_0 \times l_1\hat{\mathbf{x}}_0 = l_1\omega_1\hat{\mathbf{y}}_0 \qquad [2]$$

Equations [1] and [2] give

$$v_C\hat{\mathbf{y}}_0 = l_1\omega_1\hat{\mathbf{y}}_0 + l_2\omega_2(\cos\theta\hat{\mathbf{y}}_0 - \sin\theta\hat{\mathbf{x}}_0)$$

which yields two scalar equations

$$0 = -l_2\omega_2\sin\theta, \qquad v_C = l_1\omega_1 + l_2\omega_2\cos\theta$$

for the solution of ω_2 and ω_1:

$$\boxed{\omega_2 = 0, \qquad \omega_1 = \frac{v_C}{l_1}\hat{\mathbf{z}}_0} \qquad [3-4]$$

Remark. We note that link 2 is *in instantaneous translation*: $\mathbf{v}_{C\in 2} = \mathbf{v}_{B\in 2} = \mathbf{v}_{P\in 2}$ for all points P of link 2. This does not imply that $\mathbf{a}_{C\in 2} = \mathbf{a}_{B\in 2}$ or that $\boldsymbol{\alpha}_2 = \mathbf{0}$ at this instant!

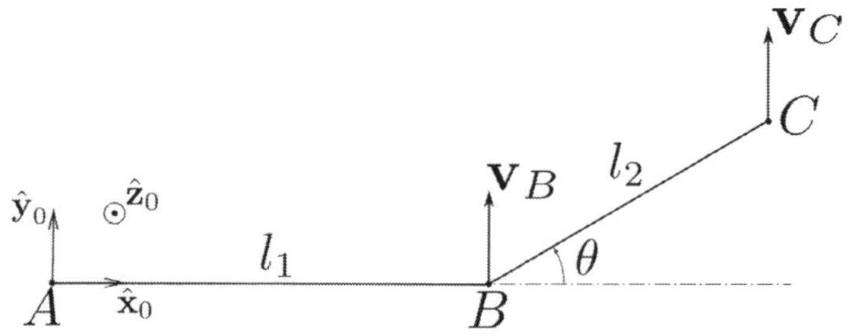

Figure 2.22 – *Problem 2.7 (Solution). The solution [3-4] can also be found by using the property of equiprojectivity of* $\mathbf{v}_{B\in 2}$ *and* $\mathbf{v}_{C\in 2}$*:* $v_C\sin\theta = v_B\sin\theta$ *leading to* $\mathbf{v}_B = \mathbf{v}_C$ *and* $\boldsymbol{\omega}_2 = \mathbf{0}$.

We cannot take the time-derivative of equations [3-4] since these equations hold only at the time of the configuration shown in 2.21. We repeat the velocity analysis for the acceleration: since $\mathbf{a}_{C\in 3} = \mathbf{0} = \mathbf{a}_{C\in 2}$ we have

$$\mathbf{0} = \mathbf{a}_{B\in 2} + \alpha_2\hat{\mathbf{z}}_0 \times \mathbf{r}_{BC} - \omega_2^2\mathbf{r}_{BC} = \mathbf{a}_{B\in 2} + l_2\alpha_2(\cos\theta\hat{\mathbf{y}}_0 - \sin\theta\hat{\mathbf{x}}_0) \qquad [5]$$

We then find $\mathbf{a}_{B\in 2} = \mathbf{a}_{B\in 1}$ by treating B as a point of body 1:

$$\mathbf{a}_{B\in 1} = \boldsymbol{\alpha}_1 \times \mathbf{r}_{AB} - \omega_1^2\mathbf{r}_{AB} = \alpha_1\hat{\mathbf{z}}_0 \times l_1\hat{\mathbf{x}}_0 - l_1\omega_1^2\hat{\mathbf{x}}_0 = l_1\alpha_1\hat{\mathbf{y}}_0 - l_1\omega_1^2\hat{\mathbf{x}}_0 \qquad [6]$$

Comparison of [5] and [6] gives

$$\mathbf{0} = l_1\alpha_1\hat{\mathbf{y}}_0 - l_1\omega_1^2\hat{\mathbf{x}}_0 + l_2\alpha_2(\cos\theta\hat{\mathbf{y}}_0 - \sin\theta\hat{\mathbf{x}}_0)$$

This gives two equations

$$0 = -l_1\omega_1^2 - l_2\alpha_2\sin\theta, \qquad 0 = l_1\alpha_1 + l_2\alpha_2\cos\theta$$

whose solution is

$$\boxed{\boldsymbol{\alpha}_2 = -\frac{v_C^2}{l_1 l_2\sin\theta}\hat{\mathbf{z}}_0, \qquad \boldsymbol{\alpha}_1 = \frac{v_C^2}{l_1^2}\cot\theta\,\hat{\mathbf{z}}_0}$$

Numerically, we find $\omega_1 = \mathbf{1.03}$ ↺ (rad/s), $\alpha_1 = \mathbf{1.83}$ ↺ (rad/s^2), $\alpha_2 = \mathbf{-4.50}$ ↺ (rad/s^2). ■

Exercise 2.8 ▪A robotic manipulator is composed of an arm 1 and forearm 2 in motion relative to a referential 0. At the instant shown, arm 1 rotates in the clockwise direction at angular velocity 5 rad/s while forearm 2 rotates relative to arm 1 in the clockwise direction at angular velocity 7.5 rad/s. At the instant shown, $\theta_1 = 50°$ and $\theta_2 = 65°$. $|OA| = 1$ m, $|AB| = 0.9$ m.

a. Find the velocity of A and B.

b. Assuming the angular velocities of both bodies constant, find the acceleration of B. ▪

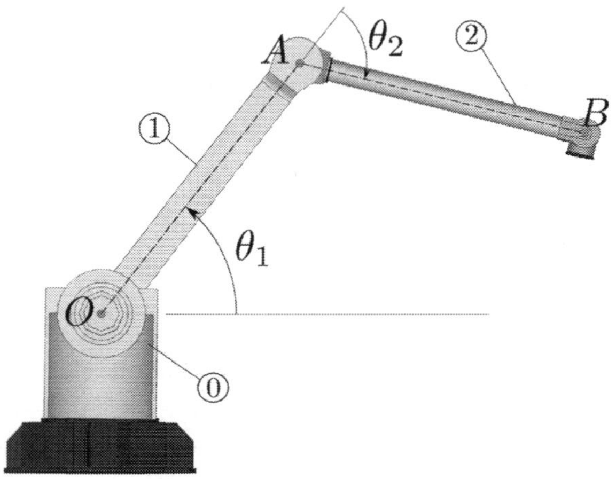

Figure 2.23 – *Problem 2.8*

SOLUTION. We first adopt a fixed set of unit vectors $(\hat{\mathbf{x}}_0, \hat{\mathbf{y}}_0, \hat{\mathbf{z}}_0)$ as shown in Figure 2.24. We express the angular velocities in terms of $\dot{\theta}_1$ and $\dot{\theta}_2$ on the chosen unit vector $\hat{\mathbf{z}}_0$:

$$\boldsymbol{\omega}_1 = \dot{\theta}_1 \hat{\mathbf{z}}_0, \qquad \boldsymbol{\omega}_2 = (\dot{\theta}_1 - \dot{\theta}_2)\hat{\mathbf{z}}_0$$

The expression of $\boldsymbol{\omega}_2$ is found by noting that the angle made by line AB relative to axis $(O, \hat{\mathbf{x}}_0)$ is $(\theta_1 - \theta_2)$ oriented in the counterclockwise direction. At the instant shown, arm 1 rotates in the clockwise direction: $\dot{\theta}_1 < 0$. Forearm 2 rotates relative to arm 1 in the clockwise direction: $\dot{\theta}_2 > 0$.

To find velocity $\mathbf{v}_A$ we treat point A as a point of body 1 and apply formula (2.9) between O and A:

$$\mathbf{v}_A = \mathbf{v}_O + \dot{\theta}_1 \hat{\mathbf{z}}_0 \times \mathbf{r}_{OA} = \dot{\theta}_1 \hat{\mathbf{z}}_0 \times L_1(\cos\theta_1 \hat{\mathbf{x}}_0 + \sin\theta_1 \hat{\mathbf{y}}_0)$$

giving

$$\boxed{\mathbf{v}_A = L_1 \dot{\theta}_1(\cos\theta_1 \hat{\mathbf{y}}_0 - \sin\theta_1 \hat{\mathbf{x}}_0)} \qquad [1]$$

To find velocity $\mathbf{v}_B$ we treat point B as a point of body 2 and apply formula (2.9) between A (since A is also a point of 2) and B:

$$\begin{aligned}
\mathbf{v}_B &= \mathbf{v}_A + (\dot{\theta}_1 - \dot{\theta}_2)\hat{\mathbf{z}}_0 \times \mathbf{r}_{AB} \\
&= L_1 \dot{\theta}_1(\cos\theta_1 \hat{\mathbf{y}}_0 - \sin\theta_1 \hat{\mathbf{x}}_0) + (\dot{\theta}_1 - \dot{\theta}_2)\hat{\mathbf{z}}_0 \times L_2(\cos(\theta_1 - \theta_2)\hat{\mathbf{x}}_0 + \sin(\theta_1 - \theta_2)\hat{\mathbf{y}}_0)
\end{aligned}$$

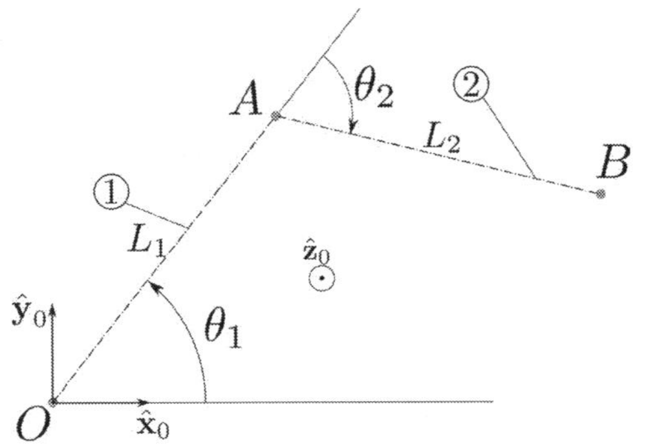

Figure 2.24 – *Problem 2.8 (Solution)*

yielding

$$\boxed{\begin{aligned}\mathbf{v}_E = \quad & \big(L_1\dot{\theta}_1 \cos\theta_1 + L_2(\dot{\theta}_1 - \dot{\theta}_2)\cos(\theta_1-\theta_2)\big)\hat{\mathbf{y}}_0 \\ & -\big(L_1\dot{\theta}_1 \sin\theta_1 + L_2(\dot{\theta}_1 - \dot{\theta}_2)\sin(\theta_1-\theta_2)\big)\hat{\mathbf{x}}_0\end{aligned}} \qquad [2]$$

To find $\mathbf{a}_B$ we may adopt the same method: relate $\mathbf{a}_B$ to $\mathbf{a}_A$ as the accelerations of two points of body 2 by using formula (2.10) after first determining $\mathbf{a}_A$ from the rotational motion of 1 about O.

A simpler method can be used by realizing that $\mathbf{v}_B$ given by [2] is valid at all times: we can obtain $\mathbf{a}_B$ by taking the time-derivative of [2]. Taking into account $\ddot{\theta}_1 = \ddot{\theta}_2 = 0$, we obtain

$$\boxed{\begin{aligned}\mathbf{a}_B = \quad & \big(-L_1\dot{\theta}_1^2 \sin\theta_1 - L_2(\dot{\theta}_1 - \dot{\theta}_2)^2\sin(\theta_1-\theta_2)\big)\hat{\mathbf{y}}_0 \\ & -\big(L_1\dot{\theta}_1^2 \cos\theta_1 + L_2(\dot{\theta}_1 - \dot{\theta}_2)^2\cos(\theta_1-\theta_2)\big)\hat{\mathbf{x}}_0\end{aligned}} \qquad [3]$$

Numerically with $\dot{\theta}_1 = -5$ (rad/s) and $\dot{\theta}_2 = 7.5$ (rad/s) we find $(v_{Ax}, v_{Ay}) = (3.8, -3.7)$ (m/s) and $(v_{Bx}, v_{By}) = (0.92, -14.1)$ (m/s), and $(a_{Bx}, a_{By}) = (-151.9, 17.2)$ (m/s^2). ■

Exercise 2.9 ■■Figure 2.25 shows a model of a hand-operated toggle press, comprised of a lever 1, connecting rod 2, toggle links 3 and 4 and of punch 5. The two toggle links 3 and 4 are characterized by $|BC| = |BD|$. When lever 1 is rotated about O at angular velocity 0.15 rad/s in the clockwise direction, punch 5 translates along the vertical direction. $|OA| = 38$ cm, $|AB| = 138$ cm, $|BC| = 78$ cm.

At the time shown, find the velocity of D by

a. using a vectorial approach connecting $\mathbf{v}_A$ to $\mathbf{v}_D$,

b. using the property of equiprojectivity of velocity vectors. ■

Solution. **a.** The vectorial solution consists of relating velocities $\mathbf{v}_D$ to $\mathbf{v}_A$ via $\mathbf{v}_B$. First we choose a basis $(\hat{\imath}, \hat{\jmath}, \hat{\mathbf{k}})$ and define the oriented angles θ_1, θ_2 and θ_3 as shown in Figure 2.26 (left). With the configuration $|BC| = |BD|$, the orientations of both links 3 and 4 are defined by the same angle θ_3. With the choice of $\hat{\mathbf{k}}$ we find the expressions of angular velocities of bodies 1, 2, 3 and 4:

$$\omega_1 = \dot{\theta}_1\hat{\mathbf{k}}, \quad \omega_2 = -\dot{\theta}_2\hat{\mathbf{k}}, \quad \omega_3 = \dot{\theta}_3\hat{\mathbf{k}}, \quad \omega_4 = -\dot{\theta}_3\hat{\mathbf{k}},$$

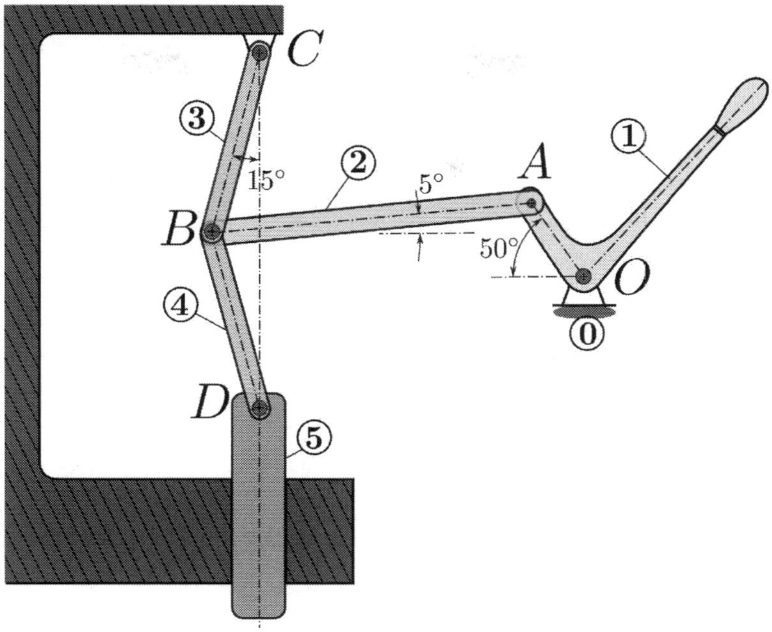

Figure 2.25 – *Problem 2.9*

We find $\mathbf{v}_A$ as a point of body 1 in rotation about O:

$$\mathbf{v}_A = \mathbf{v}_{A\in 1} = \boldsymbol{\omega}_1 \times \mathbf{r}_{OA} = \dot{\theta}_1 \hat{\mathbf{k}} \times l_1(-\cos\theta_1 \hat{\imath} - \sin\theta_1 \hat{\jmath}) = l_1\dot{\theta}_1(-\cos\theta_1 \hat{\jmath} + \sin\theta_1 \hat{\imath}) \quad [1]$$

Next we find $\mathbf{v}_B$ as a point of body 3 in rotation about C:

$$\mathbf{v}_B = \mathbf{v}_{B\in 3} = \boldsymbol{\omega}_3 \times \mathbf{r}_{CB} = \dot{\theta}_3 \hat{\mathbf{k}} \times l_3(\cos\theta_3 \hat{\jmath} - \sin\theta_3 \hat{\imath}) = l_3\dot{\theta}_3(-\cos\theta_3 \hat{\imath} - \sin\theta_3 \hat{\jmath}) \quad [2]$$

We can relate $\mathbf{v}_B = \mathbf{v}_{B\in 3} = \mathbf{v}_{B\in 2}$ to $\mathbf{v}_A = \mathbf{v}_{A\in 1} = \mathbf{v}_{A\in 2}$ since both points are attached to body 2:

$$\mathbf{v}_B = \mathbf{v}_A + \boldsymbol{\omega}_2 \times \mathbf{r}_{AB} = \mathbf{v}_A - \dot{\theta}_2 \hat{\mathbf{k}} \times l_2(-\cos\theta_2 \hat{\imath} + \sin\theta_2 \hat{\jmath}) = \mathbf{v}_A + l_2\dot{\theta}_2(\cos\theta_2 \hat{\jmath} + \sin\theta_2 \hat{\imath})$$

With the expressions [1-2] we find two equations

$$-l_3\dot{\theta}_3 \cos\theta_3 = l_2\dot{\theta}_2 \sin\theta_2 + l_1\dot{\theta}_1 \sin\theta_1 \quad [3]$$

$$-l_3\dot{\theta}_3 \sin\theta_3 = l_2\dot{\theta}_2 \cos\theta_2 - l_1\dot{\theta}_1 \cos\theta_1 \quad [4]$$

Finally we find $\mathbf{v}_D = \mathbf{v}_{D\in 5} = \mathbf{v}_{D\in 4}$ from $\mathbf{v}_B = \mathbf{v}_{B\in 4}$ as follows

$$\mathbf{v}_D = \mathbf{v}_B + \boldsymbol{\omega}_4 \times \mathbf{r}_{BD} = \mathbf{v}_B - \dot{\theta}_3 \hat{\mathbf{k}} \times l_3(\cos\theta_3 \hat{\jmath} + \sin\theta_3 \hat{\imath}) = \mathbf{v}_B + l_3\dot{\theta}_3(\cos\theta_3 \hat{\imath} - \sin\theta_3 \hat{\jmath})$$

Using [2] we obtain

$$\mathbf{v}_D = -2l_3\dot{\theta}_3 \sin\theta_3 \hat{\jmath}$$

We need $\dot{\theta}_3$ which is found from [3-4] by eliminating $\dot{\theta}_2$:

$$-l_3\dot{\theta}_3(\cos\theta_3 \cos\theta_2 - \sin\theta_3 \sin\theta_2) = l_1\dot{\theta}_1(\sin\theta_1 \cos\theta_2 + \cos\theta_1 \sin\theta_2)$$

yielding

$$l_3\dot{\theta}_3 = -l_1\dot{\theta}_1 \sin(\theta_1 + \theta_2)\cos(\theta_3 + \theta_2)$$

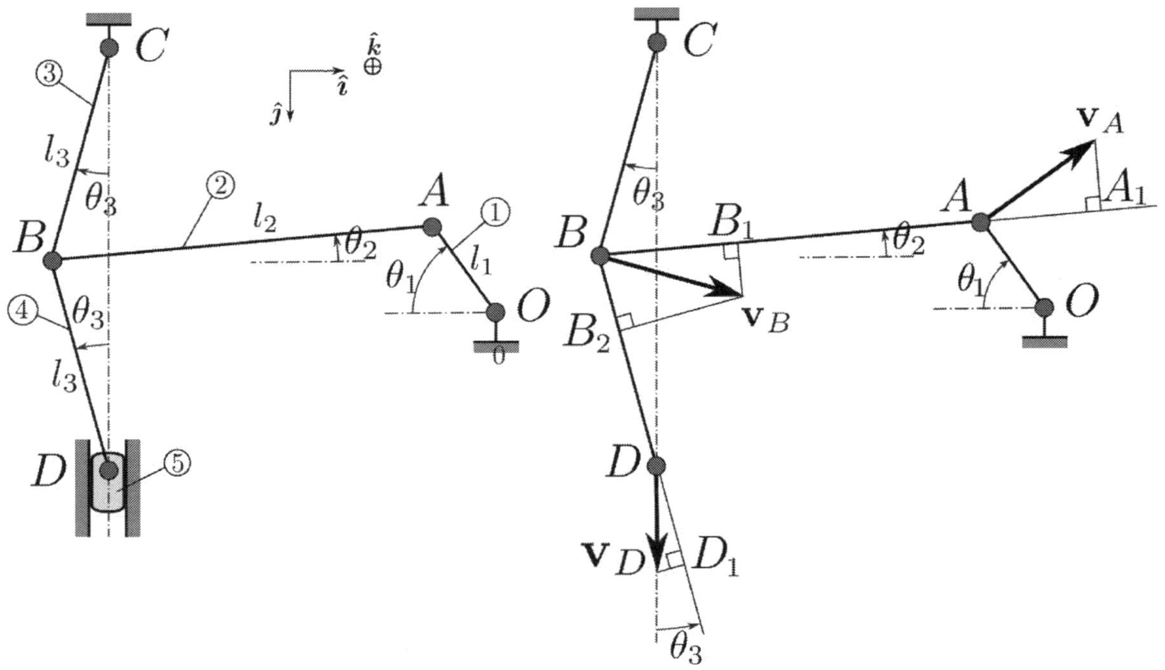

Figure 2.26 – *Problem 2.9 (Solution)*

Finally

$$\boxed{\mathbf{v}_D = 2l_1\dot{\theta}_1\frac{\sin(\theta_1+\theta_2)}{\cos(\theta_3+\theta_2)}\sin\theta_3\,\hat{\jmath}}$$

With $l_1 = 0.38$ m, $\dot{\theta}_1 = 0.15$ rad/s, $\theta_1+\theta_2 = 55°$, $\theta_3+\theta_2 = 20°$, $\theta_3 = 15°$, we find $\mathbf{v}_D = \mathbf{2.57}$ cm/s.

b. Here we use the equiprojectivity of the velocities of link 2: the projections of $\mathbf{v}_A$ and $\mathbf{v}_B$ on line AB must be identical. Referring to Figure 2.26 (right), this implies that $\overline{AA_1} = \overline{BB_1}$ or

$$v_A\cos\left(\frac{\pi}{2}-\theta_1-\theta_2\right) = v_B\cos(\theta_2+\theta_3)$$

with $v_A = l_1\dot{\theta}_1$ (A rotates about O) and $v_B = -l_3\dot{\theta}_3$ (B rotates about C with $\dot{\theta}_3 < 0$). This gives

$$l_3\dot{\theta}_3 = -l_1\dot{\theta}_1\frac{\sin(\theta_1+\theta_2)}{\cos(\theta_3+\theta_2)}$$

Next we use the equiprojectivity of the velocities of link 4: the projections of $\mathbf{v}_B$ and $\mathbf{v}_D$ on line BD must be identical, i.e. $\overline{DD_1} = \overline{BB_2}$:

$$v_D\cos\theta_3 = v_B\cos\left(\frac{\pi}{2}-2\theta_3\right) = v_B\sin 2\theta_3$$

This gives $v_D = 2v_E\sin\theta_3 = -2l_3\dot{\theta}_3\sin\theta_3$ or

$$\boxed{v_D = 2l_1\dot{\theta}_1\frac{\sin(\theta_1+\theta_2)}{\cos(\theta_3+\theta_2)}\sin\theta_3}$$

The direction of $\mathbf{v}_D$ shown in 2.26 (right) shows that $\mathbf{v}_D$ points in the $+\hat{\jmath}$ direction: the result is identical to that found in a). ■

2.6 Rolling Motion

CONSIDER two rigid bodies 1 and 2 in relative planar motion with a common contact point I (or contact line). We know from Section 2.5 that the kinematics at I is characterized by the difference of velocities $\mathbf{v}_{I\in 1/0} - \mathbf{v}_{I\in 2/0}$ which must remain in the direction of the tangent line $\mathcal{L}$ at I to the boundaries of bodies 1 and 2. This velocity difference quantifies the slip of body 2 relative body 1 and further characterizes the kinematics between the two bodies.

DEFINITION (ROLLING WITHOUT SLIPPING)

Two bodies 1 and 2 in contact at a point I are said to *roll without slipping* (over some time interval) if the *slip velocity* $\mathbf{v}_{I\in 2/0} - \mathbf{v}_{I\in 1/0}$ between the two bodies vanishes:

$$\mathbf{v}_{I\in 1/0} = \mathbf{v}_{I\in 2/0} \tag{2.13}$$

If this condition is not satisfied, then the bodies necessarily slip relative to one another.

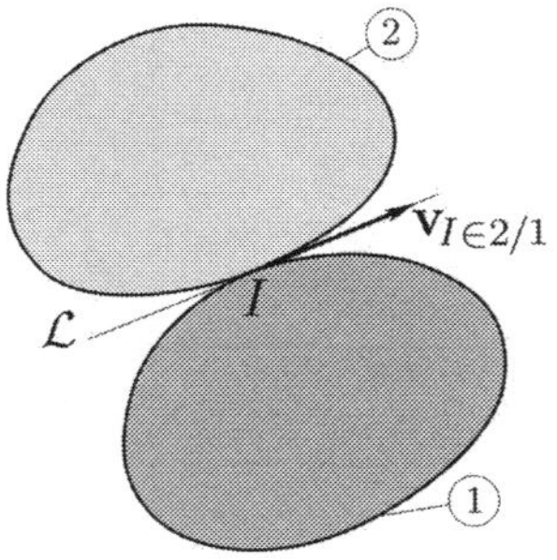

Figure 2.27 – *Body 2 rolls and slips relative to body 1.*

Remark 12. We will learn in Section 2.9 that the slip velocity actually takes the expression $\mathbf{v}_{I\in 2/1}$: it depends only on the relative motion between the two bodies, as expected.

Remark 13. If the no-slip condition at I is imposed by writing $\mathbf{v}_{I\in 1/0} = \mathbf{v}_{I\in 2/0}$, then the contact condition (2.12) is automatically satisfied.

NOTE that in case of no-slip, equation (2.13) holds at all time and can be time-differentiated. However, it is not equivalent to setting $\mathbf{a}_{I\in 1/0} = \mathbf{a}_{I\in 2/0}$. It turns out that $\mathbf{a}_{I\in 1/0}$ **is never equal to** $\mathbf{a}_{I\in 2/0}$. We can demonstrate this important fact by examining the case of a sphere 1 (of center C and radius r) rolling without slipping along a fixed horizontal support Ox of a referential 0 as shown in Figure 2.28.
Since the sphere is in contact with a fixed support, the no-slip condition (2.13) at contact point I can be written as:

$$\mathbf{v}_{I\in 1/0} = \mathbf{0} = \mathbf{v}_C + \boldsymbol{\omega}_1 \times \mathbf{r}_{CI}$$

With $\boldsymbol{\omega}_1 = \omega\hat{\mathbf{k}}$ and $\mathbf{r}_{IC} = r\hat{\boldsymbol{\jmath}}$ we find $\mathbf{v}_C = v_C\hat{\boldsymbol{\imath}} = \omega\hat{\mathbf{k}} \times r\hat{\boldsymbol{\jmath}} = -r\omega\hat{\boldsymbol{\imath}}$. Since this result holds at all time, we also have $\mathbf{a}_C = -r\alpha\hat{\boldsymbol{\imath}}$ with $\alpha = \dot{\omega}$. Now we can relate $\mathbf{a}_{I\in 1/0}$ to $\mathbf{a}_C$ by using (2.10) as the acceleration of two points of the same body:

$$\mathbf{a}_{I\in 1/0} = \mathbf{a}_C + \boldsymbol{\alpha} \times \mathbf{r}_{CI} - \omega^2\mathbf{r}_{CI} = -r\alpha\hat{\boldsymbol{\imath}} + \alpha\hat{\mathbf{k}} \times (-r\hat{\boldsymbol{\jmath}}) - \omega^2(-r\hat{\boldsymbol{\jmath}}) = -r\alpha\hat{\boldsymbol{\imath}} + r\alpha\hat{\boldsymbol{\imath}} + r\omega^2\hat{\boldsymbol{\jmath}}$$

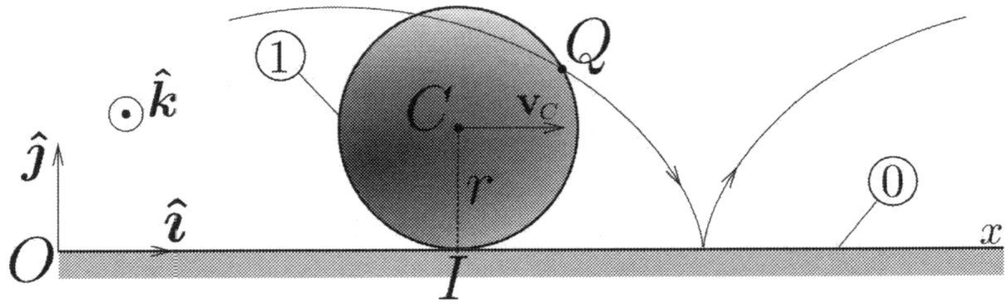

Figure 2.28 – *Sphere 1 rolling without slipping on a support Ox of referential 0. No slip imposes* $\mathbf{v}_{I\in 1/0} = \mathbf{0}$. *However* $\mathbf{v}_{I/0} \neq \mathbf{0}$. *Similarly,* $\mathbf{a}_{I\in 1/0} \neq \mathbf{0} \neq \mathbf{a}_{I/0}$.

We finally obtain

$$\mathbf{a}_{I\in 1/0} = r\omega^2 \hat{\boldsymbol{\jmath}}$$

This shows that $\mathbf{a}_{I\in 1/0} \neq 0$. *Hence we cannot write the time-derivative of (2.13) in the form* $\mathbf{a}_{I\in 1/0} = \mathbf{a}_{I\in 2/0}$. The reason for this is that point I is a point neither attached to body 1 nor to body 2 at all time. In the case of the rolling sphere, we can see that the contact point I keeps changing relative to $\mathcal{B}$: a particular point Q of $\mathcal{B}$ comes in contact with the support at some time t_1. At some later time $t_2 > t_1$, a different point of $\mathcal{B}$ comes in contact with the support. This explains why the notation $\mathbf{v}_{I\in 1/0}$ is necessary: in the example of the rolling sphere, the velocity $\mathbf{v}_{I/0}$ is found to be

$$\mathbf{v}_{I/0} = \frac{d}{dt}\mathbf{r}_{OI} = \frac{d}{dt}\mathbf{r}_{OC} = v_C \hat{\boldsymbol{\imath}} = -r\omega \hat{\boldsymbol{\imath}}$$

whereas $\mathbf{v}_{I\in 1/0} = \mathbf{0}$. We see that, in the case of point I, $\mathbf{v}_{I/0} \neq \mathbf{v}_{I\in 1/0}$.

Remark 14. We will learn in Section 2.9 that $\mathbf{v}_{I/0} = \mathbf{v}_{I/1} + \mathbf{v}_{I\in 1/0}$.

PROPERTIES

①. The no-slip condition (2.13) at a point I will yield at most two equations. However the normal component of the slip velocity will often automatically vanish whenever the condition of no-loss of contact is satisfied. In this case, (2.13) will yield a single equation (along the tangent direction at I).

②. In practice rolling without slipping can be achieved by using friction wheels in direct contact. The no-slip condition is maintained as long as the pressure exerted is sufficiently high. The higher the pressure, the higher the rolling friction, and the quicker the damage to the materials. The performance of the friction wheels will gradually degrade. Another solution is to use geared wheels (spur gears or helical gears): two meshing gears effectively behave as if two fictitious circles called *pitch circles* roll without slipping. Figure 2.29 shows a kinematic sketch of two meshing gears and their corresponding pitch circles of radii r_1 and r_2: it can be shown that the no-slip condition leads to the ratio $\omega_2/\omega_1 = r_1/r_2 = N_1/N_2$ where N_i is the number of teeth of the ith gear.

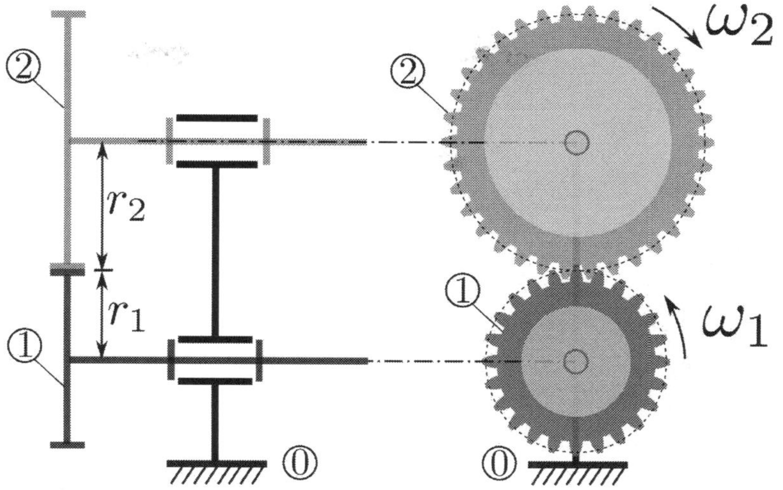

Figure 2.29 – *Kinematic sketch of meshing gears: side view (left), front view (right).*

EXAMPLE PROBLEMS

General Procedure: Kinematics of Rolling Motions

To analyze the kinematics of systems in rolling motion, follow these guidelines:

① To derive the no-slip condition (2.13) at a point I, the velocities $\mathbf{v}_{I\in 1/0}$ or $\mathbf{v}_{I\in 2/0}$ must be determined by using formula (2.9), that is, by relating I to another point attached to body 1 or 2:

$$\mathbf{v}_{I\in 1/0} = \mathbf{v}_{A\in 1/0} + \boldsymbol{\omega}_1 \times \mathbf{r}_{AI}, \qquad \mathbf{v}_{I\in 2/0} = \mathbf{v}_{B\in 2/0} + \boldsymbol{\omega}_2 \times \mathbf{r}_{BI}$$

where the points A and B are chosen so that their velocities are readily found.

② If body 1 rolls without slipping relative to a fixed support of referential 0, then the point of contact satisfies $\mathbf{v}_{I\in 1/0} = 0$ and the velocity of any point of body 1 can be found according to

$$\mathbf{v}_{P\in 1/0} = \boldsymbol{\omega}_{1/0} \times \mathbf{r}_{IP}$$

Hence the velocities of points of body 1 are found as if the body were in rotation about I.

Exercise 2.10 ▪In this mechanism, the plunger 1 is in translation at constant speed $\dot{y} = v_B$. It sets in motion the arm 3 by the use of roller 2 which is pin-connected at its center A to arm 3. The roller rolls without slipping relative to the plunger at contact point I.
Find the angular velocity $\boldsymbol{\omega}_2$ and $\boldsymbol{\omega}_3$ of the roller and arm as a function of v_B and the geometric parameters r, l and angle θ. ▪

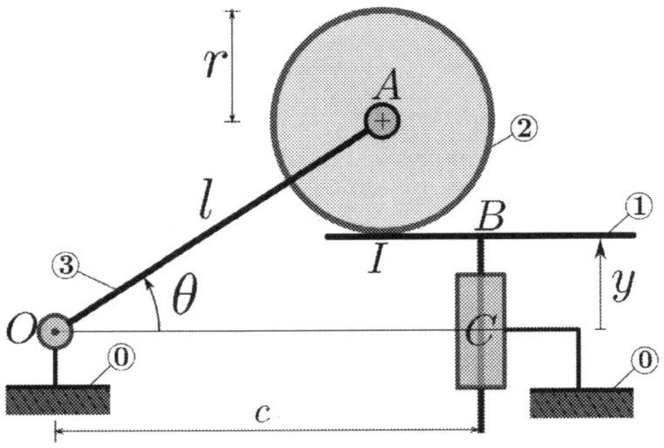

SOLUTION. We choose a basis of fixed unit vectors $(\hat{\mathbf{x}}_0, \hat{\mathbf{y}}_0, \hat{\mathbf{z}}_0)$ as shown in Figure 2.30. To find angular velocities $\boldsymbol{\omega}_2 = \omega_2\hat{\mathbf{z}}_0$ and $\boldsymbol{\omega}_3 = \omega_3\hat{\mathbf{z}}_0$ of bodies 2 and 3, we connect the motions of the 3 bodies through the no-slip condition at point I which can be stated as

$$\mathbf{v}_{I\in 1/0} = \mathbf{v}_{I\in 2/0} \qquad [1]$$

To find $\mathbf{v}_{I\in 1/0}$ we must view I as instantaneously attached to plunger 1: since the plunger is in translational motion along the $\hat{\mathbf{y}}_0$ direction relative to referential 0, its velocity field satisfies $\mathbf{v}_{P\in 1/0} = v_B\hat{\mathbf{y}}_0$ for all points P. This gives $\mathbf{v}_{I\in 1/0} = v_B\hat{\mathbf{y}}_0$.

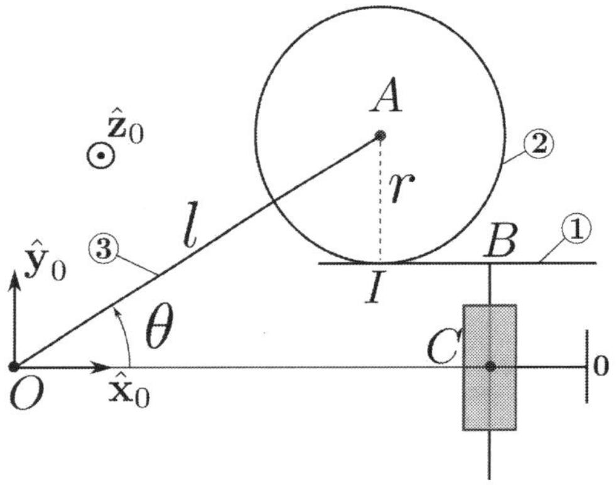

Figure 2.30 – *Problem 2.10*

To find $\mathbf{v}_{I\in 2/0}$ we must view I as instantaneously attached to roller 2 which is in general planar motion: we must relate $\mathbf{v}_{I\in 2/0}$ to $\mathbf{v}_{A\in 2/0}$

$$\mathbf{v}_{I\in 2/0} = \mathbf{v}_{A\in 2/0} + \omega_2\hat{\mathbf{z}}_0 \times \mathbf{r}_{AI} = \mathbf{v}_{A\in 2/0} + r\omega_2\hat{\mathbf{x}}_0$$

Finally we find $\mathbf{v}_{A\in 2/0} = \mathbf{v}_{A\in 3/0}$ by finding the velocity of A as a point of arm 3 in rotation about O:

$$\mathbf{v}_{A\in 3/0} = \omega_3\hat{\mathbf{z}}_0 \times \mathbf{r}_{OA} = l\omega_3 \times (\cos\theta\hat{\mathbf{x}}_0 + \sin\theta\hat{\mathbf{y}}_0) = l\omega_3(\cos\theta\hat{\mathbf{y}}_0 - \sin\theta\hat{\mathbf{x}}_0)$$

Now equation [1] becomes:

$$v_B\hat{\mathbf{y}}_0 = l\omega_3(\cos\theta\hat{\mathbf{y}}_0 - \sin\theta\hat{\mathbf{x}}_0) + r\omega_2\hat{\mathbf{x}}_0$$

which gives to equations

$$0 = -l\omega_3\sin\theta + r\omega_2, \qquad v_B = l\omega_3\cos\theta$$

to solve for ω_2 and ω_3 (note that $\omega_3 = \dot{\theta}$)

$$\boxed{\omega_2 = \frac{v_B}{r}\tan\theta\,\hat{\mathbf{z}}_0, \qquad \omega_3 = \frac{v_B}{l\cos\theta}\,\hat{\mathbf{z}}_0}$$

■

Exercise 2.11 ■Rigid bodies 1 and 2 are two identical cylinders of radius R. The bottom cylinder rolls outward without slipping at constant velocity v_0. The top cylinder rolls without slipping relative to the bottom cylinder while remaining in contact with the vertical wall. The line joining their centers A and B makes the angle φ relative to the vertical.
Using a vectorial method, find the angular velocities of the two cylinders, the velocity of B, and the slip velocity of the top cylinder relative to the vertical wall in terms of v_0, R and angle φ. Find these quantities for $v_0 = 0.5$ m/s, $R = 0.1$ m, and $\varphi = \pi/4$.
■

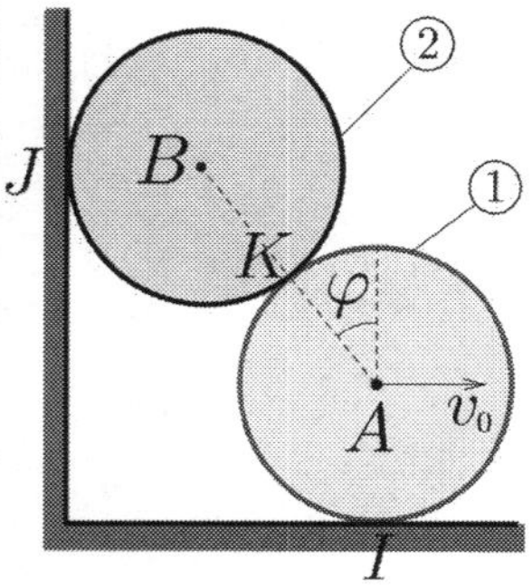

SOLUTION. First we choose unit vectors $(\hat{\imath}, \hat{\jmath}, \hat{\mathbf{k}})$ as shown in Figure 2.31. Call I, J and K the contact points. The velocity of center A is known: $\mathbf{v}_A = v_0\hat{\imath}$. First we impose the no-slip condition at contact point I by writing $\mathbf{v}_{I\in 1/0} = \mathbf{0}$: this gives

$$\mathbf{v}_A = \mathbf{v}_{I\in 1/0} + \boldsymbol{\omega}_1 \times \mathbf{r}_{IA} = \omega_1\hat{\mathbf{k}} \times R\hat{\jmath} = -R\omega_1\hat{\imath} = v_0\hat{\imath}$$

This gives the angular velocity of the bottom cylinder

$$\boxed{\boldsymbol{\omega}_1 = -\frac{v_0}{R}\,\hat{\mathbf{k}}}$$

Next we find $\mathbf{v}_{K\in 1/0}$ for contact point K between 1 and 2:

$$\mathbf{v}_{K\in 1/0} = \mathbf{v}_A + \boldsymbol{\omega}_1 \times \mathbf{r}_{AK} = v_0\hat{\imath} + \omega_1\hat{\mathbf{k}} \times R(\cos\varphi\hat{\jmath} - \sin\varphi\hat{\imath})$$

This gives

$$\mathbf{v}_{K\in 1/0} = (v_0 + v_0\cos\varphi)\hat{\imath} + v_0\sin\varphi\hat{\jmath}$$

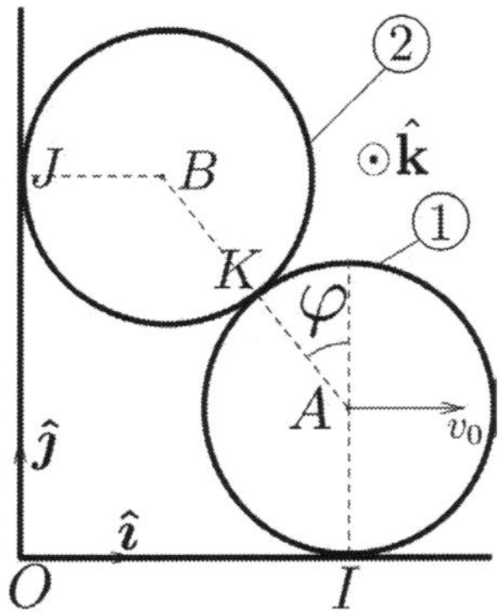

Figure 2.31 – *Problem 2.11 (Solution)*

Next we impose the no-slip condition at K: $\mathbf{v}_{K\in 1/0} = \mathbf{v}_{K\in 2/0}$. This enables us to find $\mathbf{v}_B$

$$\mathbf{v}_B = \mathbf{v}_{K\in 2/0} + \boldsymbol{\omega}_2 \times \mathbf{r}_{KB} = (v_0 + v_0 \cos\varphi)\hat{\boldsymbol{\imath}} + v_0 \sin\varphi\hat{\boldsymbol{\jmath}} + \omega_2\hat{\mathbf{k}} \times R(\cos\varphi\hat{\boldsymbol{\jmath}} - \sin\varphi\hat{\boldsymbol{\imath}})$$

or

$$\mathbf{v}_B = (v_0 + v_0 \cos\varphi)\hat{\boldsymbol{\imath}} + v_0 \sin\varphi\hat{\boldsymbol{\jmath}} - R\omega_2(\cos\varphi\hat{\boldsymbol{\imath}} + \sin\varphi\hat{\boldsymbol{\jmath}}) \qquad [1]$$

Finally we observe that center B remains at a constant distance R from the vertical axis: thus $\mathbf{v}_B = v_B\hat{\boldsymbol{\jmath}}$. Now equation [1] can be written as

$$v_B\hat{\boldsymbol{\jmath}} = (v_0 + v_0 \cos\varphi)\hat{\boldsymbol{\imath}} + v_0 \sin\varphi\hat{\boldsymbol{\jmath}} - R\omega_2(\cos\varphi\hat{\boldsymbol{\imath}} + \sin\varphi\hat{\boldsymbol{\jmath}})$$

which can be solved to give the two unknowns ω_2 and v_B:

$$v_B = v_0 \sin\varphi - R\omega_2 \sin\varphi, \qquad 0 = (v_0 + v_0\cos\varphi) - R\omega_2\cos\varphi$$

giving

$$\boxed{\boldsymbol{\omega}_2 = \frac{v_0}{R}\frac{1+\cos\varphi}{\cos\varphi}\hat{\mathbf{k}}} \qquad \boxed{\mathbf{v}_B = -v_0 \tan\varphi\,\hat{\boldsymbol{\jmath}}}$$

The slip velocity of the top cylinder relative to the vertical wall is the velocity $\mathbf{v}_{J\in 2/0}$:

$$\mathbf{v}_{J\in 2/0} = \mathbf{v}_B + \boldsymbol{\omega}_2 \times \mathbf{r}_{BJ} = -v_0 \tan\varphi\,\hat{\boldsymbol{\jmath}} + \frac{v_0}{R}\frac{1+\cos\varphi}{\cos\varphi}\hat{\mathbf{k}} \times (-R\hat{\boldsymbol{\imath}})$$

$$\boxed{\mathbf{v}_{J\in 2/0} = -v_0\frac{1+\sin\varphi+\cos\varphi}{\cos\varphi}\hat{\boldsymbol{\jmath}}}$$

For $\varphi = \pi/4$, $v_0 = 0.5$ (m/s) and $R = 0.1$ (m), we find $\omega_1 = \mathbf{5}$ (rad/s) ↺, $\omega_2 = \mathbf{12.1}$ (rad/s) ↻ and $v_B = \mathbf{0.5}$ (m/s) ↓. ■

Exercise 2.12 ■■Figure 2.32 shows a model of a ball bearing comprised of
- inner ring 1 of radius r_1 in rotation about O,
- outer ring 2 of inner radius r_2 in rotation about O,
- spherical ball 3 of center C and diameter $r_2 - r_1$ in rolling motion without slipping at contact points I and J with the rings,
- cage 4 used to secure the balls.

The inner and outer rings can rotate independently about O at angular velocity $\boldsymbol{\omega}_1$ and $\boldsymbol{\omega}_2$.

a. Write the no-slip condition at I ($|OI| = r_1$).
b. Write the no-slip condition at J ($|OJ| = r_2$).
c. From a) and b) find the angular velocity $\boldsymbol{\omega}_3$ of ball 3 and the velocity of center C in terms of the angular velocities $\boldsymbol{\omega}_1$ and $\boldsymbol{\omega}_2$. ■

Solution. **a.** First we introduce the basis of unit vectors $(\hat{\mathbf{u}}, \hat{\mathbf{v}}, \hat{\mathbf{k}})$ as shown in Figure 2.33 such that $\mathbf{r}_{OC} = \frac{r_1+r_2}{2}\hat{\mathbf{u}}$ and $\frac{d\hat{\mathbf{u}}}{dt} = \dot{\theta}\hat{\mathbf{v}}$. The no-slip condition at contact point I between bodies 1 and 3 can be stated as

$$\mathbf{v}_{I\in 1/0} = \mathbf{v}_{I\in 3/0} \qquad [1]$$

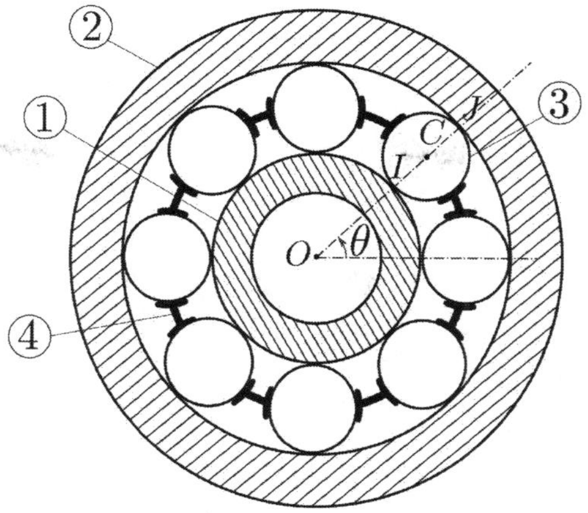

Figure 2.32 – *Problem 2.12*

Since body 1 is in rotation about O we can find $\mathbf{v}_{I\in 1/0}$ as

$$\mathbf{v}_{I\in 1/0} = \boldsymbol{\omega}_1 \times \mathbf{r}_{OI} = \omega_1 \hat{\mathbf{k}} \times r_1 \hat{\mathbf{u}} = r_1\omega_1\hat{\mathbf{v}}$$

To find velocity $\mathbf{v}_{I\in 3}$ we must relate I to another point attached to body 3: choosing C we find

$$\mathbf{v}_{I\in 3/0} = \mathbf{v}_C + \boldsymbol{\omega}_3 \times \mathbf{r}_{CI} = \mathbf{v}_C + \omega_3 \hat{\mathbf{k}} \times \frac{r_2 - r_1}{2}(-\hat{\mathbf{u}})$$

We find $\mathbf{v}_C$ by taking the time-derivative of $\mathbf{r}_{OC}$ to find

$$\mathbf{v}_C = \frac{d}{dt}\left(\frac{r_1 + r_2}{2}\hat{\mathbf{u}}\right) = \frac{r_1 + r_2}{2}\dot{\theta}\hat{\mathbf{v}}$$

This gives

$$\mathbf{v}_{I\in 3/0} = \left(\frac{r_1 + r_2}{2}\dot{\theta} - \frac{r_2 - r_1}{2}\omega_3\right)\hat{\mathbf{v}}$$

Equation [1] now gives

$$r_1\omega_1 = \frac{r_1 + r_2}{2}\dot{\theta} - \frac{r_2 - r_1}{2}\omega_3 \qquad [2]$$

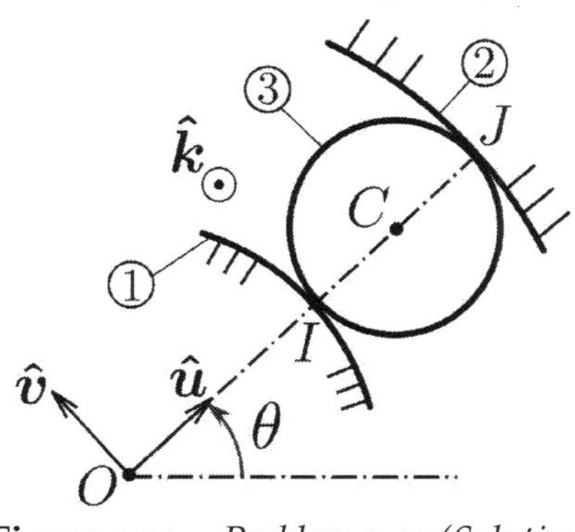

Figure 2.33 – *Problem 2.12 (Solution)*

b. We can repeat the same procedure at point J between bodies 2 and 3: we need to impose the condition $\mathbf{v}_{J\in 2/0} = \mathbf{v}_{J\in 3/0}$, with

$$\mathbf{v}_{J\in 2/0} = \boldsymbol{\omega}_2 \times \mathbf{r}_{OJ} = \omega_2 \hat{\mathbf{k}} \times r_2 \hat{\mathbf{u}} = r_2\omega_2\hat{\mathbf{v}}$$

and

$$\mathbf{v}_{J\in 3/0} = \mathbf{v}_C + \boldsymbol{\omega}_3 \times \mathbf{r}_{CJ} = \frac{r_1 + r_2}{2}\dot{\theta}\hat{\mathbf{v}} + \frac{r_2 - r_1}{2}\omega_3\hat{\mathbf{v}}$$

This gives the equation

$$r_2\omega_2 = \frac{r_1 + r_2}{2}\dot{\theta} + \frac{r_2 - r_1}{2}\omega_3 \qquad [3]$$

c. Given ω_1 and ω_2 we can find the angular velocities ω_3 and $\dot{\theta}$ by solving [2-3]: we obtain

$$\boxed{\boldsymbol{\omega}_3 = \frac{r_2\omega_2 - r_1\omega_1}{r_2 - r_1}\hat{\mathbf{k}}} \qquad \dot{\theta} = \frac{r_2\omega_2 + r_1\omega_1}{r_2 + r_1}$$

From the expression of $\dot{\theta}$, we can then find the velocity of C:

$$\boxed{\mathbf{v}_C = \frac{1}{2}(r_1 + r_2)\dot{\theta}\hat{\mathbf{v}} = \frac{1}{2}(r_2\omega_2 + r_1\omega_1)\hat{\mathbf{v}}}$$

Note that the expressions of ω_3 and $\dot{\theta}$ are valid at all time and can be time-differentiated to obtain $\boldsymbol{\alpha}_3$ and $\ddot{\theta}$. ■

2.7 Instantaneous Center of Rotation

We now present another method, of graphical nature, for the solution of planar kinematics problems. It hinges on the concept of *instantaneous center of rotation*.

Instantaneous Center of Rotation

At any given time, a rigid body $\mathcal{B}$ in general planar motion appears to be in rotation about a point I called *instantaneous center of rotation* (IC), that is, for any point P attached to $\mathcal{B}$ we have

$$\mathbf{v}_P = \boldsymbol{\omega}_\mathcal{B} \times \mathbf{r}_{IP} \qquad (2.14)$$

as long as $\boldsymbol{\omega}_\mathcal{B} \neq 0$.

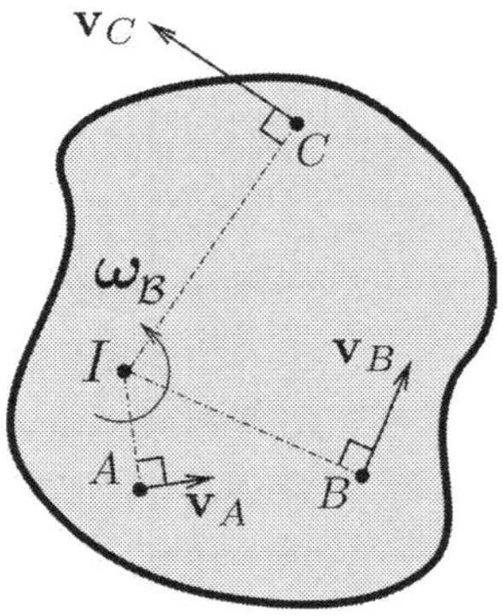

Figure 2.34 – *Instantaneous center I: body $\mathcal{B}$ appears to be in rotation about I.*

Proof. The existence of I is easily proved. Suppose that at a given instant we know the velocity $\mathbf{v}_A$ of a point A of $\mathcal{B}$ and the angular velocity $\boldsymbol{\omega}_{\mathcal{B}}$ *assumed non-zero*, then I is solution of the equation $\mathbf{v}_A = \boldsymbol{\omega}_{\mathcal{B}} \times \mathbf{r}_{IA}$. We can take the cross-product with $\boldsymbol{\omega}_{\mathcal{B}}$ of both sides of this equation. We use the formula $\mathbf{A} \times (\mathbf{B} \times \mathbf{C}) = (\mathbf{A} \cdot \mathbf{C})\mathbf{B} - (\mathbf{A} \cdot \mathbf{B})\mathbf{C}$ and obtain

$$\boldsymbol{\omega}_{\mathcal{B}} \times \mathbf{v}_A = \boldsymbol{\omega}_{\mathcal{B}} \times (\boldsymbol{\omega}_{\mathcal{B}} \times \mathbf{r}_{IA}) = (\boldsymbol{\omega}_{\mathcal{B}} \cdot \mathbf{r}_{IA})\boldsymbol{\omega}_{\mathcal{B}} - \omega_{\mathcal{B}}^2 \mathbf{r}_{IA}$$

If I lies in the same plane as A, we have $\boldsymbol{\omega}_{\mathcal{B}} \cdot \mathbf{r}_{IA} = 0$. This gives the position of I relative to A ($\omega_{\mathcal{B}} \neq 0$):

$$\mathbf{r}_{AI} = \frac{\boldsymbol{\omega}_{\mathcal{B}} \times \mathbf{v}_A}{\omega_{\mathcal{B}}^2} \tag{2.15}$$

In practice, this formula is not used since the determination of I is done graphically. ∎

Remark 15. Point I must be considered as instantaneously attached to $\mathcal{B}$ (or its rigid extension). Otherwise $\mathcal{B}$ is in pure rotation about I.

Remark 16. Equation (2.14) gives in modulus $v_P = \omega_{\mathcal{B}}|IP|$ with $v_P = |\mathbf{v}_P|$, $\omega_{\mathcal{B}} = |\boldsymbol{\omega}_{\mathcal{B}}|$ and $|IP| = |\mathbf{r}_{IP}|$.

GRAPHICAL DETERMINATION OF I. Two cases must be considered.

CASE 1: The velocities of two points A and B of $\mathcal{B}$ are known *in direction*. Then I is located at the intersection of the normals based at A and B to $\mathbf{v}_A$ and $\mathbf{v}_B$, respectively. This is a direct consequence of $\mathbf{v}_A = \boldsymbol{\omega}_{\mathcal{B}} \times \mathbf{r}_{IA}$ and $\mathbf{v}_B = \boldsymbol{\omega}_{\mathcal{B}} \times \mathbf{r}_{IB}$. See Figure 2.35 (left).

Remark 17. If the two normals do not intersect, then necessarily $\mathbf{v}_A = \mathbf{v}_B$ and $\boldsymbol{\omega}_{\mathcal{B}} = \mathbf{0}$: in this case $\mathcal{B}$ is said to be *in instantaneous translation* (I is located at infinity).

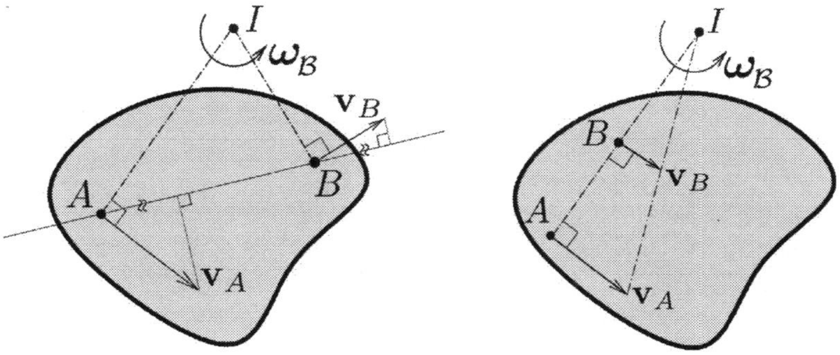

Figure 2.35 – *Graphical determination of I: Case 1 (left) and Case 2 (right). Note the equiprojectivity of velocities: the projections of $\mathbf{v}_A$ and $\mathbf{v}_B$ on line AB are equal.*

CASE 2: If $\mathbf{v}_A$ and $\mathbf{v}_B$ are parallel and normal to line AB, then I lies on AB and is found according to the velocity triangle construction of Section 2.3:

$$\frac{v_A}{|IA|} = \frac{v_B}{|IB|}$$

See Figure 2.35 (right). In this case, the precise location of I requires that velocities $\mathbf{v}_A$ and $\mathbf{v}_B$ be known in both direction and magnitude.

Remark 18. In case 2, if $\mathbf{v}_A = \mathbf{v}_B$, then necessarily $\omega_\mathcal{B} = \mathbf{0}$.

PROPERTIES

①. Body $\mathcal{B}$ is in instantaneous rotation about I as far as velocities are concerned: however I cannot be used for accelerations, that is, $\mathbf{a}_{I\in\mathcal{B}} \neq \mathbf{0}$. In fact it can be shown that there exists an *instantaneous center of acceleration* J, that is, a point satisfying $\mathbf{a}_{J\in\mathcal{B}} = \mathbf{0}$ which never coincides with I. See Section 2.11. Point I is often called *instantaneous center of zero velocity* to emphasize that I is not the instantaneous center of zero acceleration.

②. If angular velocity $\omega_\mathcal{B}$ has been found to be zero (at a particular configuration), it does not follow that the angular acceleration $\alpha_\mathcal{B}$ vanishes.

③. As time evolves, the position of I changes both relative to the fixed referential $\mathcal{E}$ and relative to body $\mathcal{B}$: the geometric point I describes two trajectories, a trajectory in $\mathcal{E}$ called *fixed centrode* $\mathcal{C}_0$ and a trajectory in $\mathcal{B}$ called *moving centrode* $\mathcal{C}_1$. *The motion of $\mathcal{B}$ relative to $\mathcal{E}$ can then be considered to be equivalent to the rolling motion of the moving centrode $\mathcal{C}_1$ without slipping on the fixed centrode $\mathcal{C}_0$*, since $\mathbf{v}_{I\in\mathcal{B}/\mathcal{E}} = \mathbf{0}$ at all time. See Figure 2.36.

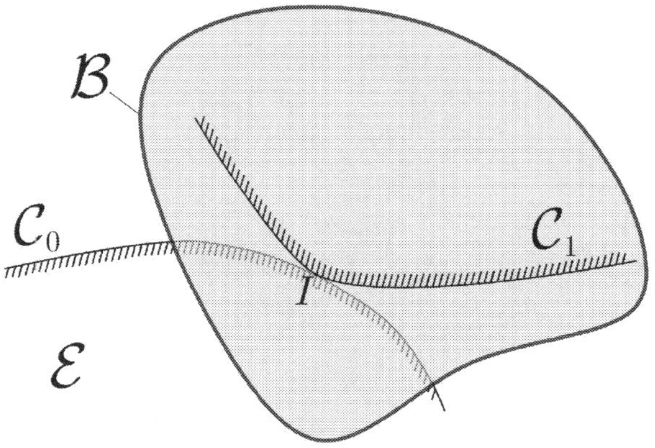

Figure 2.36 – *Fixed and moving centrodes: $\mathcal{C}_1$ rolls without slipping on $\mathcal{C}_0$.*

EXAMPLE PROBLEMS

GENERAL PROCEDURE: GRAPHICAL KINEMATIC ANALYSIS

① This method does not use vectors: it is based on a graphical construction.

② It hinges on finding graphically the position of the IC of each body of the mechanism at a given instant, by typically using the direction of the velocities of two points of the same body: see Figure 2.35 (left).

③ Once the IC $I_\mathcal{B}$ of body $\mathcal{B}$ is found, kinematic input/output relations are found by applying $v_{P\in\mathcal{B}} = \omega_\mathcal{B}|I_\mathcal{B}P|$ to two points P attached (possibly instantaneously) to $\mathcal{B}$.

④ Since this method gives velocities (linear or angular) in magnitude, the direction of these quantities is indicated in the graphical construction.

⑤ If the configuration of the mechanism is general (that is, not specific to a particular instant), then accelerations can be obtained by taking the time-derivative of velocity equations.

Exercise 2.13 ▪Consider the mechanism of problem 2.7. Find the angular velocities of links 1 and 2 by the method of instantaneous centers. Can the angular accelerations of links 1 and 2 be found from these results? ■

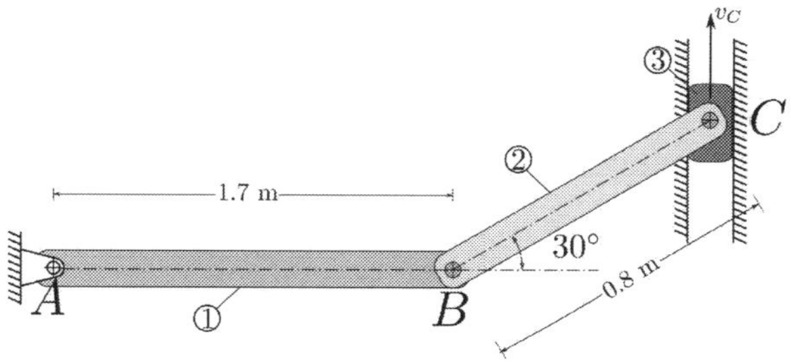

Figure 2.37 – *Problem 2.13*

SOLUTION. First we find the location of the IC of link 2: we use the direction of velocities $\mathbf{v}_C$ of C and $\mathbf{v}_B$ of B. Both points are attached to 2. Velocity $\mathbf{v}_C$ is imposed. Velocity $\mathbf{v}_B$ is obtained as that of a point of link 1 since $\mathbf{v}_{B\in 1} = \mathbf{v}_{B\in 2}$. Since link 1 rotates about A, $\mathbf{v}_B$ is normal to line AB. We conclude that both velocities $\mathbf{v}_C$ and $\mathbf{v}_B$ are parallel but are not normal to line BC: the line perpendicular to $\mathbf{v}_C$ at C is parallel to the line perpendicular to $\mathbf{v}_B$ at B. We conclude that $\boxed{\omega_2 = 0}$ and $\mathbf{v}_B = \mathbf{v}_C$. Link 2 is in *instantaneous translation* at this instant.

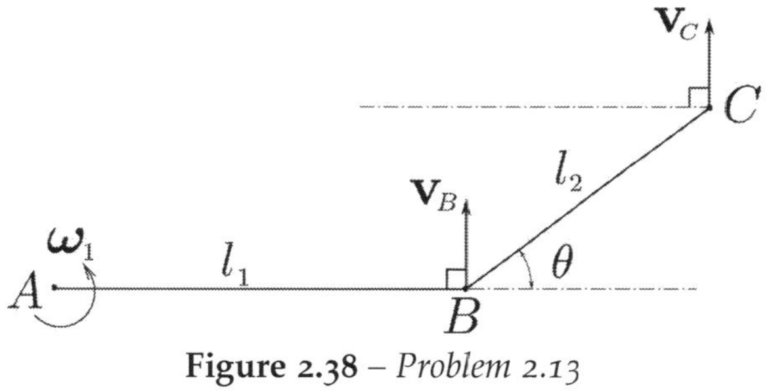

Figure 2.38 – *Problem 2.13*

To find ω_1 of link 1, we use $v_B = \omega_1 |AB| = l_1\omega_1 = v_C$ giving $\boxed{\omega_1 = v_C/l_1}$ directed in the anticlockwise direction. The construction is shown in Figure 2.38. We cannot find the angular accelerations of links 1 and 2 since the results found by this method are only valid at the time of the configuration. ■

Exercise 2.14 ▪A rod 1 is connected to two sliders at pins A and B (with $|AB| = L$) which are constrained to move along perpendicular directions. The slider at A moves at constant speed v_A as shown.
a. Find the location of the IC of rod 1. Then find its angular velocity ω_1 as a function of speed v_A and angle θ.
b. Find the corresponding fixed and moving centrodes, that is, the trajectories of I viewed by observers attached to referentials 0 and 1. Conclude that the motion of 1 relative to 0 is equivalent to a gear motion. ▪

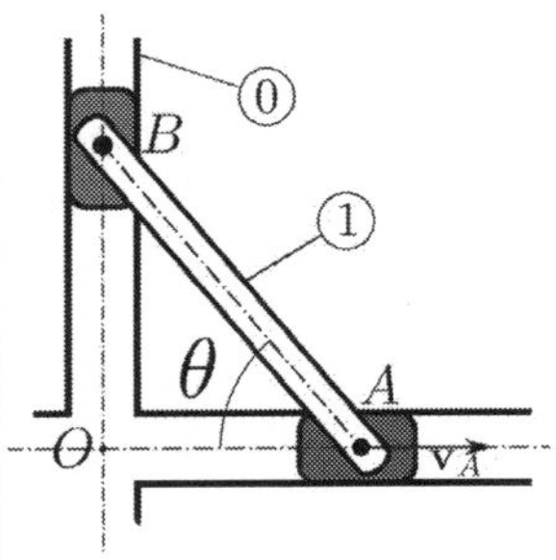

SOLUTION. **a.** The IC of rod 1 is found from the direction of $\mathbf{v}_A$ and $\mathbf{v}_B$: I is located at the intersection of lines Ay and line Bx: hence the figure $OAIB$ forms a rectangle. The angular velocity ω_1 satisfies

$$v_A = \omega_1 |IA|$$

with $|IA| = L \sin\theta$. This gives

$$\boxed{\omega_1 = \frac{v_A}{L \sin\theta} \quad \circlearrowleft}$$

directed in the counterclockwise direction. See Figure 2.39 (left). The Cartesian coordinates of I on axes Oxy are easily found to be

$$x_I = L\cos\theta, \qquad y_I = L\sin\theta$$

These coordinates define the fixed centrode $\mathcal{C}_0$, that is, the trajectory of I in the fixed referential. $\mathcal{C}_0$ is described by the equation $x_I^2 + y_I^2 = L^2$ (by eliminating angle θ): it is the circle of center O and radius L. The fixed centrode must be viewed as a subset of the fixed referential 0.

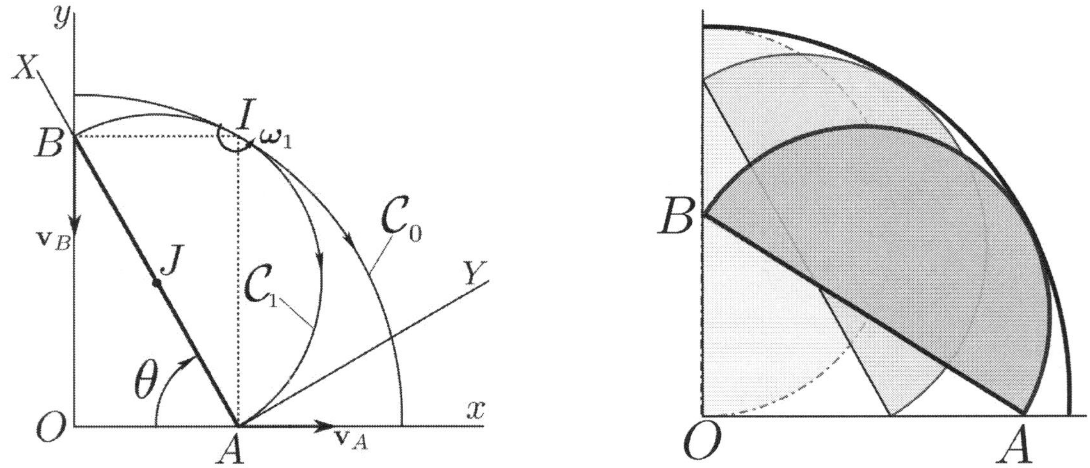

Figure 2.39 – *Problem 2.14 (Solution): construction of IC I (left), and sketch of the centrodes. One can easily verify that the two centrodes are tangential at I.*

Point I also moves relative to an observer attached to rod 1: during the motion of the rod, I moves along the moving centrode $\mathcal{C}_1$. $\mathcal{C}_1$ is found by determining the coordinates (X_I, Y_I) of I in the Cartesian coordinate axes AXY (with axis AX directed along AB).

We find

$$X_I = |AI| \sin\theta = L \sin^2\theta, \qquad Y_I = |AI| \cos\theta = L \sin\theta \cos\theta$$

or

$$X_I = \frac{L}{2}(1 - \cos 2\theta), \qquad Y_I = \frac{L}{2} \sin 2\theta$$

Eliminating θ gives the equation $(X_I - L/2)^2 + Y_I^2 = L^2/4$: this is the equation of the circle of center J midpoint of line segment AB and radius $L/2$ (passing through both points A and B). As rod 1 moves, centrode $\mathcal{C}_1$ (attached to the rod) rolls without slipping on centrode $\mathcal{C}_0$ (attached to fixed support) at contact point I. Hence the motion 1/0 is equivalent to a geared mechanism. Figure 2.39 (right) shows a few snapshots of the moving centrode. ■

Exercise 2.15 ▪Consider the mechanism of problem 2.10. Arm 3 is set in rotation by the translation of plunger 1 and by the coupling of 1 and 3 with roller 2.
By using the method of instantaneous centers, find the angular velocity ω_2 and ω_3 of the roller and arm as a function of v_B and the geometric parameters. ■

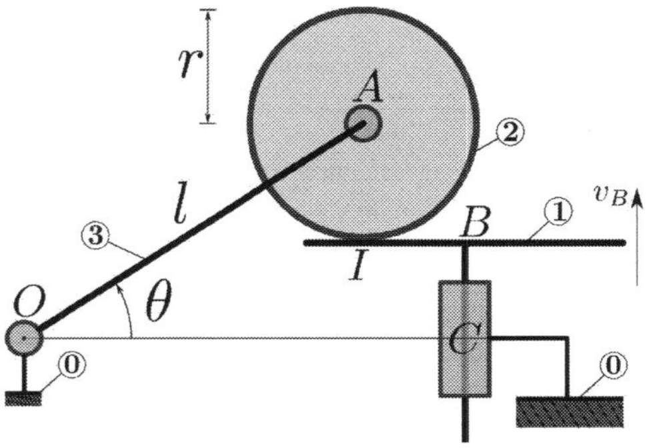

SOLUTION. The solution hinges on the determination of the IC I_2 of roller 2. It is found by using the knowledge of the direction of $\mathbf{v}_A$ (normal to line OA since point A rotates about O) and of $\mathbf{v}_{I\in 2} = \mathbf{v}_{I\in 1} = \mathbf{v}_B$ (since the roller does not slip relative to the plunger). Hence point I_2 is located at the intersection of line OA (the perpendicular at A to $\mathbf{v}_A$) and the perpendicular at I to $\mathbf{v}_B$. See Figure 2.40 for a construction. Once I_2 is obtained, we can sketch the direction of angular velocity ω_2 (counterclockwise) and that of angular velocity ω_3 (counterclockwise).

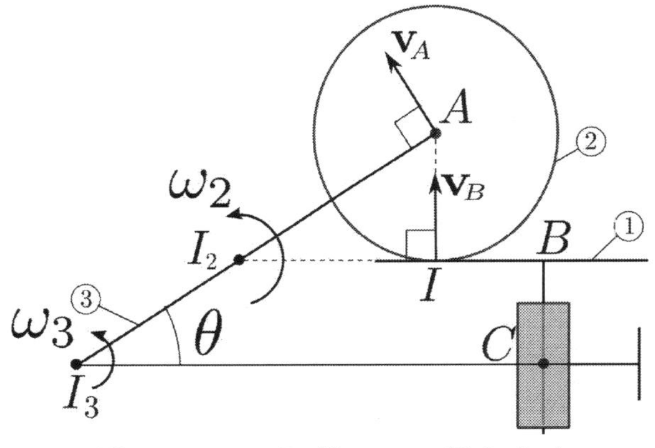

Figure 2.40 – *Problem 2.15 (Solution)*

The angular speeds are obtained by using the relationship $v_{P\in 2} = \omega_2 |I_2 P|$ for $P = I$ and $P = A$:

$$v_B = \omega_2 |I_2 I|, \qquad v_A = \omega_2 |I_2 A| = \omega_3 |I_3 A| \qquad [1-2]$$

We find the lengths $|I_2 I|$ and $|I_2 A|$ in triangle IAI_2

$$|I_2 I| = r \cot\theta, \qquad |I_2 A| = \frac{r}{\sin\theta} \qquad [3-4]$$

Using [3-4], equations [1-2] give the final result:

$$\boxed{\omega_2 = \frac{v_B}{r}\tan\theta \quad \circlearrowleft} \qquad \boxed{\omega_3 = \frac{v_B}{l\cos\theta} \quad \circlearrowleft}$$

■

Exercise 2.16 ■■ In the mechanism displayed in Figure 2.41, the disk 1 rolls without slipping on the horizontal support 0 so that center C travels to the right at constant speed v_C. Slider 3 travels horizontally at height $H = r + L/2$ from the horizontal support. The connecting rod 2 is pinned to disk at center C.
a. In the position $\theta = 0$, show that the angular velocity of the connecting rod is zero.
b. In the position $\theta = \pi/2$, find the speed ratio v_B/v_C. ■

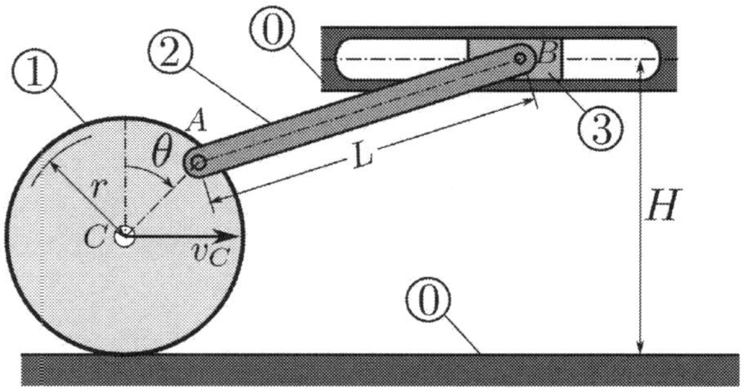

Figure 2.41 – *Problem 2.16*

Solution. **a.** The contact point I of the disk with the support is the IC of the disk. Hence when $\theta = 0$ the velocity of A is parallel to $\mathbf{v}_C$ and $v_A = 2v_C$. Since $\mathbf{v}_B$ is parallel to $\mathbf{v}_A$ the angular velocity of connecting rod 2 is necessarily zero

$$\boxed{\omega_2 = 0, \qquad \mathbf{v}_B = \mathbf{v}_A = 2\mathbf{v}_C}$$

Note that $\boldsymbol{\omega}_2 = \mathbf{0}$ does not imply $\boldsymbol{\alpha}_2 = \mathbf{0}$ at the position $\theta = 0$. Rod 2 is in instantaneous translation: $\mathbf{v}_{P\in 2/0} = \mathbf{v}_A$ for all points P. The construction is summarized in Figure 2.42.

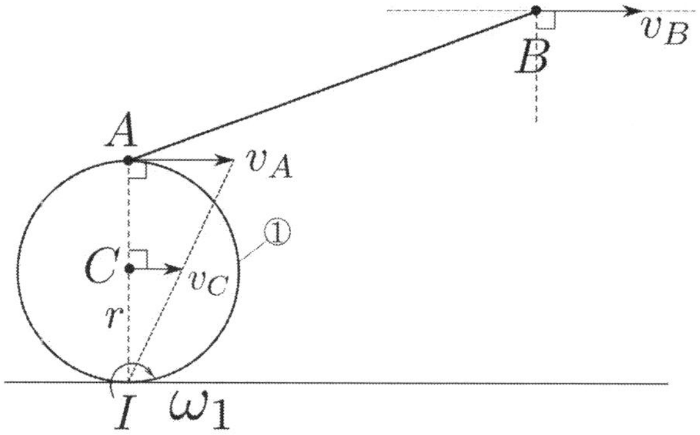

Figure 2.42 – *Problem 2.16 (Solution). $\theta = 0$.*

b. We need to build the location of the IC (denoted K) of rod 2. First we find the direction of $\mathbf{v}_A$ from the knowledge of the IC of the disk (point I): K lies on line IA.

From the knowledge of the direction of $\mathbf{v}_B$, K also lies on the normal to $\mathbf{v}_B$ at B. The intersection of these two lines gives the location of K. See Figure 2.43.
From the knowledge of K we can find ω_2 and speed v_B: between points A, B and K (3 points of rod 2) we can write

$$v_A = \sqrt{2} v_C = |AK|\,\omega_2, \qquad v_B = |BK|\,\omega_2 \qquad [1-2]$$

We need to find the lengths $|AK|$ and $|BK|$: in triangle ABK we find (using the fact that $H = r + L/2$) according to the law of sines

$$\frac{|AK|}{\sin 120°} = \frac{L}{\sin 45°}, \qquad \frac{|BK|}{\sin 15°} = \frac{L}{\sin 45°}$$

or

$$\frac{|BK|}{|AK|} = \frac{\sin 120°}{\sin 15°}$$

Then from equations [1-2] we find

$$\boxed{\frac{v_B}{v_C} = \sqrt{2}\frac{\sin 120°}{\sin 15°} = 3 + \sqrt{3} \approx 4.73}$$

We cannot find $\mathbf{a}_B$ from this result even if $\mathbf{a}_C$ is given: the expression of $\mathbf{v}_B$ is only valid at $\theta = \pi/2$. ∎

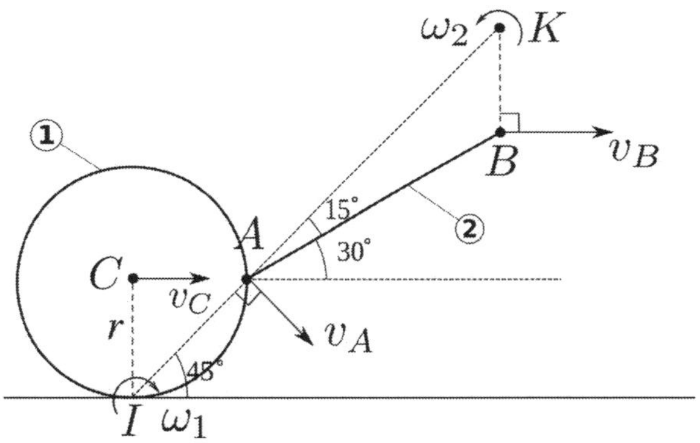

Figure 2.43 – *Problem 2.16 (Solution). Construction of IC of rod 2 for $\theta = \pi/2$ (denoted K).*

Exercise 2.17 ▪▪ Consider the two cylinders of problem 2.11. Recall that cylinder 1 rolls without slipping at both contact points I and K.

Assuming that center A moves to the right at speed v_0, find the angular velocities of the cylinders and the velocity of center B in terms of v_0 and angle φ using the method of instantaneous center of rotation. ∎

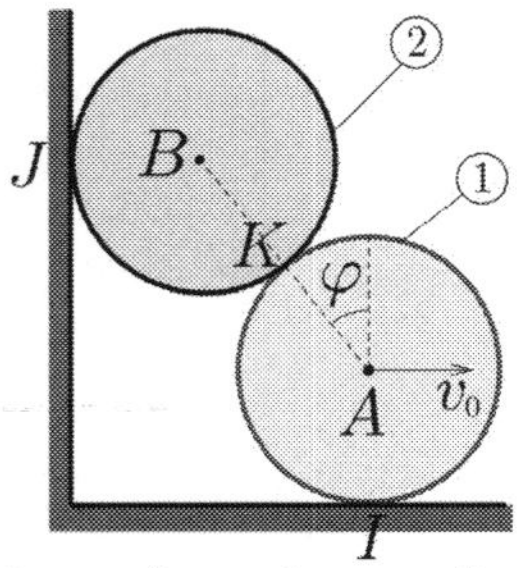

Solution. The IC of cylinder 1 is located at contact point I since there is no-slip of 1 with the fixed support. We can then write (in magnitude):

$$v_A = |IA|\,\omega_1, \qquad v_{K\in 1} = |IK|\,\omega_1$$

with $v_A = v_0$, $|IA| = R$ and $|IK| = 2R\cos(\varphi/2)$. This gives

$$\boxed{\omega_1 = \frac{v_0}{R}} \quad \circlearrowright$$

and

$$v_{K\in 1} = 2v_0 \cos\frac{\varphi}{2}$$

The no-slip condition at K can be stated as $v_{K\in 1} = v_{K\in 2} = v_K$. With the knowledge of direction of $\mathbf{v}_{K\in 2}$ we find that the instantaneous center Q of cylinder 2 is located on line IK. We need the velocity of another point of 2 to locate Q: we can use that of B since B is constrained to move vertically. This shows that Q is at the intersection of lines IK and BJ. See the construction in Figure 2.44.

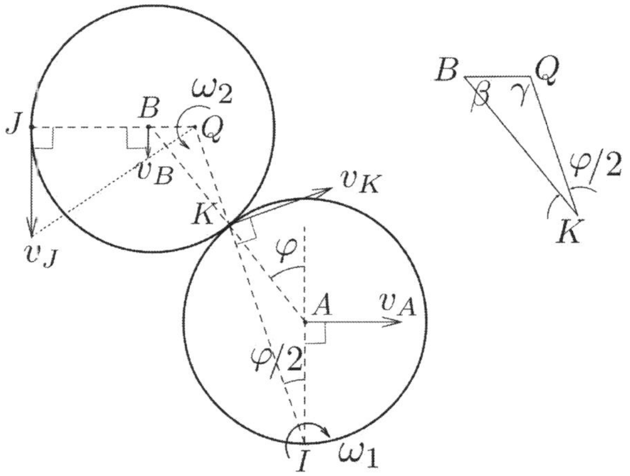

Figure 2.44 – *Problem 2.17 (Solution)*

We can then write

$$v_{K\in 2} = |QK|\,\omega_2, \qquad v_B = |QB|\,\omega_2$$

We need to find the lengths $|QK|$ and $|QB|$: in triangle QBK we find

$$\frac{|QK|}{\sin\beta} = \frac{R}{\sin\gamma}, \qquad \frac{|QB|}{\sin\frac{\varphi}{2}} = \frac{R}{\sin\gamma}$$

according to the law of sines. In triangle QBK we also find the angles $\beta = \pi/2 - \varphi$ and $\gamma = (\pi + \varphi)/2$. This gives:

$$|QK| = R\frac{\cos\varphi}{\cos\frac{\varphi}{2}}$$

and

$$|QB| = R\tan\frac{\varphi}{2}$$

To find ω_2 we use $v_{K\in 1} = 2v_0\cos\frac{\varphi}{2} = v_{K\in 2} = |QK|\,\omega_2$. Using the expression of length $|QK|$, this gives

$$\boxed{\omega_2 = \frac{2v_0}{R}\frac{\cos^2\frac{\varphi}{2}}{\cos\varphi} = \frac{v_0}{R}\frac{1+\cos\varphi}{\cos\varphi}} \quad \circlearrowleft$$

Finally we find v_B with the knowledge of ω_2 and length $|QB|$

$$\boxed{v_B = |QB|\,\omega_2 = v_0 \tan\varphi} \qquad \downarrow$$

The directions of these vectors are indicated in the construction of Figure 2.44. ■

Exercise 2.18 ■■ A bulldozer travels at constant speed v_C along a straight path. It is driven by two tracks mounted on four identical wheels of radius R. The tracks do not slip relative to the ground and relative to the rotating wheels.
a. Find the angular velocity of the wheels 1 and 2 and the velocity of the upper section 4 of the tracks.
b. Generalize the previous results assuming the front wheels are larger than the back wheels. Denote by R_1 and R_2 the wheels' radii and α the angle between the upper and lower sections of the tracks. ■

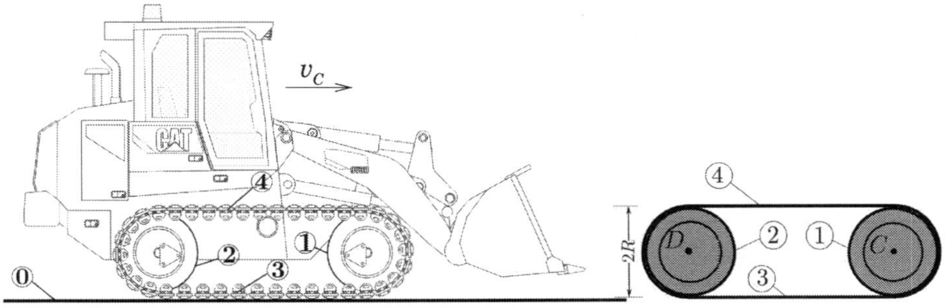

Figure 2.45 – *Problem 2.18*

SOLUTION **a.** We use the method of instantaneous of centers and solve this problem graphically (see Figure 2.46). All points P of track section 3 in contact with the ground have zero velocity: since the track does not slip (w.r.t. the ground and the wheels), point I is the IC of the wheel: $\mathbf{v}_{I\in 3} = \mathbf{0} = \mathbf{v}_{I\in 1}$. Furthermore, center C of wheel 1 has a rectilinear motion at speed v_C which is related to the wheel's angular velocity ω_1 by the relation $v_C = R\omega_1$:

$$\boxed{\omega_1 = \frac{v_C}{R} \qquad \circlearrowright}$$

Point J of the upper track section 4 has speed $v_{J\in 4} = v_{J\in 1} = 2R\omega_1 = 2v_C$:

$$\boxed{v_J = 2v_C} \qquad \rightarrow$$

All points Q of 4 of the track have the same velocity $v_{Q\in 4} = 2v_C$. Front and rear wheels rotate at the same angular velocity $\omega_2 = \omega_1$.
b. If the front and rear wheels have different radii, then they will rotate at different angular velocities to guarantee the no-slip conditions. However their centers C and D have to move at the same speed v_C. The IC of front wheel 1 (of radius R_1) is point I and the angular velocity ω_1 is given by $\omega_1 = \frac{v_C}{R_1}$. This gives the speed of point $J \in 1$: $v_{J\in 1} = |IJ|\,\omega_1 = v_{J\in 4}$. In triangle ICJ we find distance $|IJ| = 2R_1\cos(\alpha/2)$.
In conclusion, we find for the front wheels

$$\boxed{\omega_1 = \frac{v_C}{R_1} \quad \circlearrowright \qquad v_{J\in 4} = 2v_C\cos(\alpha/2)}$$

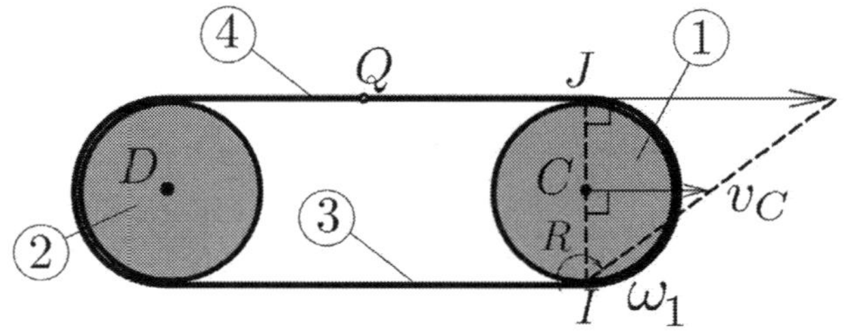

Figure 2.46 – *Problem 2.18 (Solution).*

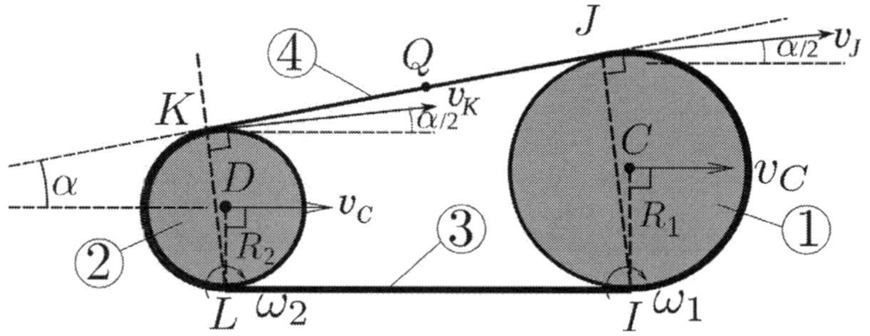

Figure 2.47 – *Problem 2.18 (Solution).*

Similarly for the rear wheel 2 (of radius R_2), we find

$$\boxed{\omega_2 = \frac{v_C}{R_2} \quad \circlearrowright \qquad v_{K\in 4} = 2v_C\cos(\alpha/2)}$$

since the same geometry is found in triangle LDK. Since $\mathbf{v}_{K\in 4} = \mathbf{v}_{J\in 4}$ we find that all points Q of the upper section 4 of the track have the same velocity of magnitude

$$v_{Q\in 4} = 2v_C\cos(\alpha/2)$$

and directed at angle $\alpha/2$ to the horizontal. See Figure 2.47.

■

RECALL from Section 2.7 that the use of the instantaneous center of rotation of a rigid body must be limited to the determination of velocities: the IC is *not* the instantaneous center of zero acceleration. This begs the question: can a graphical method based on the determination of ICs be used to find acceleration information? The answer is yes, as long as the configuration is not specific to a particular configuration. If an angular velocity has been found to be valid at all time, then it can be used to find the corresponding angular acceleration by a simple time-differentiation.
The following problems illustrate this point.

Exercise 2.19 ▪Consider the mechanism of problem 2.7. It is driven by the slider 3 at constant velocity $v_C = 1.75$ m/s in the upward direction. The geometry is given by $|AB| = 1.7$ m, $|BC| = 0.8$ m and the angle of 30° between line AB and BC. Find the angular accelerations of links 1 and 2 by the method of instantaneous centers. ■

SOLUTION. We already solved this problem by the method of instantaneous centers. However the solution found in problem 2.13 is only valid at the time of the given configuration and cannot possibly give the angular accelerations. Instead we must modify the configuration of the system so that it becomes valid at all time: it is sufficient to modify the orientation of link 1 and introduce two oriented angles ϕ and θ to define the orientation of the two links. We will recover the original configuration by setting $\phi = 0$ and $\theta = \theta_0 = 30°$. See Figure 2.48.

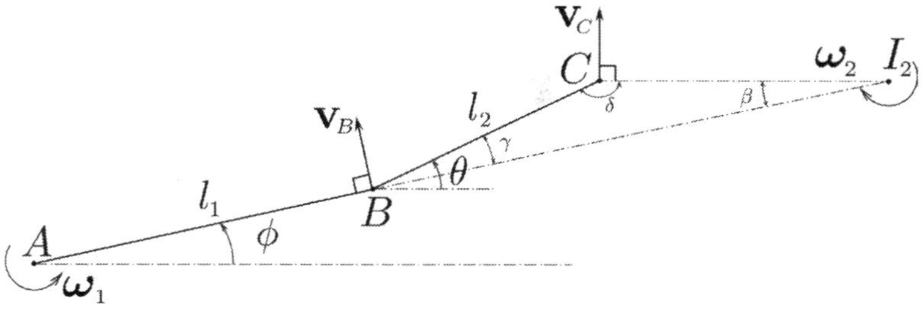

Figure 2.48 – *Problem 2.19*

First we find the location of the IC denoted I_2 of link 2. Using the direction of $\mathbf{v}_C$ and $\mathbf{v}_B$ we find I_2 at the intersection of line AB and the line normal to $\mathbf{v}_C$ at C. See Figure 2.48 for the graphical construction. Since we need the distances $|I_2B|$ and $|I_2C|$ we apply the law of sines in triangle BCI_2 after introducing the angles β, γ and δ in the construction:

$$\frac{|I_2B|}{\sin\delta} = \frac{|I_2C|}{\sin\gamma} = \frac{|BC|}{\sin\beta}$$

with $|BC| = l_2$, $\gamma = \theta - \phi$, $\beta = \phi$ and $\delta = \pi - \theta$. Then we relate the speeds of B and C (two points of link 2) to angular speed ω_2:

$$v_C = \omega_2|I_2C| = l_2\omega_2\frac{\sin(\theta-\phi)}{\sin\phi}, \qquad v_B = \omega_2|I_2B| = l_2\omega_2\frac{\sin\theta}{\sin\phi} \qquad [1-2]$$

Finally we relate the speed of B to angular speed ω_1 since $v_B = v_{B\in 1} = v_{B\in 2}$:

$$v_B = l_1\omega_1 \qquad [3]$$

We can solve for ω_1 and ω_2 in terms of v_C:

$$\boxed{\omega_1 = \frac{v_C}{l_1}\frac{\sin\theta}{\sin(\theta-\phi)} \circlearrowleft, \quad \omega_2 = \frac{v_C}{l_2}\frac{\sin\phi}{\sin(\theta-\phi)} \circlearrowleft} \qquad [4-5]$$

We recover the expressions $\omega_1 = v_C/l_1$ and $\omega_2 = 0$ found in problem 2.13 by setting $\phi = 0$ in [4-5]. Equations [4-5] are valid at all time and can be time-differentiated to obtain angular accelerations (recall that v_C is constant):

$$\alpha_1 = \frac{v_C}{l_1}\frac{\dot\theta\cos\theta\sin(\theta-\phi) - (\dot\theta-\dot\phi)\sin\theta\cos(\theta-\phi)}{\sin^2(\theta-\phi)}$$

$$\alpha_2 = \frac{v_C}{l_2}\frac{\dot\phi\cos\phi\sin(\theta-\phi) - (\dot\theta-\dot\phi)\sin\phi\cos(\theta-\phi)}{\sin^2(\theta-\phi)}$$

Since $\dot\theta < 0$ and $\dot\phi > 0$, we have $\dot\phi = \omega_1$ and $\dot\theta = -\omega_2$:

$$\alpha_1 = \frac{v_C}{l_1}\frac{-\omega_2\cos\theta\sin(\theta-\phi) + (\omega_2+\omega_1)\sin\theta\cos(\theta-\phi)}{\sin^2(\theta-\phi)}$$

$$\alpha_2 = \frac{v_C}{l_2} \frac{\omega_1 \cos\phi \sin(\theta - \phi) + (\omega_2 + \omega_1) \sin\phi \cos(\theta - \phi)}{\sin^2(\theta - \phi)}$$

Now we can set $\phi = 0$ and $\theta = \theta_0$ to obtain the angular accelerations of the two links at the given configuration:

$$\boxed{\alpha_1 = \frac{v_C^2}{l_1^2} \cot\theta_0 \quad \circlearrowleft, \qquad \alpha_2 = \frac{v_C^2}{l_1 l_2 \sin\theta_0} \quad \circlearrowright}$$

We recover the results of problem 2.7. ■

Exercise 2.20 ▪▪A four-bar mechanism with pin joints at O, A, B and C is driven by link 1 at constant angular velocity $\omega_1 = 1.2$ rad/s in the clockwise direction. By finding the location of the IC of link 2, find the angular velocity and angular acceleration of link 3. Use $|OA| = 100$ cm, $|AB| = 130$ cm, $|BC| = 85$ cm. ■

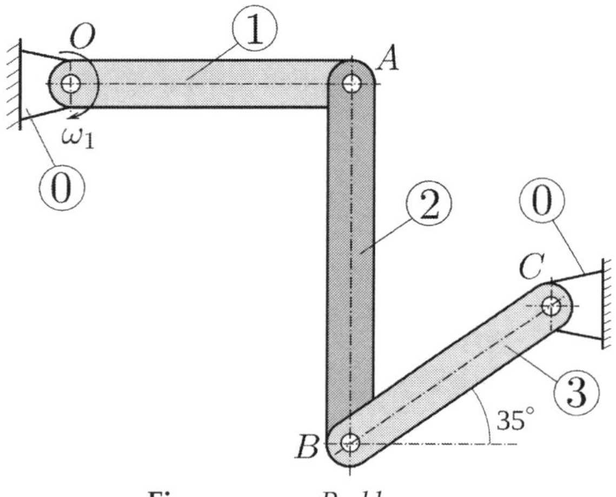

Figure 2.49 – *Problem 2.20*

SOLUTION. If we find the location of the IC (denoted K) of link 2 based on the configuration of Figure 2.49, we will not be able to find the angular acceleration of link 3 from its angular velocity since the expression of the latter would only be valid at a specific time. Instead we must adopt the general configuration shown in Figure 2.50 parametrized by the angles θ_1, θ_2 and θ_3. Figure 2.49 corresponds to $\theta_1 = \theta_2 = 0$. We then find the expression of angular velocity ω_3 from the location of IC K of link 2 as a function of θ_1, θ_2 and θ_3 and $\dot{\theta}_1 = \omega_1$. With the direction of velocities $\mathbf{v}_A$ (A rotates about O) and $\mathbf{v}_B$ (B rotates about C), we find the location of K as the intersection of line OA and line BC. With the knowledge of K we can find v_A and v_B as the speeds of two points of link 2 (see Figure 2.50):

$$v_A = |KA|\,\omega_2 = l_1\omega_1, \qquad v_B = |KB|\,\omega_2 = l_3\omega_3$$

where we denote $l_1 = |OA|$, $l_2 = |AB|$, and $l_3 = |BC|$. This gives angular velocities ω_2 and ω_3 in term of ω_1:

$$\omega_2 = \omega_1 \frac{l_1}{|KA|}, \qquad \omega_3 = \omega_1 \frac{l_1}{l_3} \frac{|KB|}{|KA|} \qquad [1-2]$$

We then apply the law of sines in triangle ABK to find

$$\frac{|KA|}{\sin\delta} = \frac{|KB|}{\sin\gamma} = \frac{l_2}{\sin(\theta_1+\theta_3)} \qquad [3-4]$$

with $\gamma = \pi/2 - \theta_1 - \theta_2$ and $\delta = \pi/2 - \theta_3 + \theta_2$. Now with equations [1-2] and [3-4] we find

$$\boxed{\omega_2 = \omega_1 \frac{l_1 \sin(\theta_1+\theta_3)}{l_2 \cos(\theta_3-\theta_2)} \circlearrowleft, \qquad \omega_3 = \omega_1 \frac{l_1 \cos(\theta_1+\theta_2)}{l_3 \cos(\theta_3-\theta_2)} \circlearrowleft} \qquad [5-6]$$

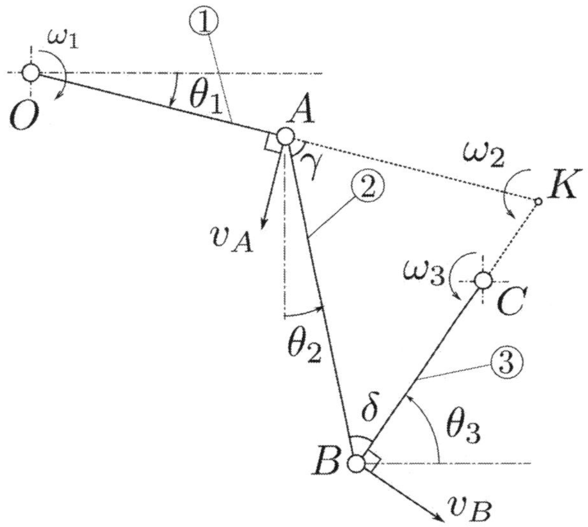

Figure 2.50 – *Problem 2.20 (Solution)*

For $\theta_1 = \theta_2 = 0$ we obtain the values reached at the time of the configuration of Figure 2.49.

$$\omega_2 = \omega_1 \frac{l_1}{l_2} \tan\theta_3 \ \circlearrowleft, \qquad \omega_3 = \omega_1 \frac{l_1}{l_3 \cos\theta_3} \ \circlearrowleft \qquad [7-8]$$

We cannot take the time-derivative of [8] to find α_3! To obtain α_3 we can take the time-derivative of ω_3 of equation [6] taking into account $\alpha_1 = 0$:

$$\alpha_3 = \omega_1 \frac{l_1}{l_3}\Big(-(\dot\theta_1+\dot\theta_2)\frac{\sin(\theta_1+\theta_2)}{\cos(\theta_3-\theta_2)} + (\dot\theta_3-\dot\theta_2)\frac{\sin(\theta_1+\theta_2)\sin(\theta_3-\theta_2)}{\cos^2(\theta_3-\theta_2)}\Big)$$

With the adopted oriented angles θ_1, θ_2 and θ_3 of Figure 2.50 we find $\dot\theta_1 = \omega_1$, $\dot\theta_2 = \omega_2$ and $\dot\theta_3 = \omega_3$. We can then substitute the expressions of ω_2 and ω_3 of [7-8] at $\theta_1 = \theta_2 = 0$ to find

$$\boxed{\alpha_3 = \omega_1^2 \frac{l_1^2}{l_3^2}\Big(1 - \frac{l_3}{l_2}\sin\theta_3\Big)\frac{\sin\theta_3}{\cos^3\theta_3} \ \circlearrowleft}$$

With $l_1 = 100$ cm, $l_2 = 130$ cm, $l_3 = 85$ cm, $\omega_1 = 1.2$ rad/s and $\theta_3 = 35°$, we find $\omega_3 = \mathbf{1.72}$ rad/s $\circlearrowleft$, and $\alpha_3 = \mathbf{1.3}$ rad/s^2 $\circlearrowleft$. ■

Exercise 2.21 ■■At a particular instant, the slider block 2 is moving at velocity $\mathbf{v}_B$ and acceleration $\mathbf{a}_B$ both pointing to the right. Determine the angular acceleration of link 1 and the acceleration of point A at this instant by using a graphical method. ■

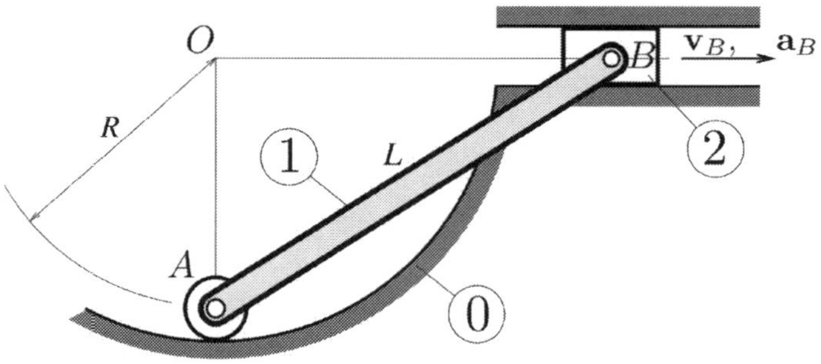

Figure 2.51 – *Problem 2.21*

Solution. It is straightforward to show that in the given configuration $\boldsymbol{\omega}_1 = \mathbf{0}$ and $\mathbf{v}_A = \mathbf{v}_B$. But these results cannot be used to find the angular acceleration $\boldsymbol{\alpha}_1$ and acceleration $\mathbf{a}_A$ of A.
We must find the IC of body 1 by defining a configuration valid at all time as shown in Figure 2.52: two angles are necessary, angle θ to define the position of A and oriented angle ϕ to define the orientation of body 1. The original configuration is obtained by setting $\theta = 0$ and $\phi = \phi_0$ with $\sin\phi_0 = R/L$. The IC is easily found at the intersection of line OA and the normal at B to $\mathbf{v}_B$. Then in triangle IAB we find

$$\frac{|IA|}{\sin(\frac{\pi}{2} - \phi)} = \frac{|IB|}{\sin(\frac{\pi}{2} + \phi - \theta)} = \frac{L}{\sin(\theta)}$$

by applying the law of sines. This leads to two equations

$$v_A = L\omega_1 \frac{\cos\phi}{\sin\theta}, \qquad v_B = L\omega_1 \frac{\cos(\phi - \theta)}{\sin\theta}$$

giving

$$\boxed{\omega_1 = \frac{v_B}{L}\frac{\sin\theta}{\cos(\phi - \theta)} \qquad v_A = v_B \frac{\cos\phi}{\cos(\phi-\theta)}} \qquad [1-2]$$

For $\theta = 0$, we recover $\omega_1 = 0$ and $v_A = v_B$. Equations [1-2] can be time-differentiated:

$$\alpha_1 = \frac{a_B}{L}\frac{\sin\theta}{\cos(\phi-\theta)} + \frac{v_B}{L}\frac{\dot{\theta}\cos\theta\cos(\phi-\theta) + (\dot{\phi} - \dot{\theta})\sin\theta\sin(\phi-\theta)}{\cos^2(\phi-\theta)}$$

For $\theta = 0$ and $\phi = \phi_0$, this expression simplifies to

$$\alpha_1 = \frac{v_B}{L}\frac{\dot{\theta}}{\cos\phi_0}$$

Since A moves along the circle of center O of radius R and angular velocity $\dot{\theta}$, its speed is given by $v_A = R\dot{\theta}$. Then using $v_A = v_B$ at $\theta = 0$ we find $v_B = R\dot{\theta}$: this allows us to find the final expression of α_1:

$$\boxed{\alpha_1 = \frac{v_B^2}{LR\cos\phi_0} \quad \circlearrowleft}$$

directed in the clockwise direction. Next we take the time-derivative of v_A:

$$\dot{v}_A = a_B \frac{\cos\phi}{\cos(\phi-\theta)} + v_B \frac{-\dot{\phi}\sin\phi\cos(\phi-\theta) + (\dot{\phi}-\dot{\theta})\cos\phi\sin(\phi-\theta)}{\cos^2(\phi-\theta)}$$

with $\dot{\phi} = -\omega_1$. At $\theta = 0$, this expression simplifies to (using $\omega_1 = 0$ and $\dot{\theta} = v_B/R$)

$$\boxed{\dot{v}_A = a_B - \frac{v_B^2}{R}\tan\phi_0}$$

Do not identify $\dot{v}_A$ with a_A! The motion of A is not rectilinear. $\dot{v}_A$ gives the tangential acceleration of A: $a_{At} = \dot{v}_A$. To complete the expression of the acceleration of A, we

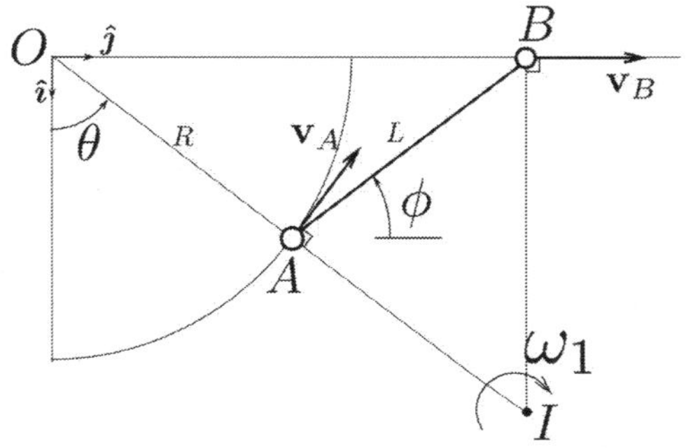

Figure 2.52 – *Problem 2.21 (Solution)*

need to find the expression of its normal acceleration

$$a_{An} = \frac{v_A^2}{R} = \frac{v_B^2}{R}$$

since A moves along a circle of radius R. Finally using the unit vectors $(\hat{\imath}, \hat{\jmath})$ defined in Figure 2.52 we find

$$\boxed{\mathbf{a}_A = (a_B - \frac{v_B^2}{R}\tan\phi_0)\hat{\jmath} - \frac{v_B^2}{R}\hat{\imath}}$$

■

2.8 Kinematics from Geometric Analysis

The kinematic analysis of mechanisms can often be done by a simple geometric method as was done for systems of particles in Section 1.8. If the mechanism is defined in an arbitrary configuration (that is, a configuration not specific to a particular instant), the method consists of finding all geometric relationships governing the mechanism whose configuration in space has been parametrized. Then time-differentiation of these equations will yield all relevant kinematic unknowns. One possible way to systematically find geometric relationships for *closed-loop* mechanisms is by writing vector loop equations, that is, equations of the type

$$\mathbf{r}_{OA_1} + \mathbf{r}_{A_1A_2} + \cdots + \mathbf{r}_{A_{N-1}A_N} + \mathbf{r}_{A_NO} = \mathbf{0} \tag{2.16}$$

where O is a fixed point and the points A_1, A_2, $\cdots$, A_N are characteristic points of the joints which interconnect pairs of links.

Note that this method is ill-suited for mechanisms which involve rigid bodies constrained to roll without slipping.

EXAMPLE PROBLEMS

General Procedure: Geometric Analysis of Mechanisms

For a kinematic analysis based on a geometric method, follow these guidelines:

① If the given configuration of the mechanism is specific to a particular time, then it must be redefined so as to make it valid at all time.

② Parametrize the position of the mechanism by defining all necessary coordinates: these are typically angles for the bodies' orientations and Cartesian/polar coordinates for critical points of the mechanism.

③ Identify as many *independent* vector loop equations: it should then be possible to express each position vector in terms of coordinates defined in step 2. Otherwise additional coordinates will need to be defined.

④ Take the time-derivative of these vector loop equations and project them on fixed unit vectors. Identify all kinematic quantities in the newly obtained equations.

⑤ Repeat step 4 to obtain acceleration information.

⑥ This method can easily be implemented for 3-dimensional mechanisms.

Exercise 2.22 ▪The slider-crank mechanism of Figure 2.53 is set in motion by the crank-arm 1. At the instant when $\theta_1 = 45°$, the angular velocity and acceleration of the crank-arm take the values $\omega_1 = 50$ rad/s and $\alpha_1 = 12.5$ rad/s^2 in the counterclockwise direction. Find the velocity $\mathbf{v}_B$ and acceleration $\mathbf{a}_B$ of slider 3 and the angular velocity $\boldsymbol{\omega}_2$ and acceleration $\boldsymbol{\alpha}_2$ of the connecting rod 2. Use $|OA| = 1.0$ m, $|AB| = 2.40$ m and $h = 0.65$ m. ▪

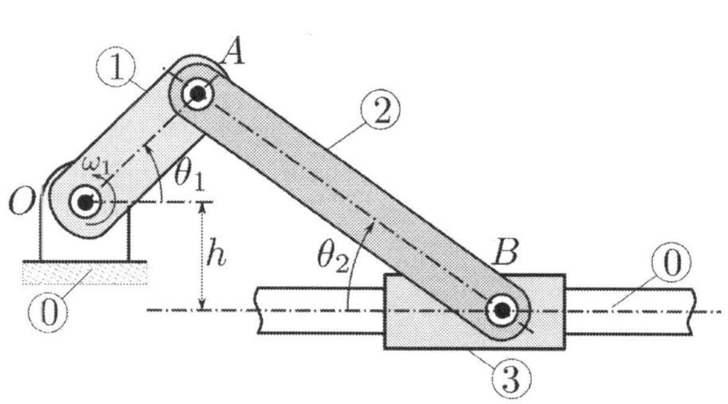

Figure 2.53 – *Problem 2.22*

Solution. We first parametrize the position of the mechanism in its general configuration. In addition to angles θ_1 and θ_2 we introduce coordinate x_B to define the position of B. We then express the vector loop $\mathbf{r}_{OA} + \mathbf{r}_{AB} + \mathbf{r}_{BO} = \mathbf{0}$ on the chosen unit vectors

$(\hat{\imath}, \hat{\jmath})$ in terms of the coordinates $(\theta_1, \theta_2, x_B)$ and constants (l_1, l_2, h) (see Figure 2.54):

$$l_1(\cos\theta_1 \hat{\imath} + \sin\theta_1 \hat{\jmath}) + l_2(\cos\theta_2 \hat{\imath} - \sin\theta_2 \hat{\jmath}) + (-x_B \hat{\imath} + h\hat{\jmath}) = \mathbf{0}$$

leading to two equations:

$$l_1 \cos\theta_1 + l_2 \cos\theta_2 - x_B = 0, \qquad l_1 \sin\theta_1 - l_2 \sin\theta_2 + h = 0 \qquad [1-2]$$

Since equations [1-2] are valid at all time, we can differentiate them w.r.t. time to

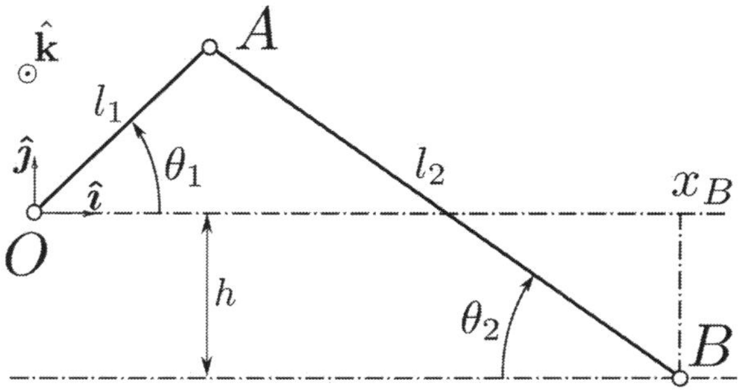

Figure 2.54 – *Problem 2.22 (Solution)*

obtain the equations

$$l_1\dot{\theta}_1 \sin\theta_1 + l_2\dot{\theta}_2 \sin\theta_2 + \dot{x}_B = 0, \qquad l_1\dot{\theta}_1 \cos\theta_1 - l_2\dot{\theta}_2 \cos\theta_2 = 0 \qquad [3-4]$$

which yield the velocity of B and the angular velocity of connecting rod 2:

$$\boxed{\mathbf{v}_B = \dot{x}_B \hat{\imath} = -l_1\omega_1 \frac{\sin(\theta_1+\theta_2)}{\cos\theta_2}\hat{\imath}} \qquad \boxed{\boldsymbol{\omega}_2 = -\dot{\theta}_2 \hat{\mathbf{k}} = -\frac{l_1}{l_2}\omega_1 \frac{\cos\theta_1}{\cos\theta_2}\hat{\mathbf{k}}} \qquad [3-4]$$

We can then differentiate [3-4] to obtain the acceleration of B and the angular acceleration $\boldsymbol{\alpha}_2$:

$$\boxed{\mathbf{a}_B = \ddot{x}_B \hat{\imath}, \quad \ddot{x}_B = -l_1\alpha_1 \frac{\sin(\theta_1+\theta_2)}{\cos\theta_2} - l_1\omega_1^2 \frac{\cos(\theta_1+\theta_2)}{\cos\theta_2} - \frac{l_1^2}{l_2}\omega_1^2 \frac{\cos^2\theta_1}{\cos^3\theta_2}} \qquad [5]$$

$$\boxed{\boldsymbol{\alpha}_2 = -\ddot{\theta}_2 \hat{\mathbf{k}}, \quad \ddot{\theta}_2 = \alpha_1 \frac{l_1 \cos\theta_1}{l_2 \cos\theta_2} + \omega_1^2 \frac{l_1 \sin\theta_1}{l_2 \cos\theta_2}\Big(-1 + \frac{l_1 \sin\theta_2 \cos^2\theta_1}{l_2 \sin\theta_1 \cos^2\theta_2}\Big)} \qquad [6]$$

With $l_1 = 1.0$ m, $l_2 = 2.4$ m, $h = 0.65$ m, and $\theta_1 = 45°$, we solve equation [2] to find angle $\theta_2 = \mathbf{34.4}°$. We then find $v_B = \mathbf{59.5}$ (m/s) $\leftarrow$, $\omega_2 = \mathbf{17.8}$ (rad/s) $\circlearrowright$, $a_B = \mathbf{1500}$ (m/s^2) $\leftarrow$, $\alpha_2 = \mathbf{670}$ (rad/s^2) $\circlearrowleft$. ■

Exercise 2.23 ■■At a particular instant, the slider block 2 is moving at speed $v_B = 1.5$ m/s and $a_B = 2.0$ m/s^2 both pointing to the right. Determine the angular acceleration of link 1 and the acceleration of point A at this instant. $|AB| = 1.2$ m and $R = 0.6$ m. ■

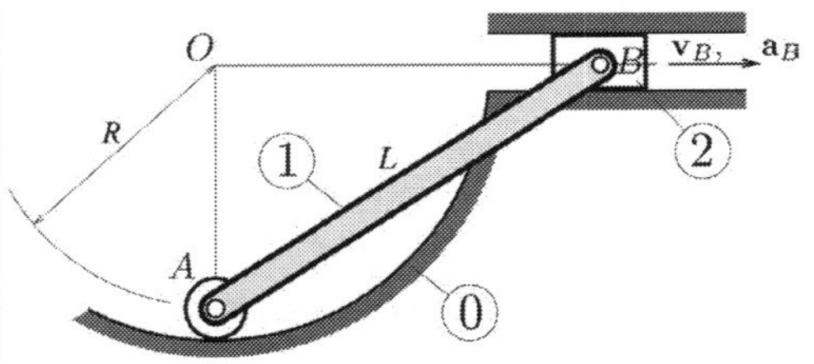

Solution. We cannot use the configuration of the Figure. Instead we modify it so

as to make the position of pin A arbitrary: see Figure 2.55. We then introduce three coordinates to parametrize the system in this configuration:
-angle θ to define the position of A on its circular path (of center O and radius R),
-oriented angle ϕ to define the orientation of line AB, leading to $\boldsymbol{\omega}_1 = \dot{\phi}\hat{\mathbf{k}}$ with the choice of basis $(\hat{\imath}, \hat{\jmath}, \hat{\mathbf{k}})$ shown in Figure 2.55,
-coordinate y_B to define the position of B (in rectilinear motion): $\mathbf{r}_{OB} = y_B\hat{\jmath}$.
The original configuration corresponds to the value $\theta = 0$ and $\phi = \phi_0$ with $\sin\phi_0 = R/L$. We then express the vector loop $\mathbf{r}_{OA} + \mathbf{r}_{AB} + \mathbf{r}_{BO} = \mathbf{0}$ on the chosen unit vectors $(\hat{\imath}, \hat{\jmath})$ in terms of the coordinates (θ, ϕ, y_B) and constants (R, L):

$$R(\cos\theta\hat{\imath} + \sin\theta\hat{\jmath}) + L(-\sin\phi\hat{\imath} + \cos\phi\hat{\jmath}) - y_B\hat{\jmath} = \mathbf{0}$$

leading to two equations

$$R\cos\theta - L\sin\phi = 0, \qquad R\sin\theta + L\cos\phi = y_B \qquad [1-2]$$

Equations [1-2] are valid at all time: we can take their time-derivative to obtain

$$R\dot{\theta}\sin\theta + L\dot{\phi}\cos\phi = 0, \qquad R\dot{\theta}\cos\theta - L\dot{\phi}\sin\phi = \dot{y}_B$$

We now have the expression of the angular velocity of rod 1:

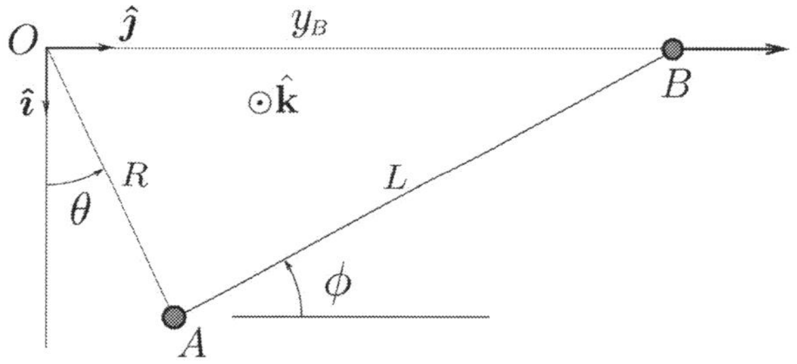

Figure 2.55 – *Problem 2.23 (Solution)*

$$\boxed{\boldsymbol{\omega}_1 = \dot{\phi}\hat{\mathbf{k}} = -\frac{\dot{y}_B}{L}\frac{\sin\theta}{\cos(\theta-\phi)}\hat{\mathbf{k}}} \qquad [3]$$

which takes the value

$$\boxed{\boldsymbol{\omega}_1 = \mathbf{0}} \qquad [3']$$

when $\theta = 0$. Point A moves along a circular path of center O and radius R at the speed

$$v_A = R\dot{\theta} = \dot{y}_B\frac{\cos\phi}{\cos(\theta-\phi)} \qquad [4]$$

The velocity of A is then given by

$$\boxed{\mathbf{v}_A = v_B\,\hat{\jmath}} \qquad [4']$$

when $\theta = 0$. These results can be recovered very simply for $\theta = 0$: in this position, the velocities of A and B are parallel, necessarily implying that $\mathbf{v}_A = \mathbf{v}_B$ and $\boldsymbol{\omega}_1 = \mathbf{0}$ (since $\phi_0 \neq 0$). However we cannot obtain $\mathbf{a}_A$ and $\boldsymbol{\alpha}_1$ from these results.

Equations [3] is valid at all time and can be differentiated to obtain:

$$\boldsymbol{\alpha}_1 = \ddot{\phi}\hat{\mathbf{k}}, \quad \ddot{\phi} = -\frac{\ddot{y}_B}{L}\frac{\sin\theta}{\cos(\theta-\phi)} - \frac{\dot{y}_B}{L}\frac{\dot{\theta}\cos\theta\cos(\theta-\phi) + (\dot{\theta}-\dot{\phi})\sin\theta\sin(\theta-\phi)}{\cos^2(\theta-\phi)}$$

When $\theta = 0$, this expression becomes (with $\dot{\phi} = \dot{\phi}_0 = 0$ and $R\dot{\theta}_0 = v_A = v_B$)

$$\boxed{\boldsymbol{\alpha}_1 = -\frac{\dot{y}_B}{L}\frac{\dot{\theta}_0}{\cos\phi_0}\hat{\mathbf{k}} = -\frac{v_B^2}{RL\cos\phi_0}\hat{\mathbf{k}}} \qquad [5]$$

Similarly, we may differentiate equation [4] to obtain:

$$\dot{v}_A = R\ddot{\theta} = \ddot{y}_B\frac{\cos\phi}{\cos(\theta-\phi)} + \dot{y}_B\frac{-\dot{\phi}\sin\phi\cos(\theta-\phi) + (\dot{\theta}-\dot{\phi})\cos\phi\sin(\theta-\phi)}{\cos^2(\theta-\phi)}$$

which becomes at $\theta = 0$:

$$\dot{v}_A = a_B - v_B\dot{\theta}\tan\phi_0 = a_B - \frac{v_B^2}{R}\tan\phi_0 \qquad [6]$$

Since the trajectory of A is circular we can identify the corresponding tangential/normal unit vectors when $\theta = 0$: $\hat{\mathbf{e}}_T = \hat{\jmath}$ and $\hat{\mathbf{e}}_N = -\hat{\imath}$. We can now write the acceleration of A in terms of its NT-components

$$\boxed{\mathbf{a}_A = \dot{v}_A\hat{\mathbf{e}}_T + \frac{v_A^2}{R}\hat{\mathbf{e}}_N = \left(a_B - \frac{v_B^2}{R}\tan\phi_0\right)\hat{\jmath} - \frac{v_B^2}{R}\hat{\imath}} \qquad [7]$$

using the expressions [4′] and [6]. For $L = 1.2$ m, $R = 0.6$ m, $v_B = 1.5$ (m/s) and $a_B = 2.0$ (m/s^2), we find $\alpha_1 = \mathbf{3.6}$ (rad/s^2) ↻ and $(a_{Ax}, a_{Ay}) = (\mathbf{-3.75}, \mathbf{-0.165})$ (m/s^2). ■

Exercise 2.24 ▪The figure shown below shows a sketch of a mechanism which transforms the continuous rotational motion of body 1 into the reciprocating rotational motion of body 4. The input crankshaft 1 transmits its motion through pin A to the connecting rod 2 which slides into the hollow shaft 3. Body 3 is guided by the ball joint 5 of center B. The output shaft of center D is driven by the connecting rod 4 which is pinned at C with body 3. It is possible to vary the output rotation by displacing the center B.
At any given time, the configuration of the mechanism is defined by the angles θ_1, θ_2 and θ_4, and the constant lengths $\lambda = |OB|$, $\mu = |BD|$, $l_1 = |OA|$, $l_4 = |CD|$.
a. Write the geometric loop equations $\mathbf{r}_{OA} + \mathbf{r}_{AB} + \mathbf{r}_{BO} = \mathbf{0}$ and $\mathbf{r}_{DC} + \mathbf{r}_{CB} + \mathbf{r}_{BD} = \mathbf{0}$ to obtain four equations. Solve these equations for $l_1 = 14$ mm, $l_4 = 36$ mm, $\lambda = 55$ mm, $\mu = 87$ mm in the rangle $0 \leq \theta_1 < 2\pi$. Then plot θ_2 and θ_4 vs θ_1. Confirm that the output motion is a reciprocating rotation.
b. Find the input/output relationship $\dot{\theta}_4$ vs $\dot{\theta}_1$. Using the data found in a), plot the ratio $\dot{\theta}_4/\dot{\theta}_1$ vs θ_1, in the range $0 \leq \theta_1 < 2\pi$. ■

SOLUTION. **a.** Before writing out the loop equations, we introduce a basis of unit vectors $(\hat{\imath}, \hat{\jmath})$ and the distances $l_2 = |AB|$ and $l_3 = |BC|$. See Figure 2.57. Note that l_2 and l_3 are time-varying. The loop equations $\mathbf{r}_{OA} + \mathbf{r}_{AB} + \mathbf{r}_{BO} = \mathbf{0}$ and $\mathbf{r}_{DC} + \mathbf{r}_{CB} + \mathbf{r}_{BD} = \mathbf{0}$

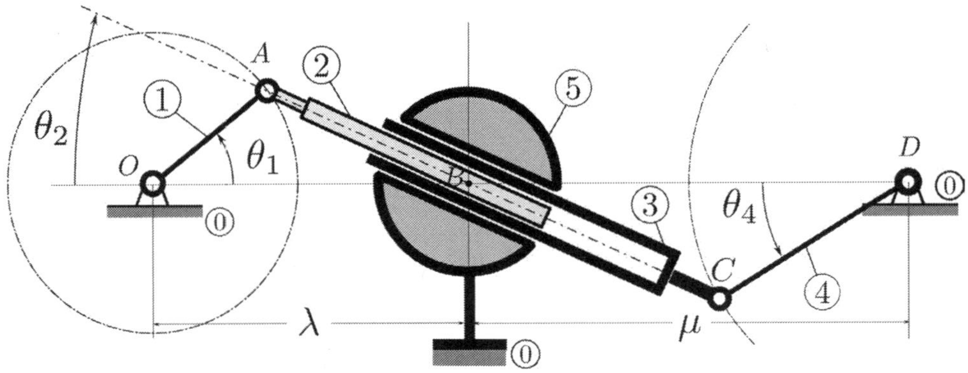

Figure 2.56 – *Problem 2.24.*

can be projected on $(\hat{\imath}, \hat{\jmath})$ to obtain four equations:

$$\lambda = l_1 \cos\theta_1 + l_2 \cos\theta_2, \qquad l_1 \sin\theta_1 = l_2 \sin\theta_2 \qquad [1-2]$$

$$\mu = l_4 \cos\theta_4 + l_3 \cos\theta_2, \qquad l_4 \sin\theta_4 = l_3 \sin\theta_2 \qquad [3-4]$$

For fixed values of the parameters (l_1, l_4, λ, μ), equations [1-4] can be solved for the four unknowns $(l_2, \theta_2, l_3, \theta_4)$ as functions of angle θ_1. These equations are solved numerically for the given data. The functions $\theta_2 = f(\theta_1)$ and $\theta_4 = g(\theta_1)$ are plotted in Figure 2.58 (left). This plot confirms that the output motion is indeed a reciprocating rotational motion: $-\theta_{4,max} \leq \theta_4 \leq \theta_{4,max}$, with $\theta_{4,max} \approx 23.2°$.

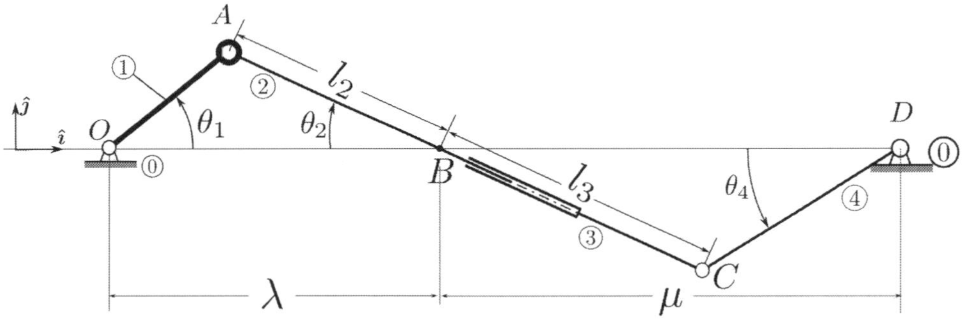

Figure 2.57 – *Problem 2.24.*

b. We now take the time-derivative of equations [1-4] to obtain

$$0 = -l_1\dot{\theta}_1 \sin\theta_1 - l_2\dot{\theta}_2 \sin\theta_2 + \dot{l}_2 \cos\theta_2, \qquad l_1\dot{\theta}_1 \cos\theta_1 = l_2\dot{\theta}_2 \cos\theta_2 + \dot{l}_2 \sin\theta_2 \quad [5-6]$$

$$0 = -l_4\dot{\theta}_4 \sin\theta_4 - l_3\dot{\theta}_2 \sin\theta_2 + \dot{l}_3 \cos\theta_2, \qquad l_4\dot{\theta}_4 \cos\theta_4 = l_3\dot{\theta}_2 \cos\theta_2 + \dot{l}_3 \sin\theta_2 \quad [7-8]$$

We eliminate $\dot{l}_2$ from [5-6], and $\dot{l}_3$ from [7-8]:

$$l_1\dot{\theta}_1 \cos(\theta_1 + \theta_2) = l_2\dot{\theta}_2, \qquad l_4\dot{\theta}_4 \cos(\theta_2 + \theta_4) = l_3\dot{\theta}_2$$

We then use [2] and [4] to eliminate l_2 and l_3:

$$\dot{\theta}_1 \cos(\theta_1 + \theta_2) \sin\theta_2 = \dot{\theta}_2 \sin\theta_1, \qquad \dot{\theta}_4 \cos(\theta_2 + \theta_4) \sin\theta_2 = \dot{\theta}_2 \sin\theta_4$$

Finally, we take the ratio of the last two equations to obtain the desired quantity:

$$\boxed{\frac{\dot{\theta}_4}{\dot{\theta}_1} = \frac{\cos(\theta_1+\theta_2)}{\cos(\theta_2+\theta_4)}\frac{\sin\theta_4}{\sin\theta_1}}$$

We can easily plot $\dot{\theta}_4/\dot{\theta}_1$ vs θ_1 using the data found in a). This is shown in Figure 2.58 (right).

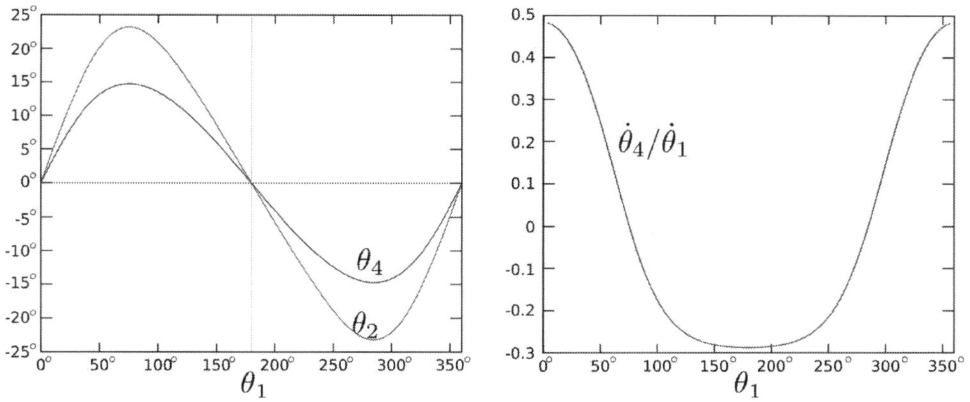

Figure 2.58 – *Problem 2.24.*

■

Exercise 2.25 ■■The mechanism shown in Figure 2.59 transforms the rotational motion of body 1 about axis $(O, \hat{\mathbf{x}}_0)$ into the rotational motion of body 3 about a concurrent perpendicular axis $(E, \hat{\mathbf{z}}_0)$. This is achieved by interconnecting body 2 between bodies 1 and 3. More specifically, body 2 can translate relative to 1 along axis $(B, \hat{\mathbf{x}}_0)$ and can translate relative to 3 along axis $(C, \hat{\mathbf{z}}_0)$. The configuration of the mechanism is defined by the two angles $\theta_1(t)$ and $\theta_3(t)$ and the parameters $l = |DC|$ and $d = |OA|$.
Find the input/output relationship between the angular velocities of 1 and 3 by expressing the vector loop equation joining points O, A, B, C, D and E. Also find the velocity $\mathbf{v}_{B\in 3/0}$. Assuming the rotation of 1 continuous, under what condition can body 3 be in alternating rotation? ■

SOLUTION. With the choice of basis $(\hat{\mathbf{x}}_0, \hat{\mathbf{y}}_0, \hat{\mathbf{z}}_0)$ defined in Figure 2.59 we need to introduce the two coordinates ρ and μ in addition to angles θ_1 and θ_3 to parametrize the position of B relative to A and C: $\mathbf{r}_{AB} = \rho(t)\hat{\mathbf{x}}_0$ and $\mathbf{r}_{BC} = \mu(t)\hat{\mathbf{z}}_0$. We can now write the vector loop equation

$$\mathbf{r}_{OA} + \mathbf{r}_{AB} + \mathbf{r}_{BC} + \mathbf{r}_{CD} + \mathbf{r}_{DE} + \mathbf{r}_{EO} = \mathbf{0} \qquad [1]$$

in terms of coordinates $(\theta_1, \theta_3, \rho, \mu)$ and the constant parameters (l, h, d, e) with $e = |OE|$ and $h = |ED|$. See Figure 2.60.
Equation [1] is then expressed on basis $(\hat{\mathbf{x}}_0, \hat{\mathbf{y}}_0, \hat{\mathbf{z}}_0)$

$$d(\cos\theta_1\hat{\mathbf{y}}_0 - \sin\theta_1\hat{\mathbf{z}}_0) + \rho(t)\hat{\mathbf{x}}_0 + \mu(t)\hat{\mathbf{z}}_0 + l(\cos\theta_3\hat{\mathbf{x}}_0 - \sin\theta_3\hat{\mathbf{y}}_0) - h\hat{\mathbf{z}}_0 - e\hat{\mathbf{x}}_0 = 0$$

We then arrive at three equations

$$\rho(t) + l\cos\theta_3 - e = 0, \quad d\cos\theta_1 - l\sin\theta_3 = 0, \quad -d\sin\theta_1 + \mu(t) - h = 0 \qquad [2-4]$$

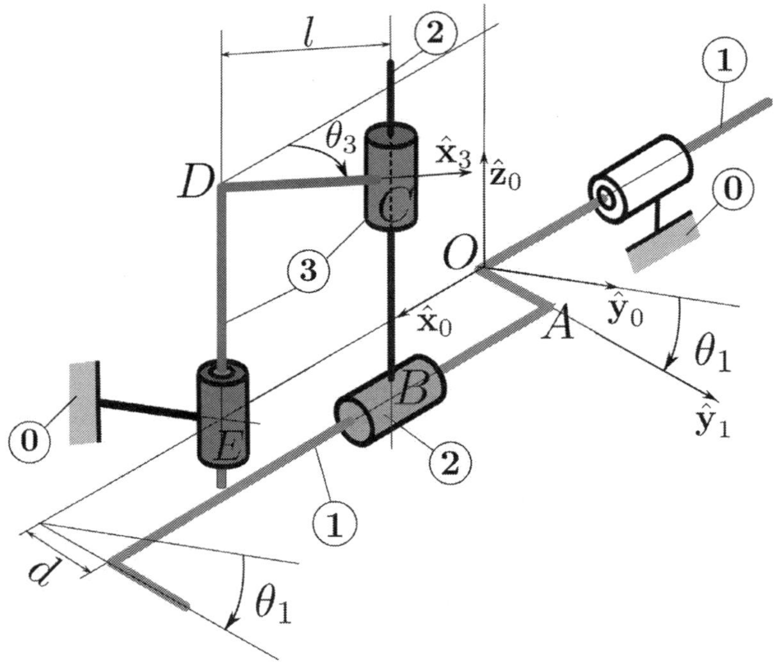

Figure 2.59 – *Problem 2.25. Note that body 2 is in translation relative to the fixed referential. However this translation is not planar.*

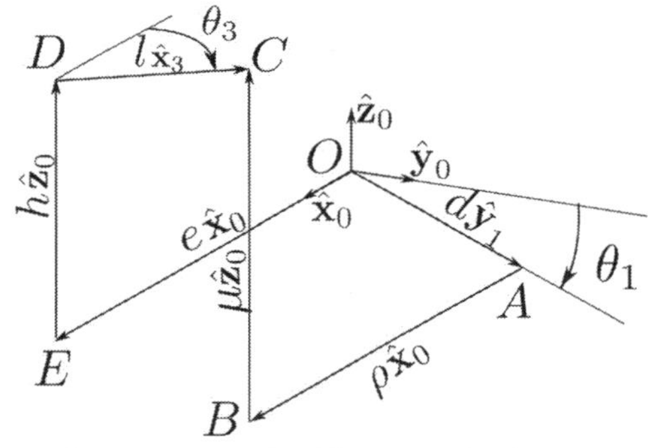

Figure 2.60 – *Problem 2.25 (Solution).*

valid at all time. We then differentiate [2-4] w.r.t. to time to obtain

$$\dot{\rho} - l\dot{\theta}_3 \sin\theta_3 = 0, \quad -d\dot{\theta}_1 \sin\theta_1 - l\dot{\theta}_3 \cos\theta_3 = 0, \quad -d\dot{\theta}_1 \cos\theta_1 + \dot{\mu} = 0 \qquad [5-7]$$

Equation [6] gives the angular velocity of body 3 in terms of the angular velocity $\boldsymbol{\omega}_1 = -\dot{\theta}_1 \hat{\mathbf{x}}_0$ of body 1:

$$\boxed{\boldsymbol{\omega}_3 = -\dot{\theta}_3 \hat{\mathbf{z}}_0, \qquad \dot{\theta}_3 = -\dot{\theta}_1 \frac{d \sin\theta_1}{l \cos\theta_3}}$$

with angle θ_3 given by [3]. To find the velocity of B we first find its position vector

$$\mathbf{r}_{OB} = d\hat{\mathbf{y}}_1 + \rho(t)\hat{\mathbf{x}}_0$$

Then we find $\mathbf{v}_B$ by differentiating $\mathbf{r}_{OB}$:

$$\mathbf{v}_B = \mathbf{v}_{B\in 3/0} = d\dot{\theta}_1(-\sin\theta_1 \hat{\mathbf{y}}_0 - \cos\theta_1 \hat{\mathbf{z}}_0) + \dot{\rho}\hat{\mathbf{x}}_0$$

which becomes after using [5]

$$\boxed{\mathbf{v}_B = -d\dot{\theta}_1 \big(\sin\theta_1 \cot\theta_3 \hat{\mathbf{x}}_0 + \sin\theta_1 \hat{\mathbf{y}}_0 + \cos\theta_1 \hat{\mathbf{z}}_0 \big)}$$

Equation [3] has a solution if $d \leq l$. If in addition $d < l$, then angle θ_3 is bounded

$$|\theta_3| \leq \theta_{3,max} \qquad \theta_{3,max} = \sin^{-1}\left(\frac{d}{l}\right)$$

In this case, the rotation of body 3 is alternating. ∎

2.9 Relative Motion Analysis

In this section, we return to particle kinematics: we seek to find out how the kinematics of a particle can be related from one referential to another. More specifically, consider a particle P in motion relative to two referentials denoted $\mathcal{E}$ and $\mathcal{F}$ in relative planar motion. The two referentials may be considered as the rigid extensions of two rigid bodies. See Figure 2.61. Assume that the motion of $\mathcal{F}$ relative to $\mathcal{E}$ is entirely known: this requires that the velocity and acceleration of a particular point A attached to $\mathcal{F}$ (relative to $\mathcal{E}$), and the angular velocity $\boldsymbol{\omega}_{\mathcal{F}} = \boldsymbol{\omega}_{\mathcal{F}/\mathcal{E}}$ and angular acceleration $\boldsymbol{\alpha}_{\mathcal{F}} = \boldsymbol{\alpha}_{\mathcal{F}/\mathcal{E}}$ of $\mathcal{F}$ (relative to $\mathcal{E}$) all be known. Recall that knowledge of these quantities implies that the velocity and acceleration of any other point B attached to $\mathcal{F}$ can be found according to the formulas:

$$\mathbf{v}_{B/\mathcal{E}} = \mathbf{v}_{A/\mathcal{E}} + \boldsymbol{\omega}_{\mathcal{F}} \times \mathbf{r}_{AB} \tag{2.17}$$

$$\mathbf{a}_{B/\mathcal{E}} = \mathbf{a}_{A/\mathcal{E}} + \boldsymbol{\alpha}_{\mathcal{F}} \times \mathbf{r}_{AB} - \omega_{\mathcal{F}}^2 \mathbf{r}_{AB} \tag{2.18}$$

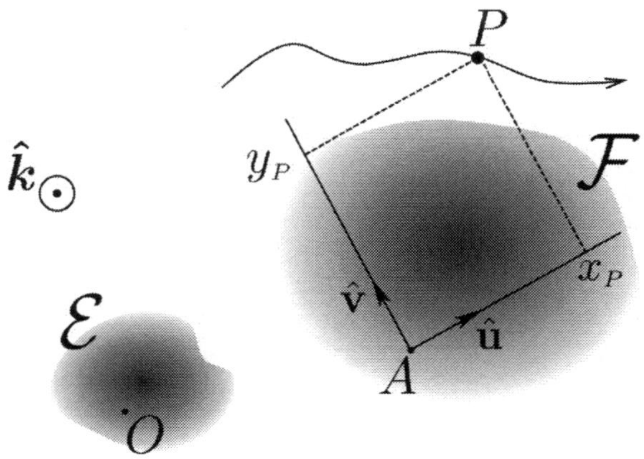

Figure 2.61 – *A particle P in motion relative to two referentials $\mathcal{E}$ and $\mathcal{F}$.*

Assuming that the velocity $\mathbf{v}_{P/\mathcal{F}}$ and acceleration $\mathbf{a}_{P/\mathcal{F}}$ of particle P relative to $\mathcal{F}$ are known, can the velocity $\mathbf{v}_{P/\mathcal{E}}$ and acceleration $\mathbf{a}_{P/\mathcal{E}}$ of P relative to $\mathcal{E}$ be found?

Remark 19. In this section, we make use of the notation $(\frac{d}{dt}\ )_{\mathcal{E}}$ to signify that the time-derivative is performed relative to referential $\mathcal{E}$: given an arbitrary vector function $\mathbf{U}(t)$ we expect that

$$\left(\frac{d\mathbf{U}}{dt}\right)_{\mathcal{E}} \neq \left(\frac{d\mathbf{U}}{dt}\right)_{\mathcal{F}}$$

1. Velocity Analysis. The starting point is the vector loop equation: $\mathbf{r}_{OP} = \mathbf{r}_{OA} + \mathbf{r}_{AP}$. We need to determine its time-derivative from the standpoint of an observer attached to $\mathcal{E}$: we obtain

$$\mathbf{v}_{P/\mathcal{E}} = \mathbf{v}_{A/\mathcal{E}} + \left(\frac{d}{dt}\mathbf{r}_{AP}\right)_{\mathcal{E}}$$

We cannot equate $\left(\frac{d}{dt}\mathbf{r}_{AP}\right)_{\mathcal{E}}$ to the velocity of P relative to $\mathcal{F}$ (since the derivative is applied relative $\mathcal{E}$) nor to the velocity of P relative to $\mathcal{E}$ (since point A is not attached to $\mathcal{E}$). To find this term, we can first write $\mathbf{r}_{AP} = x_P(t)\hat{\mathbf{u}} + y_P(t)\hat{\mathbf{v}}$ where $(\hat{\mathbf{u}}, \hat{\mathbf{v}})$ are two units vectors attached to $\mathcal{F}$. Then we obtain:

$$\left(\frac{d}{dt}\mathbf{r}_{AP}\right)_{\mathcal{E}} = \dot{x}_P\hat{\mathbf{u}} + \dot{y}_P\hat{\mathbf{v}} + x_P\frac{d\hat{\mathbf{u}}}{dt} + y_P\frac{d\hat{\mathbf{v}}}{dt}$$

Now we recognize the sum $\dot{x}_P\hat{\mathbf{u}} + \dot{y}_P\hat{\mathbf{v}}$ as the velocity $\mathbf{v}_{P/\mathcal{F}}$. We can also write

$$\frac{d\hat{\mathbf{u}}}{dt} = \boldsymbol{\omega}_{\mathcal{F}} \times \hat{\mathbf{u}}, \qquad \frac{d\hat{\mathbf{v}}}{dt} = \boldsymbol{\omega}_{\mathcal{F}} \times \hat{\mathbf{v}}$$

which then leads to

$$x_P\frac{d\hat{\mathbf{u}}}{dt} + y_P\frac{d\hat{\mathbf{v}}}{dt} = \boldsymbol{\omega}_{\mathcal{F}} \times (x_P\hat{\mathbf{u}} + y_P\hat{\mathbf{v}}) = \boldsymbol{\omega}_{\mathcal{F}} \times \mathbf{r}_{AP}$$

Finally, regrouping all the terms, we obtain:

$$\mathbf{v}_{P/\mathcal{E}} = \mathbf{v}_{P/\mathcal{F}} + (\mathbf{v}_{A/\mathcal{E}} + \boldsymbol{\omega}_{\mathcal{F}} \times \mathbf{r}_{AP})$$

Upon inspection of formula (2.17), we recognize that the sum $\mathbf{v}_{A/\mathcal{E}} + \boldsymbol{\omega}_{\mathcal{F}} \times \mathbf{r}_{AP}$ can be viewed as the velocity of P as if P were attached to referential (rigid body) $\mathcal{F}$: now we cannot write this velocity as $\mathbf{v}_{P/\mathcal{E}}$ since it is not the velocity of particle P. Instead we may denote it as $\mathbf{v}_{P\in\mathcal{F}/\mathcal{E}}$: the notation $P \in \mathcal{F}$ makes it clear that we interpret this velocity as that of a particular point of $\mathcal{F}$ relative to $\mathcal{E}$.[1] This point of $\mathcal{F}$ happens to coincide (at any given time) with the position of particle P in the plane of motion.

Change of Referential (Velocity)

Given two referentials $\mathcal{E}$ and $\mathcal{F}$ in relative motion, the velocity $\mathbf{v}_{P/\mathcal{E}}$ can be found from velocity $\mathbf{v}_{P/\mathcal{F}}$ according to the formula

$$\mathbf{v}_{P/\mathcal{E}} = \mathbf{v}_{P/\mathcal{F}} + \mathbf{v}_{P\in\mathcal{F}/\mathcal{E}} \tag{2.19}$$

where $\mathbf{v}_{P\in\mathcal{F}/\mathcal{E}}$ is known as the *transport velocity* of P by $\mathcal{F}$ relative to $\mathcal{E}$. The velocities vectors $\mathbf{v}_{P/\mathcal{E}}$ and $\mathbf{v}_{P/\mathcal{F}}$ are commonly called *absolute velocity* and *relative velocity*,

[1] We are familiar with the notation $\mathbf{v}_{P\in\mathcal{F}/\mathcal{E}}$. It was first introduced in Section 2.3.

respectively.

This formula establishes the relationship between the velocities of particle P relative to two referentials $\mathcal{E}$ and $\mathcal{F}$. The transport velocity $\mathbf{v}_{P\in\mathcal{F}/\mathcal{E}}$ is found in practice by applying formula (2.17) with a judicious choice of point A.

2. Acceleration Analysis. We can derive a similar relationship between the accelerations of particle P relative to $\mathcal{E}$ and $\mathcal{F}$. We start by taking the time derivative (relative to $\mathcal{E}$) of equation (2.19) to obtain

$$\mathbf{a}_{P/\mathcal{E}} = \left(\frac{d}{dt}\mathbf{v}_{P/\mathcal{F}}\right)_{\mathcal{E}} + \left(\frac{d}{dt}\mathbf{v}_{P\in\mathcal{F}/\mathcal{E}}\right)_{\mathcal{E}}$$

As in the previous derivation, the first term $\left(\frac{d}{dt}\mathbf{v}_{P/\mathcal{F}}\right)_{\mathcal{E}}$ is neither the acceleration of P relative to $\mathcal{E}$ nor the acceleration of P relative to $\mathcal{F}$: instead we write

$$\left(\frac{d}{dt}\mathbf{v}_{P/\mathcal{F}}\right)_{\mathcal{E}} = \left(\frac{d}{dt}(\dot{x}_P\hat{\mathbf{u}} + \dot{y}_P\hat{\mathbf{v}})\right)_{\mathcal{E}} = (\ddot{x}_P\hat{\mathbf{u}} + \ddot{y}_P\hat{\mathbf{v}}) + (\dot{x}_P\boldsymbol{\omega}_{\mathcal{F}} \times \hat{\mathbf{u}} + \dot{y}_P\boldsymbol{\omega}_{\mathcal{F}} \times \hat{\mathbf{v}})$$

The first sum $(\ddot{x}_P\hat{\mathbf{u}} + \ddot{y}_P\hat{\mathbf{v}})$ can now be recognized as the acceleration $\mathbf{a}_{P/\mathcal{F}}$ (it is the time-derivative of $\mathbf{v}_{P/\mathcal{F}}$ from the standpoint of an observer attached to $\mathcal{F}$). The second term $(\dot{x}_P\boldsymbol{\omega}_{\mathcal{F}} \times \hat{\mathbf{u}} + \dot{y}_P\boldsymbol{\omega}_{\mathcal{F}} \times \hat{\mathbf{v}})$ can be expressed as $\boldsymbol{\omega}_{\mathcal{F}} \times \mathbf{v}_{P/\mathcal{F}}$. So we have obtained

$$\left(\frac{d}{dt}\mathbf{v}_{P/\mathcal{F}}\right)_{\mathcal{E}} = \mathbf{a}_{P/\mathcal{F}} + \boldsymbol{\omega}_{\mathcal{F}} \times \mathbf{v}_{P/\mathcal{F}}$$

Now to find the second term $\left(\frac{d}{dt}\mathbf{v}_{P\in\mathcal{F}/\mathcal{E}}\right)_{\mathcal{E}}$ it is tempting to equate it to $\mathbf{a}_{P\in\mathcal{F}/\mathcal{E}}$. But this is incorrect (as we have already encountered in previous sections). Instead, we use formula (2.17) (using point A):

$$\left(\frac{d}{dt}\mathbf{v}_{P\in\mathcal{F}/\mathcal{E}}\right)_{\mathcal{E}} = \left(\frac{d}{dt}(\mathbf{v}_{A/\mathcal{E}} + \boldsymbol{\omega}_{\mathcal{F}} \times \mathbf{r}_{AP})\right)_{\mathcal{E}} = \mathbf{a}_{A/\mathcal{E}} + \boldsymbol{\alpha}_{\mathcal{F}} \times \mathbf{r}_{AP} + \boldsymbol{\omega}_{\mathcal{F}} \times \left(\frac{d}{dt}\mathbf{r}_{AP}\right)_{\mathcal{E}}$$

The expression of the last term $\left(\frac{d}{dt}\mathbf{r}_{AP}\right)_{\mathcal{E}}$ was found in the derivation of (2.19)

$$\left(\frac{d}{dt}\mathbf{r}_{AP}\right)_{\mathcal{E}} = \mathbf{v}_{P/\mathcal{E}} - \mathbf{v}_{A/\mathcal{E}} = \mathbf{v}_{P/\mathcal{F}} + \boldsymbol{\omega}_{\mathcal{F}} \times \mathbf{r}_{AP}$$

We can now regroup all the terms to find the following expression:

$$\mathbf{a}_{P/\mathcal{E}} = \mathbf{a}_{P/\mathcal{F}} + \boldsymbol{\omega}_{\mathcal{F}} \times \mathbf{v}_{P/\mathcal{F}} + \mathbf{a}_{A/\mathcal{E}} + \boldsymbol{\alpha}_{\mathcal{F}} \times \mathbf{r}_{AP} + \boldsymbol{\omega}_{\mathcal{F}} \times (\mathbf{v}_{P/\mathcal{F}} + \boldsymbol{\omega}_{\mathcal{F}} \times \mathbf{r}_{AP})$$

or

$$\mathbf{a}_{P/\mathcal{E}} = \mathbf{a}_{P/\mathcal{F}} + \left(\mathbf{a}_{A/\mathcal{E}} + \boldsymbol{\alpha}_{\mathcal{F}} \times \mathbf{r}_{AP} - \omega_{\mathcal{F}}^2\mathbf{r}_{AP}\right) + 2\boldsymbol{\omega}_{\mathcal{F}} \times \mathbf{v}_{P/\mathcal{F}}$$

As in the previous section, we recognize the sum $\left(\mathbf{a}_{A/\mathcal{E}} + \boldsymbol{\alpha}_{\mathcal{F}} \times \mathbf{r}_{AP} - \omega_{\mathcal{F}}^2\mathbf{r}_{AP}\right)$ upon inspection of formula (2.18) as the acceleration of P as if P is considered attached to referential (rigid body) $\mathcal{F}$: we denote this acceleration as $\mathbf{a}_{P\in\mathcal{F}/\mathcal{E}}$. Once again, the notation $P \in \mathcal{F}$ makes it clear that we must interpret this acceleration as that of a particular point (instantaneously) attached to $\mathcal{F}$ (and thus in motion with $\mathcal{F}$ relative to $\mathcal{E}$).

Change of Referential (Acceleration)

Given two referentials $\mathcal{E}$ and $\mathcal{F}$ in relative motion, the acceleration $\mathbf{a}_{P/\mathcal{E}}$ of P relative to $\mathcal{E}$ can be found from the acceleration $\mathbf{a}_{P/\mathcal{F}}$ of P relative to $\mathcal{F}$ according to the formula

$$\mathbf{a}_{P/\mathcal{E}} = \mathbf{a}_{P/\mathcal{F}} + \mathbf{a}_{P\in\mathcal{F}/\mathcal{E}} + 2\boldsymbol{\omega}_{\mathcal{F}} \times \mathbf{v}_{P/\mathcal{F}} \tag{2.20}$$

where
(i) $\mathbf{a}_{P\in\mathcal{F}/\mathcal{E}}$ is referred to as the *transport acceleration* of P by $\mathcal{F}$ relative to $\mathcal{E}$,
(ii) $2\boldsymbol{\omega}_{\mathcal{F}} \times \mathbf{v}_{P/\mathcal{F}}$ is called *Coriolis acceleration* of P.
(iii) the accelerations vectors $\mathbf{a}_{P/\mathcal{E}}$ and $\mathbf{a}_{P/\mathcal{F}}$ are commonly called *absolute acceleration* and *relative acceleration*, respectively.

Remark 20. We will learn in Part III that, when Newton's Second Law is applied to a non-Newtonian referential, the transport and Coriolis accelerations are the cause of two inertial forces.

Next we consider the two special cases of formulas (2.19) and (2.20) corresponding to the translational and rotational motion of $\mathcal{F}$ relative to $\mathcal{E}$.

3. Special Cases

Translation. Recall that if $\mathcal{F}$ is in translation relative to $\mathcal{E}$, its angular velocity and acceleration vanish at all time: $\boldsymbol{\omega}_{\mathcal{F}} = \mathbf{0}$, and $\boldsymbol{\alpha}_{\mathcal{F}} = \mathbf{0}$. This implies that the transport velocity and acceleration of P take the same expression irrespective of the motion of particle P:

$$\mathbf{v}_{P\in\mathcal{F}/\mathcal{E}} = \mathbf{v}_{A/\mathcal{E}}, \qquad \mathbf{a}_{P\in\mathcal{F}/\mathcal{E}} = \mathbf{a}_{A/\mathcal{E}}$$

Furthermore, the Coriolis acceleration of particle P is necessarily zero. With these simplifications, the change of referential formulas (2.19) and (2.20) take the following form

$$\mathbf{v}_{P/\mathcal{E}} = \mathbf{v}_{P/\mathcal{F}} + \mathbf{v}_{A/\mathcal{E}}$$

$$\mathbf{a}_{P/\mathcal{E}} = \mathbf{a}_{P/\mathcal{F}} + \mathbf{a}_{A/\mathcal{E}}$$

Rotation. If $\mathcal{F}$ is in rotation relative to $\mathcal{E}$ about axis $\Delta(O, \hat{\mathbf{k}})$ (let point A of $\mathcal{F}$ coincide with point O of $\mathcal{E}$), then the transport velocity and accelerations take the simple expressions

$$\mathbf{v}_{P\in\mathcal{F}/\mathcal{E}} = \omega_{\mathcal{F}}\hat{\mathbf{k}} \times \mathbf{r}_{OP}, \qquad \mathbf{a}_{P\in\mathcal{F}/\mathcal{E}} = \alpha_{\mathcal{F}}\hat{\mathbf{k}} \times \mathbf{r}_{OP} - \omega_{\mathcal{F}}^2\mathbf{r}_{OP}$$

leading to the following expressions of formulas (2.19) and (2.20)

$$\mathbf{v}_{P/\mathcal{E}} = \mathbf{v}_{P/\mathcal{F}} + \omega_{\mathcal{F}}\hat{\mathbf{k}} \times \mathbf{r}_{OP}$$

$$\mathbf{a}_{P/\mathcal{E}} = \mathbf{a}_{P/\mathcal{F}} + \alpha_{\mathcal{F}}\hat{\mathbf{k}} \times \mathbf{r}_{OP} - \omega_{\mathcal{F}}^2\mathbf{r}_{OP} + 2\omega_{\mathcal{F}}\hat{\mathbf{k}} \times \mathbf{v}_{P/\mathcal{F}}$$

4. Application to Rigid Body Motion. We learned in Section 2.4 that, given a rigid body $\mathcal{B}$ in motion relative to a referential $\mathcal{E}$, the velocities $\mathbf{v}_{P\in\mathcal{B}/\mathcal{E}}$ and $\mathbf{v}_{P/\mathcal{E}}$ may not always be the same quantity. We can now find out under what condition this situation occurs with the use of fundamental formula (2.19):

$$\mathbf{v}_{P/\mathcal{E}} = \mathbf{v}_{P\in\mathcal{B}/\mathcal{E}} + \mathbf{v}_{P/\mathcal{B}} \tag{2.21}$$

Equation (2.21) shows that the velocity difference $\mathbf{v}_{P/\mathcal{E}} - \mathbf{v}_{P\in\mathcal{B}/\mathcal{E}}$ is in fact equal to the velocity $\mathbf{v}_{P/\mathcal{B}}$: hence if P is not fixed relative to $\mathcal{B}$, then the two velocities $\mathbf{v}_{P\in\mathcal{B}/\mathcal{E}}$ and $\mathbf{v}_{P/\mathcal{E}}$ are distinct quantities:
- velocity $\mathbf{v}_{P/\mathcal{E}}$ is found by determining the rate of change (w.r.t. an observer attached to $\mathcal{E}$) of a position vector $\mathbf{r}_{OP}$ where O is an arbitrary fixed point of $\mathcal{E}$,
- velocity $\mathbf{v}_{P\in\mathcal{B}/\mathcal{E}}$ is typically found by relating it to the velocity of a point, say A, attached to $\mathcal{B}$ *at all time*:

$$\mathbf{v}_{P\in\mathcal{B}/\mathcal{E}} = \mathbf{v}_{A/\mathcal{E}} + \boldsymbol{\omega}_{\mathcal{B}} \times \mathbf{r}_{AP}$$

Remark 21. We also see that, whenever point P is attached to $\mathcal{B}$ *at all time*, then there is no need to use the notation $\mathbf{v}_{P\in\mathcal{B}/\mathcal{E}}$:

$$\mathbf{v}_{P/\mathcal{E}} = \mathbf{v}_{P\in\mathcal{B}/\mathcal{E}}$$

Kinematics of Point-Contact Between Rigid Bodies. We now re-examine the kinematics of two bodies 1 and 2 in point contact. See Figure 2.62 (left). We know from Section 2.6 that contact kinematics is characterized by the slip velocity at the contact point I, defined as $\mathbf{v}_{I\in 2/1}$. With the use of (2.21), we can write two expressions of this velocity:

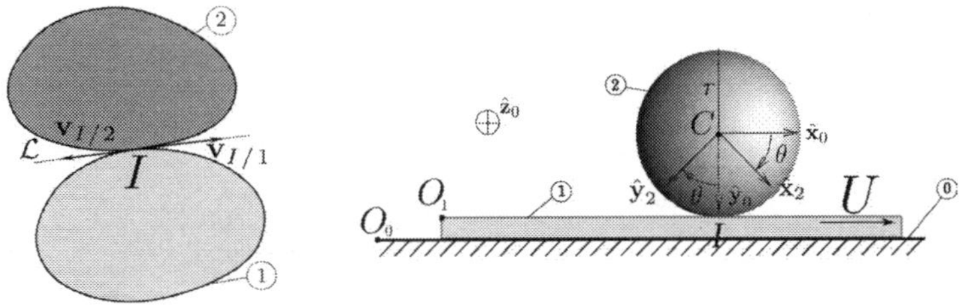

Figure 2.62 – *Two rigid bodies in contact motion (left). Sphere 2 rolls and slips on moving support 1 (right).*

$$\mathbf{v}_{I\in 2/1} = \mathbf{v}_{I/1} - \mathbf{v}_{I/2} \tag{2.22}$$

$$\mathbf{v}_{I\in 2/1} = \mathbf{v}_{I\in 2/0} - \mathbf{v}_{I\in 1/0} \tag{2.23}$$

Formula (2.22) is a straightforward application of (2.21). As the two bodies roll relative to one another, contact point I is in motion relative to both bodies at velocity $\mathbf{v}_{I/1}$ and $\mathbf{v}_{I/2}$, respectively, along trajectories which are easily identified as the bodies' boundaries. Consequently, both velocities $\mathbf{v}_{I/1}$ and $\mathbf{v}_{I/2}$, and hence $\mathbf{v}_{I\in 2/1}$ are *necessarily tangent at I to the bodies*.
A more practical way of determining the slip velocity is to use formula (2.23) which, again can be proved by the use of (2.21). Consider the point $I \in 2$ (that is, instantaneously attached

to body 2) and let us relate the ("absolute") velocity of $I \in 2$ relative to referential 0 and to its ("relative") velocity relative to referential 1. This can be done by determining the transport velocity $\mathbf{v}_{I\in 1/0}$: $\mathbf{v}_{I\in 2/0} = \mathbf{v}_{I\in 2/1} + \mathbf{v}_{I\in 1/0}$.

Application. As an illustrative application of formulas (2.22) and (2.23), consider the motion of sphere 2 (of center C and radius r, basis $(\hat{\mathbf{x}}_2, \hat{\mathbf{y}}_2)$) rolling and slipping along a straight support $1(O_1, \hat{\mathbf{x}}_0, \hat{\mathbf{y}}_0)$ which itself translates at velocity $U\hat{\mathbf{x}}_1$ relative to a referential 0 (of origin O_0). We wish to find the slip velocity $\mathbf{v}_{I\in 2/1}$. See Figure 2.62 (right). The sphere is observed relative to referential 0 to rotate at angular velocity $\boldsymbol{\omega}_{2/0} = \dot{\theta}\hat{\mathbf{z}}_0$ and to translate at velocity $\mathbf{v}_{C/0} = v_C\hat{\mathbf{x}}_0$. With formula (2.22), we need to evaluate the velocities $\mathbf{v}_{I/2}$ and $\mathbf{v}_{I/1}$:

$$\mathbf{v}_{I/1} = \Big(\frac{d\mathbf{r}_{O_1 I}}{dt}\Big)_1 = \Big(\frac{\mathbf{r}_{O_1 O_0} + \mathbf{r}_{O_0 I}}{dt}\Big)_1 = v_C\hat{\mathbf{x}}_0 - U\hat{\mathbf{x}}_0$$

$$\mathbf{v}_{I/2} = \Big(\frac{d\mathbf{r}_{CI}}{dt}\Big)_2 = \Big(\frac{r\hat{\mathbf{y}}_0}{dt}\Big)_2 = r\Big(\frac{d}{dt}(\cos\theta\hat{\mathbf{y}}_2 + \sin\theta\hat{\mathbf{x}}_2)\Big)_1 = r\dot{\theta}(-\sin\theta\hat{\mathbf{y}}_2 + \cos\theta\hat{\mathbf{x}}_2) = r\dot{\theta}\hat{\mathbf{x}}_0$$

This leads to the following expression of the slip velocity

$$\mathbf{v}_{I\in 2/1} = \mathbf{v}_{I/1} - \mathbf{v}_{I/2} = (v_C - r\dot{\theta} - U)\hat{\mathbf{x}}_0$$

Now with use of (2.23), we find

$$\mathbf{v}_{I\in 2/1} = \underbrace{\mathbf{v}_{I\in 2/0}}_{\mathbf{v}_{C\in 2/0} + \boldsymbol{\omega}_{2/0}\times\mathbf{r}_{CI}} - \mathbf{v}_{I\in 1/0} = (v_C\hat{\mathbf{x}}_0 - r\dot{\theta}\hat{\mathbf{x}}_0) - U\hat{\mathbf{x}}_0$$

Although the same expressions are found, the second method is preferable.

Application: Gravity Assistance of Space Probes. Gravity assistance has been instrumental to the success of interplanetary probes for the exploration of the solar system. A prove moves through space almost exclusively by exploiting the gravitation attraction of the Sun and planets. The strategy for the planning of probe's path is based on taking advantage of the attraction of one or more celestial bodies to control the probe's velocity in both magnitude and direction. Occasionally, the probe's motors are used to perform small changes to its the orbital plane or orientation. Without spending additional energy, it is possible to alter a probe trajectory, speed and orientation by using a mechanism known as *gravity assistance* or *slingshot effect*. When the probe enters a planet's sphere of influence (the so-called Hill's sphere), it is "captured" by the planet until it reaches its perigee (the point of closest approach) with maximum speed. The probe then bounces back to escape the sphere of influence of the planet. Clearly, its path is altered. But how does it achieve an increase (or decrease) in speed?

If we study the probe in the geocentric referential $\mathcal{G}$ of the planet (relative to which the planet's center is fixed), the probe enters Hill's sphere with a sufficiently large velocity $\mathbf{v}_{\infty/\mathcal{G}}$ to follow a hyperbolic trajectory and to finally exit Hill's sphere with a velocity $\mathbf{v}'_{\infty/\mathcal{G}}$. Application of the principle of energy conservation in $\mathcal{G}$ shows that the velocities $\mathbf{v}_{\infty/\mathcal{G}}$ and $\mathbf{v}'_{\infty/\mathcal{G}}$ are identical in magnitude. The probe reaches a maximum speed v_B when it reaches its perigee at B, but this acceleration over a short span cannot possibly result in a significant gain in time over large interplanetary distances. Hence, the study in the geocentric referential cannot predict the mechanism of gravity assistance.

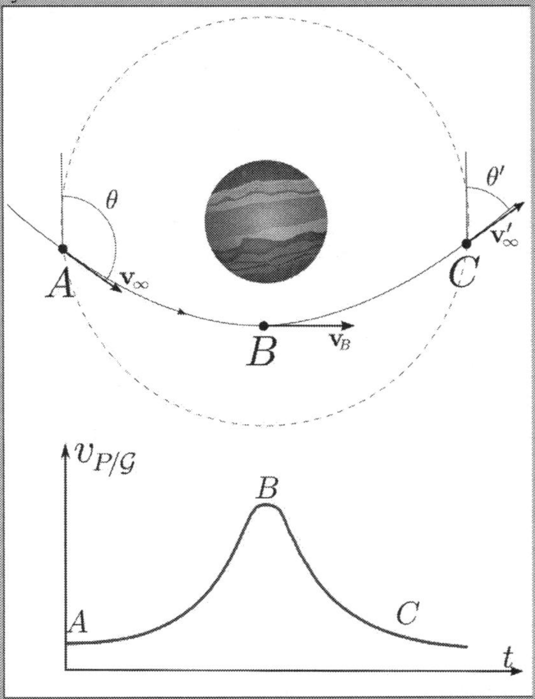

Figure 2.63 – *Trajectory of a probe P relative to geocentric referential.*

In fact, the proper way to explain gravity assistance is to view the motion of the probe relative to the heliocentric referential $\mathcal{H}$ (relative to which the Sun's center is fixed). We can assume that the time of residence of the probe in Hill's sphere is small compared to the period of revolution of the planet. We can then consider that the planet center has a uniform rectilinear motion relative to $\mathcal{H}$ at velocity $\mathbf{U}$. The velocity of the probe before and after entering Hill's sphere is then $\mathbf{v}_{\infty/\mathcal{G}} + \mathbf{U}$ and $\mathbf{v}'_{\infty/\mathcal{G}} + \mathbf{U}$ with $|\mathbf{v}_{\infty/\mathcal{G}}| = |\mathbf{v}'_{\infty/\mathcal{G}}| = v_\infty$. The gain (or loss) in speed in then predicted by the relationship

$$|\mathbf{v}'_{\infty/\mathcal{G}} + \mathbf{U}|^2 - |\mathbf{v}_{\infty/\mathcal{G}} + \mathbf{U}|^2 = 2\mathbf{U} \cdot (\mathbf{v}'_{\infty/\mathcal{G}} - \mathbf{v}_{\infty/\mathcal{G}})$$
$$= 2Uv_\infty(\cos\theta' - \cos\theta)$$

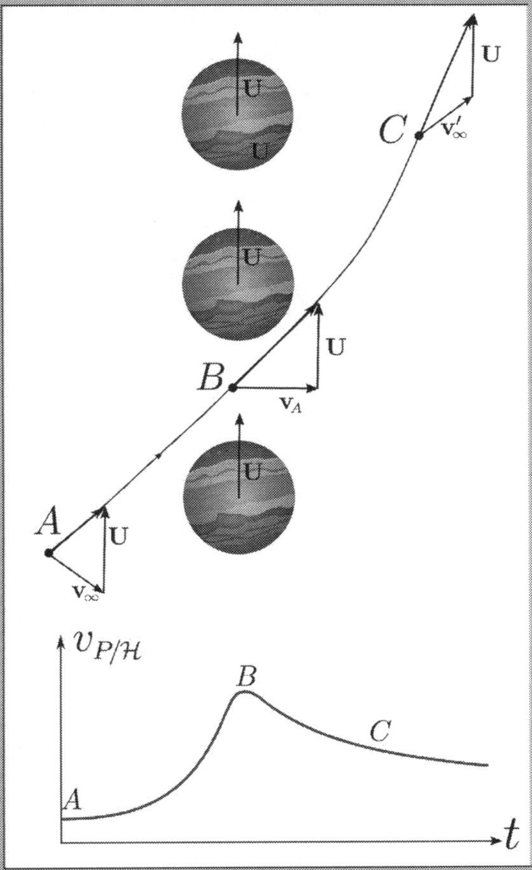

Figure 2.64 – *Trajectory of a probe P relative to the heliocentric referential.*

We then see that, if the probe passes behind the planet, $\cos\theta' > \cos\theta$ leading to $|\mathbf{v}'_{\infty/\mathcal{H}}| > |\mathbf{v}_{\infty/\mathcal{H}}|$. Since v_∞ and U are of the same order of magnitude, a significant gain in speed is obtained. With the fly-bys of several planets, a probe can exhibit accelerations large enough to reach speeds sufficient to reach large distances over shorter time spans. Note that, if the probe passes in front of the planet, it can be slowed down in the heliocentric referential.

The most striking early examples of gravity assistance are the two Voyager probes launched in 1977 during their long voyage to the planets of the outer Solar System. Voyager 2 would have taken around 30 years without gravitational assistance to reach Neptune. It took less than 12 years with the assistance of Jupiter, Saturne and Uranus, along with a favorable configuration of the planets taking place only every 176 years. More recently, the Cassini-Huygens probe was launched in 1997 with the mission of exploring Saturn. To reach Saturn, it first underwent two flybys of Venus in 1998 and 1999, a flyby of Earth in 1999, and a final gravitational assistance of Jupiter in 2000 to finally reach Saturn in 2004.

EXAMPLE PROBLEMS

General Procedure: Relative Kinematic Analysis of Particles

To solve a kinematics particle problem by a relative motion analysis, follow these guidelines:

① Clearly identify the "fixed" referential $\mathcal{E}$ and the "moving" referential $\mathcal{F}$. Define the corresponding "fixed" unit vectors and the "moving" unit vectors (if needed).

② To find the transport velocity $\mathbf{v}_{P\in\mathcal{F}/\mathcal{E}}$ and transport acceleration $\mathbf{a}_{P\in\mathcal{F}/\mathcal{E}}$ of particle P, find the velocity and acceleration of a judiciously chosen point attached to $\mathcal{F}$ at all time, then apply formulas (2.17) and (2.18).

③ It can be useful to use a graphical construction of the velocity formula (2.19). See Figure 2.65.

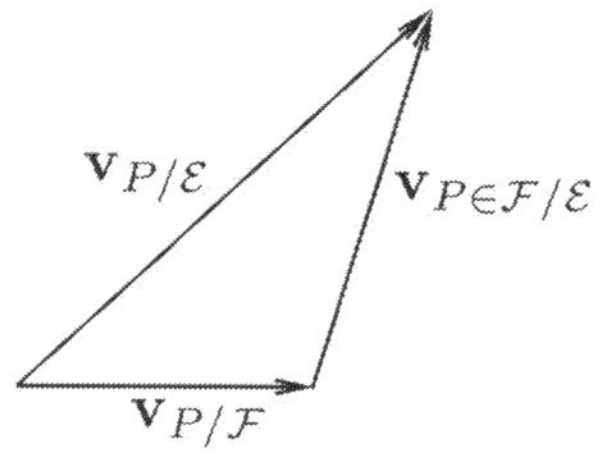

Figure 2.65 – *Construction of formula (2.19).*

Exercise 2.26 ■ An ant A "walks" along a diameter of a turntable $\mathcal{F}$ which is in rotation about an axis $(O, \hat{\mathbf{k}})$ of a fixed referential $\mathcal{E}$. Denote by $\omega\hat{\mathbf{k}}$ and $\alpha\hat{\mathbf{k}}$ the angular velocity and acceleration of $\mathcal{F}$ relative to $\mathcal{E}$.
Find the "absolute" velocity and acceleration of particle A from its "relative" velocity and acceleration, given that $\mathbf{r}_{OA} = \rho(t)\hat{\mathbf{u}}$ with $\hat{\mathbf{u}}$ a unit vector attached to $\mathcal{F}$. ■

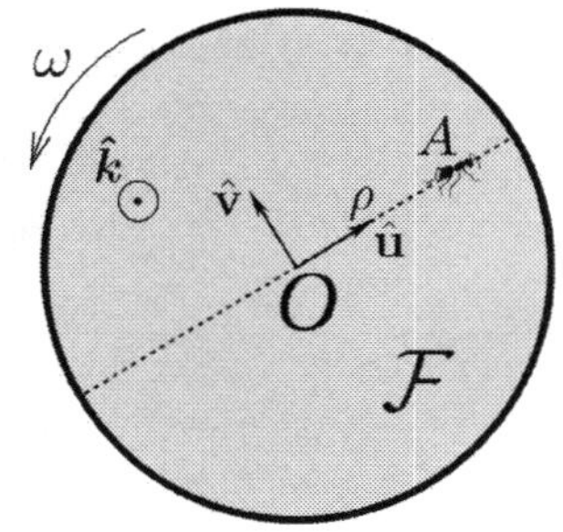

SOLUTION. The motion of $\mathcal{F}$ relative to $\mathcal{E}$ is a rotation about axis $(O, \hat{\mathbf{k}})$ at angular velocity $\boldsymbol{\omega}_{\mathcal{F}} = \omega(t)\hat{\mathbf{k}}$ and angular acceleration $\boldsymbol{\alpha}_{\mathcal{F}} = \alpha(t)\hat{\mathbf{k}}$. Define unit vector $\hat{\mathbf{v}} = \hat{\mathbf{k}} \times \hat{\mathbf{u}}$. The relative velocity and acceleration of particle A, that is relative to $\mathcal{F}$, are easily found to be

$$\mathbf{v}_{A/\mathcal{F}} = \left(\frac{d}{dt}\mathbf{r}_{OA}\right)_{\mathcal{F}} = \dot{\rho}\,\hat{\mathbf{u}}, \qquad \mathbf{a}_{A/\mathcal{F}} = \left(\frac{d}{dt}\mathbf{v}_{A/\mathcal{F}}\right)_{\mathcal{F}} = \ddot{\rho}\,\hat{\mathbf{u}}$$

since A has a rectilinear motion relative to $\mathcal{F}$. The transport velocity and acceleration are found by considering the point $A(t)$ which coincides with a point of $\mathcal{F}$, that is, a point instantaneously moving with $\mathcal{F}$:

$$\mathbf{v}_{A\in\mathcal{F}/\mathcal{E}} = \boldsymbol{\omega}_{\mathcal{F}} \times \mathbf{r}_{OA} = \omega\hat{\mathbf{k}} \times \rho\hat{\mathbf{u}} = \rho\omega\hat{\mathbf{v}}$$
$$\mathbf{a}_{A\in\mathcal{F}/\mathcal{E}} = \boldsymbol{\alpha}_{\mathcal{F}} \times \mathbf{r}_{OA} - \omega^2\mathbf{r}_{OA} = \rho\alpha\hat{\mathbf{v}} - \rho\omega^2\hat{\mathbf{u}}$$

The Coriolis acceleration of A is given by

$$2\boldsymbol{\omega}_{\mathcal{F}} \times \mathbf{v}_{A/\mathcal{F}} = 2\omega\hat{\mathbf{k}} \times \dot{\rho}\,\hat{\mathbf{u}} = 2\omega\dot{\rho}\hat{\mathbf{v}}$$

We can now apply formulas (2.19) and (2.20) to find the absolute velocity and acceleration of A:

$$\boxed{\mathbf{v}_{A/\mathcal{E}} = \mathbf{v}_{A/\mathcal{F}} + \mathbf{v}_{A\in\mathcal{F}/\mathcal{E}} = \dot{\rho}\,\hat{\mathbf{u}} + \rho\omega\hat{\mathbf{v}}}$$

$$\mathbf{a}_{A/\mathcal{E}} = \mathbf{a}_{A/\mathcal{F}} + \mathbf{a}_{A\in\mathcal{F}/\mathcal{E}} + 2\boldsymbol{\omega}_{\mathcal{F}/\mathcal{E}} \times \mathbf{v}_{A/\mathcal{F}} = \ddot{\rho}\,\hat{\mathbf{u}} + (\rho\alpha\hat{\mathbf{v}} - \rho\omega^2\hat{\mathbf{u}}) + 2\omega\dot{\rho}\hat{\mathbf{v}}$$

or

$$\boxed{\mathbf{a}_{A/\mathcal{E}} = (\ddot{\rho} - \rho\omega^2)\hat{\mathbf{u}} + (\rho\alpha + 2\omega\dot{\rho})\hat{\mathbf{v}}}$$

We recognize that the same answer can be obtained by the use of polar coordinates (r, θ) of particle P, recognizing that $\omega = \dot{\theta}$, $\alpha = \ddot{\theta}$, $\hat{\mathbf{u}} = \hat{\mathbf{u}}_r$ and $\hat{\mathbf{v}} = \hat{\mathbf{u}}_\theta$. ■

Exercise 2.27 ■■ A competition consists of crossing a river of width D which flows at constant speed v. A contestant starts at A and must reach C straight across. He can swim from A to B by keeping a constant direction and run from B to C. His speed is V in still water and $U > V$ on land.

a. Find the path ABC which minimizes the time T_{ABC} taken to reach C.

b. Find T_{ABC} for $D = 100$ m, $v = 0.5$ m/s, $V = 0.5$ m/s and $U = 3.5$ m/s. ■

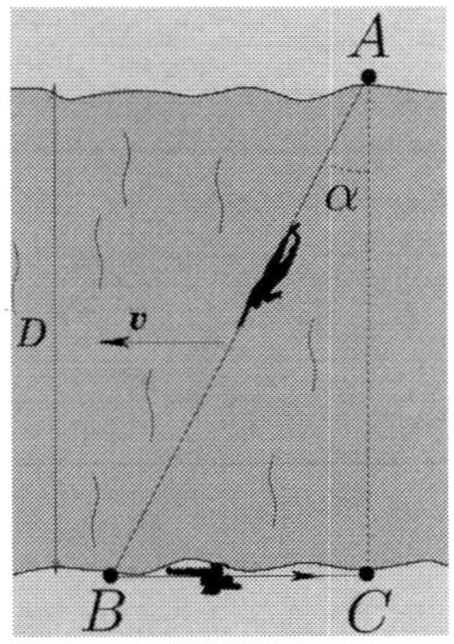

SOLUTION. **a.** Two referentials can be defined: referential $\mathcal{E}$ attached to the ground, and referential $\mathcal{F}$ moving with the river and in uniform rectilinear translation relative to $\mathcal{E}$ at velocity $v\hat{\imath}$. The rectilinear path AB is directed along the absolute velocity of the swimmer P whose magnitude and direction are unknown. This velocity can be written on unit vectors $(\hat{\imath}, \hat{\jmath})$ in the following form:

$$\mathbf{v}_{P/\mathcal{E}} = u(\sin\alpha\hat{\imath} + \cos\alpha\hat{\jmath})$$

It is related to the velocity of P relative to the flowing river according to

$$\mathbf{v}_{P/\mathcal{E}} = \mathbf{v}_{P\in\mathcal{F}/\mathcal{E}} + \mathbf{v}_{P/\mathcal{F}} = v\hat{\imath} + \mathbf{v}_{P/\mathcal{F}} \qquad [1]$$

where we have identified the transport velocity as $v\hat{\imath}$. The relative velocity $\mathbf{v}_{P/\mathcal{F}}$ is known in magnitude (the speed V of the swimmer in still water) and is unknown in direction. Since it is expected that this velocity must be directed upstream to arrive as close to C as possible we write $\mathbf{v}_{P/\mathcal{F}}$ as follows

$$\mathbf{v}_{P/\mathcal{F}} = V(-\sin\beta\hat{\imath} + \cos\beta\hat{\jmath})$$

where angle β is unknown. The graphical construction of equation [1] is shown in Figure 2.66. Equation [1] can now be written as

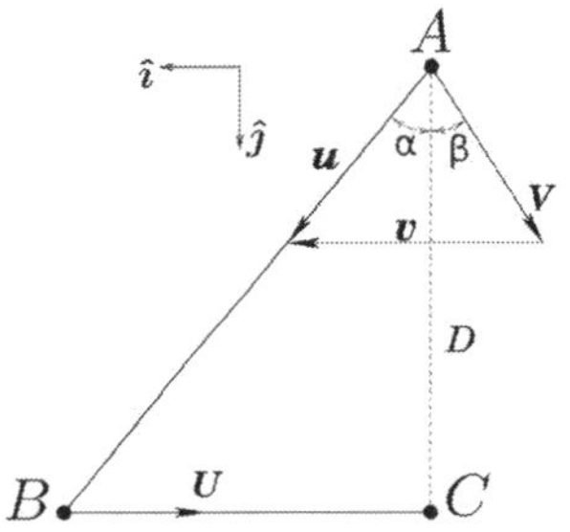

Figure 2.66 – *Problem 2.27 (Solution)*

$$u(\sin\alpha\hat{\imath} + \cos\alpha\hat{\jmath}) = v\hat{\imath} + V(-\sin\beta\hat{\imath} + \cos\beta\hat{\jmath})$$

This gives two scalar equations

$$u\sin\alpha = v - V\sin\beta, \qquad u\cos\alpha = V\cos\beta \qquad [2-3]$$

The time taken for the swimmer to cross the river is then given by

$$T_{AB} = \frac{L_{AB}}{u} = \frac{D}{u\cos\alpha} = \frac{D}{V\cos\beta}$$

where we have used equation [3] in the last equality. Similarly the time taken to travel along BC at speed U is given by

$$T_{BC} = \frac{L_{BC}}{U} = \frac{D\tan\alpha}{U}$$

The total time T_{ABC} is then given by

$$T_{ABC} = \frac{D}{V\cos\beta} + \frac{D\tan\alpha}{U}$$

We need to minimize time T_{ABC} w.r.t. all possible paths, that is, w.r.t. angle α or β:

indeed these two angles are not independent as seen by equations [2-3]. For instance we can express $\tan\alpha$ in terms of β by eliminating speed u:

$$\tan\alpha = \frac{v - V\sin\beta}{V\cos\beta} \qquad [4]$$

We can now express T_{ABC} in terms of variable β:

$$\boxed{T_{ABC} = \frac{D}{V\cos\beta} + \frac{D}{U}\frac{v - V\sin\beta}{V\cos\beta}} \qquad [5]$$

To minimize T_{ABC} , we consider the function

$$\tau(\beta) = \frac{U + v - V\sin\beta}{U\cos\beta}$$

The first term $1/\cos\beta$ attains its minimum at $\beta = 0$. The second term shifts this minimum to a value $\beta^* > 0$. To see this, we determine $d\tau/d\beta$:

$$\frac{d\tau}{d\beta} = \frac{(U+v)\sin\beta - V}{U\cos^2\beta}$$

We see that $d\tau/d\beta$ vanishes at the angle β^* given by

$$\boxed{\sin\beta^* = \frac{V}{U+v}} \qquad [6]$$

The angle α^* corresponding to the optimal path is then found from equation [4].
b. For $v = 0.5$ (m/s), $V = 0.5$ (m/s), and $U = 3.5$ (m/s), we find $\beta^* =$ **7.2**° using equation [6] and $\alpha^* =$ **80.5**°. The corresponding minimal time is T_{ABC} =3 min 47 s using $D = 100$ m. ■

Exercise 2.28 ▪▪ An aircraft must travel along a straight path between two positions A and B separated by a distance D. Due to a horizontal wind blowing opposite to the path of the aircraft at a speed w and angle α relative to line AB, the pilot must fly the aircraft at an angle β into the wind. The aircraft flies at constant speed v relative to the air.
a. Find the angle β that the aircraft must make relative to line AB so as to travel along the intended path. Under what condition is the trajectory realizable? Find angle β for $v = 100$ km/h, $D = 500$ km, $\alpha = 30°$ and $w = 50$ km/h.
b. The aircraft must travel from A to B then return to A. Find the corresponding time T_{AB} taken by the aircraft. Compare with the time t_{AB} in the absence of wind. Can T_{AB} ever be smaller than t_{AB}? ■

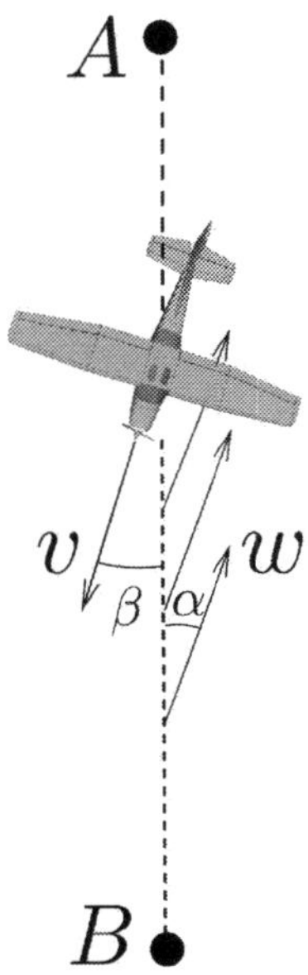

SOLUTION. **a.** Two referentials can be defined: referential $\mathcal{E}$ attached to the ground, and referential $\mathcal{F}$ moving with the air and in translation relative to $\mathcal{E}$ at velocity

$w(-\cos\alpha\hat{\imath} + \sin\alpha\hat{\jmath})$. The relative velocity $\mathbf{v}_{P/\mathcal{F}}$ of the airplane (defined as P) is known in magnitude (the speed v of the airplane in quiescent air) and is unknown in direction defined by angle β:

$$\mathbf{v}_{P/\mathcal{F}} = v(\cos\beta\hat{\imath} - \sin\beta\hat{\jmath})$$

These two expressions add up to give the absolute velocity of P as shown Figure 2.67 (left)

$$\mathbf{v}_{P/\mathcal{E}} = \mathbf{v}_{P\in\mathcal{F}/\mathcal{E}} + \mathbf{v}_{P/\mathcal{F}} = w(-\cos\alpha\hat{\imath} + \sin\alpha\hat{\jmath}) + v(\cos\beta\hat{\imath} - \sin\beta\hat{\jmath})$$

If the trajectory of P is to be line AB, velocity $\mathbf{v}_{P/\mathcal{E}}$ must be directed along $\hat{\imath}$ (the desired direction of travel): this imposes the equation

$$w\sin\alpha = v\sin\beta$$

This gives the necessary angle β

$$\boxed{\sin\beta = \frac{w}{v}\sin\alpha} \qquad [1]$$

Assuming $0 \leq \alpha < \pi/2$, a solution β of [1] exists if the condition $\boxed{w < v}$ is satisfied. If $v < w$, P cannot travel from A to B. For $v = 100$ km/h, $w = 50$ km/h and $\alpha = 30°$ we find $\beta = \mathbf{14.5}°$.

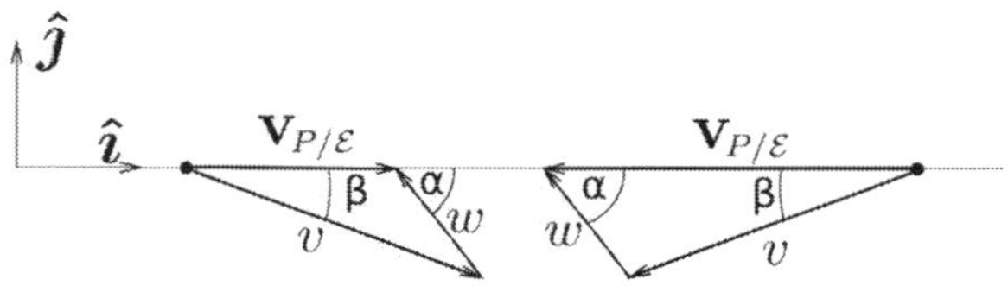

Figure 2.67 – *Problem 2.28 (Solution)*

b. Along the path from A to B the velocity of P is

$$\mathbf{v}_{P/\mathcal{E}} = (-w\cos\alpha + v\cos\beta)\hat{\imath}$$

with angle β given by [1]. Along the return trip from B to A, the velocity of P is now (see Figure 2.67, right)

$$\mathbf{v}_{P/\mathcal{E}} = -(w\cos\alpha + v\cos\beta)\hat{\imath}$$

where angle β is still given by [1]. Then time T_{AB} to complete the round trip is given by

$$\boxed{T_{AB} = \frac{D}{(-w\cos\alpha + v\cos\beta)} + \frac{D}{(w\cos\alpha + v\cos\beta)}}$$

In the absence of wind, this time would be $t_{AB} = 2D/v$. For $v = 100$ km/h, $w = 50$ km/h and $\alpha = 30°$, $\beta = 14.5°$ and $D = 500$ km, we find $T_{AB} = \mathbf{12.9}$ hours and $t_{AB} = \mathbf{10}$ hours. We can show that $T_{AB} > t_{AB}$ for all values of $k = w/v < 1$: Taking into account [1], the ratio T_{AB}/t_{AB} takes the expression

$$\frac{T_{AB}}{t_{AB}} = \frac{(1 - k^2\sin^2\alpha)^{1/2}}{1 - k^2}$$

If is easy to prove that $T_{AB}/t_{AB} > 1$ for all values of $0 < k < 1$ and all angles α in the interval $0 \leq \alpha \leq \pi/2$. ■

2.10 General Method of Kinematic Analysis

We wish to present in this section a new method for the kinematic analysis of mechanisms. This method is based on simple and intuitive formulas and is applicable to all mechanisms. To prove these formulas, consider three rigid bodies 1, 2 and 3 in relative planar motion as shown in Figure 2.68. We can show that their kinematics are coupled to each other in pairwise manner. As shown in the figure, we define the oriented angles θ_1, θ_2 and θ_3 which satisfy $\theta_1 + \theta_2 = \theta_3$ at all time. Each of these angles defines an angular velocity in agreement with equation (2.3):

$$\boldsymbol{\omega}_{2/1} = \dot{\theta}_1 \hat{\mathbf{k}}, \quad \boldsymbol{\omega}_{3/2} = \dot{\theta}_2 \hat{\mathbf{k}}, \quad \boldsymbol{\omega}_{3/1} = \dot{\theta}_3 \hat{\mathbf{k}}$$

with the choice of unit vector $\hat{\mathbf{k}}$ as shown in Figure 2.68. Since $\dot{\theta}_1 + \dot{\theta}_2 = \dot{\theta}_3$ we arrive at the *angular velocity loop equation*:

$$\boldsymbol{\omega}_{3/2} + \boldsymbol{\omega}_{2/1} = \boldsymbol{\omega}_{3/1}$$

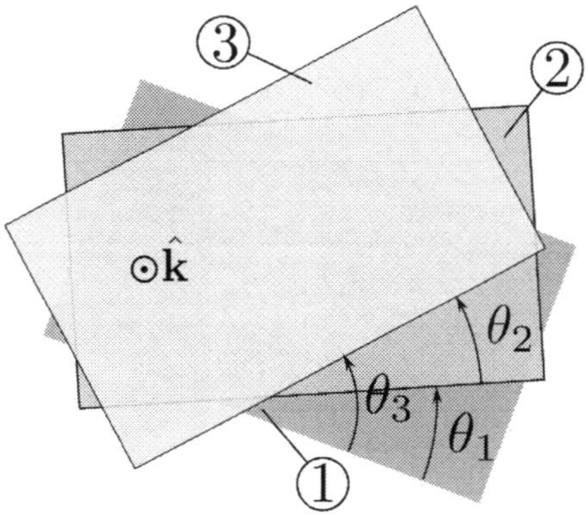

Figure 2.68 – *Three rigid bodies in planar motion.*

Remark 22. Assume that bodies 1 and 3 are rigidly connected so that $\boldsymbol{\omega}_{3/1} = \mathbf{0}$: this gives

$$\boldsymbol{\omega}_{1/2} = -\boldsymbol{\omega}_{2/1}$$

This result is also shown by realizing that angle θ_1 oriented in the opposite direction defines the orientation of body 1 relative to body 2.

Another loop equation can be derived by considering an arbitrary point P in the plane of motion: assume that P is instantaneously attached to body 3 and let us relate its velocity $\mathbf{v}_{P\in 3/1}$ relative to body 1 to its velocity $\mathbf{v}_{P\in 3/2}$ relative to body 2. According to equation (2.21), these two velocities are related through the transport velocity $\mathbf{v}_{P\in 2/1}$ in the following form

$$\mathbf{v}_{P\in 3/1} = \mathbf{v}_{P\in 3/2} + \mathbf{v}_{P\in 2/1}$$

We summarize these results as follows:

Kinematic Loop Equations

Given three rigid bodies 1, 2 and 3 in relative planar motion, the following *kinematic loop equations* can be written

$$\omega_{3/2} + \omega_{2/1} + \omega_{1/3} = \mathbf{0} \tag{2.24}$$

$$\mathbf{v}_{P\in 3/2} + \mathbf{v}_{P\in 2/1} + \mathbf{v}_{P\in 1/3} = \mathbf{0} \tag{2.25}$$

at any time, and for any point P.

Properties

1. In the three velocity terms of (2.25), point P may need to be viewed as instantaneously attached to bodies 1, 2 and 3, that is, $\mathbf{v}_{P\in i/j} \neq \mathbf{v}_{P/j}$.
2. Formulas (2.24) and (2.25) can easily be generalized to a system of N interconnected rigid bodies 1, 2, ..., N in motion relative to a referential 0:

$$\omega_{N/N-1} + \omega_{N-1/N-2} + \cdots + \omega_{1/0} + \omega_{0/N} = \mathbf{0}$$

$$\mathbf{v}_{P\in N/N-1} + \mathbf{v}_{P\in N-1/N-2} + \cdots + \mathbf{v}_{P\in 1/0} + \mathbf{v}_{P\in 0/N} = \mathbf{0}$$

3. Formula (2.25) cannot be extended to accelerations due to the presence of Coriolis terms.

The use of kinematic loop formulas is the basis for a method of kinematic analysis of mechanisms. They provide a way to interconnect the motion of the rigid bodies of a system and hence offers a practical method to find input/output kinematic relationships. This method of analysis is very powerful. It is applicable to all possible types of mechanisms. It is based on the following procedure:

General Procedure: Kinematic Analysis of Mechanisms

1. Given a kinematic diagram of the mechanism, describe the mechanism by sketching a connectivity diagram, that is, a graph of all pairwise joints between the bodies of the mechanism. Carefully identity the nature of all joints connecting pairs (i, j). Here is a graph of a mechanism comprised of 6 rigid bodies $1, 2, \cdots, 6$ in motion in referential 0.

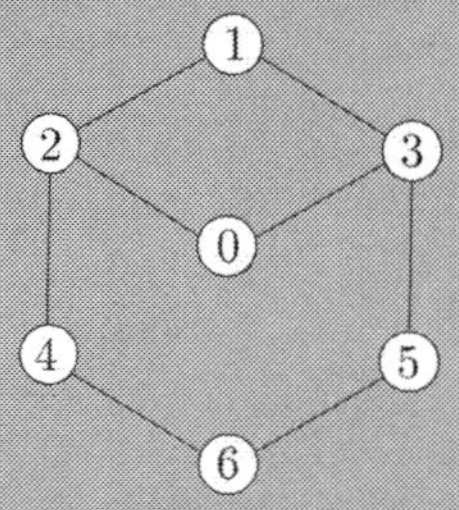

2. Identify all *independent* closed loops from the connectivity diagram found in step 1. In the graph shown above, two independent loops can be found.

3. For each loop found in step 2, derive the kinematic loop equations based on kinematic loop formulas (2.24-2.25): for each closed loop $i-j-k\cdots-n-p-i$, the closure of the loop can expressed by

$$\omega_{i/j}+\omega_{j/k}+\cdots+\omega_{n/p}+\omega_{p/i}=\mathbf{0} \quad (2.26)$$

$$\mathbf{v}_{Q\in i/j}+\mathbf{v}_{Q\in j/k}+\cdots+\mathbf{v}_{Q\in n/p}+\mathbf{v}_{Q\in p/i}=\mathbf{0} \quad (2.27)$$

In these equations, the terms $(\omega_{i/j}, \mathbf{v}_{Q\in i/j})$ are determined by taking into account the nature of the joint connecting the pair (i,j), without taking into account any other connections with bodies of the mechanism. An adequate point Q must be identified. Typically the chosen point plays a key role in the geometry of the loop. For instance, if the joint i/j is a pivot of axis $(A,\hat{\mathbf{k}})$, then $\mathbf{v}_{Q\in i/j}=\omega_{i/j}\hat{\mathbf{k}}\times\mathbf{r}_{AQ}$.
4. Project each loop equation on an adequate set of unit vectors.
5. Solve for all unknowns. In particular, determine the kinematic input/output law which governs the mechanism. If the configuration of the mechanism is valid at all time, the loop equations can be time-differentiated to obtain acceleration information.

The following problems demonstrate this procedure.

EXAMPLE PROBLEMS

Exercise 2.29 ▪ This mechanism is driven by the crank-arm 1 which rotates at constant angular velocity ω_1 in the clockwise direction. The slider 2 is pinned to the crank-arm and is confined to move along the slot of arm 3.
At the instant shown, lines OA and OB are perpendicular and the geometry is defined by the constant length r and h.
At the instant shown, find the angular velocity of arm 3 and the slip velocity $v_{2/3}$ of the slider as a function of ω_1, r and h.
Can the acceleration of arm 3 be found? ▪

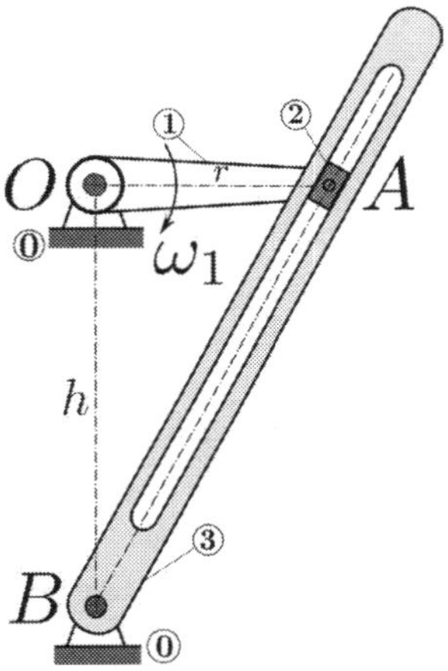

Solution. We cannot analyze this mechanism with a standard vectorial method. This is due to the slider joint between bodies 2 and 3. The mechanism can be viewed as the closed loop $0-1-2-3-0$ to which corresponds a kinematic loop equation. We can write this equation at point A as follows:

$$\mathbf{v}_{A\in 1/0}+\mathbf{v}_{A\in 2/1}+\mathbf{v}_{A\in 3/2}+\mathbf{v}_{A\in 0/3}=\mathbf{0} \quad [1]$$

First, we choose a basis of unit vectors $(\hat{\imath},\hat{\jmath},\hat{\mathbf{k}})$ and we introduce angle θ and unit vector $\hat{\mathbf{u}}$ as shown in Figure 2.69. We then find each velocity term of equation [1] by identifying the nature of each joint at play:
(i) the joint $1-0$ is a pivot of center O leading to:

$$\mathbf{v}_{A\in 1/0} = \boldsymbol{\omega}_{1/0} \times \mathbf{r}_{OA} = \omega_1 \hat{\mathbf{k}} \times r\hat{\imath} = r\omega_1 \hat{\jmath}$$

(ii) the joint $2-1$ is a pivot of center A leading to: $\mathbf{v}_{A\in 2/1} = \mathbf{0}$.
(iii) the joint $3-2$ is a slider along direction $\hat{\mathbf{u}}$ leading to: $\mathbf{v}_{A\in 3/2} = v_{3/2}\hat{\mathbf{u}}$, where $v_{3/2}$ is the slip velocity $v_{2/3}$ of the slider.
(iv) finally, the joint $3-0$ is a pivot of center B:

$$\mathbf{v}_{A\in 3/0} = -\mathbf{v}_{A\in 0/3} = \boldsymbol{\omega}_{3/0} \times \mathbf{r}_{BA} = \omega_3 \hat{\mathbf{k}} \times (-h\hat{\jmath} + r\hat{\imath}) = \omega_3(h\hat{\imath} + r\hat{\jmath})$$

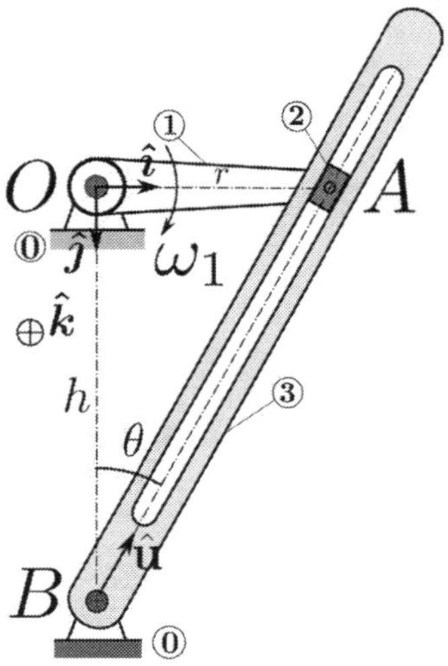

Figure 2.69 – *Problem 2.29*

Now we project equation [1] on $(\hat{\imath}, \hat{\jmath})$ to obtain the following two equations:

$$h\omega_3 = v_{3/2}\sin\theta, \qquad r\omega_3 = -v_{3/2}\cos\theta + r\omega_1$$

which lead to

$$\boxed{\boldsymbol{\omega}_3 = \omega_1 \frac{r^2}{r^2+h^2}\hat{\mathbf{k}}, \qquad v_{3/2} = \omega_1 \frac{rh}{\sqrt{r^2+h^2}}} \qquad [2]$$

Note that the angular velocity loop equation yields $\boldsymbol{\omega}_{2/1} = (\omega_3 - \omega_1)\hat{\mathbf{k}}$. We cannot obtain the angular acceleration $\boldsymbol{\alpha}_3$ from [2] since $\boldsymbol{\omega}_3$ is not valid at all time (and cannot be time-differentiated). Moreover, we cannot write an acceleration loop equation. The solution is to redo this analysis by starting with a general configuration of the mechanism. ■

Exercise 2.30 ▪ In this mechanism, the plunger 1 is in translation at constant speed $v_B = \dot{y}_1$. It sets in motion the arm 3 by the use of roller 2 which is pin-connected at its center A to arm 1. The roller rolls without slipping relative to the plunger at contact point I.
1. In the loop ⓪-①-②-③-⓪, briefly describe the nature of each joint between pairs i/j of rigid bodies.
2. Write the corresponding kinematic loop equation at a point of your choice to find the angular velocities $\boldsymbol{\omega}_{3/0}$, $\boldsymbol{\omega}_{3/2}$ and $\boldsymbol{\omega}_{2/1}$ in terms of $v_B = \dot{y}_1$, angle θ_3 and the geometric parameters.
3. Find the angular acceleration $\boldsymbol{\alpha}_{3/0}$ of arm 3. ■

SOLUTION. **a.** This mechanism is a single-loop mechanism. In the loop ⓪-①-②-③-⓪,

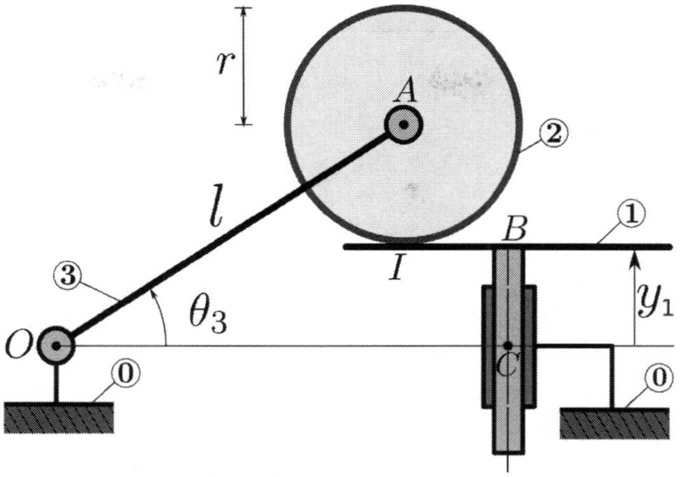

(i) joint 3/0 is a pivot centered at O,
(ii) joint 3/2 is a pivot centered at A,
(iii) joint 2/1 is no-slip point-contact at I,
(iv) joint 1/0 is slider along $(C, \hat{\mathbf{y}}_0)$.
b. We write the kinematic loop equation (2.25) at point I (we can also choose point A):

$$\mathbf{v}_{I\in 3/2} + \mathbf{v}_{I\in 2/1} + \mathbf{v}_{I\in 1/0} + \mathbf{v}_{I\in 0/3} = \mathbf{0} \qquad [1]$$

with
$\mathbf{v}_{I\in 0/3} = -\mathbf{v}_{I\in 3/0} = -\omega_{3/0}\hat{\mathbf{z}}_0 \times \mathbf{r}_{OI} = -\omega_{3/0}\hat{\mathbf{z}}_0 \times (l\hat{\mathbf{x}}_3 - r\hat{\mathbf{y}}_0)$ (pivot about O)
$\mathbf{v}_{I\in 3/2} = \omega_{3/2}\hat{\mathbf{z}}_0 \times \mathbf{r}_{AI} = r\omega_{3/2}\hat{\mathbf{x}}_0$ (pivot about A)
$\mathbf{v}_{I\in 2/1} = \mathbf{0}$ (contact at I without slip),
$\mathbf{v}_{I\in 1/0} = v_B\hat{\mathbf{y}}_0$ (slider along $\hat{\mathbf{y}}_0$).
Equation [1] becomes

$$\omega_{3/0}(l\hat{\mathbf{y}}_3 + r\hat{\mathbf{x}}_0) = r\omega_{3/2}\hat{\mathbf{x}}_0 + v_B\hat{\mathbf{y}}_0 \qquad [2]$$

We project [2] on $(\hat{\mathbf{x}}_0, \hat{\mathbf{y}}_0)$ to obtain

$$(-l\sin\theta_3 + r)\omega_{3/0} = r\omega_{3/2}, \qquad l\omega_{3/0}\cos\theta_3 = v_B \qquad [3-4]$$

Next we write kinematic loop equation (2.24): $\boldsymbol{\omega}_{3/2} + \boldsymbol{\omega}_{2/1} + \boldsymbol{\omega}_{1/0} + \boldsymbol{\omega}_{0/3} = \mathbf{0}$:

$$\omega_{3/2} + \omega_{2/1} = \omega_{3/0} \qquad [5]$$

using the fact that $\boldsymbol{\omega}_{1/0} = \mathbf{0}$. Equations [3-5] solve for the 3 unknowns $(\omega_{3/0}, \omega_{3/2}, \omega_{2/1})$:

$$\boxed{\omega_{3/0} = \frac{v_B}{l\cos\theta_3}, \quad \omega_{3/2} = \frac{v_B}{r}\frac{r - l\sin\theta_3}{l\cos\theta_3}, \quad \omega_{2/1} = \frac{v_B}{r}\tan\theta_3} \qquad [6-8]$$

Note that $\omega_{2/0} = \omega_{2/1}$.
3. We can find the angular acceleration $\boldsymbol{\alpha}_{3/0}$ by taking the time-derivative of [6] since [6] holds at all time

$$\boxed{\boldsymbol{\alpha}_{3/0} = \frac{v_B}{l}\frac{\dot{\theta}_3\sin\theta_3}{\cos^2\theta_3}\hat{\mathbf{z}}_0 = \frac{v_B^2}{l^2}\frac{\sin\theta_3}{\cos^3\theta_3}\hat{\mathbf{z}}_0}$$

where we used $\dot{\theta}_3 = \omega_{3/0}$. ■

Exercise 2.31 ■■ The Geneva drive shown in Figure 2.70 is a mechanism which transforms the continuous rotation of drive wheel 1 at constant angular velocity ω_1 into the

intermittent rotation of a slotted wheel 2. Whenever pin A of 1 is engaged into one of the four slots of 2, body 1 drives the rotation of body 2 by a quarter turn. The drive wheel 1 is also configured to block the motion of 2 whenever pin A is freed from the slots. Denote $|OA| = R$ and $|OB| = b$ and assume $R/b < 1$.
a. Use the kinematic loop equation at point A to find the angular velocity ω_2 as a function of ω_1 and position θ_1 of A.
b. Study the angular position of 2 as a function of θ_1 in the interval $-\pi \leq \theta_1 \leq \pi$. Conclude that the ratio b/R must take the value $\sqrt{2}$. Plot the angular position of 2 during a complete revolution of 1. ■

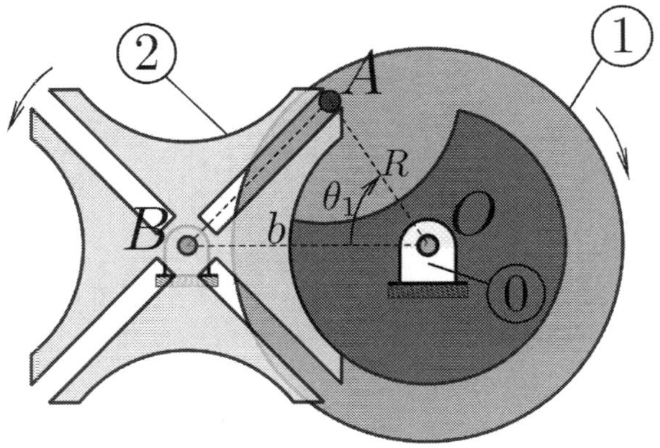

Figure 2.70 – *Problem 2.31*

SOLUTION. **a.** First we adopt the fixed unit vectors $(\hat{\imath}, \hat{\jmath}, \hat{\mathbf{k}})$ and introduce angle θ_2 as shown in Figure 2.71 to define the orientation of body 2. The mechanism is described by a single closed loop 0-1-2-0 with the following joints:
- joint 0/1 is a pivot of axis $(O, \hat{\mathbf{k}})$,
- joint 1/2 is a slider along line AB,
- joint 2/0 is a pivot of axis $(B, \hat{\mathbf{k}})$.
In order to couple the motion of 2 to that of 1, we write the kinematic loop equation

$$\mathbf{v}_{A\in 2/0} = \mathbf{v}_{A\in 2/1} + \mathbf{v}_{A\in 1/0} \qquad [1]$$

at point A. We need to find the expression of each term of [1]. First we find $\mathbf{v}_{A\in 1/0}$ by taking into account that joint 0/1is a pivot:

$$\mathbf{v}_{A\in 1/0} = \boldsymbol{\omega}_1 \times \mathbf{r}_{OA} = -\dot{\theta}_1 \hat{\mathbf{k}} \times R(-\cos\theta_1 \hat{\imath} + \sin\theta_1 \hat{\jmath}) = R\dot{\theta}_1(\cos\theta_1 \hat{\jmath} + \sin\theta_1 \hat{\imath})$$

with $R = |OA|$. Next velocity $\mathbf{v}_{A\in 2/0}$ is found as the velocity of a point of 2 in rotation about B with body 2:

$$\mathbf{v}_{A\in 2/0} = \boldsymbol{\omega}_2 \times \mathbf{r}_{BA} = \dot{\theta}_2 \hat{\mathbf{k}} \times \rho(\cos\theta_2 \hat{\imath} + \sin\theta_2 \hat{\jmath}) = \rho\dot{\theta}_2(\cos\theta_2 \hat{\jmath} - \sin\theta_2 \hat{\imath})$$

with $\rho = |AB|$.
Finally velocity $\mathbf{v}_{A\in 1/2} = -\mathbf{v}_{A\in 2/1}$ is found by realizing that pin $A \in 1$ has a rectilinear motion relative to 2 along line AB (joint 1/2 is a slider):

$$\mathbf{v}_{A\in 1/2} = v_{1/2}(\cos\theta_2 \hat{\imath} + \sin\theta_2 \hat{\jmath})$$

Now equation [1] becomes

$$\rho\dot{\theta}_2(\cos\theta_2\hat{\jmath} - \sin\theta_2\hat{\imath}) = R\dot{\theta}_1(\cos\theta_1\hat{\jmath} + \sin\theta_1\hat{\imath}) - v_{1/2}(\cos\theta_2\hat{\imath} + \sin\theta_2\hat{\jmath})$$

We arrive at two equations

$$-\rho\dot{\theta}_2 \sin\theta_2 = R\dot{\theta}_1 \sin\theta_1 - v_{1/2}\cos\theta_2 \quad [2]$$

$$\rho\dot{\theta}_2 \cos\theta_2 = R\dot{\theta}_1 \cos\theta_1 - v_{1/2l}\sin\theta_2 \quad [3]$$

which can be solved to give $(\dot{\theta}_2, v_{1/2})$:

$$\dot{\theta}_2 = \dot{\theta}_1 \frac{R}{\rho}\cos(\theta_1 + \theta_2)$$

$$v_{1/2} = R\dot{\theta}_1 \sin(\theta_1 + \theta_2)$$

We need to find ρ (a variable quantity): in triangle OAB we find two geometric equations

$$\rho\sin\theta_2 = R\sin\theta_1, \qquad b\sin\theta_2 = R\sin(\theta_1+\theta_2) \quad [4-5]$$

Using the expression of ρ we find the angular velocity of slotted wheel 2

$$\boxed{\boldsymbol{\omega}_2 = \dot{\theta}_2\hat{\mathbf{k}}, \quad \dot{\theta}_2 = \dot{\theta}_1\frac{\sin\theta_2}{\sin\theta_1}\cos(\theta_1+\theta_2)} \quad [6]$$

where angle θ_2 is found by using [5]. Finally the velocity $\mathbf{v}_{A\in 1/2}$ is given by

$$\boxed{\mathbf{v}_{A\in 1/2} = v_{1/2}\hat{\mathbf{u}}_2 = R\dot{\theta}_1\sin(\theta_1+\theta_2)\hat{\mathbf{u}}_2, \quad \hat{\mathbf{u}}_2 = (\cos\theta_2\hat{\imath} + \sin\theta_2\hat{\jmath})} \quad [7]$$

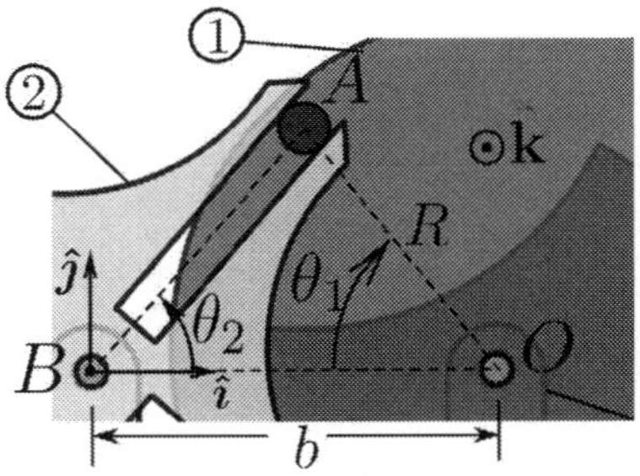

Figure 2.71 – *Problem 2.31 (Solution).*

b. We can express angle θ_2 in terms of θ_1 by using [5]:

$$\theta_2 = \tan^{-1}\left(\frac{\sin\theta_1}{k - \cos\theta_1}\right)$$

by denoting $k = b/R$. Since $k > 1$ is assumed, θ_2 is defined for all angles in the interval $-\pi \le \theta_1 \le \pi$. To find the range of values taken by θ_2 we find $d(\tan\theta_2)/d\theta_1$:

$$\frac{d\tan\theta_2}{d\theta_1} = \frac{k\cos\theta_1 - 1}{(k-\cos\theta_1)^2}$$

We see that θ_2 takes its extremal values for the angle θ_1 satisfying $\cos\theta_1 = 1/k$ and the corresponding position of 2 is given by $\tan\theta_2 = 1/\sqrt{1-k^2}$. A quarter rotation of 2 implies that angle θ_2 must vary in the range $-\pi/4 \le \theta_2 \le \pi/4$: this leads to $1/\sqrt{1-k^2} = 1$ or $\boxed{k = \sqrt{2}}$. For $k = \sqrt{2}$, the clockwise rotation of 1 in the interval $-\pi/4 \le \theta_1 \le \pi/4$ corresponds to a quarter turn of 2 in the counterclockwise sense.

In the intervals $-\pi < \theta_1 < -\pi/4$ and $\pi/4 < \theta_1 < \pi$, the rotation of 2 is blocked until pin A re-engages into the next slot.

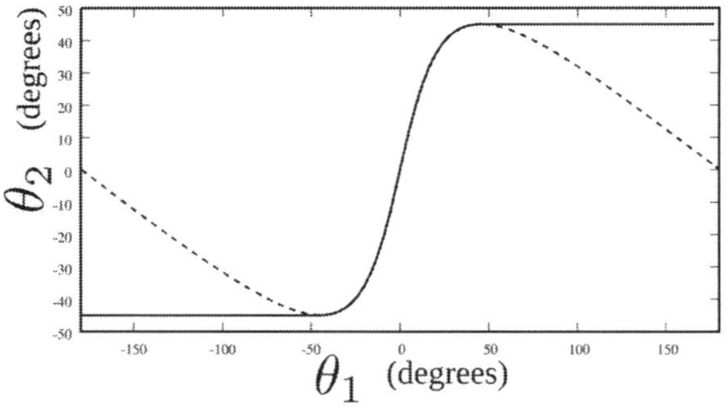

Figure 2.72 – *Problem 2.31 (Solution). Angle $\theta_2 = \tan^{-1}(\sin\theta_1/(k - \cos\theta_1))$ versus θ_1 for $b/R = \sqrt{2}$. The portion represented by the dashed line must be replaced by $\theta_2 = \pm\pi/4$ corresponding to the stationary phase of the slotted wheel.*

■

Exercise 2.32 ■■ In the epicyclic gear train of Figure 2.73, sun gear 1 is coupled to the outer ring gear 3 by the double satellite gear 2 which is mounted on the carrier 4. The latter is coaxial with the sun and ring gears. The geometry of the gear train is defined by the parameters r_1, r_2', r_2'' and r_3. **a.** After identifying two independent closed loops, write the corresponding kinematic loop equations.
b. From the equations derived in a) show that the angular velocities $\omega_{1/0}$, $\omega_{3/0}$ and $\omega_{4/0}$ satisfy the equation

$$\omega_{1/0} - \lambda\omega_{3/0} + (\lambda - 1)\omega_{4/0} = 0$$

where λ is a constant to be found. ■

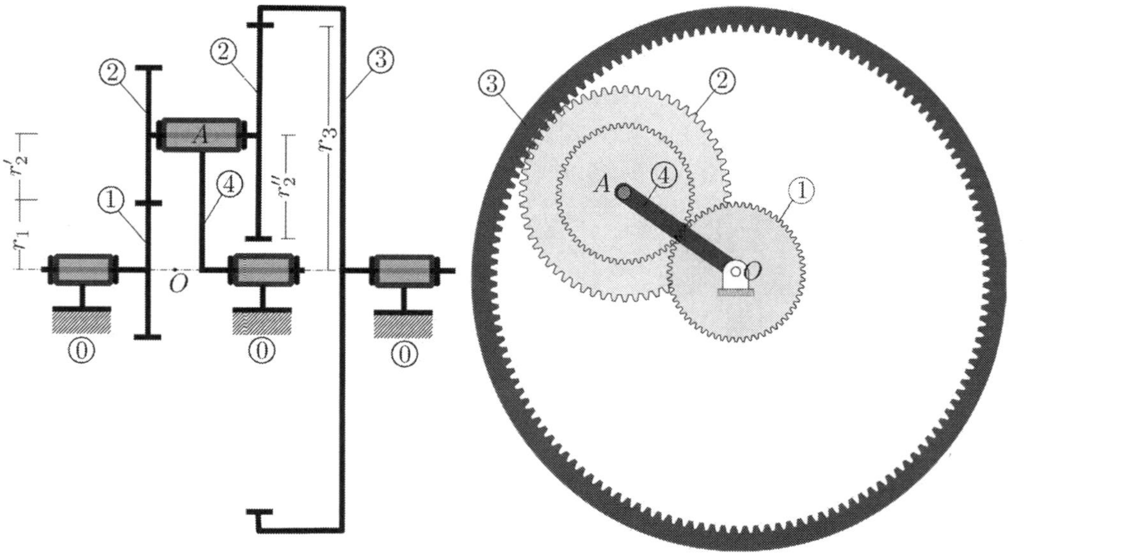

Figure 2.73 – *Problem 2.32*

SOLUTION. **a.** We re-sketch the mechanism in Figure 2.74 and define the basis of unit vectors $(\hat{\mathbf{u}}, \hat{\mathbf{v}}, \hat{\mathbf{k}})$ (with $\hat{\mathbf{k}}$ along the axis of rotation). The point I is the contact point between gears 1 and 2. J is the contact point between gears 2 and 3.

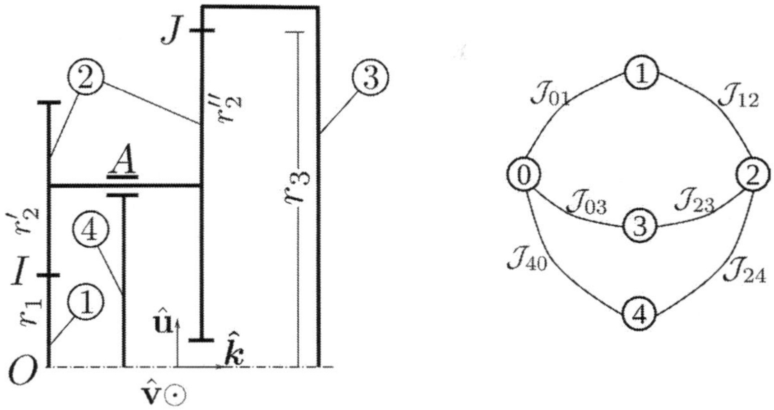

Figure 2.74 – *Problem 2.32 (Solution). In connectivity diagram (right), 2 closed loops are identified.*

Then we sketch in Figure 2.74 (right) a graph displaying the interconnections of the mechanism where the joints between the pairs of interconnected bodies are as follows:
-$\mathcal{J}_{01}$ is a pivot of axis $(O, \hat{\mathbf{k}})$
-$\mathcal{J}_{12}$ is point contact at I (without slip)
-$\mathcal{J}_{24}$ is a pivot of axis $(A, \hat{\mathbf{k}})$
-$\mathcal{J}_{23}$ is point contact at J (without slip)
-$\mathcal{J}_{03}$ is a pivot of axis $(O, \hat{\mathbf{k}})$
-$\mathcal{J}_{04}$ is a pivot of axis $(O, \hat{\mathbf{k}})$
Two closed loops can be chosen: 0-1-2-4-0 and 0-3-2-4-0. Then we can write the corresponding kinematic loop equations:

$$\mathbf{v}_{P\in 1/0} + \mathbf{v}_{P\in 2/1} + \mathbf{v}_{P\in 4/2} + \mathbf{v}_{P\in 0/4} = \mathbf{0} \quad [1]$$

$$\mathbf{v}_{P\in 3/0} + \mathbf{v}_{P\in 2/3} + \mathbf{v}_{P\in 4/2} + \mathbf{v}_{P\in 0/4} = \mathbf{0} \quad [2]$$

where point P needs to be identified in [1] and [2]. For equation [1] we choose point I due to the no-slip condition: $\mathbf{v}_{I\in 2/1} = \mathbf{0}$. We can find the expressions of the 3 remaining terms:

$$\mathbf{v}_{I\in 1/0} = \boldsymbol{\omega}_{1/0} \times \mathbf{r}_{OI} = \omega_{1/0}\hat{\mathbf{k}} \times r_1\hat{\mathbf{u}} = r_1\omega_{1/0}\hat{\mathbf{v}}$$

$$\mathbf{v}_{I\in 4/2} = \boldsymbol{\omega}_{4/2} \times \mathbf{r}_{AI} = -r_2'\omega_{4/2}\hat{\mathbf{v}}$$

$$\mathbf{v}_{I\in 0/4} = -\mathbf{v}_{I\in 4/0} = -\boldsymbol{\omega}_{4/0} \times \mathbf{r}_{OI} = -r_1\omega_4\hat{\mathbf{v}}$$

Then equation [1] becomes

$$r_1\omega_{1/0} - r_2'\omega_{4/2} - r_1\omega_{4/0} = 0 \quad [1']$$

We proceed the same way for the second loop: we choose point J and take into account $\mathbf{v}_{J\in 2/3} = \mathbf{0}$. We then find

$$\mathbf{v}_{J\in 3/0} = r_3\omega_{3/0}\hat{\mathbf{v}}, \quad \mathbf{v}_{J\in 4/2} = r_2''\omega_{4/2}\hat{\mathbf{v}}, \quad \mathbf{v}_{J\in 0/4} = -r_3\omega_{4/0}\hat{\mathbf{v}}$$

Then equation [2] becomes

$$r_3\omega_{3/0} + r_2''\omega_{4/2} - r_3\omega_{4/0} = 0 \qquad [2']$$

b. We have obtained 2 equations governing 4 unknowns: $\omega_{1/0}$, $\omega_{4/2}$, $\omega_{3/0}$ and $\omega_{4/0}$. If we eliminate $\omega_{2/4}$ we obtain an equation governing $\omega_{1/0}$, $\omega_{3/0}$ and $\omega_{4/0}$. We multiply [1′] by r_2'', [2′] by r_2' then add $r_2'\times$[2′] and $r_2''\times$[1′] to obtain:

$$r_1r_2''\omega_{1/0} - r_2'r_2''\omega_{4/2} - r_1r_2''\omega_{4/0} + (r_3r_2'\omega_{3/0} + r_2''r_2'\omega_{4/2} - r_3r_2'\omega_{4/0}) = 0$$

leading to

$$r_1r_2''\omega_{1/0} + (-r_3r_2' - r_1r_2'')\omega_{4/0} + r_3r_2'\omega_{3/0} = 0$$

We then divide the last equation by r_1r_2'' and obtain

$$\boxed{\omega_{1/0} + (-\frac{r_3r_2'}{r_1r_2''} - 1)\omega_{4/0} + \frac{r_3r_2'}{r_1r_2''}\omega_{3/0} = 0}$$

We identify λ as $\boxed{\lambda = -r_3r_2'/r_1r_2''}$. Note that λ represents the ratio $\omega_{1/0}/\omega_{3/0}$ when the carrier 4 is stationary. ■

Exercise 2.33 ▪▪ The figure shows a schematic model of a hydraulic pump whose working principle is to transform the rotational motion of an eccentric 1 into the linear reciprocating motion of a piston 2. The eccentric has center C and radius R and rotates about O at constant angular velocity ω_1. The position of C relative to O is defined by the eccentricity e. The contact point A of the piston with the eccentric is defined by $\mathbf{r}_{OA} = \lambda(t)\hat{\mathbf{x}}_0 + y_2(t)\hat{\mathbf{y}}_0$. A spring (not shown) guarantees that contact at A is achieved at all time.
a. Method 1: Find a relationship between the input variable $\theta(t)$ and the output variable $y_2(t)$ (and the parameters e and R). Deduce the velocity of the piston as a function of time.
b. Method 2: Write the loop equation $\mathbf{v}_{A\in 2/0} = \mathbf{v}_{A\in 2/1} + \mathbf{v}_{A\in 1/0}$ at contact point A, recover the relationship between y_2 and θ. Find the slip velocity at A. ■

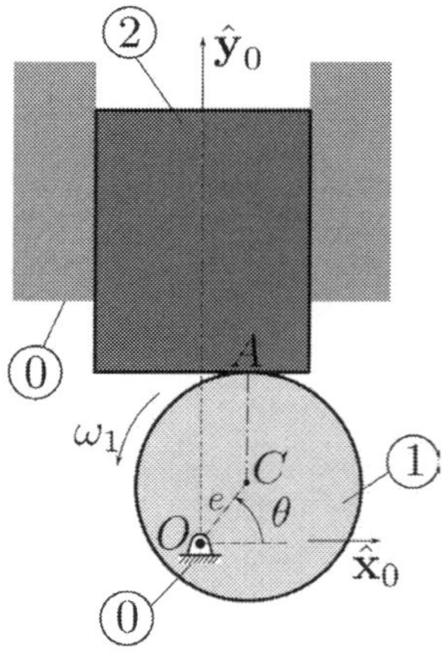

Solution **a.** We can write the vector loop equation $\mathbf{r}_{OA} + \mathbf{r}_{AC} + \mathbf{r}_{CO} = \mathbf{0}$ between points O, A and C on $(\hat{\mathbf{x}}_0, \hat{\mathbf{y}}_0)$ in terms of y_2, θ and parameters e, R:

$$(\lambda\hat{\mathbf{x}}_0 + y_2\hat{\mathbf{y}}_0) - R\hat{\mathbf{y}}_0 - e(\cos\theta\hat{\mathbf{x}}_0 + \sin\theta\hat{\mathbf{y}}_0) = \mathbf{0}$$

On unit vector $\hat{\mathbf{y}}_0$ we obtain the input-output relationship

$$\boxed{y_2 = R + e\sin\theta, \qquad \theta = \omega_1 t} \qquad [1]$$

The velocity of piston 2 (which is in translation) is then given by $\mathbf{v}_{A\in 2/0} = \dot{y}_2\hat{\mathbf{y}}_0$

$$\boxed{\mathbf{v}_{A\in 2/0} = e\omega_1\cos(\omega_1 t)} \qquad [2]$$

Remark. We can find the flow rate Q of the pump from the expression of $\dot{y}_2$: $Q = \mathcal{A}\dot{y}_2 = e\mathcal{A}\omega_1 \cos(\theta)$, for $\dot{y}_2 > 0$ (during the upward motion of the piston), where $\mathcal{A}$ is the piston's cross-sectional area.

b. In order to write the kinematic loop equation $\mathbf{v}_{A\in 2/0} = \mathbf{v}_{A\in 2/1} + \mathbf{v}_{A\in 1/0}$ we need to determine the expression of each velocity vector $\mathbf{v}_{A\in 2/0}$, $\mathbf{v}_{A\in 2/1}$ and $\mathbf{v}_{A\in 1/0}$. The piston is in translation along axis $(O, \hat{\mathbf{y}}_2)$: this implies that

$$\mathbf{v}_{A\in 2/0} = \dot{y}_2 \hat{\mathbf{y}}_0$$

The eccentric is in rotation about O, thus leading to

$$\mathbf{v}_{A\in 1/0} = \dot{\theta}\hat{\mathbf{z}}_0 \times \mathbf{r}_{OA} = \dot{\theta}\hat{\mathbf{z}}_0 \times (e\cos\theta\hat{\mathbf{x}}_0 + (R + e\sin\theta)\hat{\mathbf{y}}_0) = e\dot{\theta}\cos\theta\hat{\mathbf{y}}_0 - (R + e\sin\theta)\hat{\mathbf{x}}_0$$

Finally, the slip velocity of the piston relative to the eccentric must not have a component along $\hat{\mathbf{y}}_0$ (otherwise contact is lost):

$$\mathbf{v}_{A\in 2/1} = v_{2/1}\hat{\mathbf{x}}_0$$

Now the loop equation can be written as

$$\dot{y}_2\hat{\mathbf{y}}_0 = e\dot{\theta}\cos\theta\hat{\mathbf{y}}_0 - (R + e\dot{\theta}\sin\theta)\hat{\mathbf{x}}_0 + v_{2/1}\hat{\mathbf{x}}_0$$

On unit vector $\hat{\mathbf{y}}_0$ we recover equation [2]:

$$\dot{y}_2 = e\dot{\theta}\cos\theta$$

which can be integrated to give

$$y_2 = e\sin\theta + \text{constant}$$

The constant is found by setting $y_2 = R$ at the value $\theta = 0$ and we recover equation [1]. The component on $\hat{\mathbf{x}}_0$ gives the slip velocity $\boxed{v_{2/1} = (R + e\dot{\theta}\sin\theta)}$. ■

Exercise 2.34 ■■■ **a.** (Kennedy's theorem) Consider three rigid bodies 1, 2 and 3 in relative planar motion as shown in Figure 2.75 (left). Denote by I_{12}, I_{23} and I_{13} the three corresponding instantaneous centers of rotation ($I_{kl} = I_{lk}$ is the IC of body k in motion relative to body l). Show that I_{12}, I_{23} and I_{13} are aligned.

b. Find the location of the six instantaneous centers of rotation I_{kl} for the four-bar mechanism of Figure 2.75 (right). From the knowledge of these centers determine the angular velocities of links 2 and 3 in terms of the angular velocity ω_1 of link 1 and angles $(\theta_1, \theta_2, \theta_3)$. ■

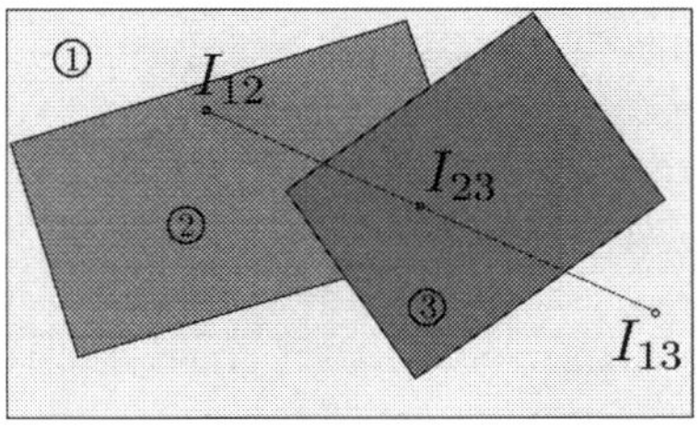

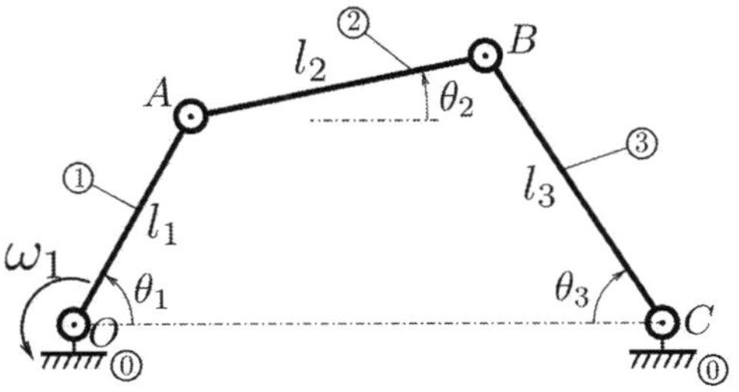

Figure 2.75 – *Problem 2.34*

SOLUTION. **a.** If we consider the closed loop 1-2-3-1, then for any point P lying in the plane of motion we can write the kinematic loop equation

$$\mathbf{v}_{P\in 3/2} + \mathbf{v}_{P\in 2/1} + \mathbf{v}_{P\in 1/3} = \mathbf{0} \qquad [1]$$

Recall that instantaneous center I_{ij} is such that $\mathbf{v}_{P\in i/j} = \boldsymbol{\omega}_{i/j} \times \mathbf{r}_{I_{ij}P}$. Using the 3 instantaneous centers I_{32}, I_{21} and I_{13}, equation [1] becomes

$$\hat{\mathbf{k}} \times (\omega_{3/2}\mathbf{r}_{I_{32}P} + \omega_{2/1}\mathbf{r}_{I_{21}P} + \omega_{1/3}\mathbf{r}_{I_{13}P}) = \mathbf{0}$$

where $\hat{\mathbf{k}}$ is a unit vector normal to the plane of motion. Since the vector $\omega_{3/2}\mathbf{r}_{I_{32}P} + \omega_{2/1}\mathbf{r}_{I_{21}P} + \omega_{1/3}\mathbf{r}_{I_{13}P}$ is not colinear to $\hat{\mathbf{k}}$, it is then necessarily zero:

$$\omega_{3/2}\mathbf{r}_{I_{32}P} + \omega_{2/1}\mathbf{r}_{I_{21}P} + \omega_{1/3}\mathbf{r}_{I_{13}P} = \mathbf{0} \qquad [2]$$

We may choose $P = I_{13}$ in equation [2] and we obtain

$$\boxed{\omega_{3/2}\mathbf{r}_{I_{32}I_{13}} + \omega_{2/1}\mathbf{r}_{I_{21}I_{13}} = \mathbf{0}} \qquad [2]$$

This shows that the three points I_{12}, I_{23} and I_{13} are colinear.
b. First we identify the centers I_{01}, I_{12}, I_{23} and I_{03} to be points O, A, B and C, respectively. To find the positions of the remaining two centers I_{13} and I_{02} we use Kennedy's theorem: point I_{13} must fall on line $I_{12}I_{23}$ but also on line $I_{01}I_{03}$, and hence is the intersection point of these two lines. Similarly, point I_{02} is the intersection point of line $I_{01}I_{12}$ and line $I_{03}I_{23}$. See Figure 2.76.

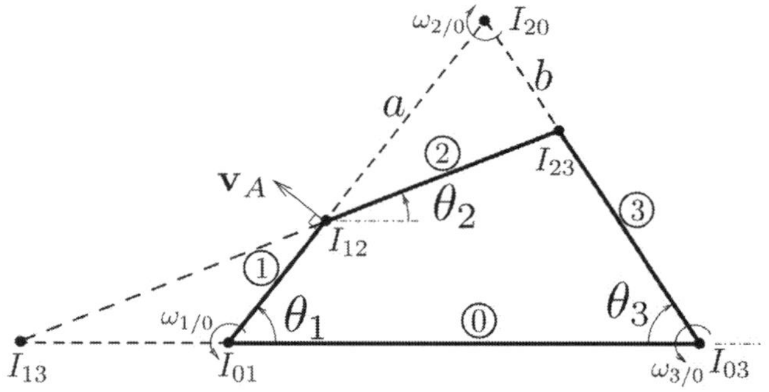

Figure 2.76 – *Problem 2.34 (Solution)*

We can relate $\omega_2 = \omega_{2/0}$ to $\omega_1 = \omega_{1/0}$ by determining the magnitude of velocity of point A first as a point in rotation about O and then as a point in rotation about I_{02}:

$$v_{A\in 1/0} = |OA|\,\omega_1 = v_{A\in 2/0} = |AI_{02}|\,\omega_2$$

with $|OA| = l_1$. The distance $a = |AI_{02}|$ is found by consideration of triangle ABI_{02} we find:

$$a = l_2 \frac{\sin(\theta_3 + \theta_2)}{\sin(\theta_3 + \theta_1)}$$

This gives

$$\boxed{\omega_2 = \omega_1 \frac{l_1 \sin(\theta_3 + \theta_1)}{l_2 \sin(\theta_3 + \theta_2)} \;\circlearrowleft} \qquad [3]$$

Similarly we can relate ω_3 to ω_2 by consideration of point B:

$$v_{B\in 3/0} = l_3\,\omega_3 = v_{B\in 2/0} = |BI_{02}|\,\omega_2$$

with

$$|BI|_{02} = b = l_2 \frac{\sin(\theta_1 - \theta_2)}{\sin(\theta_3 + \theta_1)}$$

This gives, after using equation [3]

$$\boxed{\omega_3 = \omega_1 \frac{l_1}{l_3} \frac{\sin(\theta_1 - \theta_2)}{\sin(\theta_3 + \theta_2)} \quad \circlearrowleft} \qquad [4]$$

■

2.11 Instantaneous Center of Acceleration

RECALL that any rigid body 1 in planar motion relative to a referential 0 admits an instantaneous center of rotation I, that is, such that $\mathbf{v}_P = \boldsymbol{\omega} \times \mathbf{r}_{IP}$ for any point P and at any given time (as long as $\boldsymbol{\omega} \equiv \boldsymbol{\omega}_{1/0} \neq \mathbf{0}$ at that time). Also recall that point I describes two curves, the *fixed centrode* relative to 0 and the *moving centrode* $\mathcal{C}_1$ relative to 1. Curve $\mathcal{C}_1$ rolls without slipping on curve $\mathcal{C}_0$. See Figure 2.77 (left).

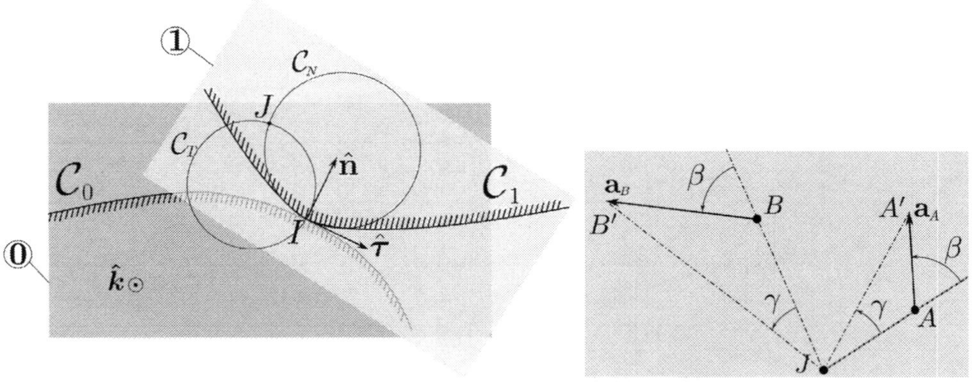

Figure 2.77 – *Center of acceleration J of body 1 relative to referential 0 (left). Graphical construction of $\mathbf{a}_B$ from $\mathbf{a}_A$: The triangles JAA' and JBB' are directly similar (right).*

Here we wish to show that the motion 1/0 can also be characterized in terms of an *instantaneous center of acceleration.* First we can express the acceleration of any point P attached to body 1 in the following way

$$\mathbf{a}_P = \frac{d}{dt}(\boldsymbol{\omega} \times \mathbf{r}_{IP}) = \boldsymbol{\alpha} \times \mathbf{r}_{IP} + \boldsymbol{\omega} \times (\mathbf{v}_P - \mathbf{v}_I) \qquad (2.28)$$

In (2.28), $\boldsymbol{\omega}$ and $\boldsymbol{\alpha}$ represent the angular velocity and acceleration of body 1, and $\mathbf{v}_I$ is the velocity of I along the fixed centrode $\mathcal{C}_0$. Do not confuse $\mathbf{v}_I \equiv \mathbf{v}_{I/0} = \mathbf{v}_{I/1}$ with $\mathbf{v}_{I\in 1/0} = \mathbf{0}$. Recall that $\mathbf{v}_I$ is necessarily tangent to $\mathcal{C}_0$ (and $\mathcal{C}_1$) at contact point I. If we define the unit tangent vector $\hat{\boldsymbol{\tau}}$ and unit normal vector $\hat{\mathbf{n}}$ at I, we can then write $\mathbf{v}_I = v_I \hat{\boldsymbol{\tau}}$. The exact sense of unit vectors $\hat{\boldsymbol{\tau}}$ and $\hat{\mathbf{n}}$ need not be known. With $\hat{\mathbf{k}} = \hat{\boldsymbol{\tau}} \times \hat{\mathbf{n}}$, we write the angular velocity $\boldsymbol{\omega} = \omega \hat{\mathbf{k}}$ and the angular acceleration $\boldsymbol{\alpha} = \alpha \hat{\mathbf{k}}$. We can now express $\mathbf{a}_P$ in the form

$$\mathbf{a}_P = \alpha \hat{\mathbf{k}} \times \mathbf{r}_{IP} - \omega^2 \mathbf{r}_{IP} + \omega v_I \hat{\mathbf{n}} \qquad (2.29)$$

where we have used $\boldsymbol{\omega} \times \mathbf{v}_P = \boldsymbol{\omega} \times (\boldsymbol{\omega} \times \mathbf{r}_{IP}) = -\omega^2 \mathbf{r}_{IP}$.

First, we ask if points of body 1 can be found with *vanishing normal acceleration*: since the tangent direction associated with the path of point P is normal to line IP, such points must satisfy $\mathbf{a}_P \cdot \mathbf{r}_{IP} = 0$. Using (2.29), we obtain

$$-\omega^2 \mathbf{r}_{IP}^2 + \omega v_I \hat{\mathbf{n}} \cdot \mathbf{r}_{IP} = 0$$

Let us define the position of P in terms of its coordinates relative to axes $(I, \hat{\boldsymbol{\tau}})$ and $(I, \hat{\mathbf{n}})$, that is, define $\mathbf{r}_{IP} = x\hat{\boldsymbol{\tau}} + y\hat{\mathbf{n}}$: then the set of points, denoted $\mathcal{C}_N$, with vanishing normal acceleration is defined by

$$(x^2 + y^2) - \frac{v_I}{\omega} y = 0 \tag{2.30}$$

Therefore, set $\mathcal{C}_N$ is the circle centered at $(0, \frac{v_I}{2\omega})$ and of radius $r_N = \frac{v_I}{2\omega}$.
Similarly, we can determine the set $\mathcal{C}_T$ of points of body 1 with *vanishing tangential acceleration*: now we set $\mathbf{a}_P \times \mathbf{r}_{IP} = \mathbf{0}$. Using (2.29) we obtain

$$(\alpha \hat{\mathbf{k}} \times \mathbf{r}_{IP}) \times \mathbf{r}_{IP} + \omega v_I \hat{\mathbf{n}} \times \mathbf{r}_{IP} = 0$$

or in terms of coordinates (x, y):

$$(x^2 + y^2) + \frac{\omega v_I}{\alpha} x = 0 \tag{2.31}$$

The set $\mathcal{C}_T$ is the circle centered at $(-\frac{\omega v_I}{2\alpha}, 0)$ and of radius $r_T = \frac{\omega v_I}{2\alpha}$.
If there exists a point J satisfying $\mathbf{v}_{J\in 1/0} = 0$ it must necessarily lie at the intersection of circles $\mathcal{C}_N$ and $\mathcal{C}_T$: ruling out the solution $x = y = 0$ (since equation (2.29) gives $\mathbf{a}_{I\in 1/0} = \omega v_I \hat{\mathbf{n}} \neq \mathbf{0}$), its coordinates (x_J, y_J) must satisfy (2.30) and (2.31) leading to $y = -\frac{\omega^2}{\alpha} x$ and the coordinates

$$x_J = -v_I \frac{\omega\alpha}{\omega^4 + \alpha^2}, \qquad y_J = v_I \frac{\omega^3}{\omega^4 + \alpha^2} \tag{2.32}$$

INSTANTANEOUS CENTER OF ACCELERATION

At any instant, the motion of body 1 relative to referential 0 can be characterized by an instantaneous center of acceleration J whose position relative to instantaneous center of rotation I is given by (2.32). Then, the acceleration of any point P attached to 1 is given by

$$\mathbf{a}_P = \boldsymbol{\alpha} \times \mathbf{r}_{JP} - \omega^2 \mathbf{r}_{JP} \tag{2.33}$$

Body 1 behaves as if in rotation about J as far as accelerations are concerned

Once its location is known, the instantaneous center of acceleration J can be used for a graphical construction of accelerations as shown in Figure 2.77 (right). If the acceleration of a point A is known, then the acceleration of any other point B can be found. As shown in section 2.3 for rotational motions, the triangles JAA' and JBB' are *directly similar*. All corresponding angles in triangles JAA' and JBB' are equal and described in the same rotational sense.

PROJECT 3: A Study of Chebyshev Linkage

A mechanical linkage is assembled by connecting three bars 1, 2 and 3 at revolute joints B and D. The linkage is then connected to the fixed referential 0 at revolute joints A and C. We denote the lengths $|AC| = L_0$, $|AB| = L_1$, $|BD| = L_2$ and $|CD| = L_3$. The orientations of the bars are defined by the angles θ_1, θ_2 and θ_3 relative to the direction of fixed line AC. Bar 1 is driven in a specified way, that is, angle θ_1 is a known function of time.

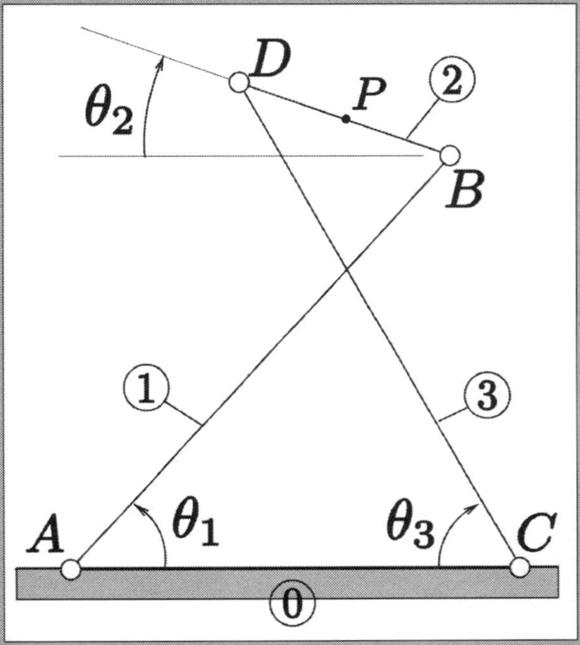

1. By the writing the vector loop equation $\mathbf{r}_{AC} = \mathbf{r}_{AB} + \mathbf{r}_{BD} + \mathbf{r}_{DC}$ on Cartesian axes, find two equations governing the unknown angles θ_2 and θ_3. Find the velocity of midpoint P of bar BD.
2. Let $L_0 = 4L, L_1 = L_3 = 5L, L_2 = 2L$ (where L is a reference length): solve the equations found in a) numerically to show that the motion of P is nearly rectilinear along the direction of AC in some interval of angle θ_1. Plot the two components of $\mathbf{v}_P$ for $\theta_1(t) = \theta_{10} + \theta_{11}\cos\omega t$, $\theta_{10} = 65^\circ$ and $\theta_{11} = 27^\circ$.
3. Let $L_2 = L_0 = l$ and $L_3 = L_1 = L > l$: find the location of instantaneous center I of bar BD. As the bar AB rotates, point I describes a fixed and moving centrodes. Find and sketch these two curves. Interpret the motion of the mechanism in terms of the motion of I on its centrodes.

PROJECT 4: Kinematic Analysis of a Piston Pump

The figure displayed below shows a sketch of a hand-operated piston pump. It is comprised of a housing 0, a piston 1, a connecting rod 2, and a lever 3. The joints at A, B and C are pivots. The joint between the housing and the piston is a slider. During the upstroke, the piston draws water, through valves a and b, into the cylinder. On the downstroke, the water is discharged into the outlet pipe. The geometry is given by $|AC| = 160$ mm, $|AD| = 490$ mm, and $|BC| = 110$ mm. At the instant shown, the angles take the values $\theta_2 = 70^\circ$ and $\theta_3 = 15^\circ$. The cylinder has diameter $d = 270$ mm, and the piston stroke length is $l = 116$ mm.

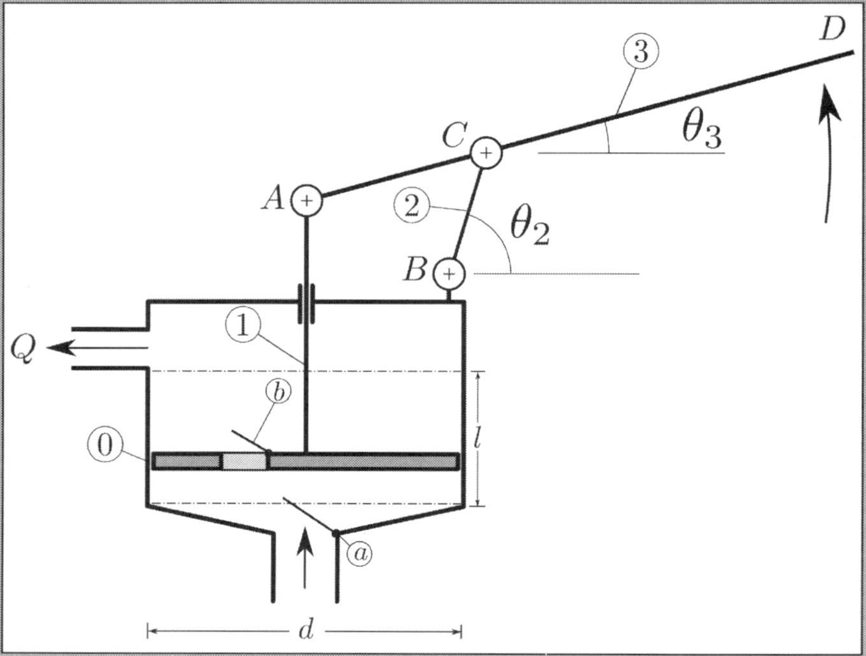

The pump is designed to yield a constant volumetric flux $Q = 400$ cm^3/s. During the downstroke, find the magnitude and direction of the velocities $\mathbf{v}_{A\in3/0}$, $\mathbf{v}_{C\in3/0}$ and $\mathbf{v}_{D\in3/0}$ by using

1. a vectorial method (choose a basis of unit vectors),
2. the method of instantaneous centers of rotation.

Plot $|\mathbf{v}_{D\in3/0}|$ versus time during a complete downstroke.

2.12 Problems

Problem 2.1 In this cross belt drive, the belt wrapped around pulleys 1 and 2 is assumed not to slip. Pulley 1 rotates at angular velocity 12.5 rev/s in the clockwise direction. The motion of the belt drives the rotation of pulley 2 about B.

Find the velocities of points I and J at any given time. Deduce the angular velocity of the driven pulley.

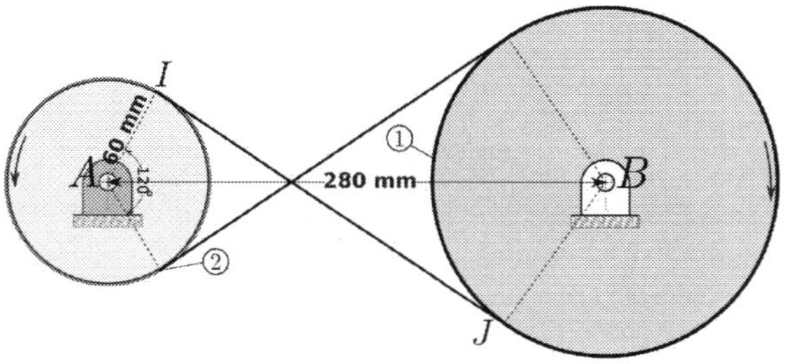

Answer: $\omega_2 = 16.7$ *rev/s (ccw).*

Problem 2.2 A slender bar AB of length 0.75 m is constrained to slide along the edges of a wedge-shaped support. Endpoint A slides at constant speed $v_A = 2.35$ m/s. At the instant shown, the bar is horizontal.

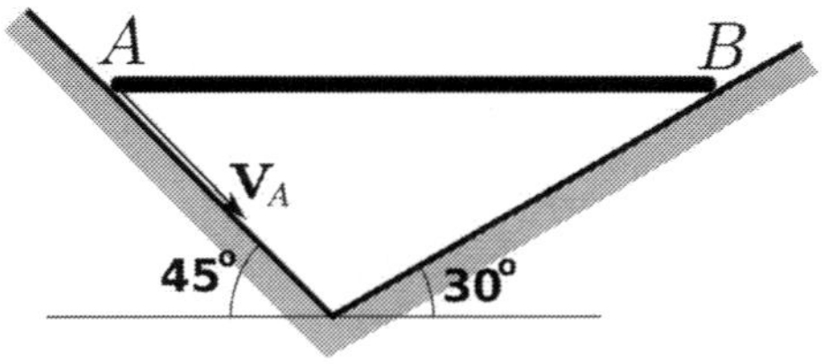

At the time of the configuration, find

a. the velocity of point B and the angular velocity of the bar,

b. the acceleration of point B and the angular acceleration of the bar.

Answer: a) $\omega = 3.49$ *rad/s (ccw), b)* $\alpha = 7.05$ *rad/s*^{2} *(cw).*

Problem 2.3 In this mechanism, the wheel 1 rotates at the constant angular velocity 12 rev/s in the clockwise direction about fixed point O. It sets in motion slider 3 by using the connecting rod 2 of length l. $r = 38$ mm, $l = 164$ mm, $h = 73$ mm.

a. Find the velocity of slider 3 and the angular velocity of rod 2 at $\phi = 0°, 90°$.

b. Find the acceleration of slider 3 and the angular acceleration of rod 2 at $\phi = 0°, 90°$.

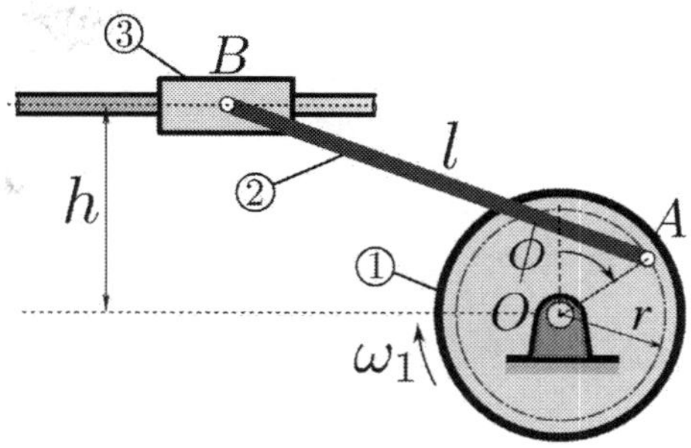

Answer: a) $\omega = 0$ $(\phi = 0°)$, *3.10 rad/s (ccw)* $(\phi = 90°)$.

Problem 2.4 At the instant shown, rod 1 has angular velocity 3 rad/s in the counterclockwise direction and angular acceleration 0.5 rad/s^2 in the clockwise direction. Body 3 is a circular slider centered at C. $|OA| = 120$ mm, $|AB| = 100$ mm, $|CB| = 130$ mm, $|AC| = 50$ mm, $\phi = 50°$. At the time of the configuration, find

a. the angular velocity of rod 2.

b. the angular acceleration of rod 2.

Answer: a) $\omega_2 = 22.6$ *rad/s (cw).*

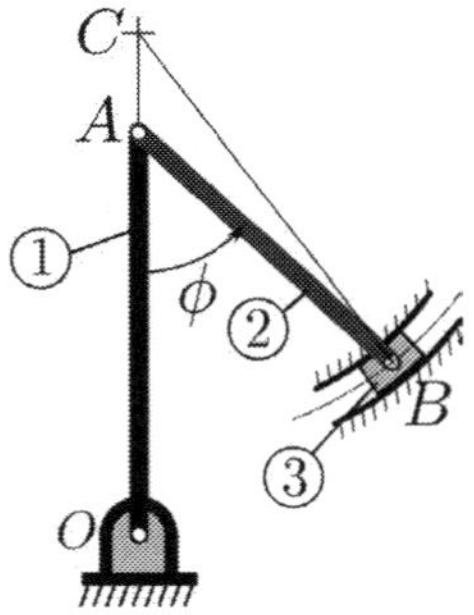

Problem 2.5 In the slider-crank mechanism, link 1 rotates at constant angular velocity ω_1 in the counterclockwise direction about O. The connecting rod 2 is pinned at A with link 1 and at B with slider 3. The orientation of link 1 is defined by angle θ_1 as shown. Define the constants $r = |OA|$ and $L = |AB|$.

Find the velocity $\mathbf{v}_{B\in 3/0}$ and acceleration $\mathbf{a}_{B\in 3/0}$ of slider 3 as a function of θ_1, ω_1, r and L.

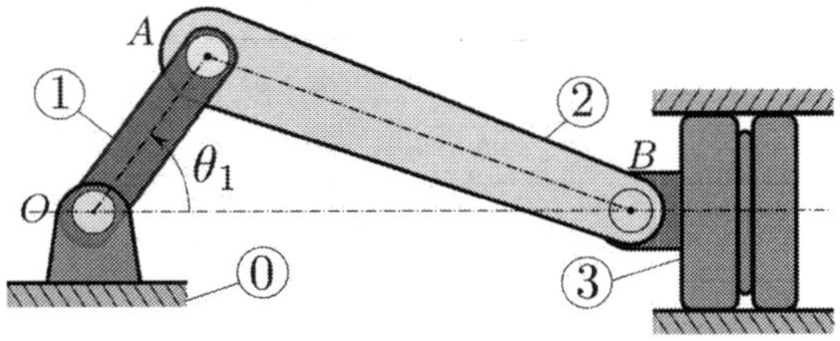

Problem 2.6 The scissor lift is a device used

to adjust a horizontal platform by mechanical means. It is designed by the assembly of 5 rigid bodies: body 0 is the fixed frame which supports the platform 3, the folding links 1 and 2, and the hydraulic cylinder assembly 4 and 5. The folding links 1 and 2 are connected by a pin joint at E midpoint of AB and CD. Link 1 is pinned with the frame 0 at A and is connected to a roller at B. Link 2 is pinned with the platform at D and is connected to a roller at C. The cylinder 4 is pinned at F with link 1, while piston 5 is pinned with link 2 at G. The geometry is characterized by $|AE| = |EB| = |CE| = |ED| = 120$ cm and $|EF| = |CG| = 45$ cm. Assuming that the piston 5 extends at speed 25 mm/s, find

a. the angular velocities ω_1 and ω_2 of folding links 1 and 2,

b. the velocities of C, E, B and D when $\theta = 60°$.

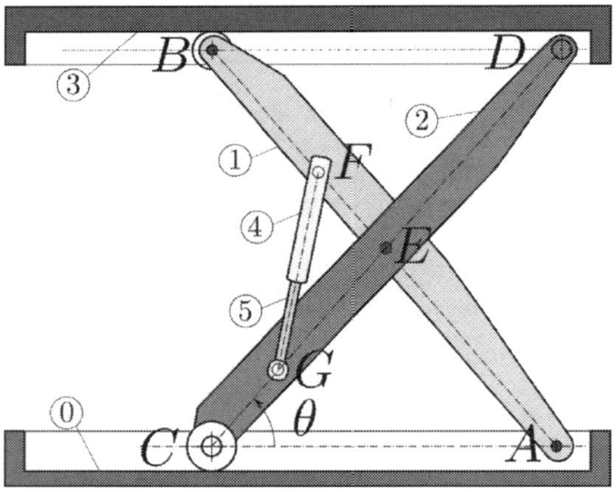

Answer: $\dot{\theta} = 0.045$ *rad/s.*

Problem 2.7 A dock leveler is a mechanism used to adjust the height of a platform 5 acting as a bridge between dock and truck. It is powered by the hydraulic cylinder 1 whose piston 2 extends at the constant velocity 5 cm/s. The piston sets in motion the two connecting rods 3 and 4 at the pivot B.

This results in the rotation of the platform 5 about D relative to the frame 0. At the time shown, the configuration of the system is described by four angles. Find the angular velocity of the platform 5. Use $|CD| = 20$ cm.

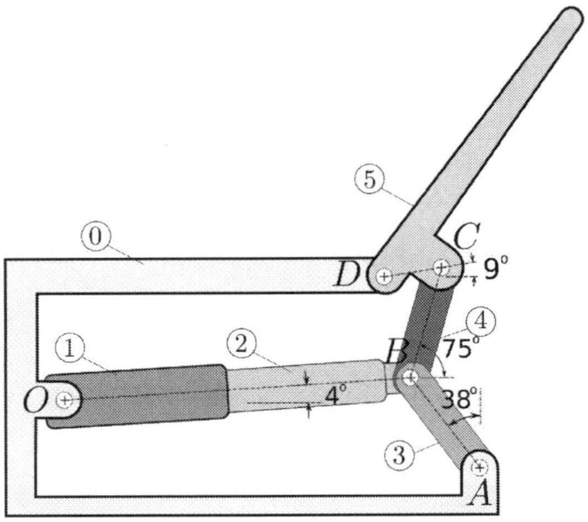

Problem 2.8 Consider three points A, B and C of a rigid body in planar motion. At the instant shown, the magnitude and direction of $\mathbf{v}_A$ ($v_A = 0.45$ m/s) are known as well as the direction of $\mathbf{v}_B$ (relative to line AB). $|AB| = 200$ mm, $|AC| = 150$ mm.

At the time shown, find the velocity of point C by

a. using a vectorial approach connecting the velocity of the three points,

b. using the property of equiprojectivity of velocity vectors.

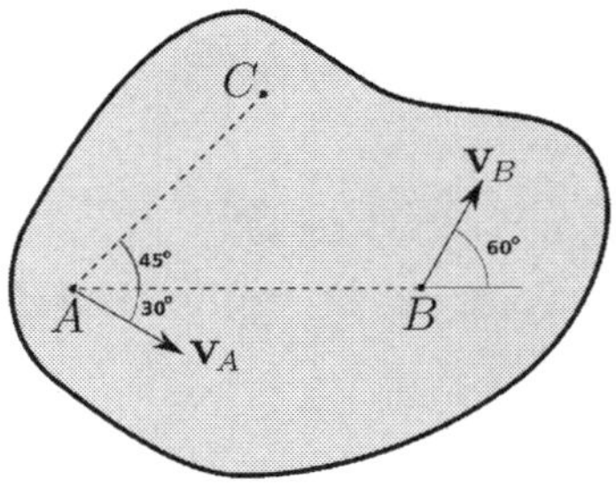

Problem 2.9 A mechanism for the automatic opening/closing of a barrier gate is composed of 3 rigid bodies. Link 1 is mounted to a driven shaft which rotates at the angular velocity $\omega_1 = 0.08$ rad/s. The gate arm is rigidly mounted to link 3. The connecting rod 2 allows the rotation of 1 to be transmitted to 3. Points O, A, B and C denote the centers of the pivot connections. The angles θ and ϕ define the orientations of links 1 and 3. The geometry is defined by $|OA| = 150$ mm, $|AB| = 75$ mm, $|BC| = 90$ mm and $|OC| = 100$ mm.

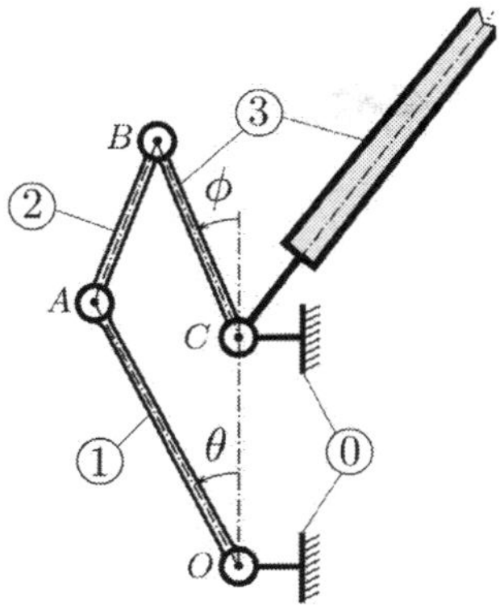

Find the angular velocity ω_3 of link 3 when $\theta = 40°$ by relating velocity $\mathbf{v}_{A\in 1}$ to velocity $\mathbf{v}_{C\in 3}$.

Problem 2.10 Two identical wheels 1 (of center A, radius r) and 2 (of center B, radius r) roll without slipping on a flat support. A rod 3 of length L is pinned to the wheels at points C and D of their rim. Center A of wheel 1 moves to the right at a given constant speed v_A. At the instant shown, the points A, B and C are aligned horizontally. The rod makes the angle ϕ with the horizontal.

a. Using a vectorial approach, find three equations which solve for the angular velocities ω_1, ω_2 and ω_3 of the three bodies in terms of speed v_A and parameters r, L and ϕ. Solve these equations. (Note that parameters r, L and ϕ are not independent).

b. Can the angular accelerations α_1, α_2 and α_3 be obtained by taking the time-derivative of the expressions you found in question a)? Explain. Find three equations which govern the unknowns α_1, α_2 and α_3.

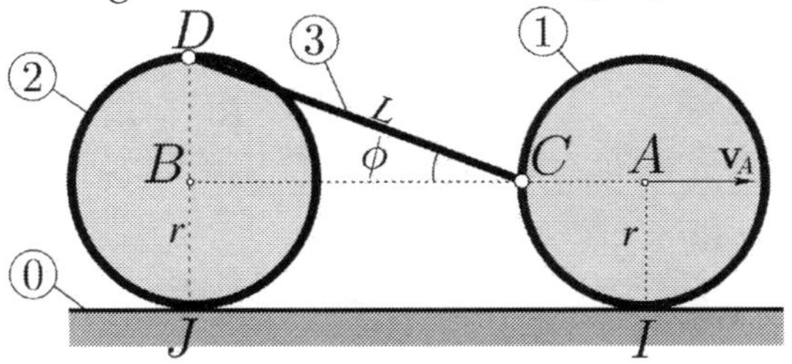

Problem 2.11 The system of problem 2.10 is modified: wheel 1 now rolls without slipping on a circular support of radius R.

a. Are the equations derived in question a) of problem 2.10 still valid? Explain.

b. Repeat question b) to find the angular accelerations α_1, α_2 and α_3.

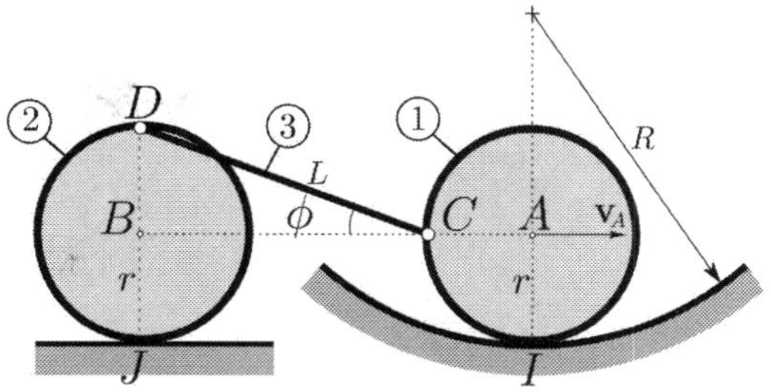

Problem 2.12 A bicycler rides his bike at the constant speed of 48 km/h. He uses the 51/14 gear ratio, where N=51 is the number of teeth in the front chainring and M=14 is the number of teeth in the rear sprocket. Determine the corresponding pedalling frequency (in rev/min) of the bicycler. Assume that the bicycle back-wheel of radius $R = 35$ cm does not slip.

Problem 2.13 A geared wheel 1 of center C and radius r is rolling in the interior of a fixed ring gear 0 of center O and radius $R > r$. The orientation of line OC is defined by angle θ.

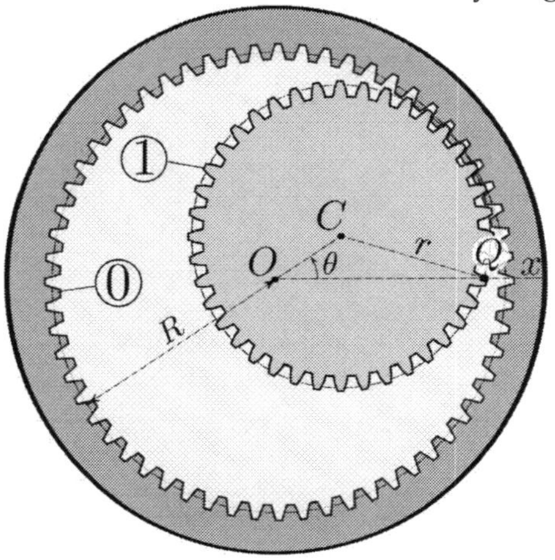

Under what condition will a point Q attached to the rim of 1 travel along the fixed axis Ox? Assuming this condition satisfied, find the velocity of center C and the angular velocity ω_1 of the wheel as a function of $\dot{\theta}$.
Describe a possible use of this mechanism.

Problem 2.14 A double gear-wheel 3 is driven in motion by the gear racks 1 and 2. The gear racks are in translation at known velocity and acceleration.

Find the angular velocity and acceleration of wheel 3 and the velocity and acceleration of its center C as a function of v_1, v_2, a_1, a_2 and the wheel radii r_1 and r_2.

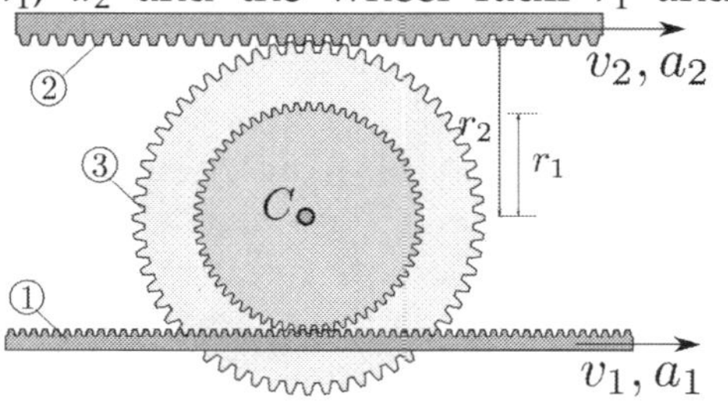

Problem 2.15 Two cylinders 1 and 2 roll without slipping on a horizontal fixed support 0. They remain in contact at point J.

a. Find the velocities of center B and the angular velocity of cylinder 2.

b. Find the slip velocity $\mathbf{v}_{J\in 2/1}$ of cylinder 2 relative to cylinder 1.

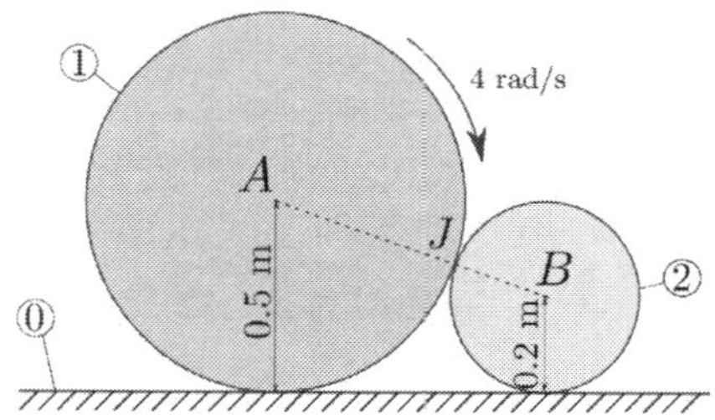

Problem 2.16 A rod 2 of length $|OA| = 105$ cm is pinned at O to a fixed support 0. It is connected at its endpoint A to a wheel 1 of center B and radius $|BI| = 50$ cm by a thin inextensible cable. The wheel rotates at constant angular velocity 0.55 rad/s about fixed point B and winds the cable around its rim, thereby setting rod 2 in motion. $|OB| = 170$ cm.

a. Find the angular velocity ω_2 of rod 2 when it reaches the position $\theta = 45°$.

b. Find the corresponding value of the angular acceleration α_2.

Hint: at any given instant, treat the section of cable AI as a rod pinned at A and rolling without slipping relative to the wheel at I.

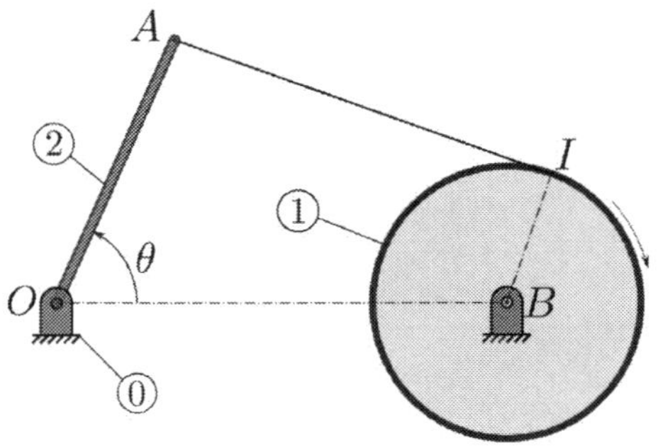

Answer: $\omega_2 = 0.307$ *rad/s (cw).*

Problem 2.17 A helicopter 1 is in rectilinear translation at constant velocity $\mathbf{v}_1 = 180$ km/h. The blades 2 rotate about center A at constant angular velocity 190 rev/min in the direction shown. At the instant shown, line BAC is perpendicular to velocity $\mathbf{v}_1$. Use $|AB| = |AC| = 5$ m.

a. Using a graphical method, sketch the velocities of the points of the blades which lie on line BC.

b. Find the velocity of the point of the blade which passes through point D. Deduce the position of instantaneous center I_2 of the blade assembly. Determine the coordinates of I_2.

c. Construct without calculation the fixed and moving centrodes.

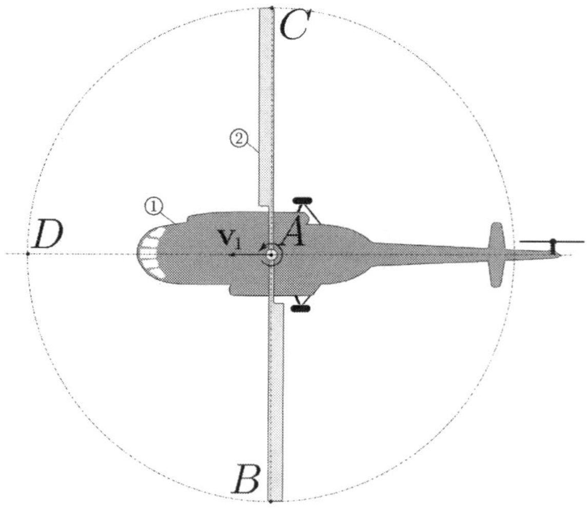

Problem 2.18 A rigid body 1 in the shape of an ellipse of center C slips along two perpendicular rigid boundaries Ox and Oy of a referential 0. A and B denote the contact points.

a. Show that the midpoint J of line segment AB, the instantaneous center I of body 1, the center C and origin O are aligned.

b. From the property found in a) show that center C describes an arc of circle.

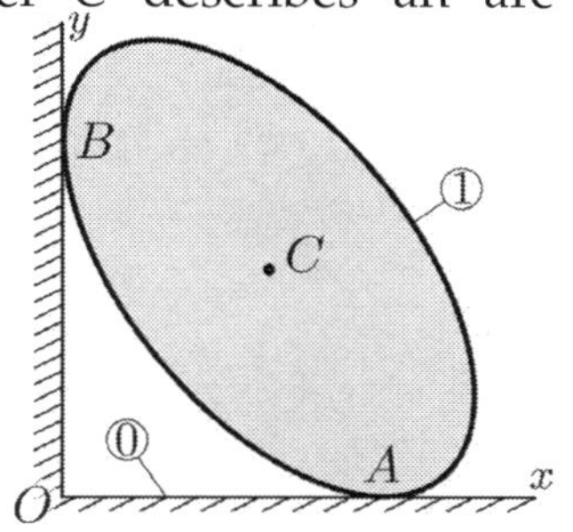

Problem 2.19 A spool 2 of center C rolls without slipping relative to a conveyor belt 1 translating at velocity $\mathbf{v}_1$ and acceleration $\mathbf{a}_1$ in the direction shown. The cable remains taut and parallel to the belt during the motion.

a. Find the velocities of points A, B and C.

b. Find the accelerations of points A, B and C.

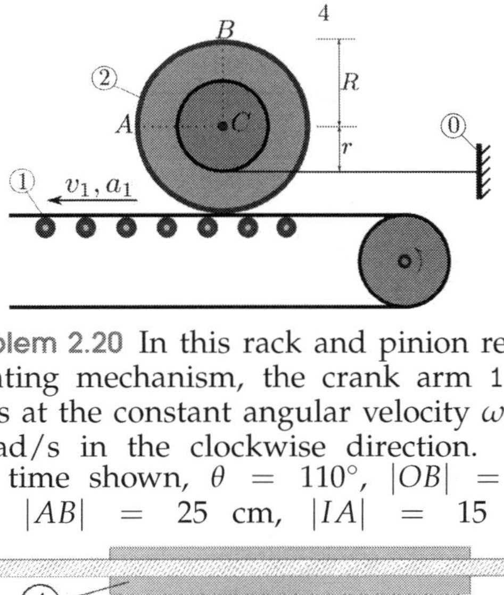

Problem 2.20 In this rack and pinion reciprocating mechanism, the crank arm 1 rotates at the constant angular velocity $\omega_1 = 5$ rad/s in the clockwise direction. At the time shown, $\theta = 110°$, $|OB| = 10$ cm, $|AB| = 25$ cm, $|IA| = 15$ cm.

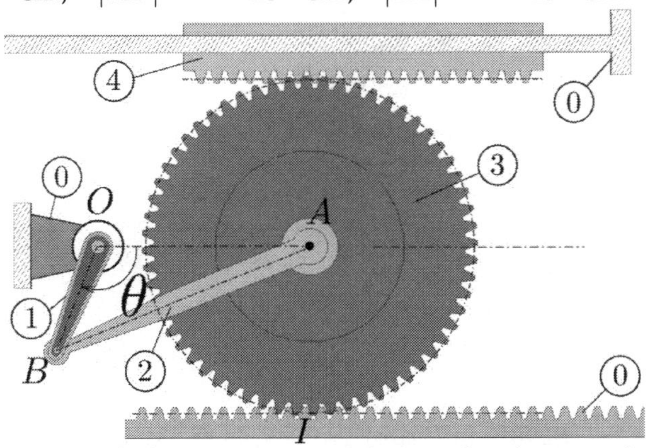

a. Find the output translational velocity of rack 4 and the angular velocities of connecting rod 2 and pinion 3.

b. Find the acceleration of rack 4.

Problem 2.21 A rod 1 is connected to two sliders at pins A and B ($|AB| = 0.75$ m). The sliders are constrained to move along directions which form the angle $\beta = 60°$. The angular velocity of rod 1 is assumed to take the constant value 0.6 rad/s in the clockwise direction.

a. Relate the velocity of B to that of A to find the speeds v_A and v_B as a function of angle θ.

b. Find the location of the IC I of rod 1. Then rederive the expressions of v_A and v_B found in a).

c. Find and plot the corresponding fixed and moving centrodes, that is, the trajectories of I viewed by observers attached to the fixed and moving frames.

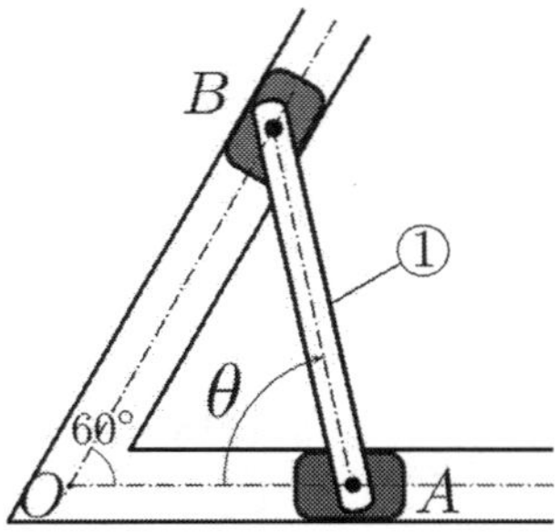

Answer: $v_A = -L\omega_1 \cos(\beta + \theta)/\sin\beta$
$v_B = L\omega \cos\theta/\sin\beta$.

Problem 2.22 Consider the system of problem 2.10. Construct the instantaneous center of zero-velocity I_3 of rod 3. Carefully indicate the directions of all (linear and angular) velocities. Find the distances from I_3 to points C and D. From this information, recover the expressions of angular velocities ω_1, ω_2 and ω_3 of the three bodies in terms of speed v_A and parameters r, L, ϕ.

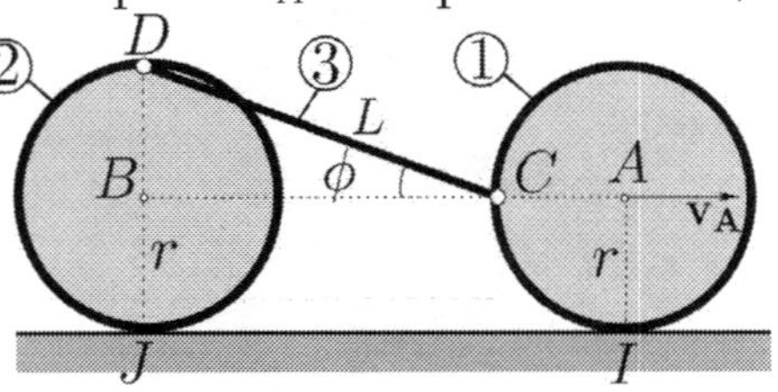

Problem 2.23 In a ball bearing, the inner ring 1 of radius r_1 rotates about O in the clockwise direction at angular velocity ω_1. The outer ring 2 of inner radius r_2 rotates about O in the clockwise direction at angular ve-

locity ω_2. Spherical ball 3 of center C and diameter $r_2 - r_1$ rolls without slipping at contact points I and J relative to the rings.

a. Find the location of the instantaneous center I_3 of ball 3 (give the distance $|OI_3|$).

b. From the knowledge of I_3 find the speed v_C of center C and the angular velocity of ball 3.

c. Under what condition is ball 3 in translational motion?

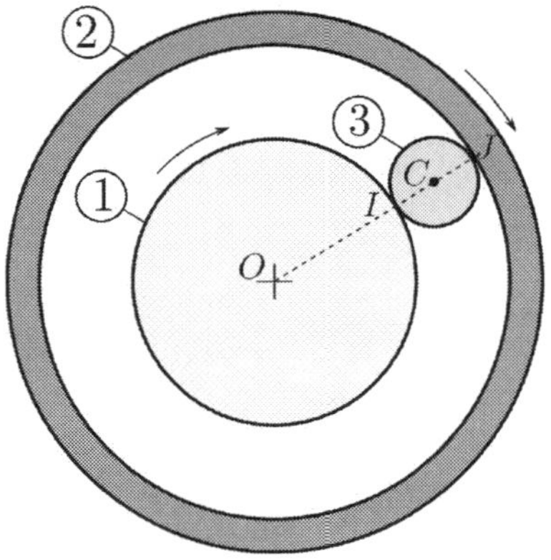

Problem 2.24 A sphere 1 of radius r rolls without slipping on a horizontal support 0 at angular velocity ω and angular acceleration α in the direction shown. Recall that contact point I is the instantaneous center of rotation of the sphere.

a. Find the set $\mathcal{C}_N$ of points P attached to 1 whose normal acceleration is zero at a given instant. *hint: impose* $\mathbf{a}_P \cdot \mathbf{r}_{IP} = 0$. *why?*

b. Find the set $\mathcal{C}_T$ of points P attached to 1 whose tangential acceleration is zero at a given instant. *hint: impose* $\mathbf{a}_P \times \mathbf{r}_{IP} = 0$.

c. Conclude that there exists a unique point J which satisfies $\mathbf{a}_{J\in 1/0} = \mathbf{0}$. Find its position relative to I. (We may call J *instantaneous center of zero acceleration.*)

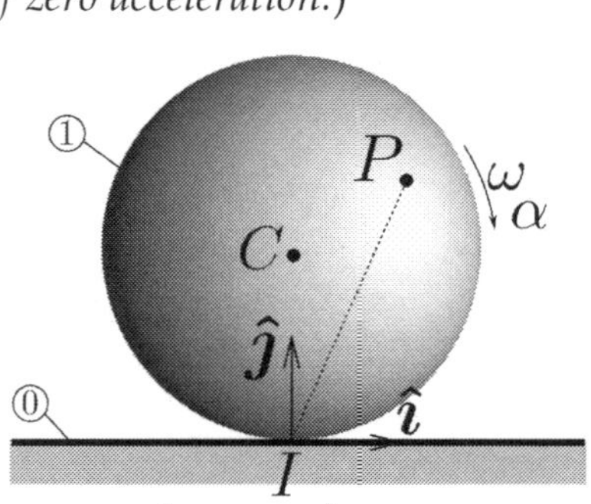

Answer: $\mathbf{r}_{IJ} = r\frac{\alpha\omega^2}{\alpha^2+\omega^4}\hat{\imath} + r\frac{\omega^4}{\alpha^2+\omega^4}\hat{\jmath}$

Problem 2.25 A rider on his Segway takes a turn of radius R. The two identical wheels 1 and 2 of radius r have centers A and B respectively. Denote by G the center of line segment AB and by e the distance between the contact points of the wheels with the road. To take the turn without skidding, show that the inner and outer wheels must rotate at different speeds ω_1 and ω_2 and find the ratio ω_1/ω_2.

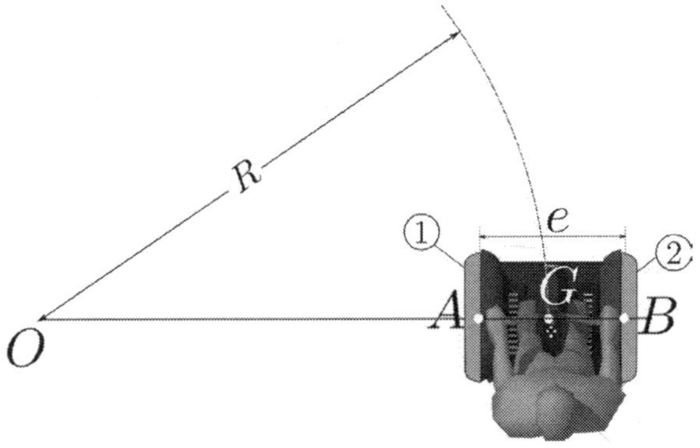

Problem 2.26 Three cylinders are in planar motion. Cylinder 1 (of center A, radius r) and cylinder 2 (of center B, radius $r/2$) roll without slipping on a horizontal fixed support 0. Cylinder 3 (of center C, radius $r/4$) rolls without slipping relative to cylinder 1 at contact point I. Center A moves at known speed v_A as shown in the figure.

a. Find the velocities of B and C.

b. Find the instantaneous center I_3 of cylinder 3. Deduce the angular velocities of the three cylinders.

c. Find the slip velocity $\mathbf{v}_{K\in 3/2}$.

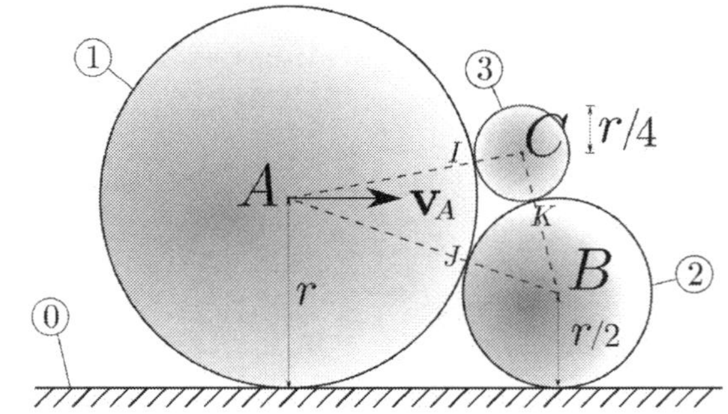

Answer: $\mathbf{v}_{K\in 3/2} = \mathbf{0}$

Problem 2.27 A rod 2 is connected to the center A of a wheel 1 of radius r and to a slider 3 at B. The wheel rolls without slipping on a fixed circular support of center O and radius $R + r$. Denote by L the length $|AB|$. The slider moves at constant speed v_B in the direction shown. The orientation of rod 2 is defined by angle ϕ.

a. Construct the location of the instantaneous centers I_1 and I_2 of bodies 1 and 2.
b. Find the velocity $\mathbf{v}_A$ of A, the angular velocities ω_1 and ω_2 of bodies 1 and 2 as a function of ϕ, v_B and the geometric parameters r, R and L.
c. Find the acceleration $\mathbf{a}_A$ of A, the angular accelerations α_1 and α_2 of bodies 1 and 2 as a function of ϕ, v_B and the geometric parameters r, R and L.

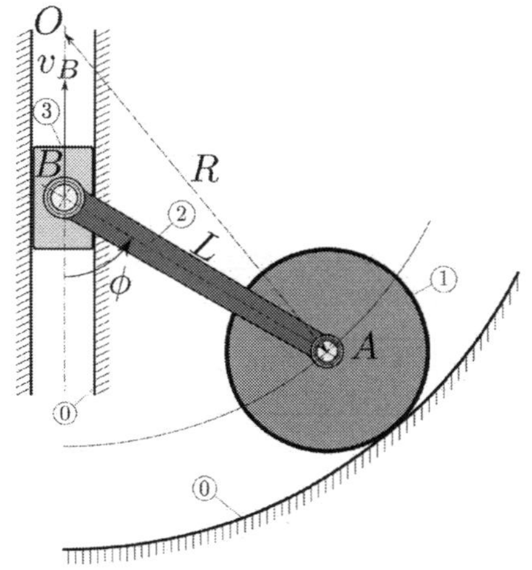

Problem 2.28 Consider the double gear-wheel/gear racks of problem 2.14. The gear racks are in translation at known velocity and acceleration in opposite direction.
Find the location of the instantaneous center of rotation I_3 of wheel 3.
From the knowledge of I_3, determine the angular velocity of wheel 3 and the velocity of its center C as a function of v_1, v_2 and the wheel radii r_1 and r_2.
Can the angular acceleration of wheel 3 and the acceleration of C be found?

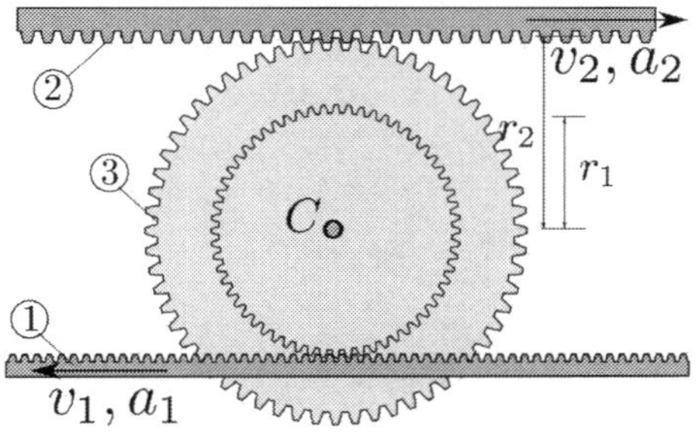

Problem 2.29 A steering mechanism ensures that a vehicle wheels of the same axle are pointing in the desired directions. The arm 1 pivots about O relative to the frame 0. It is actuated by a hydraulic cylinder (not shown). The tie rod 2 connects the arm 1 to the steering arm 3 which pivots about the kingpin C relative to the frame. Only the right-hand side of the mechanism is shown. The configuration of the mechanism is defined by the three angles θ_1, θ_2 and θ_3 and the fixed lengths $l_1 = |OA|$, $l_2 = |AB|$ and $l_3 = |BC|$.

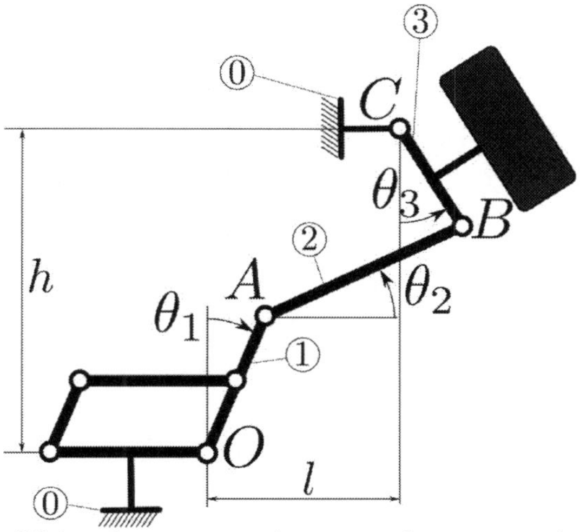

a. Write the geometric vector loop equation between O, A, B and C to find 2 scalar equations governing θ_1, θ_2 and θ_3.
b. Assume that at $\theta_1 = 0$, we have $\theta_2 = \theta_2^0$ and $\theta_3 = 0$ and that for $\theta_1 \neq 0$, $\theta_2 = \theta_2^0 + \Delta\theta_2$, with $\Delta\theta_2 \ll 1$. Show that angle θ_3 is governed by the equation

$$c_2 \tan^2 \frac{\theta_3}{2} + c_1 \tan \frac{\theta_3}{2} + c_0 = 0$$

Express c_0, c_1 and c_0 in terms of θ_1, and the constants θ_2^0, l_1, l_2 and l_3. Plot θ_3 vs θ_1 for $\theta_2^0 = 25°$, $l_1 = 0.5$ m, $l_2 = 0.8$ m and $l_3 = 0.4$ m for $-30° < \theta_1 < 30°$.

Problem 2.30 A laboratory agitator is based on a four-bar linkage. The drive link 1 and the output link 3 are interconnected by the coupler link 2. Link 1 is in rotation about fixed point O. Link 3 which holds the agitator vessel is in rotation about fixed point C. Denote $l_1 = |OA|$, $l_2 = |AB|$, $l_3 = |BC|$. The constants H and D define the distances between fixed points O and C.
a. Write the geometric loop equation between O, A, B and C to find an equation between angles θ_1, θ_3 and the geometric parameters l_1, l_2, l_3, D and H.
b. From the equation found in a) determine the relationship between $\dot{\theta}_1$ and $\dot{\theta}_3$.

c. Assume that the geometry of the mechanism is such that the angle θ_3 remains small: using the approximation $\sin\theta_3 \approx 0$ and $\cos\theta_3 \approx 1$ show that $\dot{\theta}_3$ is given by

$$\dot{\theta}_3 = \frac{l_1}{l_3}\dot{\theta}_1 \frac{(D - l_3)\sin\theta_1 + H\cos\theta_1}{H + l_1 \sin\theta_1}$$

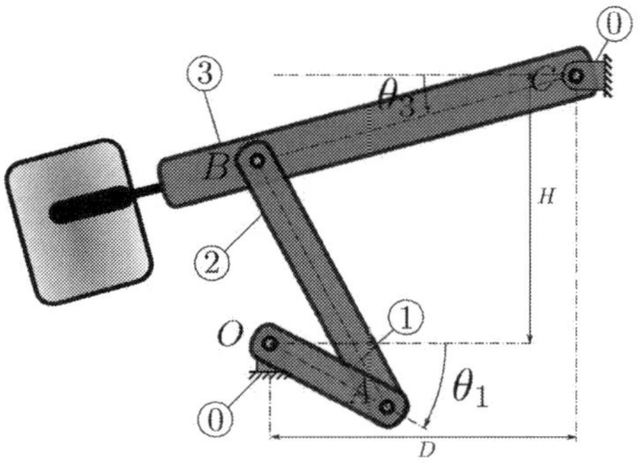

Problem 2.31 A small mass P can slide along a rod $\mathcal{F}$ which rotates about referential $\mathcal{E}$ of Cartesian axes $Oxyz$ about axis Oz with constant angular velocity $\dot{\theta} = \omega$. Rod $\mathcal{F}$ passes through O and maintains a constant angle α relative to axis Oz.

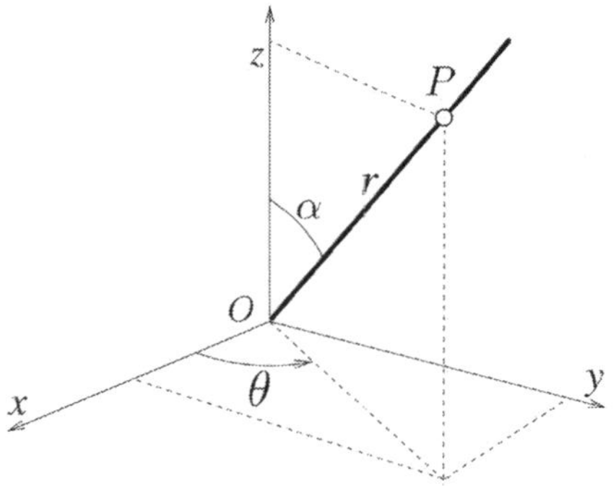

The position of P relative to $\mathcal{F}$ is defined by the distance $r(t) = |OP|$.

a. Find the velocity of P relative to $\mathcal{F}$ in terms of r and $\dot{r}$. Deduce the velocity $\mathbf{v}_{P/\mathcal{E}}$.

b. Repeat for the acceleration of P: find $\mathbf{a}_{P/\mathcal{E}}$ from $\mathbf{a}_{P/\mathcal{F}}$.

Problem 2.32 A probe enters the gravitational sphere of influence of Earth to gain speed by gravitational assistance. Relative to the geocentric referential Cxy attached to Earth, the probe approaches Earth along the Cx axis with speed $v_\infty = 5$ km/s. It then escapes Earth's gravitational field with speed $v_\infty = 5$ km/s along the hyperpolic trajectory which makes the angle $\theta = 45°$ with axis Cx. In the heliocentric referential $\mathcal{H}$, the orbital velocity of Earth is $V_C = 30$ km/s and is assumed directed along axis Cy.

Find the gain in speed of the probe relative to the Heliocentric referential $\mathcal{H}$ due to its travel in Earth's gravitational field.

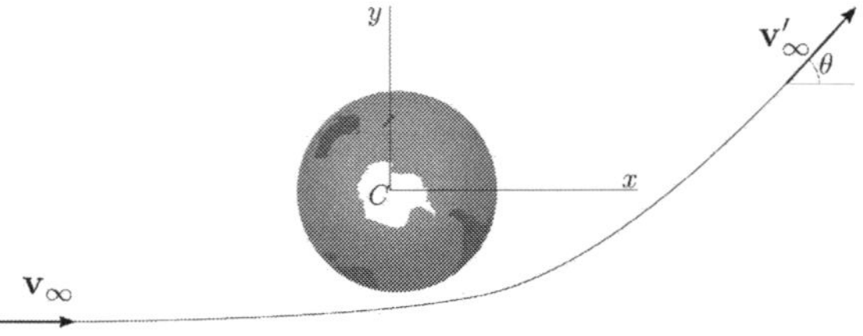

Answer: $\Delta v = 3.3$ km/s.

Problem 2.33 A small ring P slides relative to a circular frame $\mathcal{C}$ of center O and radius $R = 0.5$ m at constant speed of 0.4 m/s. Frame $\mathcal{C}$ rotates about its diameter Oz at constant angular velocity $\omega = 1$ rad/s relative to referential $\mathcal{E}$. At time $t = 0$, $\mathcal{C}$ lies in the plane Oxz of $\mathcal{E}$ and P is at point A on axis Ox.

a. Find the velocity and acceleration of P relative to $\mathcal{C}$.

b. Using the quantities found in a) find the velocity and acceleration of P relative to $\mathcal{E}$.

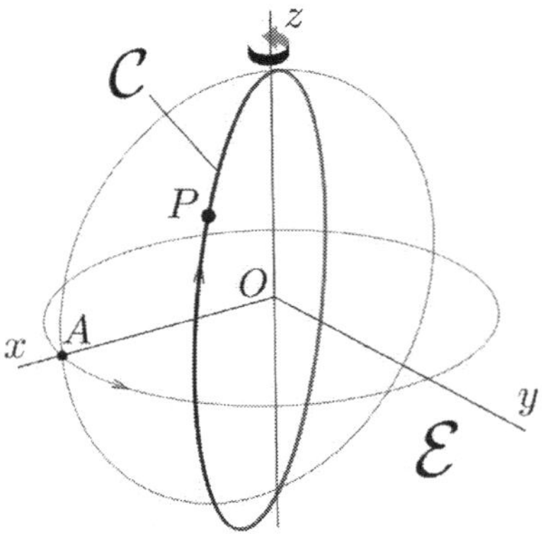

Problem 2.34 A spherical ball A is connected at pin B to a vertical shaft by a slender rod of length l. The shaft is set in rotation at constant angular velocity ω relative to the support $\mathcal{E}$. Neglect the distance from B to the axis of the shaft.

a. Find the velocity of A relative to the shaft, then relative to the support in terms of l, ω, θ and $\dot{\theta}$. Find $|\mathbf{v}_A|$ for $\omega = 2$ rev/s, $l = 25$ cm, $\theta = 30°$ and $\dot{\theta} = 2.5$ rad/s.

b. Repeat question a) to find the acceleration of A in terms of l, ω, θ, $\dot{\theta}$ and $\ddot{\theta}$. Find $|\mathbf{a}_A|$ using the data of a) and $\ddot{\theta} = -0.5$ rad/s^2.

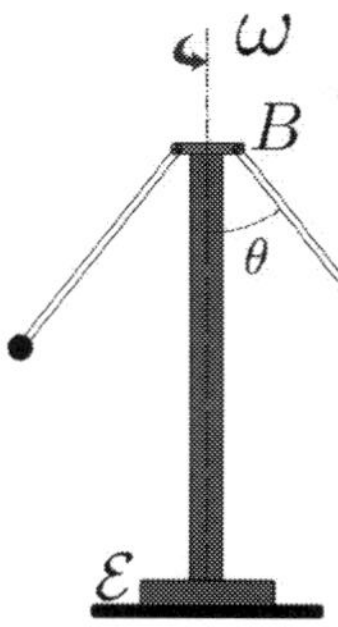

Problem 2.35 Four particles A, B, C and D initially located at the corners of a square travel at the same constant speed v along the sides, A toward B, B toward C, C toward D and D toward A as shown. The square is in rectilinear translation relative to a fixed referential 0 at constant speed v.

Show that the absolute trajectories of the four particles, that is, viewed by an observer fixed in referential 0, intersect at a point I.

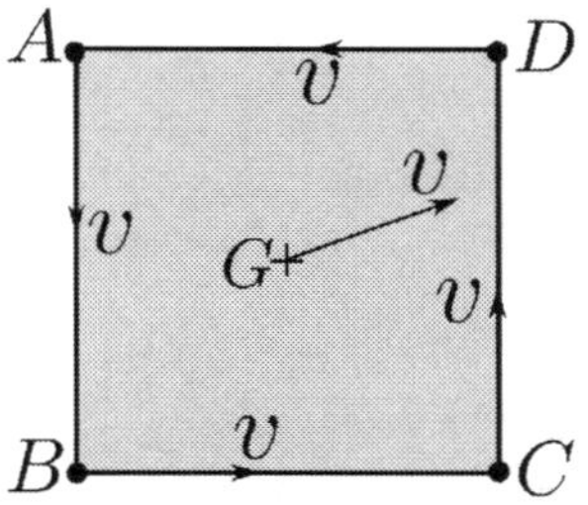

Problem 2.36 An angler attempts to cross a river of width D whose banks are parallel straight lines. The speed of his boat in calm water is the constant v. The current velocity relative to the ground is of constant magnitude u and in the direction of the river's banks. The angler has two choices: cross the river in the shortest path, or cross in the minimum time.

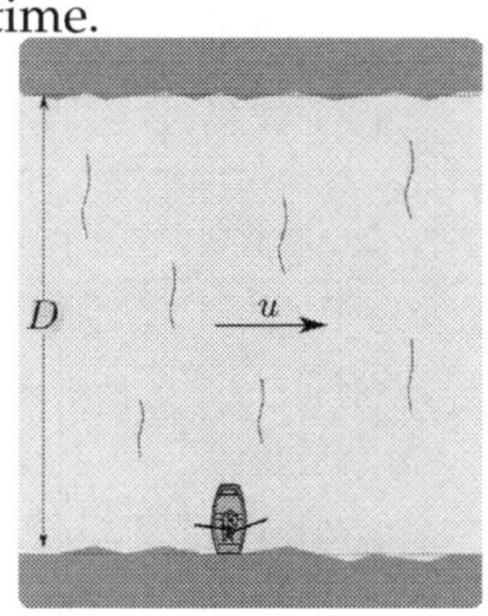

a. Which direction should the angler give his boat so that the shortest path is traveled? Find the duration T_1 of the crossing. (*hint: two cases must be considered: $u > v$ and $u < v$*)

b. Which direction should the boat take for the shortest duration T_2 of the crossing? Sketch T_1/T_2 versus the ratio $k = u/v$.

Answer: $T_1/T_2 = 1/\sqrt{1-k^2}, \quad (v > u)$

Problem 2.37 A sailboat P travels in a current so that a buoy located at a fixed point O always stays directly abeam to P, that is, the line connecting O and P stays perpendicular to the boat's relative velocity. The boat's speed in still water is the constant v. The current has a constant velocity $\mathbf{u}$ (relative to the ground). Denote by $\hat{\mathbf{e}}_r$ the unit vector directed from O to P, and by $r(t)$ the radial distance. With axis Ox defined as the fixed line perpendicular to $\mathbf{u}$, $\theta(t)$ is the angle made by line OP with axis Ox.

a. Find the absolute velocity of P, that is, measured relative to the ground on the polar basis $(\hat{\mathbf{e}}_r, \hat{\mathbf{e}}_\theta)$.

b. Find the trajectory $r(\theta)$ of P assuming that at time $t = 0$, $\theta = 0$ and $r = r_0$. Characterize this trajectory according to the ratio $e = u/v$. Plot the trajectories of P for $e = 0.5, 1, 2$.

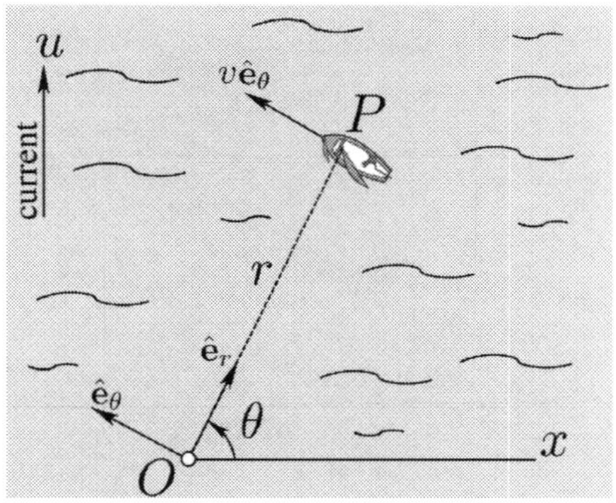

Answer: $r = r_0(e+1)/(e\cos\theta + 1)$

Problem 2.38 A disk-shaped eccentric cam 1 is driven in rotation about O at angular velocity 1300 rev/min in the counterclockwise direction. It sets follower 2 in translation due by a point contact joint at A and a prismatic joint with the support 0. The geometry is characterized by $|OC| = 32$ mm and $|CA| = 46$ mm.

By writing the kinematic loop equation ⓪-①-②-⓪ at A, find at the position shown

a. the slip velocity $\mathbf{v}_{A\in 2/1}$ between the cam and the follower,

b. the velocity of the follower $\mathbf{v}_{A\in 2/0}$.

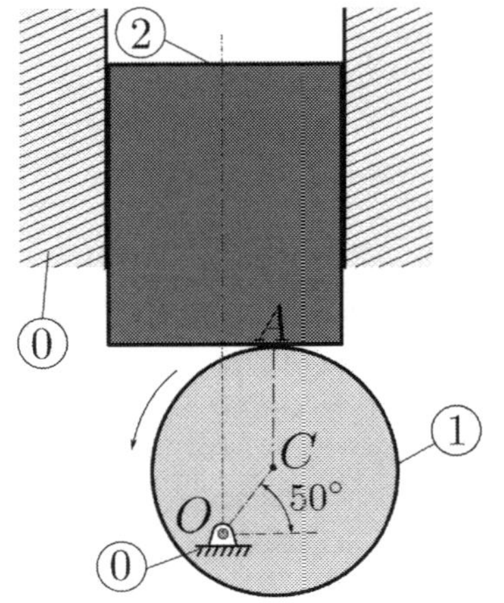

Problem 2.39 Two cams 1 and 2 are connected to a fixed support 0 by pivots centered at A and B, respectively. The rotation of cam 1 sets cam 2 in rotation due to a point contact joint at J. At the instant shown, cam 1 rotates at the angular velocity of 0.45 rad/s in the counterclockwise direction and the geometry is given by $|AB| = 350$ mm and $|AJ| = 200$ mm.

a. Construct the location of the instantaneous center of rotation I_{12} of cam 2 relative to cam 1.

b. Using I_{12} find the angular velocity of cam 2 and the slip velocity of cam 2 relative to cam 1.

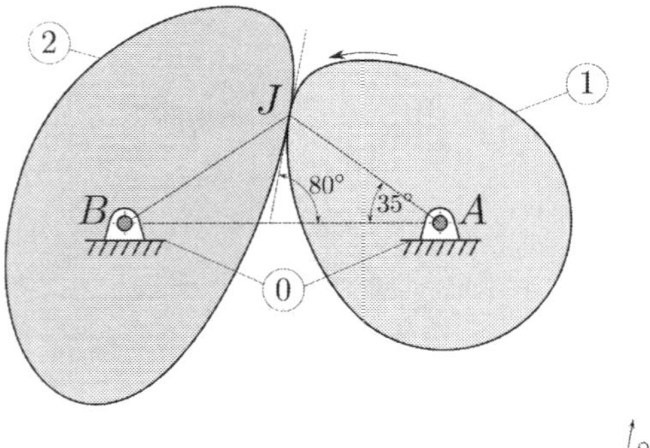

Answer: $\omega_2 = 0.262$ *rad/s ccw,* $v_{J\in 2/1} = 0.124$ *m/s* $\angle 80°$

Problem 2.40 A swing-arm quick-return mechanism consists of three rigid bodies in motion relative to a referential 0. Drive link 1 rotates uniformly at angular velocity ω_1 about fixed point C in the counterclockwise direction. It is connected at pin B to slider 2 which is constrained to translate along the slotted arm 3. Arm 3 is constrained to rotate about fixed point A. The orientation of link 1 is defined by angle θ_1 and the geometry is defined by the constant parameters $r = |CB|$ and $l = |AC|$ with $r < l$. We also denote $\lambda(t) = |AB|$.

a. Write the kinematic loop equation ⓪-①-②-⓪ at point B to find the slip velocity $\mathbf{v}_{B\in 2/3}$ and the angular velocity $\boldsymbol{\omega}_3$ of slotted arm 3 (as a function of θ_1, θ_3, r and l and λ).

b. Find two geometric relationships which relate λ and θ_3 to θ_1, r and l. Deduce that the angular velocity ω_3 (defined in the counterclockwise direction) is given by

$$\omega_3 = r\omega_1 \frac{r + l\sin\theta_1}{r^2 + l^2 + 2rl\sin\theta_1}$$

c. From the result of b) conclude that the output motion is a reciprocating motion. Find the ratio r/l so that arm 3 oscillates between the values $\theta_3 = \pm\pi/3$. Plot the corresponding function ω_3/ω_1 vs θ_1 for $0 \leq \theta_1 < 2\pi$ and explain why the mechanism is characterized as "quick-return".

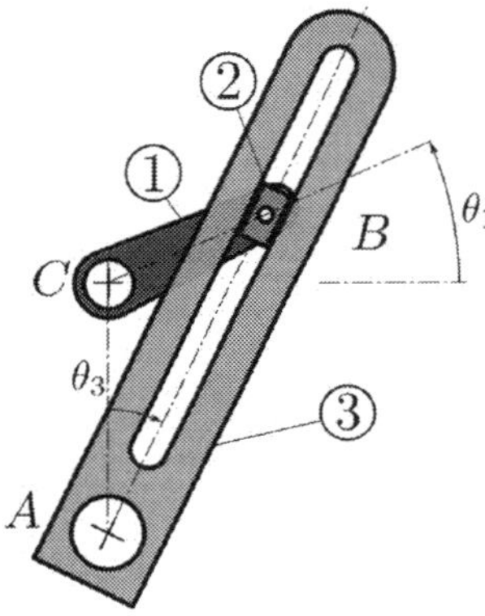

Problem 2.41 In this mechanism, the motion of the hydraulic cylinder/piston assembly 1-2 controls the motion of plunger 5. The cylinder/piston assembly 1-2 is pinned at point O fixed in referential 0 and at point A with lever 3 itself pinned at B to referential 0. Lever 3 supports a roller 4 of center C. Roller 4 is in contact with the plunger 5. The goal is find the relationship between the plunger's velocity v_5 and the piston's extensional velocity $v_{2/1}$ (measured positively in the directions shown). At the instant shown, the configuration is given by angles $\theta_1 = 25°$ and $\theta_3 = 20°$ and the geometry is defined by $|BA| = 150$ mm and $|BC| = 350$ mm.

a. From the equation $\mathbf{v}_{C\in 5/0} = \mathbf{v}_{C\in 5/4} + \mathbf{v}_{C\in 4/3} + \mathbf{v}_{C\in 3/0}$, find the relationship between velocity v_5 and angular velocity $\omega_{3/0}$.

b. From the equation $\mathbf{v}_{A\in 3/0} = \mathbf{v}_{A\in 3/2} + \mathbf{v}_{A\in 2/1} + \mathbf{v}_{A\in 1/0}$, find the relationship be-

tween velocity $v_{2/1}$ and angular velocity $\omega_{3/0}$.
c. Using the results of a) and b) find the ratio $v_{2/1}/v_5$.

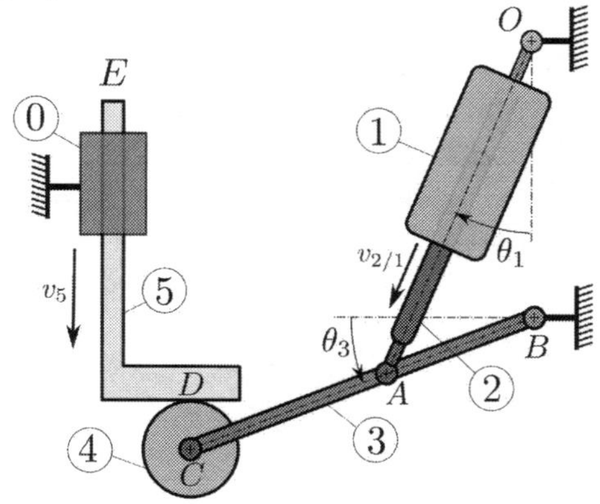

Answer: $v_{2/1}/v_5 = 0.322$.

Problem 2.42 In this epicyclic gear train the carrier 1 supports the double satellite gear 2 which is meshed with two ring gears 3 and 4. The geometry is defined by the radii r_1, r_2', r_2''.

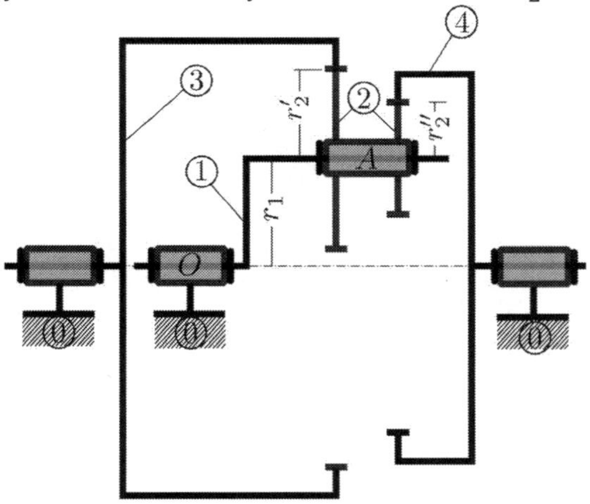

a. Identify one (or more) kinematic loop equation(s). Deduce relationship(s) between the angular velocities ω_1, ω_2, ω_3 and ω_4.
b. Assume that ring gear 3 is stationary. Find the ratio ω_1/ω_4 for $r_2'' = 100$ mm, $r_2' = 102.5$ mm, $r_3 = r_1 + r_2' = 210$ mm and $r_4 = r_1 + r_2'' = 215.5$ mm. Comment about the possible use of this mechanism.
Answer: $\omega_1/\omega_4 = 862$

Problem 2.43 A disk 2 of center C and radius R is set in motion by a rod 1 which rotates about pivot O at constant angular velocity ω_1 in the clockwise direction. The disk is in contact with the rod at point I and with the fixed support 1 at point J. We are interested in predicting the kinematics of the disk. (Neglect the thickness of the rod).
Part I. *Assume that there is no slip at I between the disk and the rod.*
1a. Using a vectorial method find the velocity $\mathbf{v}_C$ of C and the angular velocity of $\boldsymbol{\omega}_2$ of the disk in terms of ω_1, θ and R. Then find the velocity $\mathbf{v}_{J\in 2/0}$ to show that there is slip at J.
2a. Find a geometric constraint between the position of C and angle θ. Then recover the relationship between v_C and ω_1 found in 1a).

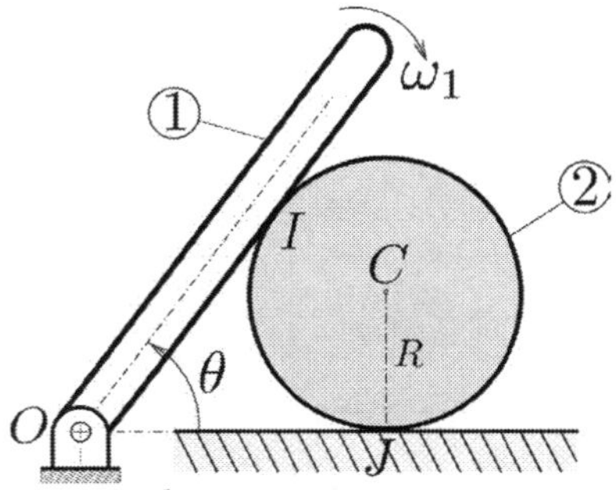

3a. Construct the instantaneous center of rotation I_2 of the disk $\mathcal{D}$. Clearly indicate on this construction the direction of all (linear or angular) velocities. Then use the knowledge of I_2 to find the angular speed ω_2 and speed v_C of C.
Part II. *Here we assume that there is no slip at contact point J.*
1b. Explain why there must be slip at I (find $\mathbf{v}_{I\in 2/1}$). Then with a method of your choice, find the velocity $\mathbf{v}_C$ of C and the angular velocity of $\boldsymbol{\omega}_2$ of the disk in terms of ω_1, θ and R.
2b. Interpret the velocity difference $\mathbf{v}_{I/0} - \mathbf{v}_{I\in 1/0}$. Then find its expression (in terms of ω_1, θ and R).
3b. With a method of your choice find the angular acceleration of the disk.

Problem 2.44 The sketch displayed below shows a model of 90° V-twin engine for a motorcycle. Body 0 represents the housing assumed fixed, 1 the crankshaft in rotation relative to the housing at angular velocity ω_1 about O, 2 and 3 the connecting rods both in rotation about pivot A relative to the crankshaft, and finally 4 and 5 the pistons in translational motion relative to the housing. The pistons are joined to the connecting rods by pivots at B and C. The axis of the pistons' cylinders make a 90 degree angle.

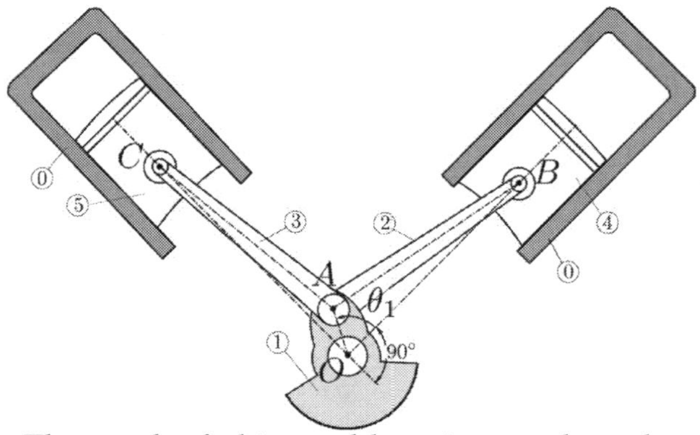

The goal of this problem is to relate the crankshaft angular velocity ω_1 to the velocity of the pistons. The crankshaft's position is defined by angle θ_1. Denote by $L_A = |OA|$, $L_B = |AB|$, $L_C = |AC|$.

a. Vectorial approach: find the velocities $\mathbf{v}_B$ and $\mathbf{v}_C$ from the velocity of A.

b. Graphical method: find the speed v_B and v_C by locating the instantaneous centers of rotation I_2 and I_3 of the connecting rods.

c. Find the speed v_B and v_C for $\omega_1 = 60$ rev/s, $\theta_1 = 30°$, $L_A = 40$ mm, $L_B = L_C = 180$ mm.

Problem 2.45 A pumpjack (a.k.a. horsehead pump) is used to mechanically pump liquid out of the oil wells. This mechanism converts the uniform rotation of a motor into a vertical reciprocating motion of the pump piston. It comprises four rigid bodies: the crank arm 1, the connecting rod 2, the I-beam 3 (which ends in a curved metal box called *horse head*) and a cable 4 connected to a piston which travels vertically through a stuffing box. During the motion, the cable wraps around the curved part of the horse head without slipping. The configuration of the system is defined by the angles θ_1, θ_2 and θ_3 as shown.

The crank arm rotates at constant angular velocity ω_1. Define $|OA| = L_1$, $|AB| = L_2$, $|BC| = L_3$. The joints at O, A, B and C are pivots. By finding the piston's velocity $\mathbf{v}_{E\in 4}$ in terms of ω_1 we can determine the flow rate of the pump.

a. Find $\mathbf{v}_{E\in 4}$ by a vectorial method in terms of the geometrical parameters θ_1, θ_2, θ_3, L_1, L_2, L_3. Does the piston travel at constant speed? Find $|\mathbf{v}_{E\in 4}|$ for the numerical values $\omega_1 = 0.025$ rev/s, $\theta_1 = 30°$, $L_1 = 5$ m, $L_2 = 14$ m, $L_3 = 7.75$ m, $H = 11.5$ m, $R = 12$ m, $D = R$, $L = 7$ m.

b. Repeat question a) by using a graphical method with the concept of instantaneous center of rotation.

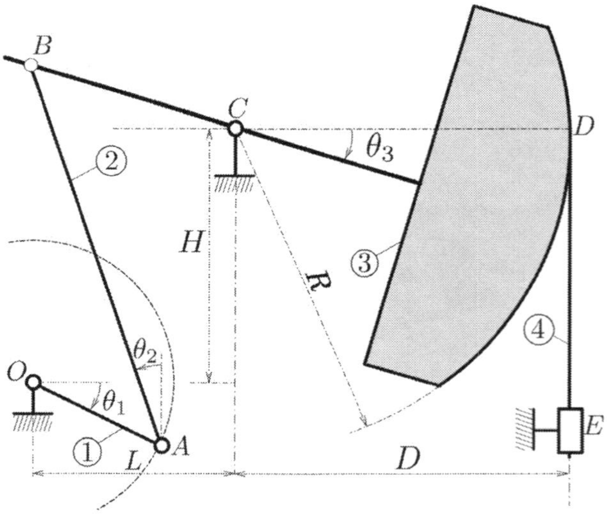

Problem 2.46 In this problem, we propose to re-examine example problem 2.24 by a graphical solution. The sketch shows a speed variator in which the rotary motion of the drive shaft 1 is turned into the oscillatory rotation of an output shaft 4 of adjustable amplitude. More specifically, the input crankshaft 1 transmits its motion through pin A to the connecting rod 2 which slides into the hollow shaft 3. Body 3 is guided by the ball joint 5 of center B. The output shaft of center D is driven by the connecting rod 4 which is pinned at C with body 3. It is possible to adjust the output rotation by displacing the center B.

At the given configuration of the mechanism, the angles take the values $\theta_1 = 45°$, $\theta_2 = 12.4°$ and $\theta_4 = 18.8°$. The geometry is defined by the constant lengths $\lambda = |OB| = 55$ mm, $\mu = |BD| = 87$ mm, $l_1 = |OA| = 14$ mm, $l_4 = |CD| = 36$ mm, $|AB| = 46$ mm, $|BC| = 54$ mm.

a. Construct the velocities $\mathbf{v}_{A\in 2/0}$ and $\mathbf{v}_{B\in 2/0}$. To justify the direction of $\mathbf{v}_{B\in 2/0}$ use the

identity $\mathbf{v}_{B\in 2/0} = \mathbf{v}_{B\in 2/3} + \mathbf{v}_{B\in 3/5} + \mathbf{v}_{B\in 5/0}$. Deduce the location of IC I_2 of body 2.
b. Sketch the direction of $\mathbf{v}_{C\in 2/0}$, $\mathbf{v}_{C\in 4/0}$, and $\mathbf{v}_{C\in 3/2}$. Then construct these velocities so that the identity $\mathbf{v}_{C\in 4/0} = \mathbf{v}_{C\in 4/3} + \mathbf{v}_{C\in 3/2} + \mathbf{v}_{C\in 2/0}$ is satisfied (Note that $\mathbf{v}_{C\in 4/3} = \mathbf{0}$).
c. Use the construction of $\mathbf{v}_{C\in 4/0}$ to determine the ratio $\dot{\theta}_4/\dot{\theta}_1$.
Answer: $\dot{\theta}_4/\dot{\theta}_1 = 0.287$

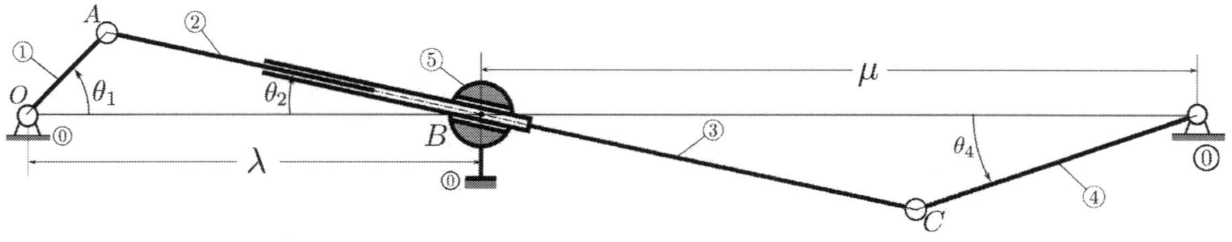

Problem 2.47 At the instant shown, camshaft 1 rotates about O at angular velocity $\omega_1 = 5$ rev/s in the counterclockwise direction. It is in contact at A with piston 2 which is (i) in translation relative to fixed support 0 and (ii) pinned at B with the connecting rod 3. Finally, connecting rod 3 is pinned at center C of geared wheel 4 which rolls without slipping at D with a fixed ring gear of center E. The geometry is defined by $|OA| = 10$ mm, $|BC| = 40$ mm, $|CD| = 15$ mm, $|DE| = 35$ mm, and the angles showed in the sketch. The goal of the problem is to find the velocity of center C and the angular velocity of geared wheel 4 at the instant shown by a graphical method.
a. From the knowledge of $\mathbf{v}_{A\in 1/0}$ and by the use of identity $\mathbf{v}_{A\in 1/0} = \mathbf{v}_{A\in 1/2} + \mathbf{v}_{A\in 2/0}$, find the direction and magnitude of $\mathbf{v}_{A\in 2/0}$.
b. From the knowledge of $\mathbf{v}_{B\in 3/0}$ and $\mathbf{v}_{C\in 3/0}$, construct the position of instantaneous center of rotation $I_{3/0}$.
c. Deduce the direction and magnitude of velocity $\mathbf{v}_{C\in 4/0}$ and angular velocity ω_4.

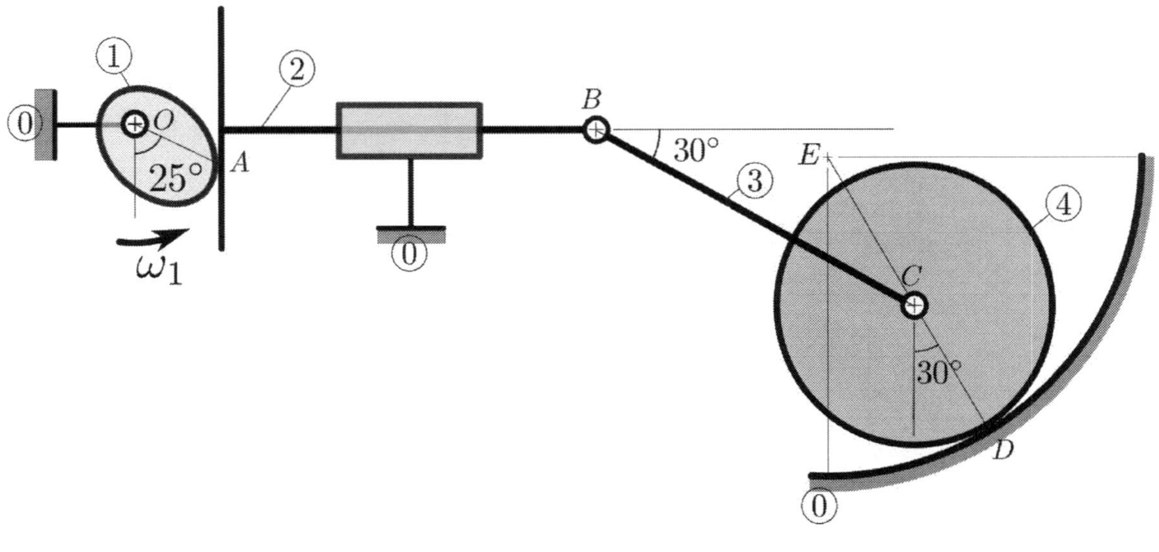

PART II: KINETICS

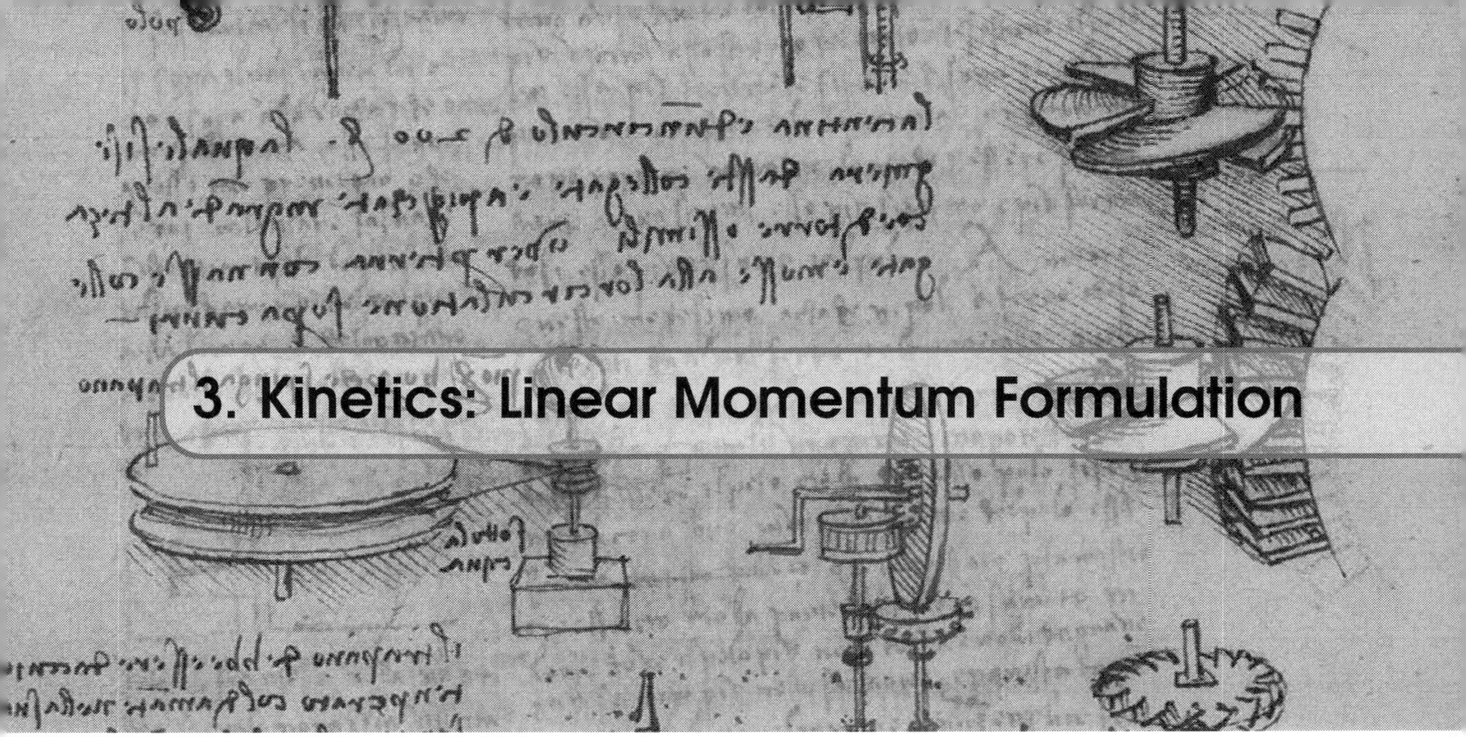

3. Kinetics: Linear Momentum Formulation

In Part II of this course, *kinetics*, we explore the relationship between the motion of material systems and the causes of their motion. In Chapter 3, the formulation of kinetics principles is achieved in terms of *linear momentum.* This formulation is applicable to particle motion and to the motion of the mass center of material systems. We start in Section 3.1 by introducing *Newton's laws of motion* which are the foundation of classical mechanics. They are applicable to point masses, and they presuppose the existence of a *Newtonian referential*. In Section 3.2, we describe the various referentials used in practice, and assess their Newtonian character. By formulating Newton's Second Law in a non-Newtonian referential, we shed light on the conditions under which a given referential can safely be assumed Newtonian. Central to classical mechanics is *Newton's Universal Law of Gravitation* (Section 3.3) which describes the effect of gravitational interaction, one of several fundamental action-at-a-distance interactions. Action-at-a-distance forces occur between two interacting systems without being in physical contact. Whenever two interacting material systems are physically in contact, the corresponding *contact forces* are more complex to model, and one must often rely on empirical laws derived from experiments. Examples of contact forces include dry frictional forces between rigid bodies, fluid-solid forces of interaction, forces due elastic springs, tensions due to ideal strings, and more. These forces play an essential role in mechanical systems. Our description is limited to forces due to dry contact, strings and springs (Sections 3.4 and 3.5). We then outline general guidelines for the solution of particle kinetics problems (Section 3.6).

Newton's Second Law relates the rate of change of *linear momentum* to the force applied to the particle, and it can be extended to other material systems, such as system of particles, rigid body or a system of rigid bodies whose study is conducted in Section 3.7. First, one can show that the linear momentum of a material system of mass m and mass center G is equivalent to that of a fictitious particle of mass m located at all time at G. Furthermore, according to *Euler's First Principle*, this particle is in motion as if acted upon by the forces external to the material system. In particular, the forces of interactions between rigid bodies or particles comprising a system become internal and drop out of the formulation. Clearly, Euler's First Principle is not sufficient to fully predict the motion of a rigid body. For a complete analysis, an angular momentum formulation will be needed. It will be described in Chapter 4.

In this chapter, we will learn

1. to characterize a Newtonian referential,
2. to sketch a free-body diagram to summarize the forces applied on a body,
3. to solve kinetics problems by using Newton's Second Law or Euler's First Principle.

3.1 Newton's Laws of Motion

Newton's laws of motion are three physical laws which describe the relationship between the forces acting upon a particle, and its motion in response to those forces.
The motion is assumed to take place in a referential $\mathcal{E}$ equipped with a clock allowing the measurements of the time of occurrence of (simultaneous) events at various locations of $\mathcal{E}$. In classical mechanics, we make the fundamental assumption that two events which occur simultaneously in a referential must occur simultaneously in any other referential $\mathcal{F}$. Similarly, we assume that the distances measured in $\mathcal{E}$ and $\mathcal{F}$ between two simultaneous events are identical. In particular, the length measured between any two points of a referential remains constant relative to any other referential.
We also assume that a *mass* can be defined for any material system, and that forces can be defined by vectors or vector fields. Newton's laws of motion can then be stated as follows.

NEWTON'S FIRST LAW

There exists at least one referential $\mathcal{E}$ relative to which any particle P is either at rest or in uniform rectilinear motion in the absence of unbalanced forces acting on P.

Such referentials are called *Newtonian* (or *inertial*, or *Galilean*). Newton's First Law is also known as Newton's law of inertia. Hence, in a Newtonian referential, a particle at rest tends to stay at rest and a particle in motion tends to stay in motion at the same velocity, unless acted upon by unbalanced forces.

NEWTON'S SECOND LAW

Relative a Newtonian referential $\mathcal{E}$, the acceleration of a particle P is proportional to the resultant force $\mathbf{F}$ exerted on P, and inversely proportional to the mass m of P:

$$m\mathbf{a}_{P/\mathcal{E}} = \mathbf{F} \tag{3.1}$$

Remark 1. The first law is just a special case of the second law when the resultant force is zero.

Remark 2. The second law shows that the larger the mass of the particle, the larger the force required to set the particle in motion with the same acceleration.

Remark 3. The S.I. unit of mass is the kilogram (kg). The S.I. unit of force is the Newton (N): 1 N = 1 kg.m/s^2.

Newton's Third Law

For every action, there is an equal and opposite reaction. More specifically, if $\mathbf{F}_{1\to 2}$ is the force caused by a particle P_1 on a particle P_2, then particle P_2 exerts a force $\mathbf{F}_{2\to 1}$ opposed to $\mathbf{F}_{1\to 2}$ and colinear to the line joining P_1 and P_2:

$$\mathbf{F}_{2\to 1} = -\mathbf{F}_{1\to 2}, \qquad \mathbf{F}_{2\to 1} \times \mathbf{r}_{P_1P_2} = \mathbf{0} \tag{3.2}$$

Remark 4. The third law is also known as the *Law of Action and Reaction*. It is independent of the chosen referential, and is applicable whether the particles are in motion or not.

Remark 5. As will be seen in the next sections, the third law has important consequences in the derivation of first principles applied to systems of particles or systems of rigid bodies.

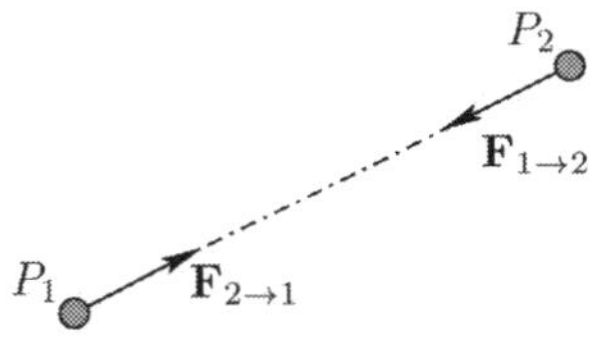

Figure 3.1 – *Newton's Third Law.*

3.2 Newtonian Referentials

No experiment has been devised to prove the existence of Newtonian referentials: for instance it is not possible to experimentally detect the uniform rectilinear motion of a body. However experiments and simple theoretical arguments can help us understand under what condition a referential can be characterized as a good approximation of a Newtonian referential.

First, we can show the following statement:

Newtonian Referentials

All Newtonian referentials are in uniform, rectilinear translation relative to one another.

Proof. This can be easily shown by using relative motion analysis of Section 2.9. Given a point P in uniform rectilinear motion relative to a Newtonian referential $\mathcal{E}$, its velocity relative to a referential $\mathcal{F}$ is given by formula (2.19)

$$\mathbf{v}_{P/\mathcal{F}} = \mathbf{v}_{P/\mathcal{E}} + \mathbf{v}_{P\in\mathcal{E}/\mathcal{F}}$$

If P is to be in uniform rectilinear motion relative to $\mathcal{F}$, then the transport velocity $\mathbf{v}_{P\in\mathcal{E}/\mathcal{F}}$ must be constant in magnitude and direction: since P is arbitrary, this implies that $\mathcal{F}$ is in uniform, rectilinear translation relative to $\mathcal{E}$. ■

Second, we can generalize Newton's Second Law (3.1) to the case of non-Newtonian referentials $\mathcal{F}$.

Law of Motion in a Non-Newtonian Referential

Newton's Second Law can be written relative to a non-Newtonian referential $\mathcal{F}$ as if $\mathcal{F}$ were Newtonian, as long as *inertial forces* $\tilde{\mathbf{F}}_i$ (proportional to the mass m of P) are added to the resultant force $\mathbf{F}$ acting on P

$$m\mathbf{a}_{P/\mathcal{F}} = \mathbf{F} + \tilde{\mathbf{F}}_i \tag{3.3}$$

where the inertial forces are of two types:
(i) *transport inertial force* $\tilde{\mathbf{F}}_{transp} = -m\mathbf{a}_{P\in\mathcal{F}/\mathcal{E}}$,
and
(ii) and *Coriolis inertial force* $\tilde{\mathbf{F}}_{cor} = -2m\boldsymbol{\omega}_{\mathcal{F}/\mathcal{E}} \times \mathbf{v}_{P/\mathcal{F}}$.
Here $\mathcal{E}$ represents a Newtonian referential relative to which $\mathcal{F}$ is in motion.

Proof. Equation (3.3) is easily shown: by applying the change of referential formula (2.20) we can substitute $\mathbf{a}_{P/\mathcal{E}} = \mathbf{a}_{P/\mathcal{F}} + \mathbf{a}_{P\in\mathcal{F}/\mathcal{E}} + 2\boldsymbol{\omega}_{\mathcal{F}/\mathcal{E}} \times \mathbf{v}_{P/\mathcal{F}}$ in Newton's Second Law (3.1) in Newtonian referential $\mathcal{E}$ to obtain (3.3) relative to non-Newtonian referential $\mathcal{F}$. ■

Remark 6. It is easy to show that both forces $\tilde{\mathbf{F}}_{transp}$ and $\tilde{\mathbf{F}}_{cor}$ vanish when the motion of $\mathcal{F}$ relative to $\mathcal{E}$ is a uniform, rectilinear translation.

Equation (3.3) plays a fundamental role in evaluating whether a referential can be considered adequately Newtonian or not: if both inertial forces $\tilde{\mathbf{F}}_{transp}$ and $\tilde{\mathbf{F}}_{cor}$ are negligible relative to resultant force $\mathbf{F}$ acting on P, then $\mathcal{F}$ can be considered Newtonian for the motion of P.

Two particular cases are often encountered in practice:

Case 1: Relative Translational Motion. If $\mathcal{F}$ is in translation relative to $\mathcal{E}$, then all points attached to $\mathcal{F}$ have the same acceleration, say $\mathbf{a}_F$, relative to $\mathcal{E}$. Since $\tilde{\mathbf{F}}_{cor} = \mathbf{0}$, we can then write (3.3) in the form

$$m\mathbf{a}_{P/\mathcal{F}} = \mathbf{F} - m\mathbf{a}_F \tag{3.4}$$

for any particle P.

Case 2: Relative Rotational Motion. If $\mathcal{F}$ is in rotation relative to $\mathcal{E}$ at constant angular velocity $\omega\hat{\mathbf{e}}_z$ about an axis $\Delta(O, \hat{\mathbf{e}}_z)$ fixed in $\mathcal{E}$, the inertial force $\tilde{\mathbf{F}}_{transp}$ is commonly referred to as the *centrifugal force*. It takes the following expression at any given time

$$\tilde{\mathbf{F}}_{transp} = -m\mathbf{a}_{P\in\mathcal{F}/\mathcal{E}} = m\omega^2\mathbf{r}_{HP}$$

where H is the projection of P onto axis Δ. Furthermore, the Coriolis force $\tilde{\mathbf{F}}_{cor} = -2m\omega\hat{\mathbf{e}}_z \times \mathbf{v}_{P/\mathcal{F}}$ is non-zero.

Remark 7. Case 2 is applicable to Earth whenever its rotational motion is taken into account.

Approximate Newtonian Referentials. Comparisons between experimental measurements and application of Newton's Second Law can indicate the extent to which a given referential can safely be assumed Newtonian. For most studies, the use of a Terrestrial referential leads to negligible errors and such a referential is an acceptable approximation of a Newtonian referential. On the other hand, for systems evolving on a large scale (such as satellites in orbit around the Earth), or evolving over long time interval, or systems evolving at very high speed (such as high speed gyroscopes), it is more appropriate to apply Newton's Second Law relative to the geocentric referential and hence accounting for Earth's rotation. Figure 3.2 displays three referentials:

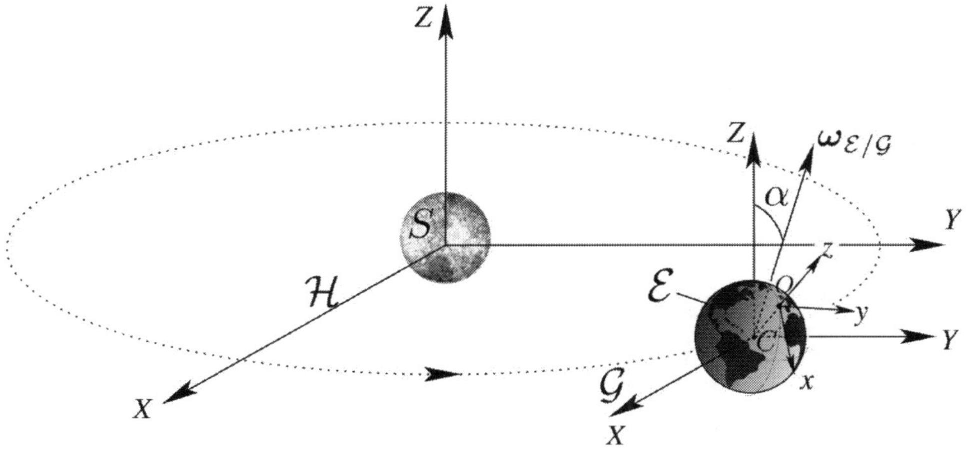

Figure 3.2 – *The heliocentric, geocentric and terrestrial referentials.*

1. The **heliocentric referential** $\mathcal{H}$ is defined by the center S of the Sun, and by three axes SX, SY and SZ pointing toward three "fixed" stars. The axes SX and SY can be chosen into the so-called *ecliptic plane* which contains the trajectory of center C of the Earth.
2. The **geocentric referential** $\mathcal{G}$ is defined by the center C of the Earth, and by axes CX, CY and CZ which remain parallel to the axes SX, SY, SZ of $\mathcal{H}$. $\mathcal{G}$ is in translation relative to $\mathcal{H}$, that is, $\boldsymbol{\omega}_{\mathcal{G}/\mathcal{H}} = \mathbf{0}$, and its motion is specified by the acceleration $\mathbf{a}_{C/\mathcal{H}}$ of center C. Hence its motion accounts for the orbital motion of Earth around the Sun. Given the small value of $|\mathbf{a}_{C/\mathcal{H}}| \approx 0.006\ \text{m/s}^2$, the non-Newtonian character of $\mathcal{G}$ can be considered quite weak for most applications.
3. The **terrestrial or laboratory referential** $\mathcal{E}$ attached to Earth with origin O taken at some fixed point of the surface of the Earth, axes Ox, Oy and Oz pointing southward, eastward, and upward, respectively. The motion of $\mathcal{E}$ relative to $\mathcal{G}$ is a rotation about the axis passing through the south and north poles, with angular velocity $\boldsymbol{\omega}_{\mathcal{E}/\mathcal{G}} = \boldsymbol{\omega}_{\mathcal{E}/\mathcal{H}}$ of magnitude $2\pi/(24 \times 3600) = 7.27 \times 10^{-5}$ rad/s, and directed from the south pole to the north pole. The direction of $\boldsymbol{\omega}_{\mathcal{E}/\mathcal{G}}$ is sensibly constant relative to the axis SZ normal to the ecliptic. Experiments show that $\mathcal{E}$ can be reasonably assumed Newtonian, as long as the extent and length of the motion are sufficiently small. *If $\mathcal{E}$ is assumed Newtonian, then any other referential $\mathcal{F}$ in uniform rectilinear translation relative to $\mathcal{E}$ can be considered Newtonian.* Consequently, any referential "in acceleration" relative to $\mathcal{E}$ cannot be considered Newtonian. One readily perceives the transport inertial force in a vehicle in acceleration or deceleration, or the centrifugal force perceived in a vehicle performing a sharp turn, or in roller coaster rides.

The famous pendulum experiment performed by Foucault in 1851, now displayed in many science museums around the world, demonstrated that the terrestrial referential is not rigorously Newtonian. This pendulum consists of a heavy mass suspended by a long cable.

It is observed that the plane of oscillations is not fixed, but exhibits a slow rotation. This can only be predicted by taking into account Earth's rotation, that is by applying Newton's Second Law relative to the geocentric referential $\mathcal{G}$. See Section 3.9 for more detail.

3.3 Gravitational Forces

THE gravitational force is a type of body forces, also known as action-at-a-distance forces which are transmitted from one continuum to another without the need of physical contact between them. Its expression is based on an empirical law known as Newton's *Law of Universal Gravitation* which can be stated as follows:

> **LAW OF UNIVERSAL GRAVITATION**
> Two particles P_1 and P_2 of mass m_1 and m_2, respectively, and at a distance r from one another, attract each other with a force proportional to their mass and inversely proportional to r^2. Hence, the force due to P_1 on P_2 can be written as
>
> $$\mathbf{F}_{1\to 2} = -G\frac{m_1 m_2}{|\mathbf{r}_{P_1P_2}|^3}\mathbf{r}_{P_1P_2} \tag{3.5}$$
>
> where $G = 6.67 \times 10^{-11}$ N.m^2/kg^2 is a universal constant, called *constant of gravitation*.

Note that $\mathbf{F}_{2\to 1} = -\mathbf{F}_{1\to 2}$ and $\mathbf{F}_{1\to 2} \times \mathbf{r}_{P_1P_2} = \mathbf{0}$ in accordance with Newton's Third Law. Every particle attracts any other particle. However, the effect of gravitational forces is only appreciable if at least one of the "particles" has a very large mass. For instance, on Earth's surface we neglect the mutual attraction of particles in comparison with Earth's gravitational pull. The spectacular success of Newton in predicting the motion of a falling body on Earth or the motion of the planets around the sun hinged on the Law of Universal Gravitation. This law is of enormous practical use for applications ranging from terrestrial to celestial mechanics.

GRAVITATIONAL FIELD DUE TO A CELESTIAL BODY. The effect of a discrete or continuous distribution Σ of mass on a particle is equivalent to a *gravitational field* $\mathbf{G}_P$ at any space location P, such that, a particle of mass m located at point P at a particular instant would be subject to the force

$$\mathbf{F}_{\Sigma\to P} = m\mathbf{G}_P \tag{3.6}$$

For a discrete set of particles $\Sigma = \{P_1, P_2, \cdots P_N\}$ of mass $m_1, m_2, \cdots, m_N$, we find

$$\mathbf{G}_P = -G\sum_{i=1}^{N}\frac{m_i\hat{\mathbf{u}}_i}{r_i^2} \tag{3.7}$$

with $r_i = |\mathbf{r}_{P_iP}|$ the distance from P_i to P, and $\hat{\mathbf{u}}_i = \mathbf{r}_{P_iP}/r_i$ the unit vector directed from P_i to P. Hence the gravitational field is obtained by a superposition of individual fields. We can extend the notion of gravitational field to the case of a continuous material system Σ:

$$\mathbf{G}_P = -G\int_\Sigma \frac{\rho\hat{\mathbf{u}}}{|\mathbf{r}_{QP}|^2}dV \tag{3.8}$$

where ρ is the mass density at a point Q of Σ, dV is an infinitesimal volume surrounding

Q, and $\hat{\mathbf{u}} = \mathbf{r}_{QP}/|\mathbf{r}_{QP}|$ the unit vector pointing from Q to P. In general, the calculation of this integral is difficult, except in the case of bodies with spherical material symmetry. Most celestial bodies such as Earth are spherical to a first order of approximation. If, in addition, we assume that the mass density ρ at point Q of body Σ is only a function of the distance from Q to the center C of the sphere, it is then clear, due to spherical symmetry, that the resulting gravitational field at point P is only a function of the distance $r = |\mathbf{r}_{CP}|$ from C to P.

Then we can show that the gravitational field of Σ at a point P *outside* Σ is given by

$$\mathbf{G}_P = -\frac{GM}{r^2}\hat{\mathbf{e}}_r \tag{3.9}$$

where $\hat{\mathbf{e}}_r = \mathbf{r}_{CP}/|\mathbf{r}_{CP}|$ is the unit vector pointing from center C to point P, and $M = \int_\Sigma \rho dV$ is the mass of Σ. Hence, *the gravitational field of a body with spherical material symmetry is equivalent to the gravitational field produced by a fictitious particle of mass M coinciding with center C*.

Remark 8. Equation (3.9) can be proved by applying Gauss's theorem for the flux of field $\mathbf{G}_P$ across the surface of the sphere of center C and radius r.

Earth's Gravitational Field. If we treat the Earth as a body of spherical material symmetry of center C and radius R, then the gravitation field at a point P located at altitude z above the Earth's surface is given by (with $r = R + z$) using (3.9)

$$\mathbf{G}_P = -G\frac{M_E}{(R+z)^2}\hat{\mathbf{e}}_r \tag{3.10}$$

where M_E is Earth's mass, and $\hat{\mathbf{e}}_r$ is the unit vector directed from C to P. If $z \ll R$, we can approximate $1/(R+z)^2$ as $(1 - 2z/R)/R^2$ to first order, and we obtain the approximation

$$\mathbf{G}_P \approx -g(1 - \frac{2z}{R})\hat{\mathbf{e}}_r \tag{3.11}$$

with

$$g = \frac{GM_E}{R^2} \tag{3.12}$$

the gravitational field at ground level. See Figure 3.3 (left). Hence, it is seen that the magnitude of $\mathbf{G}$ decreases with altitude, though the variation is very weak: with $R = 6,400$ km, we find a decrease of 1% at $z = 32$ km. Its direction also varies weakly: two points on the surface of the Earth whose corresponding unit vectors $\hat{\mathbf{e}}_r$ make an angle of 1° are separated by a distance of 112 km. Consequently, for most terrestrial applications, we may treat the gravitational field as a constant vector in magnitude and direction

$$\mathbf{G}_P = -g\hat{\mathbf{e}}_r \tag{3.13}$$

With $M_E = 6 \times 10^{24}$ kg, we find $g \approx 9.8$ m/s^2.

Weight of a Particle and Gravitational Acceleration. Consider in a referential $\mathcal{E}$ a plumb line experiment, whereby a mass particle P of mass m is suspended by a string attached at a point Q fixed in $\mathcal{E}$. If P is at rest in $\mathcal{E}$, we call *vertical* in $\mathcal{E}$ the direction taken by the plumb line. See Figure 3.3 (right).

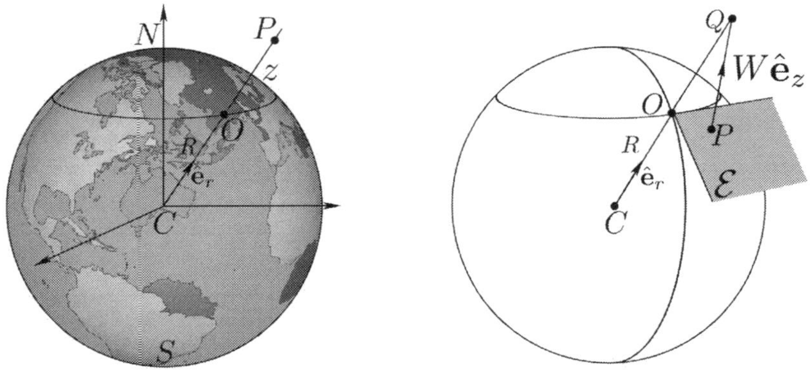

Figure 3.3 – *Earth's gravitational field (left). Plumb line experiment (right).*

Definition (Weight of a particle)

We define the *weight* $\mathbf{W}$ of particle P of mass m in referential $\mathcal{E}$ the force which balances the tension in the string (in the absence of any other external perturbing influence). If the weight of P is written as

$$\mathbf{W} = m\mathbf{g} \tag{3.14}$$

then, the vector $\mathbf{g}$ is called the *gravitational acceleration* at the location of P in referential $\mathcal{E}$.

Hence the weight of a particle and the corresponding gravitational acceleration depend on the local value of the gravitational field of one or more celestial bodies and on the "transport" acceleration of referential $\mathcal{E}$ in motion relative to some Newtonian referential. The same particle of mass m will have a different weight whether the plumb line experiment is done on Earth, on the Moon, in the space station, or in an elevator in acceleration. See problem 3.8 for the effect of Earth's rotation on the gravitational acceleration.

In terrestrial applications, $\mathbf{g}$ is mostly a function of Earth's gravity field and to a much lesser extent a function of Earth's rotation (more specifically, the motion of the terrestrial frame $\mathcal{E}$ relative to the geocentric referential $\mathcal{G}$). Consider the plumb line experiment done in the terrestrial referential $\mathcal{E}$ attached to Earth. Let us approximate the Earth's gravitational field as done previously

$$\mathbf{G}_P = -g\hat{\mathbf{e}}_r, \qquad g = \frac{GM_E}{R^2}$$

treating the Earth as a body of spherical material symmetry, and neglecting the effect of altitude ($z \approx 0$). Denote by $\hat{\mathbf{e}}_z$ the unit vector along the string from P to the attachment point Q, and by $\hat{\mathbf{e}}_r$ the unit vector directed from Earth's center C to P or to any "nearby" point such as origin O. See Figure 3.3 (right). The string tension is given by $W\hat{\mathbf{e}}_z$, denoting by W the weight of particle P in $\mathcal{E}$. If we treat $\mathcal{E}$ as a Newtonian referential then the equilibrium of P in $\mathcal{E}$ can be stated as

$$m\mathbf{a}_{P/\mathcal{E}} \equiv \mathbf{0} \approx -mg\hat{\mathbf{e}}_r + W\hat{\mathbf{e}}_z$$

This shows that in referential $\mathcal{E}$ the vertical is given by $\hat{\mathbf{e}}_z \approx \hat{\mathbf{e}}_r$ and that the weight of P is $W \approx mg$.

3.4 Contact Forces

MOST technological applications involve rigid bodies in contact with one another. Here we consider the case of a small body treated as a particle P constrained to slide along a rigid support. This precludes the possibility of this body to tip over the surface. The support may be the surface of a rigid body: the particle may not penetrate this rigid surface, but it may lose contact with it due to the effect of applied forces. In such a case, the constraint between P and the rigid body is said to be *unilateral*. See Figure 3.4 (left).

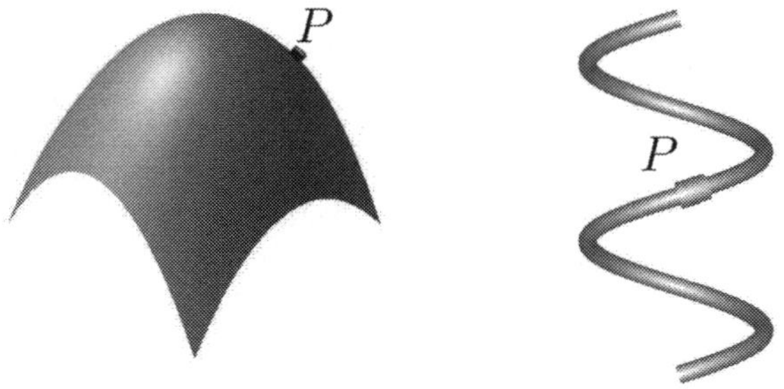

Figure 3.4 – *Small object in unilateral contact (left), in bilateral contact (right).*

The constraint of P with its support may also be *bilateral*. This may be realized by constraining the particle to slide between two parallel surfaces. Another bilateral constraint may be achieved by forcing P to slide along a curve, such as the case of a small ring sliding on a rigid straight or curved tube, or the case of a small object sliding inside a hollow tube of small cross-section. In a bilateral constraint, the particle is not allowed to leave the surface or the curve along which it moves. Note that, in the case of a constraint with a curve, the trajectory taken by P is known. See Figure 3.4 (right).
The contact of a particle P with a rigid support necessarily involves an additional force which modifies the trajectory that P would otherwise take under the action of a known applied force $\mathbf{F}_a$ and in the absence of the support. We call this force the *reaction of the support* on P, and we denote it by $\mathbf{R}$. Relative to a Newtonian referential $\mathcal{E}$, the dynamics of P is governed by the equation

$$m\mathbf{a}_P = \mathbf{F}_a + \mathbf{R} \tag{3.15}$$

according to Newton's Second Law. The reaction force $\mathbf{R}$ generally leads to three unknown scalar quantities in this equation in addition to the independent coordinates which parametrize the motion of P. If P is constrained to a surface, P has two degrees of freedom. If P is constrained to a curve, P has one degree of freedom. In both cases, the problem at hand is thus *indeterminate* since fewer equations than unknowns are available. To fix this indetermination, we need to come up with empirical constitutive laws governing the reaction force. When the contact between the rigid bodies is dry, these laws are known as *Coulomb laws*. The simplest modeling assumption corresponds to *frictionless contact*.

FRICTIONLESS CONTACT. The contact of P with its support $\mathcal{S}$ is said to be *frictionless* if the reaction force $\mathbf{R}$ remains perpendicular to the support. Since the velocity $\mathbf{v}_{P/\mathcal{S}}$ of P is always tangential to the support, frictionless contact implies $\mathbf{R} \cdot \mathbf{v}_{P/\mathcal{S}} = 0$. In the case of a particle constrained to move along a smooth curve $\mathcal{C}$ without friction, the reaction force must lie in the plane perpendicular to curve $\mathcal{C}$ at P. At any location P of $\mathcal{C}$, we can uniquely define the Frenet basis, comprised of the unit vectors $\hat{\mathbf{e}}_T$ (tangential), $\hat{\mathbf{e}}_N$ (principal normal) and

$\hat{\mathbf{e}}_B = \hat{\mathbf{e}}_T \times \hat{\mathbf{e}}_N$ (binormal). The reaction force must take the form $\mathbf{R} = R_N\hat{\mathbf{e}}_N + R_B\hat{\mathbf{e}}_B$, that is, the component $R_T = \mathbf{R} \cdot \hat{\mathbf{e}}_T$ must be zero at all time. The single degree of freedom which characterizes the motion of P can be taken as $s(t)$, the arc-length measured along $\mathcal{C}$ from some origin P_0. Then the motion of P along curve $\mathcal{C}$ is obtained by projecting (3.15) on $(\hat{\mathbf{e}}_T, \hat{\mathbf{e}}_N, \hat{\mathbf{e}}_B)$:

$$\begin{aligned} m\ddot{s} &= F_{aT} \\ m\frac{\dot{s}^2}{\rho} &= F_{aN} + R_N \\ 0 &= F_{aB} + R_B \end{aligned} \tag{3.16}$$

where ρ denotes the radius of curvature of curve $\mathcal{C}$ at point P, and F_{aT}, F_{aN}, F_{aB} are the components of the known applied force $\mathbf{F}_a$ on basis $(\hat{\mathbf{e}}_T, \hat{\mathbf{e}}_N, \hat{\mathbf{e}}_B)$. The problem is now well-posed: we have three unknowns $s(t)$, R_N and R_B solutions of three equations (3.16). A similar approach would show that, if P is in constrained to move along a surface $\mathcal{S}$, the problem is fully resolved in terms of the coordinates $(q_1(t), q_2(t))$ which parametrize the motion of P, and the sole component $R_N = \mathbf{R} \cdot \hat{\mathbf{e}}_N$ of the reaction along the unit normal vector to surface $\mathcal{S}$.

Properties

①. If the constraint of P with $\mathcal{S}$ is *bilateral*, then the normal components of $\mathbf{R}$ can change sign during the motion.

②. If, however, the constraint between P and $\mathcal{S}$ is *unilateral*, that is, if P is free to lose contact with surface $\mathcal{S}$, then the Cartesian coordinates of P satisfy an inequality of the type

$$f(x, y, z) \geq 0$$

the equality being satisfied as long as the contact is maintained. When contact is realized, we must impose the condition $R_N > 0$, where $R_N = \mathbf{R} \cdot \hat{\mathbf{e}}_N$ is the normal component of $\mathbf{R}$ measured along the unit *outward* normal $\hat{\mathbf{e}}_N$ to $\mathcal{S}$. As soon as $R_N \leq 0$ along the motion of P, particle P leaves the support $\mathcal{S}$ and in this case $\mathbf{R}$ vanishes.

③. The condition $\mathbf{R} \cdot \mathbf{v}_{P/\mathcal{S}} = 0$ implies that the reaction force $\mathbf{R}$ does not work relative to $\mathcal{S}$. Note that the support $\mathcal{S}$ may be in motion relative to referential $\mathcal{E}$: in this case the velocity $\mathbf{v}_{P/\mathcal{E}}$ is not necessarily tangent to the surface $\mathcal{S}$ or curve $\mathcal{C}$: $\mathbf{R}$ will work relative to $\mathcal{E}$! See Chapter 5 for more detail.

Finally, note the fundamental difference between the frictionless contact with a surface $\mathcal{S}$ and that with a curve $\mathcal{C}$:

1. In the case of frictionless contact with surface $\mathcal{S}$, for a given position of P on $\mathcal{S}$, the direction of reaction force $\mathbf{R}$ is perfectly known.
2. In the case of frictionless contact with curve $\mathcal{C}$, the path taken by P is known on $\mathcal{C}$, yet the direction of $\mathbf{R}$ is unknown.

Contact with Friction. Experiments show that frictionless contact is an idealized model, and that the reaction force $\mathbf{R}$ has a component which tends to slow down the motion of particle P, that is, which is opposed to the velocity $\mathbf{v}_{P/\mathcal{S}}$ of P relative to the support $\mathcal{S}$. In the case of a surface $\mathcal{S}$ of contact, the reaction force $\mathbf{R}$ is decomposed into a component N along unit outward normal $\hat{\mathbf{e}}_N$ and a tangential component $\mathbf{F}$, called *friction force*, perpendicular to

$\hat{\mathbf{e}}_N$, hence lying in tangential plane Π to surface $\mathcal{S}$ at P (See Figure 3.5):

$$\mathbf{R} = \mathbf{F} + N\hat{\mathbf{e}}_N$$

Definition (Coulomb laws of friction)

Assuming *dry contact*, that is, in the absence of lubrication, the empirical Coulomb laws state the existence of coefficients μ_s and μ_k, called *coefficients of static and kinetic friction*, which characterize the reaction force exerted on P in the following way:

Case 1: if P is *fixed relative* to $\mathcal{S}$, then $\mathbf{R}$ satisfies the inequality:

$$|\mathbf{F}| \leq \mu_s |N| \tag{3.17}$$

The reaction force lies inside the friction cone whose half-angle ϕ satisfies $\tan\phi = \mu_s$. See Figure 3.5 (left).

Case 2: if P is *in motion relative* to $\mathcal{S}$, then $\mathbf{R}$ satisfies the equality:

$$|\mathbf{F}| = \mu_k |N| \tag{3.18}$$

Furthermore, $\mathbf{F}$ is colinear and opposed to the velocity $\mathbf{v}_{P/\mathcal{S}}$ of P relative to its support. See Figure 3.5 (right).

Remark 9. Angle ϕ is the maximum inclination reached by the reaction force $\mathbf{R}$ relative to the normal direction $\hat{\mathbf{n}}$.

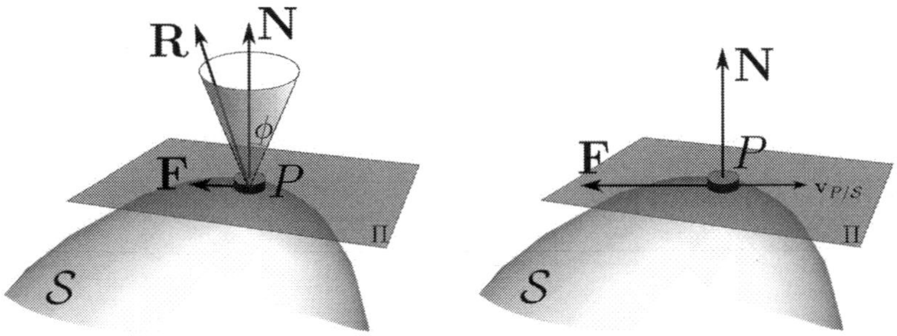

Figure 3.5 – *Reaction force* $\mathbf{R} = \mathbf{F} + \mathbf{N}$ *of support* $\mathcal{S}$ *on particle* P. Left: *P is in equilibrium relative to $\mathcal{S}$. $\mathbf{R}$ lies inside the friction cone* ($\tan\phi = \mu_s$). Right: *P slips relative to $\mathcal{S}$. The friction force $\mathbf{F}$ opposes velocity $\mathbf{v}_{P/\mathcal{S}}$ and its magnitude satisfies* $|\mathbf{F}| = \mu_k |\mathbf{N}|$.

Remark 10. The value of μ_s is in general larger than that of μ_k for given surfaces in contact. To simplify the analysis, coefficients μ_s and μ_k are often assumed to be equal. The value of these coefficients depends on the nature of the surfaces in contact. For instance, $\mu_s \approx 0.2$ for steel on steel contact, and $\mu_s \approx 0.6$ for rubber (car tire) in contact with a surface coated with bituminous asphalt (road). [1]

Remark 11. In the case of a guiding curve $\mathcal{C}$, the laws stated above are identical, with component $\mathbf{F}$ always tangential to $\mathcal{C}$. In addition, if P moves relative to $\mathcal{C}$, then $\mathbf{F}$ is always

[1] The study of friction forces is the domain of *tribology*. It is the subject of active research activities with huge economic stakes.

opposed to $\mathbf{v}_{P/C}$.

3.5 Force exerted by strings and springs

Here we examine the special case of two contact forces which often come into play in mechanics:

FORCE EXERTED BY AN IDEAL STRING. A particle P is attached to a support by an inextensible, massless string at A. See Figure 3.6 (left). If the string is taut, it takes the shape of a straight line AP and it exerts on P the force $\mathbf{T} = -T\hat{\mathbf{u}}$ where $\hat{\mathbf{u}}$ is the unit vector pointing from A to P. T is the *tension* in the string: it satisfies $T > 0$ as long as the string is taut, and becomes $T = 0$ as soon as the string is slack. In general, the tension $\mathbf{T}$ is unknown in both magnitude and direction. Note that the attachment point A is not necessarily fixed.

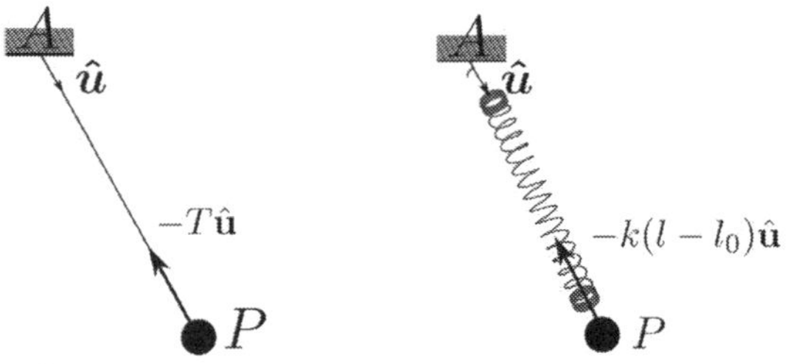

Figure 3.6 – *Force exerted by a string on particle P (left). Force exerted by a spring on particle P (right).*

FORCE EXERTED BY A LINEAR SPRING. A particle P is attached to a linear spring of constant k and unstretched length l_0. See Figure 3.6 (right). The spring is attached at point A of a support. If its length at any time is denoted l, the force $\mathbf{F}$ exerted by the spring on P is proportional to the *stretch* $\delta = l - l_0$. More specifically, $\mathbf{F}$ takes the expression

$$\mathbf{F} = -k(l - l_0)\hat{\mathbf{u}}$$

where $\hat{\mathbf{u}}$ is the unit vector pointing from A to P. Again the attachment point A is not necessarily fixed.

3.6 Solution Method of Particle Kinetics Problems

Now that we have described the laws governing motion, and the forces responsible for these motions, we are ready to solve particle kinetics problems. Such problems require an organized method of solution. To successfully attack each problem, the solution should ideally be broken down into the following steps:

> GENERAL PROCEDURE: PARTICLE KINETICS PROBLEMS
>
> ① Read the problem carefully, identify the knowns and unknowns. Define all relevant symbols. Formulate in a few words the main steps your solution of a specific problem will hinge upon.
>
> ② Define the system modeled as a particle P and the referential $\mathcal{E}$ relative to which the motion of P is studied. Identify whether referential $\mathcal{E}$ is Newtonian or not.

③ Define a coordinate system in $\mathcal{E}$ most suited for the motion of P and all relevant fixed or moving unit vectors associated with the chosen coordinate system. This is best done in a sketch.

④ Once the coordinate system is defined, determine the kinematic quantities relevant to the solution: this amounts to determining the expressions of $\mathbf{v}_P$ and $\mathbf{a}_P$ on the relevant unit vectors of the chosen coordinate system. If needed, define all initial conditions (initial position and initial velocity).

⑤ In problems involving systems of interconnected particles, determine the geometric/kinematic constraint equations.

⑥ Describe all forces exerted on P: this should involve only physical forces which will be sketched in a free-body diagram. The free-body diagram will typically include gravitational and contact forces.

⑦ Once the acceleration of P and the forces exerted on P have been defined, we can know apply Newton's Second Law which will be projected on 3 unit vectors to obtain 3 scalar equations (2 scalar equations for a two-dimensional problem).

⑧ At this point, we must compare the number of unknowns with the number of equations: if these two numbers match, the problem can be solved. Otherwise, some information is missing (typically an equation is missing or too many unknowns have been defined): return to steps 4 and 5 and fix this problem. We are then typically confronted with two types of problems:

> TYPE I: The problem involves solving for an unknown force and/or unknown acceleration at a given instant. The solution amounts to solving a simple (linear) algebraic problem.
>
> TYPE II: The problem involves solving for the position/velocity of P at a later time, given initial conditions. This problem is more challenging since it involves solving the *equations of motion*, that is, the ordinary differential equations which govern the motion of P. These equations of motion are found by decoupling the unknown motion coordinates from the unknown forces. Most likely, the equations of motion can be solved analytically (review Appendix C for basic techniques of integration). Otherwise, one must resort to a numerical solution (usually when frictional effects are taken into account).
>
> ⑨ Check the dimensional consistency of the results and finally find their numerical values.

Remark 12. In Step 2, do not use a non-Newtonian referential unless the effect of Earth's rotation is taken into account.

Remark 13. For instance, in Step 3, if the motion of P is rectilinear, define a Cartesian coordinate system Oxy, with Ox along the path of P. If the motion of P is circular, then the polar or NT coordinate system should be used.

Remark 14. Step 5 is often neglected: this leads to situations of unsolvability since equations are missing to complete the solution.

Remark 15. In Step 8, mistakes are often due to the improper use of Coulomb laws whenever frictional effects are taken into account.

Remark 16. For Type II problems, we will see in Chapter 5 that problems can often be solved in a more convenient manner by using an energy method, whenever only one degree of freedom is involved.

Remark 17. For Step 9, review Appendix A: checking the dimensionality is an easy way to find mistakes.

EXAMPLE PROBLEMS

Exercise 3.1 ▪ A small cube P of mass m is set in motion on a plane inclined at angle θ with the horizontal. It departs from point O with initial velocity $\mathbf{v}_0$. Under the effect of gravity and friction forces, it reaches the highest position A and then slides back toward point O. The times to travel from O to A and from A to O are measured to be T_1 and T_2, respectively. Show that the knowledge of T_1 and T_2 gives an expression of the coefficient of kinetic friction μ_k. Find μ_k for $m = 125$ g, $v_0 = 1.25$ m/s, $T_1 = 5.2$ s, $T_2 = 9.8$ s, and $\theta = 30°$. ▪

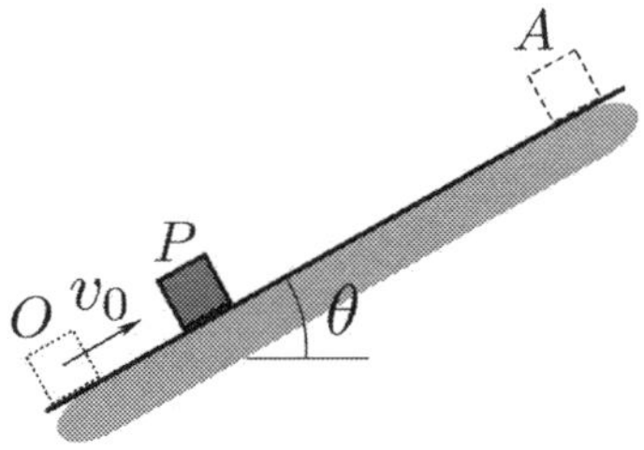

Solution. *Method: We determine the time T_1 taken by P to travel from O to A then the time T_2 to return from A to O. This leads to two equations for the unknowns $L = |OA|$ and μ_k. Eliminating L yields a relationship between μ_k and the ratio T_2/T_1.*

Step 1. We adopt a Cartesian coordinate system with the axis Ox directed along the incline, O being the initial position of particle P. Let $x(t)$ be the coordinate of P on axis Ox with $x(0) = 0$ and $\dot{x}(0) = v_0$ (initial conditions). See Figure 3.7 (left). The velocity and acceleration of P at time t take the expression:

$$\mathbf{v}_P = \dot{x}\hat{\mathbf{e}}_x, \qquad \mathbf{a}_P = \ddot{x}\hat{\mathbf{e}}_x$$

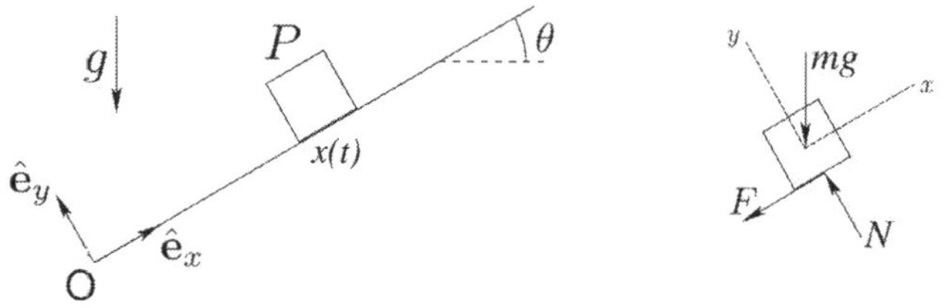

Figure 3.7 – *Problem 3.1 (Solution).*

Step 2. We sketch a free-body diagram in the upward phase of motion: P is subject to the gravitational force $m\mathbf{g}$ and the reaction of the support $\mathbf{R} = N\hat{\mathbf{e}}_y - F\hat{\mathbf{e}}_x$ with $F = \mu_k N$. See Figure 3.7 (right). Note that the friction force F must oppose the velocity of P.

Step 3. Upward motion from O to A: we apply Newton's Second Law: $m\mathbf{a}_P = m\mathbf{g} + N\hat{\mathbf{e}}_y - F\hat{\mathbf{e}}_x$ and project on the unit vectors $(\hat{\mathbf{e}}_x, \hat{\mathbf{e}}_y)$ to obtain two scalar equations

$$m\ddot{x} = -mg\sin\theta - F, \qquad 0 = N - mg\cos\theta$$

which yields $N = mg\cos\theta$ and $F = mg\mu_k\cos\theta$. We then obtain the (constant) acceleration of P:

$$\ddot{x} = -g(\sin\theta + \mu_k\cos\theta)$$

which gives by integration (with $\dot{x} = v_0$ at $t = 0$)

$$\dot{x} = -g(\sin\theta + \mu_k\cos\theta)t + v_0$$

At point A, $\dot{x} = 0$: this yields the time T_1 to reach A

$$g(\sin\theta + \mu_k\cos\theta)T_1 = v_0$$

Remark. You may ask at this point under what condition particle P slides back down. At point A, P is momentarily at rest: the reaction force is given by $F = mg\sin\theta$ and $N = mg\cos\theta$. Particle P remains at A as long as $\tan\theta \leq \mu_s$. To slide back down, the friction coefficients must satisfy $\mu_k < \mu_s < \tan\theta$. We must expect that as $\mu_k \to \tan\theta$ (from below), the time to travel from A back to O becomes increasingly large.

Note that P travels a maximum distance

$$L = -\frac{1}{2}g(\sin\theta + \mu_k\cos\theta)T_1^2 + v_0T_1 = \frac{1}{2}g(\sin\theta + \mu_k\cos\theta)T_1^2 \qquad [1]$$

Step 4. Downward motion from A to O: on the downward motion, the friction force is now directed along $+\hat{\mathbf{e}}_x$:

$$m\ddot{x} = -mg\sin\theta + F, \qquad 0 = N - mg\cos\theta$$

which now gives

$$\ddot{x} = g(-\sin\theta + \mu_k\cos\theta)$$

and by integration (reset $t = 0$ at A with $x(0) = L$ and $\dot{x}(0) = 0$ at A):

$$\dot{x}(t) = g(-\sin\theta + \mu_k\cos\theta)t, \qquad x(t) = \frac{1}{2}g(-\sin\theta + \mu_k\cos\theta)t^2 + L$$

To find the time taken to reach O we must impose $x(T_2) = 0$:

$$\frac{1}{2}g(-\sin\theta + \mu_k\cos\theta)T_2^2 + L = 0 \qquad [2]$$

Step 5. We can now solve for unknown μ_k by using equations [1-2] to find

$$\boxed{\mu_k = \frac{\tau^2 - 1}{\tau^2 + 1}\tan\theta, \qquad \tau = \frac{T_2}{T_1}}$$

Note that $\mu_k = 0$ for $\tau = 1$ as expected! Also $\mu_k \to \tan\theta$ as $\tau \to \infty$ (note that time T_2 is necessarily larger than time T_1). Numerically we find $\mu_k = \mathbf{0.32}$ with $\tau = 1.88$ and $\tan\theta = 1/\sqrt{3}$. ■

Exercise 3.2 ■■ A block P of mass $2m$ is suspended by a cable of length l to a fixed horizontal support at O. A second cable connects P to block Q of mass m via a small frictionless pulley fixed at point A ($|OA| = l$). The system is released from rest in the position shown. Neglect the size of the pulley.
Find the acceleration of P and Q immediately after release.
Are there configurations for which the system stays at rest upon release? ■

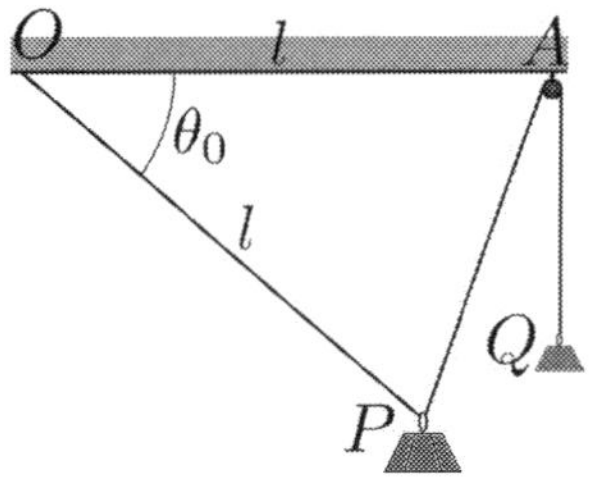

SOLUTION. *Method: The motion of Q is constrained to that of P. The accelerations of both P and Q are functions of the value of $\ddot{\theta}$ at $t = 0^+$. Also we note that the motion of P is circular. We find the acceleration of P by applying Newton's Second Law to Q then P: this should yield 3 equations. This is a Type I problem: we find $\ddot{\theta}$ by solving a system of linear equations.*

STEP 1. We choose the Newtonian referential $\mathcal{E}(O, \hat{\mathbf{e}}_x, \hat{\mathbf{e}}_y)$ attached to the horizontal support. See Figure 3.8 (left). We also define the polar unit vectors $(\hat{\mathbf{e}}_r, \hat{\mathbf{e}}_\theta)$ associated with the motion of block P: P has a circular path centered at O and of radius l. The initial conditions are given by $\theta = \theta_0$ and $\dot{\theta} = 0$. Angle θ_0 defines the initial configuration of the system: since triangle OPA is isosceles, we find angle $\phi = \pi/2 - \theta/2$. Its acceleration (in polar coordinates) is given by

$$\mathbf{a}_P = l(\ddot{\theta}\hat{\mathbf{e}}_\theta - \dot{\theta}^2\hat{\mathbf{e}}_r) = l\ddot{\theta}_0\hat{\mathbf{e}}_\theta \qquad [1]$$

since at time $t = 0^+$ we set $\dot{\theta} = 0$ (the system is released from rest). Finally block Q is in rectilinear motion along line $(A, \hat{\mathbf{e}}_y)$ and its acceleration is given by $\mathbf{a}_Q = \ddot{y}_Q\hat{\mathbf{e}}_y$. We can relate coordinate y_Q to angle θ by guaranteeing that the length of cable PAQ remains constant along the motion: with $|AP| = 2l\sin(\theta/2)$ we find

$$y_Q + 2l\sin\frac{\theta}{2} = Cst$$

Then two time-derivatives give

$$\ddot{y}_Q = -l\ddot{\theta}_0\cos\frac{\theta_0}{2} \qquad [2]$$

after taking account $\dot{\theta} = 0$.

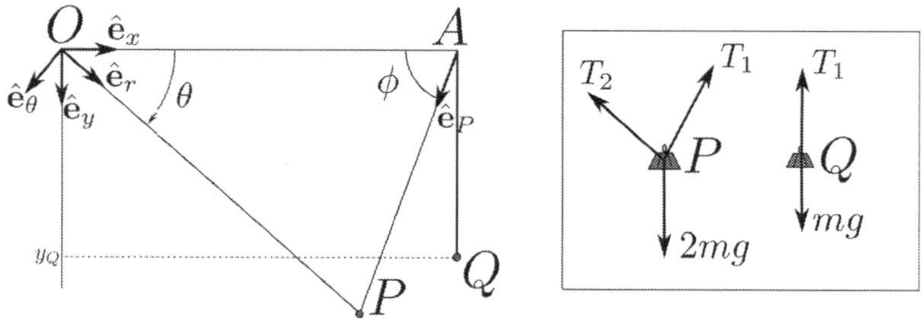

Figure 3.8 – *Problem 3.2 (Solution). Definition of the geometry (left). Free-body diagrams (right).*

STEP 2. Next we sketch free-body diagrams for block P, then for block Q (see Figure 3.8, right). Application of Newton's Second Law to P and Q gives the equations

$$2ml\ddot{\theta}_0\hat{\mathbf{e}}_\theta = 2mg\hat{\mathbf{e}}_y - T_2\hat{\mathbf{e}}_r - T_1\hat{\mathbf{e}}_P, \qquad m\ddot{y}_Q\hat{\mathbf{e}}_y = (mg - T_1)\hat{\mathbf{e}}_y \qquad [3-4]$$

We project equation [3] on $(\hat{\mathbf{e}}_r, \hat{\mathbf{e}}_\theta)$ to find

$$0 = 2mg\sin\theta_0 - T_2 - T_1\sin\frac{\theta_0}{2}, \qquad 2ml\ddot{\theta}_0 = 2mg\cos\theta_0 - T_1\cos\frac{\theta_0}{2} \qquad [5-6]$$

and [4] becomes after using [2]

$$-ml\ddot{\theta}_0\sin\frac{\theta_0}{2} = mg - T_1 \qquad [7]$$

We can solve [6-7] to find $\ddot{\theta}_0$:

$$\boxed{\ddot{\theta}_0 = \frac{2g}{l}\frac{2\cos\theta_0 - \cos\frac{\theta_0}{2}}{4+\sin\theta_0}} \qquad [8]$$

Once $\ddot{\theta}_0$ has been found, we obtain the acceleration of P and Q from [1] and [2].

Remark. Note that equation [8] does not give the equation of motion. [8] is only valid at $t = 0^+$. To find the equation of motion, equation [3-4] must be modified to account for $\dot{\theta} \neq 0$.

Step 3. We notice that $\ddot{\theta}_0$ can vanish if the condition $2\cos\theta_0 = \cos\frac{\theta_0}{2}$ is satisfied. If a solution of this equation is found, it must correspond to an equilibrium position. With $\cos\theta = 2\cos^2\frac{\theta}{2} - 1$ we find the quadratic equation

$$4\cos^2\frac{\theta_0}{2} - \cos\frac{\theta_0}{2} - 2 = 0$$

We find two roots $\cos\frac{\theta_0}{2} = (1 \pm \sqrt{33})/8$. We keep only the positive root and find the equilibrium angle $\theta_{eq} = \mathbf{65}^\circ$. If $\theta_0 < \theta_{eq}$ then angle θ will increase toward the value θ_{eq}. If $\theta_0 > \theta_{eq}$ then angle θ will decrease toward the value θ_{eq}. The system will likely oscillate about this equilibrium.

Remark. After finding the equation of motion, the reader can check that the period of small amplitude oscillations is $8.33(l/g)^{1/2}$.

■

Exercise 3.3 ■■ A pendulum of mass m and length l is attached in the interior of a car. The car is initially at rest and the pendulum's inclination angle with the vertical is $\theta = 0$. The car is then set in motion at a constant acceleration a_C along a fixed horizontal direction.
Find the tension in the string and the angular acceleration $\ddot{\theta}$ as a function of angle θ. Show that the pendulum can reach an equilibrium angle θ_e. If the value $\theta_e = 15^\circ$ is measured, find the corresponding acceleration a_C. ■

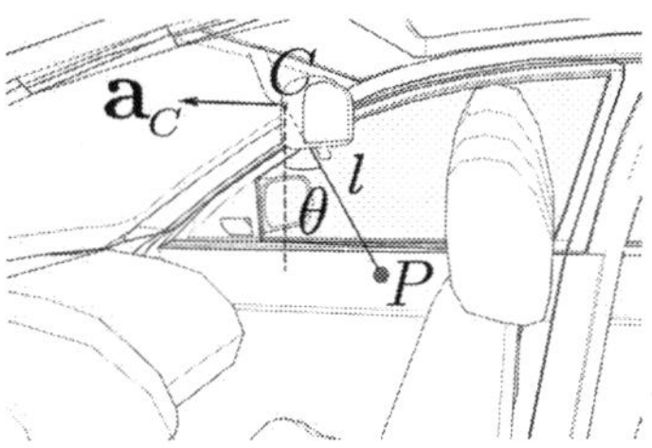

Solution. We cannot apply Newton's Second Law relative to the car: being in acceleration, it does not define a Newtonian referential. Instead we must apply the Second Law in the referential $\mathcal{E}$ relative to which the motion of the car is observed (acceleration $\mathbf{a}_C$ is measured relative to $\mathcal{E}$). We choose unit vectors $\hat{\mathbf{e}}_x$, $\hat{\mathbf{e}}_y$, $\hat{\mathbf{e}}_r$, $\hat{\mathbf{e}}_\theta$ and an arbitrary origin O attached to $\mathcal{E}$ as shown in Figure 3.9 (left). To find $\mathbf{a}_P$, we define Cartesian coordinate x_C so as to be able to write the position vector of P:

$\mathbf{r}_{OP} = x_C\hat{\mathbf{e}}_x + l\hat{\mathbf{e}}_r$ (with $\ddot{x}_C = a_C$). After two time-derivatives, we find $\mathbf{v}_P = v_C\hat{\mathbf{e}}_x + l\dot{\theta}\hat{\mathbf{e}}_\theta$ and

$$\mathbf{a}_P = a_C\hat{\mathbf{e}}_x + l(\ddot{\theta}\hat{\mathbf{e}}_\theta - \dot{\theta}^2\hat{\mathbf{e}}_r)$$

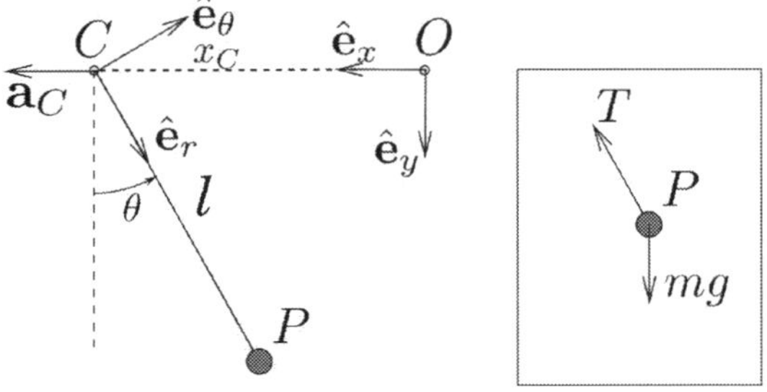

Figure 3.9 – *Problem 3.3*

After sketching a free-body diagram in Figure 3.9 (right) (P is subject to gravity mg and to the tension T of the string), we apply Newton's Second Law:

$$m\mathbf{a}_P = -T\hat{\mathbf{e}}_r + mg\hat{\mathbf{e}}_y$$

which we project on $(\hat{\mathbf{e}}_r, \hat{\mathbf{e}}_\theta)$ to obtain

$$-T + mg\cos\theta = m(-a_C\sin\theta - l\dot{\theta}^2), \qquad -mg\sin\theta = m(-a_C\cos\theta + l\ddot{\theta}) \qquad [1-2]$$

Equation [2] gives the angular acceleration

$$\ddot{\theta} = \frac{a_C}{l}\cos\theta - \frac{g}{l}\sin\theta \qquad [3]$$

Equation [3] is actually the equation of motion of P: given initial conditions θ_0 and $\dot{\theta}_0$ at $t = 0$ we can find a first integral of motion $\frac{1}{2}\dot{\theta}^2 = \frac{a_C}{l}\sin\theta + \frac{g}{l}\cos\theta + Cst$. See Appendix C for a review of ODEs. The equilibrium position $\theta = \theta_{eq}$ is obtained by setting $\ddot{\theta} = 0$:

$$\boxed{\tan\theta_e = \frac{a_C}{g}}$$

Numerically, we find $a_C = \mathbf{0.268}g = \mathbf{2.6}\ \mathrm{m/s^2}$. ■

Exercise 3.4 ▪ A small object P of mass m is placed on top of a horizontal oscillating platform set in motion by a crankshaft which rotates uniformly at angular velocity ω. The crankshaft geometry is defined by $|OA| = a$ and $|AB| = b$ with $b > a$. Under what condition can P travel along vertical axis Oz without losing contact with the platform? ▪

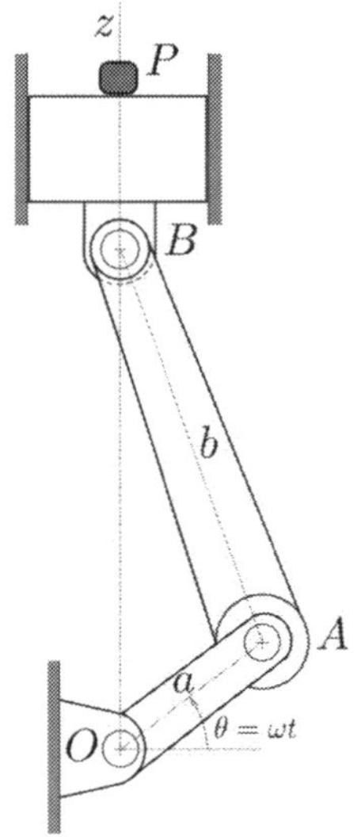

SOLUTION. *Method: We find the minimum value taken by the normal reaction N of the support on P. The needed condition is then found by imposing* $\min N > 0$.

STEP 1. We adopt a vertical Cartesian coordinate axis Oz attached to the Newtonian referential relative to which the system is in motion. P is in rectilinear motion along Oz: let $z(t)$ be the coordinate of P on axis Oz.

We can express z in terms of a, b and the angles θ and ϕ defined in Figure 3.10 (left):

$$z = a\sin\theta + b\cos\phi + c$$

where c is a constant and where angle ϕ is related to angle $\theta = \omega t$ according to

$$a\cos\theta = b\sin\phi$$

The velocity and acceleration of P at time t take the expression:

$$\mathbf{v}_P = \dot{z}\hat{\mathbf{e}}_z, \quad \dot{z} = a\omega(\cos\theta + \sin\theta\tan\phi)$$

(using $b\dot{\phi}\cos\phi = -a\omega\sin\theta$)

$$\mathbf{a}_P = \ddot{z}\hat{\mathbf{e}}_z, \quad \ddot{z} = a\omega^2(-\sin\theta + \cos\theta\tan\phi - \frac{a}{b}\frac{\sin^2\theta}{\cos^3\phi}) \qquad [1]$$

STEP 2. Particle P is subject to gravity $-mg\hat{\mathbf{e}}_z$ and to the normal reaction $N\hat{\mathbf{e}}_z$ due to the oscillating support. See 3.10 (center).
STEP 3. We can now apply Newton's Second Law which we project on $\hat{\mathbf{e}}_z$ to obtain

$$m\ddot{z} = N - mg$$

leading to the expression of N:

$$N = mg + m\ddot{z}$$

To guarantee that P never loses contact during the motion we need to impose $N > 0$ at all time. It is sufficient to impose $\min N > 0$. It is equivalent to find the position of the crankshaft for which $\ddot{z}$ is minimum. It is easy to see that each of the three terms which make up the expression of $\ddot{z}$ takes its minimum for $\theta = \pi/2$ for which $\phi = 0$ (it is easy to see that $-\pi/2 < \phi < -\pi/2$ during a full rotation of arm OA): then we obtain

$$\min N = N_{\theta=\pi/2} = mg + ma\omega^2(-1 - \frac{a}{b})$$

This is also confirmed graphically by plotting N vs θ in Figure 3.10 (left). Then for P to never lose contact we must impose the condition

$$\boxed{g > a\omega^2(1 + \frac{a}{b})}$$

■

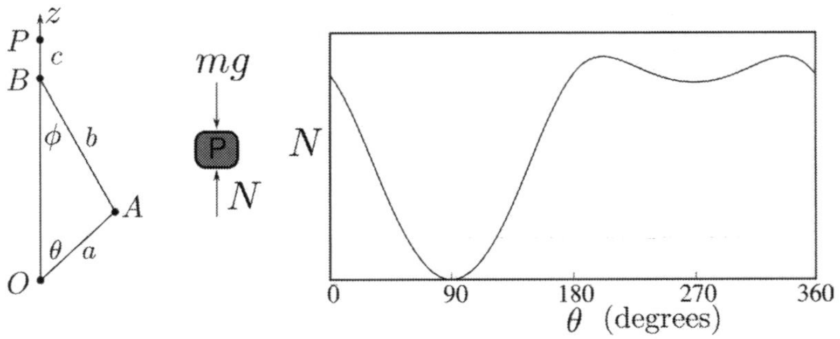

Figure 3.10 – *Problem 3.4 (Solution). Left: definition of geometry. Center: free-body diagram. Right: normal reaction force N vs θ for a/b = 1/2.*

Exercise 3.5 ▪ To measure the coefficient of static friction μ_s between two surfaces, the following experiment is conducted: a coin P of mass m treated as a particle is placed on a turntable at a distance d from its center. The turntable is set in motion at constant angular velocity Ω. At some critical value Ω_c, the coin is observed to slip. Show that knowledge of Ω_c gives the value of μ_s. Find μ_s for $m = 10$ g, $\Omega_c = 1.2$ rev/s, and $d = 12$ cm. ▪

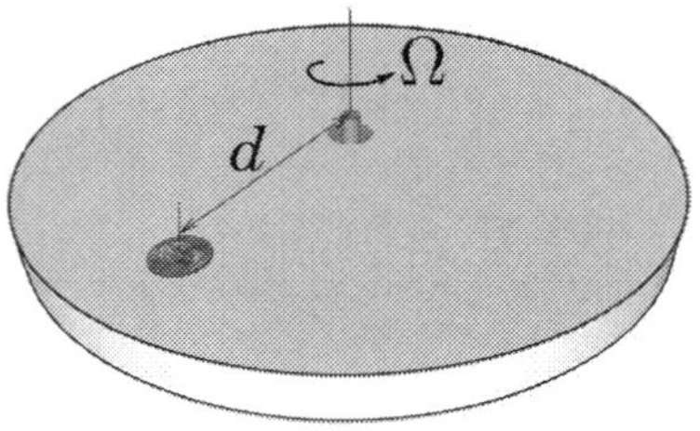

SOLUTION. *Method: as long as the coin does not slip, it is in equilibrium relative to the turntable and its trajectory is circular. Application of Newton's Second Law in polar coordinates will yield the reaction force. Then by imposing $|F| < \mu_s N$ we should find the expression of Ω_c vs μ_s.*

STEP 1. Assume that P is at rest relative to the turntable which rotates at constant angular velocity Ω relative to the referential $\mathcal{E}$ assumed Newtonian. The trajectory of P relative to $\mathcal{E}$ is a circle of center O and radius d. We adopt a polar coordinate system to define the position of P: the polar coordinates of P are $(r = d, \theta)$ with $\dot{d} = 0$ and $\dot{\theta} = \Omega$. We can then express the kinematics of P on the unit vectors $(\hat{\mathbf{e}}_r, \hat{\mathbf{e}}_\theta)$

$$\mathbf{v}_P = d\Omega\hat{\mathbf{e}}_\theta, \qquad \mathbf{a}_P = -d\Omega^2\hat{\mathbf{e}}_r$$

where we used the fact that Ω is constant. Also denote $\hat{\mathbf{e}}_z = \hat{\mathbf{e}}_r \times \hat{\mathbf{e}}_\theta$ (it would be equivalent to use the NT coordinate system since $\hat{\mathbf{e}}_N = -\hat{\mathbf{e}}_r$ and $\hat{\mathbf{e}}_T = \hat{\mathbf{e}}_\theta$).

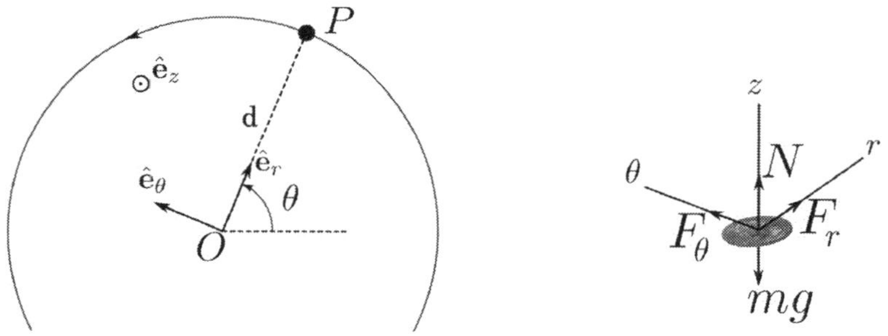

Figure 3.11 – *Problem 3.5 (Solution). Definition of geometry (left). Free-body diagram (right).*

STEP 2. Next we sketch a free-body diagram for P (see Figure 3.11, right): P is subject to the reaction force $\mathbf{R} = F_r\hat{\mathbf{e}}_r + F_\theta\hat{\mathbf{e}}_\theta + N\hat{\mathbf{e}}_z$ and to gravity $m\mathbf{g} = -mg\hat{\mathbf{e}}_z$. Since the friction force must lie in the plane of motion, it must have two components: $\mathbf{F} = F_r\hat{\mathbf{e}}_r + F_\theta\hat{\mathbf{e}}_\theta$. As long as P is not slipping relative to the turntable we must have the condition

$$|\mathbf{F}| < \mu_s|N| \qquad [1]$$

STEP 3. We now apply Newton's Second Law $m\mathbf{a}_P = \mathbf{R} + m\mathbf{g}$ to obtain 3 equations by projection on $(\hat{\mathbf{e}}_r, \hat{\mathbf{e}}_\theta, \hat{\mathbf{e}}_z)$:

$$\begin{aligned} -md\Omega^2 &= F_r \\ 0 &= F_\theta \\ 0 &= N - mg \end{aligned}$$

This gives the 3 components (F_r, F_θ, N). We now impose condition [1] to obtain

$$\boxed{\Omega^2 < \Omega_c^2 = \frac{\mu_s g}{d}}$$

Note that the fact that $F_\theta = 0$ stems from the assumption $\Omega = Cst$. With $\Omega_c = 1.2$ rev/s and $d = 12$ cm, we find $\mu_s \approx \mathbf{0.7}$. ■

Exercise 3.6 ▪▪ Two particles P of mass m and Q of mass M are joined by an inextensible massless string. P is constrained to move without friction on a horizontal support. The string can slide without friction through a small hole located at point O. Q is constrained to move along vertical axis Oz. At time $t = 0$, P is at a distance $r = r_0$ from O and is given a velocity $\mathbf{v}_0$ perpendicular to line OP. Show that Q stays in equilibrium if the initial conditions (r_0, v_0) satisfy a condition to be found. ■

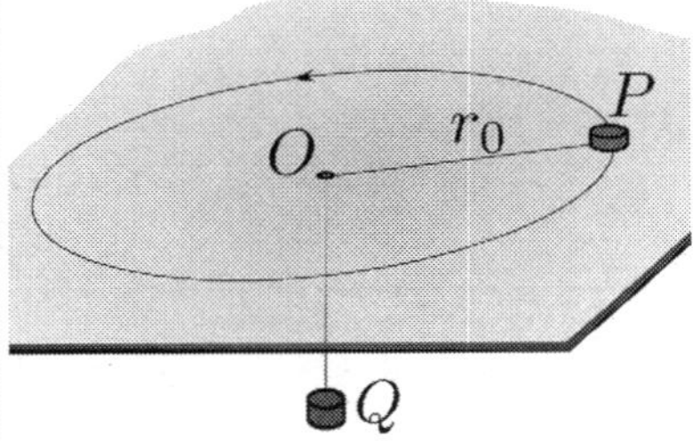

SOLUTION. STEP 1. If Q is in equilibrium, then P moves along a circle of center O and radius r_0. We can use the NT coordinate system to represent the kinematics of P.

$$\mathbf{a}_P = \dot{v}_P \hat{\mathbf{e}}_T + \frac{v_P^2}{r_0} \hat{\mathbf{e}}_N$$

We also define a unit vector $\hat{\mathbf{e}}_z$ directed downward.
STEP 2. We sketch the free-body diagrams of P and Q in Figure 3.12. Particle P is subject to the following forces:
- the gravitational force $m\mathbf{g} = mg\hat{\mathbf{e}}_z$,
- the reaction force $\mathbf{R} = -N\hat{\mathbf{e}}_z$ $(N > 0)$ of the horizontal support: it is normal to the surface of contact since friction is neglected.
- the tension $T\hat{\mathbf{e}}_N$ $(T > 0)$.

Particle Q is subject to the following forces:
- the gravitational force $M\mathbf{g} = Mg\hat{\mathbf{e}}_z$,
- the tension $\mathbf{T} = -T\hat{\mathbf{e}}_z$ $(T > 0)$, neglecting the friction of the sliding string through the hole at O.

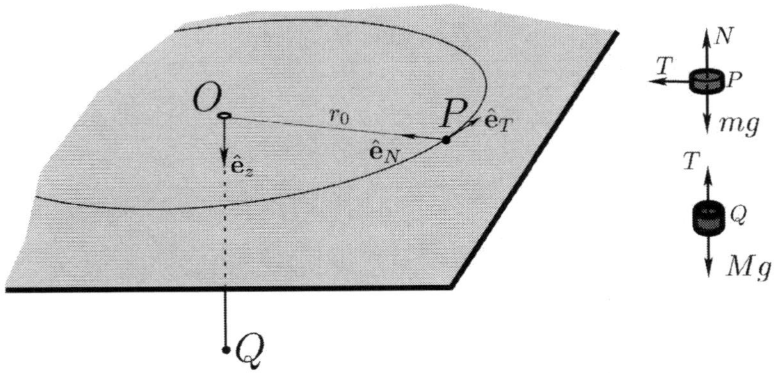

Figure 3.12 – *Problem 3.6 (Solution).*

STEP 3. We now apply Newton's Second Law (i) for P: $m\mathbf{a}_P = m\mathbf{g} + \mathbf{R} + T\hat{\mathbf{e}}_N$ and (ii) for Q: $\mathbf{0} = M\mathbf{g} - T\hat{\mathbf{e}}_z$ and project on basis $(\hat{\mathbf{e}}_N, \hat{\mathbf{e}}_T, \hat{\mathbf{e}}_z)$ to obtain four scalar equations for the three unknowns v_P, N, T:

$$m\frac{v_P^2}{r_0} = T, \quad m\dot{v}_P = 0, \quad 0 = mg - N, \quad Mg - T = 0 \qquad [1-4]$$

Equation [2] shows that the velocity of P stays constant: $v_P = v_0$. Then equations [1] and [4] lead to $m\frac{v_0^2}{r_0} = Mg$. This leads to the condition

$$\boxed{\frac{mv_0^2}{Mgr_0} = 1}$$

for the circular motion of P to exist. Note that this simple approach does not tell us whether this circular motion is stable or not. See Problem 5.11 for a more complete solution. ■

Exercise 3.7 ▪▪▪ A small ball P of mass m is in motion in a vertical plane by staying in contact with a bowl-like support in the shape of an arc of cycloid parametrized as

$$\begin{aligned} x &= R(\phi + \sin\phi) \\ y &= R(1 - \cos\phi) \end{aligned} \qquad (-\pi < \phi < \pi)$$

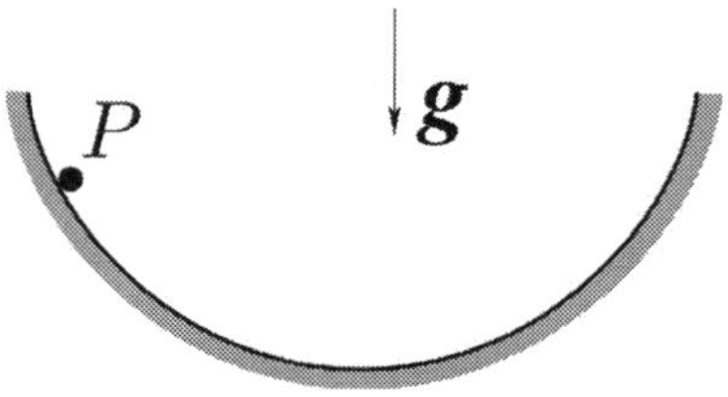

Upon release from rest, it oscillates along this support under the effect of gravity. The contact is assumed frictionless.
Find the equation of motion in terms of the arc-length s (set $s = 0$ at $\phi = 0$). Then find the period of oscillations of P: show that the result is independent of the amplitude of the oscillation! Compare this result with the case of a circular support:

$$x = R\sin\phi, \qquad y = R(1 - \cos\phi)$$

■

Solution. The problem statement suggests the use of the NT coordinate system to find the equation of motion: we can justify this choice by examining Newton's Second Law:

$$m\mathbf{a}_P = m\mathbf{g} + N\hat{\mathbf{e}}_N \qquad [1]$$

given that P is subject to gravity and the reaction N of the support. The latter is normal to the path since the contact is frictionless. Hence to eliminate N we need to project [1] onto the unit tangential vector $\hat{\mathbf{e}}_T$ (See Figure 3.13):

$$\mathbf{a}_P \cdot \hat{\mathbf{e}}_T = \mathbf{g} \cdot \hat{\mathbf{e}}_T \qquad [2]$$

Step 1. We find the tangent unit vector $\hat{\mathbf{e}}_T$ along the path by determining the velocity of P:

$$\mathbf{v}_P = \dot{x}\hat{\mathbf{e}}_x + \dot{y}\hat{\mathbf{e}}_y = R\dot{\phi}[(1+\cos\phi)\hat{\mathbf{e}}_x + \sin\phi\hat{\mathbf{e}}_y] = 2R\dot{\phi}\cos\frac{\phi}{2}(\cos\frac{\phi}{2}\hat{\mathbf{e}}_x + \sin\frac{\phi}{2}\hat{\mathbf{e}}_y) = \dot{s}\hat{\mathbf{e}}_T$$

leading to

$$\hat{\mathbf{e}}_T = \cos\frac{\phi}{2}\hat{\mathbf{e}}_x + \sin\frac{\phi}{2}\hat{\mathbf{e}}_y \qquad \dot{s} = 2R\dot{\phi}\cos\frac{\phi}{2}$$

where s denotes the arc-length measured on the trajectory of P.

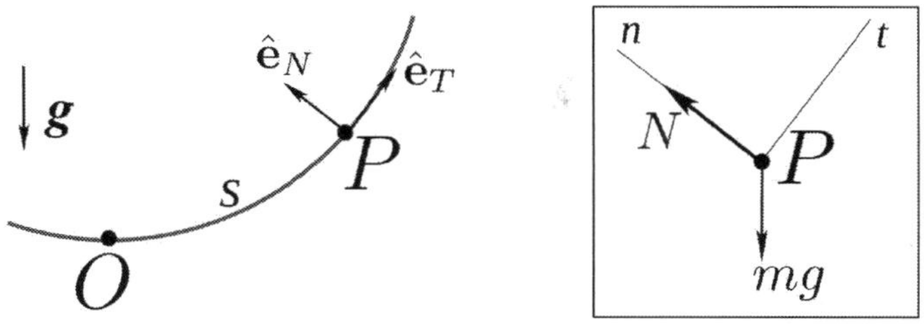

Figure 3.13 – *Problem 3.7 (Solution)*

STEP 2. We then obtain the r.h.s. of [2]:

$$\mathbf{g} \cdot \hat{\mathbf{e}}_T = -g\hat{\mathbf{e}}_y \cdot \hat{\mathbf{e}}_T = -g \sin \frac{\phi}{2}$$

Recall that the tangential acceleration is equal to the rate of change of speed v_P: $\mathbf{a}_P \cdot \hat{\mathbf{e}}_T = \dot{v}_P = \ddot{s}$. So equation [2] becomes

$$\ddot{s} = -g \sin \frac{\phi}{2} \qquad [3]$$

STEP 3. Next we must relate s and ϕ: we use the speed $v_P = \dot{s} = 2R\dot{\phi} \cos \frac{\phi}{2}$ and this gives

$$\frac{ds}{d\phi} = 2R \cos \frac{\phi}{2}$$

We can integrate this equation to find $s = 4R \sin \frac{\phi}{2}$ assuming $s = 0$ at $\phi = 0$ as suggested. Now the equation of motion [3] becomes

$$\boxed{\ddot{s} = -\frac{g}{4R} s} \qquad [4]$$

We recognize a pendulum equation: the general solution of linear ODE [4] is $s(t) = s_0 \cos(\omega t + \theta_0)$ where s_0 and θ_0 are two constants of integration, and $\omega^2 = g/4R$. The period of oscillations

$$\boxed{T = \frac{2\pi}{\omega} = 4\pi \sqrt{\frac{R}{g}}} \qquad [5]$$

is a *constant independent of the amplitude* s_0, however large s_0 (actually s_0 must be less than $4R$). This is rather remarkable and unexpected: *no matter where P starts on the support, P will arrive at the bottom of the support in the same amount of time!*

Remark. The oscillations of P are said to be *isochronous*.

If we replace the cycloidal support by the circular support

$$x = R \sin \phi, \qquad y = R(1 - \cos \phi)$$

we should not expect the same property. It is easy to show that [4] now becomes

$$\ddot{s} = -g \sin(\frac{s}{R}) \qquad [4']$$

Comparing [4′] with [4] we find that the ODE [4′] is nonlinear. We can show that this causes the period to be a function of amplitude. It is possible to find an expression of the period of oscillations by first writing [4′] in the following non-dimensional form:

$$\frac{d^2\theta}{d\tau^2} = -\sin\theta$$

with $s = R\theta$, $t = \Omega\tau$ and $\Omega^2 = g/R$. We can integrate this equation by first pre-multiplying both sides by $d\theta/d\tau$ to find (see Appendix C)

$$\frac{d\theta}{d\tau} = \sqrt{2(\cos\theta_0 - \cos\theta)}$$

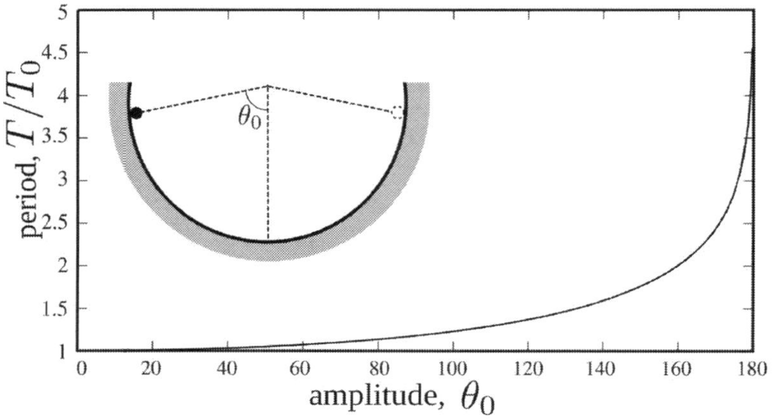

Figure 3.14 – *Problem 3.7 (Solution). Period of oscillation vs amplitude for a circular support (assuming no-loss of contact).*

where we have imposed $d\theta/d\tau = 0$ at $\theta = \theta_0$. This gives the period T versus the amplitude θ_0 (returning to dimensional variables)

$$T = \frac{4}{\Omega}\int_0^{\theta_0} \frac{d\theta}{\sqrt{2(\cos\theta_0 - \cos\theta)}}$$

With the help of a numerical solver, we compute T/T_0 ($T_0 = 2\pi/\Omega$) for the amplitude in the range $0 \leq \theta_0 < \pi$ and plot the results in Figure 3.14.

Remark. If the amplitude is small, then $\frac{d^2\theta}{d\tau^2} \approx -\Omega^2\theta$: the period is a constant to a first order of approximation.

■

In the following problems, we consider examples which illustrate the effect of Earth's rotation on simple static or dynamic experiments.

Exercise 3.8 ■■ The goal of this problem is to find the effect of Earth's rotation on the gravitational acceleration at a location of colatitude ϕ on Earth $\mathcal{E}$. Consider a particle P of mass m suspended by a string to a fixed point Q. By expressing the equilibrium of P in $\mathcal{E}$ under the effect of the gravitational force $m\mathbf{g}_0$ (pointing toward the Earth's center C), the tension $\mathbf{W}$ in the string (the weight of P) and the centrifugal inertial force $mR\omega_E^2 \sin\phi \hat{\mathbf{u}}$ due to Earth's rotation, find the weight W of P, the corresponding gravitational $\mathbf{g}$ acceleration and the angle θ defining the direction of the vertical $\hat{\mathbf{e}}_z$. Use $\phi = 45°$, $g_0 = 9.8$ m/s^2, $R = 6400$ km, and $\omega_E = 1$ rev/day.
What would the Earth angular velocity have to be to make the gravitational acceleration vanish at the equator? ■

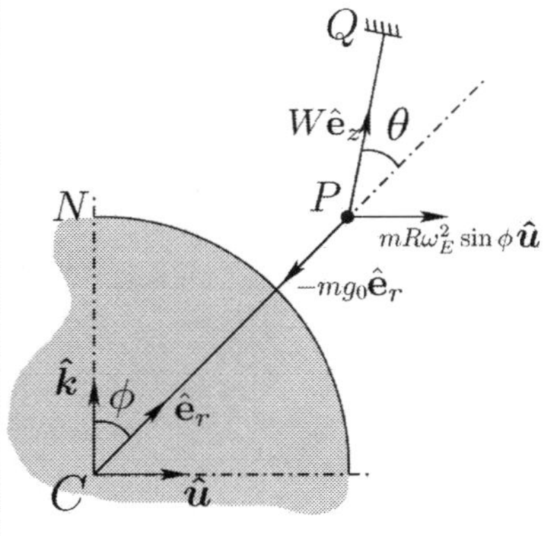

Solution. If we account for the rotation of the Earth, then $\mathcal{E}$ cannot be considered Newtonian. The equilibrium of P is now governed by the equation

$$m\mathbf{a}_{P/\mathcal{G}} = -mg_0\hat{\mathbf{e}}_r + W\hat{\mathbf{e}}_z$$

where $\mathcal{G}$ is the geocentric referential defined by Earth's axis of rotation SN, with two additional axes pointing to "fixed" stars, and about which referential $\mathcal{E}$ (Earth) has angular velocity $\boldsymbol{\omega}_{\mathcal{E}/\mathcal{G}} = \omega_E \hat{\mathbf{k}}$, with $\omega_E = 2\pi$ rad/day. $\hat{\mathbf{k}}$ is the unit vector directed from the south pole S to the north pole N. The acceleration of P in $\mathcal{G}$ is determined as

$$\mathbf{a}_{P/\mathcal{G}} = \mathbf{a}_{P\in\mathcal{E}/\mathcal{G}} = \omega_E\hat{\mathbf{k}} \times (\omega_E\hat{\mathbf{k}} \times \mathbf{r}_{CP}) = -R\omega_E^2 \sin\phi\hat{\mathbf{u}}$$

since P is fixed in $\mathcal{E}$, and the motion of $\mathcal{E}$ relative to $\mathcal{G}$ is a uniform rotation about axis $(C, \hat{\mathbf{k}})$. Unit vector $\hat{\mathbf{u}}$ is defined as the vector perpendicular to $\hat{\mathbf{k}}$ lying in the plane containing axis $(C, \hat{\mathbf{k}})$ and point P. Hence the equilibrium equation of P in $\mathcal{E}$ can be stated as

$$-mg_0\hat{\mathbf{e}}_r + W\hat{\mathbf{e}}_z + mR\omega_E^2 \sin\phi\ \hat{\mathbf{u}} = \mathbf{0}$$

Let us define the ratio of accelerations ϵ by

$$\epsilon = \frac{R\omega_E^2}{g_0}$$

With $R = 6400$ km, $\omega_E = \frac{2\pi}{24\times3600}$ rad/s, and $g_0 = 9.8$ m/s^2, we obtain a very small value $\epsilon = 0.0034$. Now from the vectorial equation

$$-\hat{\mathbf{e}}_r + \frac{W}{mg_0}\hat{\mathbf{e}}_z + \epsilon\sin\phi\hat{\mathbf{u}} = \mathbf{0} \qquad [1]$$

we can determine the direction of the vertical $\hat{\mathbf{e}}_z$ and the weight W of P in referential $\mathcal{E}$. Denote by θ the angle $(\hat{\mathbf{e}}_r, \hat{\mathbf{e}}_z)$. To eliminate W from equation [1], we take the cross-product with $\hat{\mathbf{e}}_z$ to obtain:

$$\tan\theta = \epsilon\frac{\sin\phi\cos\phi}{1-\epsilon\sin^2\phi} \qquad [2]$$

Since $\epsilon \ll 1$, the angle θ is small and can be approximated by (recall that $\tan\theta \approx \theta$

when $\theta \to 0$).

$$\boxed{\theta = \epsilon \sin\phi \cos\phi}$$

to first order in ϵ. To obtain the weight W and the corresponding gravitational acceleration $g = W/m$ in $\mathcal{E}$, we take the scalar product of equation [1] with $\hat{\mathbf{e}}_r$ to get

$$\boxed{W = mg(1 - \epsilon \sin^2\phi), \qquad g = g_0(1 - \epsilon \sin^2\phi)}$$

By setting $\epsilon = 0$, we recover the results obtained by treating $\mathcal{E}$ as a Newtonian referential. At colatitude $\phi = \pi/4$ we find

$$\theta = 0.2^\circ, \qquad g = 0.997\, g_0$$

These results show that considering $\mathcal{E}$ as a Newtonian referential for a body at rest is a valid approximation: the error incurred by neglecting the Earth's rotation is very small.
To obtain $g = 0$, we need to set $\epsilon = 1$: This gives ω_E 17 times larger than the value of 1 rev/day. ■

Exercise 3.9 ■■■ Consider a particle P of mass m in motion relative to a terrestrial referential $\mathcal{E}$ of Cartesian coordinate axes $Oxyz$ where axis Ox points to the South, axis Oy points to the East, and axis Oz points upward. P moves under the sole effect of a constant gravitational acceleration $\mathbf{g} = -g\hat{\mathbf{e}}_z$ and of Coriolis inertial force due to Earth's rotation.

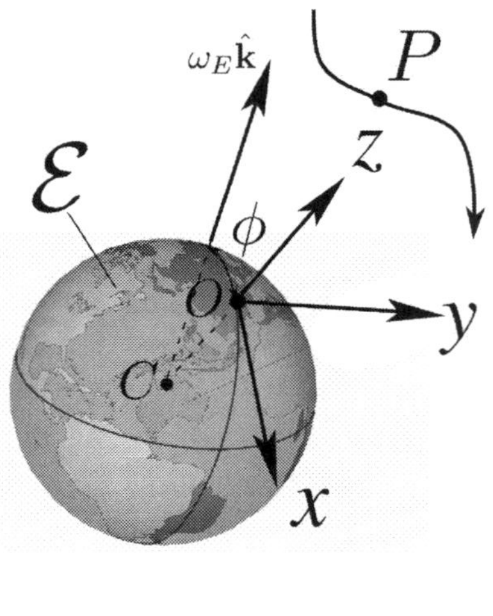

a. Show that the position vector $\mathbf{r} = \mathbf{r}_{OP}$ of P is solution of the equation

$$\frac{d\mathbf{r}}{dt} = \mathbf{v}_0 + \mathbf{g}t + 2\omega_E \hat{\mathbf{k}} \times (\mathbf{r}(0) - \mathbf{r}(t)) \qquad [1]$$

where $\mathbf{r}(0)$ is the initial value of $\mathbf{r}$, and $\mathbf{v}_0$ is the initial velocity of P relative to $\mathcal{E}$.

b. In order to find of solution of equation [1] we can take advantage of the small value of ω_E and seek an approximate solution in the form

$$\mathbf{r}(t) = \mathbf{r}_0(t) + \omega_E \mathbf{r}_1(t) + \omega_E^2 \mathbf{r}_2(t) + \cdots$$

Find the expressions of $\mathbf{r}_0(t)$, $\mathbf{r}_1(t)$ and $\mathbf{r}_2(t)$ as functions of time.

c. Assume that P is dropped without initial speed from a height h directly above O. Find the eastward and southward deviation from O of the impact location of P into the horizontal plane Oxy. Numerically, find the deviations of P for $h = 200$ m at colatitude $\phi = 45^\circ$. ■

SOLUTION. **a.** We apply Newton's Second Law in non-Newtonian referential $\mathcal{E}$ by accounting for the Coriolis inertial force (and neglecting the centrifugal force) to find

$$\frac{d\mathbf{v}_P}{dt} = \mathbf{g} - 2\omega_E \hat{\mathbf{k}} \times \mathbf{v}_P$$

Since $\hat{\mathbf{k}}$ is fixed relative to $\mathcal{E}$ we can integrate this equation to find

$$\mathbf{v}_P = \frac{d\mathbf{r}}{dt} = \mathbf{v}_0 + \mathbf{g}t - 2\omega_E \hat{\mathbf{k}} \times \big(\mathbf{r}(t) - \mathbf{r}(0)\big)$$

taking into account the initial conditions $\mathbf{v}_P = \mathbf{v}_0$ at $t = 0$.
b. An exact solution of equation [1] is possible. But it is more practical to solve it in terms of an expansion in powers of ω_E:

$$\mathbf{r}(t) = \mathbf{r}_0(t) + \omega_E \mathbf{r}_1(t) + \omega_E^2 \mathbf{r}_2(t) + \cdots \qquad [2]$$

The leading term $\mathbf{r}_0$ is found by substituting [2] in [1] and collecting all terms in $\omega_E^0 = 1$: we find

$$\frac{d\mathbf{r}_0}{dt} = \mathbf{v}_0 + \mathbf{g}t$$

whose integration gives

$$\boxed{\mathbf{r}_0 = \mathbf{r}(0) + \mathbf{v}_0 t + \frac{1}{2}\mathbf{g}t^2}$$

This solution corresponds to neglecting Coriolis force, and would be the exact parabolic trajectory if referential $\mathcal{E}$ is assumed Newtonian.
At the next order, we collect all terms in powers of ω_E and find

$$\frac{d\mathbf{r}_1}{dt} = -2\hat{\mathbf{k}} \times [\mathbf{r}_0(t) - \mathbf{r}(0)] = -2\hat{\mathbf{k}} \times [\mathbf{v}_0 t + \frac{1}{2}\mathbf{g}t^2]$$

whose integration gives

$$\boxed{\mathbf{r}_1 = -\hat{\mathbf{k}} \times [\mathbf{v}_0 t^2 + \frac{1}{3}\mathbf{g}t^3]}$$

since $\mathbf{r}_1(0) = \mathbf{0}$. Finally at order 2, we find

$$\frac{d\mathbf{r}_2}{dt} = -2\hat{\mathbf{k}} \times \mathbf{r}_1(t) = 2\hat{\mathbf{k}} \times \left[\hat{\mathbf{k}} \times (\mathbf{v}_0 t^2 + \frac{1}{3}\mathbf{g}t^3)\right]$$

whose integration gives

$$\boxed{\mathbf{r}_2 = \frac{2}{3}\hat{\mathbf{k}} \times \left[\hat{\mathbf{k}} \times (\mathbf{v}_0 t^3 + \frac{1}{4}\mathbf{g}t^4)\right]}$$

c. With $\mathbf{v}_0 = \mathbf{0}$ and $\mathbf{r}(0) = h\hat{\mathbf{e}}_z$, we find from the results of b)

$$\mathbf{r}(t) = h\hat{\mathbf{e}}_z + \frac{1}{2}\mathbf{g}t^2 - \frac{1}{3}\omega_E \hat{\mathbf{k}} \times \mathbf{g}t^3 + \frac{1}{6}\omega_E^2 \hat{\mathbf{k}} \times (\hat{\mathbf{k}} \times \mathbf{g})t^4 + O(\omega_E^3)$$

We may project $\mathbf{r}$ on axes $Oxyz$ to find the Cartesian coordinates

$$x(t) = \frac{1}{6}g\omega_E^2(\sin\phi\cos\phi)t^4, \quad y(t) = \frac{1}{3}g\omega_E(\sin\phi)t^3, \quad z(t) = h - \frac{1}{2}gt^2 - \frac{1}{6}g\omega_E^2(\sin^2\phi)t^4$$

The expressions of $x(t)$ and $y(t)$ show that P is deflected toward the east and south directions, the southward deflection being much smaller by the eastward deflection.

We find the time t_i of impact by neglecting the ω^2 term in the expression of $z(t)$:

$$t_i = \sqrt{\frac{2h}{g}}$$

Then southward and eastward deflections at impact are given by

$$\boxed{x_i = \frac{2h}{3g}\omega_E^2 \sin\phi\cos\phi, \qquad y_i = \frac{2}{3}\omega_E \sin\phi\sqrt{\frac{2h^3}{g}}}$$

For $g = 9.8$ m/s^2, $\phi = 45°$ and $h = 200$ m, we find the values $x_i \approx \mathbf{3.6 \times 10^{-8}}$ m and $y_i \approx \mathbf{0.044}$ m. Only the eastward deviation would be measurable experimentally. ■

3.7 Euler's First Principle

WE seek here to extend Newton's Second Law to more general material systems. Consider first the case of a system Σ of N particles P_1 of mass m_1, P_2 of mass m_2, ..., P_N of mass m_N in motion in a Newtonian referential $\mathcal{E}$. See Figure 3.15 (left). First we define the notion of mass center of Σ.

DEFINITION (MASS CENTER)

The *mass center* G of a system Σ of N particles is the point defined by

$$\mathbf{r}_{AG} = \frac{1}{m}(m_1\mathbf{r}_{AP_1} + \cdots + m_N\mathbf{r}_{AP_N}) \tag{3.19}$$

where A is an arbitrary point and $m = m_1 + \cdots + m_N$ is the mass of Σ. Furthermore, G satisfies the equation

$$m_1\mathbf{r}_{GP_1} + \cdots + m_N\mathbf{r}_{GP_N} = \mathbf{0} \tag{3.20}$$

Denote by $\mathbf{v}_i$ the velocity of particle P_i and by $\mathbf{a}_i$ its acceleration relative to $\mathcal{E}$.

DEFINITION (LINEAR MOMENTUM)

The quantity $m_i\mathbf{v}_i$ is called *linear momentum of particle* P_i. The sum of the N linear momenta $(m_1\mathbf{v}_1 + m_2\mathbf{v}_2 + \cdots + m_N\mathbf{v}_N)$ is the *linear momentum of system* Σ, denoted $\mathbf{L}_\Sigma$.

We can then prove the following statement:

THEOREM OF THE MASS CENTER

The linear momentum of system Σ is equal to the momentum of a fictitious particle of mass m located at the mass center G of Σ:

$$\mathbf{L}_\Sigma = m\mathbf{v}_G \tag{3.21}$$

Proof. This result is easily obtained: if point A in equation (3.19) is chosen to be fixed, then time-differentiation of (3.19) gives the expression of the velocity of mass center G

$$\mathbf{v}_G = \frac{1}{m}(m_1\mathbf{v}_1 + m_2\mathbf{v}_2 + \cdots + m_N\mathbf{v}_N)$$

∎

Next we describe the forces acting on the system.

External and internal forces. We denote by $\overline{\Sigma}$ the system's exterior (the complement of Σ). When particle P_i is viewed as an element of system Σ, the total force $\mathbf{F}_i$ acting on particle P_i can be considered to be the sum of two contributions:

(i) a contribution emanating from the exterior $\overline{\Sigma}$ denoted as $\mathbf{F}_{ext\to i}$ and called *external force* acting on P_i,

(ii) contributions emanating from the other $N-1$ particles denoted $\mathbf{F}_{j\to i}$, $j = 1, ..., N$ $(j \neq i)$ and called *internal forces*.

Hence we can write the resultant force acting on particle P_i as $\mathbf{F}_i = \mathbf{F}_{ext\to i} + \sum_{j=1}^N \mathbf{F}_{j\to i}$ (with the adopted convention $\mathbf{F}_{i\to i} = \mathbf{0}$). Then the resultant force acting on system Σ will be viewed as the sum of the N forces $\mathbf{F}_i$, $i = 1, ..., N$:

$$\sum_{i=1}^N \mathbf{F}_i = \sum_{i=1}^N \mathbf{F}_{ext\to i} + \sum_{i=1}^N \sum_{j=1}^N \mathbf{F}_{j\to i}$$

The first term $\sum_{i=1}^N \mathbf{F}_{ext\to i}$ is the resultant external force acting on Σ, and is denoted $\mathbf{F}_{ext\to\Sigma}$. The second term $\sum_{i=1}^N \sum_{j=1}^N \mathbf{F}_{j\to i}$ *vanishes* in view of Newton's Third Law: $\mathbf{F}_{i\to j} + \mathbf{F}_{j\to i} = \mathbf{0}$.

Resultant Force acting on a System

The resultant force acting on system Σ is the external force $\mathbf{F}_{ext\to\Sigma}$. The sum of the internal forces vanishes.

In order to extend Newton's Second Law to the case of a system of particles, we sum the following N equations written relative to a Newtonian referential

$$m_i\mathbf{a}_i = \mathbf{F}_i$$

Since $\sum_{i=1}^N \mathbf{F}_i = \mathbf{F}_{ext\to\Sigma}$ and $\sum_{i=1}^N m_i\mathbf{a}_i = m\mathbf{a}_G$. This leads to

Euler's First Principle: System of Particles

In a Newtonian referential, the mass center G of a system of particles of total mass m has the same motion as that of a fictitious particle G of mass m acted upon by the resultant external force

$$m\mathbf{a}_G = \mathbf{F}_{ext\to\Sigma} \tag{3.22}$$

Let us now consider the case of a *rigid body* $\mathcal{B}$. It is modeled as a continuum of volumetric mass density ρ: at any location P an infinitesimal volume dV has mass $dm = \rho dV$. Newton's

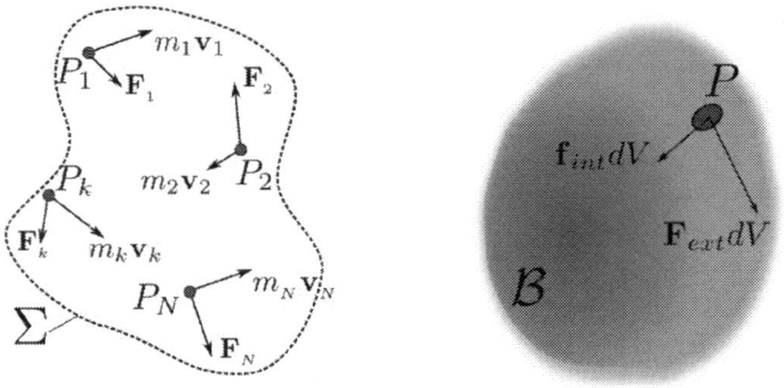

Figure 3.15 – *System of particles (left). Rigid body (right)*

Second Law applied to particle P of mass dm relative to a Newtonian referential $\mathcal{E}$ can be stated as

$$\mathbf{a}_{P/\mathcal{E}}dm = (\mathbf{F}_{ext} + \mathbf{f}_{int})dV \tag{3.23}$$

where $\mathbf{F}_{ext}$ and $\mathbf{f}_{int}$ are the external and internal forces (per unit volume) acting on P. See Figure 3.15 (right).
Integration over $\mathcal{B}$ gives

$$\int_{\mathcal{B}} \mathbf{a}_{P/\mathcal{E}}dm = \int_{\mathcal{B}} (\mathbf{F}_{ext} + \mathbf{f}_{int})dV \tag{3.24}$$

Two simplifications can be made at this point. First, the contribution of the internal forces vanishes due to Newton's Third Law: $\int_{\mathcal{B}} \mathbf{f}_{int}dV = 0$. Second, the integral $\int_{\mathcal{B}} \mathbf{a}_{P/\mathcal{E}}dm$ can be evaluated as a direct consequence of the definition of the mass center.

DEFINITION (MASS CENTER OF A RIGID BODY)

There exists a unique point G of $\mathcal{B}$, called *mass center*, such that

$$\int_{\mathcal{B}} \mathbf{r}_{GP}\, dm = \mathbf{0} \tag{3.25}$$

Then, for any point A (fixed or not) the position of G is given by:

$$\mathbf{r}_{AG} = \frac{1}{m}\int_{\mathcal{B}} \mathbf{r}_{AP}\, dm \tag{3.26}$$

where m is the mass of the body.

Proof. Equation (3.26) is a consequence of (3.25):

$$\int_{\mathcal{B}} \mathbf{r}_{GP}\, dm = \int_{\mathcal{B}} (\mathbf{r}_{GA} + \mathbf{r}_{AP})dm = m\mathbf{r}_{GA} + \int_{\mathcal{B}} \mathbf{r}_{AP}dm = 0$$

■

As a consequence of this definition, the linear momentum of the body can be obtained by time-differentiation of (3.26) (by choosing A fixed in $\mathcal{E}$):

LINEAR MOMENTUM OF A RIGID BODY

The linear momentum of rigid body $\mathcal{B}$ of mass m relative to referential $\mathcal{E}$ is equal to the linear momentum of its mass center G with mass m

$$\int_{\mathcal{B}} \mathbf{v}_{P/\mathcal{E}}\, dm = m\mathbf{v}_{G/\mathcal{E}} \tag{3.27}$$

Finally time-differentiation of (3.27) gives

$$\int_{\mathcal{B}} \mathbf{a}_{P/\mathcal{E}}\, dm = m\mathbf{a}_{G/\mathcal{E}} \tag{3.28}$$

Now equation (3.24) leads to the expression of Euler's First Principle.

EULER'S FIRST PRINCIPLE (RIGID BODY)

Relative to a Newtonian referential $\mathcal{E}$, the mass center G of a rigid body $\mathcal{B}$ of mass m has the same motion as that of a fictitious particle G of mass m acted upon by the resultant external force $\mathbf{F}_{ext\to\mathcal{B}}$ acting on $\mathcal{B}$

$$m\mathbf{a}_{G/\mathcal{E}} = \mathbf{F}_{ext\to\mathcal{B}} \tag{3.29}$$

Euler's First Principle can be generalized to systems of rigid bodies: consider a system Σ of two rigid bodies 1 (of mass m_1 and mass center G_1) and 2 (of mass m_2 and mass center G_2). Two types of forces are acting on bodies 1 and 2: forces external to the system $\mathbf{F}_{ext\to 1}$ and $\mathbf{F}_{ext\to 2}$ and forces of interactions $\mathbf{F}_{1\to 2}$ and $\mathbf{F}_{2\to 1}$ between the two bodies. Application of (3.29) to body 1 and body 2 gives

$$m_1\mathbf{a}_{G_1/\mathcal{E}} = \mathbf{F}_{ext\to 1} + \mathbf{F}_{2\to 1}$$

$$m_2\mathbf{a}_{G_2/\mathcal{E}} = \mathbf{F}_{ext\to 2} + \mathbf{F}_{1\to 2}$$

Adding these two equations gives

$$m_1\mathbf{a}_{G_1/\mathcal{E}} + m_2\mathbf{a}_{G_2/\mathcal{E}} = \mathbf{F}_{ext\to 1} + \mathbf{F}_{2\to 1} + \mathbf{F}_{ext\to 2} + \mathbf{F}_{1\to 2} = \mathbf{F}_{ext\to\Sigma}$$

where we have used $\mathbf{F}_{1\to 2} = -\mathbf{F}_{2\to 1}$ according to Newton's Third Law. This leads to:

EULER'S FIRST PRINCIPLE (SYSTEM OF RIGID BODIES)

Relative to a Newtonian referential $\mathcal{E}$, the mass center G of a system $\Sigma = 1,2$ of two rigid bodies of total mass m has the same motion as that of a fictitious particle G of mass m acted upon by the resultant *external* force acting on the system

$$\underbrace{m\mathbf{a}_{G/\mathcal{E}}}_{m_1\mathbf{a}_{G_1/\mathcal{E}} + m_2\mathbf{a}_{G_2/\mathcal{E}}} = \mathbf{F}_{ext\to\Sigma} \tag{3.30}$$

Remark 18. Euler's First Principle is in fact applicable to any material system (such as a fluid), as long as the mass of the system is conserved during the motion.

Remark 19. Euler's First Principle justifies the study of particles in mechanics, since the motion of the mass center of a body is equivalent to the motion of a particle.

PROPERTIES

Conservation laws can be obtained from Euler's First Principle under certain conditions:

①. If the resultant external force $\mathbf{F}_{ext\to\Sigma}$ vanishes at all time, then the linear momentum $m\mathbf{v}_G$ of the system (of particles or rigid bodies) is conserved, leading to

$$m_1\mathbf{v}_1 + m_2\mathbf{v}_2 + \cdots + m_N\mathbf{v}_N = Cst \tag{3.31}$$

where $m_i\mathbf{v}_i$ is the linear momentum of particle P_i or the ith body.

②. If the component of resultant external force $\mathbf{F}_{ext\to\Sigma}$ along a *fixed* direction $\hat{\mathbf{u}}$ vanishes at all time, then the component of the linear momentum $m\mathbf{v}_G$ of the system along $\hat{\mathbf{u}}$ is conserved, leading to

$$m_1\mathbf{v}_1\cdot\hat{\mathbf{u}} + m_2\mathbf{v}_2\cdot\hat{\mathbf{u}} + \cdots + m_N\mathbf{v}_N\cdot\hat{\mathbf{u}} = Cst \tag{3.32}$$

These equations have important consequences on the motion of the interacting bodies. They are called *first integrals of motion*, since they can be obtained by a single integration of the equations of motion.

Remark 20. Equation (3.32) is only obtained if $\hat{\mathbf{u}}$ is a constant unit vector: $\mathbf{F}_{ext\to\Sigma}\cdot\hat{\mathbf{u}} = 0$ implies $m\mathbf{a}_G\cdot\hat{\mathbf{u}} = 0 = d(m\mathbf{v}_G\cdot\hat{\mathbf{u}})/dt$ if $d\hat{\mathbf{u}}/dt = 0$.

EXAMPLE PROBLEMS

GENERAL PROCEDURE: EULER'S FIRST PRINCIPLE FOR THE SOLUTION OF KINETICS PROBLEMS

Proceed as described in Section 3.6:

① In Steps 3 and 4, parametrize the position of each particle of the system. Find the velocity and acceleration of each particle/mass center.

② In Step 6, draw a free-body diagram for the system, thereby eliminating the contributions of internal forces.

③ In Step 7, first apply Euler's First Principle for the system: inspect your free-body diagram to see if the system's linear momentum is conserved along a particular direction along which $\mathbf{F}_{ext\to\Sigma}$ vanishes. For a complete solution, return to the Newton's Second Law applied to each particle and extract particular equations to solve for the desired unknowns.

Exercise 3.10 ■ A "double" pendulum is assembled by connecting a small body P of mass m to a slider Q of mass M by a massless rod of length l. The rod can rotate freely about a pivot attached to the cart. The slider can slide without friction along the horizontal direction Ox of a Newtonian referential $\mathcal{E}$. The motion of the system lies in a vertical plane. At time $t = t_0$ the system is released without initial velocity, line QP making an angle $\theta = \theta_0$ with the vertical.
a. Describe the motion of the mass center G of the system. Deduce a first integral of motion. Integrate this equation.
b. Show that the path of P is elliptical. ■

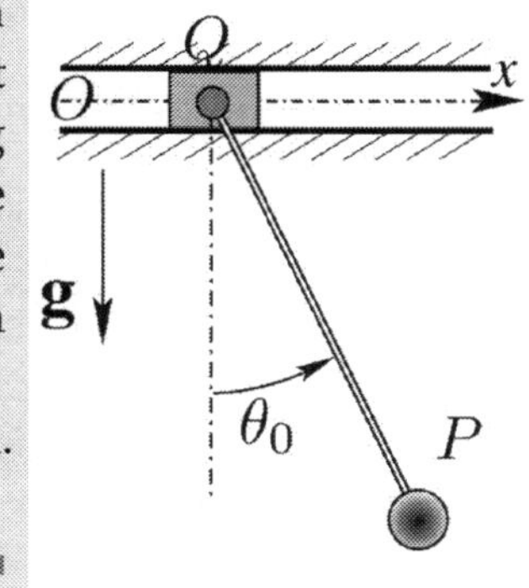

SOLUTION. **a.** We first need to define the problem's geometry: let $(\hat{\mathbf{e}}_x, \hat{\mathbf{e}}_z)$ a set of fixed unit vectors in the plane of motion with $\hat{\mathbf{e}}_x$ along horizontal axis Ox and $\hat{\mathbf{e}}_z$ along vertical downward. Define $\hat{\mathbf{e}}_y = \hat{\mathbf{e}}_z \times \hat{\mathbf{e}}_x$.

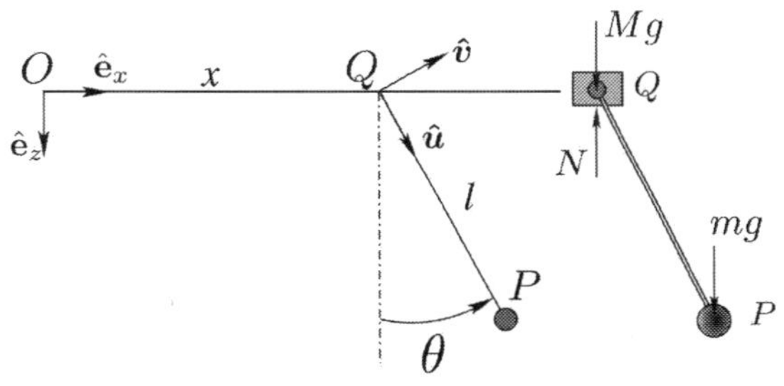

Figure 3.16 – *Problem 3.10 (Solution)*

The position of the system is defined by coordinate $x(t)$ of Q along axis Ox and by the oriented angle $\theta(t)$ which line QP makes with the vertical. See Figure 3.16. Then introduce unit vector $\hat{\mathbf{u}}$ along line QP directed from Q to P, and $\hat{\mathbf{v}} = \hat{\mathbf{e}}_y \times \hat{\mathbf{u}}$. The position vectors of particles P and Q are then given by $\mathbf{r}_{OQ} = x\hat{\mathbf{e}}_x$ and $\mathbf{r}_{OP} = x\hat{\mathbf{e}}_x + l\hat{\mathbf{u}}$. Then the velocities of particles P and Q are given by

$$\mathbf{v}_Q = \dot{x}\,\hat{\mathbf{e}}_x, \qquad \mathbf{v}_P = \dot{x}\,\hat{\mathbf{e}}_x + l\dot{\theta}\hat{\mathbf{v}}.$$

At time $t = 0$, we have $\theta = \theta_0$, and $\dot{x} = 0$ and $\dot{\theta} = 0$ since the system is released without initial velocity.
Denote by Σ the system $\{P, Q, \text{rod}\}$ and by G its mass center. The external forces acting on Σ are the weights $M\mathbf{g}$ and $m\mathbf{g}$ of each body, and the reaction force $N\hat{\mathbf{e}}_z$ directed along $\hat{\mathbf{e}}_z$ only in the absence of friction. Hence the resultant force $\mathbf{F}_{ext\to\Sigma}$ on Σ is vertical: $\mathbf{F}_{ext\to\Sigma} \cdot \hat{\mathbf{e}}_x = 0$. This implies that $(m+M)\mathbf{a}_G \cdot \hat{\mathbf{e}}_x = 0$ and hence by integration that ($\hat{\mathbf{e}}_x$ is a constant unit vector)

$$(m+M)\mathbf{v}_G \cdot \hat{\mathbf{e}}_x = (m+M)\dot{x}_G = Cst = 0 \qquad [1]$$

The constant of integration is zero because the system is released without initial speed. This shows that *the mass center G of the system moves along a vertical line.* Integration of [1] gives $x_G = Cst = 0$, assuming that the origin O is located directly above G at time $t = 0$.
Equation [1] can also be interpreted by stating that *the linear momentum of Σ along Ox*

is conserved. With $(M+m)\mathbf{v}_G = M\mathbf{v}_Q + m\mathbf{v}_P$, we obtain

$$(M\mathbf{v}_Q + m\mathbf{v}_P)\cdot\hat{\mathbf{e}}_x = \boxed{(M+m)\dot{x} + ml\dot{\theta}\cos\theta = 0} \qquad [2]$$

Integration of [2] gives

$$\boxed{(M+m)x + ml\sin\theta = 0} \qquad [3]$$

since $x_G = 0$ is chosen at $t = 0$.

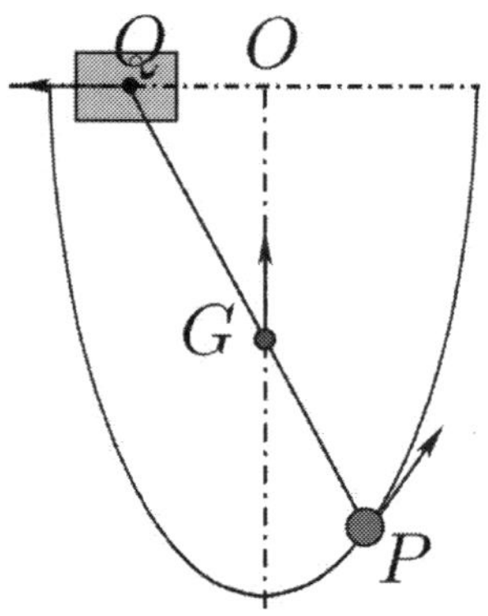

Figure 3.17 – *Problem 3.10*

b. The position of particle P is given by $\mathbf{r}_{OP} = (x + l\sin\theta)\hat{\mathbf{e}}_x + l\cos\theta\hat{\mathbf{e}}_z$, where $x = -\frac{m}{m+M}l\sin\theta$ using equation [3]. The coordinates of P are then given by

$$x_P = \frac{M}{M+m}l\sin\theta, \qquad z_P = l\cos\theta$$

Eliminating θ gives the Cartesian equation of the trajectory of P:

$$\boxed{\left(\frac{x_P}{l(1+\frac{m}{M})}\right)^2 + \left(\frac{z_P}{l}\right)^2 = 1}$$

This equation is that of an ellipse (see Figure 3.17). ■

Exercise 3.11 ▪▪ The pendulum of example 3.10 is configured so that the slider Q is initially set in contact with a vertical backstop. The pendulum is initially displaced at some angle $\theta_0 < 0$ as shown and released without initial speed. Angle θ is measured positively is the counterclockwise direction. Find the relationship between the slider's velocity and the angular velocity $\dot{\theta}$. Is the situation identical to that of example 3.10? Carefully describe the motion. ■

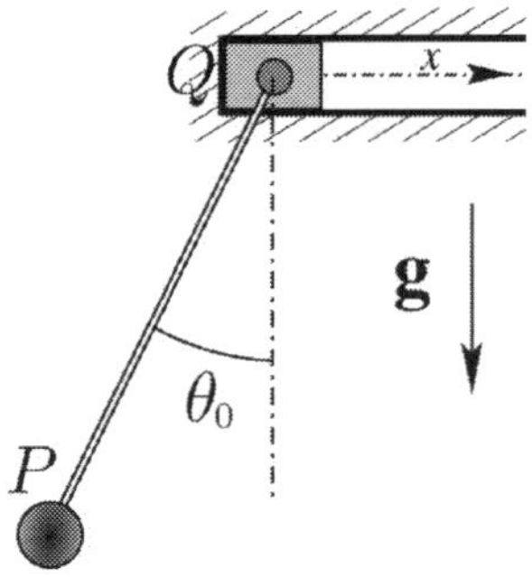

SOLUTION. We use the notations of the solution of example 3.10. If the slider is initially in contact with the backstop, we can expect an initial phase during which the slider remains in contact with the backstop: indeed as the angle increases from the initial value $\theta_0 < 0$, the slider will be pushed backward as the pendulum swings forward. Hence, we assume that there exists an initial phase during which $\dot{x} = 0$ and the external force exerted to system Σ is augmented by a horizontal reaction $\mathbf{R} = R\hat{\mathbf{e}}_x$ due to the backstop. As long as $R > 0$, contact is maintained. If $R(t)$ vanishes at some time $t_1 > 0$, then the slider ceases to be in contact with the backstop. During the first phase, the motion of Σ along the x-axis is governed by Euler's First Principle on the x-axis

$$R = (m+M)\mathbf{a}_{G_\Sigma} \cdot \hat{\mathbf{e}}_x = \frac{d}{dt}[(m+M)\dot{x} + ml\dot{\theta}\cos\theta] = ml(\ddot{\theta}\cos\theta - \dot{\theta}^2\sin\theta) \qquad (3.33)$$

We can predict the sign of reaction R without having to find the angular acceleration $\ddot{\theta}$ and angular velocity $\dot{\theta}$ as a function of θ. Indeed the angular acceleration $\ddot{\theta}$ is positive ($\dot{\theta}$ increases) in the range $\theta_0 \leq \theta < 0$, becomes zero for $\theta = 0$, and is negative for $\theta > 0$. Hence $\text{sign}(\ddot{\theta}) = -\text{sign}(\sin\theta)$. This shows that

$$\begin{cases} R > 0, & \theta_0 \leq \theta < 0 \\ R = 0, & \theta = 0 \\ R < 0, & \theta > 0 \end{cases}$$

Consequently, *the slider ceases to be in contact with the backstop the moment the pendulum crosses the vertical in the forward direction.* The time t_1 at which this occurs is of course a function of the initial angle θ_0.

In the second phase ($t \geq t_1$), once contact with the backstop ceases, we are in a similar situation as in example 3.10: the linear momentum of Σ along axis Ox is conserved. In this case, however, *the linear momentum of Σ on axis Ox is not zero,* since

$$(m+M)\mathbf{v}_{G_\Sigma} \cdot \hat{\mathbf{e}}_x = Cst = m\mathbf{v}_G(t_1) \cdot \hat{\mathbf{e}}_x = ml\dot{\theta}_1$$

where the constant is obtained at time $t = t_1$ when $\dot{\theta} = \dot{\theta}_1$ and $\dot{x} = 0$. Hence in the second phase, *the mass center of Σ moves uniformly in the direction of axis Ox.* Furthermore, the slider's velocity $\dot{x}$ is given by

$$\dot{x} = \frac{m}{M+m} l(\dot{\theta}_1 - \dot{\theta}\cos\theta) \qquad (3.34)$$

We expect in phase 2 that $\dot{\theta} \leq \dot{\theta}_1$ and that $\dot{\theta} = \dot{\theta}_1 > 0$ at $\theta = 0$. This shows that the slider's velocity remains positive and becomes zero whenever the pendulum crosses the vertical with a positive velocity (in the counterclockwise direction). Hence *the slider always moves forward in a non-uniform way in the x-direction, coming to a stop whenever the pendulum crosses the vertical in the forward direction. The slider never comes back toward the backstop.* The situation is thus quite different from that of example 3.10. ■

Exercise 3.12 ■■ A cube P of mass m is released without initial velocity on a smooth wedge Q of mass M. The wedge is free to slide without friction on a horizontal surface. Assume that the motions take place in a plane and denote by α the angle of the wedge.
Find the acceleration of particle P relative to the wedge. Then find the acceleration of P relative to the fixed support. Examine the limiting cases $M \gg m$ and $M \ll m$. ■

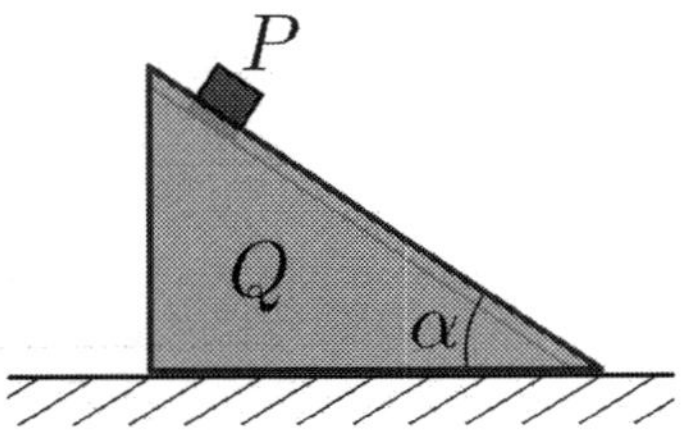

SOLUTION. Introduce the fixed unit vectors $(\hat{\mathbf{e}}_x, \hat{\mathbf{e}}_z)$ with $\hat{\mathbf{e}}_x$ along the horizontal support and $\hat{\mathbf{e}}_z$ directed along the vertical upward. The position of wedge Q is defined by coordinate $x(t)$ along axis $(O, \hat{\mathbf{e}}_x)$: its velocity and acceleration are then given by

$$\mathbf{v}_Q = \dot{x}\,\hat{\mathbf{e}}_x, \qquad \mathbf{a}_Q = \ddot{x}\,\hat{\mathbf{e}}_x$$

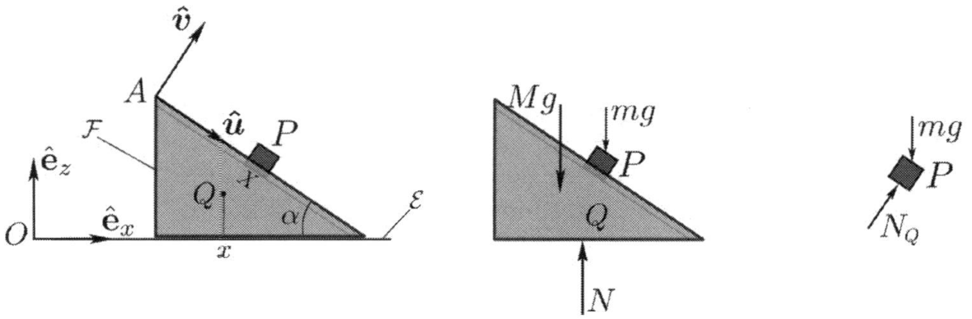

Figure 3.18 – *Problem 3.12 (Solution).*

Next we introduce the unit vectors $(\hat{\mathbf{u}}, \hat{\mathbf{v}})$ attached to the wedge, with $\hat{\mathbf{u}}$ directed along the inclined edge of the wedge, and $\hat{\mathbf{v}}$ perpendicular to this edge. See Figure 3.18. The position of cube P relative to the wedge can be defined by coordinate X along axis $(A, \hat{\mathbf{u}})$: $\mathbf{r}_{AP} = X(t)\hat{\mathbf{u}}$. Note that the orientation of $(\hat{\mathbf{u}}, \hat{\mathbf{v}})$ relative to $(\hat{\mathbf{e}}_x, \hat{\mathbf{e}}_z)$ does not change: basis $(\hat{\mathbf{u}}, \hat{\mathbf{v}})$ is also fixed relative to the support $\mathcal{E}$. If we denote by $\mathcal{F}$ the referential attached to the wedge, the relative velocity and acceleration of P are given by

$$\mathbf{v}_{P/\mathcal{F}} = \dot{X}\hat{\mathbf{u}}, \qquad \mathbf{a}_{P/\mathcal{F}} = \ddot{X}\hat{\mathbf{u}}.$$

The change of referential formulas give the absolute velocity $\mathbf{v}_P = \mathbf{v}_{P/\mathcal{E}}$ and acceleration $\mathbf{a}_P = \mathbf{a}_{P/\mathcal{E}}$ of P:

$$\mathbf{v}_P = \mathbf{v}_{P/\mathcal{F}} + \mathbf{v}_{P\in\mathcal{F}/\mathcal{E}} = \dot{X}\hat{\mathbf{u}} + \dot{x}\hat{\mathbf{e}}_x$$

$$\mathbf{a}_P = \mathbf{a}_{P/\mathcal{F}} + \mathbf{a}_{P\in\mathcal{F}/\mathcal{E}} = \ddot{X}\hat{\mathbf{u}} + \ddot{x}\hat{\mathbf{e}}_x$$

Denote by Σ the system wedge/cube. The external forces acting on Σ are the weights $M\mathbf{g}$ and $m\mathbf{g}$ of both bodies, and the reaction force $N\hat{\mathbf{e}}_z$ directed along $\hat{\mathbf{e}}_z$ in the absence of friction. Hence the resultant external force $\mathbf{F}_{ext\to\Sigma}$ acting on Σ is vertical: $\mathbf{F}_{ext\to\Sigma} \cdot \hat{\mathbf{e}}_x = 0$. This implies that $(M\mathbf{a}_Q + m\mathbf{a}_P) \cdot \hat{\mathbf{e}}_x = 0$ according to Euler's first principle

$$(M + m)\ddot{x} + m\ddot{X}\cos\alpha = 0 \qquad [1]$$

This equation is readily integrated

$$\boxed{(M\mathbf{v}_Q + m\mathbf{v}_P) \cdot \hat{\mathbf{e}}_x = (M + m)\dot{x} + m\dot{X}\cos\alpha = \text{constant}}$$

which shows that *the linear momentum of Σ on axis Ox is conserved.* As the cube slides downward to the right ($\dot{X} > 0$), the wedge slides backward to the left ($\dot{x} < 0$).
To obtain another equation, we apply Newton's Second Law for particle P along axis $(A, \hat{\mathbf{u}})$ in order to eliminate the normal reaction force of the wedge: we project equation $m\mathbf{a}_P = N_Q\hat{\mathbf{v}} + m\mathbf{g}$ on $\hat{\mathbf{u}}$ to obtain

$$\ddot{x}\cos\alpha + \ddot{X} = g\sin\alpha \qquad [2]$$

We now solve equations [1-2] for $\ddot{x}$ and $\ddot{X}$:

$$\boxed{\ddot{x} = -g\frac{\sin\alpha\cos\alpha}{k + \sin^2\alpha}, \qquad \ddot{X} = g(1 + k)\frac{\sin\alpha}{k + \sin^2\alpha}} \qquad [3]$$

denoting $k = M/m$. Two limiting cases can be examined:
- if $M \gg m$, then $\ddot{x} \to 0$ and $\ddot{X} \to g \sin\alpha$: the cube slides down on a stationary wedge.
- if $M \ll m$, then $\ddot{x} \to -g\cot\alpha$ and $\ddot{X} \to g/\sin\alpha$: in this case the wedge slides backward so as to maintain the (absolute) path of P in a vertical downward direction.
■

As an application of Euler's First Principle, we consider in the next section the case of *closed systems*.

3.8 Dynamics of a Closed System of Two Particles

DEFINITION (CLOSED SYSTEM)

We call *closed system* (or *isolated system*) of two particles, a system $\Sigma = \{P_1, P_2\}$ in motion in a Newtonian referential $\mathcal{E}$ acted upon by external forces whose resultant is zero at all time.

Since the resultant external force $\mathbf{F}_{ext\to\Sigma}$ is zero, Euler's First Principle gives $m\mathbf{a}_{G/\mathcal{E}} = \mathbf{0}$. Consequently, *the linear momentum of the system*

$$m\mathbf{v}_{G/\mathcal{E}} = Cst \tag{3.35}$$

is conserved during the motion. We can then state that:

CENTROIDAL REFERENTIAL

The mass center of a closed system Σ is in rectilinear uniform motion in $\mathcal{E}$. The *centroidal referential* $\mathcal{E}^*$, defined as the referential in translational motion relative to $\mathcal{E}$ at the velocity $\mathbf{v}_{G/\mathcal{E}}$ of the mass center G of Σ, is a Newtonian referential.

Indeed, we know from Section 3.2 that, if $\mathcal{E}$ is Newtonian, then any referential (such as $\mathcal{E}^*$) in rectilinear uniform translation relative to $\mathcal{E}$ is also Newtonian. We can also write:

$$\mathbf{v}_{P_i/\mathcal{E}^*} = \mathbf{v}_{P_i/\mathcal{E}}, \qquad \mathbf{a}_{P_i/\mathcal{E}^*} = \mathbf{a}_{P_i/\mathcal{E}} \qquad (i = 1, 2) \tag{3.36}$$

Since we know the motion of G in $\mathcal{E}$, it is sufficient to study the motion of P_1 and P_2 relative to the centroidal referential $\mathcal{E}^*$. Application of Newton's Second Law to P_1 and P_2 in $\mathcal{E}^*$ gives

$$m_1 \frac{d^2}{dt^2}\mathbf{r}_{GP_1} = \mathbf{F}_{2\to1} \tag{3.37}$$

$$m_2 \frac{d^2}{dt^2}\mathbf{r}_{GP_2} = \mathbf{F}_{1\to2} \tag{3.38}$$

denoting $\mathbf{F}_{i\to j}$ the forces of interaction between P_1 and P_2. We can then exploit the fact that $\mathbf{F}_{2\to1} = -\mathbf{F}_{1\to2}$ (Newton's Third Law) to show that it is sufficient to study the motion of a single (fictitious) particle. First denote $\mathbf{r}_{12} = \mathbf{r}_{P_1P_2}$. We can express the position of P_1 and P_2 in $\mathcal{E}^*$ in terms of $\mathbf{r}_{12}$ by taking into account the equation $m_1\mathbf{r}_{GP_1} + m_2\mathbf{r}_{GP_2} = \mathbf{0}$: starting with

$\mathbf{r}_{GP_1} = \mathbf{r}_{GP_2} - \mathbf{r}_{12} = -\frac{m_1}{m_2}\mathbf{r}_{GP_1} - \mathbf{r}_{12}$, we find

$$\mathbf{r}_{GP_1} = -\frac{m_2}{m_1+m_2}\mathbf{r}_{12}, \qquad \mathbf{r}_{GP_2} = \frac{m_1}{m_1+m_2}\mathbf{r}_{12} \tag{3.39}$$

Now equations (3.37-3.38) become

$$-\frac{m_1 m_2}{m_1+m_2}\frac{d^2}{dt^2}\mathbf{r}_{12} = \mathbf{F}_{2\to 1}, \qquad \frac{m_1 m_2}{m_1+m_2}\frac{d^2}{dt^2}\mathbf{r}_{12} = \mathbf{F}_{1\to 2} \tag{3.40}$$

These two equations are identical. It is then possible to state:

Motion of the Reduced Closed System

In a closed system of two particles, it is sufficient to study the motion of fictitious particle P defined by $\mathbf{r}_{GP} = \mathbf{r}_{12}$ of "reduced" mass $m = \frac{m_1 m_2}{m_1+m_2}$ acted upon by force $\mathbf{F}_{2\to 1}$:

$$m\mathbf{a}_{P/\mathcal{E}^*} = \mathbf{F}_{1\to 2} \tag{3.41}$$

Remark 21. This shows that a two-body problem can be converted to a one-body problem.

Remark 22. If $m_1 \gg m_2$, then G coincides with P_1 and fictitious particle P of mass $m \approx m_2$ coincides with P_2. Such is the case for the closed system $\{P_1$= Sun, P_2 =Earth$\}$ or $\{P_1$ = Earth, P_2 = satellite$\}$.

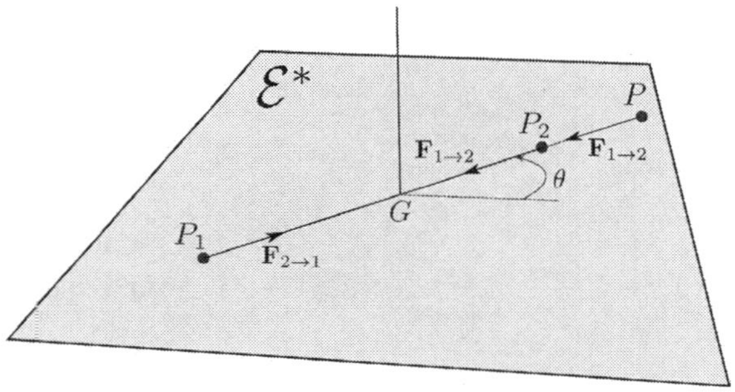

Figure 3.19 – *Motion of P_1, P_2 and fictitious particle P in the centroidal referential $\mathcal{E}^*$. The trajectories lie in a plane of $\mathcal{E}^*$.*

Fictitious particle P has a central force motion in $\mathcal{E}^*$ since force $\mathbf{F}_{1\to 2}$ is always colinear to line GP: it is possible to show that the motion of P (and hence of P_1 and P_2) is planar and characterized by the first integral $r^2\dot{\theta} = Cst$ if $(r = |GP|, \theta)$ are the polar coordinates of P in $\mathcal{E}^*$. For a given interaction law, the motion of P can be entirely predicted: for instance, for the $1/r^2$ gravitational attraction, the motion is a conic section (see Appendix D). Once the position of P is known, then the positions of P_1 and P_2 relative to the centroidal referential $\mathcal{E}^*$ can be obtained by a simple scaling of $\mathbf{r}_{12}$, according to (3.39).

Remark 23. See Section 3.10 for more detail on central force motion.

The following problem demonstrates the applicability of this approach on a simple mechanical system.

Exercise 3.13 ■■■ Two cubes P_1 of mass $2m$ and P_2 of mass m are connected by a linear spring of constant k and unstretched length l_0. The system rests on a smooth horizontal support and the spring is unstretched when cube P_2 is set in motion in a direction perpendicular to line P_1P_2 with speed v_0.
a. Show that the system is equivalent to a closed system. Describe the motion of the mass center G of the system.
b. Show that the length l of the spring satisfies the equation

$$\ddot{l} - \frac{l_0^2 v_0^2}{l^3} = -\frac{3k}{2m}(l - l_0)$$

with $\dot{l} = 0$ and $l = l_0$ at time $t = 0$. Find the value l/l_0 when l is maximum for $\frac{mv_0^2}{kl_0^2} = 1$. ■

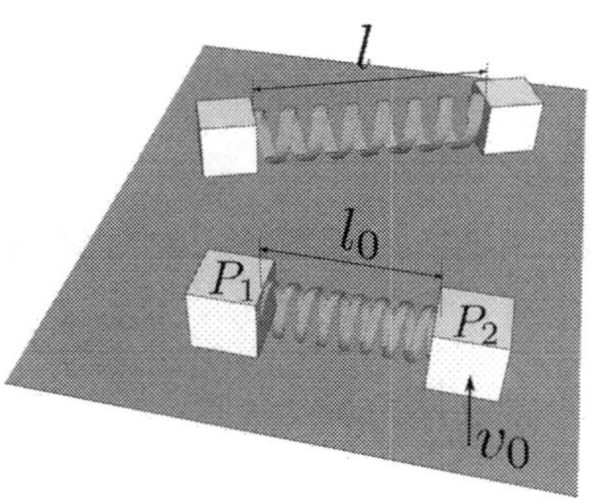

SOLUTION. *This system has 4 degrees of freedom and would require 4 ODEs to solve for the motions of P_1 and P_2. It can be reduced to a single ODE by treating it as a closed system.*
a. First, we define basis $(\hat{\mathbf{e}}_x, \hat{\mathbf{e}}_y)$ in the plane of motion, with $\hat{\mathbf{e}}_x$ directed along the initial position of line P_1P_2. Let origin O coincide with the initial position of the mass center G of system $\Sigma = \{P_1, \text{spring}, P_2\}$. See Figure 3.20.
The external forces (the cubes' weights and the support's reaction) acting on the system are normal to the plane of motion: therefore the forces in the plane of motion satisfy $\mathbf{F}_{ext \to \Sigma} = \mathbf{0}$. The system behaves like a closed system in the plane of motion. Euler's First Principle implies $3m\mathbf{a}_G = \mathbf{0}$, leading to, by integration,

$$3m\mathbf{v}_G = 2m\mathbf{v}_{P_1} + m\mathbf{v}_{P_2} = Cst = mv_0\hat{\mathbf{e}}_y \qquad [1]$$

Hence, *the mass center G of the system is in rectilinear uniform motion at speed $v_0/3$ along axis Oy.*

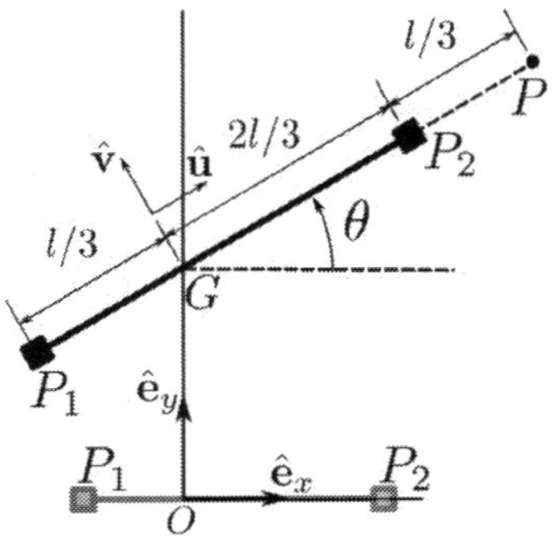

Figure 3.20 – *Problem 3.13*

b. We know that we can predict the motion of P_1 and P_2 in the centroidal referential $\mathcal{E}^*(G, \hat{\mathbf{e}}_x, \hat{\mathbf{e}}_y)$ by studying the motion of the fictitious particle P of mass $m_1m_2/(m_1 + m_2) = 2m/3$ and of position $\mathbf{r}_{GP} = \mathbf{r}_{P_1P_2} = l\hat{\mathbf{u}}$ recognizing the length l of the spring as the distance $|P_1P_2|$. Recall that the position of P_1 and P_2 relative to P are given by

$$\mathbf{r}_{GP_1} = -\frac{1}{3}\mathbf{r}_{GP}, \qquad \mathbf{r}_{GP_2} = \frac{2}{3}\mathbf{r}_{GP} \qquad [2]$$

Then the motion of P is governed by

$$\frac{2m}{3}\frac{d^2}{dt^2}\mathbf{r}_{GP} = \mathbf{F}_{1\to 2} = -k(l-l_0)\hat{\mathbf{u}} \qquad [3]$$

We can express the acceleration of P in terms of its polar coordinates (l, θ)

$$\mathbf{a}_{P/\mathcal{E}^*} = (\ddot{l} - l\dot{\theta}^2)\hat{\mathbf{u}} + (2\dot{l}\dot{\theta} + l\ddot{\theta})\hat{\mathbf{v}}$$

We can then project [2] on the polar unit vectors $(\hat{\mathbf{u}}, \hat{\mathbf{v}})$ to obtain:

$$\frac{2m}{3}(\ddot{l} - l\dot{\theta}^2) = -k(l-l_0), \qquad \frac{2m}{3}(l\ddot{\theta} + 2\dot{l}\dot{\theta}) = 0 \qquad [4-5]$$

We know that equation [5] leads to a first integral

$$l^2\dot{\theta} = Cst = l_0^2\dot{\theta}_0 \qquad [6]$$

We need to express $\dot{\theta}_0$ in terms of v_0:

$$\mathbf{v}_{P_2} = \mathbf{v}_G + \frac{d}{dt}(\frac{2l}{3}\hat{\mathbf{u}}) = \frac{v_0}{3}\hat{\mathbf{e}}_y + \frac{2}{3}(\dot{l}\hat{\mathbf{u}} + l\dot{\theta}\hat{\mathbf{v}})$$

This gives at $t = 0$:

$$\dot{l} = 0, \qquad v_0 = l_0\dot{\theta}_0$$

We can now express [4] in terms of l and its derivatives by using [6]:

$$\boxed{\ddot{l} - \frac{l_0^2 v_0^2}{l^3} = -\frac{3k}{2m}(l-l_0)}$$

We can integrate this equation by pre-multiplying by $\dot{l}$: we find

$$\frac{\dot{l}^2}{2} + \frac{l_0^2 v_0^2}{2l^2} = -\frac{3k}{4m}(l-l_0)^2 + C$$

To find constant C we impose $\dot{l} = 0$ when $l = l_0$: $C = v_0^2/2$ This gives

$$\boxed{\frac{\dot{l}^2}{2} + \frac{v_0^2}{2}\left(\frac{l_0^2}{l^2} - 1\right) = -\frac{3k}{4m}(l-l_0)^2}$$

When $\dot{l} = 0$ we find the algebraic equation

$$\frac{x-1}{x+1}x^2 = \frac{2}{3}\kappa$$

in unknown $x = l/l_0$ with parameter $\kappa = \frac{mv_0^2}{kl_0^2}$. This equation is solved numerically to find $l/l_0 = \mathbf{1.65}$ for $\kappa = 1$. ■

3.9 Foucault's Pendulum

Foucault's pendulum experiment was first demonstrated in 1851 by hanging a heavy mass from the Panthéon in Paris, located at latitude $\lambda = \pi/2 - \phi = 48.8°$. The pendulum's length was chosen to be $L = 67$ m, the mass was released without initial speed at a distance of 3 m from its equilibrium position. The experiments revealed that the oscillations do not take place in a fixed vertical plane of the terrestrial referential $\mathcal{E}$, as it should, if the only forces acting on P are the string's tension and the gravitational force , both contained in the plane defined by points O, P_0 the initial position of P, and Q the attachment point. Rather, P exhibits a more complex motion, whereby the plane of oscillations rotates slowly about the vertical axis Oz with a period of 31 hours and 47 minutes. These results cannot be explained if $\mathcal{E}$ is assumed Newtonian. Hence Foucault's pendulum demonstrated the non-Newtonian character of referential $\mathcal{E}$. This experiment is now replicated in many science museums around the world where the motion of the pendulum is maintained indefinitely by electromagnetic forces to compensate its gradual slowing down due to frictional forces.

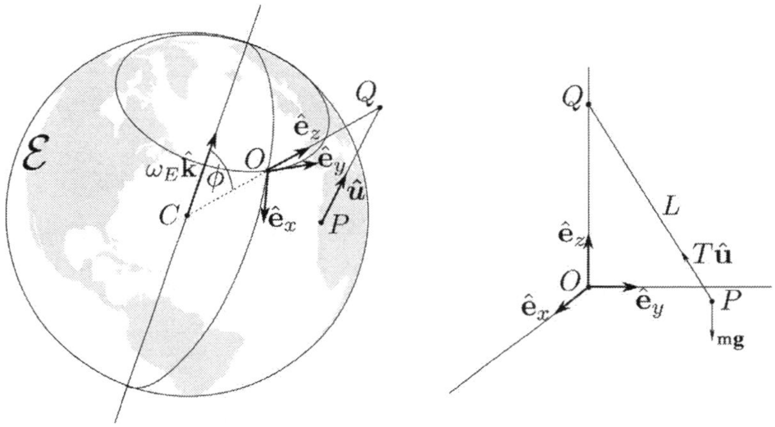

Figure 3.21 – *Foucault's pendulum in the terrestrial referential $\mathcal{E}$ with origin O at colatitude ϕ, and axes Ox, Oy and Oz pointing southward, eastward, and upward, respectively.*

To predict the motion of P we use Newton's Second Law relative to $\mathcal{E}$ by adding Coriolis inertial force to account for Earth's rotation at a colatitude ϕ

$$m\mathbf{a}_{P/\mathcal{E}} = m\mathbf{g} + \mathbf{T} - 2m\omega_E \hat{\mathbf{k}} \times \mathbf{v}_{P/\mathcal{E}} \tag{3.42}$$

where $\mathbf{T}$ is the string's tension, and where we treat the gravitational acceleration as the constant vector $\mathbf{g} \approx -g\hat{\mathbf{e}}_z$ parallel to axis Oz with $g = 9.8$ m/s^2. In equation (3.42), we have neglected the effect of the centrifugal force whose magnitude $mR\omega_E^2 \sin\phi$ is negligible compared to that of the other forces acting on P. We have also neglected all possible frictional forces. We can write express (3.42) in terms of the Cartesian coordinates $(x(t), y(t), z(t))$ of P relative to axes $Oxyz$:

$$m\ddot{x} = T_x + 2m\omega_E \dot{y} \cos\phi \tag{3.43}$$

$$m\ddot{y} = T_y - 2m\omega_E (\dot{x} \cos\phi + \dot{z} \sin\phi) \tag{3.44}$$

$$m\ddot{z} = T_x - mg + 2m\omega_E \dot{y} \sin\phi \tag{3.45}$$

where (T_x, T_y, T_z) are the components of $\mathbf{T}$ on $(\hat{\mathbf{e}}_x, \hat{\mathbf{e}}_y, \hat{\mathbf{e}}_z)$. Assume that the coordinates of attachment point Q are $(0, 0, L)$, and that the coordinates of initial position P_0 are $(0, y_0, z_0)$

with $z_0 \ll L$. Denote by $\hat{\mathbf{u}}$ the unit vector pointing from P to Q:

$$\hat{\mathbf{u}} = \frac{1}{L}\mathbf{r}_{PQ} = -\frac{x}{L}\hat{\mathbf{e}}_x - \frac{y}{L}\hat{\mathbf{e}}_y - \frac{z-L}{L}\hat{\mathbf{e}}_z$$

Now equations (3.43-3.45) become

$$m\ddot{x} = -T\frac{x}{L} + 2m\omega_E \dot{y}\cos\phi \tag{3.46}$$

$$m\ddot{y} = -T\frac{y}{L} - 2m\omega_E(\dot{x}\cos\phi + \dot{z}\sin\phi) \tag{3.47}$$

$$m\ddot{z} = -T\frac{z-L}{L} - mg + 2m\omega_E \dot{y}\sin\phi \tag{3.48}$$

To simplify these equations we note that the motion is essentially contained in a horizontal plane, given the small initial deviation of P from its equilibrium position. Hence we have $z \ll L$ and $\ddot{z} - 2\omega_E \dot{y}\sin\phi \ll g$, which shows that the tension remains sensibly constant: equation (3.48) becomes

$$T \approx mg,$$

leading to $z = z_0 \approx 0$. Furthermore, if point O is located sufficiently far from the equator ($\phi \neq \pi/2$) then we expect $\dot{x}\cos\phi \approx \dot{y}\cos\phi \gg \dot{z}\sin\phi$. This leads to the simplified system of linear homogeneous ODEs

$$\ddot{x} + \omega_0^2 x = 2\omega_E \dot{y}\cos\phi \tag{3.49}$$

$$\ddot{y} + \omega_0^2 y = -2\omega_E \dot{x}\cos\phi \tag{3.50}$$

where we have denoted $\omega_0^2 = g/L$. Note that ω_0 is the frequency of oscillations of the pendulum in the absence of Coriolis forces. We expect $\omega_E \ll \omega_0$. This system of equations is easily integrated by introducing the complex variable $Z(t) = x(t) + iy(t)$. Then we have

$$\ddot{Z} = \ddot{x} + i\ddot{y} = -\omega_0^2(x+iy) + 2\omega_E\cos\phi(\dot{y} + i\dot{x}) = -\omega_0^2 Z - 2i(\omega_E\cos\phi)\dot{Z}$$

This ODE admits 2 solutions of the type $\exp(st)$ with s solution of the quadratic equation $s^2 + 2i(\omega_E\cos\phi)s + \omega_0^2 = 0$:

$$s_{1,2} = -i\omega_E\cos\phi \pm i(\omega_0^2 + \omega_E^2\cos^2\phi) \approx i(-\omega_E\cos\phi \pm \omega_0)$$

Hence the solution is of the form

$$Z(t) = x(t) + iy(t) = e^{-i(\omega_E\cos\phi)t}\left(ae^{i\omega_0 t} + be^{-i\omega_0 t}\right)$$

where the constants of integration are found by imposing the initial conditions: at $t = 0$, we impose $\dot{Z} = 0$ leading to

$$\frac{b}{a} = \frac{\omega_0 - \omega_E\cos\phi}{\omega_0 + \omega_E\cos\phi} \approx 1$$

and $Z = iy_0$ leading to $a = b = iy_0/2$. The final solution is then given by

$$Z(t) = x(t) + iy(t) = iy_0\cos(\omega_0 t)e^{-i(\omega_E\cos\phi)t}$$

or

$$\boxed{\begin{aligned} x(t) &= y_0 \cos(\omega_0 t) \sin(\omega_E t \cos\phi) \\ y(t) &= y_0 \cos(\omega_0 t) \cos(\omega_E t \cos\phi) \end{aligned}} \tag{3.51}$$

From this result, we can conclude that
(i) over a short time scale $T_0 = 2\pi/\omega_0$, the effect of Coriolis forces on the motion of P is negligible: indeed

$$Z(t + T_0) = Z(t)e^{-i(\omega_E \cos\phi)T_0} \approx Z(t)$$

since $\omega_E T_0$ is a very small number. Hence, over a time scale T_0, the pendulum oscillates in a vertical plane fixed in $\mathcal{E}$.
(ii) over a much larger time scale $T_E = 2\pi/\omega_E \gg T_0$, the effect of Coriolis forces is seen to cause a rotation of the plane of oscillations about vertical axis Oz with constant angular velocity $\omega = -\omega_E \cos\phi$: this plane completes a full rotation over a period $T = T_E/\cos\phi$. This rotation is clockwise when seen by an observer looking down on plane Oxy in the northern hemisphere. The rotation is anticlockwise in the southern hemisphere.
Numerically, we find $T_0 = 16.4$ seconds, and $T = 31$ hours 47 min, in total agreement with the experimental measurements. This agreement supports the hypothesis that the geocentric referential $\mathcal{G}$ is a better approximation of a Newtonian referential than the terrestrial referential $\mathcal{E}$. However, for motions taking place during sufficiently short time intervals (particularly near the equator), $\mathcal{E}$ may still be considered as a satisfactory approximation to a Newtonian referential. Furthermore, these conclusions do not mean that referential $\mathcal{G}$ is always a good approximation of a Newtonian referential.

3.10 Central Force Motion

Such motions are encountered in celestial mechanics where the resultant force is the gravitational attraction due a body at O.

Definition (Central Force Motions)

A particle P is said to be in *central force motion* if it is acted upon by a resultant force which remains colinear to line OP, relative to a fixed point O, called *pole*, of some Newtonian referential $\mathcal{E}$.

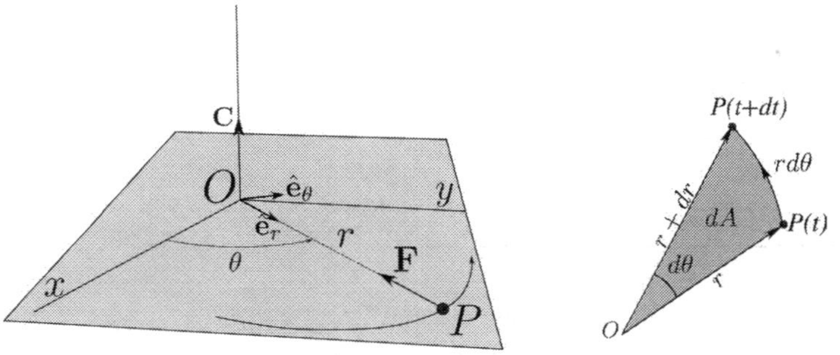

Figure 3.22 – *Central force motion (left). Area swept by the position vector* $\mathbf{r}_{OP}$ *during time dt (right).*

Since the resultant force $\mathbf{F}$ is colinear to $\mathbf{r}_{OP}$, we can show that the vector $\mathbf{r}_{OP} \times m\mathbf{v}_P$ (the angular momentum about O) remains constant:

$$\frac{d}{dt}(\mathbf{r}_{OP} \times m\mathbf{v}_P) = \mathbf{v}_P \times m\mathbf{v}_P + \mathbf{r}_{OP} \times m\mathbf{a}_P = \mathbf{r}_{OP} \times \mathbf{F} = \mathbf{0}$$

Let us denote by $m\mathbf{C}$ this constant vector, and call Oz the axis of $\mathcal{E}$ parallel to $\mathbf{C}$. With $\mathbf{r}_{OP} \times \mathbf{v}_P = \mathbf{C}$, we see that both vectors $\mathbf{r}_{OP}$ and $\mathbf{v}_P$ remain perpendicular to the same constant vector: *the particle P is necessarily in motion in a plane perpendicular to* $\mathbf{C}$. Denote by Oxy this plane. See Figure 3.22 (left). It is defined by the initial position vector $\mathbf{r}_{OP_0}$ and the initial velocity $\mathbf{v}_0$ of P. If we adopt the polar coordinates (r, θ) to define the position of P relative to axis Ox, the constant vector $\mathbf{C} = C\hat{\mathbf{e}}_z$ is given by

$$\mathbf{C} = r\hat{\mathbf{e}}_r \times (\dot{r}\hat{\mathbf{e}}_r + r\dot{\theta}\hat{\mathbf{e}}_\theta) = r^2\dot{\theta}\hat{\mathbf{e}}_z$$

which shows that the constant C is given by

$$C = r^2\dot{\theta} \tag{3.52}$$

Remark 24. The result $r^2\dot{\theta} = Cst$ can be obtained by imposing $a_\theta = r\ddot{\theta} + 2\dot{r}\dot{\theta} = \frac{1}{r}\frac{d}{dt}(r^2\dot{\theta}) = 0$. There is an important geometric interpretation of equation (3.52): as the position vector $\mathbf{r}_{OP}$ varies in the plane Oxy, it sweeps an area $A(t)$ which varies according to (see Figure 3.22, right)

$$\frac{dA}{dt} = \frac{1}{2}r^2\dot{\theta} = \frac{C}{2} \tag{3.53}$$

We can then state:

Invariance of the Areal Velocity

In a central force motion, the *areal velocity dA/dt* of point P with respect to pole O remains constant: *the position vector sweeps equal areas during equal time intervals.*

Remark 25. This result is known as *Kepler's Second Law*: in 1602, Kepler formulated the law, based on Ticho Brahe's measurements, that planets of our solar system sweep out equal areas in equal times. His next empirical breakthrough came in 1605 with the discovery that the planets have elliptical path, with the sun located at one focus (Kepler's First Law).

Two useful formulas, known as *Binet's formulas*, can be derived for the speed v_P and the acceleration $\mathbf{a}_P$ of a particle in central force motion. By taking into account the constant $C = r^2\dot{\theta}$, we can relate time derivative to derivative with respect to angle θ according to:

$$\frac{d}{dt} = \frac{C}{r^2}\frac{d}{d\theta}$$

We start with v_P^2 in polar coordinates:

$$v_P^2 = \dot{r}^2 + r^2\dot{\theta}^2 = \frac{C^2}{r^4}\left(\frac{dr}{d\theta}\right)^2 + r^2\frac{C^2}{r^4}$$

We now denote $u = 1/r$ and use the identity $dr/d\theta = -r^2(du/d\theta)$. We arrive at *Binet's first formula*

$$v_P^2 = C^2\Big(\Big(\frac{du}{d\theta}\Big)^2 + u^2\Big) \qquad (u = 1/r)$$

Similarly, for the radial acceleration (recall that $a_\theta = 0$)

$$a_r = (\ddot{r} - r\dot{\theta}^2) = \dot{\theta}\frac{d}{d\theta}(\dot{\theta}\frac{dr}{d\theta}) - r\frac{C^2}{r^4} = \frac{C}{r^2}\frac{d}{d\theta}(\frac{C}{r^2}\frac{dr}{d\theta}) - \frac{C^2}{r^3} = C^2\Big(\frac{1}{r^2}\frac{d}{d\theta}(-\frac{d}{d\theta}(\frac{1}{r})) - \frac{1}{r^3}\Big)$$

leading to *Binet's second formula*

$$a_r = -C^2u^2\Big(\frac{d^2u}{d\theta^2} + u\Big) \qquad (u = 1/r)$$

Binet's Formulas

In a central force motion, the velocity and radial acceleration of a particle P can be expressed in polar coordinates in the form

$$v_P^2 = C^2\Big(\Big(\frac{du}{d\theta}\Big)^2 + u^2\Big) \tag{3.54}$$

and

$$a_r = -C^2u^2\Big(\frac{d^2u}{d\theta^2} + u\Big) \tag{3.55}$$

where $u = 1/r$.

These formulas are especially useful for central forces which are functions of $u = 1/r$, such as in the case of gravitational attraction. This is illustrated with the following problem.

Exercise 3.14 Kepler's Third Law. Assume that a particle P of mass m is subject to a central force due to the gravitational body of mass M located at point O:

$$\mathbf{F} = -G\frac{mM}{r^2}\hat{\mathbf{e}}_r$$

Show that the trajectory of P is a conic section.
In the case of an elliptical orbit, show that the square of the motion's period T is proportional to the cube of the semiaxis major of the ellipse, the constant of proportionality being a function of constant GM only (Kepler's Third Law). ■

The equation in polar coordinates of P's trajectory can easily be found by integrating $u(\theta)$ using Binet's second formula (3.55):

$$\frac{d^2u}{d\theta^2} + u = \frac{1}{p}$$

with $1/p = GM/C^2$. The solution of this second-order linear ODE is given by

$$u = \frac{1}{p} + A\cos\theta + B\sin\theta = \frac{1}{p}(1 + e\cos(\theta - \phi))$$

where the constants of integration e and ϕ are found from the initial conditions satisfied by P. Finally we obtain

$$\boxed{r = \frac{p}{1 + e\cos(\theta - \phi)}}$$

which is recognized as the equation in polar coordinate of a *conic section* of *semi-latus-rectum* p, *eccentricity* e and *focus* O. The orbit is a *hyperbola* for $e > 1$, a *parabola* for $e = 1$ and an *ellipse* for $0 \le e < 1$ (a circle for $e = 0$). For a review of conic sections, see Appendix D.

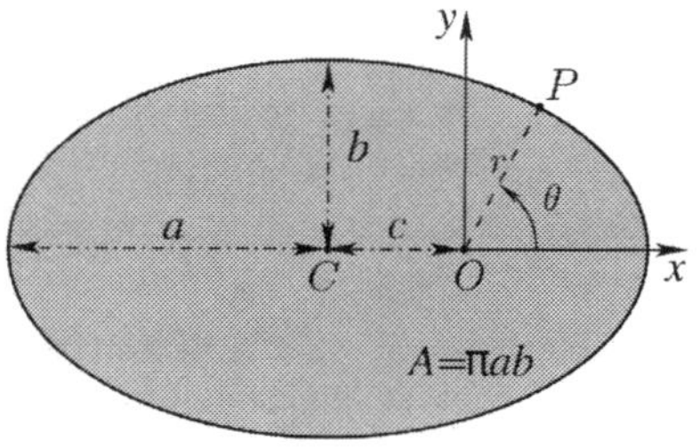

Figure 3.23 – *Problem 3.14: Elliptic trajectory* ($0 \le e < 1$.

Assuming $e < 1$ and $\phi = 0$, the semiaxis major a and semiaxis minor b of the ellipse can be found by expressing the trajectory in Cartesian coordinates

$$\left(x - \frac{ep}{1-e^2}\right)^2 + \frac{y^2}{1-e^2} = \frac{p^2}{(1-e^2)^2}$$

from which we immediately obtain

$$a = \frac{p}{1-e^2}, \qquad b = \frac{p}{\sqrt{1-e^2}}$$

To find the period T of the motion of P, we know that the total area $A(T) = \pi ab$ of the ellipse is swept by the position vector $\mathbf{r}_{OP}$ in time T. Furthermore, the areal speed dA/dt is the constant $C/2$: hence $A(t) = Ct/2$, and $A(T) = CT/2 = \pi ab$, leading to (using $b^2 = pa$)

$$T^2 = \frac{4\pi^2}{C^2}a^2b^2 = \frac{4\pi^2 p}{C^2}\,a^3$$

or (using $p/C^2 = 1/GM$)

$$\boxed{T^2 = \frac{4\pi^2}{GM}a^3}$$

Note that we could have expressed the period T in terms of the semiaxis minor b: $T = \frac{2\pi GM}{C^3}b^3$. However the constant of proportionality is a function of the initial conditions of P. *In Kepler's third law,* T^2/a^3 *is a constant, identical for all planets of the solar system.*

■

PROJECT 5: Optimal Performance of a Crossbow

A projectile P of mass m is placed inside the tube AB of a crossbow. An elastic string is connected to the arms of the crossbow at C and D. The tube AB is capped at A, and open at B. It possesses lateral slots, so as to allow the free passage of the string. The projectile is launched by bringing it to position A. It is then released without initial speed. We set $|OB| = |OC| = |OD| = l$ and $|OA| = 4l/3$, where point O is the center of segment CD. In its unstretched configuration, the string has length $|CD|$. The elasticity of the arms is neglected. When the string is stretched, each strand exerts the force $k(|CP| - |OC|)/|OC|$ on P where k is a constant. The tube exerts a friction force characterized by a (static and kinetic) friction coefficient μ. The crossbow is positioned in a vertical plane, making the inclination angle θ with the vertical, as shown.

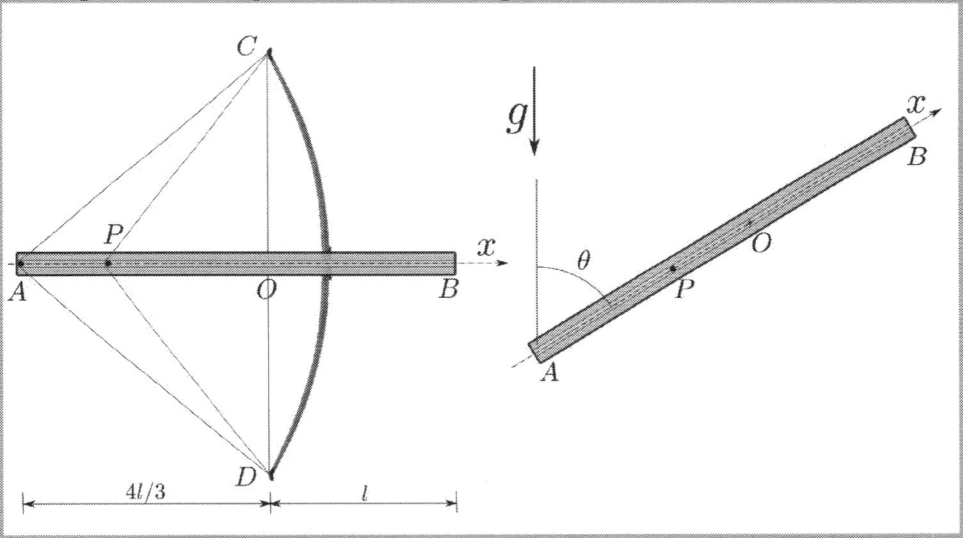

1. To measure k and μ, the crossbow is set in the vertical direction ($\theta = 0$) with A at the lowest position. With a mass $m = 112$ g, it is observed that P reaches the position of A under the sole effect of gravity and the string forces (that is, without being in contact with the bottom cap). It is also observed that equilibrium is lost at A, as soon as the inclination angle θ exceeds the value $\theta = \pi/2$. Deduce the values of k and μ.

2. With the values of k and μ found in 1), we are interested in predicting the performance of the crossbox for varying values of m and θ by determining the range of the projectile from point B.

2a. Find the ODE governing the position x of P for $-4l/3 \leq x \leq 0$. Deduce the speed v_1 at $x = 0$.

2b. Find the ODE governing the position x of P for $0 \leq x \leq l$. Deduce the exit speed v_2 at $x = l$.

2c. Find the range reached by the projectile as a function of m and θ. Show that for $m > m_{max}$, the crossbow cannot be used. For $m < m_{max}$, find the angle θ which maximizes the range.

PROJECT 6: GRAVITATIONAL SPHERE OF INFLUENCE

The goal of this project is to study the concept of the sphere of influence (SOI) of a celestial body. The SOI is the region of space in which the gravitation of a celestial body on the motion of a spacecraft dominates that of other bodies. This simplifies considerably the study of the spacecraft in the SOI. On the interplanetary scale, the Sun is the dominant attractor, except inside the SOI of one of the planets.

Consider three bodies P_0 (the reference body of mass m_0), P_1 (the perturbing body of mass m_1) and P_2 (a body of negligible mass m_2 moving in the gravitational fields of P_0 and P_1). Recall, that relative to a Newtonian referential of origin O, the motion of a body P_i of mass m_i in the presence of n bodies is governed by

$$\frac{d^2\mathbf{r}_{OP_i}}{dt^2} = \sum_{j=0, j\neq i}^{n} \frac{Gm_j}{r_{ij}^3}\mathbf{r}_{ij}, \qquad \mathbf{r}_{ij} = \mathbf{r}_{P_iP_j} \qquad [1]$$

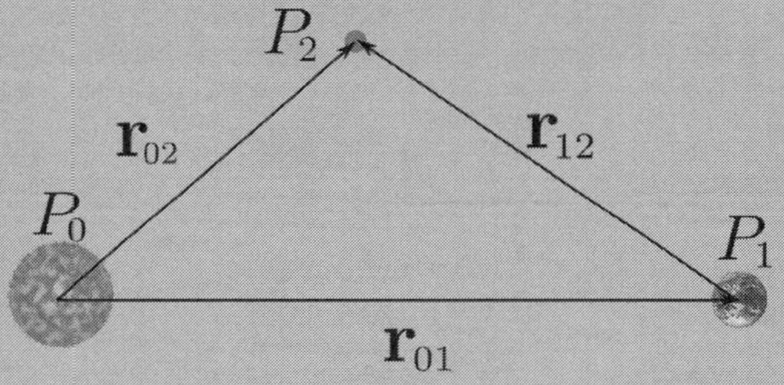

1. Write equation [1] for $i = 0$, $i = 1$ and $i = 2$ to obtain equations [2-4]. By subtracting equation [2] from equation [4], find the equation

$$\frac{d^2\mathbf{r}_{02}}{dt^2} + \underbrace{G\frac{m_0+m_2}{r_{02}^3}\mathbf{r}_{02}}_{\mathbf{A}_0} = \underbrace{-Gm_1\left(\frac{\mathbf{r}_{12}}{r_{12}^3} + \frac{\mathbf{r}_{01}}{r_{01}^3}\right)}_{\mathbf{B}_1}$$

This equation describes the motion of P_2 relative to P_0 under the attraction of P_0 (term $\mathbf{A}_0$) and perturbed by the presence of P_1 (term $\mathbf{B}_1$). Similarly, derive the equation of motion of P_2 about P_1 under the attraction of P_1 (term $\mathbf{A}_1$) perturbed by the presence of P_0 (term $\mathbf{B}_0$):

$$\frac{d^2\mathbf{r}_{12}}{dt^2} + \mathbf{A}_1 = \mathbf{B}_0$$

2. In order to evaluate the effect of the perturbation of body P_1 on the gravitational field of body P_0, we introduce the ratio $\rho_1 = |\mathbf{B}_1|/|\mathbf{A}_0|$. Similarly, we define the ratio $\rho_0 = |\mathbf{B}_0|/|\mathbf{A}_1|$. Express ρ_0 and ρ_1 as a function of $x = r_{12}/D$ (with $D = r_{01}$), and the ratio $\epsilon = \sqrt{m_1/m_0}$ when P_0, P_1 and P_2 are aligned. Sketch these two functions vs $0 \leq x \leq 1$ to show that there exists a value x^* such that $\rho_0 = \rho_1$. This value defines two spheres, the SOIs of bodies P_0 and P_1.

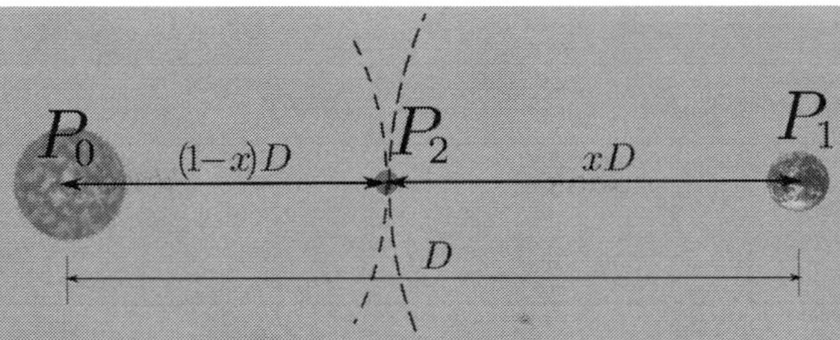

3. Application 1: for the system Earth-Moon, find x^* for $m_1/m_0 = \epsilon^2 = 1/81$.
4. Application 2: for the system Sun-Earth, find x^* for $m_1/m_0 = \epsilon^2 = 1/333{,}000$.
5. From these results, assuming that body P_0 is massive compared to body P_1 ($\epsilon \ll 1$), the solution x^* of the equation $\rho_0 = \rho_1$ is a small number (compared to 1). In this case, show that the radius of the SOI of smaller body P_1 is given by

$$x^* = \frac{r_{SOI}}{D} = \kappa \left(\frac{m_1}{m_0}\right)^{2/5}$$

Find κ. Using this formula, compare the values of x^* for the Earth/Moon and the Sun-Earth systems found in q. 3) and 4) and evaluate the relative errors of this approximation.

3.11 Problems

Problem 3.1 The trajectory of Earth (of mass 6×10^{24} kg) around the Sun (of mass 2×10^{30} kg) in the Heliocentric referential can be approximated as a circle of radius $R = 1.5 \times 10^8$ km under the sole effect of the Sun's gravitational attraction.
a. Show that the motion of Earth is uniform.
b. Find the orbital speed of Earth around the Sun using the gravitational constant $G = 6.67 \times 10^{-11}$ N·m²·kg⁻².

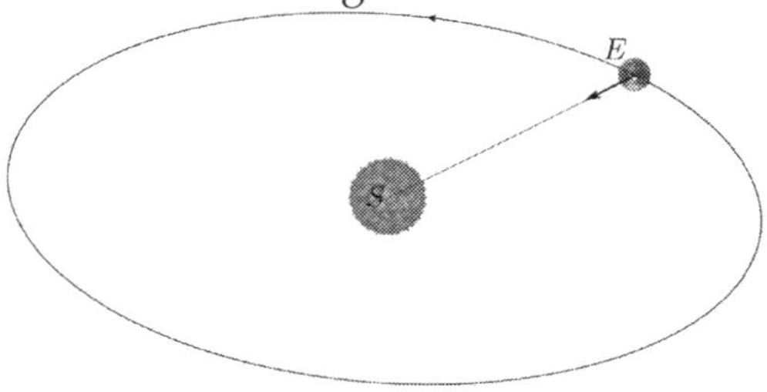

Answer: b) $v = 30$ km/s

Problem 3.2 A spherical ball of mass $m = 150$ g is initially suspended by two identical strings of length $l = 45$ cm attached at A and B. The string at B is then cut and the ball swings about A without initial speed.
a. Find the acceleration of P and the string's tension immediately after P is set in motion.
b. Derive the equation of motion of P and find the speed of P when the string reaches the vertical position.

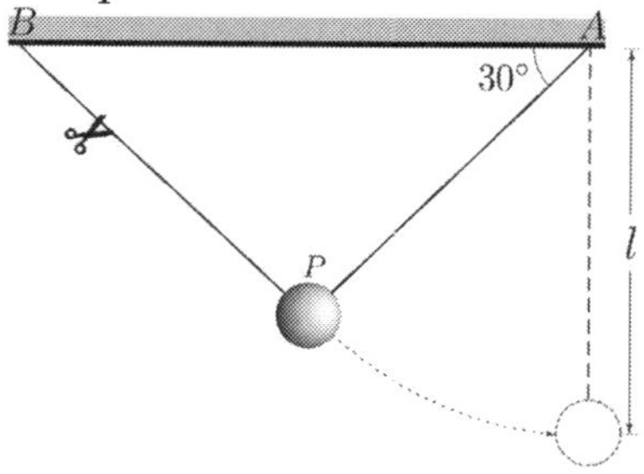

Answer: b) $v_1 = 2.1$ m/s

Problem 3.3 A body of mass $m = 2.5$ kg is shot vertically with the initial speed $v_0 = 26$ m/s. The body is known to reach the terminal speed $v_\infty = 52$ m/s ($\dot{v} \to 0$ as $v \to v_\infty$) due to the air drag modeled as the force mkv^2 (where v is the speed of P). What is the maximal height reached by P? Use $g = 9.8$ m/s²).
Answer: $H = 30.8$ m

Problem 3.4 A bouncy ball is thrown from O at speed v_0 and launch angle α. It bounces on the vertical wall at A to rebound on horizontal surface at B. The impacts are perfectly elastic. Find the minimum speed v_0, and the corresponding angle α, so that the ball returns to O.

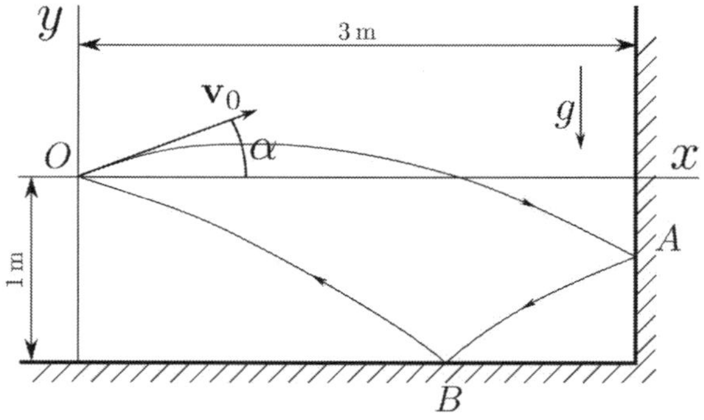

Answer: $\alpha = 30°$

Problem 3.5 A block A of mass 0.5 kg is connected to a spring of constant $k = 4$ N/m. Its contact with the horizontal support is characterized by a coefficient of kinetic friction $\mu_A = 0.1$. Block A supports a block B of mass 0.3 kg whose contact with A is characterized by a coefficient of static friction $\mu_{AB} = 0.15$. The system is pulled to the right and released from rest.

What is the maximum stretch δ which can be given to the spring so that B does not slip relative to A upon release?

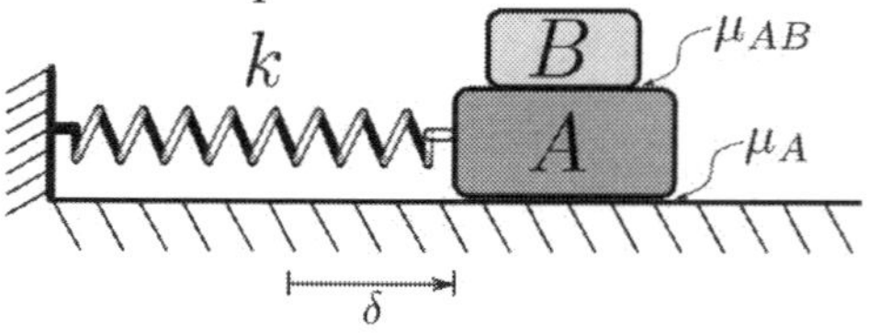

Answer: $\delta = 0.49$ m

Problem 3.6 A small object P of mass m is spun inside a centrifuge which rotates at constant angular velocity Ω about a vertical axis. The contact is characterized by the coefficient of static friction μ_s. If P is to remain at a constant distance r from the axis of rotation without slipping relative to the centrifuge, show that Ω must satisfy

$$\Omega_1 < \Omega < \Omega_2$$

Find the values of Ω_1 and Ω_2 for $r = 60$ cm, $m = 150$ g, $\alpha = 70°$, $\mu_s = 0.3$.

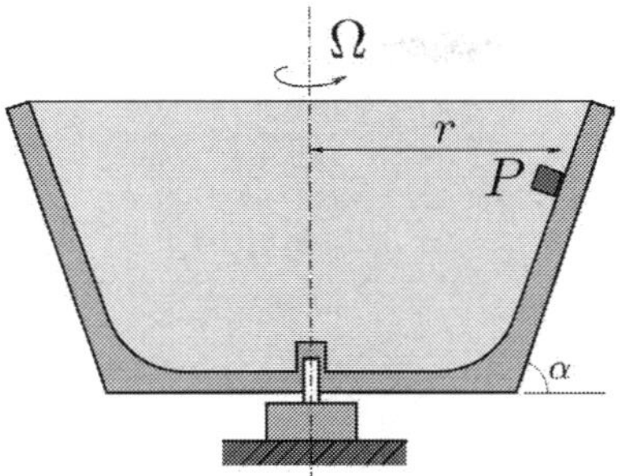

Answer: $\Omega_1 = 0.74$ *rev/s*, $\Omega_2 = 2.68$ *rev/s*

Problem 3.7 A small cube P of mass m can slide along a straight, smooth groove directed along a diameter of a disk. The disk, of radius R, is in rotation about a vertical axis at the constant angular velocity of ω. At $t = 0$, the position of P is released without initial speed (relative to the disk) from the distance $r = R/2$.
Find the time taken by P to reach $r = R$. Find the reaction force of the disk on P at this position.

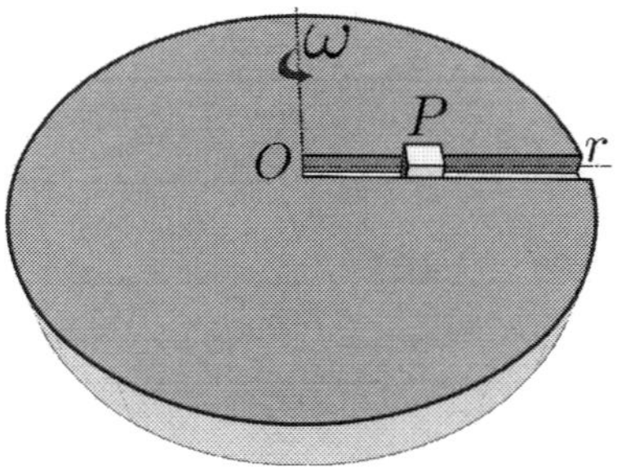

Answer: $\omega T = 1.32$

Problem 3.8 Two blocks A and B of mass m_A and m_B lie on a inclined plane. They are connected by a cord which passes through a pulley attached to the plane. Both strands of the rope are parallel to the steepest line. The coefficients of (static and kinetic) friction between the plane and the blocks are μ_A and μ_B. The mass of the cord and pulley is neglected. Assume $m_A > m_B$.

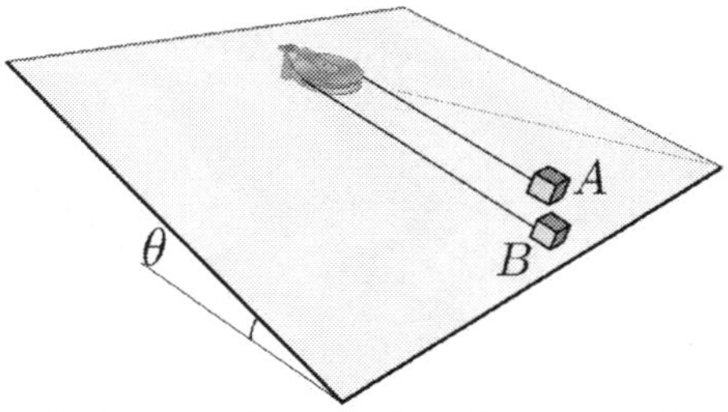

a. Find the condition for the two blocks to remain in equilibrium relative to the plane.
b. The two blocks are slipping. Find their accelerations for $\theta = 30°$, $\mu_A = \mu_B = 0.05$, and $m_A/m_B = 1.25$.
Answer: b) $a_A/g = 0.012$

Problem 3.9 An inextensible, massless rope is supported by a pulley of negligible mass. Its end points are attached to two blocks P of mass m and Q of mass M resting on a horizontal support. The pulley is held at a height such that the two branches of the rope are vertical and taut. The pulley is suddenly pulled upward with a force $\mathbf{F}$. Assume that $m < M$.
Describe the impending motion of P and Q. In each case give the acceleration of P and Q.
Hint: 3 cases must be considered: (i) $F < 2mg$, *(ii)* $2mg < F < 2Mg$ *and (iii)* $F > 2Mg$.

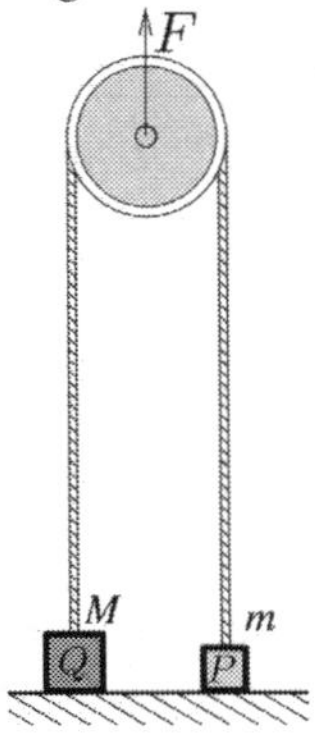

Problem 3.10 Three blocks A, B and C are connected in a vertical plane by two cables AOB and BQC which pass through frictionless pulleys at O and Q. Assume that the cables and pulleys are of negligible mass. A and B have identical mass $2m$. C is of mass m. The system is initially held in the configuration shown (defined by angles α and β) and released from rest.
a. Show that there is a configuration (α_e, β_e) corresponding to equilibrium.
b. Assuming $(\alpha, \beta) \neq (\alpha_e, \beta_e)$, find the initial acceleration of A, B and C when $\alpha = \beta = 45°$.

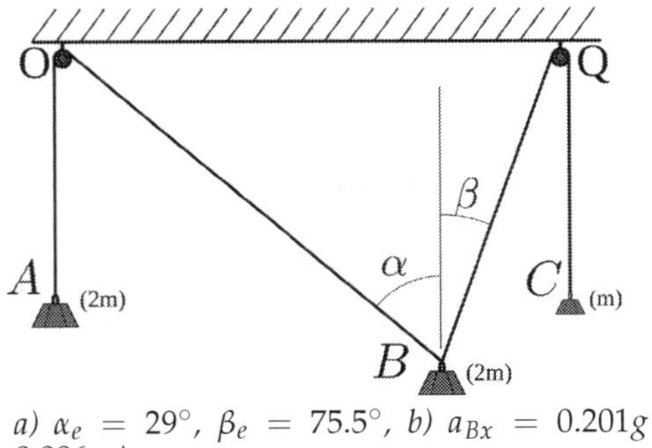

Ans.: a) $\alpha_e = 29°$, $\beta_e = 75.5°$, *b)* $a_{Bx} = 0.201g \leftarrow$, $a_{By} = 0.006g \uparrow$

Problem 3.11 The rolling trolley C of a tower crane moves horizontally at the acceleration a_C. A hook P is suspended by a cable of length l to the trolley. It is observed to exhibit small amplitude oscillations about an equilibrium angle θ_e.
Find angle θ_e and the period of these oscillations for $a_C = 5.1$ m/s^2, $l = 20$ m.
Hint: introduce angle $\epsilon = \theta - \theta_e$ *with* $\epsilon \ll 1$, *then use the approximations* $\sin\epsilon \approx \epsilon$ *and* $\cos\epsilon \approx 1$.

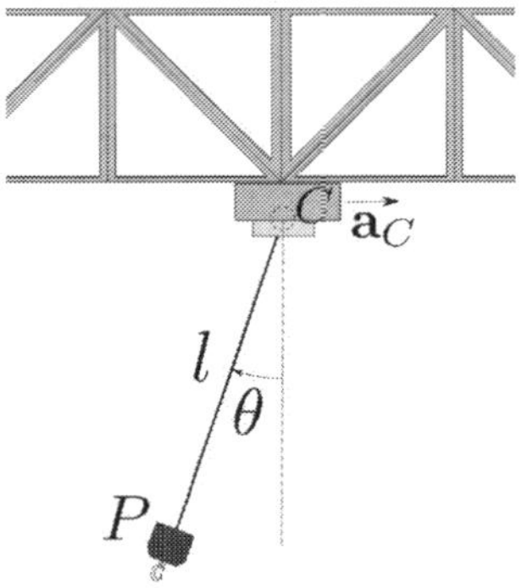

Answer: $\theta_e = 27.5°$, $T = 8.4$ s

Problem 3.12 A projectile P is fired from a point O with initial speed $v_0 = 40$ m/s at angle $\alpha = 45°$ above a horizontal firing range. In addition to gravity, P is subject to the force $\mathbf{F} = -m\lambda\mathbf{v}_P$ due to air friction ($\lambda > 0$).
a. Show that the path described by P is given by

$$\mathbf{r}_{OP} = \frac{\lambda\mathbf{v}_0 - \mathbf{g}}{\lambda^2}(1 - e^{-\lambda t}) + \frac{\mathbf{g}t}{\lambda}$$

with $\mathbf{v}_0 = v_0(\cos\alpha\hat{\mathbf{e}}_x + \sin\alpha\hat{\mathbf{e}}_z)$ and $\mathbf{g} = -g\hat{\mathbf{e}}_z$. Show that the parabolic path is recovered in the limit $\lambda \to 0$. Also show that $x(t)$ tends to a finite limit when $t \to \infty$.
b. Plot the paths for $\lambda = 0.1$ s^{-1} and $\lambda = 0$. Then estimate the difference $R_0 - R_{0.1}$ between the corresponding ranges.

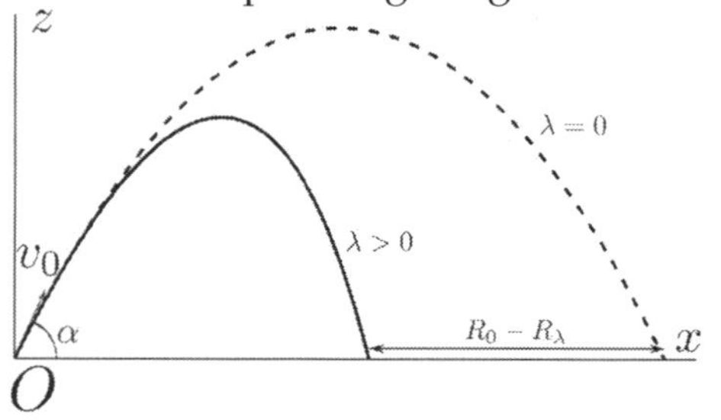

Answer: a). $x_\infty = (v_0\cos\alpha)/\lambda$. *b).* $R_0 - R_{0.1} = 44$ *m*

Problem 3.13 Consider in a vertical plane a stationary circular disk of center O and radius r whose surface has been roughened. A small body P of mass m, modeled as a particle, is set without initial speed on the edge of the disk in the vicinity of the highest point. We denote by θ_0 the initial angle between the vertical and the line OP and by μ the coefficient of static and kinetic friction between the disk and P.

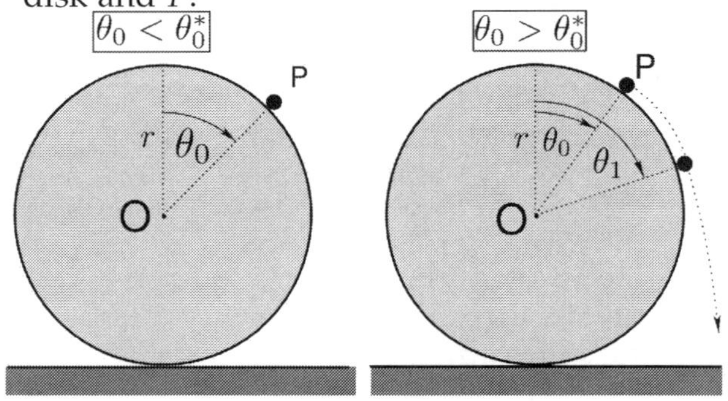

a. Find the minimum value θ_0^* of angle θ_0 above which particle P ceases to be in equilibrium.

b. Assume that $\theta_0 > \theta_0^*$. P slides along the edge of the disk. Show that the position of P is given by angle θ solution of the ODE

$$\ddot{\theta} - \mu\dot{\theta}^2 = \frac{g}{r}(\sin\theta - \mu\cos\theta) \qquad [1]$$

with $\theta = \theta_0$ and $\dot{\theta} = 0$ at $t = 0$.

c. Define $u = \dot{\theta}^2$. Show that [1] leads to the first-order ODE

$$\frac{1}{2}\frac{du}{d\theta} - \mu u = \frac{g}{r}(\sin\theta - \mu\cos\theta) \qquad [2]$$

Solve equation [2] by imposing $u = 0$ at $\theta = \theta_0$. Then find the angle θ_1 corresponding to the loss of contact of P with the disk.

Problem 3.14 A particle P of mass m is constrained to move along a horizontal circle of center C and radius R without friction. In addition to the support's reaction N, P is subject to a force $-f(r)\hat{\mathbf{u}}$ which always points to point O of the circle (with $r = |OP|$). The goal of this problem is to find the dependence of $f(r)$ upon r so that N remains constant.

a. Apply Newton's Second Law in a polar coordinate system centered at C.

b. Use a) to find N and impose $dN/dt = 0$ to find an equation governing $f(r)/r$. Conclude that $f(r) = -c/r^3$ where c is a constant.

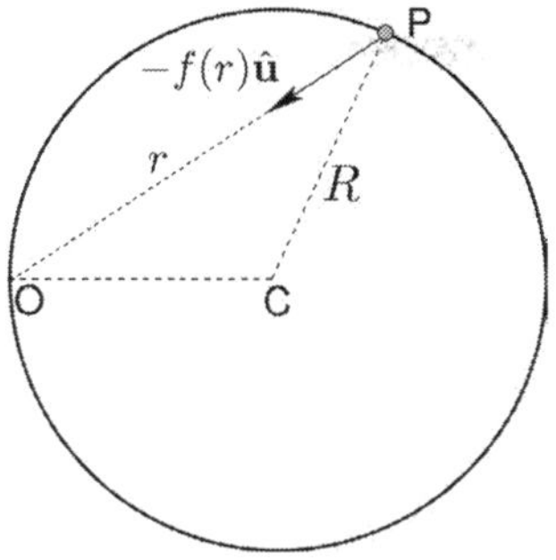

Problem 3.15 A glider of mass m is in a planar motion at speed v. In addition to gravity, the glider is subject to the drag force $R_x = C_x \rho S v^2/2$ and to the lift force $R_z = C_z \rho S v^2/2$, where ρ is the air mass density, A is the area of the wings, and (C_z, C_x) are lift and drag coefficients.

a. Find the ODEs which govern the glider's path.

b. Show that there exists an angle θ for which the glider descends at constant speed.

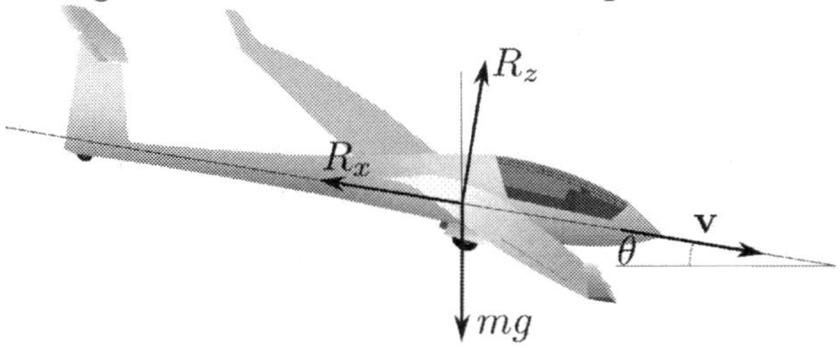

Problem 3.16 A small spherical ball of mass m is in motion under the effect of gravity in the interior of a cone of vertical axis Oz and half-angle α. The contact is frictionless. Its initial height is $z = h > 0$, and its initial velocity is of magnitude v_0, tangent to the cone, and perpendicular to axis Oz. Hence $\mathbf{v}_0$ lies in the horizontal plane $z = h$. Assume $\theta = 0$ at $t = 0$.

a. Apply Newton's Second Law to find three equations in cylindrical coordinates (r, θ, z) and the normal reaction force N.

b. Show that the motion satisfies the first integral $r^2 \dot{\theta} = Cst = h v_0 \tan\alpha$.

c. Use the result of b) to find the ODE governing z. Show that this equation leads to another first integral.

d. Under which condition is the motion of P circular? Is this motion stable?

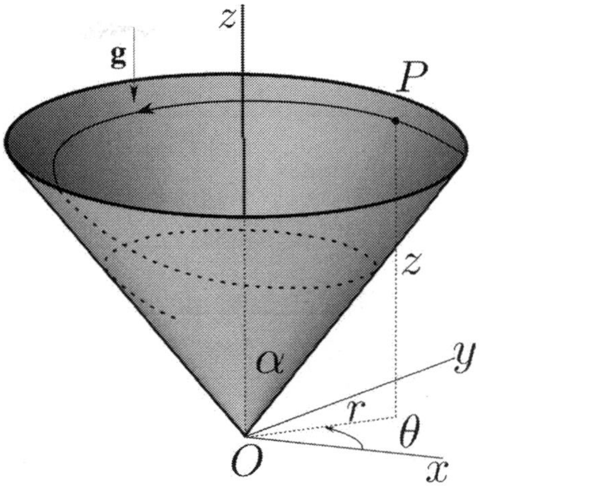

Problem 3.17 Two small spheres P_1 of mass $m_1 = 50$ g and $P_2 = 110$ g of mass m_2 are connected by an inextensible string. They are placed on a vertical support which takes the shape of a cycloid:

$$x(\theta) = R(\theta - \sin\theta), \qquad y(\theta) = R(1 - \cos\theta)$$

with $0 \le \theta \le 2\pi$ and $R = 1$ m. The string is initially taut and the system is released without initial speed. Friction is neglected. The spheres then slip under the effect of gravity. At $t = 0$, the initial positions of P_1 and P_2 are defined by $\theta_1 = 90°$ and $\theta_2 = 200°$.

Find the accelerations of the spheres immediately upon release.

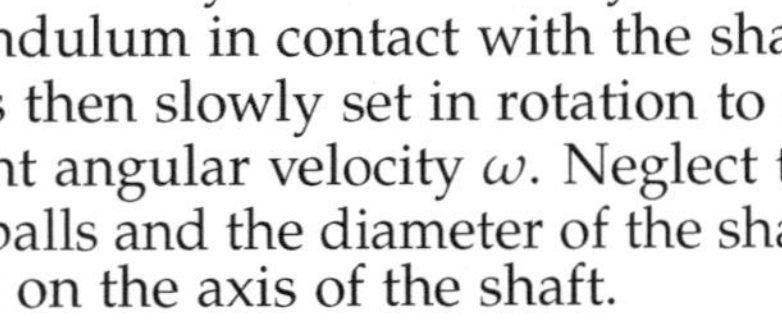

Problem 3.18 A small ball A of mass m is connected to a vertical shaft at O by a string of length l. The system is initially at rest, with the pendulum in contact with the shaft. The shaft is then slowly set in rotation to reach a constant angular velocity ω. Neglect the size of the balls and the diameter of the shaft, that is, O is on the axis of the shaft.

a. Show that the line OA describes a cone of half-angle ϕ only if ω exceeds a minimum value ω_1. Find angle ϕ as a function of ω. What happens if $\omega < \omega_1$?

b. The string is known to break if its tension reaches the value $20mg$. Find the value ω_2 of angular velocity ω when the string breaks.

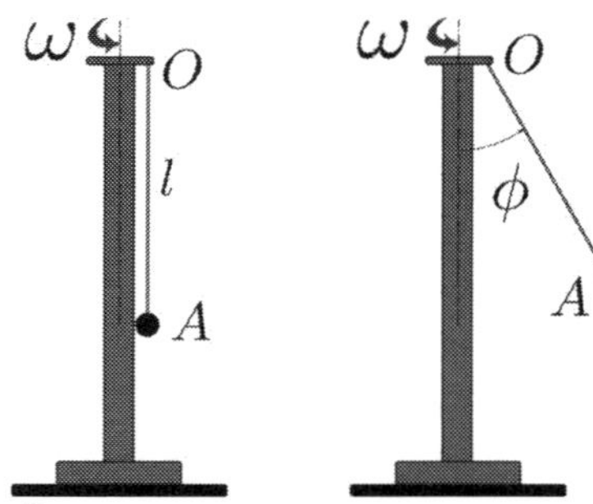

Answer: $\omega_2^2 = 20g/l$

Problem 3.19 A small ring P of mass m is constrained to slide along a vertical circular support of center O and radius R without friction. It is connected to point C of the support by an elastic string which exerts a force $-mkr\hat{\mathbf{u}}$ always pointing to point C of the circle (with $r = |CP|$, k constant). P is released at $t = 0$ at the position $\theta = \theta_0$ without initial speed.

a. By application of Newton's Second Law, find the ODE governing angle θ and the reaction force of the support as a function of θ. Show that P behaves like a pendulum of length R with an effective gravitational acceleration $g - kR$. Conclude that P can oscillate about B or C and find the period of the small amplitude oscillations.

b. Find the speed of P as a function of θ. Find the reaction force of the support as a function of θ.

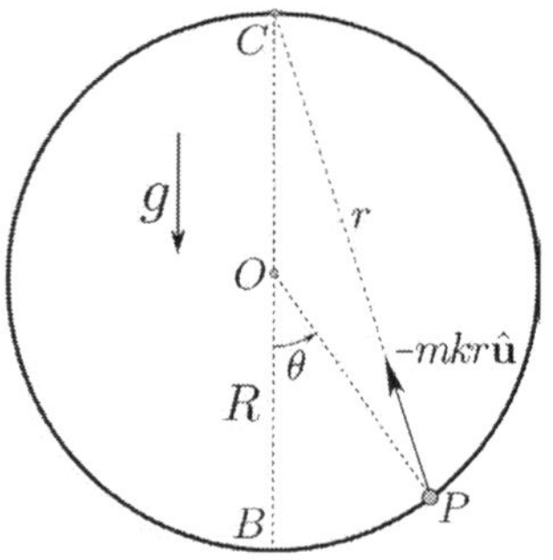

Problem 3.20 A small ring P of mass m is constrained to slide along a vertical circular support of center O and radius R without friction. It is connected to point B of the support by an elastic string which exerts the force of attraction force $-mkr\hat{\mathbf{u}}$ (with $r = |BP|$, k constant). At $t = 0$, P is located at A and is given a speed v_0 (the diameter AB is horizontal). Find the speed of P as a function of angle θ. Deduce that P admits two equilibrium positions given by $\tan 2\theta = g/kR$. Discuss the stability of these points.

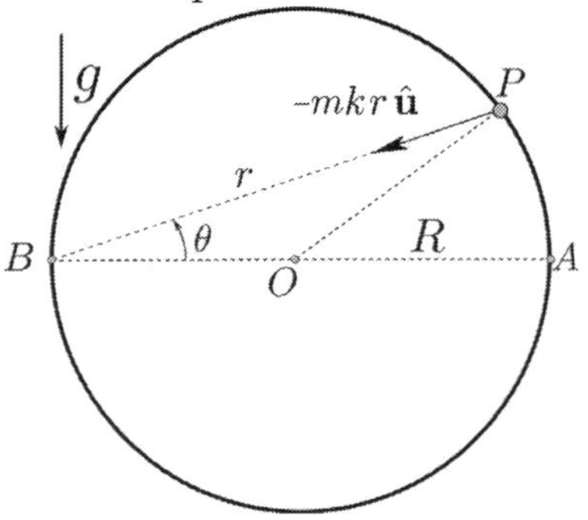

Problem 3.21 A particle P of mass m is in motion along a circle of center C and radius R under the action of an external resultant force $\mathbf{F}$ which is constantly directed toward a point O fixed on the circle.

a. Show that the magnitude of $\mathbf{F}$ must be proportional to $1/r^5$.

b. Show that the angle between line CO and line CP satisfies the equation $\theta - \sin\theta = kt + k'$, where k and k' are constants to be determined. Is the motion of P periodic?

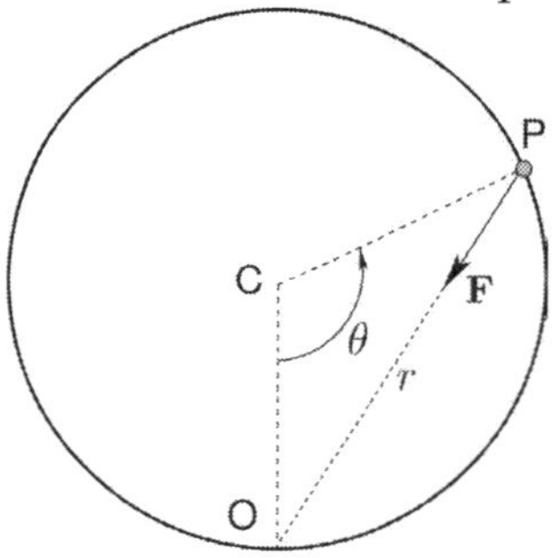

Problem 3.22 A heavy object P of mass $m = 450$ g is attached to the rim of a fixed cylinder of radius $R = 1$ m by an inextensible string of length $l = 2\pi$ m. At $t = 0$, the string is taut and P is released without initial speed. P is then set into a swinging motion in a vertical plane, while the string winds around the cylinder.

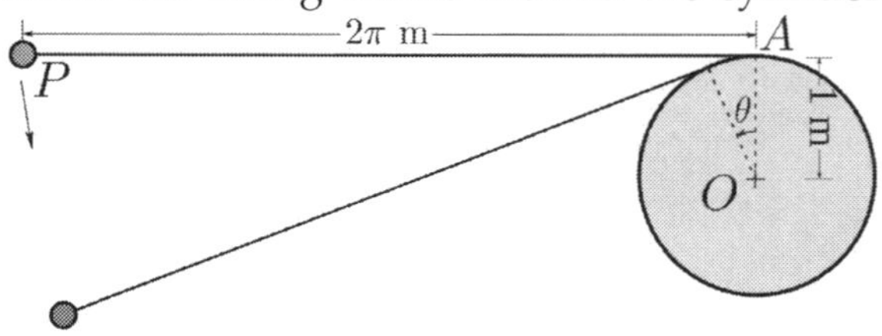

a. Apply Newton's Second Law to find two scalar equations. Show that one of these equations leads to a first integral.

b. Find the unwound length l of the string,

the speed v of P, and the tension in the string T as function of angle θ. Show that the string becomes slack at some angle $\theta = \theta_*$.

Problem 3.23 A particle P of mass m is attached to an inextensible string whose other extremity Q_0 is attached to the rim of a fixed cylinder of vertical axis. The particle is set on a smooth horizontal plane. The string length is initially l_0. At $t = 0$, the string is taut and P is given a velocity $\mathbf{v}_0$ perpendicular to Q_0P_0. P is then set in motion, while the string winds around the cylinder.

a. Apply Newton's Second Law to find three scalar equations. Show that one of these equations leads to a first integral.

b. Using the results of b), find the unwound length $l(t)$ of the string, the tension in the string $T(t)$ as functions of time. Show that the time at which $l = 0$ is given by $t_* = l_0^2/2Rv_0$. Find the corresponding value of tension T.

c. Now assume that the contact is characterized by a kinetic friction coefficient μ_k. Modify the three equations found in a) to find $l(t)$ versus time.

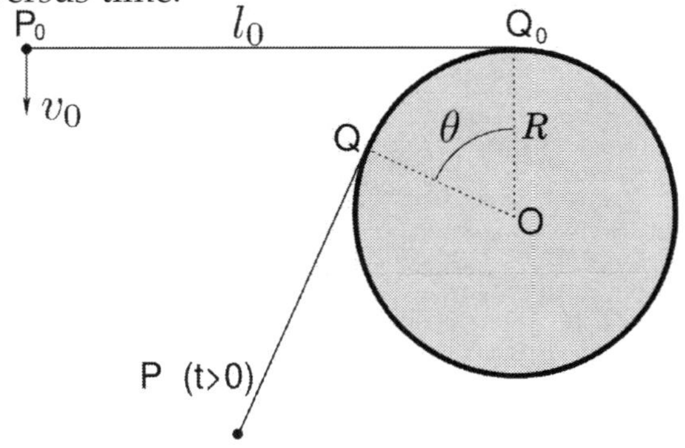

Problem 3.24 A small parcel P slides down a helical chute of vertical axis Oz under the effect of gravity. P travels along the path given by the following equations in Cartesian coordinates

$$x = R\cos\theta, \qquad y = R\sin\theta, \qquad z = p\frac{\theta}{2\pi}$$

where p and R are constant parameters. The contact of P with the chute is frictionless. P starts at position A without initial speed and ends at position B dropping from a height H.

a. Show that angle θ is governed by the following equation

$$\ddot{\theta} = -\frac{g}{R}\sin\beta\cos\beta$$

where angle β is defined by $\tan\beta = p/(2\pi R)$. Then find θ versus time t.

b. Find the reaction force of the chute on P at any position on the path.

c. Find the speed of P at position B and the time t_B taken to reach this position.

d. Find v_B, t_B and $\dot{\theta}$ at B for $R = 1$ m, $H = 10$ m, and the value of p corresponding to a drop with 6 revolutions.

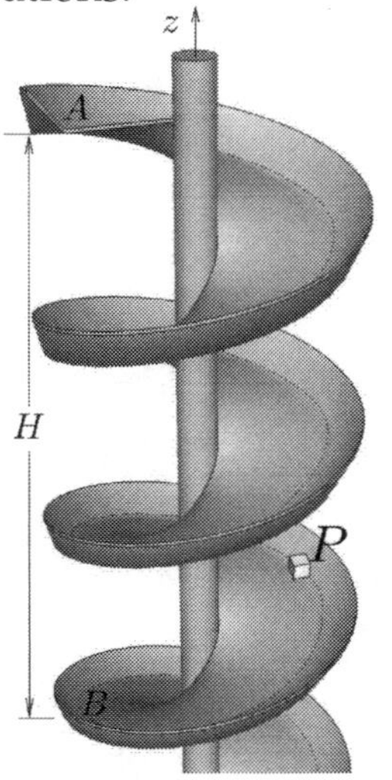

Problem 3.25 Two particles P_1 (of mass m_1) and P_2 (of mass m_2) form a closed system of mass center G such that the corresponding fictitious particle P of reduced mass $m_1m_2/(m_1+m_2)$ defined by $\mathbf{r}_{GP} = \mathbf{r}_{P_1P_2}$ travels along the conic section

$$r = \frac{p}{1 + e\cos\theta}$$

in the centroidal referential $\mathcal{E}^*$.
Find the force of interaction $\mathbf{F}_{1\to 2}$ between P_1 and P_2.

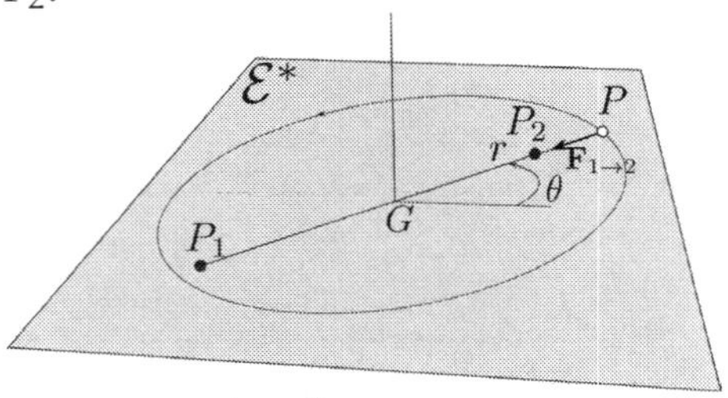

Answer: $\mathbf{F}_{1\to 2} = -(K/r^2)\hat{\mathbf{e}}_r$

Problem 3.26 Approximately half of the stars in our galaxy are members of so-called bi-

nary star systems. Such systems consist of two stars A and B orbiting about their mass center G. Assume that both stars have a circular motion centered about G relative to the centroidal referential under the effect of their mutual gravitational attraction. Three quantities are measured: (i) the distance of 3.55 billion km of separation between A and B, (ii) the ratio $r_B/r_A = 3$ of the radii of their circular paths, (iii) their common orbital period $T = 88$ years.
Show these quantities allow for the determination of the masses m_A and m_B of stars A and B.
Hint: Show that the given data allows for the determination of sum and ratio of the masses of A and B.

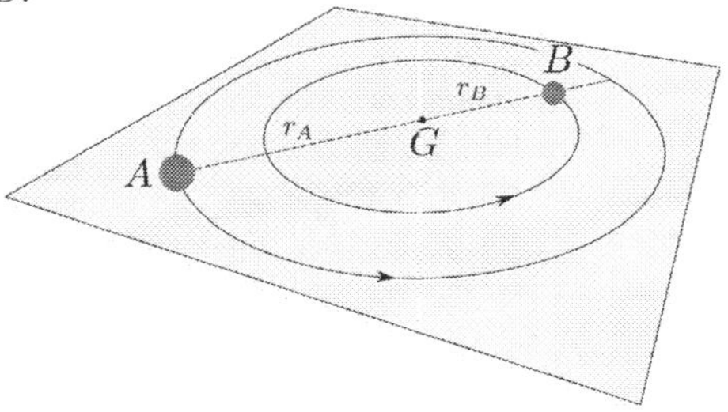

Answer: $m_A = 1.29M$, $m_B = 0.432M$, $M = 1.988 \times 10^{30}$ *kg*

Problem 3.27 Three celestial bodies P_1, P_2 and P_3 of mass m_1, m_2 and m_3 respectively are in gravitational interaction. They move in a plane of the corresponding centroidal referential $\mathcal{E}^*$ and remain at a constant distance from each other ($r_{ij} = Cst$). Recall that mass center G is defined by $m_1\mathbf{r}_1 + m_2\mathbf{r}_2 + m_3\mathbf{r}_3 = \mathbf{0}$.
a. Show that P_1, P_2 and P_3 are in uniform circular motion about G.
b. Apply Newton's Second Law to P_1. Then use $\mathbf{a}_1 = -\omega_1^2\mathbf{r}_1$ to show that $r_{12} = r_{13}$.
c. Conclude that if P_1, P_2 and P_3 are not aligned, then they must form an equilateral triangle.

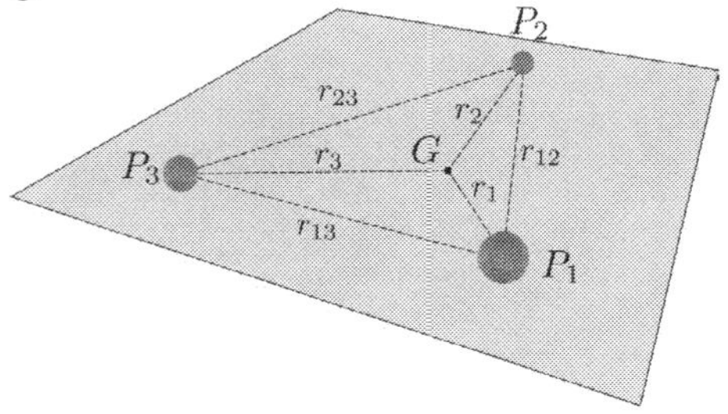

Problem 3.28 Two crates A and B of mass m_A and m_B ($m_A > m_B$) are set on top of the two sides of a wedge-shaped body C of mass m_C. The two crates are connected by a rope passing over a pulley attached to body C. All sources of friction are neglected. Body C may slide freely relative to a horizontal support. The bodies are released without initial speed. Find the acceleration of body C, and the acceleration of the two crates relative to body C.

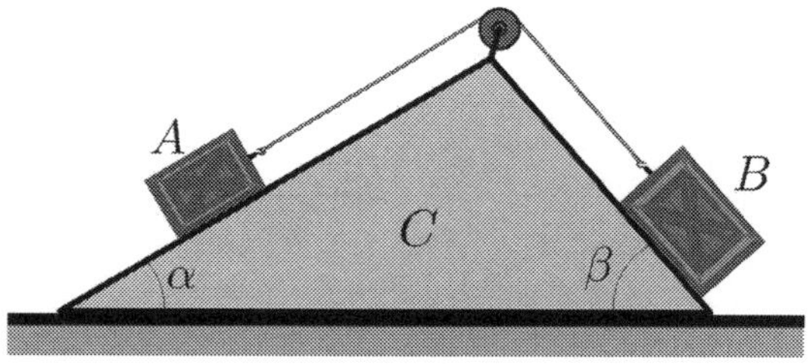

Problem 3.29 A block B of mass m_B is set inside the track formed along the sloped edge of a wedge-shaped body A of mass m_A and angle α. Body B is attached by a cord passing through the pulley P to a vertical wall. Body B is released without initial speed, and is observed to slip along the track. All sources of friction are neglected.
Find the acceleration of body A.

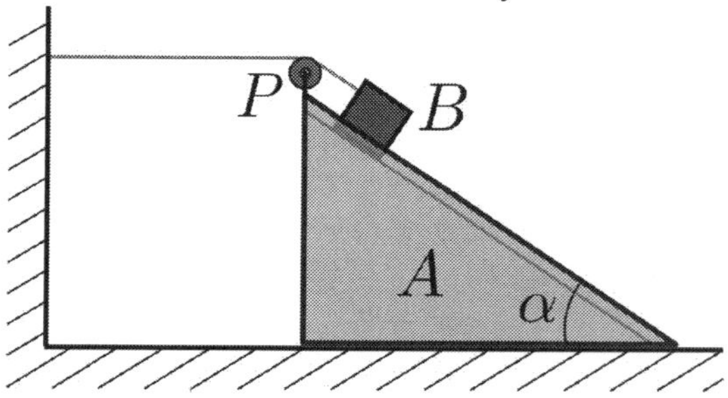

Problem 3.30 Four blocks are stacked pairwise, as shown in the figure.

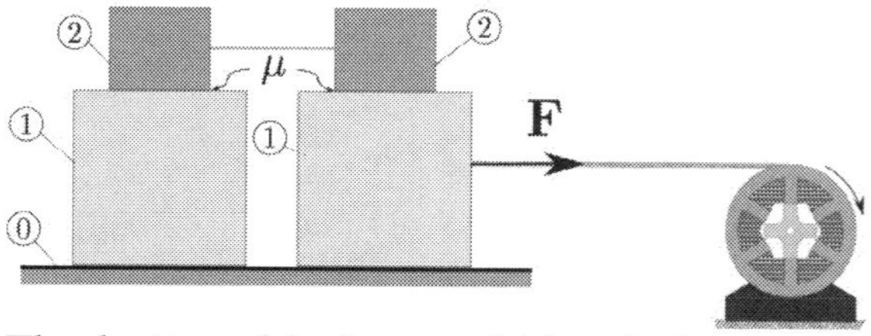

The bottom blocks 1, of identical mass M, are resting on a smooth surface (friction is neglected). The top blocks 2, of identical mass m, are connected by a taut cord and are resting on top of the bottom blocks. The friction coefficients between the blocks are $\mu_s \approx \mu_k = \mu \neq 0$. The system is at rest when the bottom block is subject to the force $\mathbf{F}$.
Find the accelerations of the blocks as a function of $\mathbf{F}$.

Problem 3.31 A cart 1 of mass M is resting on a horizontal support 0. Its frictionless rollers have negligible mass. Hence, it is free to move along the horizontal direction. It supports a hemispherical container of radius R. A small spherical body 2 of mass m is released without initial speed from the position shown, the cart being in contact with a vertical wall. Neglect friction between the two bodies.
Describe the motion of the cart. In particular, find its maximum speed.

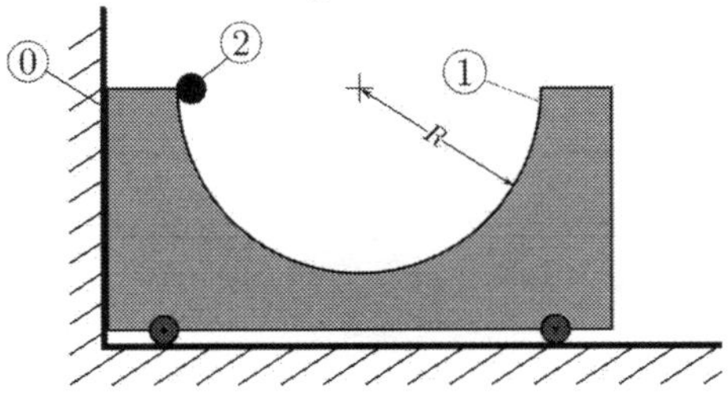

Problem 3.32 Two groups of people are standing still on a flatcar separated by a distance L. The first group has mass m_1, the second group mass m_2. The flatcar of mass M is initially at rest. The two groups then trade places and stand still again. Neglect the resistance of the tracks on the flatcar.
Find the total displacement the flatcar is observed to make.

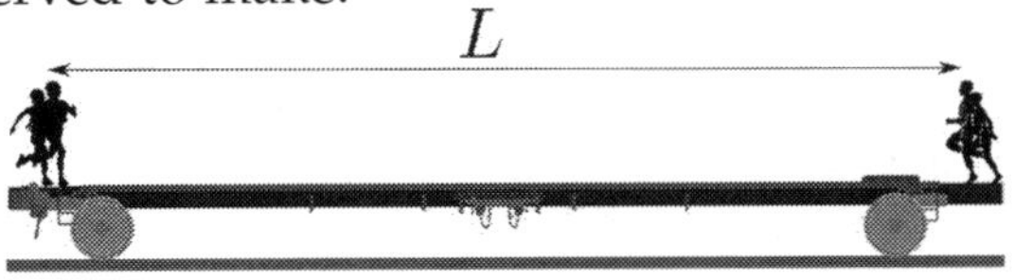

Problem 3.33 Two stars A of mass m_A and B of mass m_B orbit about their mass center G along circular trajectory of radius r_A and r_B, respectively, under the effect of their mutual gravitational attraction.

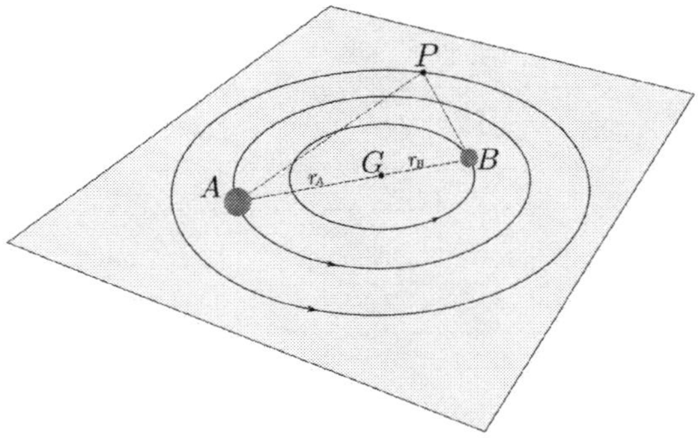

a. Find the angular velocity ω of the line AB joining the two stars, as a function of m_A, m_B, r_A, r_B and G (the universal gravitational constant).

a. A small body P of mass m ($m \ll m_A$, $m \ll m_B$) rotates about G along a circular trajectory under the effect of the gravitational attraction of both stars. Show that, for P to remain stationary relative to both A and B, it is necessary to have $|AP| = |BP| = r_A + r_B$. Deduce that triangle APB is equilateral.

Problem 3.34 Two parallel rods support two small sliding collars, each of identical mass m. The collars are connected by a massless rod of length l, making the constant angle φ relative to the rods. The contact at A is frictionless. The contact at B is characterized by the friction coefficients $\mu_s \approx \mu_k = \mu$. The effect of gravity on both A and B is neglected. A constant force F is exerted on A.
a. Show that if $\mu > \mu_1$, no motion is possible. Find μ_1.
b. Assuming that $\mu < \mu_1$, show that two possible motions are possible whether $\mu < \mu_2$ or $\mu_2 < \mu < \mu_1$. Find μ_2 and describe the motions.

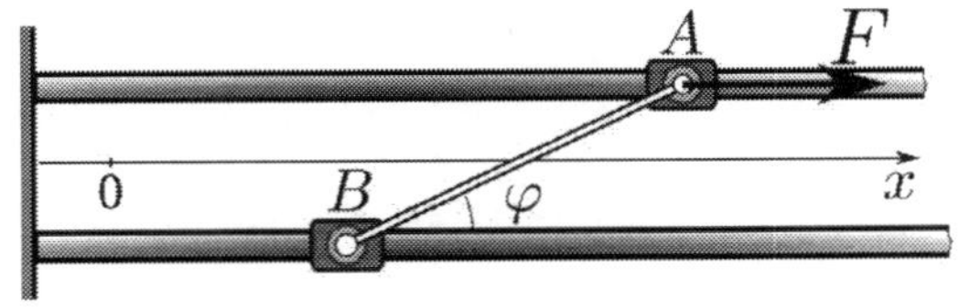

Problem 3.35 A platform $\mathcal{F}$ rotates relative to a Newtonian referential at constant angular velocity Ω about a fixed vertical axis $(O, \hat{\mathbf{k}})$.

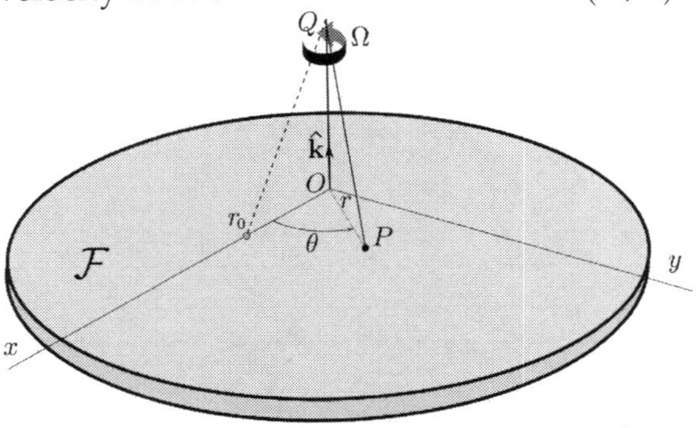

A pendulum P of mass m and connected to a point Q of axis $(O, \hat{\mathbf{k}})$ is installed in $\mathcal{F}$. When undisturbed, P coincides with O. The length $l = |PQ|$ is assumed large compared to the amplitude of oscillation so that P can be considered to move in the plane Oxy attached to $\mathcal{F}$. At $t = 0$, P is located on axis Ox at a distance r_0 from O. Its position is defined by

its polar coordinates (r, θ) in plane Oxy.
a. Show that the equation of motion of P relative to $\mathcal{F}$ is given by

$$\frac{d^2\mathbf{r}_{OP}}{dt^2} + 2\Omega\hat{\mathbf{k}} \times \frac{d\mathbf{r}_{OP}}{dt} + (\omega_0^2 - \Omega^2)\mathbf{r}_{OP} = \mathbf{0}$$

where the derivatives are performed relative to $\mathcal{F}$. Identify the value of ω_0 and describe the physical meaning of each term of this equation.

b. Express the equation of motion in polar coordinates (r, θ) to obtain two ODEs. Show that a particular solution is given by

$$r(t) = r_0 \cos(\omega_0 t), \qquad \theta(t) = -\Omega t$$

c. Plot the solution in the xy-plane for $r_0 = 0.1$ m, $\omega_0 = 3$ s^{-1}, and $\Omega = 0.6$ s^{-1}.

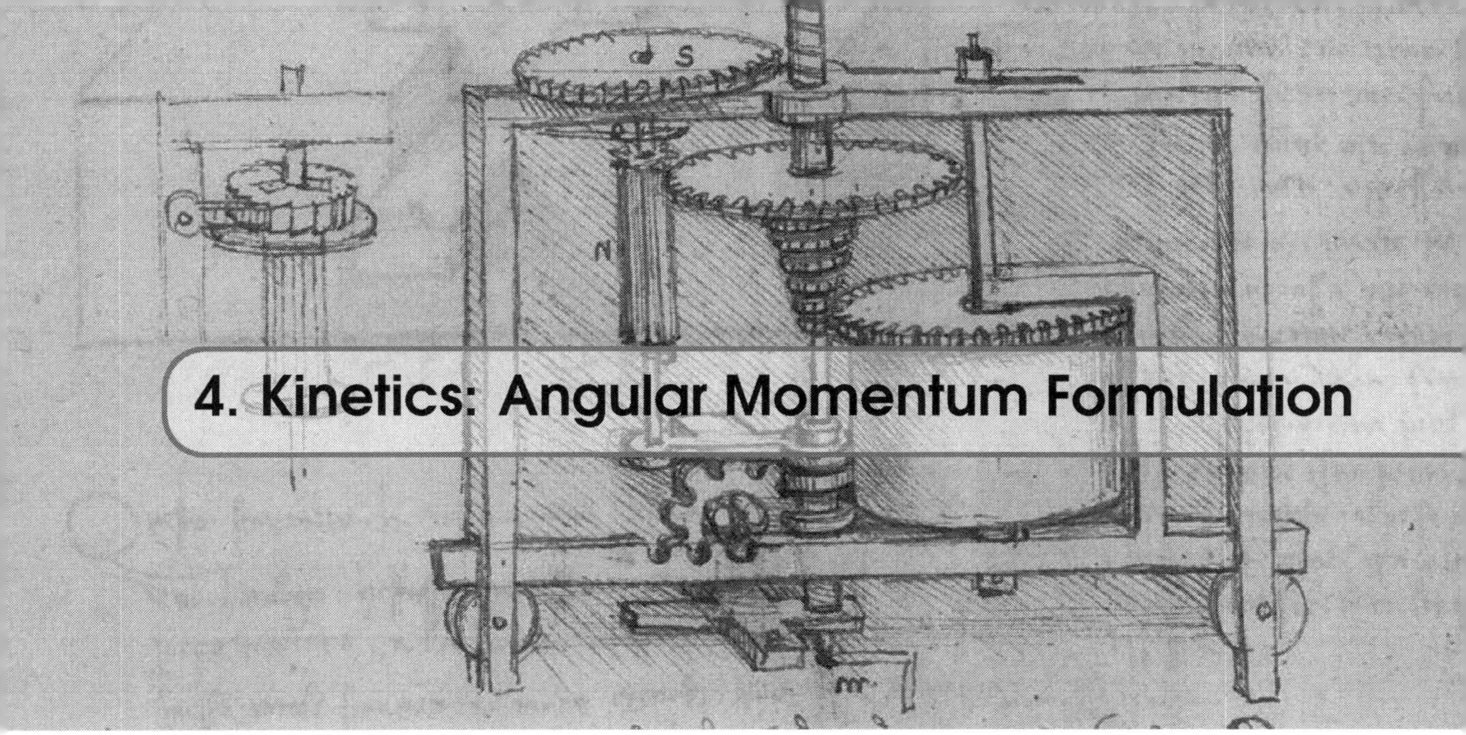

4. Kinetics: Angular Momentum Formulation

IN THIS CHAPTER, we derive new kinetics laws formulated in terms of *angular momentum*. On one hand, for a single particle, the angular momentum form of Newton's Second Law (Section 4.1) can lead to fewer equations, and its application adds nothing fundamentally new. On the other hand, for systems of particles or rigid bodies, Euler's First Principle only predicts the motion of the system's or body's mass center, as was shown in Chapter 3, and it is thus not sufficient for a complete analysis. In such cases, the angular momentum formulation will lead to a new kinetics principle called *Euler's Second Principle*. We first derive and illustrate this principle for systems of particles in Section 4.2. In Section 4.3, we learn how to determine the angular momentum of a rigid body in planar motion. We show that the angular momentum can be calculated quite simply from the body's *mass moment and products of inertia*. In Section 4.4, we discuss the properties of these quantities. In particular, we tabulate the moments/products of inertia of simple shapes which can be used to find those of more complex shapes. We then derive the expression of *Euler's Second Principle* for a rigid body (Section 4.5). This principle is critical to predict the body's rotation motion (or absence of rotational motion). Prior to illustrating the application of Euler's Principles to rigid body motion, we review in Section 4.6 how contact forces can be modeled by Coulomb laws and, in particular, learn about the specificities of *frictionless joints* interconnecting rigid bodies (Section 4.7). Finally, we outline general guidelines for the solution of rigid body kinetics problems (Section 4.8).
In Section 4.9, we examine the challenging case of *systems of rigid bodies* and outline strategies to derive equations of motion (ODEs) or *first integrals of motion*. The latter translate into conservation of linear or angular momentum. As a consequence of Newton's Third Law, the contribution of internal mechanical actions disappears from the formulation of Euler's principles, whether applied to systems of particles or systems

of rigid bodies. Finally, in Section 4.10, we examine a special application of rigid body kinetics: the *dynamic balancing* of bodies in rotational motion.

In this chapter, we will learn how to

1. determine the moments and products of inertia of a rigid body,
2. determine the angular momentum of a particle, system of particle or rigid body about an arbitrary point,
3. apply Euler's principles to the kinetic analysis of one or more rigid bodies.

4.1 Angular Momentum Form of Newton's Second Law

THE angular momentum form of Newton's Second Law is obtained by taking the cross-product of both sides of equation (3.1) with a particular position vector. This leads us to define two quantities as moments of vectors bound at P: (i) the angular momentum, and (ii) the moment of the force applied at P.

DEFINITION (ANGULAR MOMENTUM)

The *angular momentum* $\mathbf{H}_A$ of particle P of mass m about point A relative to referential $\mathcal{E}$ is the vector defined by

$$\mathbf{H}_A = \mathbf{r}_{AP} \times m\mathbf{v}_P \tag{4.1}$$

Hence, $\mathbf{H}_A$ can be interpreted as the moment about A of the linear momentum $m\mathbf{v}_P$ treated as a vector bound at P. It is easy to verify that the angular momentum about another point B can be obtained from that about A according to the formula

$$\mathbf{H}_B = \mathbf{H}_A + \mathbf{r}_{BA} \times m\mathbf{v}_P \tag{4.2}$$

Remark 1. Note that the angular momentum is a function of the choice of referential $\mathcal{E}$ and should be denoted $\mathbf{H}_{A/\mathcal{E}}$.

Remark 2. The angular momentum of P can be defined about a point A either fixed or in motion in a referential $\mathcal{E}$.

Likewise, by considering the force $\mathbf{F}$ acting on particle P as a vector bound at P, the moment of $\mathbf{F}$ about point A is the vector defined by

$$\mathbf{M}_A = \mathbf{r}_{AP} \times \mathbf{F} \tag{4.3}$$

Then the moment of $\mathbf{F}$ about any other point is found to be

$$\mathbf{M}_B = \mathbf{M}_A + \mathbf{r}_{BA} \times \mathbf{F}$$

Now we can derive the angular momentum form of Newton's Second Law: we take the cross-product of both sides of $m\mathbf{a}_P = \mathbf{F}$ with position vector $\mathbf{r}_{OP}$:

$$\mathbf{r}_{OP} \times m\mathbf{a}_P = \mathbf{r}_{OP} \times \mathbf{F}$$

If point O is fixed in $\mathcal{E}$, then we can write

$$\mathbf{r}_{OP} \times m\mathbf{a}_P = \frac{d}{dt}(\mathbf{r}_{OP} \times m\mathbf{v}_P) - \mathbf{v}_P \times m\mathbf{v}_P = \frac{d\mathbf{H}_O}{dt}$$

Furthermore the term $\mathbf{r}_{OP} \times \mathbf{F}$ is recognized as the moment $\mathbf{M}_O$ about O of the resultant force acting on P. We have obtained the angular momentum form of Newton's Second Law, which can be generalized as follows.

Newton's Second Law: Angular Momentum form

The time rate of change of the angular momentum of particle P about an arbitrary point A of velocity $\mathbf{v}_A$ is related to the moment about A of the resultant force acting on P according to

$$\frac{d\mathbf{H}_A}{dt} + \mathbf{v}_A \times m\mathbf{v}_P = \mathbf{M}_A \tag{4.4}$$

relative to a Newtonian referential. About a fixed point O of the referential, equation (4.4) takes the form

$$\frac{d\mathbf{H}_O}{dt} = \mathbf{M}_O \tag{4.5}$$

Proof. By time-differentiating the equation $\mathbf{H}_A = \mathbf{H}_O + \mathbf{r}_{AO} \times m\mathbf{v}_P$, we find

$$\frac{d\mathbf{H}_A}{dt} = \frac{d\mathbf{H}_O}{dt} - \mathbf{v}_A \times m\mathbf{v}_P + \mathbf{r}_{AO} \times m\mathbf{a}_P$$

where we can replace $d\mathbf{H}_O/dt$ by $\mathbf{M}_O$ and $m\mathbf{a}_P$ by $\mathbf{F}$

$$\frac{d\mathbf{H}_A}{dt} = \mathbf{M}_O - \mathbf{v}_A \times m\mathbf{v}_P + \mathbf{r}_{AO} \times \mathbf{F}$$

We then recognize the sum $\mathbf{M}_O + \mathbf{r}_{AO} \times \mathbf{F}$ as the moment $\mathbf{M}_A$ about A of the resultant force to obtain (4.4). ■

Properties

①. The angular momentum form of Newton's Second Law may yield less information. For instance, if the motion of P evolves in plane Oxy, then both $\mathbf{H}_O$ and $\mathbf{M}_O$ are normal to this plane: only one equation is obtained from (4.5).
②. The angular momentum form of Newton's law may yield the equation of motion or a first integral of motion (conservation of angular momentum) in a more efficient manner.

These two points are illustrated in the following examples.

Example 1: Pendulum Motion. Let P of mass m be connected to fixed point O by an inextensible string of length l. See Figure 4.1 (left). We apply equation (4.5) about O by using the polar unit vectors $(\hat{\mathbf{e}}_r, \hat{\mathbf{e}}_\theta)$: from the expression $\mathbf{v}_P = l\dot{\theta}\hat{\mathbf{e}}_\theta$ of the velocity of P we obtain the angular momentum of P about O

$$\mathbf{H}_O = l\hat{\mathbf{e}}_r \times ml\dot{\theta}\hat{\mathbf{e}}_\theta = ml^2\dot{\theta}\hat{\mathbf{e}}_z$$

with $\hat{\mathbf{e}}_z = \hat{\mathbf{e}}_r \times \hat{\mathbf{e}}_\theta$. From the resultant force $\mathbf{F} = (-T + mg\cos\theta)\hat{\mathbf{e}}_r - mg\sin\theta\hat{\mathbf{e}}_\theta$, we obtain the total moment about O as

$$\mathbf{M}_O = l\hat{\mathbf{e}}_r \times \mathbf{F} = -mgl\sin\theta\hat{\mathbf{e}}_z$$

Then equation (4.5) yields the equation of motion of P (the pendulum equation):

$$ml^2\ddot{\theta} + mgl\sin\theta = 0$$

Note that only one equation is obtained! If tension T is needed, one must return to Newton's Second Law.

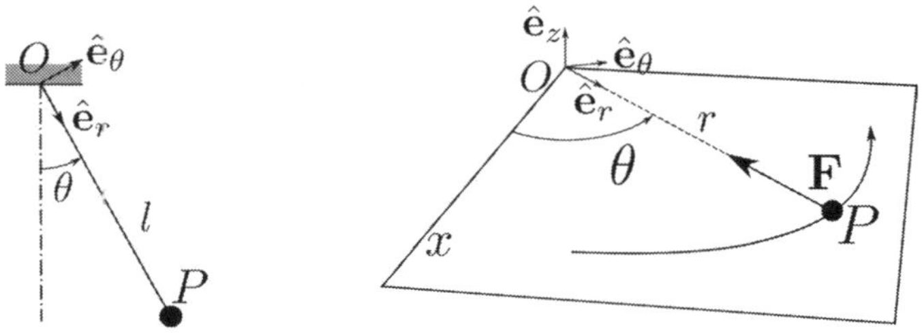

Figure 4.1 – *Pendulum motion (left). Central force motion (right).*

Example 2: Central Force Motion. Let P of mass m be acted upon by a central force $\mathbf{F}$ in plane Oxy relative to pole O. See Figure 4.1 (right). P is defined by its polar coordinates (r, θ). We apply equation (4.5) about O by using the polar unit vectors $(\hat{\mathbf{e}}_r, \hat{\mathbf{e}}_\theta)$. From the expression $\mathbf{v}_P = \dot{r}\hat{\mathbf{e}}_r + r\dot{\theta}\hat{\mathbf{e}}_\theta$ of the velocity of P, we obtain the angular momentum of P about O

$$\mathbf{H}_O = r\hat{\mathbf{e}}_r \times m(\dot{r}\hat{\mathbf{e}}_r + r\dot{\theta}\hat{\mathbf{e}}_\theta) = mr^2\dot{\theta}\hat{\mathbf{e}}_z$$

with $\hat{\mathbf{e}}_z = \hat{\mathbf{e}}_r \times \hat{\mathbf{e}}_\theta$. Since $\mathbf{F}$ points toward O, its moment about O is zero: we obtain $d\mathbf{H}_O/dt = \mathbf{0}$. Hence the angular momentum of P about O is conserved. This gives the first integral

$$mr^2\dot{\theta} = Cst$$

which expresses the conservation of angular momentum of P about O.

4.2 Euler's Second Principle for Systems of Particles

First, we generalize the concept of angular momentum for a system Σ of N particles P_1, P_2,..., P_N of mass $m_1, m_2, \ldots, m_N$, respectively.

DEFINITION (ANGULAR MOMENTUM OF A SYSTEM)

The *angular momentum* of a system Σ about point A relative to referential $\mathcal{E}$ is defined as the sum of the angular momenta about A of the N particles

$$\mathbf{H}_A = \mathbf{r}_{AP_1} \times m_1 \mathbf{v}_{P_1} + \cdots + \mathbf{r}_{AP_N} \times m_N \mathbf{v}_{P_N} \tag{4.6}$$

Again, we can relate the angular momentum about two distinct points according to

$$\mathbf{H}_B = \mathbf{H}_A + \mathbf{r}_{BA} \times m\mathbf{v}_G \tag{4.7}$$

where $m\mathbf{v}_G$ is the linear momentum of the system.
Next we wish to generalize the angular momentum form (4.5) to the system Σ about a fixed point O of a Newtonian referential. To simplify the analysis, we consider the case of $N = 2$ particles. We first write (4.5) for each particle:

$$\frac{d}{dt}(\mathbf{r}_{OP_1} \times m_1 \mathbf{v}_{P_1}) = \mathbf{r}_{OP_1} \times (\mathbf{F}_{ext \to 1} + \mathbf{F}_{2 \to 1})$$

$$\frac{d}{dt}(\mathbf{r}_{OP_2} \times m_2 \mathbf{v}_{P_2}) = \mathbf{r}_{OP_2} \times (\mathbf{F}_{ext \to 2} + \mathbf{F}_{1 \to 2})$$

where the forces $\mathbf{F}_{ext \to 1}$ and $\mathbf{F}_{ext \to 2}$ are external to the system and $\mathbf{F}_{2 \to 1} = -\mathbf{F}_{1 \to 2}$ is the (internal) force of interaction between P_1 and P_2. We then add these two equations to obtain

$$\frac{d}{dt}\mathbf{H}_O = \mathbf{r}_{OP_1} \times \mathbf{F}_{ext \to 1} + \mathbf{r}_{OP_2} \times \mathbf{F}_{ext \to 2} + \mathbf{r}_{OP_1} \times \mathbf{F}_{2 \to 1} + \mathbf{r}_{OP_2} \times \mathbf{F}_{1 \to 2}$$

where $\mathbf{H}_O$ is the angular momentum of a system Σ about point O. The sum of the first two terms $\mathbf{r}_{OP_1} \times \mathbf{F}_{ext \to 1}$ and $\mathbf{r}_{OP_2} \times \mathbf{F}_{ext \to 2}$ are moments of external forces which we denote $\mathbf{M}_{O,ext \to \Sigma}$. The sum of the two terms $\mathbf{r}_{OP_1} \times \mathbf{F}_{2 \to 1}$ and $\mathbf{r}_{OP_2} \times \mathbf{F}_{1 \to 2}$ cancel each other out in view of Newton's Third Law:

$$\mathbf{r}_{OP_1} \times \mathbf{F}_{2 \to 1} + \mathbf{r}_{OP_2} \times \mathbf{F}_{1 \to 2} = (-\mathbf{r}_{OP_1} + \mathbf{r}_{OP_2}) \times \mathbf{F}_{1 \to 2} = \mathbf{r}_{P_1 P_2} \times \mathbf{F}_{1 \to 2} = \mathbf{0}$$

since the two vectors $\mathbf{r}_{P_1 P_2}$ and $\mathbf{F}_{1 \to 2}$ are colinear. The final result is known as *Euler's Second Principle*:

EULER'S SECOND PRINCIPLE

Relative to a Newtonian referential, the time rate of change of the angular momentum of a system of particles about a fixed point O is equal to the moment about O of the external forces acting on the system

$$\frac{d\mathbf{H}_O}{dt} = \mathbf{M}_{O,ext \to \Sigma} \tag{4.8}$$

Furthermore, the same equation holds about the mass center G of the system:

$$\frac{d\mathbf{H}_G}{dt} = \mathbf{M}_{G,ext \to \Sigma} \tag{4.9}$$

Proof. To prove equation (4.9), we take the cross-product of Newton's Second Law with the vector $\mathbf{r}_{GP_1}$ for P_1 and $\mathbf{r}_{GP_2}$ for P_2:

$$\mathbf{r}_{GP_1} \times m_1\mathbf{a}_{P_1} = \mathbf{r}_{GP_1} \times (\mathbf{F}_{ext\to1} + \mathbf{F}_{2\to1})$$

$$\mathbf{r}_{GP_2} \times m_2\mathbf{a}_{P_2} = \mathbf{r}_{GP_2} \times (\mathbf{F}_{ext\to2} + \mathbf{F}_{1\to2})$$

By summing these two equations, we again eliminate the contribution of internal forces:

$$\mathbf{r}_{GP_1} \times m_1\mathbf{a}_{P_1} + \mathbf{r}_{GP_2} \times m_2\mathbf{a}_{P_2} = \mathbf{M}_{G,ext\to\Sigma}$$

We can relate the left-hand side of this equation to the angular momentum of the system about G: these terms can be written as

$$\frac{d}{dt}(\mathbf{r}_{GP_1} \times m_1\mathbf{v}_{P_1} + \mathbf{r}_{GP_2} \times m_2\mathbf{v}_{P_2}) - \mathbf{v}_G \times (m_1\mathbf{v}_{P_1} + m_2\mathbf{v}_{P_2}) = \frac{d\mathbf{H}_G}{dt}$$

where we have used $m_1\mathbf{v}_{P_1} + m_2\mathbf{v}_{P_2} = m\mathbf{v}_G$. This gives (4.9). ■

PROPERTIES

①. In both Euler's First and Second Principles, the contribution of internal forces must not be taken into account.

②. Euler's Second Principle takes the form $d\mathbf{H}_Q/dt = \mathbf{M}_{Q,ext\to\Sigma}$ only in two cases: (i) if point Q is fixed, (ii) if Q is the mass center of the system.

③. For a single particle, the angular momentum formulation does not provide new information vs. the linear momentum formulation of Newton's Second Law. This is not the case for a system of particles. The linear momentum formulation of Euler's First Principle allows the prediction of the motion of the mass center G of the system. The angular momentum formulation of Euler's Second Principle predicts the behavior of the system about the mass center G.

This last point is demonstrated by examining the case of a closed system of particles.

CLOSED SYSTEM OF TWO PARTICLES. Recall that the system is said to be *closed* if the external forces acting on P_1 (of mass m_1) and P_2 (of mass m_2) satisfy $\mathbf{F}_{ext\to1} = \mathbf{F}_{ext\to2} = \mathbf{0}$. In Section 3.8, we found by applying Euler's First Principle that the mass center G of system Σ has a uniform rectilinear motion, a consequence of the fact that $\mathbf{F}_{ext\to\Sigma} = \mathbf{F}_{ext\to1} + \mathbf{F}_{ext\to2} = \mathbf{0}$. Let us now apply Euler's Second Principle (4.9) about G in the centroidal referential $\mathcal{E}^*$ (which is Newtonian): since the external forces acting on the system satisfy $\mathbf{M}_{G,ext\to\Sigma} = \mathbf{0}$, we find that the angular momentum $\mathbf{H}_{G/\mathcal{E}^*}$ of the system is conserved, that is,

$$\mathbf{r}_{GP_1} \times m_1\mathbf{v}_{P_1/\mathcal{E}^*} + \mathbf{r}_{GP_2} \times m_2\mathbf{v}_{P_2/\mathcal{E}^*} = Cst \qquad (4.10)$$

We can express this equation in terms of the fictitious particle P of mass $m_1m_2/(m_1+m_2)$ defined by $\mathbf{r}_{GP} = \mathbf{r}_{P_1P_2}$ (see equation 3.39):

$$\mathbf{r}_{GP_1} = -\frac{m_2}{m_1+m_2}\mathbf{r}_{GP}, \qquad \mathbf{r}_{GP_2} = \frac{m_1}{m_1+m_2}\mathbf{r}_{GP}$$

$$\mathbf{v}_{P_1/\mathcal{E}^*} = -\frac{m_2}{m_1 + m_2}\mathbf{v}_{P/\mathcal{E}^*} \qquad \mathbf{v}_{P_2/\mathcal{E}^*} = \frac{m_1}{m_1 + m_2}\mathbf{v}_{P/\mathcal{E}^*}$$

Then equation (4.10) becomes

$$\mathbf{r}_{GP} \times \frac{m_1 m_2}{m_1 + m_2}\mathbf{v}_{P/\mathcal{E}^*} = Cst \tag{4.11}$$

This shows that *the angular momentum of P about G (relative to $\mathcal{E}^*$) is conserved.* This is equivalent to the first integral $r^2\dot{\theta} = Cst$ mentioned in Section 3.8. Consequently, the application of Euler's Second Principle predicts the behavior of P (and hence of P_1 and P_2) about G.

EXAMPLE PROBLEMS

General Procedure: Kinetics of Systems of Particles

To study the kinetics of systems of particles, apply Euler's Principles by following these guidelines:

① Define the system Σ and the Newtonian referential $\mathcal{E}$ relative to which the system is studied. Define all necessary unit vectors.

② After parametrizing the position of particle A of Σ relative to $\mathcal{E}$, find its linear momentum $m_A\mathbf{v}_A$ and angular momentum $\mathbf{H}_Q = \mathbf{r}_{QA} \times m_A\mathbf{v}_A$ (about a point Q of interest). Repeat for the other particles.

③ Draw a free-body diagram for the system (do not take into account the internal forces).

④ Apply Euler's Principles as needed. Carefully choose the point Q about which the moments are evaluated. Ideally, point Q is chosen so as to make the moment of an *unknown, external force* zero.

Exercise 4.1 ▪ Two monkeys M_1 and M_2 of identical mass m are hanging to the two ends of a massless, inextensible rope which passes over a pulley of negligible mass. The pulley (of radius r) can rotate without friction about a horizontal axis passing through point O.
The system pulley-rope-monkeys is maintained initially at rest. The two monkeys are initially at the same level relative to the ground.
a. As soon as the system is released, monkey M_1 decides to climb up the rope, while monkey M_2 holds tightly to the rope. Describe the vertical motions of the monkeys relative to the ground. Which monkey will arrive first at the level of the pulley?
b. Now assume that monkey M_2 is heavier than monkey M_1. Is it possible for the lighter monkey to raise the heavier monkey? ▪

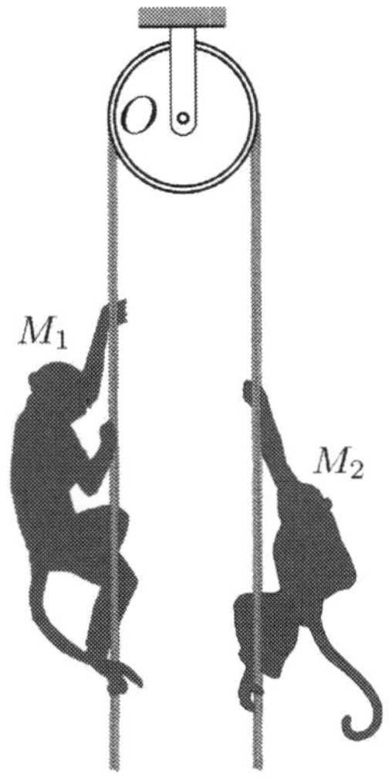

SOLUTION. **a.** CASE 1: $m_1 = m_2 = m$. We define an axis By directed upward and a basis $(\hat{\boldsymbol{\imath}}, \hat{\boldsymbol{\jmath}}, \hat{\mathbf{k}})$ as shown in Figure 4.2. The positions of the monkeys are defined by coordinates y_1 and y_2: Their motions are rectilinear, with velocities $\mathbf{v}_1 = \dot{y}_1\hat{\boldsymbol{\jmath}}$ and $\mathbf{v}_2 = \dot{y}_2\hat{\boldsymbol{\jmath}}$. We sketch a free-body diagram for the system $\Sigma = \{pulley, M_1, M_2\}$ pulley-rope-monkeys. All forces are directed downward. We find $\mathbf{M}_{O,ext\to\Sigma} = 0$ for a frictionless pulley and equal masses $m_1 = m_2$:

$$\mathbf{M}_{O,ext\to\Sigma} = \mathbf{r}_{OM_1} \times m\mathbf{g} + \mathbf{r}_{OM_2} \times m\mathbf{g} = \mathbf{0}$$

This necessarily implies that $\mathbf{H}_{O,\Sigma} = Cst = 0$: The angular momentum of the system about O is conserved. We find $\mathbf{H}_O$ as follows:

$$\mathbf{H}_{O,\Sigma} = \mathbf{r}_{OM_1} \times m\mathbf{v}_1 + \mathbf{r}_{OM_2} \times m\mathbf{v}_2 = mr(\dot{y}_2 - \dot{y}_1)\hat{\mathbf{k}} \Longrightarrow \boxed{\dot{y}_1 = \dot{y}_2}$$

This shows that the monkeys climb up the rope at the same rate. They both arrive at the same time at the level of O.

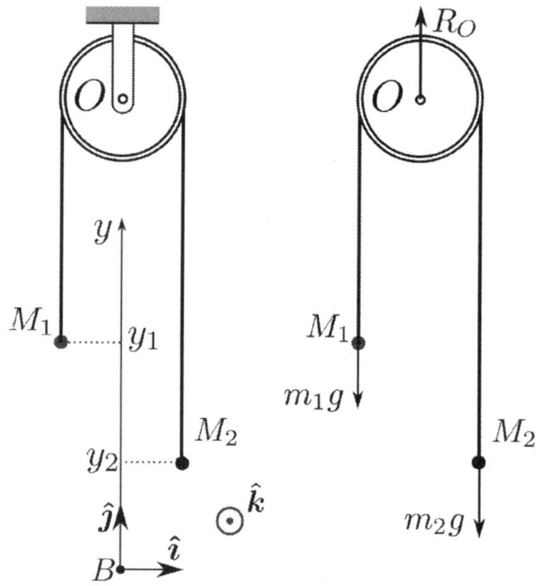

Figure 4.2 – *Problem 4.1 (Solution)*

b. CASE 2: $m_2 > m_1$. Now $\mathbf{M}_{O,ext\to\Sigma} = (m_1 - m_2)gr\hat{\mathbf{k}}$. Now the angular momentum of the system is not conserved.

$$(m_1 - m_2)gr\hat{\mathbf{k}} = \frac{d}{dt}\mathbf{H}_{O,\Sigma} = r(m_2\ddot{y}_2 - m_1\ddot{y}_1)\hat{\mathbf{k}}$$

We integrate this equation, using $\dot{y}_1 = \dot{y}_2 = 0$ at $t = 0$:

$$-(m_2 - m_1)gt = m_2\dot{y}_2 - m_1\dot{y}_1 \Longrightarrow \boxed{m_2\dot{y}_2 = m_1\dot{y}_1 - (m_2 - m_1)gt}$$

If at any given time $\dot{y}_1 > (1 - m_1/m_2)gt$ then $\dot{y}_2 > 0$. *If the lighter monkey climbs up the rope at an acceleration exceeding the value $(1 - m_1/m_2)g$, then he will be able to raise the heavier monkey.*

■

Exercise 4.2 ▪ A pendulum is assembled by connecting a particle P of mass m to a fixed point O by a string of length l. P is set in motion at speed v so that its path is a circle in a horizontal plane while the string makes the constant angle α with the vertical.
a. Find the angular momentum $\mathbf{H}_O$ of P about O.
b. From the time-rate of change of $\mathbf{H}_O$ determine the speed v to give P to guarantee its motion. ▪

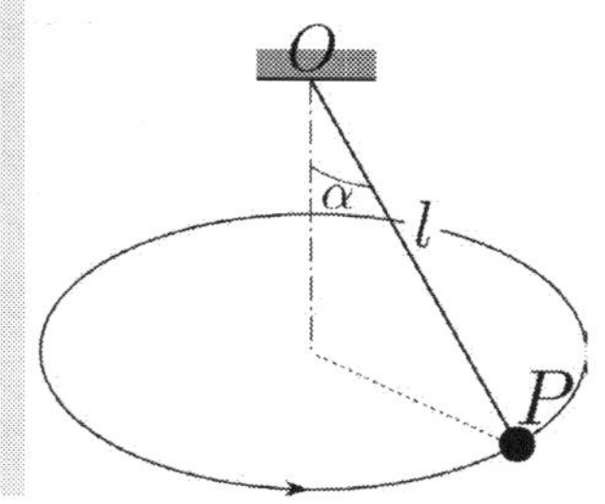

SOLUTION. **a.** Since P moves along a circular path, it is convenient to introduce the unit vectors $(\hat{\mathbf{e}}_T, \hat{\mathbf{e}}_N, \hat{\mathbf{e}}_z)$ as shown in Figure 4.3. Note that $\hat{\mathbf{e}}_T \times \hat{\mathbf{e}}_N = \hat{\mathbf{e}}_z$. On this basis of unit vectors, the velocity of P takes the expression $\mathbf{v}_P = v\hat{\mathbf{e}}_T$. With $\mathbf{r}_{OP} = -l(\cos\alpha\hat{\mathbf{e}}_z + \sin\alpha\hat{\mathbf{e}}_N)$ we can find the expression of angular momentum $\mathbf{H}_O$ as follows

$$\mathbf{H}_O = -l(\cos\alpha\hat{\mathbf{e}}_z + \sin\alpha\hat{\mathbf{e}}_N) \times mv\hat{\mathbf{e}}_T = mlv(-\cos\alpha\hat{\mathbf{e}}_N + \sin\alpha\hat{\mathbf{e}}_z)$$

b. P is subject to the following forces:
(i) the gravitational force $m\mathbf{g} = -mg\hat{\mathbf{e}}_z$ whose moment about O is $\mathbf{r}_{OP} \times (-mg\hat{\mathbf{e}}_z) = mgl\sin\alpha\hat{\mathbf{e}}_T$,
(ii) the tension $\mathbf{T}$ of the string whose moment about O is $\mathbf{r}_{OP} \times \mathbf{T} = \mathbf{0}$ since $\mathbf{T}$ is colinear to $\mathbf{r}_{OP}$.

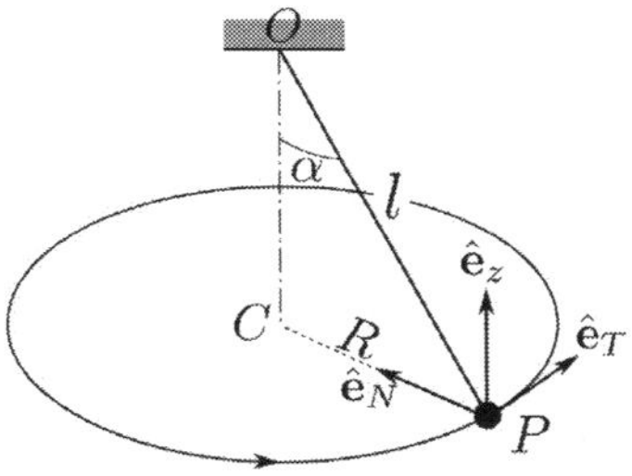

Figure 4.3 – *Problem 4.2 (Solution)*

Application of the angular momentum form of Newton's Second Law about point O is convenient since it eliminates the contribution of unknown tension $\mathbf{T}$:
$d\mathbf{H}_O/dt = \mathbf{M}_O$ gives

$$\frac{d}{dt}\left(mlv(-\cos\alpha\hat{\mathbf{e}}_N + \sin\alpha\hat{\mathbf{e}}_z)\right) = mgl\sin\alpha\hat{\mathbf{e}}_T$$

Since angle α is constant and $d\hat{\mathbf{e}}_N/dt = -\omega\hat{\mathbf{e}}_T = -(v/R)\hat{\mathbf{e}}_T$ we find

$$\dot{v}(-\cos\alpha\hat{\mathbf{e}}_N \sin\alpha\hat{\mathbf{e}}_z) + \frac{v^2}{R}\cos\alpha\hat{\mathbf{e}}_T = g\sin\alpha\hat{\mathbf{e}}_T$$

This leads to $\dot{v} = 0$ on $\hat{\mathbf{e}}_N$, that is, $v = \textit{constant}$ as expected. The value of v is found from the contribution on $\hat{\mathbf{e}}_T$:

$$ml\cos\alpha\frac{v^2}{R} = mgl\sin\alpha$$

with $R = l\sin\alpha$ we find

$$\boxed{v^2 = gl\frac{\sin^2\alpha}{\cos\alpha}}$$

▪

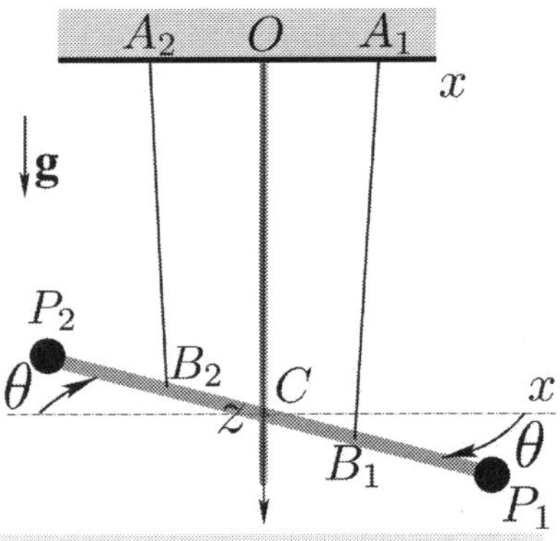

Exercise 4.3 ▪▪ A bar of negligible mass and length $4l$ supports two identical particles P_1 and P_2 of mass m. A vertical guiding rod Oz passes through a small hole of the rod located at midpoint C: the bar can slide without friction along Oz. Two elastic strings OA_1B_1 and OA_2B_2 of unstretched length l and elastic constant k connect point O to points B_1 and B_2 of the rod by passing without friction through small stationary rings A_1 and A_2 located on horizontal axis Ox with $|OA_1| = |OA_2| = |CB_1| = |CB_2| = l$.

We consider the motions of P_1 and P_2 such that rod P_1P_2 remains horizontal under the effect of the strings' elastic forces and gravity. Denote by z the coordinate of center C on axis Oz and by θ the oriented angle θ as shown in the figure. Assume that at time $t = 0$, the system is released from at $z = z_0$ and $\theta = \theta_0$.

a. Express the velocity of P_1 and P_2 in terms of l, z and θ.

b. Find the expression of the tension $\mathbf{T}_1$ that elastic string OA_1B_1 exerts at B_1 in terms of l, z and θ and unit vectors of your choice. Repeat for tension $\mathbf{T}_2$ that elastic string OA_2B_2 exerts at B_2.

c. Find the equations of motion governing the coordinates z and θ. Then find the period of small oscillations of the system. ▪

SOLUTION. **a.** The system has two degrees of freedom described by the cylindrical coordinates (θ, z) which define the position of P_1 and P_2:

$$\mathbf{r}_{OP_1} = z\hat{\mathbf{e}}_z + 2l\hat{\mathbf{e}}_r, \qquad \mathbf{r}_{OP_2} = z\hat{\mathbf{e}}_z - 2l\hat{\mathbf{e}}_r$$

See Figure 4.4 This leads to the velocities of P_1 and P_2 on basis $(\hat{\mathbf{e}}_r, \hat{\mathbf{e}}_\theta, \hat{\mathbf{e}}_z)$

$$\mathbf{v}_{P_1} = \dot{z}\hat{\mathbf{e}}_z + 2l\dot{\theta}\hat{\mathbf{e}}_\theta, \quad \mathbf{v}_{P_2} = \dot{z}\hat{\mathbf{e}}_z - 2l\dot{\theta}\hat{\mathbf{e}}_\theta \qquad [1]$$

b. The force of elastic string OA_1B_1 exerted on the rod at point B_1 is directed along vector $\mathbf{r}_{B_1A_1}$:

$$\mathbf{T}_1 = k(L + l - l)\frac{1}{L}\mathbf{r}_{B_1A_1} = k\mathbf{r}_{B_1A_1} = k(l\hat{\mathbf{e}}_x - z\hat{\mathbf{e}}_z - l\hat{\mathbf{e}}_r) \qquad [2]$$

denoting by $L = |B_1A_1|$. Similarly the tension of string OA_2B_2 exerted on the rod at point B_2 is given by

$$\mathbf{T}_2 = k\mathbf{r}_{B_2A_2} = k(-l\hat{\mathbf{e}}_x - z\hat{\mathbf{e}}_z + l\hat{\mathbf{e}}_r). \qquad [3]$$

c. The equations of motion are found by applying Euler's First and Second Principles to system Σ consisting of particles P_1 and P_2 and rod P_1P_2. First we list the external forces acting on Σ:
(i) the gravitational force $2mg\hat{\mathbf{e}}_z$ acting through mass center C,
(ii) the elastic forces $\mathbf{T}_1$ and $\mathbf{T}_2$ acting at B_1 and B_2,
Given that C is the mass center of the system, Euler's First Principle can be written as

$$2m\mathbf{a}_C = \mathbf{T}_1 + \mathbf{T}_2 + 2mg\hat{\mathbf{e}}_z$$

With $\mathbf{a}_C = \ddot{z}\hat{\mathbf{e}}_z$ and $\mathbf{T}_1 + \mathbf{T}_2 = -2kz\hat{\mathbf{e}}_z$ (from equations [2-3]) we find (by projection

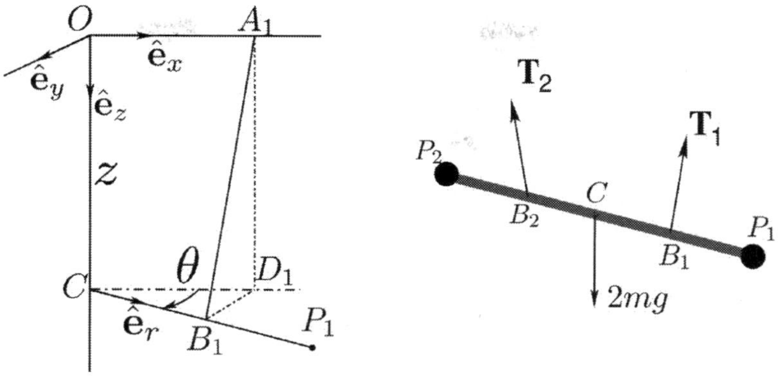

Figure 4.4 – *Problem 4.3*

along $\hat{\mathbf{e}}_z$)

$$\boxed{2m\ddot{z} + 2kz = 2mg} \qquad [4]$$

Before applying the Second Principle about mass center C, we need to find the system's angular momentum about C:

$$\mathbf{H}_C = \mathbf{r}_{CP_1} \times m\mathbf{v}_{P_1} + \mathbf{r}_{CP_2} \times m\mathbf{v}_{P_2} = 2l\hat{\mathbf{e}}_r \times m(\dot{z}\hat{\mathbf{e}}_z + 2l\dot{\theta}\hat{\mathbf{e}}_\theta) - 2l\hat{\mathbf{e}}_r \times m(\dot{z}\hat{\mathbf{e}}_z - 2l\dot{\theta}\hat{\mathbf{e}}_\theta)$$

which simplifies to

$$\mathbf{H}_C = 8ml^2\dot{\theta}\hat{\mathbf{e}}_z$$

Then by using equation (4.9) we find

$$\frac{d\mathbf{H}_C}{dt} = \mathbf{M}_{C,ext\to\Sigma} = \mathbf{r}_{CB_1} \times \mathbf{T}_1 + \mathbf{r}_{CB_2} \times \mathbf{T}_2 = l\hat{\mathbf{e}}_r \times (\mathbf{T}_1 - \mathbf{T}_2)$$

With $l\hat{\mathbf{e}}_r \times (\mathbf{T}_1 - \mathbf{T}_2) = -2kl^2 \sin\theta\hat{\mathbf{e}}_z$, we find

$$\boxed{\ddot{\theta} + \frac{k}{4m}\sin\theta = 0} \qquad [5]$$

We notice from equations [4-5] that the two motions are independent of each other. The vertical motion exhibits linear oscillation of period

$$\boxed{T_z = 2\pi\sqrt{\frac{m}{k}}}$$

For small initial angle θ_0, the rotational motion exhibits oscillations of period

$$\boxed{T_\theta = 2\pi\sqrt{\frac{4m}{k}} = 2T_z}$$

We see in this example that the First Principle predicts the motion of the mass center C along axis Oz, while the Second Principle predicts the rotation of line P_1P_2 about C. ∎

4.3 Angular Momentum of a Rigid Body

IN VIEW of deriving Euler's Second Principle, we wish here to determine a general expression of the angular momentum of a rigid body $\mathcal{B}$. We assume that body $\mathcal{B}$ is in planar motion, and we denote by G its mass center. We begin with the definition of the angular momentum about a point A. It can be defined as a sum of infinitesimal angular momenta $\mathbf{r}_{AP} \times \mathbf{v}_P \, dm$.

DEFINITION (ANGULAR MOMENTUM OF A RIGID BODY)

The angular momentum of rigid body $\mathcal{B}$ about an arbitrary point A is the vector defined as

$$\mathbf{H}_A = \int_{\mathcal{B}} \mathbf{r}_{AP} \times \mathbf{v}_P \, dm \tag{4.12}$$

where point A is not necessarily fixed in the referential of interest. Furthermore, angular momenta can be obtained from one point to another according to

$$\mathbf{H}_A = \mathbf{H}_B + \mathbf{r}_{AB} \times m\mathbf{v}_G \tag{4.13}$$

where $m\mathbf{v}_G$ is the body's linear momentum.

Proof. Equation (4.13) is straightforward to prove: write $\mathbf{r}_{AP} = \mathbf{r}_{AB} + \mathbf{r}_{BP}$ in the expression of $\mathbf{H}_A$ and the two terms $\mathbf{H}_B$ and $\mathbf{r}_{AB} \times m\mathbf{v}_G$ appear after using the property (3.26) of the mass center. ■

Remark 3. When multiple bodies are involved, we will use the notation $\mathbf{H}_{A,1}$, $\mathbf{H}_{A,2}$, ... for bodies 1, 2, ...

Equation (4.13) shows that, if the expression of the angular momentum about a particular point is known, the angular momentum about any other point can be obtained. However the integral expression of (4.12) to find a particular value of $\mathbf{H}_A$ is not useful in practice. To derive a tractable expression, we exploit the fact that (i) the velocity field $P \in \mathcal{B} \mapsto \mathbf{v}_P$ satisfies the fundamental property

$$\mathbf{v}_P = \mathbf{v}_{A \in \mathcal{B}} + \boldsymbol{\omega}_{\mathcal{B}} \times \mathbf{r}_{AP} \tag{4.14}$$

and (ii) that the body $\mathcal{B}$ is in planar motion. Recall that point A must be viewed as a point attached to $\mathcal{B}$, at least in an instantaneous way. Referring to Figure 4.5, we introduce three unit vectors $(\hat{\mathbf{x}}, \hat{\mathbf{y}}, \hat{\mathbf{z}})$ with $(\hat{\mathbf{x}}, \hat{\mathbf{y}})$ in the plane of motion. Then we define the Cartesian coordinates (x, y, z) of generic point P of $\mathcal{B}$ on the three axes $(A, \hat{\mathbf{x}})$, $(A, \hat{\mathbf{y}})$, and $(A, \hat{\mathbf{z}})$, that is

$$\mathbf{r}_{AP} = x\hat{\mathbf{x}} + y\hat{\mathbf{y}} + z\hat{\mathbf{z}} \tag{4.15}$$

The angular velocity of $\mathcal{B}$ can be expressed as $\boldsymbol{\omega}_{\mathcal{B}} = \omega_{\mathcal{B}}\hat{\mathbf{z}}$. Using (4.14) and (4.15), we can now determine the integrand of (4.12):

$$\mathbf{r}_{AP} \times \mathbf{v}_P = \mathbf{r}_{AP} \times \mathbf{v}_{A \in \mathcal{B}} + \mathbf{r}_{AP} \times (\omega_{\mathcal{B}}\hat{\mathbf{z}} \times \mathbf{r}_{AP})$$

with

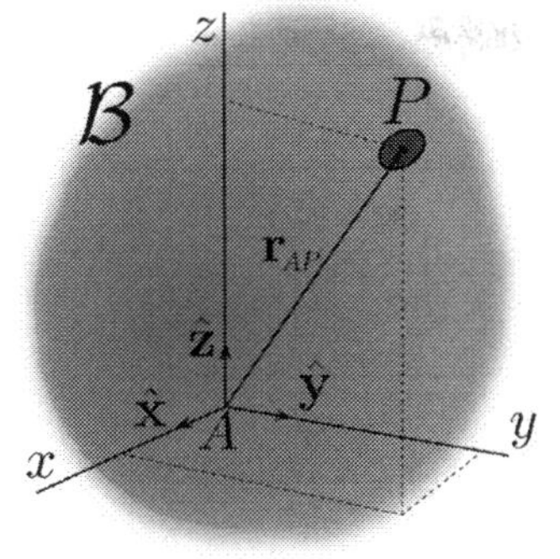

Figure 4.5 – *Coordinate system Axyz attached to rigid body $\mathcal{B}$.*

$$\begin{aligned}\mathbf{r}_{AP} \times (\hat{\mathbf{z}} \times \mathbf{r}_{AP}) &= \mathbf{r}_{AP} \times \big(\hat{\mathbf{z}} \times (x\hat{\mathbf{x}} + y\hat{\mathbf{y}} + z\hat{\mathbf{z}})\big) = (x\hat{\mathbf{x}} + y\hat{\mathbf{y}} + z\hat{\mathbf{z}}) \times (x\hat{\mathbf{y}} - y\hat{\mathbf{x}}) \\ &= -xz\hat{\mathbf{x}} - yz\hat{\mathbf{y}} + (x^2 + y^2)\hat{\mathbf{z}}\end{aligned}$$

This yields the following expression of the angular momentum $\mathbf{H}_A$:

$$\mathbf{H}_A = \mathbf{r}_{AG} \times m\mathbf{v}_{A\in\mathcal{B}} - \omega_{\mathcal{B}}\hat{\mathbf{x}} \int_{\mathcal{B}} xz\, dm - \omega_{\mathcal{B}}\hat{\mathbf{y}} \int_{\mathcal{B}} yz\, dm + \omega_{\mathcal{B}}\hat{\mathbf{z}} \int_{\mathcal{B}} (x^2 + y^2)\, dm$$

where we have used $\int_{\mathcal{B}} \mathbf{r}_{AP}\, dm = m\mathbf{r}_{AG}$ according to (3.26). This last expression leads to the following definitions.

Definition (Moment and Products of Inertia)

The inertia characteristics of body $\mathcal{B}$ about axes $Axyz$ are defined by the integrals:

$$I_{Azz} = \int_{\mathcal{B}} (x^2 + y^2)\, dm \tag{4.16}$$

$$I_{Axz} = -\int_{\mathcal{B}} xz\, dm, \qquad I_{Ayz} = -\int_{\mathcal{B}} yz\, dm \tag{4.17}$$

where:
(i) I_{Azz} is called *moment of inertia about axis Az*,
(ii) I_{Axz} is called *product of inertia about axes Ax and Az*,
(iii) I_{Ayz} is called *product of inertia about axes Ay and Az*.
Then, the angular momentum of body $\mathcal{B}$ about A is given by

$$\mathbf{H}_A = \mathbf{r}_{AG} \times m\mathbf{v}_{A\in\mathcal{B}} + I_{Axz}\omega_{\mathcal{B}}\hat{\mathbf{x}} + I_{Ayz}\omega_{\mathcal{B}}\hat{\mathbf{y}} + I_{Azz}\omega_{\mathcal{B}}\hat{\mathbf{z}} \tag{4.18}$$

Properties

①. The three moments of inertia I_{Axx}, I_{Ayy}, I_{Azz} and three products of inertia I_{Axz}, I_{Ayz}, I_{Axy} are measures of the *mass distribution* of the body about A. The S.I. unit of moment and product of inertia is kg.m^2.

②. The first term $\mathbf{r}_{AG} \times m\mathbf{v}_{A\in\mathcal{B}}$ vanishes in two important cases:
(i) if point A coincides with the mass center G of the body,

(ii) if point A is a fixed point of $\mathcal{B}$, or if A is the instantaneous center of rotation of $\mathcal{B}$ (in both cases $\mathbf{v}_{A\in\mathcal{B}} = \mathbf{0}$).

③. The moment I_{Azz} is specific to the axis Az: indeed the integrand $x^2 + y^2$ represents the square of the distance of generic point P to axis Az. For bodies of non-standard shape or for which mass is not distributed uniformly, it is convenient to define the *radius of gyration* k_Δ relative to an axis Δ: it is the distance measured from axis Δ to a fictitious point about which all the mass of the body would be concentrated to give the same moment of inertia of the body about Δ:

$$I_\Delta = mk_\Delta^2$$

④. Given three rectangular axes $Axyz$, two additional moments of inertia I_{Axx} and I_{Ayy} and one additional product I_{Axy} can be defined. Their expressions are given by

$$I_{Axx} = \int_{\mathcal{B}} (y^2 + z^2)\, dm, \quad I_{Ayy} = \int_{\mathcal{B}} (x^2 + z^2)\, dm, \quad I_{Axy} = -\int_{\mathcal{B}} xy\, dm$$

⑤. No geometric interpretation can be given for the products of inertia I_{Axz} and I_{Ayz}: they reflect the lack of (material) symmetry of the body w.r.t. to planes Axz. In fact, it is possible to show that there always exist axes $AXYZ$ called *principal axes of inertia* relative to which all products of inertia vanish, whatever choice of point A.

4.4 Determination of Moments and Products of Inertia

THE determination of the moments and products of inertia of a rigid body involves the calculation of volume integrals, and hence, constitutes more a problem of calculus than of mechanics. If one dimension of the body is much smaller than the other two, then the mass of $\mathcal{B}$ is effectively modeled as distributed along a surface, and the calculation of moments and products of inertia amounts to determining surface integrals. If the body possesses two negligible dimensions so as to be modeled as a one-dimensional object, then the calculation of moments and products of inertia amounts to simple line integrals.

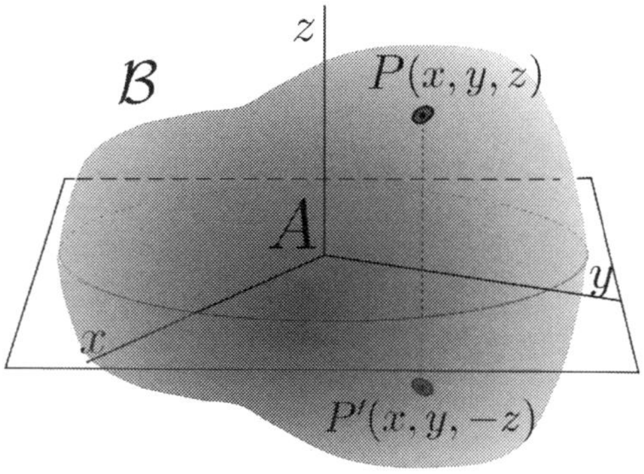

Figure 4.6 – *Body $\mathcal{B}$ with material plane of symmetry Axy.*

PLANE MATERIAL SYMMETRY. Given three axes Ax, Ay and Az intersecting at a particular point A, the products of inertia can be found to vanish if the body admits certain *material* symmetries. More specifically, consider a body $\mathcal{B}$ of mass density ρ whose axes Ax, Ay and Az admits the following property: for every point $P \in \mathcal{B}$ of coordinates (x, y, z) there exists a point $P' \in \mathcal{B}$ of coordinates $(x, y, -z)$ such that

$$\rho(x, y, z) = \rho(x, y, -z)$$

Then, plane Axy is said to be a *plane of material symmetry*. See Figure 4.6. It is then readily shown that both products of inertia I_{Axz} and I_{Ayz} vanish:

$$I_{Axz} = -\int_{\mathcal{B}} xz\,\rho\, dxdydz = -\int_{\mathcal{B}^+} xz\,\rho\, dxdydz - \int_{\mathcal{B}^-} xz\,\rho\, dxdydz = 0$$

where $\mathcal{B}^+$ and $\mathcal{B}^-$ are the partitions of $\mathcal{B}$ corresponding to $z \geq 0$ and $z \leq 0$, respectively. The change of variable $z \to -z$ shows that $\int_{\mathcal{B}^-} xzdm = \int_{\mathcal{B}^+} x(-z)dm$. Likewise, $I_{Ayz} = 0$.

AXISYMMETRY. If the body admits two planes of material symmetry, say Axy and Axz, then all products of inertia must vanish: $I_{Axz} = I_{Ayz} = I_{Axy} = 0$. For instance this occurs if the body is axisymmetric about axis Az and if mass density ρ is only a function of the distance r of point P to axis Az. In that case, the axes Ax and Ay play the same role so that $I_{Axx} = I_{Ayy} = J$ and this implies that

$$I_{Azz} = 2J - 2\int_{\mathcal{B}} z^2\, dm$$

PLATES. If the body is a *plate* lying in plane Axy (that is, its thickness is negligible), then $z = 0$ for all points: this leads to

$$I_{Axz} = I_{Ayz} = 0, \qquad I_{Azz} = I_{Axx} + I_{Ayy}$$

since $\int_{\mathcal{B}} y^2 dm = I_{Axx}$ and $\int_{\mathcal{B}} x^2 dm = I_{Ayy}$.

We now list the moments and products of inertia of standard shapes. In all cases, a particular set of axes $Axyz$ is chosen so as to take into account the body's symmetries. The mass density is assumed uniform throughout the body (the body is then said to be *homogeneous*). Point G denotes the mass center of the body, m its mass. Only the non-zero products are given.

ONE-DIMENSIONAL BODIES. We first list one-dimensional bodies: two dimensions of the (wire-like) body are negligible and the mass is assumed to be distributed along a line/curve.

Slender rod of length l

$I_{Gxx} = I_{Gyy} = \frac{1}{12} m l^2$

$I_{Gzz} = 0$

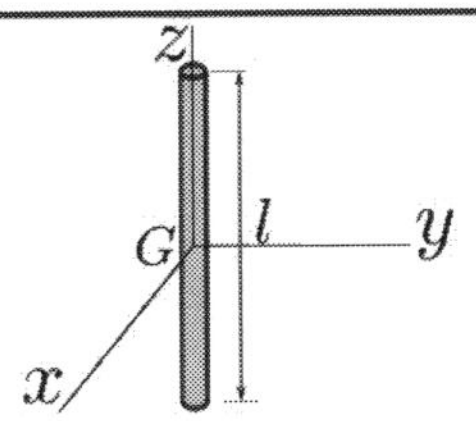

Circle of radius R

$I_{Gxx} = I_{Gyy} = \frac{1}{2} m R^2$

$I_{Gzz} = \ mR^2$

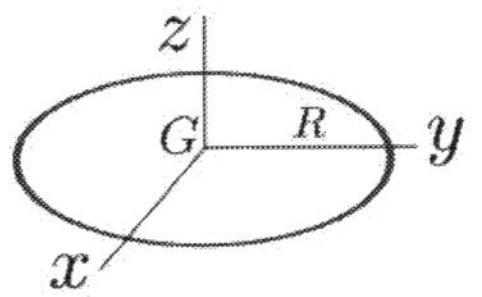

Two-Dimensional Bodies. Next we list two-dimensional bodies (plates or shells) for which one dimension of the body is negligible and the mass is assumed to be uniformly distributed on a surface. Four shapes are listed: spherical shell, disk, rectangular plate, and right triangular plate.

Spherical shell of radius R

$$I_{Gxx} = I_{Gyy} = I_{Gzz} = \tfrac{2}{3}mR^2$$

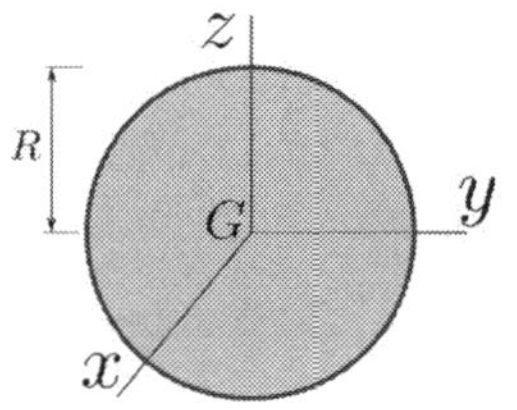

Disk of radius R

$$I_{Gxx} = I_{Gyy} = \tfrac{1}{4}mR^2$$

$$I_{Gzz} = \tfrac{1}{2}mR^2$$

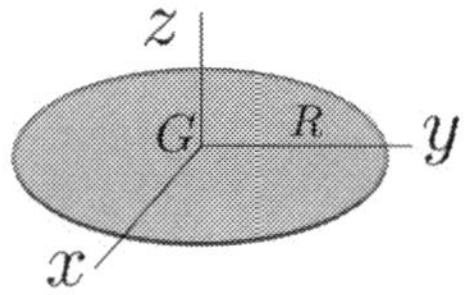

Rectangular plate of side lengths (a, b)

$$I_{Gxx} = \tfrac{1}{12}mb^2$$

$$I_{Gyy} = \tfrac{1}{12}ma^2$$

$$I_{Gzz} = \tfrac{1}{12}m(a^2 + b^2)$$

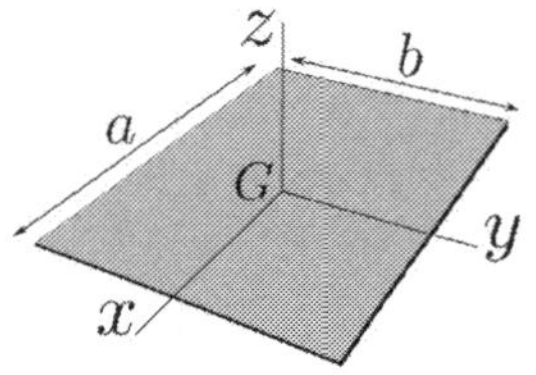

Right triangular plate

$$I_{Oxx} = \tfrac{1}{6}mb^2$$

$$I_{Oyy} = \tfrac{1}{6}ma^2$$

$$I_{Ozz} = \tfrac{1}{6}m(a^2 + b^2)$$

$$I_{Oxy} = -\tfrac{1}{12}mab$$

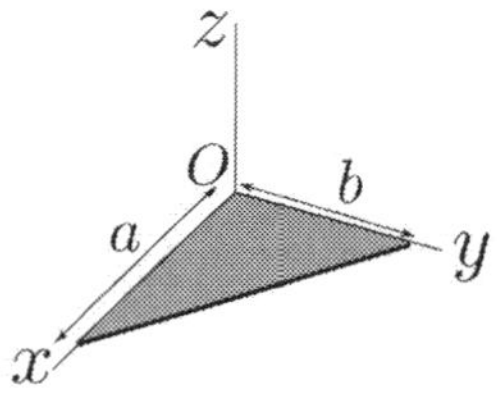

Remark 4. The mass center G of a right-triangular plate is located at $(a/3, b/3, 0)$.

Three-Dimensional Bodies. Finally, we list three-dimensional bodies: full sphere, hollow sphere, cuboid, full cylinder, cone, hollow cylinder, and torus.

Sphere of radius R

$$I_{Gxx} = I_{Gyy} = I_{Gzz} = \tfrac{2}{5} mR^2$$

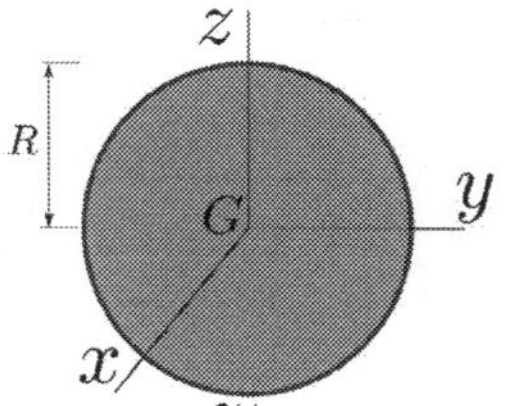

Hollow sphere of inner radius r
outer radius R

$$I_{Gxx} = I_{Gyy} = I_{Gzz} = \tfrac{2}{5} mR^2 \frac{1-\frac{r^5}{R^5}}{1-\frac{r^3}{R^3}}$$

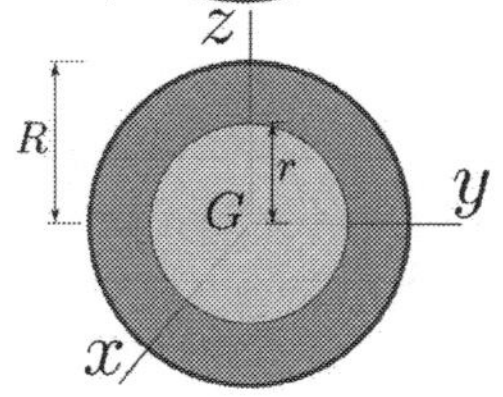

Cuboid of side lengths (a, b, c)

$$I_{Gxx} = \tfrac{1}{12} m(b^2 + c^2)$$
$$I_{Gyy} = \tfrac{1}{12} m(a^2 + c^2)$$
$$I_{Gzz} = \tfrac{1}{12} m(a^2 + b^2)$$

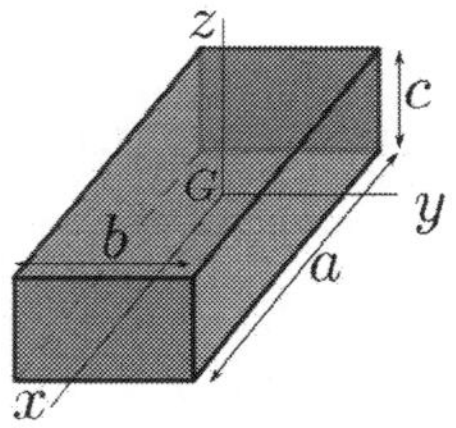

Cylinder of radius R and height H

$$I_{Gxx} = I_{Gyy} = \tfrac{1}{4} mR^2 + \tfrac{1}{12} mH^2$$
$$I_{Gzz} = \tfrac{1}{2} mR^2$$

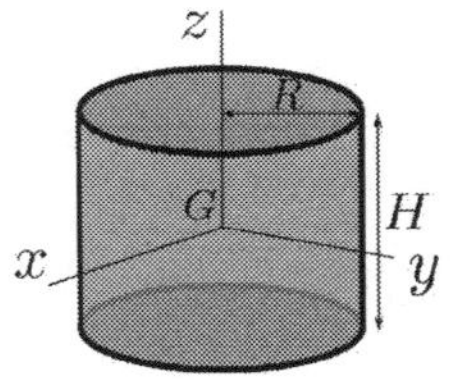

Cone of height H and base radius R

$$I_{Oxx} = I_{Oyy} = \tfrac{3}{10} m(2H^2 + \tfrac{1}{2} R^2)$$
$$I_{Ozz} = \tfrac{3}{10} mR^2$$

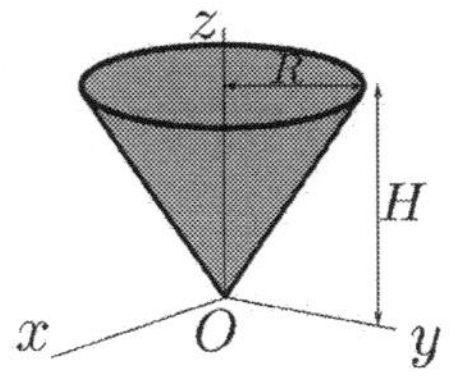

Hollow cylinder of inner radius r outer radius R
height H

$$I_{Gxx} = I_{Gyy} = \tfrac{1}{4} m(r^2 + R^2) + \tfrac{1}{12} mH^2$$
$$I_{Gzz} = \tfrac{1}{2} m(R^2 + r^2)$$

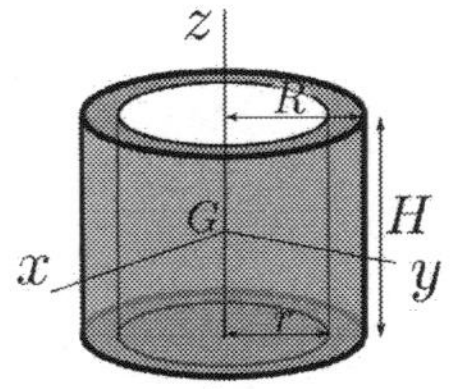

Torus of radii (r, R)

$$I_{Gxx} = I_{Gyy} = m(R^2 + \tfrac{5}{8} r^2)$$
$$I_{Gzz} = m(R^2 + \tfrac{3}{4} r^2)$$

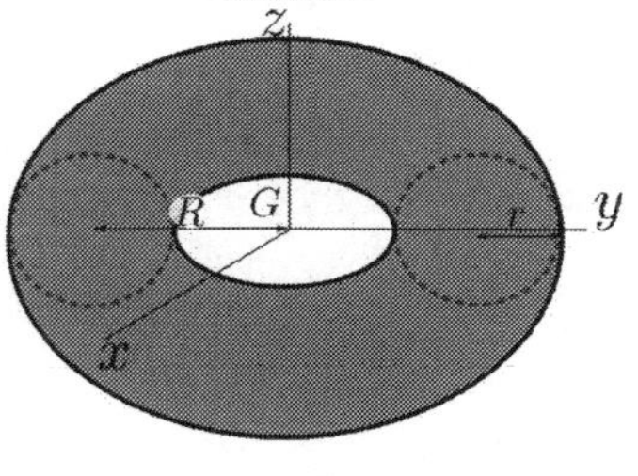

With the use of this data, we can find the inertia characteristics of more complex shapes by using two important results: (i) the extensive property of inertia moments/products, (ii) the parallel axis theorem.

Extensive Property of Moments & Products of Inertia. Consider a rigid body $\mathcal{B}$ whose union with body $\mathcal{B}_2$ is equal (identical) to the union of two bodies $\mathcal{B}_0$ and $\mathcal{B}_1$: we obtain a composite, hollowed out body $\mathcal{B} = (\mathcal{B}_0 \bigcup \mathcal{B}_1) \setminus \mathcal{B}_2$ as shown in Figure 4.7. Assuming that the three bodies $\mathcal{B}_0$, $\mathcal{B}_1$ and $\mathcal{B}_2$ are disjoint, it is possible to obtain the moments and products of inertia of $\mathcal{B}$ from those of $\mathcal{B}_0$, $\mathcal{B}_1$ and $\mathcal{B}_2$. Consider an arbitrary set of orthogonal axes $Axyz$. Then we can write, as a consequence of the addition property of multiple integrals over disjoint sets:

$$\int_{\mathcal{B}} (x^2 + y^2) dm + \int_{\mathcal{B}_2} (x^2 + y^2) dm = \int_{\mathcal{B}_0} (x^2 + y^2) dm + \int_{\mathcal{B}_1} (x^2 + y^2) dm$$

This leads to the moment relationship

$$I_{Azz} = I_{Azz,0} + I_{Azz,1} - I_{Azz,2} \tag{4.19}$$

and similarly for the products

$$I_{Axz} = I_{Axz,0} + I_{Axz,1} - I_{Axz,2}. \tag{4.20}$$

This property is useful whenever the moments and products (about the same axes) of the disjointed parts of $\mathcal{B}$ are known.

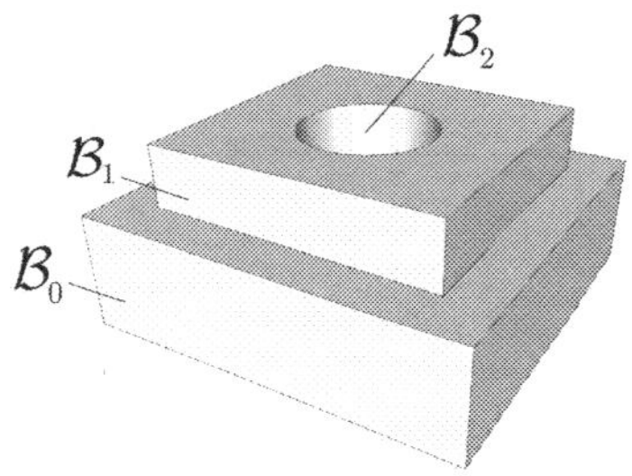

Figure 4.7 – *Composite body* $\mathcal{B} = (\mathcal{B}_0 \bigcup \mathcal{B}_1) \setminus \mathcal{B}_2$.

The Parallel Axis Theorem. The inertia information given in the previous tables have a limitation: the data is restricted to a particular choice of axes $Axyz$. This limitation is overcome by the *Parallel Axis Theorem.* Consider in Figure 4.8 a body $\mathcal{B}$ of mass center G and mass m for which two sets of parallel axes $Gxyz$ and $Ax'y'z'$ are defined. For any point P of $\mathcal{B}$, we define the coordinates (x, y, z) and (x', y', z') of P

$$\mathbf{r}_{GP} = x\hat{\mathbf{x}} + y\hat{\mathbf{y}} + z\hat{\mathbf{z}}, \qquad \mathbf{r}_{AP} = x'\hat{\mathbf{x}} + y'\hat{\mathbf{y}} + z'\hat{\mathbf{z}}$$

and the coordinates $(\bar{x}, \bar{y}, \bar{z})$ of A

$$\mathbf{r}_{GA} = \bar{x}\hat{\mathbf{x}} + \bar{y}\hat{\mathbf{y}} + \bar{z}\hat{\mathbf{z}}$$

The vector loop equation $\mathbf{r}_{GP} = \mathbf{r}_{GA} + \mathbf{r}_{AP}$ gives the equations

$$x = x' + \bar{x}, \qquad y = y' + \bar{y}, \qquad z = z' + \bar{z}$$

We can relate the moment of inertia I_{Azz} to the moment of inertia I_{Gzz} as follows:

$$\begin{aligned} I_{Azz} &= \int_{\mathcal{B}} (x'^2 + y'^2) dm = \int_{\mathcal{B}} ((x - \bar{x})^2 + (y - \bar{y})^2) dm \\ &= \int_{\mathcal{B}} (x^2 + y^2) dm - 2 \int_{\mathcal{B}} (\bar{x}x + \bar{y}y) dm + m(\bar{x}^2 + \bar{y}^2) \\ &= I_{Gzz} + m(\bar{x}^2 + \bar{y}^2) \end{aligned}$$

Similarly, for the products of inertia:

$$\begin{aligned} I_{Axz} &= -\int_{\mathcal{B}} x'z' dm = -\int_{\mathcal{B}} (x - \bar{x})(z - \bar{z}) dm \\ &= -\int_{\mathcal{B}} xz dm + \int_{\mathcal{B}} (\bar{z}x + \bar{x}z) dm - m\bar{x}\bar{z} \\ &= I_{Gxz} - m\bar{x}\bar{z} \end{aligned}$$

where we have used $\int_{\mathcal{B}} x dm = \int_{\mathcal{B}} y dm = \int_{\mathcal{B}} z dm = 0$ in accordance with equation (3.25).

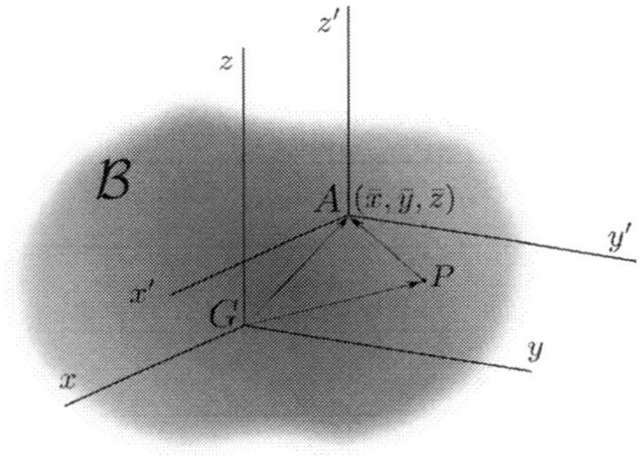

Figure 4.8 – *Parallel axes $Gxyz$ and $Ax'y'z'$ to body $\mathcal{B}$.*

We summarize these results with the following theorem.

Parallel Axis Theorem

Given two sets of parallel rectangular axes $Gxyz$ through mass center G and $Ax'y'z'$ through arbitrary point A, the corresponding moments and products of inertia are related according to

$$I_{Axx} = I_{Gxx} + m(\bar{y}^2 + \bar{z}^2), \quad I_{Ayy} = I_{Gyy} + m(\bar{x}^2 + \bar{z}^2), \quad I_{Azz} = I_{Gzz} + m(\bar{x}^2 + \bar{y}^2) \quad (4.21)$$

$$I_{Axz} = I_{Gxz} - m\bar{x}\bar{z}, \quad I_{Ayz} = I_{Gyz} - m\bar{y}\bar{z}, \quad I_{Axy} = I_{Gxy} - m\bar{x}\bar{y} \quad (4.22)$$

where $(\bar{x}, \bar{y}, \bar{z})$ are the coordinates of A on the axes $Gxyz$.

Remark 5. Note that to obtain I_{Axx} from I_{Gxx} the correction term $m(\bar{y}^2 + \bar{z}^2)$ involves the square of the distance between the two axes Ax and Gx.

PROPERTIES

①. The Parallel Axis Theorem requires that moment of inertia I_{Azz} be obtained from I_{Gzz}, that is, about an axis parallel to Az and passing through the body's mass center. However any two moments I_{Azz} and I_{Bzz} about two parallel axes Az and Bz can be related by applying (4.21) twice, first about point A, then about point B, and subsequently taking the difference of the two resulting equations.

②. Equations (4.21) show that the smallest possible values taken by the moments of inertia of a rigid body are obtained about axes passing through its mass center.

③. For an axisymmetric body of axis Gz, the inertia properties about G (or any point of axis Gz) are invariant by rotation about Gz: hence the chosen axes Gx and Gy need not be attached to the body.

We will illustrate the calculation of moments and products of inertia after deriving Euler's Second Principle.

4.5 Euler's Second Principle

Now that we have found the general expression of the angular momentum of a rigid body $\mathcal{B}$, we seek to derive Euler's Second Principle which, we surmise, is pivotal to describe the rotational motion of $\mathcal{B}$. We start with the form of this principle about the body's mass center by finding the time rate-of-change (relative to a Newtonian referential $\mathcal{E}$) of angular momentum $\mathbf{H}_G$ according to equation (4.12):

$$\frac{d}{dt}\mathbf{H}_G = \int_{\mathcal{B}} \frac{d}{dt}(\mathbf{r}_{GP} \times \mathbf{v}_P)dm = \int_{\mathcal{B}} ((\mathbf{v}_P - \mathbf{v}_G) \times \mathbf{v}_P + \mathbf{r}_{GP} \times \mathbf{a}_P)dm = \int_{\mathcal{B}} \mathbf{r}_{GP} \times \mathbf{a}_P dm$$

where we have used $\int_{\mathcal{B}} \mathbf{v}_P dm = m\mathbf{v}_G$. Next, we use Newton's Second Law (3.23) applied to P of mass dm to find

$$\frac{d}{dt}\mathbf{H}_G = \int_{\mathcal{B}} \mathbf{r}_{GP} \times (\mathbf{F}_{ext} + \mathbf{f}_{int})dV = \int_{\mathcal{B}} \mathbf{r}_{GP} \times \mathbf{F}_{ext} dV$$

where we have set the resultant moment of internal forces $\int_{\mathcal{B}} \mathbf{r}_{GP} \times \mathbf{f}_{int} dV$ to zero due to Newton's Third Law. We are left with the resultant moment of external forces which we denote $\mathbf{M}_{G,ext\to\mathcal{B}}$. This gives the mass center form of Euler's Second principle:

EULER'S SECOND PRINCIPLE: MASS CENTER FORM

With respect to a Newtonian referential, the time rate-of-change of the angular momentum of a rigid body about its mass center G is equal to the resultant moment of the external forces acting on the body:

$$\frac{d}{dt}\mathbf{H}_G = \mathbf{M}_{G,ext\to\mathcal{B}} \tag{4.23}$$

Equations (4.23) can be conveniently expressed on a basis of unit vectors $(\hat{\mathbf{x}}, \hat{\mathbf{y}}, \hat{\mathbf{z}})$ *attached* to $\mathcal{B}$ (with $\hat{\mathbf{z}}$ normal to the plane of motion). See Figure 4.9. Using the expression (4.18) and recalling that $\frac{d\hat{\mathbf{x}}}{dt} = \omega_{\mathcal{B}}\hat{\mathbf{y}}$ and $\frac{d\hat{\mathbf{y}}}{dt} = -\omega_{\mathcal{B}}\hat{\mathbf{x}}$ we find

$$\frac{d}{dt}\mathbf{H}_G = (I_{Gxz}\alpha_{\mathcal{B}} - I_{Gyz}\omega_{\mathcal{B}}^2)\hat{\mathbf{x}} + (I_{Gyz}\alpha_{\mathcal{B}} + I_{Gxz}\omega_{\mathcal{B}}^2)\hat{\mathbf{y}} + I_{Gzz}\alpha_{\mathcal{B}}\hat{\mathbf{z}}$$

denoting $\omega_{\mathcal{B}} = \boldsymbol{\omega}_{\mathcal{B}} \cdot \hat{\mathbf{z}}$ and $\alpha_{\mathcal{B}} = \boldsymbol{\alpha}_{\mathcal{B}} \cdot \hat{\mathbf{z}}$. This gives three equations by projection onto $(\hat{\mathbf{x}}, \hat{\mathbf{y}}, \hat{\mathbf{z}})$

$$I_{Gxz}\alpha_{\mathcal{B}} - I_{Gyz}\omega_{\mathcal{B}}^2 = M_{Gx} \tag{4.24}$$

$$I_{Gyz}\alpha_{\mathcal{B}} + I_{Gxz}\omega_{\mathcal{B}}^2 = M_{Gy} \tag{4.25}$$

$$I_{Gzz}\alpha_{\mathcal{B}} = M_{Gz} \tag{4.26}$$

where (M_{Gx}, M_{Gy}, M_{Gz}) denote the components of moment $\mathbf{M}_{G,ext\to\mathcal{B}}$ on basis $(\hat{\mathbf{x}}, \hat{\mathbf{y}}, \hat{\mathbf{z}})$.

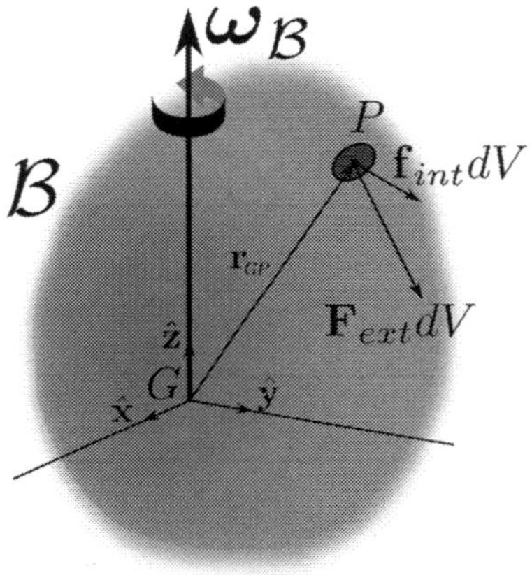

Figure 4.9 – *Basis* $(\hat{\mathbf{x}}, \hat{\mathbf{y}}, \hat{\mathbf{z}})$ *is attached to body* $\mathcal{B}$.

It is also of interest to obtain a version of Euler's Second Principle applicable about an arbitrary point A. A first method consists of relating the moment of external forces about A to that about G:

$$\mathbf{M}_{A,ext\to\mathcal{B}} = \int_{\mathcal{B}} \mathbf{r}_{AP} \times \mathbf{F}_{ext} dV = \int_{\mathcal{B}} (\mathbf{r}_{AG} + \mathbf{r}_{GP}) \times \mathbf{F}_{ext} dV = \mathbf{r}_{AG} \times \mathbf{F}_{ext\to\mathcal{B}} + \mathbf{M}_{G,ext\to\mathcal{B}}$$

where we have used $\int_{\mathcal{B}} \mathbf{F}_{ext} dV = \mathbf{F}_{ext\to\mathcal{B}}$. We can then use Euler's First Principle to identify the resultant external force $\mathbf{F}_{ext\to\mathcal{B}}$ with $m\mathbf{a}_G$: this gives

$$\mathbf{M}_{A,ext\to\mathcal{B}} = \mathbf{r}_{AG} \times m\mathbf{a}_G + \mathbf{M}_{G,ext\to\mathcal{B}} = \mathbf{r}_{AG} \times m\mathbf{a}_G + \frac{d}{dt}\mathbf{H}_G$$

Another version involves the angular momentum about A which can be related to $\mathbf{H}_G$ according to:

$$\mathbf{H}_A = \int_{\mathcal{B}} (\mathbf{r}_{AG} + \mathbf{r}_{GP}) \times \mathbf{v}_P dm = \mathbf{r}_{AG} \times m\mathbf{v}_G + \mathbf{H}_G$$

Then time-differentiation gives:

$$\frac{d}{dt}\mathbf{H}_A = (\mathbf{v}_G - \mathbf{v}_A) \times m\mathbf{v}_G + \mathbf{r}_{AG} \times m\mathbf{a}_G + \frac{d}{dt}\mathbf{H}_G$$

Since $\frac{d}{dt}\mathbf{H}_G = \mathbf{M}_{A,ext\to\mathcal{B}} - \mathbf{r}_{AG} \times m\mathbf{a}_G$ we obtain the equation

$$\frac{d}{dt}\mathbf{H}_A + \mathbf{v}_A \times m\mathbf{v}_G = \mathbf{M}_{A,ext\to\mathcal{B}}$$

This is summarized as follows:

Euler's Second Principle: General Forms

With respect to a Newtonian referential, Euler's Second Principle about an arbitrary point A can take two forms given by

$$\frac{d}{dt}\mathbf{H}_G + \mathbf{r}_{AG} \times m\mathbf{a}_G = \mathbf{M}_{A,ext\to\mathcal{B}} \tag{4.27}$$

or

$$\frac{d}{dt}\mathbf{H}_A + \mathbf{v}_A \times m\mathbf{v}_G = \mathbf{M}_{A,ext\to\mathcal{B}} \tag{4.28}$$

where $\mathbf{M}_{A,ext\to\mathcal{B}}$ is the resultant moment of external forces about A.

Remark 6. The term $\mathbf{r}_{AG} \times m\mathbf{a}_G$ is often referred to as the *kinetic moment about A*.

Remark 7. We recover (4.23) when we set $A = G$ in (4.27) or (4.28).

The second version given by (4.28) is useful whenever the body admits a point O fixed in the referential at all time (as occurs when body $\mathcal{B}$ is in rotation about axis Oz): $\mathbf{v}_{O\in\mathcal{B}} = \mathbf{v}_O = \mathbf{0}$. Then, with $\mathbf{H}_O = I_{Oxz}\omega_{\mathcal{B}}\hat{\mathbf{x}} + I_{Oyz}\omega_{\mathcal{B}}\hat{\mathbf{y}} + I_{Ozz}\omega_{\mathcal{B}}\hat{\mathbf{z}}$ according to (4.18), we arrive at a set of equations which take the same form as that obtained about the mass center:

$$I_{Oxz}\alpha_{\mathcal{B}} - I_{Oyz}\omega_{\mathcal{B}}^2 = M_{Ox} \tag{4.29}$$

$$I_{Oyz}\alpha_{\mathcal{B}} + I_{Oxz}\omega_{\mathcal{B}}^2 = M_{Oy} \tag{4.30}$$

$$I_{Ozz}\alpha_{\mathcal{B}} = M_{Oz} \tag{4.31}$$

where (M_{Ox}, M_{Oy}, M_{Oz}) denote the components of external moment $\mathbf{M}_{O,ext\to\mathcal{B}}$ on basis $(\hat{\mathbf{x}}, \hat{\mathbf{y}}, \hat{\mathbf{z}})$.

Properties

①. In many problems of planar kinetics, application of Euler's Second Principle (4.23) yields only one equation along unit vector $\hat{\mathbf{z}}$, that is, both sides of (4.24) and (4.25) are identically zero. This occurs whenever the following conditions are satisfied:
(i) both products of inertia I_{Gxz} and I_{Gyz} vanish because the plane of motion Gxy is a *plane of symmetry*, and
(ii) the lines of action of all external forces lie in the plane of motion of G.

②. If the body is in *translation* relative to referential $\mathcal{E}$, then Euler's Second Principle takes the form

$$\mathbf{M}_{G,ext\to\mathcal{B}} = \mathbf{0}, \qquad \text{or} \qquad \mathbf{M}_{A,ext\to\mathcal{B}} = \mathbf{r}_{AG} \times m\mathbf{a}_G$$

since in this case $\mathbf{H}_G = \mathbf{0}$.

③. The form (4.28) of Euler's Second Principle can be useful when applied at the instantaneous center of rotation J of the body since the corresponding angular momentum is given by $\mathbf{H}_J \cdot \hat{\mathbf{e}}_z = I_{Jzz}\omega$. However the term $\mathbf{v}_J \times m\mathbf{v}_G$ must be taken into account since in general $\mathbf{v}_J \neq \mathbf{0}$ (whereas $\mathbf{v}_{J\in\mathcal{B}} = \mathbf{0}$).

EXAMPLE PROBLEMS

General Procedure: Calculation of Moment/Product of Inertia

To calculate the moments and products of inertia of a rigid body, follow these guidelines:

① To find the moments and products of inertia of a rigid body relative to a particular set of orthogonal axes $Axyz$, use a combination of the extensive property of the moments/products of inertia and the Parallel Axis Theorem. Identify all possible plane symmetries.

② Recall that the moment of inertia comes into play in the component of angular momentum along the normal to the plane of motion.

③ We will outline in Section 4.8 a general procedure for the solution of rigid body kinetics problems.

Exercise 4.4 ▪ Figure 4.10 shows a cylinder of axis Oz and radius $4R$ and possessing a cylindrical hole of radius R and axis Cz located at a distance $2R$ from axis Oz. Denoting by m the mass of $\mathcal{B}$, find the moment and products of inertia I_{Ozz}, I_{Oxz} and I_{Oyz} in terms of geometric parameters H and R. If the body is set in rotation about axis Oz, find the corresponding angular momentum $\mathbf{H}_O$ about O. ▪

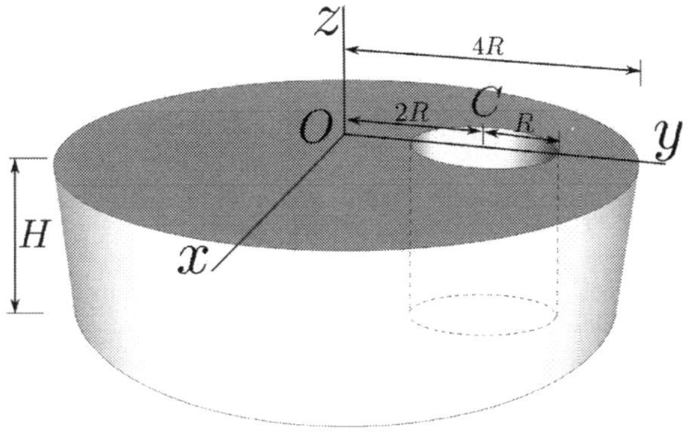

Figure 4.10 – *Problem 4.4*

Solution. Body $\mathcal{B}$ can be viewed as the removal of body 1 (the cylinder of mass m_1, radius R, and axis Cz) from body 2 (the cylinder of mass m_2, axis Oz, and radius $4R$). This leads to the equations:

$$I_{Ozz} = I_{Ozz,2} - I_{Ozz,1}, \quad I_{Oxz} = I_{Oxz,2} - I_{Oxz,1}, \quad I_{Oyz} = I_{Oyz,2} - I_{Oyz,1}$$

We note that body $\mathcal{B}$ is symmetric w.r.t. plane Oyz: this should lead to $I_{Oxz} = 0$. For body 2 we have

$$I_{Ozz,2} = \frac{1}{2} m_2 (4R)^2, \quad I_{Oxz,2} = 0, \quad I_{Oyz,2} = 0$$

For body 1 we use the Parallel Axis Theorem

$$I_{Ozz,1} = I_{Czz,1} + m_1 d^2 = \frac{1}{2} m_1 R^2 + m_1 (2R)^2, \quad I_{Oxz,1} = 0, \quad I_{Oyz,1} = -m_1 (2R)(-\frac{H}{2})$$

Next we relate the masses of body 1 and body 2 to mass m:

$$m_1 = \frac{1}{15} m, \quad m_2 = \frac{16}{15} m$$

Finally we find

$$\boxed{I_{Ozz} = \frac{247}{30} m R^2, \quad I_{Oxz} = 0, \quad I_{Oyz} = -\frac{m}{15} RH}$$

To find $\mathbf{H}_O$, we use formula (4.18): $\mathbf{H}_O = \omega_{\mathcal{B}} (I_{Oxz} \hat{\mathbf{x}} + I_{Oyz} \hat{\mathbf{y}} + I_{Ozz} \hat{\mathbf{z}})$

$$\boxed{\mathbf{H}_O = m R \omega_{\mathcal{B}} (-\frac{1}{15} H \hat{\mathbf{y}} + \frac{247}{30} R \hat{\mathbf{z}})}$$

where $\hat{\mathbf{y}}$ and $\hat{\mathbf{z}}$ are the unit vectors directed along axis Oy and Oz, respectively. ■

Exercise 4.5 ■■ Consider an arbitrary triangular plate ABC of side lengths a, b and c and mass center G. Show that its moment of inertia about axis Gz is given by

$$I_{Gzz} = \frac{m}{36} (a^2 + b^2 + c^2)$$

■

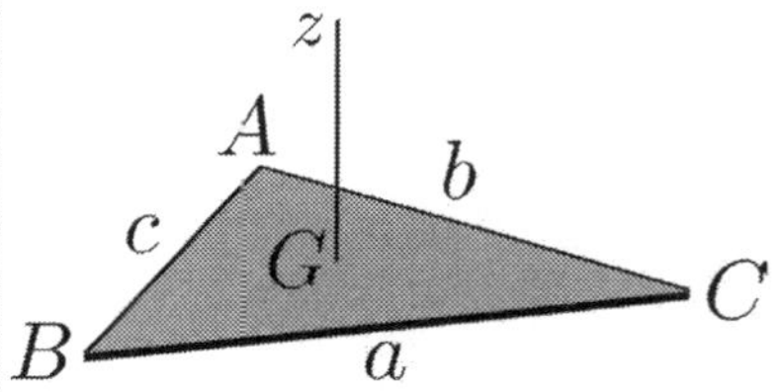

Solution. We will use a combination of the extensive property and the Parallel Axis Theorem by splitting the triangle into 2 right-triangular plates of mass center G_1 and G_2 and mass $m_1 = m a_1 / a$ and $m_2 = m a_2 / a$, respectively. See Figure 4.11. We assume that A can be projected onto side BC to obtain point H. We know the moment of inertia of each right triangle AHB (partition 1) and AHC (partition 2) about H:

$$I_{Hzz,1} = \frac{1}{6} m_1 (a_1^2 + h^2), \qquad I_{Hzz,2} = \frac{1}{6} m_2 (a_2^2 + h^2)$$

Then we use the Parallel Axis Theorem to find $I_{G_1zz,1}$ and $I_{G_2zz,2}$

$$\begin{aligned} I_{G_1zz,1} &= I_{Hzz,1} - m_1 |HG_1|^2 = \frac{1}{6} m_1 (a_1^2 + h^2) - \frac{1}{9} m_1 (a_1^2 + h^2) = \frac{1}{18} m_1 (a_1^2 + h^2) \\ &= \frac{1}{18} m_1 c^2 \end{aligned}$$

Similarly $I_{G_2zz,2} = \frac{1}{18} m_2 b^2$. Next we apply again the Parallel Axis Theorem to find

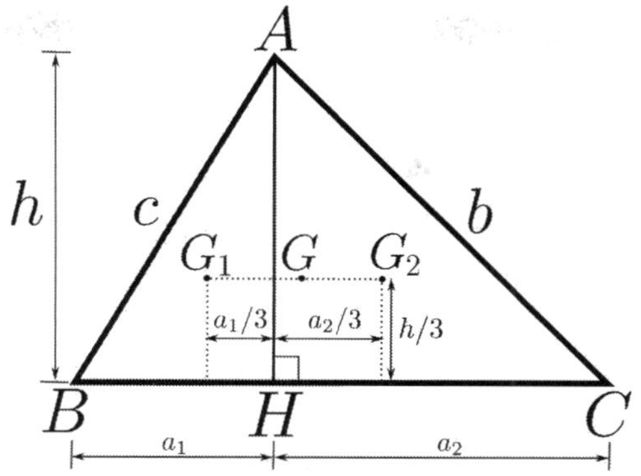

Figure 4.11 – *Problem 4.5 (Solution)*

$I_{Gzz,1}$ and $I_{Gzz,2}$ by using

$$|GG_1| = \frac{m_2}{m}|G_1G_2| = \frac{a_2}{3}, \qquad |GG_2| = \frac{m_1}{m}|G_1G_2| = \frac{a_1}{3}$$

leading to

$$I_{Gzz,1} = I_{G_1zz,1} + \frac{1}{9}m_1a_2^2 = \frac{1}{18}m_1(c^2 + 2a_2^2), \qquad I_{Gzz,2} = \frac{1}{18}m_2(b^2 + 2a_1^2)$$

Now we can obtain I_{Gzz}:

$$I_{Gzz} = I_{Gzz,1} + I_{Gzz,2} = \frac{1}{18}m_1(c^2 + 2a_2^2) + \frac{1}{18}m_2(b^2 + 2a_1^2)$$

With $m_1 = ma_1/a$ and $m_2 = ma_2/a$ we find

$$I_{Gzz} = \frac{m}{18}\frac{a_1c^2 + 2aa_1a_2 + a_2b^2}{a} \qquad [1]$$

Finally we need to express a_1 and a_2 in terms of (a, b, c): in triangle ABC the law of cosines gives (denoting by θ the angle at B)

$$b^2 = c^2 + a^2 - 2ac\cos\theta = c^2 + a^2 - 2aa_1$$

giving

$$a_1 = \frac{c^2 - b^2 + a^2}{2a} \qquad [2]$$

Similarly we find

$$a_2 = \frac{b^2 - c^2 + a^2}{2a} \qquad [3]$$

Finally we substitute (a_1, a_2) of [2-3] into [1]:

$$I_{Gzz} = \frac{m}{36a^2}\left(c^2(c^2 - b^2 + a^2) + (c^2 - b^2 + a^2)(b^2 - c^2 + a^2) + b^2(b^2 - c^2 + a^2)\right)$$

After simplifications, we find

$$\boxed{I_{Gzz} = \frac{m}{36}(a^2 + b^2 + c^2)}$$

■

Exercise 4.6 ▪ A rigid body is assembled by connecting two identical spherical shells of mass m and radius R to an axle Oz of negligible mass by massless rods AB and CD.
a. Find the moment of inertia I_Δ of a spherical shell about one of its diameter Δ without performing an integration.
b. Find the angular momentum of the body about O if the body is rotating about axis Oz at angular velocity $\omega_{\mathcal{B}}$. ▪

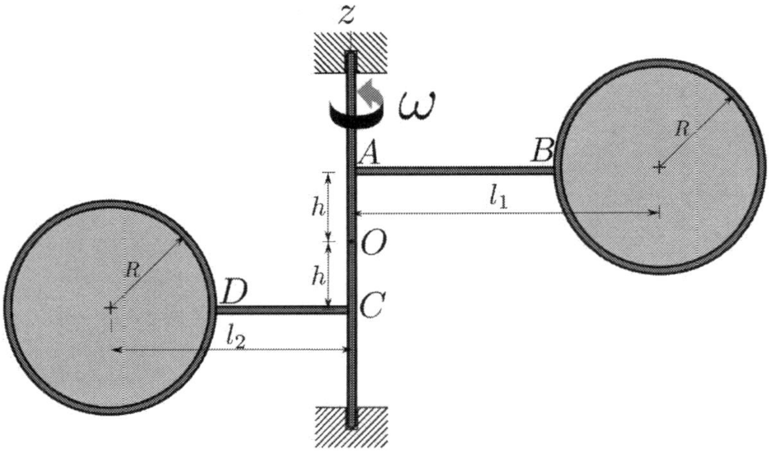

Figure 4.12 – *Problem 4.6*

Solution. **a.** Consider a set of axis $Gxyz$ centered about the mass center G of a spherical shell: by symmetry we can state that

$$\int x^2 dm = \int y^2 dm = \int z^2 dm = J$$

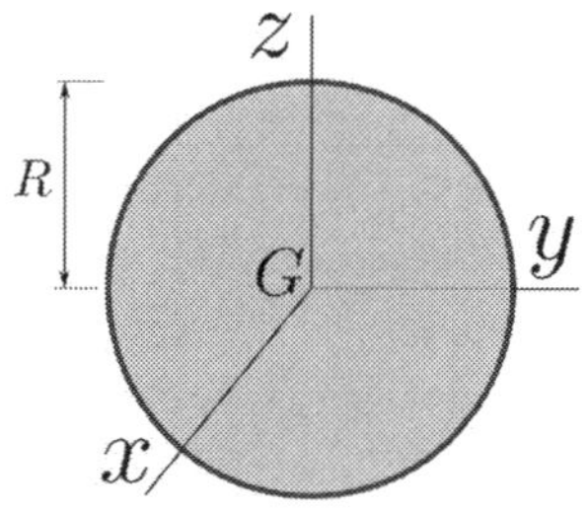

Figure 4.13 – *Problem 4.6*

We call J the common value of these three integrals. The value of J can be found as follows

$$\int (x^2 + y^2 + z^2) dm = 3J = \int r^2 dm = mR^2$$

since all points of the shell satisfy $r^2 = x^2 + y^2 + z^2 = R^2$. This gives the value $J = \frac{1}{3} mR^2$. Now we can find the moment of inertia about any diameter, for instance by choosing axis Gz for Δ

$$I_\Delta = \int (x^2 + y^2) dm = 2J$$

giving

$$\boxed{I_\Delta = \frac{2}{3} m R^2}$$

b. Given unit vectors $(\hat{\mathbf{x}}, \hat{\mathbf{y}}, \hat{\mathbf{z}})$ attached to the body as shown in Figure 4.14, recall that the angular momentum about O is given by

$$\mathbf{H}_O = \omega(I_{Oxz}\hat{\mathbf{x}} + I_{Oyz}\hat{\mathbf{y}} + I_{Ozz}\hat{\mathbf{z}})$$

since O is a fixed point of the body. Thus we need to determine the expression of the moment I_{Ozz} and products I_{Oxz} and I_{Oyz} of inertia.

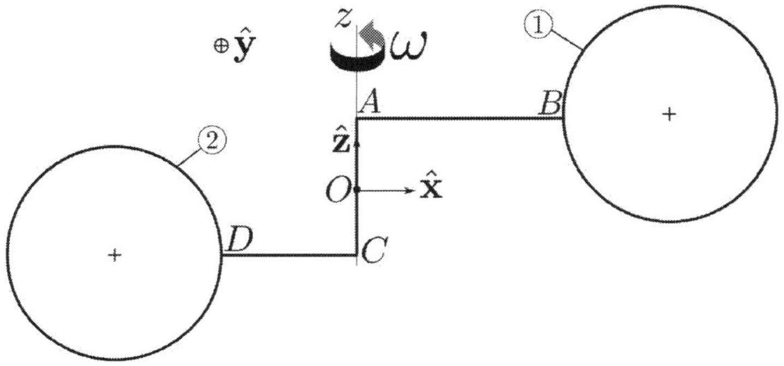

Figure 4.14 – *Problem 4.6 (Solution)*

We find by splitting the body into its two parts 1 and 2 and by application of the Parallel Axis Theorem

$$I_{Ozz} = I_{Ozz,1} + I_{Ozz,2} = (J_\Delta + m l_1^2) + (J_\Delta + m l_2^2) = m(\frac{4}{3}R^2 + l_1^2 + l_2^2)$$

$$I_{Oxz} = I_{Oxz,1} + I_{Oxz,2} = -mhl_1 - mhl_2 = -mh(l_1 + l_2)$$

$$I_{Oyz} = 0$$

This last result is due to the symmetry of the body about plane Oxz.

$$\boxed{\mathbf{H}_O = -m\omega h(l_1 + l_2)\hat{\mathbf{x}} + m\omega(\frac{4}{3}R^2 + l_1^2 + l_2^2)\hat{\mathbf{z}}}$$

■

Exercise 4.7 ■ A twirling baton is thrown with an initial angular velocity $\omega_0 = 3$ rad/s from a position O to a position A a distance 5 m away. Assuming that it travels in a vertical plane under the sole effect of gravity, find the initial velocity which must be given to its mass center G so that the baton describes a complete rotation when it reaches A. ■

SOLUTION. First, we adopt a set of fixed unit vectors $(\hat{\imath}, \hat{\jmath}, \hat{\mathbf{k}})$ with $\hat{\imath}$ along the horizontal, $\hat{\jmath}$ along the vertical directed upward and $\hat{\mathbf{k}} = \hat{\imath} \times \hat{\jmath}$. Figure 4.16. We denote the initial velocity of G as $\mathbf{v}_0 = v_0(\cos\theta_0\hat{\imath} + \sin\theta_0\hat{\jmath})$ and the initial angular velocity as $\boldsymbol{\omega}_0 = -\omega_0\hat{\mathbf{k}}$. We use Cartesian coordinates (x_G, y_G) to define the position of G which satisfy $x_G = y_G = 0$ and $\dot{x}_G = v_0\cos\theta_0$, $\dot{y}_G = v_0\sin\theta_0$ at time $t = 0$. Since the baton is assumed to be subject to the sole action of gravity, we have $\mathbf{F}_{ext\to\mathcal{B}} = -mg\hat{\jmath}$

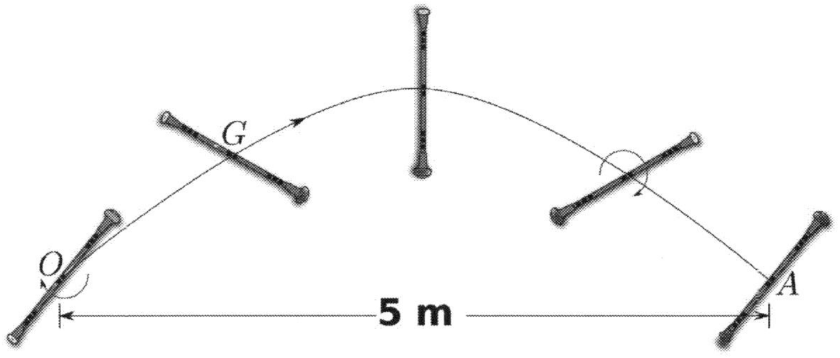

Figure 4.15 – *Problem 4.7*

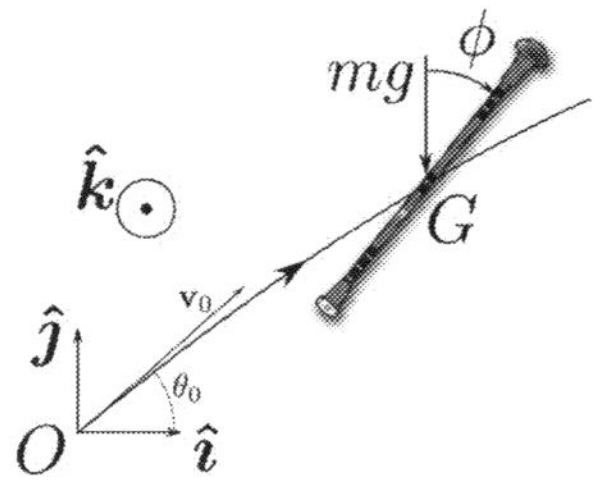

Figure 4.16 – *Problem 4.7 (Solution)*

and $\mathbf{M}_{G,ext\to\mathcal{B}} = \mathbf{0}$. Application of the Euler's First Principle $m\mathbf{a}_G = -mg\hat{\jmath}$ gives by projection on $(\hat{\imath}, \hat{\jmath})$:

$$m\ddot{x}_G = 0, \qquad m\ddot{y}_G = -mg \qquad [1-2]$$

while application of the Second Principle about G gives on $\hat{\mathbf{k}}$ (with $\boldsymbol{\omega}_{\mathcal{B}} = -\omega\hat{\mathbf{k}}$):

$$\frac{d}{dt}(\hat{\mathbf{k}} \cdot \mathbf{H}_G) = 0 = -I_{Gzz}\dot{\omega} \qquad [3]$$

Integration of equations [1-3] is straightforward

$$x_G = (v_0 \cos\theta_0)t, \qquad y_G = (v_0 \sin\theta_0)t - \frac{g}{2}t^2, \qquad \phi = \omega_0 t + \phi_0$$

since $\omega = Cst = \omega_0$. Angle ϕ defines the baton's orientation as shown in Figure 4.16. Now we can impose the requirements on the motion of the baton: (i) $\Delta\phi = 2\pi$ (complete rotation) at $t = t_A$, (ii) $x_G = D$ (given) and $y_G = 0$ at $t = t_A$ (G reaches A).

$$2\pi = \omega_0 t_A, \qquad (v_0 \cos\theta_0)t_A = D, \qquad (v_0 \sin\theta_0)t_A - \frac{g}{2}t_A^2 = 0$$

These equations are easily solved to find v_0 and θ_0:

$$\boxed{\tan\theta_0 = \frac{2\pi^2 g}{D\omega_0^2}, \qquad v_0 = \sqrt{\frac{D^2\omega_0^2}{4\pi^2} + \frac{\pi^2 g^2}{\omega_0^2}}}$$

With $\omega_0 = 3$ (rad/s) and $D = 5$ m, we find $\theta_0 = \mathbf{77}^\circ$ and $v_0 = \mathbf{10.5}$ (m/s). ■

Exercise 4.8 ▪ A milk carton of mass m and dimensions $b \times b \times h$ is placed on the conveyor

belt of a cash register. It is at rest when the belt suddenly starts up at constant acceleration a.
Find the conditions which must be satisfied by acceleration a for the milk carton to neither slip nor tip. Treat the milk carton as a uniform rectangular cuboid. ■

Figure 4.17 – *Problem 4.8*

SOLUTION. We assume that the milk carton $\mathcal{B}$ neither slips nor tips. We will find conditions to be satisfied by acceleration a to guarantee that $\mathcal{B}$ is in translation at acceleration $a\hat{\imath}$. The basis $(\hat{\imath}, \hat{\jmath}, \hat{\mathbf{k}})$ is defined in Figure 4.18. Note that the angular velocity of $\mathcal{B}$ satisfies $\omega = \mathbf{0}$ during the motion.

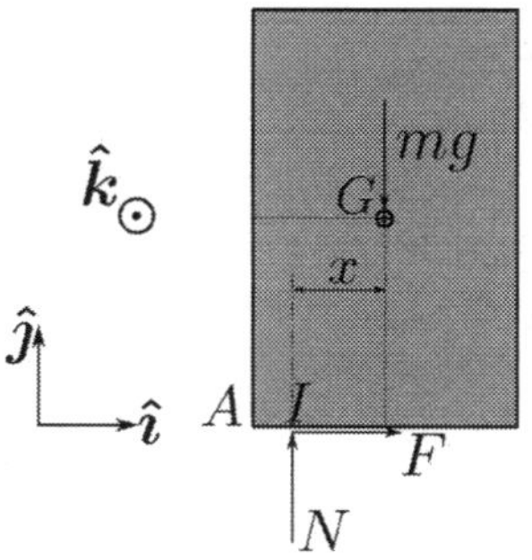

Figure 4.18 – *Problem 4.8 (Solution)*

Next we sketch a free-body diagram to show all external forces acting on $\mathcal{B}$: see Figure 4.18. The body is acted upon by
(i) gravitational forces with resultant $(-mg\hat{\jmath})$ applied at mass center G,
(ii) the contact forces exerted by the belt on $\mathcal{B}$ whose resultant must lie in the plane of motion of G at some unknown point I. The position of I is defined by (unknown) distance x. We write the resultant as $\mathbf{R} = N\hat{\jmath} + F\hat{\imath}$ where N is the normal reaction and F the friction force.
Next we apply Euler's First Principle $m\mathbf{a}_G = \mathbf{F}_{ext\to\mathcal{B}} = F\hat{\imath} + (N - mg)\hat{\jmath}$ to get two equations

$$ma = F, \qquad N - mg = 0 \qquad [1-2]$$

We also need to apply Euler's Second Principle $\frac{d}{dt}\mathbf{H}_G = \mathbf{M}_{G,ext\to\mathcal{B}} = \mathbf{0}$ taking into account the fact that $\mathbf{H}_G = 0$ for a body in translation. This leads to a single equation (the moments are directed along $\hat{\mathbf{k}}$):

$$\frac{h}{2}F - xN = 0 \qquad [3]$$

Equations [1-3] allow us to solve for the unknowns x, F, N:

$$F = ma, \qquad N = mg, \qquad x = \frac{h}{2}\frac{a}{g}$$

At this point we need to impose two conditions: $|F| < \mu_s N$ to guarantee that $\mathcal{B}$ moves without slipping and $x < \frac{b}{2}$ to guarantee that the point I of application of the contact forces does not fall beyond the corner point A. When I coincides with A the body necessarily tips about A. These two conditions give

$$ma < \mu_s mg, \qquad \frac{h}{2}\frac{a}{g} < \frac{b}{2}$$

leading to

$$\boxed{a \leq a_{max} = \kappa g, \qquad \kappa = \min(\mu_s, \frac{b}{h})}$$

Acceleration a must not exceed the upper bound a_{max}. ■

4.6 Contact Forces

We have already seen in Section 3.4 that contact forces lead to indeterminations in the solution of particle kinetics problems unless constitutive laws are introduced. Here we wish to complement these laws to include the case of rigid bodies in point contact.
Consider bodies 1 and 2 in relative planar motion. See Figure 4.19. Assume that the bodies are in contact at a single point I. Assume that at least one of the bodies possesses a smooth surface, and consider the common tangent line $\mathcal{L}$ to the bodies' boundaries at point I. [1] Denote by $\hat{\mathbf{n}}_{12}$ the unit vector normal to $\mathcal{L}$ pointing from 1 and 2. In general, the presence of body 1 creates a reaction force $\mathbf{R}_{1\to2}$ on body 2 whose line of action is assumed to lie in the plane of motion of I and to pass through I:

$$\mathbf{R}_{1\to2} = N_{12}\hat{\mathbf{n}}_{12} + \mathbf{F}_{12}, \qquad \mathbf{M}_{I,1\to2} = \mathbf{0} \qquad (4.32)$$

where

1. the component N_{12} is called the *normal reaction force* at I. It must satisfy the inequality $N_{12} \geq 0$ at all times if the point-contact is *unilateral*, that is, if the rigid bodies are free to lose contact from one another under the action of other applied forces.
2. the vector $\mathbf{F}_{12}$ is the *tangential reaction force* or *frictional force* at I. It is a measure of *resistance to the sliding motion* of body 2 relative to body 1.
3. setting $\mathbf{M}_{I,1\to2} = \mathbf{0}$ is equivalent to *neglecting pivoting and rolling friction.*

Remark 8. If the equations of motion predict that at some time t^*, $N_{12} = 0$ and subsequently $N_{12} < 0$ for $t > t^*$, then the contact at I is lost and necessarily $\mathbf{R}_{1\to2} = \mathbf{0}$ for $t > t^*$.

[1] If one of the boundaries is not smooth, then define line $\mathcal{L}$ as the tangent line to the other (smooth) boundary.

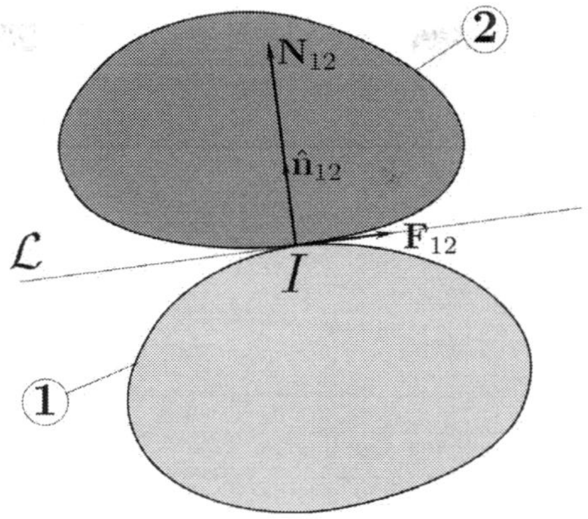

Figure 4.19 – *Reaction force* $\mathbf{R}_{1\to2} = N_{12}\hat{\mathbf{n}}_{12} + \mathbf{F}_{12}$ *of body 1 on body 2.*

Hence, in general, point-contact between two rigid bodies in planar motion results in two additional unknowns. The most drastic simplification amounts to neglecting friction:

Definition (Frictionless Contact)

The contact at point I is said to be *frictionless* if the reaction force $\mathbf{R}_{1\to2}$ is assumed normal to the surfaces in contact and applied at I, that is,

$$\mathbf{R}_{1\to2} = N_{12}\hat{\mathbf{n}}_{12} \quad \text{and} \quad \mathbf{M}_{I,1\to2} = \mathbf{0} \tag{4.33}$$

If friction must be taken into account, then we rely on phenomenological laws governing the sliding motion of body 2 relative to body 1 under dry contact conditions, that is, in the absence of lubrication. It is necessary to distinguish two kinematic cases whether there is slip or not.

Definition (Coulomb Laws)

The contact between bodies 1 and 2 is characterized by two dimensionless coefficients μ_s and μ_k, called *coefficients of static and kinetic friction*. These coefficients characterize the reaction force $\mathbf{R}_{1\to2}$ in the following manner:

Case 1: if body 2 slips relative to body 1, i.e. $\mathbf{v}_{I\in2/1} \neq 0$, then

$$|\mathbf{F}_{12}| = \mu_k |N_{12}|, \tag{4.34}$$

and the tangential component $\mathbf{F}_{12}$ must be colinear and opposed to the slip velocity $\mathbf{v}_{I\in2/1}$:

$$\mathbf{F}_{12} \times \mathbf{v}_{I\in2/1} = \mathbf{0} \qquad \text{and} \qquad \mathbf{F}_{12} \cdot \mathbf{v}_{I\in2/1} < 0 \tag{4.35}$$

Case 2: if body 2 rolls without slipping on body 1 about I, $\mathbf{v}_{I\in2/1} = \mathbf{0}$, then the inequality

$$|\mathbf{F}_{12}| \leq \mu_s |N_{12}| \tag{4.36}$$

must be satisfied.

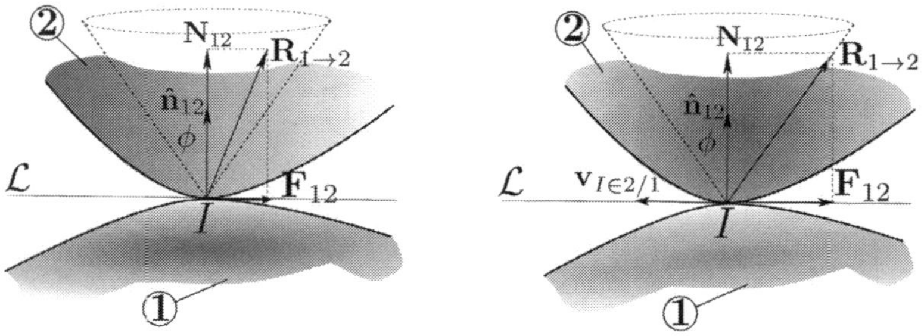

Figure 4.20 – *Coulomb laws: case 1 (left)* $\mathbf{v}_{I\in2/1} = \mathbf{0}$*, case 2 (right)* $\mathbf{v}_{I\in2/1} \neq 0$.

PROPERTIES

①. We can define a unique angle ϕ by the relation $\tan\phi = \mu_s$. The inequality (4.36) can be stated as

$$\frac{|\mathbf{F}_{12}|}{|N_{12}|} \leq \tan\phi$$

which shows that the contact force $\mathbf{R}_{1\to2}$ must lie inside the cone (known as the *friction cone*) of axis $(I, \hat{\mathbf{n}}_{12})$ and half-angle ϕ. Angle ϕ is the maximum inclination reached by $\mathbf{R}_{1\to2}$ relative to the normal $\hat{\mathbf{n}}_{12}$. See Figure 4.20.

②. In both cases stated above, one additional scalar equation is provided. This is generally sufficient to resolve the indetermination.

③. We will show in Section 5.8 that in the case of no-slip (CASE 2) the friction force is non-dissipative: the work done by $\mathbf{F}_{12}$ *relative to body 1* vanishes.

④. It is known that the value of coefficients of friction depends on the nature of the surfaces in contact (such as the type of materials, the state of roughness, etc) and on the magnitude of the slip velocity. Various empirical laws can be found in the literature. It can be shown that $\mu_k < \mu_s$. However, it is often assumed that $\mu_k = \mu_s$ since the values taken by μ_s and μ_k are generally close to one another for a given pair of surfaces. These values may range from 0.05 for smooth surfaces to as much as 1.2 for very rough surfaces.

⑤. These laws extend to the case of line contact between two cylindrical surfaces. The line of action of the resultant contact force $\mathbf{R}_{1\to2}$ necessarily lies in the plane of motion of the mass centers of the bodies.

4.7 Frictionless Joints

When two rigid bodies 1 and 2 are constrained by joints other than point contact, the laws of friction defined in the previous section may be difficult to apply unless friction is neglected. This leads to the concept of frictionless joints.

DEFINITION (FRICTIONLESS JOINT)

A joint between two rigid bodies 1 and 2 is said to be *frictionless* if the elementary contact force $d\mathbf{R}_{1\to2}$ acting on an element of area dA is directed along the local unit normal vector $\hat{\mathbf{n}}_{12}$ to the contact surface $\mathcal{S}$ (see Figure 4.21)

$$d\mathbf{R}_{1\to 2} = N_{12}\hat{\mathbf{n}}_{12}\, dA \tag{4.37}$$

at any point Q of $\mathcal{S}$.

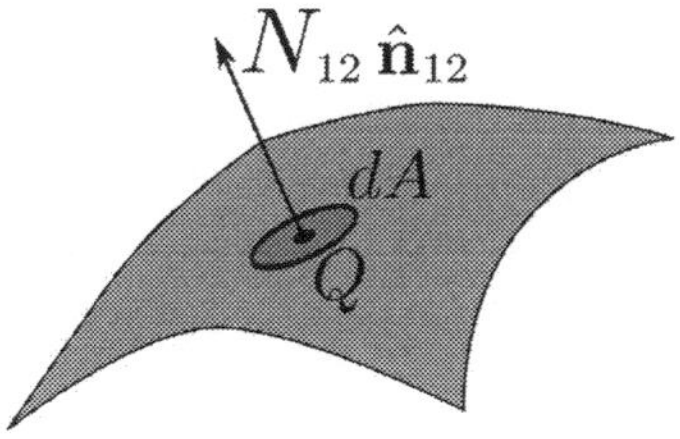

Figure 4.21 – *Elementary contact force.*

In practice, we are not interested in these local forces, but rather on their global effect in terms of a resultant force $\mathbf{R}_{1\to 2} = \int_{\mathcal{S}} N_{12}\hat{\mathbf{n}}_{12}\, dA$ and a resultant moment $\mathbf{M}_{A,1\to 2} = \int_{\mathcal{S}} \mathbf{r}_{AQ} \times N_{12}\hat{\mathbf{n}}_{12}\, dA$ about a particular point A. What is the consequence of property (4.37) on these quantities? We examine four cases of interest in planar kinetics (see Figure 4.22):

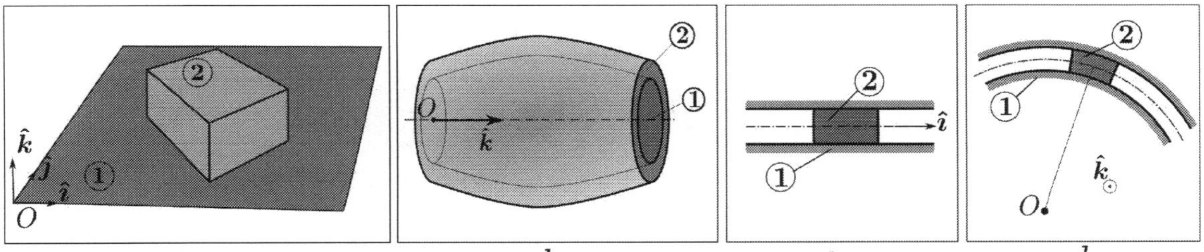

Figure 4.22 – *Frictionless Joints: a) planar joint, b) pivot, c) slider, d) circular slider.*

General Planar Joint. Assume that a planar surface of body 2 coincides with a planar surface of body 1. Define unit vectors $(\hat{\imath}, \hat{\jmath})$ in the plane of motion and $\hat{\mathbf{k}}$ normal to the plane of motion. Then, the action of body 1 on body 2 satisfies

$$\mathbf{R}_{1\to 2}\cdot\hat{\imath} = \mathbf{R}_{1\to 2}\cdot\hat{\jmath} = 0, \qquad \mathbf{M}_{A,1\to 2}\cdot\hat{\mathbf{k}} = 0 \tag{4.38}$$

about any point A. This result is readily seen from $\mathbf{R}_{1\to 2} = \int_{\mathcal{S}} N_{12}\hat{\mathbf{k}}\, dA$ and $\mathbf{M}_{A,1\to 2} = \int_{\mathcal{S}} \mathbf{r}_{AQ} \times N_{12}\hat{\mathbf{k}}\, dA$ since the unit normal $\hat{\mathbf{n}}_{12}$ is equal to $\hat{\mathbf{k}}$ for all points of contact Q.

Revolute Joint. Assume that body 2 is constrained to rotate about an axis $(O, \hat{\mathbf{k}})$ relative to body 1. Assume that an axisymmetric surface of body 1 is in contact with an axisymmetric surface of body 2. Then the action of body 1 on body 2 satisfies

$$\mathbf{M}_{A,1\to 2}\cdot\hat{\mathbf{k}} = \int_S \hat{\mathbf{k}}\cdot(\mathbf{r}_{AQ} \times N_{12}\hat{\mathbf{n}}_{12})dA = 0$$

for any point A of axis $(O, \hat{\mathbf{k}})$. Note, however, that in general $\mathbf{M}_{A,1\to 2} = M_{Ax}\hat{\imath} + M_{Ay}\hat{\jmath} \neq \mathbf{0}$ if $(\hat{\imath}, \hat{\jmath})$ are two arbitrary unit vectors normal to $\hat{\mathbf{k}}$.

SLIDER. Assume that body 2 is constrained to translate relative to body 1along the direction of unit vector $\hat{\imath}$. Then the action of body 1 on body 2 satisfies

$$\mathbf{R}_{1\to 2} \cdot \hat{\imath} = 0 \tag{4.39}$$

This is a consequence of the fact that the local force $d\mathbf{R}_{1\to 2}$ is everywhere normal to $\hat{\imath}$. Note, however, that in general $\mathbf{M}_{A,1\to 2} \neq \mathbf{0}$ for any choice of point A.

CIRCULAR SLIDER. Finally assume that body 2 is constrained to slide along a circular groove of body 1 centered about O. Then the action of body 1 on body 2 satisfies

$$\hat{\mathbf{k}} \cdot \mathbf{M}_{O,1\to 2} = 0 \tag{4.40}$$

We leave it to the reader to prove this last result. See Section 5.8 for a possible answer using energy arguments.

PROPERTIES

As expected, frictionless joints are *non-dissipative*, that is, the power dissipated by all local contact forces $d\mathbf{R}_{1\to 2}$ vanishes for the standpoint of an observer attached to body 1. Indeed, we can show that these forces satisfy

$$\int_S \mathbf{v}_{Q\in 2/1} \cdot d\mathbf{R}_{1\to 2} = \int_S \mathbf{v}_{Q\in 2/1} \cdot N_{12}\hat{\mathbf{n}}_{12}\, dA = 0 \tag{4.41}$$

since $\mathbf{v}_{Q\in 2/1}$ is necessarily normal to $\hat{\mathbf{n}}_{12}$.

4.8 Solution Method of Rigid Body Kinetics Problems

We have already outlined in Section 3.6 guidelines for the solution of particle kinetics problems. Most are applicable for the solution of rigid body kinetics problems.

PRACTICAL GUIDELINES

①. Read the problem carefully, identify the knowns and unknowns. Define all relevant symbols. Formulate in a few words the main steps of a solution.

②. Identify the motion of each rigid body of the system: is the body in translation, rotation or general planar motion? This step is important to parametrize the motion and to facilitate the body's kinematic analysis.

③. Define a set of unit vectors $(\hat{\imath}, \hat{\jmath}, \hat{\mathbf{k}})$ attached to the Newtonian referential $\mathcal{E}$ relative to which the system is studied. Choose $\hat{\mathbf{k}} = \hat{\imath} \times \hat{\jmath}$ normal to the plane of motion. It may also be useful to define unit vectors attached to a particular rigid body. If needed, parametrize the position of each rigid body: define (at most) 2 coordinates to define the position of the body's mass center G and an oriented angle to define the body's orientation. Summarize this step by defining all these quantities (unit vectors, coordinates) in a sketch.

④. At this stage, express the kinematic quantities $\mathbf{v}_G$ and $\boldsymbol{\omega}_{\mathcal{B}}$ on the chosen basis of unit vectors (possibly in terms of the coordinates defined in step 3. Time-differentiate $\mathbf{v}_G$ to find $\mathbf{a}_G$. The solution of the problem may also hinge on the determination of a slip velocity: the direction of $\mathbf{v}_{I\in 2/1} \neq \mathbf{0}$, if known, will give the direction of the friction force $\mathbf{F}_{1\to 2}$ exerted at contact point I by body 1 on body 2. If the condition $\mathbf{v}_{I\in 2/1} = \mathbf{0}$ must be imposed, then an important constraint equation is obtained.

⑤. If needed, define all initial conditions (initial position and initial velocity).

⑥. Describe all forces exerted on each body by sketching free-body diagrams. Each diagram must clearly indicate the line of action of each force in the plane of motion of the body's mass center.

⑦. Once the body's kinematics and the forces and moments exerted on the body have been properly defined,

we can now apply Euler's First and Second Principles which will be projected on 3 unit vectors to obtain at most 6 scalar equations, and most often just 3 equations (2 force equations in the plane of motion, 1 moment equation normal to the plane of motion).

⑧. At this point, compare the number of unknowns with the number of equations: if these two numbers match, the problem can be solved. Otherwise, some information is missing (typically an equation is missing or too many unknowns have been defined): return to steps 4 and 6 to fix this problem. Often the Coulomb laws presented in Section 4.6 have not been applied properly.
We are then typically confronted with two types of problems:

Type I: The problem involves solving for an unknown force/moment or unknown acceleration at a given instant. The solution amounts to solving a simple (linear) algebraic problem.

Type II: The problem involves solving for the position/velocity of the body at a later time, given initial conditions. This problem is more challenging since it involves solving the *equations of motion*, that is, the ordinary differential equations which govern one or more coordinates. These equations of motion are found by decoupling the unknown motion coordinates from the unknown forces. Most likely, the equations of motions can be solved analytically for simple ODEs (see Appendix C). Otherwise one must resort to a numerical solution (usually when frictional effects are taken into account).

⑨. Check the dimensional consistency of your results, then find their numerical values.

EXAMPLE PROBLEMS

Exercise 4.9 ▪ A mirror of mass 3 kg is attached by two identical strings to a fixed point O. It is at rest when the string OB snaps, thereby releasing the mirror.

Find the angular acceleration of the mirror and the tension in the cord OA immediately after the string OB snaps (treat the body as a uniform rectangular plate).

Compare with the tension at equilibrium.

Treat the mirror as a rectangular plate. ▪

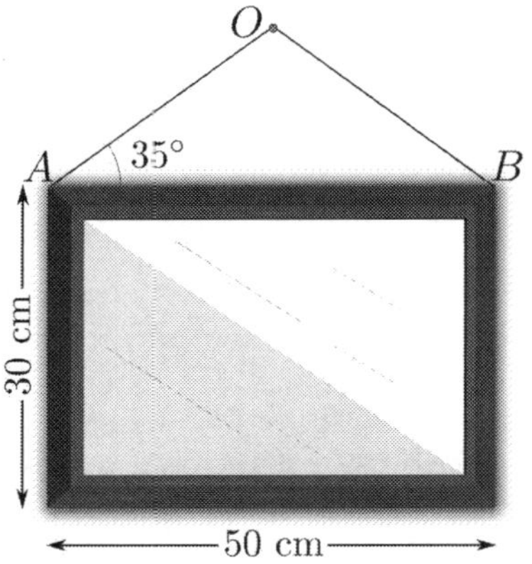

Solution. We introduce the geometric parameters b, c and θ and define the fixed unit vectors $(\hat{\imath}, \hat{\jmath}, \hat{\mathbf{k}})$ as shown in Figure 4.23 (left).

Step 1. There is no need to parametrize the motion: this is a "type 1" problem which involves solving for a force and angular acceleration at time $t = 0^+$. However we need to relate the acceleration of mass center G to that of corner point A. We note that A is in circular motion: its acceleration can be written in terms of its tangential and normal components as follows:

$$\mathbf{a}_A = \mathbf{a}_{A_T} + \mathbf{a}_{A_N} = \mathbf{a}_{A_T} = \dot{v}_A(\cos\theta\hat{\imath} + \sin\theta\hat{\jmath}) \quad [1]$$

where we have set $\mathbf{a}_{A_N} = \mathbf{0}$ since the initial speed of A is zero. We then find $\mathbf{a}_G$ as

$$\mathbf{a}_G = \mathbf{a}_A + \alpha_0\hat{\mathbf{k}} \times \mathbf{r}_{AG} - \omega_0^2\mathbf{r}_{AG} = \mathbf{a}_A + \frac{\alpha_0}{2}(b\hat{\jmath} - c\hat{\imath}) \quad [2]$$

setting $\boldsymbol{\alpha} = \alpha_0\hat{\mathbf{k}}$ and $\boldsymbol{\omega} = \omega_0\hat{\mathbf{k}} = \mathbf{0}$ at $t = 0^+$. We expect $\alpha_0 < 0$.

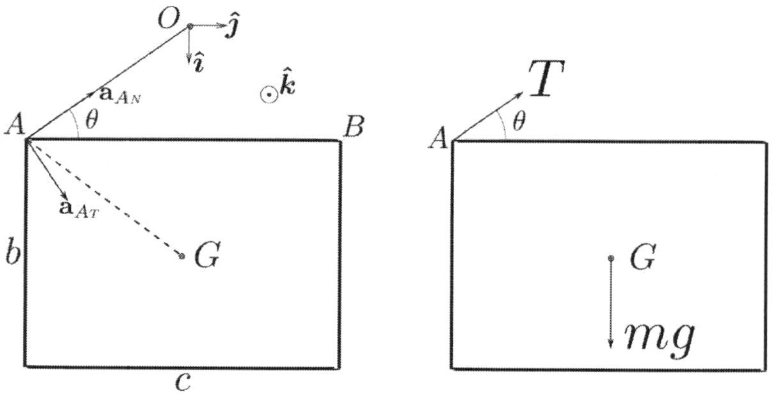

Figure 4.23 – *Problem 4.9 (Solution).*

Step 2. The next step consists of sketching a free-body diagram: see Figure 4.23 (right). The body is subject to the tension $T(\cos\theta\hat{\jmath} - \sin\theta\hat{\imath})$ ($T > 0$) acting at A and to the gravitational force $mg\hat{\imath}$ acting at G. Euler's First Principle $m\mathbf{a}_G = \mathbf{F}_{ext\to\mathcal{B}}$ projected on $(\hat{\imath}, \hat{\jmath})$ gives 2 equations using expressions [1-2] to find $\mathbf{a}_G$

$$m(\dot{v}_A\cos\theta - \frac{c}{2}\alpha_0) = -T\sin\theta + mg \quad [3]$$

$$m(\dot{v}_A\sin\theta + \frac{b}{2}\alpha_0) = T\cos\theta \quad [4]$$

Euler's Second principle $\frac{d}{dt}\mathbf{H}_G = \mathbf{M}_{G,ext\to\mathcal{B}}$ about mass center G gives

$$\frac{m}{12}(b^2+c^2)\alpha_0\hat{\mathbf{k}} = \mathbf{r}_{GA}\times T(\cos\theta\hat{\jmath}-\sin\theta\hat{\imath}) = -\frac{T}{2}(b\cos\theta + c\sin\theta)\hat{\mathbf{k}} \qquad [5]$$

using $I_{Gzz} = \frac{m}{12}(b^2+c^2)$.

Remark. It is tempting to apply Euler's Second principle about A in the form $I_{Azz}\boldsymbol{\alpha} = \mathbf{r}_{AG}\times m\mathbf{g}$ since $\mathbf{v}_G = \mathbf{v}_A = \mathbf{0}$ at $t=0^+$ eliminating the contribution of the tension. But this is incorrect! Why?

STEP 3. We have found 3 equations [3-5] governing 3 unknowns $(\dot{v}_A, \alpha_0, T)$: eliminating $\dot{v}_A$ from [3-4] gives

$$T = mg\sin\theta + \frac{m}{2}\alpha_0(b\cos\theta + c\sin\theta)$$

Then substitution of T in [5] gives

$$\boxed{\alpha_0 = -6g\sin\theta\frac{b\cos\theta+c\sin\theta}{b^2+c^2+3(b\cos\theta+c\sin\theta)^2}} \quad \boxed{T = mg\sin\theta\frac{b^2+c^2}{b^2+c^2+3(b\cos\theta+c\sin\theta)^2}}$$

With $b=0.3$ m, $c=0.5$, $m=3$ kg, $\theta=35°$, we find $\alpha_0 = \mathbf{-15.1}$ rad/s^2 and $T=\mathbf{4.8}$ N. At equilibrium $T_0 = mg/2\sin\theta = \mathbf{25.6}$ N. ∎

Exercise 4.10 ▪ A uniform slender bar of mass m and length $2l$ is supported by strings OB and QA in a vertical plane. String QA is cut, releasing the bar from rest.
Find the bar's angular acceleration and the string's tension immediately after release. ▪

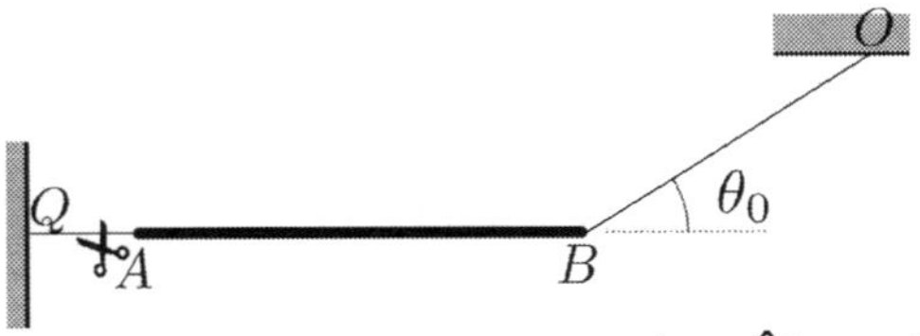

SOLUTION. STEP 1. We start by defining a fixed basis of unit vectors $(\hat{\imath},\hat{\jmath},\hat{\mathbf{k}})$ and the basis $(\hat{\mathbf{u}},\hat{\mathbf{u}},\hat{\mathbf{k}})$ such that $\mathbf{r}_{OB} = b\hat{\mathbf{u}}$. See Figure 4.24 (left). We need to define the acceleration of the mass center G at time $t=0^+$. At this instant, we have $\mathbf{v}_G=\mathbf{0}$ and $\omega = \mathbf{0}$. If we introduce the angular acceleration of the bar as $\boldsymbol{\alpha} = \alpha_0\hat{\mathbf{k}}$, we find

$$\mathbf{a}_G = \mathbf{a}_B + \boldsymbol{\alpha}\times\mathbf{r}_{BG} - \omega^2\mathbf{r}_{BG} = \mathbf{a}_B + l\alpha_0\hat{\jmath}$$

To find $\mathbf{a}_B$, we recognize that the trajectory of B is circular: using NT coordinates, we find

$$\mathbf{a}_B = \dot{v}_B\hat{\mathbf{v}} + \frac{v_B^2}{b}(-\hat{\mathbf{u}}) = \dot{v}_B\hat{\mathbf{v}}$$

using $v_B = 0$ at $t=0^+$. Finally, we obtain the expression

$$\mathbf{a}_G = \dot{v}_B\hat{\mathbf{v}} + l\alpha_0\hat{\jmath} \qquad [1]$$

STEP 2. The next step consists of sketching a free-body diagram: See Figure 4.24 (right). The bar is subject to gravity and the string's tension: $\mathbf{F}_{ext\to bar} = mg\hat{\jmath} - T\hat{\mathbf{u}}$.
STEP 3. We now apply Euler's 1st principle: $m\mathbf{a}_G = -T\hat{\mathbf{u}} + mg\hat{\jmath}$ to obtain 2 equations (using [1])

$$-T\cos\theta_0 = -m\dot{v}_B\sin\theta_0, \qquad -T\sin\theta_0 + mg = m(\dot{v}_B\cos\theta_0 + l\alpha_0) \qquad [2-3]$$

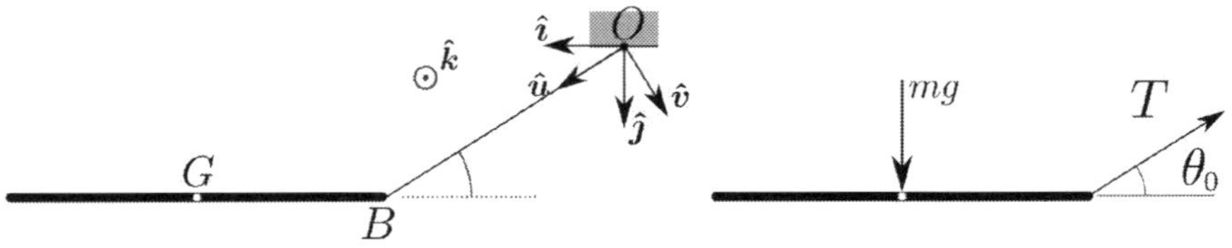

Figure 4.24 – *Problem 4.10 (Solution).*

and Euler's 2nd principle: $d\mathbf{H}_G/dt = \mathbf{M}_{G,ext\to bar}$ to obtain 1 equation on $\hat{\mathbf{k}}$

$$m\frac{l^2}{3}\alpha_0 = lT\sin\theta_0 \qquad [4]$$

STEP 4. We can now solve equations [2-4] for the 3 unknowns $(\alpha_0, \dot{v}_B, T)$. We find

$$\boxed{\alpha_0 = \frac{3g}{l}\frac{\sin^2\theta_0}{1+3\sin^2\theta_0}, \qquad T = mg\frac{\sin\theta_0}{1+3\sin^2\theta_0}}$$

Note that these results are not functions of the length b of the string. ■

Exercise 4.11 ■■ A cylindrical chemical drum is initially at rest on a flatbed truck as shown in Figure 4.25. The truck then starts up at constant acceleration a_T. The drum is observed to roll without slipping on the flatbed. The contact is characterized by a coefficient of static friction μ_s. The drum is of mass m, radius R and moment of inertia $\frac{1}{2}mR^2$ about its axis. Denote by L the maximum distance traveled by the drum on the flatbed.

a. Find the distance traveled by the truck when the drum falls off the flatbed.

b. Find the condition satisfied by a_T to guarantee rolling without slipping. ■

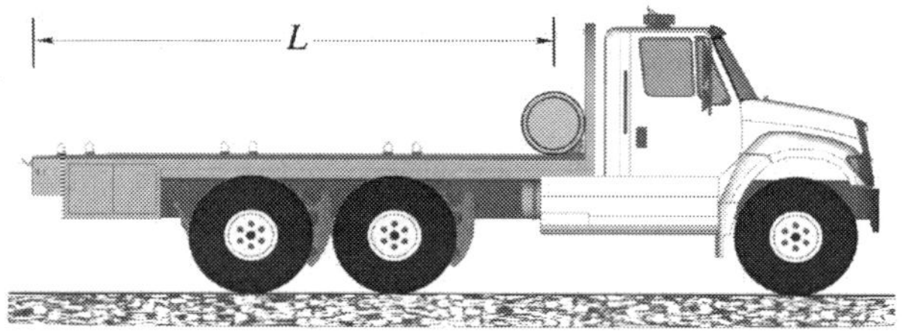

Figure 4.25 – *Problem 4.11*

SOLUTION. **a.** First we note that this is a "type 2" kinetics problem: given initial conditions, find the state of the system at a later time. Define O as a fixed origin and Q the point of the flatbed coinciding with the initial position of the contact point I of the drum. Define the position of Q by x_Q and the position of mass center G relative to the truck by X_G on an axis QX attached to the truck. See Figure 4.26 (left). The position of G is then given by $\mathbf{r}_{OG} = (x_Q - X_G)\hat{\imath} + R\hat{\jmath}$. By taking two time-derivatives we find the acceleration of G

$$\mathbf{a}_G = (a_T - \ddot{X}_G)\hat{\imath}$$

since $\ddot{x}_Q = a_T$. This result can also be found by applying equation (2.20):

$$\mathbf{a}_{G/\mathcal{E}} = \mathbf{a}_{G/truck} + \mathbf{a}_{G\in truck/\mathcal{E}}$$

If the angular velocity of the drum is written as $\boldsymbol{\omega} = \omega\hat{\mathbf{k}}$ (with $\hat{\mathbf{k}} = \hat{\imath}\times\hat{\jmath}$), then the no-slip condition can be stated as

$$\mathbf{v}_{G/truck} = -\dot{X}_G\hat{\boldsymbol{\imath}} = \omega\hat{\mathbf{k}} \times \mathbf{r}_{IG} = -R\omega\hat{\boldsymbol{\imath}}$$

leading to $\dot{X}_G = R\omega$ and by time-differentiation

$$\boxed{\ddot{X}_G = R\dot{\omega}} \qquad [1]$$

At $t = 0$, $\dot{X}_G = \dot{x}_Q = 0$. We also note that, when $X_G = L$, the drum falls off the truck.

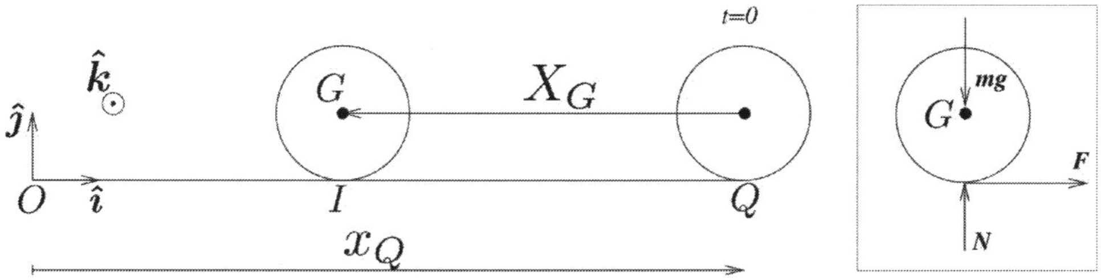

Figure 4.26 – *Problem 4.11 (Solution).*

However we need to find X_G as a function of time.
We sketch a free-body diagram for the drum in Figure 4.26 (right) and apply Euler's principles to obtain 3 equations:
(i) equation $m\mathbf{a}_G = \mathbf{F}_{ext\rightarrow drum} = F\hat{\boldsymbol{\imath}} + (N - mg)\hat{\boldsymbol{\jmath}}$ gives

$$N = mg, \qquad m(a_T - \ddot{X}_G) = F \qquad [2-3]$$

(ii) equation $I_{Gzz}\boldsymbol{\alpha} = \mathbf{M}_{G,ext\rightarrow drum}$

$$\tfrac{1}{2}mR^2\dot{\omega} = RF \qquad [4]$$

We can solve [1-4] to find $\ddot{X}_G$ versus a_T:

$$\boxed{\ddot{X}_G = \frac{2}{3}a_T}$$

By two time-integrations, we obtain $X_G(t) = \frac{1}{3}a_T t^2$ (set $X_G = 0$ at time $t = 0$). When $X_G = L = \frac{1}{3}a_T\tau^2$, the drum falls off the truck. This gives $a_T\tau^2 = 3L$. During this time the truck travels the distance $\Delta x_Q = \frac{1}{2}a_T\tau^2$:

$$\boxed{\Delta x_Q = \frac{3}{2}L}$$

b. According to Coulomb laws, the drum rolls without slipping as long as the reaction force of the flatbed on the drum satisfies $|F| < \mu_s N$. According to [1-2] the components of the reaction force are given by

$$F = m(a_T - \ddot{X}_G) = \frac{1}{3}ma_T, \qquad N = mg$$

This leads to the condition

$$\boxed{a_T < 3\mu_s g}$$

■

Exercise 4.12 ■■ Two uniform disks 1 (of mass center O, radius R_1 and mass m_1) and 2 (of mass center A, radius R_2 and mass m_2) rotate about their frictionless axes at angular velocity Ω_1 and Ω_2, respectively. The two disks are then brought in contact keeping their axes parallel and horizontal (the distance between the centers O and A is maintained at the value $R_1 + R_2$).
Assume that the disks slip under the effect of dry friction (characterized by coefficient of kinetic friction μ).
Show that the disks eventually stop slipping relative to one another and find their final angular velocities. Comment your result when $R_1\Omega_1 + R_2\Omega_2 = 0$. ■

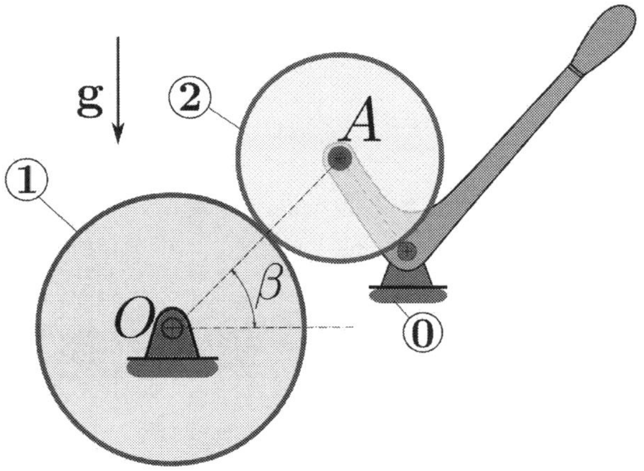

Figure 4.27 – *Problem 4.12*

SOLUTION. First we choose a basis of fixed unit vectors $(\hat{\imath}, \hat{\jmath}, \hat{\mathbf{k}})$ as shown in Figure 4.28 (left), with $\mathbf{r}_{OA} = (R_1 + R_2)\hat{\imath}$. We define the angular velocities $\boldsymbol{\omega}_1 = \omega_1\hat{\mathbf{k}}$ and $\boldsymbol{\omega}_2 = \omega_2\hat{\mathbf{k}}$ and assume $\omega_1 > 0$ and $\omega_2 > 0$. At time $t = 0$, we have $\omega_1 = \Omega_1$ and $\omega_2 = \Omega_2$. We call I the point of contact between the two bodies. Since the solution hinges on finding the friction force $\mathbf{F}_{12}$ which body 1 exerts on body 2, we need to find the slip velocity $\mathbf{v}_{I\in 2/1} = \mathbf{v}_{I\in 2/0} - \mathbf{v}_{I\in 1/0}$:

$$\mathbf{v}_{I\in 2/1} = \boldsymbol{\omega}_2 \times \mathbf{r}_{AI} - \boldsymbol{\omega}_1 \times \mathbf{r}_{OI} = \omega_2\hat{\mathbf{k}} \times (-R_2\hat{\imath}) - \omega_1\hat{\mathbf{k}} \times (R_1\hat{\imath}) = -(R_2\omega_2 + R_1\omega_1)\hat{\jmath} \quad [1]$$

This shows that the slip velocity is directed along $(-\hat{\jmath})$. Hence the friction force is necessarily directed in the direction $(+\hat{\jmath})$. We can now sketch a free-body diagram for bodies 1 and 2. See Figure 4.28 (right). Body 2 is subject to
-the reaction force $\mathbf{R}_{1\to 2} = F_{12}\hat{\jmath} + N_{12}\hat{\imath}$, with $F_{12} = \mu N_{12} > 0$,
-the gravitational force $m_2\mathbf{g}$,
-the reaction $\mathbf{R}_A$ (passing through A since the pin is frictionless).
Body 1 is subject to
-the reaction force $\mathbf{R}_{2\to 1} = -F_{12}\hat{\jmath} - N_{12}\hat{\imath}$,
-the gravitational force $m_1\mathbf{g}$,
-the reaction $\mathbf{R}_O$ (passing through O since the pin is frictionless).
Since both bodies are in rotation, we apply Euler's Second principle about their center of rotation, thereby eliminating the contribution of unknown reactions $\mathbf{R}_O$ and $\mathbf{R}_A$:

$$\frac{d}{dt}\mathbf{H}_{O,1} = \mathbf{M}_{O,ext\to 1}, \qquad \frac{d}{dt}\mathbf{H}_{A,2} = \mathbf{M}_{A,ext\to 2}$$

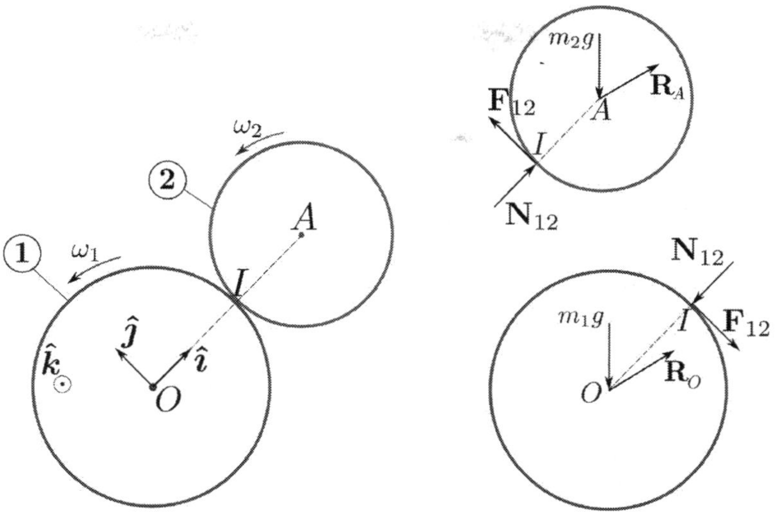

Figure 4.28 – *Problem 4.12*

with

$$\mathbf{H}_{O,1} = \frac{1}{2}m_1R_1^2\omega_1\hat{\mathbf{k}}, \qquad \mathbf{H}_{A,2} = \frac{1}{2}m_2R_2^2\omega_2\hat{\mathbf{k}}$$

and

$$\mathbf{M}_{O,ext\to 1} = \mathbf{r}_{OI} \times (-F_{12}\hat{\jmath} - N_{12}\hat{\imath}) = -R_1F_{12}\hat{\mathbf{k}}, \quad \mathbf{M}_{A,ext\to 2} = \mathbf{r}_{AI} \times (F_{12}\hat{\jmath} + N_{12}\hat{\imath}) = -R_2F_{12}\hat{\mathbf{k}}$$

This leads to the equations

$$\frac{1}{2}m_1R_1^2\dot{\omega}_1 = -R_1F_{12}, \qquad \frac{1}{2}m_2R_2^2\dot{\omega}_2 = -R_2F_{12} \qquad [2-3]$$

We can eliminate F_{12} between [2] and [3] to find

$$m_1R_1\dot{\omega}_1 = m_2R_2\dot{\omega}_2$$

which can be integrated to give

$$\boxed{m_1R_1(\omega_1 - \Omega_1) = m_2R_2(\omega_2 - \Omega_2)} \qquad [4]$$

To show that the disks eventually stop slipping relative to one another, we find the time rate-of-change of the slip velocity $\mathbf{v}_{I\in 2/1}$ according to [1]

$$\frac{d}{dt}\mathbf{v}_{I\in 2/1} = -(R_2\dot{\omega}_2 + R_1\dot{\omega}_1)\hat{\jmath} = 2F_{12}(\frac{1}{m_1} + \frac{1}{m_2})\hat{\jmath}$$

after using [2-3]. Since $F_{12} > 0$, the slip velocity's component on $\hat{\jmath}$ will increase from the negative value $-(R_2\Omega_2 + R_1\Omega_1)$ at $t = 0$ to eventually reach the value 0 at some time t^*: *this shows that the two bodies eventually stop slipping relative to one another*. This is not surprising since the role of the friction force is to oppose the slip. At time $t = t^*$ the angular velocities take the values $\omega_1 = \omega_1^*$ and $\omega_2 = \omega_2^*$ which satisfy $\mathbf{v}_{I\in 2/1} = \mathbf{0}$ and equation [4]:

$$R_2\omega_2^* + R_1\omega_1^* = 0, \qquad m_1R_1(\omega_1^* - \Omega_1) = m_2R_2(\omega_2^* - \Omega_2) \qquad [5-6]$$

We can solve [5-6] to find

$$\boxed{\begin{aligned}\omega_1^* &= \frac{1}{R_1}\frac{m_1R_1\Omega_1 - m_2R_2\Omega_2}{m_1+m_2}\\ \omega_2^* &= \frac{1}{R_2}\frac{m_2R_2\Omega_2 - m_1R_1\Omega_1}{m_1+m_2}\end{aligned}}$$

If the initial angular velocities Ω_1 and Ω_2 satisfy the condition $R_1\Omega_1 + R_2\Omega_2 = 0$, the slip velocity is initially zero and the two disks should continue to rotate without slipping relative to one another, that is, we should expect $\omega_1^* = \Omega_1$ and $\omega_2^* = \Omega_2$. We can verify that this is indeed the case. ■

Exercise 4.13 ▪▪ A spherical ball of mass m and radius r is thrown with backspin on a horizontal straight support Ox. Its mass center G has initial speed v_0 and its initial (counterclockwise) angular velocity is ω_0. The contact is characterized by the coefficient of kinetic friction μ. Rolling friction is neglected.

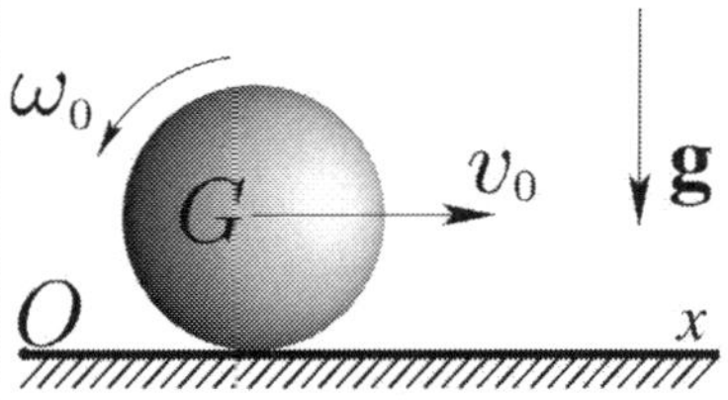

a. Show that the ball initially slips. Find the slip velocity in terms of the speed $v(t)$ of mass center G and the angular velocity $\omega(t)$.
b. Apply Euler's principles to find $v(t)$ and $\omega(t)$ as functions of time. Then show that the ball will cease to slip at some time t_1. Find time t_1 and the corresponding distance L_1 traveled by the ball.
c. What happens at time $t = t_1$? Does the ball roll forward?
d. Find the velocity of G and the angular velocity of the ball at time $t \geq t_1$. Find the friction force and describe the motion of the ball. ■

Solution **a.** First we choose Cartesian unit vectors $(\hat{\mathbf{e}}_x, \hat{\mathbf{e}}_y, \hat{\mathbf{e}}_z)$ as shown in Figure 4.29. The kinematics of the sphere is entirely specified by the velocity $\mathbf{v}_G = v(t)\hat{\mathbf{e}}_x$ of its mass center G, and by its angular velocity $\boldsymbol{\omega} = \omega(t)\hat{\mathbf{e}}_z$. Then the slip velocity of the ball is given by

$$\mathbf{v}_{I\in\mathcal{B}} = \mathbf{v}_G + \boldsymbol{\omega}\times\mathbf{r}_{GI} = (v + r\omega)\hat{\mathbf{e}}_x$$

At time $t = 0$, the slip velocity takes the expression $\mathbf{v}_{I\in\mathcal{B}} = (v_0 + r\omega_0)\hat{\mathbf{e}}_x$. Since $(v_0 + r\omega_0) > 0$, the ball initially slips, and we can assume that the initial phase of motion is also characterized by slipping.

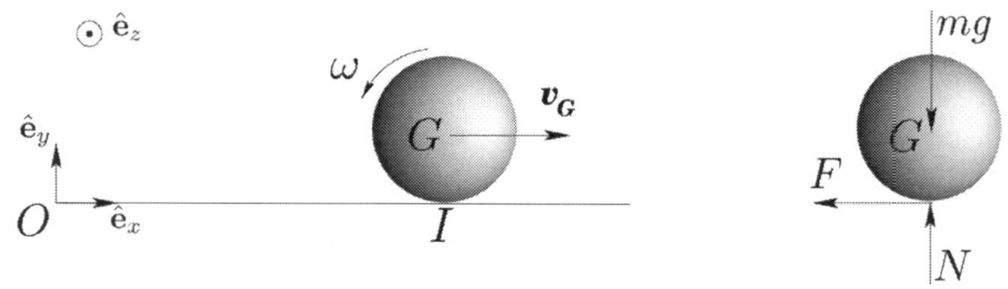

Figure 4.29 – *Problem 4.13 (Solution)*

b. The external forces acting on the ball are (i) gravity $m\mathbf{g}$ applied at mass center G, and (ii) the contact force $N\hat{\mathbf{e}}_y - F\hat{\mathbf{e}}_x$ exerted at the contact point I: note that $F > 0$ so as to oppose the slip velocity which points in the $+\hat{\mathbf{e}}_x$ direction. Application of Euler's First Principle $m\mathbf{a}_G = m\mathbf{g} - F\hat{\mathbf{e}}_x + N\hat{\mathbf{e}}_y$ gives by projection on $(\hat{\mathbf{e}}_x, \hat{\mathbf{e}}_y)$

$$m\dot{v} = -F, \qquad N = mg \tag{1-2}$$

Then application of the Second principle $\frac{d}{dt}\mathbf{H}_G = \mathbf{M}_{G,ext\to\mathcal{B}}$ about mass center G gives on unit vector $\hat{\mathbf{e}}_z$

$$\frac{2}{5}mr^2\dot{\omega} = -rF \tag{3}$$

using $I_{Gzz} = \frac{2}{5}mr^2$ for a full uniform sphere. We have obtained 3 equations solving for 4 unknowns v, ω, F, N. One more equation is given by Coulomb Laws which specify that

$$|F| = F = \mu N = \mu mg \tag{4}$$

since the ball is slipping. Now we can solve equations [1-4] to obtain

$$\dot{\omega} = -\frac{5\mu g}{2r}, \qquad \dot{v} = -\mu g$$

showing that both v and ω are (linearly) decreasing functions of time. Eventually the slip velocity $(v + r\omega)\hat{\mathbf{e}}_x$ will reach the value $\mathbf{0}$. We obtain by integration (taking into account $v = v_0$ and $\omega = \omega_0$ at $t = 0$)

$$\boxed{\omega = \omega_0 - \frac{5g}{2r}\mu t, \qquad v = v_0 - \mu g t} \tag{5-6}$$

leading to the slip velocity

$$\mathbf{v}_{I\in\mathcal{B}} = (v_0 + r\omega_0 - \frac{7\mu}{2r}gt)\hat{\mathbf{e}}_x.$$

The slip ceases at time

$$\boxed{t_1 = \frac{2}{7\mu g}(v_0 + r\omega_0)} \tag{7}$$

The length traveled by G is obtained by integrating $\dot{x} = v_0 - \mu g t$ from $t = 0$ to $t = t_1$:

$$L_1 = \Delta x = v_0 t_1 - \frac{1}{2}\mu g t_1^2$$

c. At time $t = t_1$, G takes the velocity

$$v_1 = v_0 - \mu g t_1 = \frac{5}{7}v_0 - \frac{2}{7}r\omega_0 \tag{8}$$

and the ball has the angular velocity

$$\omega_1 = \omega_0 - \frac{5g}{2r}\mu t_1 = -\frac{v_1}{r} \tag{9}$$

At this point, *the ball will roll forward if* $\boxed{5v_0 > 2r\omega_0}$, *stop if* $5v_0 = 2r\omega_0$ *or roll backward if* $5v_0 < 2r\omega_0$.

d. For $t \geq t_1$ the ball will roll without slipping: once it stops slipping, it continues to roll without slipping. Its translational and rotational motions are now coupled: setting

$\mathbf{v}_{I \in \mathcal{B}} = \mathbf{0}$ gives

$$v_G = -r\omega, \qquad t \geq t_1$$

Assuming that the ball rolls forward (that is, assuming $5v_0 > 2r\omega_0$): $v_G > 0$ and $\omega < 0$. The motion is still governed by the equations [1-3]. We find

$$-F = m\dot{v}_G = -mr\dot{\omega} = \frac{2}{5}mr\dot{\omega}$$

This shows that

$$m\dot{v}_G = -mr\dot{\omega} = -F = 0$$

The ball rolls indefinitely at the constant angular velocity ω_1 while its mass center G moves at the constant speed v_1. The condition $|F| \leq \mu_s N$ as specified by Coulomb laws is of course satisfied. Addition of rolling friction in the model would "fix" this unphysical situation. ■

4.9 Kinetics for Multi-Body Systems

We have already learned in Section 3.7 how to extend Euler's First Principle to systems of rigid bodies. As a consequence of the Law of Action and Reaction, the forces of interactions between rigid bodies, the so-called *internal* forces, drop out of the principle's formulation. Our first task in this section is to extend Euler's Second Principle (4.27) or (4.28) to systems of rigid bodies. We proceed as in Section 3.7. Consider a system Σ comprised of rigid body 1 (of mass m_1 and mass center G_1) and rigid body 2 (of mass m_2 and mass center G_2). As was done for the derivation Euler's First Principle, we separate actions internal to the system to those external to the system. Given an arbitrary point A, application of (4.28) to each body gives

$$\frac{d}{dt}\mathbf{H}_{A,1} + \mathbf{v}_A \times m_1\mathbf{v}_{G_1} = \underbrace{\mathbf{M}_{A,ext\to 1}}_{\text{external}} + \underbrace{\mathbf{M}_{A,2\to 1}}_{\text{internal}}$$

$$\frac{d}{dt}\mathbf{H}_{A,2} + \mathbf{v}_A \times m_2\mathbf{v}_{G_2} = \underbrace{\mathbf{M}_{A,ext\to 2}}_{\text{external}} + \underbrace{\mathbf{M}_{A,1\to 2}}_{\text{internal}}$$

Adding these two equations gives

$$\frac{d}{dt}(\mathbf{H}_{A,1} + \mathbf{H}_{A,2}) + \mathbf{v}_A \times (m_1\mathbf{v}_{G_1} + m_2\mathbf{v}_{G_2}) = \underbrace{\mathbf{M}_{A,ext\to 1} + \mathbf{M}_{A,ext\to 2}}_{\text{external}} + \underbrace{\mathbf{M}_{A,2\to 1} + \mathbf{M}_{A,1\to 2}}_{\text{internal}}$$

This last equation can be simplified by noting that

-the sum $\mathbf{H}_{A,1} + \mathbf{H}_{A,2}$ is the system's angular momentum $\mathbf{H}_{A,\Sigma}$ about A,
-the sum $m_1\mathbf{v}_{G_1} + m_2\mathbf{v}_{G_2}$ is the system's linear momentum $m\mathbf{v}_G$,
-the sum $\mathbf{M}_{A,ext\to 1} + \mathbf{M}_{A,ext\to 2}$ is the resultant moment $\mathbf{M}_{A,ext\to\Sigma}$ about A of all external actions on Σ,
-the sum $\mathbf{M}_{A,2\to 1} + \mathbf{M}_{A,1\to 2}$ vanishes according to the Law of Action and Reaction.

We can now state Euler's Second Principle for system Σ.

Euler's Second Principle: System of Rigid Bodies

With respect to a Newtonian referential, Euler's Second Principle applied to a system Σ of mass m and mass center G takes the form

$$\frac{d}{dt}\mathbf{H}_{A,\Sigma} + \mathbf{v}_A \times m\mathbf{v}_G = \mathbf{M}_{A,ext\to\Sigma} \tag{4.43}$$

where $\mathbf{M}_{A,ext\to\Sigma}$ is the resultant moment of external actions on Σ about arbitrary point A.

Remark 9. Another version of Euler's Second Principle is based on (4.27): the left-hand-side of (4.43) is obtained by summing $\frac{d}{dt}\mathbf{H}_{G_i,i} + \mathbf{r}_{AG_i} \times m_i\mathbf{a}_G = \mathbf{M}_{A,ext\to i}$

$$\frac{d}{dt}\mathbf{H}_{G,\Sigma} + \mathbf{r}_{AG_1} \times m_1\mathbf{a}_{G_1} + \mathbf{r}_{AG_2} \times m_2\mathbf{a}_{G_2} = \mathbf{M}_{A,ext\to\Sigma}$$

Solution Method for Multi-Body Kinetics Problems. When faced with a multi-body kinetics problem, the solution will seem daunting unless a methodical approach is adopted. Given a system Σ of N interconnected rigid bodies `1,2,...,N` the first step consists of identifying all interconnections and to list all unknowns. These unknowns are typically of two types: (i) unknown motion variables which define the configuration of system Σ at any given time, (ii) unknown contact forces or moments internal or external to the system. Ideally, a free-body diagram for *each* body of the system should be sketched. Whenever the various interconnections are modeled as frictionless joints, it is vital to identify the specificities of each joint which can be later exploited in the solution. The next step is to identify which of the unknowns listed above are asked for. Ideally a solution method which eliminates the unwanted unknowns should be identified.

At this point, two strategies can be employed:

Strategy #1. Apply Euler's First Principle to a *subset* Σ' of Σ along a direction $\hat{\mathbf{u}}$ to obtain

$$\mathbf{F}_{ext\to\Sigma'} \cdot \hat{\mathbf{u}} = m'\mathbf{a}_{G'} \cdot \hat{\mathbf{u}} \tag{4.44}$$

where G' and m' are the mass center and mass of system Σ'. Typically Σ' is chosen so as to make one or more particular forces *internal* to Σ'. Unit vector $\hat{\mathbf{u}}$ is chosen normal to particular forces *external* to Σ'. It may then happen that $\mathbf{F}_{ext\to\Sigma'} \cdot \hat{\mathbf{u}} = 0$, that is, the resultant external force acting on Σ' has a vanishing component along $\hat{\mathbf{u}}$. Then *if unit vector* $\hat{\mathbf{u}}$ *is fixed*, equation (4.44) necessarily leads to

$$m'\mathbf{v}_{G'} \cdot \hat{\mathbf{u}} = (m_1\mathbf{v}_{G_1} + m_2\mathbf{v}_{G_2} + \cdots) \cdot \hat{\mathbf{u}} = \text{Constant} \tag{4.45}$$

where bodies `1,2,...` belong to system Σ'. Equation (4.45) expresses the **conservation of the linear momentum of system Σ' along direction $\hat{\mathbf{u}}$.**

STRATEGY #2. Apply Euler's Second Principle to subset Σ'' of Σ about a point A to obtain

$$\mathbf{M}_{A,ext\to\Sigma''}\cdot\hat{\mathbf{k}} = (\frac{d}{dt}\mathbf{H}_{A,\Sigma''} + \mathbf{v}_A \times m''\mathbf{v}_{G''})\cdot\hat{\mathbf{k}} \qquad (4.46)$$

where $\hat{\mathbf{k}}$ is a unit vector normal to the plane of motion, m'' and G'' are the mass and mass center of Σ''. Again system Σ'' is chosen so as to make particular action (forces/couples) *internal* to Σ''. Point A is chosen so as to make the line of action of a particular *external* force pass through A. It may then happen that $\mathbf{M}_{A,ext\to\Sigma''}\cdot\hat{\mathbf{k}} = 0$, that is, the resultant external moment about A acting on Σ'' is vanishing along $\hat{\mathbf{k}}$. Then, *if point A is fixed*, equation (4.46) necessarily leads to

$$\mathbf{H}_{A,\Sigma''}\cdot\hat{\mathbf{k}} = \mathbf{H}_{A,1}\cdot\hat{\mathbf{k}} + \mathbf{H}_{A,2}\cdot\hat{\mathbf{k}} + \cdots = \text{Constant} \qquad (4.47)$$

where bodies 1, 2, ... belong to system Σ''. Equation (4.47) expresses the **conservation of the angular momentum of system Σ'' about axis Az.**

Remark 10. The assumption that A be fixed can sometimes be removed whenever A is such that the term $\mathbf{v}_A \times m''\mathbf{v}_{G''}$ vanishes: this requires that velocities $\mathbf{v}_A$ and $\mathbf{v}_{G''}$ be colinear.

We illustrate these strategies with the following problems.

EXAMPLE PROBLEMS

Exercise 4.14 ▪ A push-pull rattle toy consists of a forked handle 1 connected to a wheel assembly 2 of negligible mass. The handle is of mass m and mass center G located at a distance h from the wheels' axle. It is then released without initial speed from the position $\theta = 0$. Denote by J the moment of inertia of body 1 about Gz.
a. Determine the trajectory of mass center G.
b. Find the equation governing angle θ. Integrate this equation to find the angular speed $\dot{\theta}$ versus θ.
c. Find the reaction force $\mathbf{R}$ of the support on each wheel as a function of θ. Show that $\mathbf{R}$ remains in the same direction. Interpret this result physically. ▪

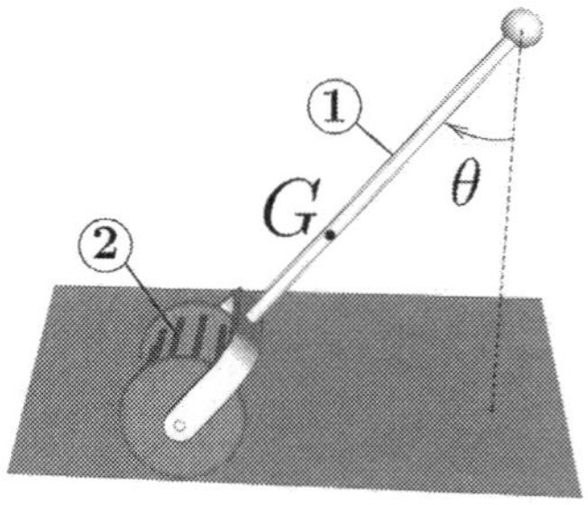

SOLUTION. **a.** As the handle is released from the vertical position, it rotates about axis Az while endpoint A translates along horizontal axis Ox: its configuration is defined by the angle θ which line AG makes with the vertical and by the coordinate x of A on axis Ox. The fixed unit vectors $(\hat{\imath}, \hat{\jmath}, \hat{\mathbf{k}})$ are defined in Figure 4.30 (left). The acceleration of mass center G is then found from that of A:

$$\mathbf{a}_G = \mathbf{a}_A - \ddot{\theta}\hat{\mathbf{k}} \times h(\sin\theta\hat{\imath} + \cos\theta\hat{\jmath}) - h\dot{\theta}^2(\sin\theta\hat{\imath} + \cos\theta\hat{\jmath})$$

$$= (\ddot{x} + h\ddot{\theta}\cos\theta - h\dot{\theta}^2\sin\theta)\hat{\imath} - h(\ddot{\theta}\sin\theta + \dot{\theta}^2\cos\theta)\hat{\jmath} \qquad ([1])$$

using $\boldsymbol{\omega}_1 = -\dot{\theta}\hat{\mathbf{k}}$ and $\boldsymbol{\alpha}_1 = -\ddot{\theta}\hat{\mathbf{k}}$.

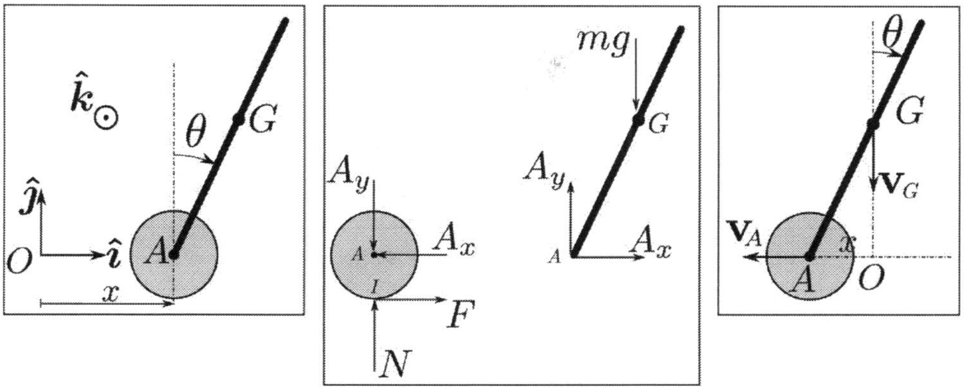

Figure 4.30 – *Problem 4.14 (Solution)*

Before applying Euler's principles we sketch the bodies' free-body diagrams in Figure 4.30 (center). Body 1 is subject to gravity at G and the reaction of body 2 at A. Body 2 is subject to the support's reaction at I and the reaction of body 1 at A. Since the motion is planar, all lines of action must lie in the plane of motion of G. For instance the reactions on the wheels are necessarily identical and add up to $F\hat{\imath} + N\hat{\jmath}$.
First we exploit the fact that body 2 is assumed massless to find 3 equations: Euler's First Principle $\mathbf{F}_{ext\to 2} = \mathbf{0}$ gives

$$F = A_x, \qquad N = A_y \qquad [2-3]$$

Euler's Second Principle $\mathbf{M}_{A,ext\to 2} = \mathbf{0}$ about point A gives

$$F = 0 \qquad [4]$$

Equation [4] shows that $A_x = 0$. Next we apply Euler's principles to body 1: the first principle $m\mathbf{a}_G = \mathbf{F}_{ext\to 1}$ gives on basis $(\hat{\imath}, \hat{\jmath})$

$$ma_{Gx} = A_x = 0, \qquad ma_{Gy} = A_y - mg \qquad [5-6]$$

the second principle $I_{Gzz}\boldsymbol{\alpha}_G = \mathbf{M}_{G,ext\to 1}$ gives on $\hat{\mathbf{k}}$:

$$J\ddot{\theta} = hN\sin\theta \qquad [7]$$

where we used $A_y = N$ from equation [3].
Equation [5] shows that $mv_{Gx} = Cst = 0$ (since the body is released without initial velocity). This shows that *there is conservation of the body's linear momentum along axis Ox. Point G moves along a vertical line.* See Figure 4.30 (right). We can now position the origin O so that line Oy represents the path of G, that is, $x_G = 0$ at all time. In the triangle OAG we easily obtain the relationship between x, y_G, h and angle θ:

$$x = -h\sin\theta, \qquad y_G = h\cos\theta \qquad [8-9]$$

b. To find the equation governing angle θ we combine equations [6] and [7] to eliminate N:

$$J\ddot{\theta} = h\sin\theta(ma_{Gy} + mg)$$

and use the component $a_{Gy} = -h(\ddot{\theta}\sin\theta + \dot{\theta}^2\cos\theta)$ from the expression of $\mathbf{a}_G$ of [1] to obtain

$$\boxed{(J + mh^2 \sin^2\theta)\ddot{\theta} + mh^2 \sin\theta\cos\theta = mgh\sin\theta} \qquad [10]$$

To integrate [10] we pre-multiply [10] by $\dot{\theta}$ and recognize that $\dot{\theta}\ddot{\theta}\sin^2\theta + \dot{\theta}\sin\theta\cos\theta = \frac{1}{2}\frac{d}{dt}(\dot{\theta}^2\sin^2\theta)$:

$$\frac{1}{2}J\dot{\theta}^2 + \frac{1}{2}mh^2\dot{\theta}^2\sin^2\theta + mgh\cos\theta = Cst = mgh$$

after using the condition $\dot{\theta} = 0$ when $\theta = 0$. This gives the angular speed $\dot{\theta}$ versus θ:

$$\boxed{\dot{\theta}^2 = 2mgh\frac{1-\cos\theta}{J + mh^2\sin^2\theta}} \qquad [11]$$

c. The reaction force $\mathbf{R} = \frac{N}{2}\hat{\jmath}$ of the support on each wheel is found from equation [7]. We need to find angular acceleration $\ddot{\theta}$ vs θ: we take the time-derivative of [11] and find

$$\ddot{\theta} = mgh\sin\theta\frac{J + mh^2(1-\cos\theta)^2}{(J + mh^2\sin^2\theta)^2}$$

This gives

$$\mathbf{R} = \frac{N}{2}\hat{\jmath} = \tfrac{1}{2}mg\frac{1+\kappa(1-\cos\theta)^2}{(1+\kappa\sin^2\theta)^2}\hat{\jmath}$$

with $\kappa = \frac{mh^2}{J}$. We can see that $N > 0$ for all values of θ: the reaction forces on the wheels remains strictly positive during the entire motion: the toy never loses contact with the horizontal support. ■

Exercise 4.15 ■■ A rigid rectangular frame ① of mass M and radius of gyration k_Δ can rotate about a vertical axis Δ. The pivots at A and B are frictionless. Two identical homogeneous cylinders ② and ③ of mass m and radius r are mounted on the vertical sides of the frame, about which they can rotate without friction. Their mass centers coincide with the centers of the bars.
Both cylinders are given the same initial angular velocity ω_0, the frame being initially at rest.

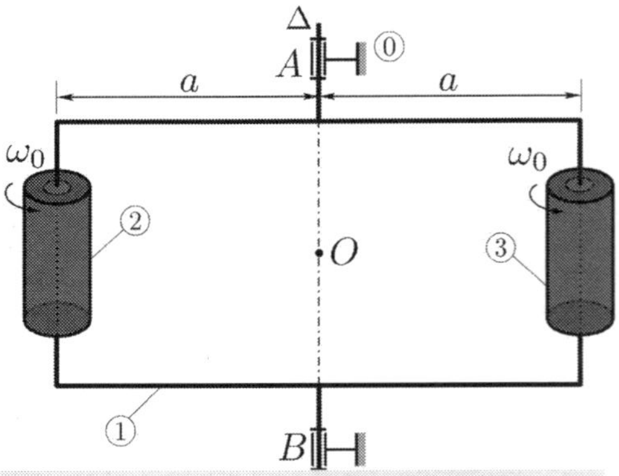

At a later time, a mechanism attached to the frame, of negligible mass, suddenly stops the rotation of cylinders ②.
Why is the frame observed to rotate? Find its angular velocity ω_1. ■

Solution.Step 1. Denote by Σ the system comprised of the frame and the two cylinders and by $(\hat{\mathbf{u}}, \hat{\mathbf{v}}, \hat{\mathbf{k}})$ a basis attached to the frame. See Figure 4.31. At any given time, the angular momentum of Σ about axis of rotation Δ (directed along unit vector $\hat{\mathbf{k}}$) is conserved since

$$\mathbf{M}_{O,ext\to\Sigma}\cdot\hat{\mathbf{k}} = \mathbf{M}_{O,joints\to 1}\cdot\hat{\mathbf{k}} + \mathbf{M}_{O,gravity\to\Sigma}\cdot\hat{\mathbf{k}} = 0$$

taking into account that the pivots at A and B are frictionless. Note that the braking mechanism is internal to Σ and that the corresponding forces play no role in Euler's second principle applied to Σ:

$$\mathbf{M}_{O,ext\to\Sigma}\cdot\hat{\mathbf{k}} = \frac{d}{dt}\mathbf{H}_{O,\Sigma}\cdot\hat{\mathbf{k}} = 0 \Longrightarrow \boxed{\mathbf{H}_{O,\Sigma}\cdot\hat{\mathbf{k}} = Cst} \quad [1]$$

STEP 2. We need to find the angular momentum $\mathbf{H}_{O,\Sigma}\cdot\hat{\mathbf{k}}$ of the system at time $t = 0$. At this instant, we have $\boldsymbol{\omega}_1 = \mathbf{0}$, $\boldsymbol{\omega}_2 = \boldsymbol{\omega}_3 = \omega_0\hat{\mathbf{k}}$.

$$\mathbf{H}_{O,1} = \mathbf{0}, \quad \mathbf{H}_{O,2} = \mathbf{H}_{O,3} = \frac{1}{2}mr^2\omega_0\hat{\mathbf{k}}$$

This gives

$$\boxed{(\mathbf{H}_{O,1} + \mathbf{H}_{O,2} + \mathbf{H}_{O,3})\cdot\hat{\mathbf{k}} = mr^2\omega_0}$$

STEP 3. Now we need to find the angular momentum $\mathbf{H}_{O,\Sigma}\cdot\hat{\mathbf{k}}$ of the system at the time t_1 when the rotation of body 2 is stopped. *Since the angular momentum of body 2 is changing, body 1 must rotate so as to conserve the value of the total angular momentum.*

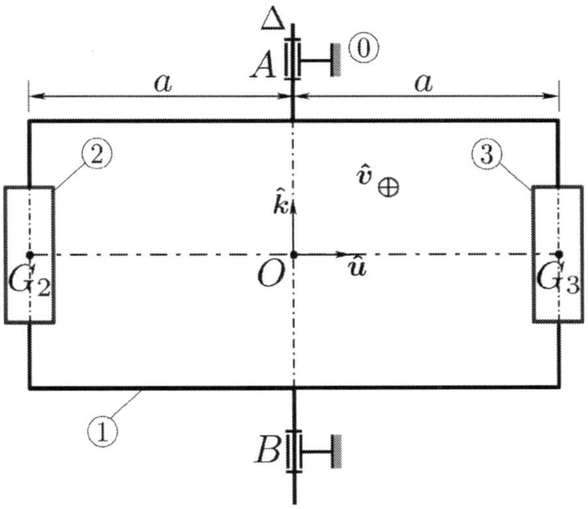

Figure 4.31 – *Problem 4.15 (Solution)*

Note that the angular momentum of body 3 about its axis is also conserved in the absence of friction on the axle of the cylinder:

$$\mathbf{H}_{G_3,3}\cdot\hat{\mathbf{k}} = Cst \Longrightarrow \boldsymbol{\omega}_3 = \omega_0\hat{\mathbf{k}}$$

This shows that the angular velocity of cylinder 3 stays constant. After cylinder 2 is stopped, let us denote $\boldsymbol{\omega}_1 = \omega_1\hat{\mathbf{k}} = \boldsymbol{\omega}_2$ the angular velocity of the frame and cylinder 2 (cylinder 2 is now rigidly attached to the frame!). Now we can find the angular momentum of the system at $t = t_1$:

$$\mathbf{H}_{O,1} = Mk_\Delta^2\omega_1\hat{\mathbf{k}}$$

$$\mathbf{H}_{O,2} = \underbrace{\mathbf{H}_{G_2,2}}_{\frac{m}{2}r^2\omega_1\hat{\mathbf{k}}} + \mathbf{r}_{OG_2}\times m\mathbf{v}_{G_2} = \frac{m}{2}r^2\omega_1\hat{\mathbf{k}} + (-a\hat{\mathbf{u}})\times(-ma\omega_1\hat{\mathbf{v}}) = \frac{m}{2}r^2\omega_1\hat{\mathbf{k}} + ma^2\omega_1\hat{\mathbf{k}}$$

$$\mathbf{H}_{O,3} = \underbrace{\mathbf{H}_{G_3,3}}_{\frac{m}{2}r^2\omega_0\hat{\mathbf{k}}} + \mathbf{r}_{OG_3} \times m\mathbf{v}_{G_3} = \frac{m}{2}r^2\omega_0\hat{\mathbf{k}} + a\hat{\mathbf{u}} \times (ma\omega_1\hat{\mathbf{v}}) = \frac{m}{2}r^2\omega_0\hat{\mathbf{k}} + ma^2\omega_1\hat{\mathbf{k}}$$

where we have use $\mathbf{v}_{G_2} = -ma\omega_1\hat{\mathbf{v}}$ and $\mathbf{v}_{G_3} = ma\omega_1\hat{\mathbf{v}}$ for the velocities of the mass centers G_2 and G_3.

Step 4. Now we can restate [1] as follows:

$$Mk_\Delta^2\omega_1 + \frac{1}{2}mr^2\omega_1 + \frac{1}{2}mr^2\omega_0 + 2ma^2\omega_1 = mr^2\omega_0$$

This equation gives the expression of ω_1

$$\boxed{\boldsymbol{\omega}_1 = \omega_1\hat{\mathbf{k}} = \frac{mr^2}{2Mk_\Delta^2 + 4ma^2 + mr^2}\omega_0\hat{\mathbf{k}}}$$

We note that when $M \gg m$, $\omega_1 \approx 0$ and when $M \ll m$, $\omega_1 \approx r^2\omega_0/(4a^2 + r^2)$. ■

Exercise 4.16 ■ A massless pulley 1 of radius R can rotate about a frictionless, horizontal axis Oz. A slender rod of mass m_1 and length $2l$ is rigidly attached to the axle of the pulley at O. The inextensible string wrapped around the pulley supports a body 2 of mass m_2. At time $t = 0$, the system is released from rest from the position $\theta = 0$. Find the ODE governing angle θ and discuss the nature of the motion. ■

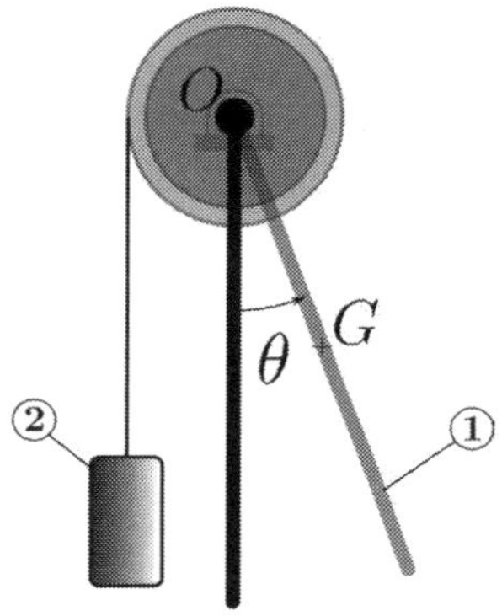

Solution. Since all is needed is the equation governing angle θ, we will apply Euler's Second Principle for the system Σ ={1,2} about point O. This eliminates the reaction of the support at O and all internal forces:

$$\frac{d}{dt}(\mathbf{H}_{O,1} + \mathbf{H}_{O,2}) = \mathbf{M}_{O,ext\to\Sigma} \qquad [1]$$

This will yield a single equation which should be sufficient since the system has one degree of freedom. Before we find the angular momentum of the system, we choose a set of fixed unit vectors $(\hat{\imath}, \hat{\jmath}, \hat{\mathbf{k}})$ and define Cartesian coordinate x in addition to θ to parametrize the system. See Figure 4.32 (left). These two coordinates are not independent: $\dot{x} = R\dot{\theta}$ (the string does not stretch). To find the angular momentum $\mathbf{H}_{O,1}$ we note that body 1 is in rotation about axis Oz at angular velocity $\boldsymbol{\omega}_1 = \dot{\theta}\hat{\mathbf{k}}$:

$$\mathbf{H}_{O,1} = I_{Ozz}\dot{\theta}\hat{\mathbf{k}} = \frac{4}{3}m_1l^2\dot{\theta}\hat{\mathbf{k}}$$

To find the angular momentum $\mathbf{H}_{O,2}$ of body 2 in rectilinear translation, we use

$$\mathbf{H}_{O,2} = \underbrace{\mathbf{H}_{G_2,2}}_{0} + \mathbf{r}_{OG_2} \times m_2\mathbf{v}_{G_2} = -R\hat{\jmath} \times m_2\dot{x}\hat{\imath} = m_2R\dot{x}_2\hat{\mathbf{k}}$$

A free-body diagram shows the forces acting on the system (see Figure 4.32, right):
-reaction $\mathbf{R}_O$ of the support which satisfies $\mathbf{M}_O \cdot \hat{\mathbf{k}} = 0$ since the pivot Oz is frictionless.

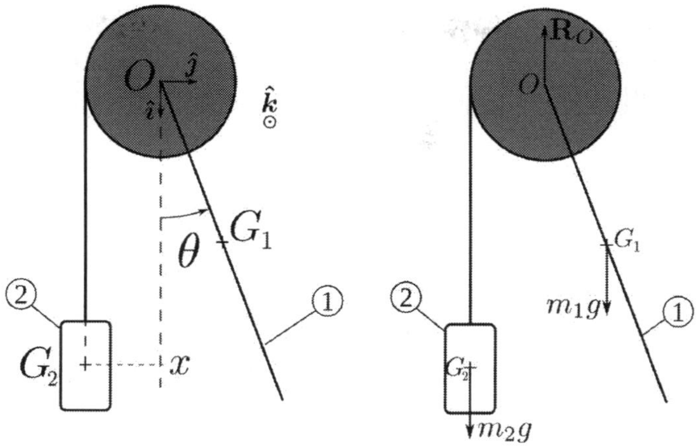

Figure 4.32 – *Problem 4.16 (Solution)*

-the effect of gravity on body 2 whose moment about O is $m_2 g R\hat{\mathbf{k}}$,
-the effect of gravity on body 1 whose moment about O is $-m_1 g l \sin\theta\hat{\mathbf{k}}$.
Equation [1] becomes

$$\boxed{(\frac{4}{3}m_1 l^2 + m_2 R^2)\ddot{\theta} = m_2 g R - m_1 g l \sin\theta} \qquad [2]$$

where we used the constraint equation $\dot{x} = R\dot{\theta}$. The motion of the system can be predicted on the basis of the right-hand-side of equation [2]: if $m_2 g R - m_1 g l \sin\theta$ remains positive for all values of θ, then $\ddot{\theta} > 0$ implies that the angular velocity increases: the pulley will rotate in the counterclockwise direction as long as there remains string wrapped around it. It is easy to see that this is guaranteed if the condition $\boxed{m_2 R > m_1 l}$ is satisfied. If $\boxed{m_2 R < m_1 l}$ is satisfied, then $\dot{\theta}$ changes sign periodically: integration of [2] gives

$$\tfrac{1}{2}\dot{\theta}^2 = g\frac{m_2 R\theta - m_1 l(1-\cos\theta)}{\frac{4}{3}m_1 l^2 + m_2 R^2}$$

after imposing $\dot{\theta} = 0$ at $\theta = 0$. Finally if $\boxed{m_2 R = m_1 l}$, then $\dot{\theta} = 0$ at all time: the system stays at rest. ■

Exercise 4.17 ■■ A circular horizontal turntable 1 of center O is free to rotate without friction about a vertical axis Oz. Denote by J its moment of inertia about Oz. A miniature circular railroad track $\mathcal{C}$, of center C and radius R, is attached to the turntable. A number n of small electric trains, each of mass m, can be set on the track. They can be turned on and off by remote control. The trains can travel at constant speed v on the track in the direction shown in Figure 4.33. Assume that $J \gg nmR^2$.

a. The center C does not coincide with center O, but O is inside the track.

1a. A single train P is set on $\mathcal{C}$. The system is initially at rest. P is then turned on. Describe the motion of the turntable.

2a. The experiment of **1a** is now repeated with n trains which cover the track entirely. Are there any qualitative changes?

b. The center C does not coincide with center O, but now O falls outside the track. Repeat experiments of **1a** and **2a**. What happens? Describe and explain the possible qualitative changes between the two cases. ■

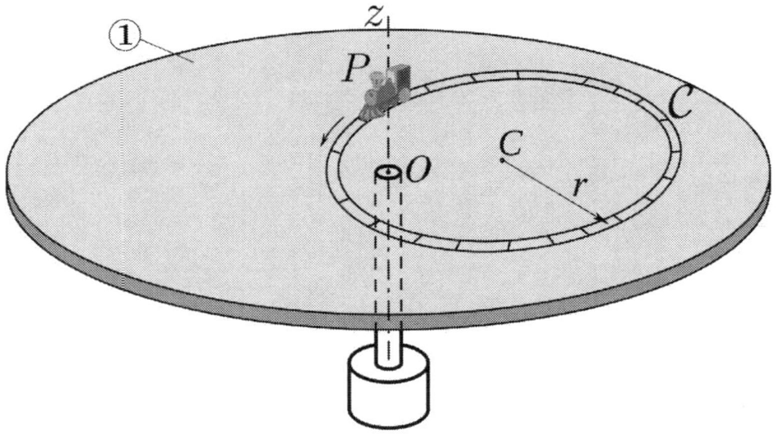

Figure 4.33 – *Problem 4.17*

SOLUTION. Intuitively we expect that the turntable 1 will rotate in response to the rotation of the trains. This can be predicted by realizing that the system's angular momentum about axis Oz is conserved. To predict the qualitative differences between the 4 cases we need to carefully determine the system's angular momentum.

We adopt a set of unit vectors $(\hat{\imath}, \hat{\jmath}, \hat{\mathbf{k}})$ attached to the turntable with $\mathbf{r}_{OC} = d\hat{\imath}$ and $\hat{\mathbf{k}}$ directed upward. Angle θ defines the position of P relative to 1. See Figure 4.34. We denote $\boldsymbol{\omega}_{1/0} = \Omega\hat{\mathbf{k}}$ the angular velocity of the turntable whose angular momentum about O satisfies:

$$\hat{\mathbf{k}} \cdot \mathbf{H}_{O,1} = J\Omega \qquad [1]$$

Then we determine the angular momentum of P about O:

$$\begin{aligned}\mathbf{H}_{O,P} &= \mathbf{r}_{OP} \times m(\mathbf{v}_{P/1} + \mathbf{v}_{P\in 1/0}) = m((d + R\cos\theta)\hat{\imath} + R\sin\theta\hat{\jmath}) \times v\hat{\mathbf{e}}_T + m\rho^2\Omega\hat{\mathbf{k}} \\ &= m(R + d\cos\theta)v\hat{\mathbf{k}} + m\rho^2\Omega\hat{\mathbf{k}} \qquad [2]\end{aligned}$$

with $\rho = |OP|$ and $d = |OC|$.

The system Σ ={1,2} is subject to the following external actions:

-the contact forces acting on the joint 0-1 which satisfy $\hat{\mathbf{k}} \cdot \mathbf{M}_{O,0\to 1} = 0$ since the joint is frictionless,

-the forces of gravity whose moment about axis Oz vanishes: $\hat{\mathbf{k}} \cdot \mathbf{M}_{O,gravity\to\Sigma} = 0$.

Application of Euler's Second Principle to system Σ about axis Oz gives

$$\hat{\mathbf{k}} \cdot \frac{d}{dt}(\mathbf{H}_{O,1} + \mathbf{H}_{O,P}) = \hat{\mathbf{k}} \cdot \mathbf{M}_{O,ext\to\Sigma} = 0$$

which shows that the angular momentum $\hat{\mathbf{k}} \cdot (\mathbf{H}_{O,1} + \mathbf{H}_{O,P})$ is conserved during the motion:

$$\hat{\mathbf{k}} \cdot (\mathbf{H}_{O,1} + \mathbf{H}_{O,P}) = Cst = 0 \qquad [3]$$

since the system is initially at rest. With the expressions [1-2] equation [3] becomes

$$(J + m\rho^2)\Omega = -m(R + d\cos\theta)v$$

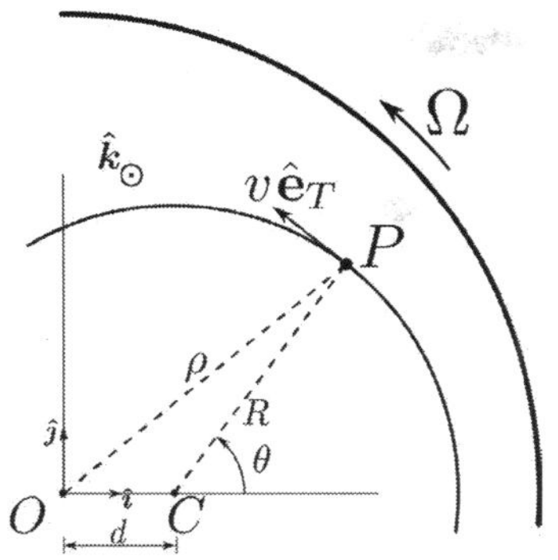

Figure 4.34 – *Problem 4.17*

Since $J \gg mR^2$, we can neglect $m\rho^2$ relative to J: this is equivalent to approximating the absolute velocity of P by its relative velocity. The motion of the turntable is slow compared to the motion of P. Finally we obtain

$$\boxed{\Omega = -\frac{mv}{J}(R + d\cos\theta)} \qquad [4]$$

We can now discuss the rotation of the turntable on the basis of result of equation [4]:
CASE 1A: if the center O is inside the track, then $0 < d < R$ and $R + d\cos\theta > 0$ for all angles $0 \leq \theta \leq 2\pi$. From [4] we conclude that *the turntable rotates continuously in the clockwise direction in a non-uniform, oscillating manner.*
CASE 2A: if the center O falls outside the track, then $d > R$ and the quantity $R + d\cos\theta$ changes sign twice per cycle of the train: we conclude that *the turntable rotates alternatively in the clockwise and anticlockwise directions, its sense of rotation changing twice per revolution of the train*. Globally, the turntable turns by the same amount $\Delta\Phi$

$$\Delta\phi = -\frac{mR^2}{J}\int_0^{2\pi}(1 + \frac{d}{R}\cos\theta)\dot{\theta} = -2\pi\frac{mR^2}{J}$$

per revolution of the train as in Case 1a.
If the track is entirely covered with n trains, we can assume that the mass is uniformly distributed over the entire track and the angular momentum of the moving trains about C is given by $\mathbf{H}_{C,trains} = nmRv\hat{\mathbf{k}}$ leading to

$$\mathbf{H}_{O,trains} = nmRv\hat{\mathbf{k}} + \mathbf{r}_{OC} \times nm\mathbf{v}_C = nm(Rv + d^2\Omega)\hat{\mathbf{k}}$$

Now equation [3] gives

$$\boxed{\Omega = -\frac{nmRv}{J}} \qquad [5]$$

after taking account the condition $J \gg nmR^2$. The *turntable turns uniformly in the clockwise direction whether O is inside the track or not* (Cases 1b and 2b). ■

Exercise 4.18 ■■ Figure 4.35 shows a front-wheel-drive automobile in planar motion along a straight horizontal direction. The mass and inertia of the wheels are neglected. The total mass of the vehicle is $M = 2100$ kg, its mass center G is located at height $h = 0.7$ m from the ground. The position of the wheels relative to G is defined by the lengths $l_1 = 1.5$ m and $l_2 = 1.9$ m. The vehicle is subject to the aerodynamic resistance $F_D = \frac{1}{2}\rho C_D S v^2$ acting through mass center G, where $\rho = 1.3$ kg/m^2 the air mass density, $C_D = 0.45$ is the drag coefficient, $S = 3.25$ m^2 is the cross-sectional area and v is the speed of the vehicle. The action of the road on the wheels is modeled by a standard dry Coulomb friction model with the coefficient of static friction $\mu_s = 0.9$. The connections between the wheels and the chassis are assumed to be frictionless pivots. Denote by Σ the entire vehicle comprised of the body 1 (including the engine) and the rear 2 and front 3 wheel assemblies. The wheels of radius r are assumed to roll without slipping relative to the road.

a. Apply Euler's Second Principle to each rear wheel to show that the friction forces acting on the rear wheels vanish. Explain why this is not the case for the friction forces acting on the front wheels.

b. Apply Euler's First Principle to Σ to relate the friction forces acting on the front wheels to the acceleration of G.

c. Apply Euler's Second Principle to Σ to relate the normal reaction forces acting on the front wheels to a_G, M, l_1, l_2, and F_D.

d. Using the previous results, find the maximum traction the car can possibly have.

e. Assume that the front wheel assembly is subject to a couple C from the engine. Find the equation which governs the speed v of the car. Assuming C constant, integrate this equation in terms of $u = v/v_{max}$ where v_{max} is the vehicle maximum speed (introduce the constant $1/\kappa = Mrv_{max}/C$). Find v_{max} for $C = 600$ N.m and $r = 0.37$ m. Finally find the maximum couple C_{max} to prevent the wheels from slipping. ■

Figure 4.35 – *Problem 4.18*

SOLUTION. **a.** The rear wheels 2 are each subject to (see Figure 4.36, left)
- the reaction force $\mathbf{R}_2 = N_2\hat{\jmath} + F_2\hat{\imath}$ of the road,
- the forces/moments due to the pivot connection with body 1 which satisfy $\mathbf{M}_{A,1\to 2} \cdot \hat{\mathbf{k}} = 0$ since the joint is assumed frictionless.

We apply Euler's Second Principle about A (which can be assumed to be the mass center) projected on $\hat{\mathbf{k}}$:

$$\hat{\mathbf{k}} \cdot \frac{d}{dt}\mathbf{H}_{A,2} = \hat{\mathbf{k}} \cdot \mathbf{M}_{A,ext\to 2} = r_2 F_2$$

Since the wheel's inertia is neglected we have $\mathbf{H}_{A,2} = \mathbf{0}$. This implies that $F_2 = 0$: the friction forces acting on the rear wheels vanish.
The friction forces F_3 acting on the front wheels cannot possibly be zero since there are

acted upon by the driving torque of the engine! ($F_3 = 0$ would not make the traction of the vehicle relative to the road possible).

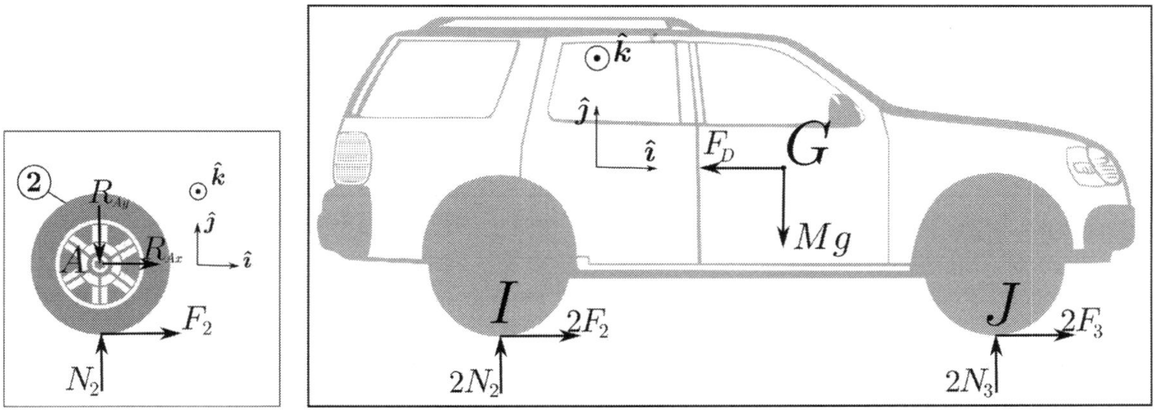

Figure 4.36 – *Problem 4.18 (Solution)*

b. The vehicle Σ is subject to the following *external forces* (see Figure 4.36, right):
- the total gravitational force $-Mg\hat{\jmath}$ acting through mass center G,
- the reaction forces $\mathbf{R}_2 = N_2\hat{\jmath}$ of the road on the rear wheels 2 taking into account that $F_2 = 0$,
- the reaction forces $\mathbf{R}_3 = N_3\hat{\jmath} + F_3\hat{\imath}$ of the road on the front wheels 3,
- the drag force $\mathbf{F}_D = -\frac{1}{2}\rho C_D S v^2\hat{\imath}$ acting through mass center G.

We apply Euler's First Principle $M\mathbf{a}_G = \mathbf{F}_{ext\to\Sigma}$ to system Σ to find two equations on $\hat{\imath}, \hat{\jmath}$:

$$Ma_G = 2F_3 - F_D, \qquad 0 = 2N_2 + 2N_3 - Mg \qquad [1-2]$$

taking into account that mass center G has a rectilinear motion along direction $\hat{\imath}$. Note that all internal actions exerted on Σ such as the torque exerted by the engine on the front wheels do not come into play when we apply Euler's First and Second Principle to system Σ.

c. To eliminate the contributions of the normal reactions N_2 on the rear wheel, we need to apply Euler's Second Principle about point I (here I is the projection of the rear wheel contact points I_l and I_r into the plane of motion of G):

$$\mathbf{M}_{I,road\to 2} = \mathbf{r}_{II_l} \times N_2\hat{\jmath} + \mathbf{r}_{II_r} \times N_2\hat{\jmath} = \mathbf{0}$$

We are left with i) the moments about I of the reaction forces on the front wheels:

$$\mathbf{M}_{I,road\to 3} = \mathbf{r}_{IJ_l} \times (N_3\hat{\jmath} + F_3\hat{\imath}) + \mathbf{r}_{IJ_r} \times (N_3\hat{\jmath} + F_3\hat{\imath}) = 2(l_1 + l_2)\hat{\imath} \times (N_3\hat{\jmath} + F_3\hat{\imath}) = 2(l_1 + l_2)N_3\hat{\mathbf{k}}$$

ii) the moment about I of $(-Mg\hat{\jmath} - F_D\hat{\imath})$:

$$\mathbf{M}_{I,ext\to 1} = \mathbf{r}_{IG} \times (-Mg\hat{\jmath} - F_D\hat{\imath}) = (-l_2 Mg + hF_D)\hat{\mathbf{k}}$$

Recall that Euler's Second Principle about point I takes the form

$$\mathbf{M}_{I,ext\to\Sigma} = \frac{d}{dt}\mathbf{H}_{G,1} + \mathbf{r}_{IG} \times M\mathbf{a}_G = -Mha_G\hat{\mathbf{k}}$$

since $\mathbf{H}_{G,1} = \mathbf{0}$ for a body in translation. Finally we obtain

$$\boxed{2(l_1 + l_2)N_3 - l_2 Mg + hF_D = -Mha_G} \qquad [3]$$

d. Since the wheels are assumed to roll without slipping, the friction forces acting on the front wheels must satisfy $|F_3| < \mu_s N_3$. The maximum value reached by F_3 is therefore $\mu_s N_3$. Using equations [1] and [3] we can find a relationship between F_3 and N_3 by eliminating $Ma_G + F_D$:

$$Ma_G + F_D = 2F_3$$

giving

$$2(l_1 + l_2)N_3 - l_2 Mg = 2F_3$$

Now with $F_3 = \mu_s N_3$ we obtain N_3:

$$N_3 = \frac{Mgl_2}{2(l_1 + l_2) + 2\mu_s h}$$

Finally the maximum possible traction is given by

$$\boxed{2F_{3,max} = \frac{\mu_s Mgl_2}{(l_1 + l_2) + \mu_s h}} \qquad [4]$$

Numerically, we find $2F_{3,max} = \mathbf{8,741}$ N.

e. Consider the front wheel assembly 3 consisting of the two front wheels and the drive shaft: in addition to the reactions $\mathbf{R}_B$ of the body on the wheels and the reactions $F_3\hat{\imath} + N_3\hat{\jmath}$ of the ground on the tires, the shaft is subject to a couple $-C\hat{\mathbf{k}}$. See Figure 4.37. Application of Euler's Second Principle about the center B of the front wheel assembly gives the equation

$$\mathbf{M}_{B,ext\to 3} \cdot \hat{\mathbf{k}} = 2rF_3 - C = 0$$

Now equation [1] becomes

$$\boxed{M\frac{dv}{dt} = \frac{C}{r} - \tfrac{1}{2}\rho C_D S v^2} \qquad [5]$$

The maximum speed is reached for $dv/dt = 0$: using [5] we find

$$\boxed{v_{max} = \sqrt{\frac{2C}{\rho C_D S r}}}$$

Now introduce $u = v/v_{max}$: the equation governed by u is

$$\frac{Mrv_{max}}{C}\frac{du}{dt} = 1 - u^2$$

With constant $1/\kappa = Mrv_{max}/C$. The unknown u is solution of $du/dt = \kappa(1-u^2)$ which is easily integrated by separation of variables $du/(1-u^2) = \kappa dt$ to give

$$\boxed{u = \frac{v}{v_{max}} = \tanh(\kappa t)}$$

For $C = 600$ N.m, $r = 0.37$ m, $\rho = 1.3$ kg/m^2, $C_D = 0.45$, $S = 3.25$ m^2, we find $v_{max} = \mathbf{150}$ km/h. The maximum couple is found by using the maximum traction $2F_{3,max}$: $C_{max} = 2rF_{3,max} = \mathbf{3200}$ N.m.

■

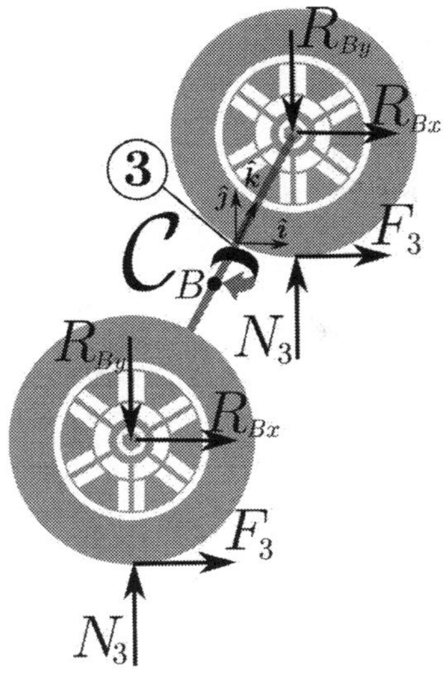

Figure 4.37 – *Problem 4.18 (Solution)*

Exercise 4.19 ■■■ A crate of mass M is pulled by an operator at constant speed $v_0\hat{\imath}$ with a horizontal force $F\hat{\imath}$ along a horizontal roller conveyor. The rollers are cylinders of mass m and radius r. The axes of two consecutive rollers are separated by a distance $d > 2r$. The rollers are at rest before coming into contact with the crate. They can rotate about their axis without friction. The coefficient of kinetic friction between the roller and the crate is μ.

a. Find the time τ taken for the rollers to stop slipping after coming into contact with the crate. Express τ in terms of v_0, μ, d, r and the ratio m/M. Then find the maximum value v_{max} of v_0 below which at most one roller slips at any given time. *Hint: guarantee that the time interval taken by the crate to come into contact with two adjacent rollers is larger than the time taken for any given roller to stop slipping.*

b. Assume that $v_0 < v_{max}$. Find the force F that the operator must exert on average to move the crate at speed v_0. Compare with the force F_1 which would be needed to pull the crate on a surface with the friction coefficient μ. Find the expression of ratio F_1/F. Find the numerical values of v_{max} and F_1/F for $\mu = 0.3$, $l = 1$ m, $r = 25$ mm, $d = 55$ mm, $M = 150$ kg, $m = 5$ kg.

c. Discuss the case $v_0 > v_{max}$. ■

SOLUTION. **a.** First, we choose the set $(\hat{\imath}, \hat{\jmath}, \hat{\mathbf{k}})$ of fixed unit vectors shown in Figure 4.39 (left) with $\hat{\jmath}$ perpendicular to the plane of motion. Consider one of the rollers which comes into contact with the crate at time $t = 0$. Denote by C its mass center and by I the contact point with the crate located in the plane of motion of G. The roller is in rotation about C with angular velocity $\omega\hat{\jmath}$. We can find the slip velocity at I as the vector given by

$$\mathbf{v}_{I\in roller/crate} = \mathbf{v}_{I\in roller/0} - \mathbf{v}_{I\in crate/0} = \mathbf{v}_C + \omega\hat{\jmath} \times r\hat{\mathbf{k}} - v_0\hat{\imath} = (r\omega - v_0)\hat{\imath}$$

Since $\omega = 0$ at $t = 0$ (the roller is at rest for $t < 0$), the slip velocity is directed in the $(-\hat{\imath})$ direction: this implies that the friction force exerted by the crate is directed in the $(+\hat{\imath})$ direction since it must be opposed to the slip velocity. To find the time τ when the slip velocity $(r\omega - v_0)\hat{\imath}$ becomes zero, we need to find ω vs t.
We can now sketch a free-body diagram for the roller in Figure 4.39 (left): it is subject

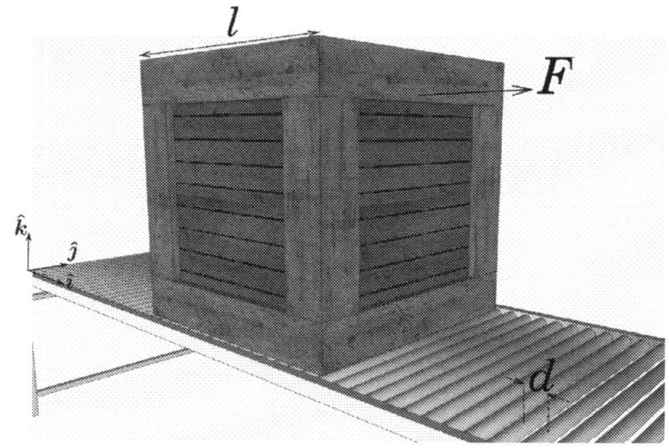

Figure 4.38 – *Problem 4.19*

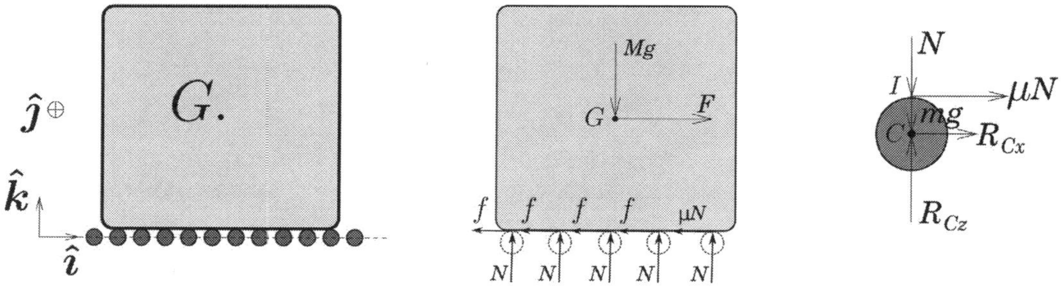

Figure 4.39 – *Problem 4.19 (Solution)*

to the gravitational force $mg(-\hat{\mathbf{k}})$, to the reaction $R_{Cx}\hat{\boldsymbol{\imath}} + R_{Cz}\hat{\mathbf{k}}$ at C, and to the reaction force $\mu N\hat{\boldsymbol{\imath}} - N\hat{\mathbf{k}}$ of the crate at I. The friction force is written $\mu N\hat{\boldsymbol{\imath}}$ since the roller is assumed in the slipping phase of its motion. Then, to predict the rotational motion of the roller, we apply Euler's Second Principle about mass center C to find

$$\tfrac{1}{2}mr^2\dot{\omega} = \mu rN \qquad [1]$$

where the moment of inertia $I_{Czz} = \frac{1}{2}mr^2$ is used. Equation [1] shows that the normal force N is needed to find $\omega(t)$. To find N, we apply Euler's First Principle $M\mathbf{a}_G = \mathbf{F}_{ext\to crate}$ to the crate (of mass M, mass center G) along the $\hat{\mathbf{k}}$-direction: after sketching a free-body diagram for the crate in Figure 4.39 (center), we find

$$nN - Mg = M\mathbf{a}_G \cdot \hat{\mathbf{k}} = 0 \qquad [2]$$

where n is the number of rollers in contact with the crate. We can find an approximation of n as

$$n = \frac{l}{d}$$

(for $d \ll l$ this is a good approximation of n). Equation [2] gives $N = dMg/l$. We can now integrate equation [1] to find ω versus t:

$$\omega = 2\mu\frac{M}{m}\frac{d}{lr}gt$$

imposing $\omega = 0$ at the instant $t = 0$ when the roller first comes in contact with the crate. Now we can set $r\omega - v_0 = 0$ to find time τ taken by the roller to stop slipping:

$$2\mu \frac{M}{m}\frac{d}{l} g\tau = v_0$$

or

$$\boxed{\tau = \frac{1}{2\mu}\frac{m}{M}\frac{l}{d}\frac{v_0}{g}}$$

If at most one roller must slip at any given time, we must guarantee that the time interval d/v_0 taken by the crate to come into contact with two adjacent rollers is larger than the time τ taken for any given roller to stop slipping: the former is the time taken to travel distance d at constant speed v_0. Therefore we must impose

$$\frac{d}{v_0} > \frac{1}{2\mu}\frac{m}{M}\frac{l}{d}\frac{v_0}{g}$$

This gives the maximum speed v_{max}:

$$\boxed{v_0 < v_{max} = d\sqrt{\frac{2\mu M g}{ml}}}$$

b. If $v_0 < v_{max}$, only one roller in contact with the crate is slipping and exerts a friction force $\mu N(-\hat{\imath})$ on the crate. The remaining $(n-1)$ rollers exert a friction force $(-f\hat{\imath})$. We complete the sketch of Figure 4.39 (center) which shows the free-body diagram for the crate.
To predict the applied horizontal force F, we apply Euler's First Principle for the crate $M\mathbf{a}_G = \mathbf{F}_{ext\to crate}$ in the $\hat{\imath}$-direction, and find

$$F - \mu N - (n-1)f = Ma_G = 0 \qquad [3]$$

since $v_G = v_0$ =constant. Recall that the normal force N is given by $N = dMg/l$. However the friction force f is still unknown. To find f, we examine a roller which does not slip: it rotates at constant angular velocity $\boldsymbol{\omega} = (v_0/r)\hat{\jmath}$ and its free-body diagram is obtained by replacing $\mu N\hat{\imath}$ by $f\hat{\imath}$. The application of Euler's Second Principle to such a roller about its mass center C gives

$$\tfrac{1}{2}mr^2\dot{\omega} = rf = 0$$

since $\omega = v_0/r = Cst$. This gives $f = 0$ and from [3]

$$\boxed{F = \mu M g \frac{d}{l}}$$

The force F_1 needed to pull the crate at constant speed on a surface of friction coefficient μ is given by

$$F_1 = \mu N_1 = \mu M g$$

This gives the ratio

$$\boxed{\frac{F_1}{F} = \frac{l}{d}}$$

For $\mu = 0.3$, $l = 1$ m, $r = 25$ mm, $d = 55$ mm, $M = 150$ kg, $m = 5$ kg, we find $v_{max} = \mathbf{0.73}$ m/s and $F_1/F = \mathbf{18}$. This last number shows the *tremendous advantage*

offered by roller conveyors to move heavy loads (in factories, distribution centers, airports, etc).
c. If $v_0 > v_{max}$, then more than one roller slips at any given time. If the condition $l/v_0 < \tau$, then all n rollers slip at any given time. This gives

$$\boxed{v_0 > v_{min} = \sqrt{\frac{2\mu Mgd}{m}}}$$

Equation [3] becomes:

$$F - n\mu N = 0$$

giving

$$\boxed{F = \mu Mg = F_1}$$

We see that in this case *the advantage of the rollers is completely lost.* ■

4.10 Dynamic Balancing of a Rigid Body in Rotation

Countless technological applications involve rigid bodies in rotational motion. The distribution of mass within a rigid body in rotation may be responsible for rapid wear and deterioration of the bearings which support the body. The object of this section is to understand this phenomenon and to show how to suppress it by a process called *dynamic balancing*.
Consider a rigid body 1 whose shaft is supported by a ball bearing attached to some Newtonian referential 0. Hence the joint between 0 and 1 is a pivot of axis $(O, \hat{\mathbf{z}}_0)$ assumed to be frictionless. See Figure 4.40. The body is assumed to be of mass m and mass center G whose location is given by $\mathbf{r}_{OG} = e\hat{\mathbf{x}}_1$ where e is a constant length and $\hat{\mathbf{x}}_1$ is a unit vector attached to the body and normal to $\hat{\mathbf{z}}_0$. Define $\hat{\mathbf{y}}_1 = \hat{\mathbf{z}}_0 \times \hat{\mathbf{x}}_1$ so that $(\hat{\mathbf{x}}_1, \hat{\mathbf{y}}_1, \hat{\mathbf{z}}_0)$ is a basis of unit vectors attached to body 1 whose orientation is defined by some oriented angle θ as shown. The plane $(O, \hat{\mathbf{x}}_1, \hat{\mathbf{y}}_1)$ is the plane of motion of G. Recall that the acceleration of G takes the expression

$$\mathbf{a}_G = e(\ddot{\theta}\hat{\mathbf{y}}_1 - \dot{\theta}^2\hat{\mathbf{x}}_1)$$

The body is also characterized by its moment of inertia I_{Ozz} and by its products of inertia I_{Oxz} and I_{Oyz} about the axes $(O, \hat{\mathbf{x}}_1, \hat{\mathbf{y}}_1, \hat{\mathbf{z}}_0)$ so that its angular momentum about O is given by (4.18)

$$\mathbf{H}_O = \dot{\theta}(I_{Oxz}\hat{\mathbf{x}}_1 + I_{Oyz}\hat{\mathbf{y}}_1 + I_{Ozz}\hat{\mathbf{z}}_0)$$

The mechanical actions exerted on the body are given by
(i) the gravitational force $-mg(\cos\theta\hat{\mathbf{x}}_1 + \sin\theta\hat{\mathbf{y}}_1)$ whose moment about G vanishes,
(ii) the contact forces exerted by the bearing on the body's shaft of resultant
$\mathbf{R} = R_x\hat{\mathbf{x}}_1 + R_y\hat{\mathbf{y}}_1 + R_z\hat{\mathbf{z}}_0$ and resultant moment $\mathbf{M}_O = M_{Ox}\hat{\mathbf{x}}_1 + M_{Oy}\hat{\mathbf{y}}_1$ (note that $M_{Oz} = 0$ since the joint is assumed frictionless),
(iii) the action of a motor equivalent to a couple $\mathcal{C}\hat{\mathbf{z}}_0$.
Other actions may be included such those due to aerodynamic and other frictional effects.

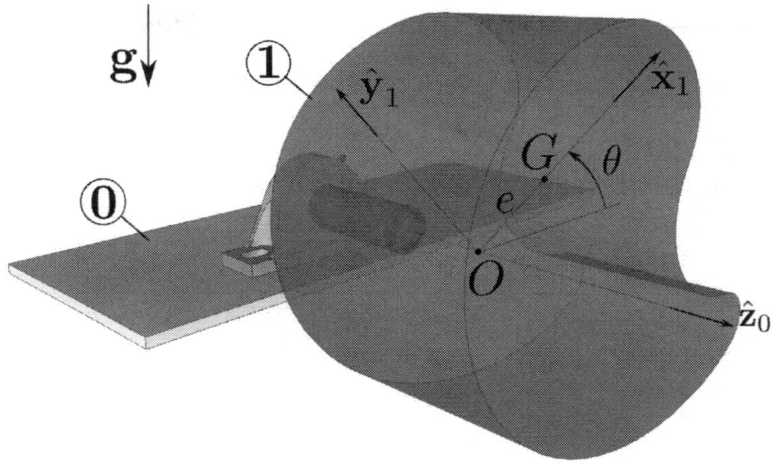

Figure 4.40 – *Rigid body 1 in rotational motion about axis Oz.*

Application of Euler's First and Second Principles can be stated as follows

$$me(\ddot{\theta}\hat{\mathbf{y}}_1 - \dot{\theta}^2\dot{x}_1) = -mg(\sin\theta\hat{\mathbf{x}}_1 + \cos\theta\hat{\mathbf{y}}_1) + R_x\hat{\mathbf{x}}_1 + R_y\hat{\mathbf{y}}_1 + R_z\hat{\mathbf{z}}_0$$

$$\frac{d}{dt}\dot{\theta}(I_{Oxz}\hat{\mathbf{x}}_1 + I_{Oyz}\hat{\mathbf{y}}_1 + I_{Ozz}\hat{\mathbf{z}}_0) = \mathbf{r}_{OG} \times m\mathbf{g} + M_{Ox}\hat{\mathbf{x}}_1 + M_{Oy}\hat{\mathbf{y}}_1 + C\hat{\mathbf{z}}_0$$

Upon projection on basis $(\hat{\mathbf{x}}_1, \hat{\mathbf{y}}_1, \hat{\mathbf{z}}_0)$ this leads to the following system of six equations

$$-me\dot{\theta}^2 = R_x - mg\sin\theta \tag{4.48}$$
$$me\ddot{\theta} = R_y - mg\cos\theta \tag{4.49}$$
$$0 = R_z \tag{4.50}$$

$$I_{Oxz}\ddot{\theta} - I_{Oyz}\dot{\theta}^2 = M_{Ox} \tag{4.51}$$
$$I_{Oxz}\dot{\theta}^2 + I_{Oyz}\ddot{\theta} = M_{Oy} \tag{4.52}$$
$$I_{Ozz}\ddot{\theta} = C - mg\sin\theta \tag{4.53}$$

These equations would allow the determination of angle $\theta(t)$ under the action of couple C and the determination of the unknown contact forces and moments $(R_x, R_y, R_z, M_{Ox}, M_{Oy})$. In particular, these equations show that the inertial terms involving $\dot{\theta}^2$ can have a significant effect on the forces and moments exerted on the bearing. For fast-rotating bodies, these inertial effects can lead to significant vibrations, noise, wear and possibly failure. It is important to minimize them. This can be achieved by *dynamic balancing*.

Definition (Static and Dynamic balancing)

A body in rotation at constant angular velocity about a fixed axis is said to be *balanced* if the bearings which support its shaft are not subject to inertial effects. The body is said to be *statically balanced* if the effects of inertial forces vanish. The body is said to be *dynamically balanced* if it is statically balanced and the effects of inertial moments vanish.

We can immediately make from equations (4.48-4.53) the following statement:

Static and Dynamic balancing

A body in rotation about an axis Oz is statically balanced when its mass center is located on its axis of rotation. Additionally, if its products of inertia I_{Oxz} and I_{Oyz} vanish, then the body is dynamically balanced:

$$e = 0, \qquad I_{Oxz} = I_{Oyz} = 0 \tag{4.54}$$

Remark 11. The condition $I_{Oxz} = I_{Oyz} = 0$ guarantees that the axis of rotation Oz is a principal axis of inertia. The Parallel Axis Theorem shows that if the body is dynamically balanced then $I_{Axz} = I_{Ayz} = 0$ for any point A of its axis of rotation.

In practice, dynamic balancing of a rigid body is achieved by altering its mass distribution, typically by adding or removing a pair of weights of mass m_1 and m_2 at particular locations P_1 and P_2 fixed relative to the body: the body is then replaced by the rigid system $\Sigma \equiv \{1, (P_1, m_1), (P_2, m_2)\}$. The masses m_1 and m_2 and the locations of P_1 and P_2 are determined by imposing conditions (4.54) for system Σ. Denote by (x_1, y_1, z_1) and (x_2, y_2, z_2) the coordinates of P_1 and P_2 on the moving axes $(O, \hat{\mathbf{x}}_1, \hat{\mathbf{y}}_1, \hat{\mathbf{z}}_0)$, that is,

$$\mathbf{r}_{OP_1} = x_1\hat{\mathbf{x}}_1 + y_1\hat{\mathbf{y}}_1 + z_1\hat{\mathbf{z}}_0, \qquad \mathbf{r}_{OP_2} = x_2\hat{\mathbf{x}}_1 + y_2\hat{\mathbf{y}}_1 + z_2\hat{\mathbf{z}}_0$$

The system's mass center and products of inertia are given by

$$(m + m_1 + m_2)\mathbf{r}_{OG_\Sigma} = m\mathbf{r}_{OG} + m_1\mathbf{r}_{OP_1} + m_2\mathbf{r}_{OP_2}$$

$$I_{Oxz,\Sigma} = I_{Oxz} - m_1x_1z_1 - m_2x_2z_2$$

$$I_{Oyz,\Sigma} = I_{Oyz} - m_1y_1z_1 - m_2y_2z_2$$

Dynamic Balancing by Two Point Masses

The conditions of dynamic balancing by two point masses P_1 and P_2 of mass m_1 and m_2 and coordinates (x_1, y_1, z_1) and (x_2, y_2, z_2), respectively, are satisfied by the equations

$$\begin{aligned} & me + m_1x_1 + m_2x_2 = 0 \\ & m_1y_1 + m_2y_2 = 0 \\ & I_{Oxz} - m_1x_1z_1 - m_2x_2z_2 = 0 \\ & I_{Oyz} - m_1y_1z_1 - m_2y_2z_2 = 0 \end{aligned} \tag{4.55}$$

Remark 12. This gives 4 equations to solve for 8 unknowns $(m_1, m_2, x_1, y_1, z_1, x_2, y_2, z_2)$: infinitely many solutions are possible.

Remark 13. Note that the addition (or removal) of a single mass does not allow dynamic balancing: setting $m_2 = 0$ in (4.55) leads to $y_1 = 0$ and $I_{Oxz} = 0$ which is not possible.

To obtain a unique solution of system of equations (4.55), constraints can be placed upon the locations P_1 and P_2 of the weights, as would be the case for the dynamic balancing of car tire/wheel assemblies. For instance, both weights are placed at the same (prescribed)

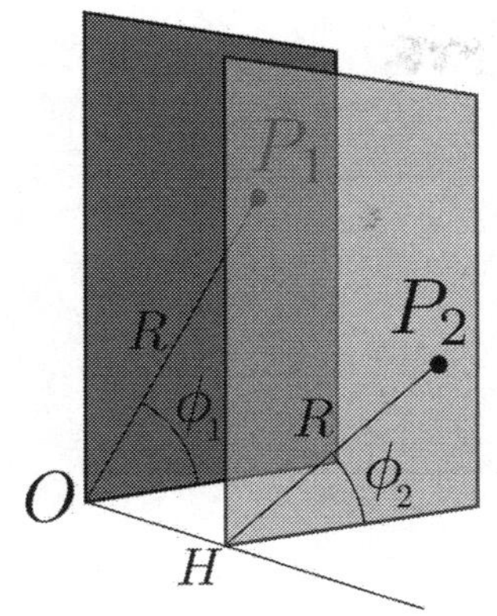

Figure 4.41 – *Dynamic balancing with weights at P_1 and P_2 located in two parallel planes $z = 0$ and $z = H$.*

distance R from the axis and in two parallel planes $z = 0$ and $z = H$. See Figure 4.41. Introducing the polar angles ϕ_1 and ϕ_2, the coordinates of P_1 and P_2 are now

$$\mathbf{r}_{OP_1} = R(\cos\phi_1 \hat{\mathbf{x}}_1 + \sin\phi_1 \hat{\mathbf{y}}_1), \qquad \mathbf{r}_{OP_2} = R(\cos\phi_2 \hat{\mathbf{x}}_1 + \sin\phi_2 \hat{\mathbf{y}}_1) + H\hat{\mathbf{z}}_0$$

Then the four unknowns $(m_1, m_2, \phi_1, \phi_2)$ are found by solving the four equations:

$$\begin{aligned}
&me + m_1 R\cos\phi_1 + m_2 R\cos\phi_2 = 0\\
&m_1 \sin\phi_1 + m_2 \sin\phi_2 = 0\\
&I_{Oxz} - m_2 HR\cos\phi_2 = 0\\
&I_{Oyz} - m_2 HR\sin\phi_2 = 0
\end{aligned}$$

A unique solution is then found.

PROJECT 7: TIMOSHENKO'S EXPERIMENT

A long plank 1 of mass m, mass center G and of negligible thickness is placed without initial speed on top of two counter-rotating cylindrical rollers 2 and 3 of radius R centered at A and B. The rollers are driven at the same constant angular velocity ω, and their axes are parallel and located in the same horizontal plane. The plank is observed to oscillate along the horizontal direction without losing contact with the rollers. The position of G is defined by coordinate $x(t)$: assume that, at $t = 0$, $x(0) = x_0$ with $0 < x_0 < L$. Denote by μ the coefficient of kinetic friction between the rollers and the plank, and by L the distance $|OA| = |OB|$ (O is the midpoint of line segment AB).

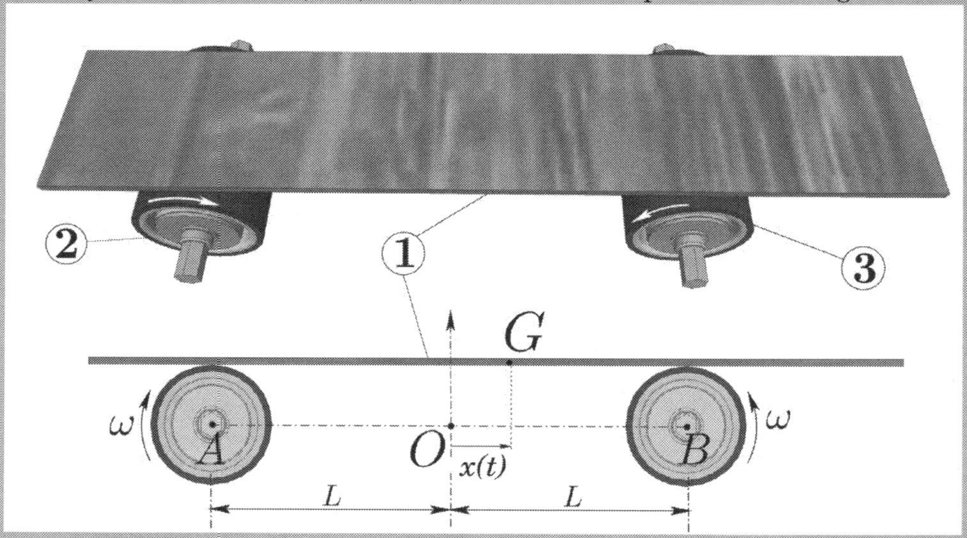

1. At $t = 0$, the plank necessarily slips relative to the rollers. Find the slip velocities of the plank relative to the rollers in terms of $x(t)$ and ω. Then sketch a free-body diagram for the plank.
2. Apply Euler's principles to find the equation of motion of the plank and the reaction forces of the rollers on the plank.
3. Show that the plank oscillates and find the period T of these oscillations (*seek a solution in the form* $x(t) = x_0 \cos \Omega t$). Show that the amplitude x_0 must satisfy a condition for these oscillations to occur. Describe the motion of the plank if this condition is not satisfied.
4. Find the friction coefficient μ corresponding to a period of oscillation $T = 2$ s, $L = R = 0.1$ m, $\omega = 0.5$ rev/s, and $x_0 = 0.05$ m.

PROJECT 8: MURPHY'S LAW

In this project, we attempt to prove that Murphy's Law is true: if toast can land butter-side down, it will do. The toast is modeled as a rectangular parallelepiped of width $2b$ and thickness $2e$, uniformly distributed mass m and mass center G. We assume that the mass center's position is given by its distance d from the corner point O of a table. Point O is at a distance h from the floor. We assume that the first phase of motion is a rotation about O due the frictional effects at point O. The orientation of the toast is defined by angle θ as shown. Assume that at $t = 0$, the toast is released from rest, corresponding to $\theta = 0$ and $\dot{\theta} = 0$.

This project is comprised of two parts:

Part I: a theoretical model and numerical prediction.

Part II: an experimental verification of the model.

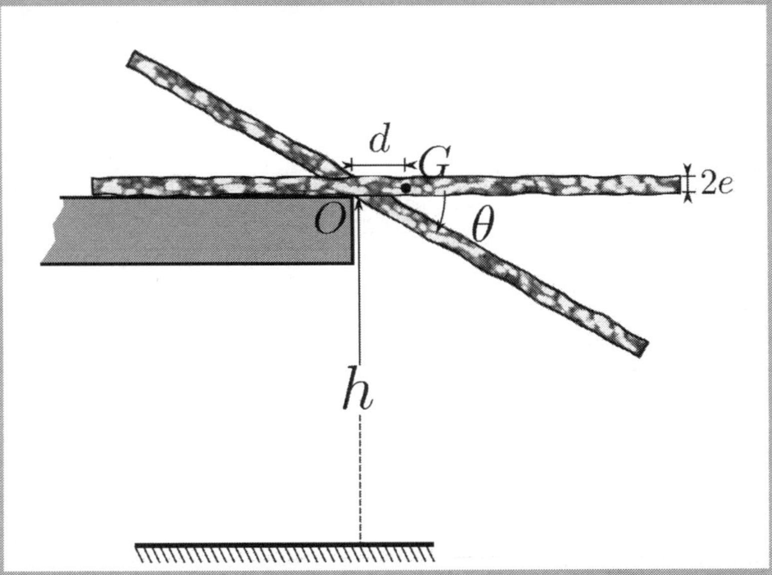

Part I: Theoretical Model & Numerical Prediction

I.1 Find $\ddot{\theta}$ and $\dot{\theta}$ as a function of θ and parameters b, e, and d. Then find the reaction force of the table on the toast as a function of angle θ.

I.2 In practice, the fall of a piece of toast is due to a slight imbalance: the coefficient d/b will not exceed 0.05. Hence, to simplify the analysis, we can assume $d = 0$. We can also assume $e \ll b$. Show that the angular velocity of the toast is given by

$$\dot{\theta}^2 = \frac{6ge}{b^2}(1 - \cos\theta)$$

Find the normal and tangential components (N, F) of the reaction force and show that the normal reaction satisfies $N < 0$ for $\theta = \pi/2$. Sketch the shape of N and F vs θ on the graph. Conclude that the toast must necessarily slip for some angle $\theta_1 < \pi/2$. Find

the expression of θ_1 in terms of the coefficient of static friction μ.
I.3 The slip of the toast relative to the corner point O is very brief for $\theta > \theta_1$: we assume that as soon as θ exceeds the value θ_1, the toast is in free fall. Find the trajectory of G and the expression of angle θ as a function of time (reset $t = 0$ when $\theta = \theta_1$).
I.4 Assuming that the toast does not turn more that one revolution, the toast will hit the ground with the buttered side up if angle θ satisfies $\theta \leq \pi/2$ or $\theta \geq 3\pi/2$ at the time $t = t_2$ taken to reach the ground. Justify this property by drawing cartoons of the toast during its free flight. Give an approximation of t_2 by assuming that $h \gg b$. Then find the value of $\theta_2 = \theta(t_2)$. For $\mu = 1$, $b = 4$ cm, $e = 4$ mm and $h = 0.75$ m, find angle θ_2. Conclude: does the toast land with the buttered side up or down? Describe the effect of each parameter (μ, b, e, h) on the outcome. For the given height $h = 0.75$ m, find out if there exist values of the parameters (μ, b, e) which guarantee that the toast lands with the buttered side up. Finally, **give a critical assessment of the proposed model.**

Part II: Experimental Work

In this part, you are expected to work freely, independently (within you group), and critically. Your ingenuity will be rewarded.
II.1 Decide on your materials: What kind of bread (e.g., white, wheat, matzo, bagel)? What kind of butter (note that it can be any kind of spread, e.g., vegan butter, cream cheese)? What kind of platform (e.g., cardboard, plywood, plastic)? You can be as creative as you want here. Once you have decided on your materials, remember to be consistent.
II.2 Conduct preliminary experiments with typical pieces of toast while varying h and observe the landings. Determine the parameters h, e, b, μ you will use for your specific experiment.
II.3 Document your chosen materials and experimental procedure in detail as if you were writing the Methods section of an academic journal article or lab report.
II.4 Compare your experimental results with your theoretical predictions. Use tables or graphs as necessary. If the experiments are not in agreement or/and the outcomes highly variable, offer some explanations: is the model acceptable? Can the model be improved? Are they uncertainties in the experiments?

PROJECT 9: "LIFE ON THE LINE"

A spectacular video entitled "Freefall Without Secured Rope" [2] has recently attracted our attention. In a carefully planned stunt to test the laws of physics, Norwegian physicist and television host Andreas Wahl is seen hanging 15 meters above the ground, attached by his harness to a rope which is allowed to slide over a horizontal, cantilevered pole. The other strand of the rope is attached horizontally to a secured hook, while supporting a weight of only a few kilograms. See Figure 4.42. Upon release of the rope from its attachment point, Wahl is seen to drop while the weight winds around the horizontal pole. After a few turns, the rope quickly tightens around the pole, decelerating the drop and leading Wahl to a complete stop, safely above the ground. The video is well worth watching, to get a sense of the experiment and to appreciate the risks involved (no safety net can be seen).

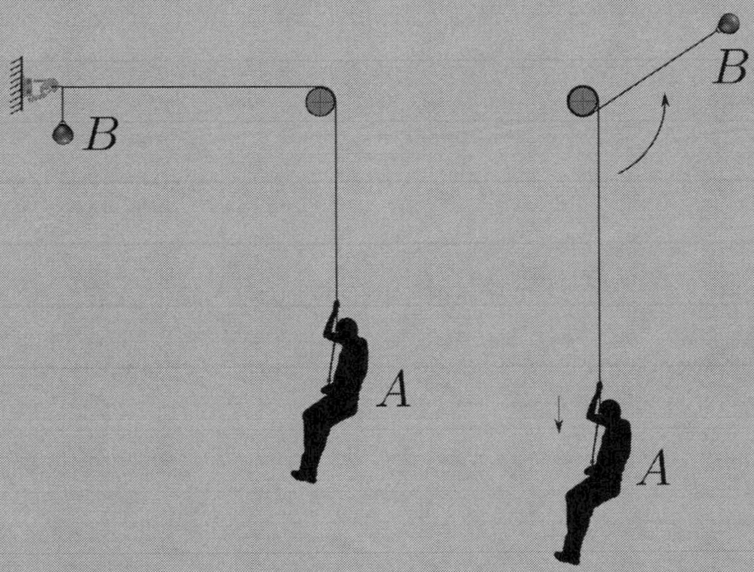

Figure 4.42 – *Freefall Without Secured Rope.*

In the first part of this problem, the goal is to propose mathematical models to predict and explain how and why things work in this experiment. In a second part, you will duplicate this experiment, and make measurements to verify your model's predictions.

Part I: Theoretical Model & Numerical Prediction

I.1. Model 1. Consider the one-particle model shown in Figure 4.43. We assume that the weight P (of mass m) is attached to a rope which does not slip relative to the pole (of radius r). The rope of length l_0 is initially taut and horizontal. P is then released without initial speed, and swings in a vertical plane, while the rope winds around the pole. Show that, during its swinging motion from the horizontal position, the rope will eventually become slack before reaching the horizontal position, at some angle $\theta = \theta_*$. Find the value of θ_* for $l_0/r = 2\pi$ and $l_0/r \to \infty$.

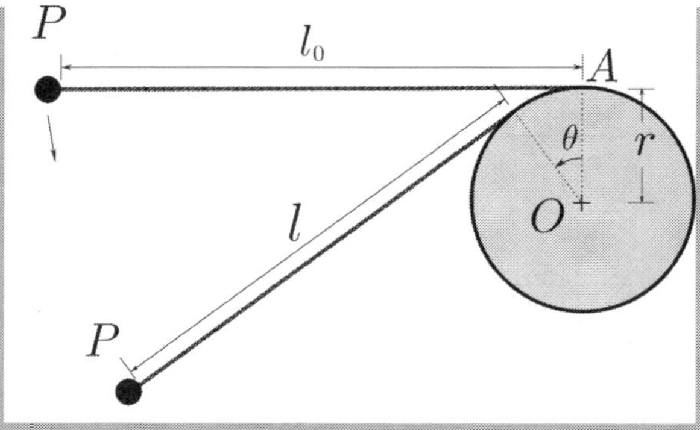

Figure 4.43 – *Model 1 (l denotes the unwound length of the rope).*

I.1. Model 2. Here we propose a better model to explain how the rope winds around the pole. Figure 4.44 shows a frictionless, two-particle model $\{P, Q\}$: particle P (of mass m) is connected to particle Q (of mass M) by a single, inextensible, massless rope. The system's configuration is defined by the coordinates (θ, z) with $(\theta, z) = (0, z_0)$ and $(\dot{\theta}, \dot{z}) = (0, 0)$ at time $t = 0$. Denote by l the unwound length of the strand of rope connected to P (with $l = l_0$ at time $t = 0$).

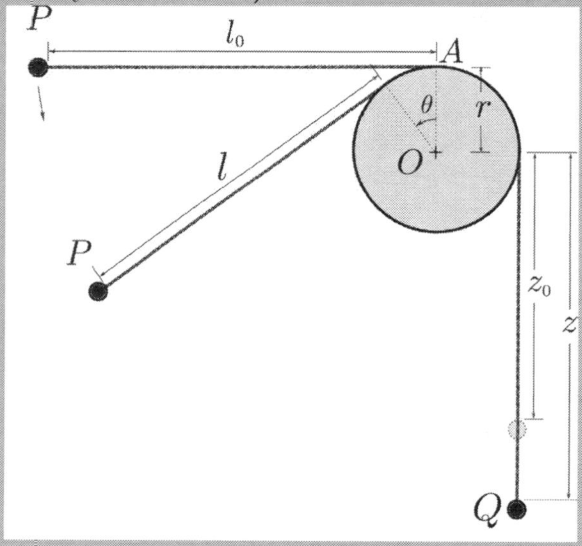

Figure 4.44 – *Model 2*

Show that the model is governed by the equations:

$$\begin{cases} m(l\ddot{\theta} - 2\dot{z}\dot{\theta} - r\dot{\theta}^2) = mg\cos\theta \\ m(\ddot{z} + l\dot{\theta}^2) = T - mg\sin\theta \\ M\ddot{z} = Mg - T \end{cases} \quad (1-3)$$

where T is the rope's tension. Write a computer code to obtain a numerical solution of equations (1-3) using reasonable values of l_0, z_0, m, M, r. Using these results, conclude your analysis by:

(i) showing that the weight P can rise above the pole by the action of the rope winding around the pole with increasing angular speed,
(ii) showing that the rope can never become slack: plot T vs t.
(iii) by finding the time $t = t_2$ and the drop exhibited by Q when the motion ends, that is, when $l = 0$. Can the number of revolutions described by P be found?
What do we learn from this model? Is it satisfactory (what happens at $t > t_2$?)?
I.3. Why does the rope appear to stop slipping relative to the pole after two or three turns? Explain why this is a key aspect to the safety of the experiment. Modify, then solve equations (1-3) to arrive at a model which would account for this effect. Obtain new quantitative information from this new model.
I.4. While the weight P is clearly seen to accelerate, the drop of Q decelerates. Why does the drop not come to an instantaneous stop? Explain why this is another vital aspect to the safety of the experiment.
I.5. Finally, **give a critical assessment of this theoretical work.**
Part II: Experimental Work
In this part, you will duplicate this experiment in the lab. Plan the experiment by carefully choosing the materials and by building an adequate setup.
II.1. Conduct experiments by testing your materials and your setup. Determine the parameters m, M, l_0, z_0, etc. Observe the phenomenon and measure the outcomes, such as the number of revolutions described by P, the total drop achieved by Q, etc. Observe the effect of parameter variations.
II.2. Compare your experimental results with your theoretical predictions. If the experiments are not in agreement or/and the outcomes highly variable, offer some explanations: is the model acceptable? Can the model be improved?

[2] "*Fritt fall uten sikret tau*" www.youtube.com/watch?v=5ZnKIPfIhAQ

4.11 Problems

Problem 4.1 A small body of mass 100 g is connected by a string of length 1 m to a point O of a vertical wall. The pendulum thus formed exhibits small oscillations by swinging about the corner point of the wall located 0.5 m immediately below O. What is the observed period of oscillation?

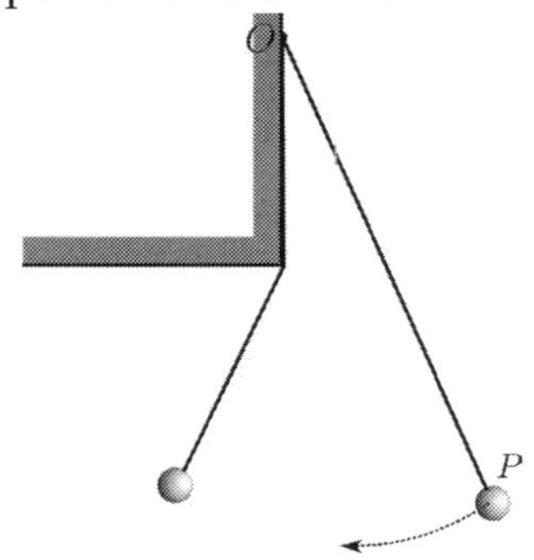

Answer: b). T = 1.7 s

Problem 4.2 A pendulum P of mass $m = 55$ g and length $l = 60$ cm is attached to a point O of plane inclined at angle $\alpha = 25°$ to the horizontal. The contact between P and the plane is assumed frictionless.

a. Find the angular momentum of P about O. Then find the moment about O of the forces acting on P.

b. From the knowledge of a) find the equation of motion of P. Deduce the period of the small amplitude oscillations.

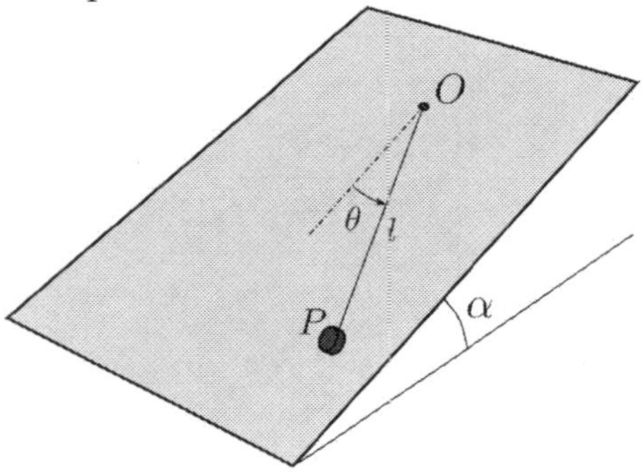

Answer: b). T = 2.4 s

Problem 4.3 A satellite of mass 1,200 kg has an elliptic orbit around Earth relative to the geocentric referential under the sole effect of Earth's gravitational attraction. At the position shown, the speed of P is 11,000 km/h.

a. Show that the angular momentum of P about the center O is conserved.

b. Find the speed of P at the positions A and B.

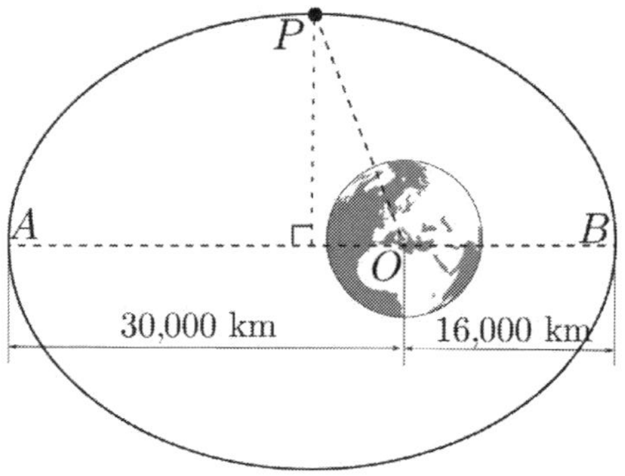

Problem 4.4 A pendulum is assembled by attaching two point masses P_1 and P_2 of identical mass m to a massless rod of length $2l$. P_1 is located at the midpoint of the rod. P_2 is located at the tip of the rod. The pendulum is then connected to a point O and is free to rotate about horizontal axis Oz without friction.

a. Apply Euler's second principle about O by finding the angular momentum of the system as the sum $\mathbf{r}_{OP_1} \times m\mathbf{v}_{P_1} + \mathbf{r}_{OP_2} \times m\mathbf{v}_{P_2}$. Deduce the equation of motion and find the period of small amplitude oscillations.

b. Repeat a) by first finding the mass center G of the system. Apply Euler's second principle about O by considering that the system is equivalent to its mass center G with mass $2m$ (in a manner similar to Euler's first principle). Are the results identical? Comment.

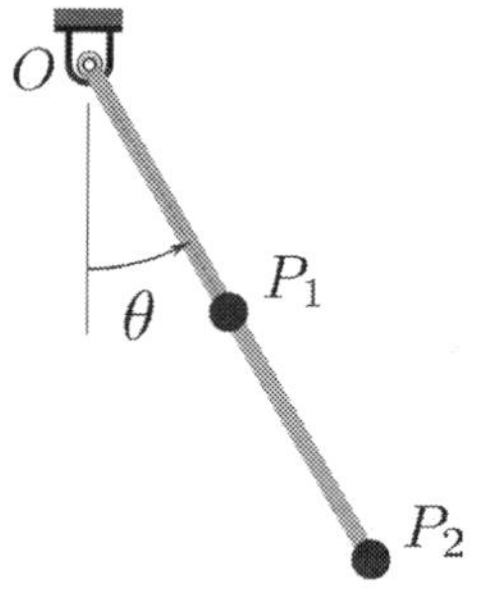

Problem 4.5 A rigid system is assembled by welding a bar of length $2l$ an angle β on the axle of a disk of radius R. The mass of this assembly is neglected compared to the two point masses P_1 and P_2 of identical mass m which are attached at the extremities of the bar. The axle of the disk is supported by bearings at A and B and can rotate about a vertical axis without friction. A string is wound around the edge of the disk. It passes through a frictionless, massless pulley at E

to support a block Q of mass M which can travel along a vertical axis.

Find the acceleration of Q upon its release.

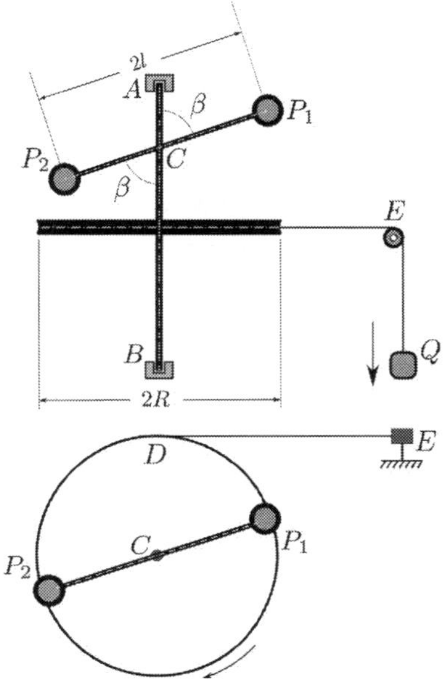

Problem 4.6 A rigid body of mass m and mass center G is assembled by rigidly connecting three slender rods as shown in the Figure. The body is in rotation about axis Gz with angular velocity ω.
In both cases, find the angular momentum of the body about G.

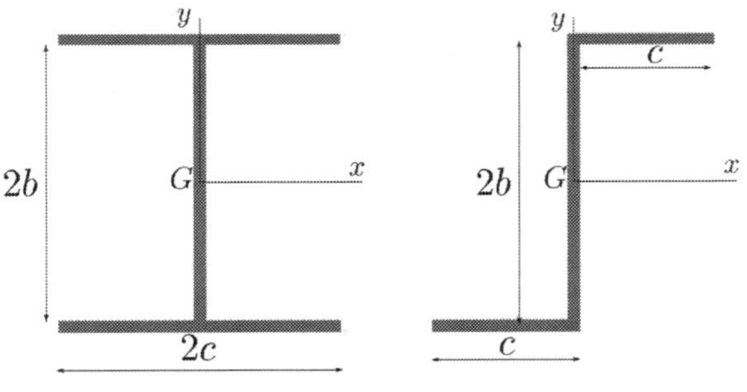

Problem 4.7 A rigid body of mass m is assembled by rigidly connecting two slender rods of length l and $2l$ as shown. The body is set in rotation about axis Oz with angular velocity ω.

a. Find the location of the body's mass center G.
b. Find the angular momentum $\mathbf{H}_O$ about O.
c. Find the angular momentum $\mathbf{H}_G$ about G.

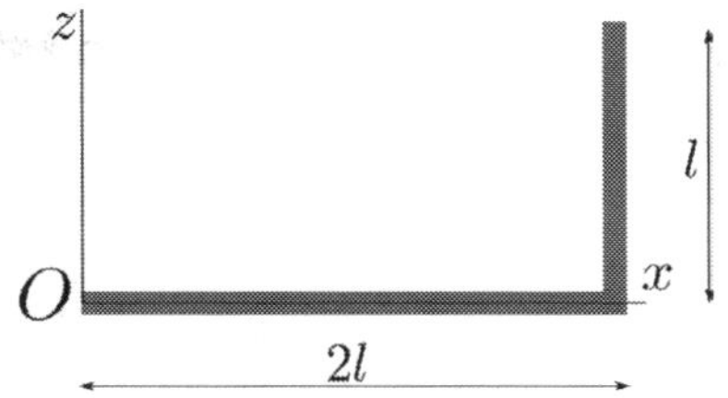

Answer: c). $\mathbf{H}_G = \frac{1}{9} ml^2\omega(-\hat{\mathbf{x}} + 4\hat{\mathbf{z}})$

Problem 4.8 Consider a uniform quarter disk of mass m and radius R.
a. Find the moments of inertia I_{Oxx}, I_{Oyy} and I_{Ozz}.
b. Find the angular momentum $\mathbf{H}_O$ if the body is set in rotation about axis Oy at angular speed ω.
c. Find the angular momentum $\mathbf{H}_O$ if the body is set in rotation about axis Oz at angular speed ω.

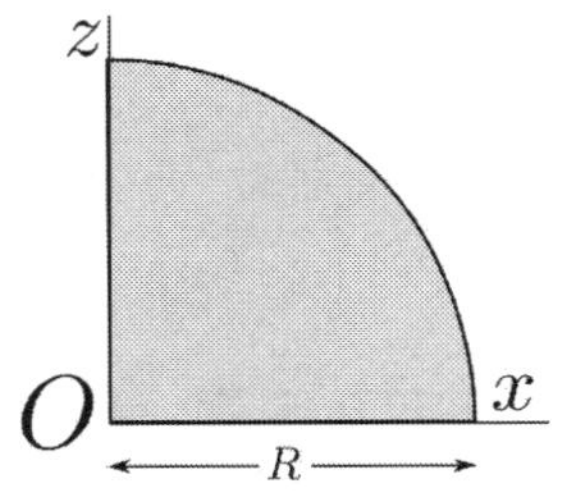

Problem 4.9 A rigid body of mass m is obtained by truncating a cone of radius R_2 and height H_2 by removing the section of cone of radius R_1 and height H_1. The body is set in rotation about axis Oz at angular velocity $\boldsymbol{\omega} = \omega\hat{\mathbf{z}}$.
Find the angular momentum $\mathbf{H}_O$ about O.

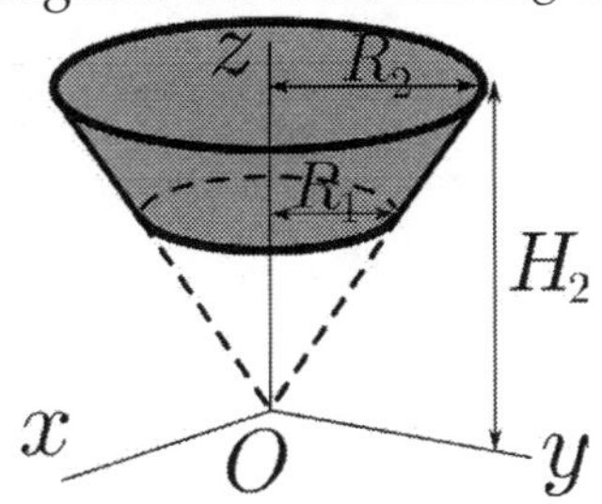

Answer: $\mathbf{H}_O = \frac{3}{10} m\omega \frac{R_2^5 - R_1^5}{R_2^3 - R_1^3}\hat{\mathbf{z}}$

Problem 4.10 A slender rod of mass m and length l is in contact with a horizontal support. The rod is slightly arched so that at any time it is contact with the support only at end point A and corner point O. The contact is assumed frictionless at O, with friction at A with coefficient of kinetic friction μ. At $t = 0$, end point B coincides with O when the body is given an initial horizontal velocity v_0.

Find the velocity v_0 so that the rod does not tip about O.

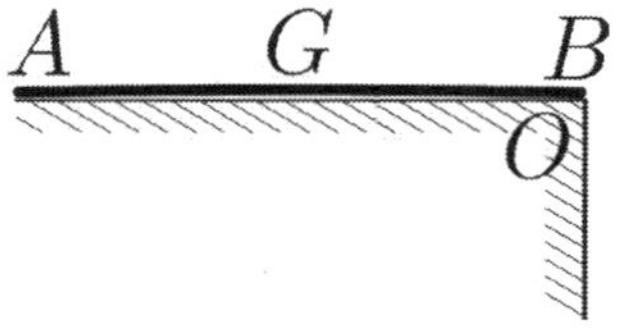

Answer: $v_0^2 = 0.307\mu l g$.

Problem 4.11 A uniform slender rod of mass m, mass center G and length $2l$ is set against a vertical wall, with its lower extremity coinciding with the corner O of the horizontal support. A small perturbation destroys this unstable equilibrium, and the rod rotates about O. The contact is assumed frictionless. The inclination of the rod with the vertical is described by angle θ.

a. Show that the motion of the rod is a rotation about point O until the angle θ reaches the value θ_1 to be found. What happens at $\theta = \theta_1$? Does the rod lose contact with the horizontal support?

b. Find the value $\dot{\theta} = \dot{\theta}_1$ when $\theta = \theta_1$ is reached.

Hint: Find and plot the two components of the reaction force at O versus angle θ.

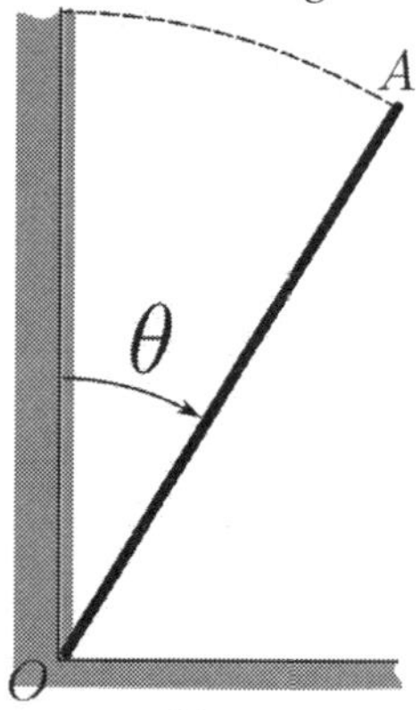

Answer: a) $\theta_1 = 48.2^\circ$, b) $\dot{\theta}_1^2 = g/2l$

Problem 4.12 A sphere 2 of mass 0.8 kg, radius 0.3 m, and mass center G is at rest on a horizontal plank 1 of mass 1.5 kg itself resting on a smooth horizontal support 0. The plank is then subject to a horizontal force of constant magnitude $F = 1.45$ N. The contact between the sphere and the plank is characterized by the coefficient of static friction $\mu_s = 0.15$.

a. Show that the sphere rolls without slipping relative to the plank.

b. Find the time taken by the sphere to fall off the plank.

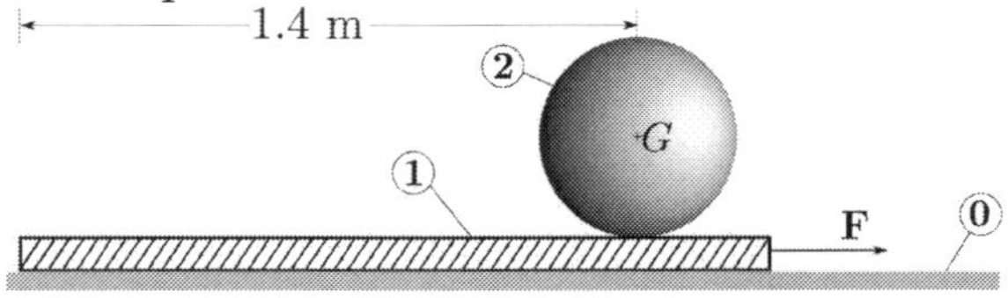

Answer: $T = 2.2$ s.

Problem 4.13 A cylinder 2 of mass m and radius R rests on a plate 1 of mass M. The plate lies on a horizontal support 0. Both contacts are characterized by the same coefficients of static and kinetic friction $\mu = \mu_s \approx \mu_k$.
The plate is then subject to a force F in the horizontal direction.
Show that the cylinder will roll without slipping if F satisfies the condition

$$F_{min} < F < F_{max}$$

Find F_{min} and F_{max}. Then find the corresponding expression of the angular acceleration of the cylinder.

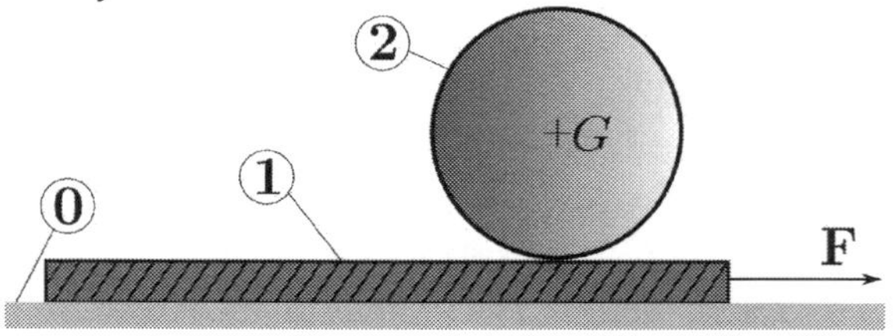

Answer: $\mu(M+m)g < F < 2\mu(2M+m)g$.

Problem 4.14 A crate of side length $2l$ and uniform mass m is set without initial speed on an inclined plane. The contact between the surfaces is characterized by coefficients of static/kinetic friction $\mu_s = \mu_k = \mu$. The angle of inclination of the plane with the horizontal is denoted by θ.

Find the condition(s) governing the parameters μ and α for the crate to immediately slip without slipping upon release.

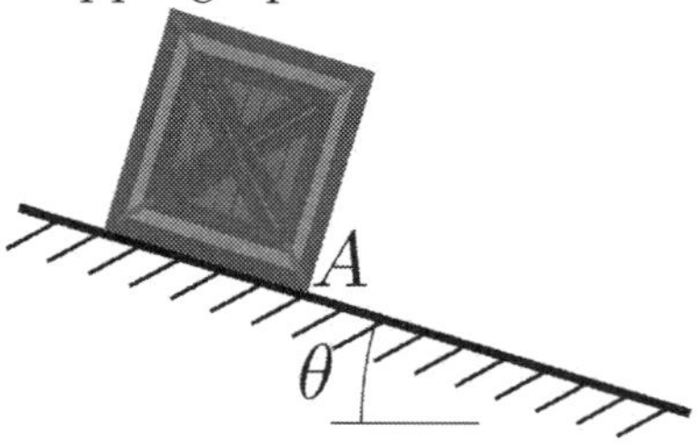

Problem 4.15 Consider the crate of problem 4.14.

a. Find the conditions which must be satisfied by the parameters μ and θ for the cube

to tip about its lower edge without slipping immediately upon release.
b. Find the angular acceleration of the cube at that instant.
Answer: $\alpha_0 = \frac{3g}{8l}(\sin\theta - \cos\theta)$.

Problem 4.16 We continue the study of the slender rod of problem 4.11 for $\theta > \theta_1$. The position of contact point B is defined by the Cartesian coordinate x.
a. Show that the motion is governed by two first integrals, that is, by equations containing only $(x, \theta, \dot{x}, \dot{\theta})$. Deduce $\dot{\theta}^2$ as a function of θ. Find $\dot{\theta}$ just before the rod hits the ground.
b. Show that once point B takes off from the wall, it cannot move back toward the wall (show that $\ddot{x} > 0$) nor can it lose contact with the ground.

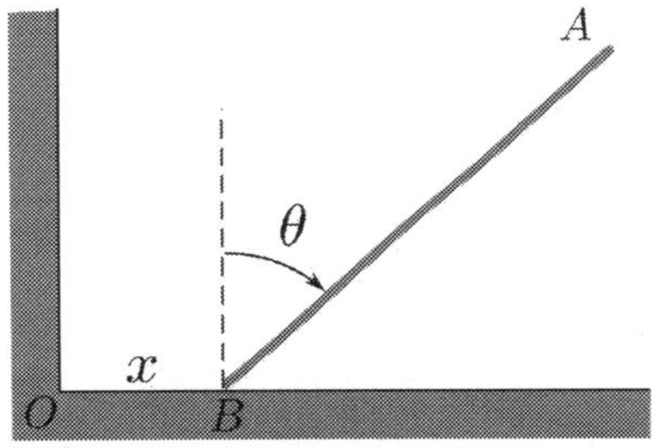

Answer: a) $\dot{\theta}^2 = 4g/3l$ *at* $\theta = \pi/2$.

Problem 4.17 Body 1 of mass 14 kg is in rectilinear translation along a rough, horizontal support 0. The corresponding coefficients of static and kinetic friction at the contact are $\mu_s = 0.18$ and $\mu_k = 0.15$. Body 3 of mass 25 kg is in rectilinear translation along the vertical direction. It is connected to body 1 by a rope which does not slip along the pulley 2. The pulley of radius $r = 0.1$ m and of mass 3 kg is free to rotate without friction about its axle Oz. Its mass moment of inertia about its axis of rotation is $J = 0.024$ kg.m^2.

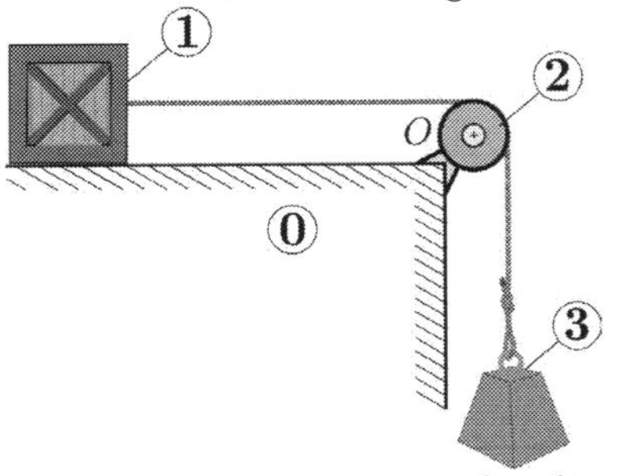

a. Show that body 1 necessarily slips when the system is released from rest.
b. Find the velocity of body 1 when body 3 has dropped 0.85 m.

Problem 4.18 A dragster of mass M and mass center G is modeled as a rear-wheel-drive vehicle. The geometry is described by the dimensions (L_1, L_2, H) as shown in the Figure and the contact between the tires and the road is characterized by coefficients of static and kinetic friction (μ_s, μ_k). The mass of the wheels is neglected.
a. Find the maximal possible acceleration a_{max} that the dragster can reach in a rectilinear translation. What is the advantage of rear-wheel-drive? (*hint: find* a_{max} *in front-wheel-drive mode*).
b. Based on the result found in a) describe ways to increase a_{max}.

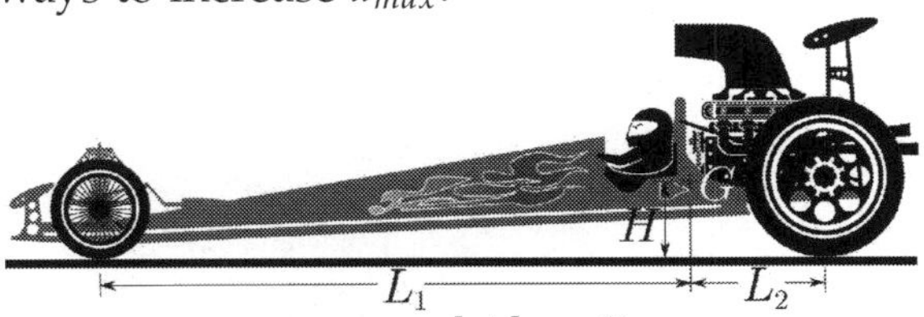

Answer: a) $a < a_{max} = g/(\frac{L_1+L_2}{\mu_s L_1} - \frac{H}{L_1})$.

Problem 4.19 A spool of mass m, radius R, mass center G, and radius of gyration k about its axis is resting on a horizontal support. A string wound around the spool's inner hub of radius r is then pulled with a force F making the angle β with the horizontal. Assume that the spool does not slip. The coefficient of static friction is μ.
a. Find the acceleration of G as a function of F, β, r and R. Conclude that the spool may move to the right or the left depending on the value of angle β.
b. Find the conditions satisfied by F to prevent the spool from losing contact or from slipping.

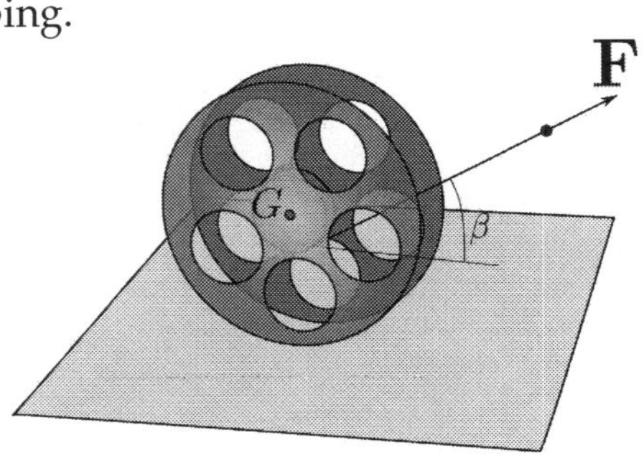

Ans: $F < mg/(\sin\beta + \frac{\kappa}{\mu}) < mg/\sin\beta$, $\kappa = (rR + k^2\cos\beta)/(R^2 + k^2)$.

Problem 4.20 A plate 1 is free to rotate about a horizontal axis Oz. It supports body 2 of negligible size located at a distance x from axis Oz. At $t = 0$, the plate is released without initial speed from its horizontal position ($\theta = 0$). Friction prevents body 2 from immediately sliding. Both bodies have equal mass $m = 0.75$ kg. The coefficients of friction are $(\mu_s, \mu_k) = (0.4, 0.35)$. The dimensions of the plate are specified by the lengths $2l = 1.2$ m and $2d = 0.5$ m

a. By considering the system $\{1,2\}$ find the angular speed $\dot{\theta}$ as a function of angle θ.

b. Find the reaction force of the plate on body 2 as a function of angle θ.

c. Show that if $x < x_{max}$ body 2 will not lose contact with the plate. Find x_{max}.

d. Assuming $x = x_{max}/2$, find the value of angle θ at which body 2 begins to slip.

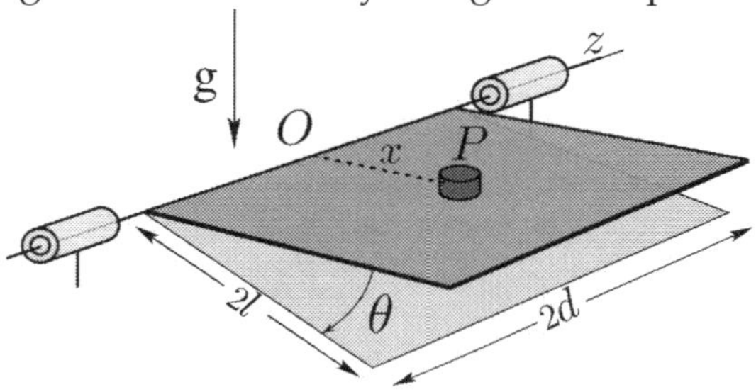

Answer: c) $x_{max} = 0.8$ *m. d)* $3.8°$

Problem 4.21 A ladder of mass m and length $2l$ is set against a smooth vertical wall at A, with its lower end B resting on a smooth horizontal support. Its inclination with the horizontal is given by angle $\theta = \theta_0 < \pi/2$. The ladder is then released without initial speed and it slips at both A and B. Treat the ladder as a uniform slender rod.

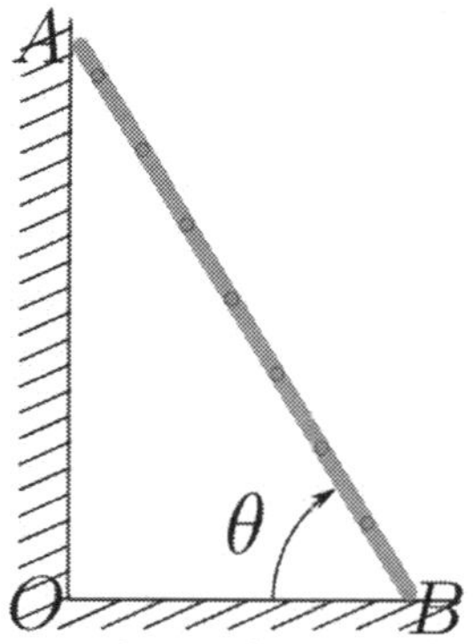

a. Find the angular velocity of the ladder as a function of angle θ.

b. Find the reactions at A and B to find out whether the ladder loses contact with its supports before the value $\theta = 0$ is reached. If so, find the value(s) of angle θ at which this takes place.

Answer: b) $N_A = 0$ *for* $\sin\theta_1 = 2\sin\theta_0$

Problem 4.22 A cylinder of mass m and radius R is connected to a fixed support by a linear spring of constant k and unstretched length l_0. The spring is stretched to length $l > l_0$ and the cylinder is released with initial speed v_0.

Find the maximum possible value taken by v_0 for the cylinder to roll without slipping on the horizontal support upon release.

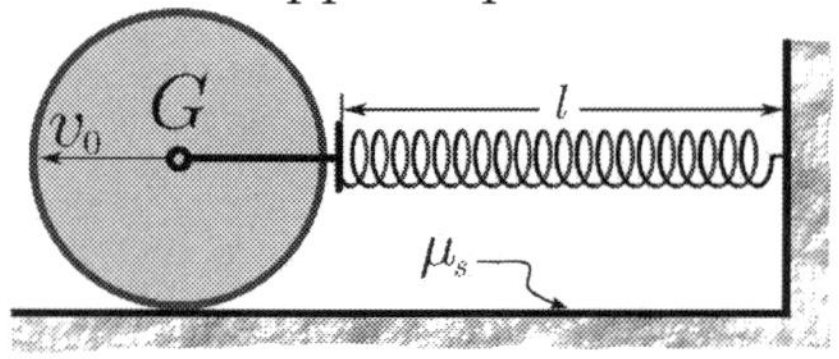

Answer: $v_0^2 < 6\mu_s^2 g^2 \frac{m}{k} - \frac{2k}{3m}(l - l_0)^2$

Problem 4.23 A uniform slender rod of mass m and length l is suspended by two strings each of length l attached to its end points A and B. It is at rest when the string connecting B is suddenly cut.

Find the ratio of the tension in the remaining string immediately after the string at B is cut to the initial equilibrium tension.

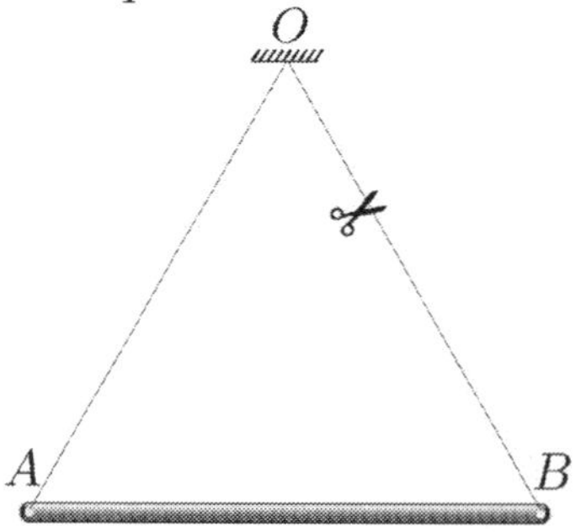

Answer: $T_1/T_0 = 6/13$

Problem 4.24 A car of mass M is at rest on a horizontal road with the passenger door left open at an angle $\theta = \theta_0$ as shown. The car then starts up at constant acceleration a_C. The door of mass m is free to rotate about frictionless vertical hinges of axis Az. The door mass moment of inertia about this axis is $I_{Azz} = mk_A^2$. The mass center G of the door is located at a distance l from axis Az.

Find the angular velocity of the door when it reaches the position $\theta = 0$. Find the corresponding time for the door to close, by making the approximation $\sin\theta \approx \theta$ during the motion.

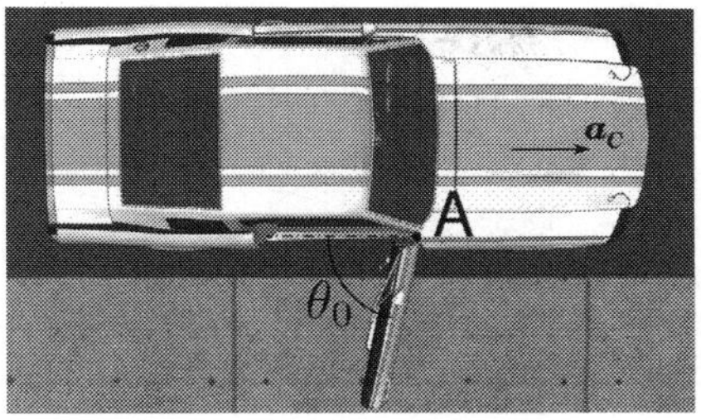

Problem 4.25 A spool of mass m, mass center G, radius of gyration k_G about axis Gz is initially at rest on a conveyor belt. The belt is then set in motion at constant acceleration a as shown. The contact between the spool and the belt is characterized by the coefficients of static and kinetic friction μ_s and μ_k. The cable is initially taut. The goal of this problem is to predict the motion of the spool depending of the value of acceleration a.

a. Assume that the spool does not slip relative to the belt. Find the spool angular acceleration, the tension in the cable and the friction force. Then find the condition satisfied by acceleration a for the no-slip motion to occur.

b. Assume that the spool slips immediately after the belt is set in motion. Find the spool angular velocity, the tension in the cable and the friction force. Then find the condition satisfied by acceleration a for this motion to occur.

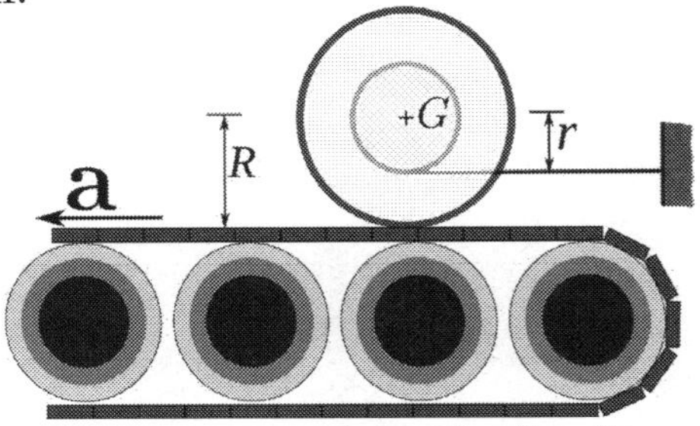

Answer: a) $a \leq a_{max} = \mu_s g(R-r)^2/(r^2+k_G^2)$

Problem 4.26 Consider a system of two identical homogeneous rods 1 and 2 moving in a vertical plane. Both rods have same mass of 0.4 kg, and same length of 1.0 m. Their end points are connected by massless inextensible strings of identical length 0.4 m. Rod 1 is free to rotate without friction about its center O. The system is initially at rest in the position shown. The string at A is then cut.

a. Find the angular accelerations $\ddot{\phi}$ and $\ddot{\theta}$ immediately after release. Do these values hold at subsequent times?

b. Find the tensions in the strings.

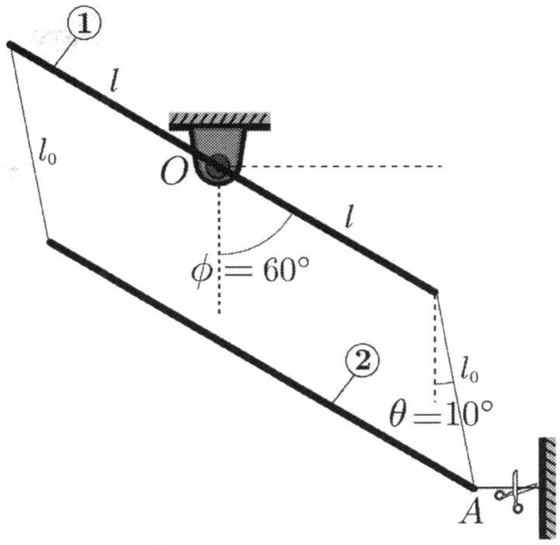

Answer: b) $T = T' = 1.93$ N

Problem 4.27 A Segway and its rider are modeled as a system $\Sigma = \{1,2,3\}$ of three interconnected rigid bodies in motion in a Newtonian referential $0(O, \hat{\mathbf{x}}_0, \hat{\mathbf{y}}_0, \hat{\mathbf{z}}_0)$. The axis $(O, \hat{\mathbf{z}}_0)$ represents the vertical directed upward. Body 1 represents the rider/powerbase/frame assembly. Its mass center is G and its mass is M. Bodies 2 and 3 are the tire/wheels, assumed of negligible mass. The wheels are in contact with the road at points I and J.

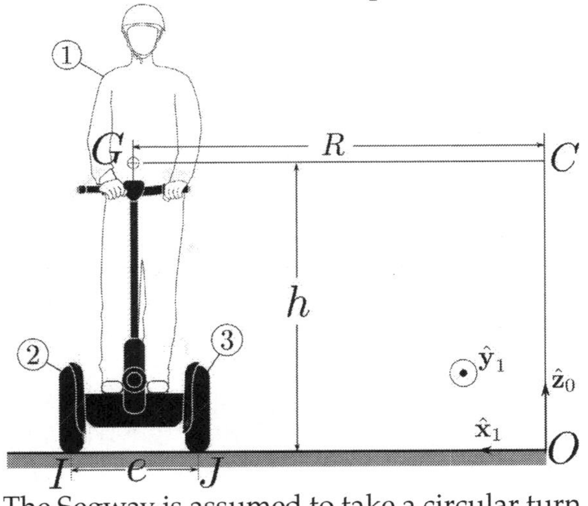

The Segway is assumed to take a circular turn, so that mass center G has a circular trajectory of center C and radius R at constant speed v_G and at a constant height h above the road.

a. Show that the referential $(G, \hat{\mathbf{x}}_1, \hat{\mathbf{y}}_1, \hat{\mathbf{z}}_0)$ has a zero angular acceleration $\boldsymbol{\alpha}_1$.

b. Apply Euler's second principle about contact point I to system Σ to find one equation governing the normal reaction force N_J acting on wheel 3, in terms of v_G and other parameters.

c. Deduce from b). that the Segway must not exceed a speed $v_{G,max}$ to take the turn without tipping.

Problem 4.28 A system is comprised of two rigid bodies, a frame 1 of mass m_1 and mass center G_1 and a homogeneous cylinder 2 of mass m_2, mass center G_2, radius r connected to frame 1 by a frictionless pivot. The frame is free to rotate about horizontal axis Oz without friction. The mass centers G_1 and G_2 are located at a distance l_1 and l_2, respectively, from axis Oz. The moment of inertia of the frame about axis Oz is denoted as I_1.

In the first experiment, the rotation of cylinder is blocked, so that the assembly makes small amplitude oscillations about the vertical axis Ox as a single rigid body.

In the second experiment, the rotation of the disk is free: the frame is released without initial speed making a small angle θ_0 with the vertical, while the cylinder rotates with initial angular velocity ω_0.

Find the periods T_1 and T_2 of small amplitude oscillations of the system in both experiments. Compare T_1 and T_2. Does the rotation of the cylinder have any influence on the period of oscillations?

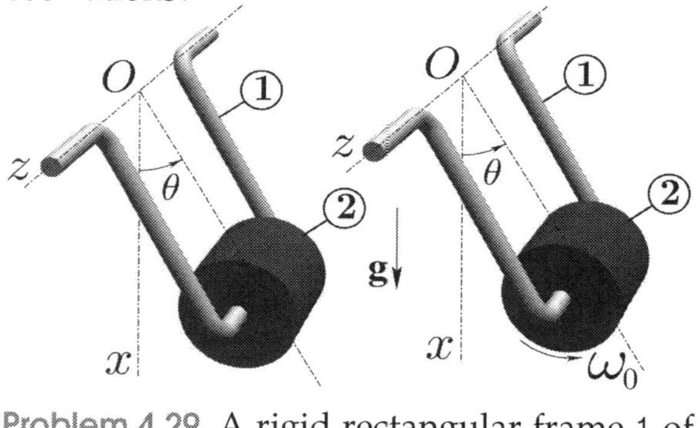

Problem 4.29 A rigid rectangular frame 1 of mass 250 g and radius of gyration $k_\Delta = 0.2$ m can rotate about a vertical axis Δ. The pivots at A and B are frictionless. Two identical homogeneous cylinders 2 and 3 of mass 500 g and radius $r = 0.1$ m are mounted on the vertical sides of the frame, about which they can rotate without friction. Their mass centers coincide with the centers of the bars.

a. Both cylinders are given the same initial angular velocity $\omega_0 = 0.45$ rev/s, the frame being initially at rest. At a later time, a mechanism attached to the frame, of negligible mass, suddenly stops the rotation of cylinders 2. Why is the frame observed to rotate? Find its angular velocity ω_1'.

b. The frame is now turning at angular velocity ω_1' with one of the cylinders immobilized, when the rotating cylinder 3 is suddenly put to a sudden complete stop relative to the frame. Find the new angular velocity ω_1'' taken then by the frame.

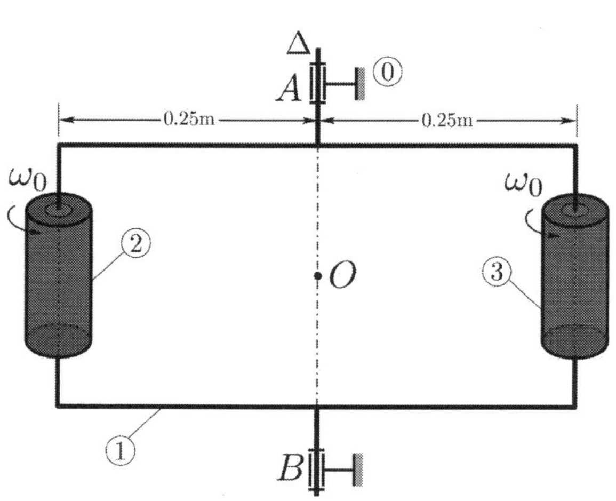

Problem 4.30 In this block and tackle configuration, pulley 1 has mass m, radius R and moment of inertia J about its axis. Its axle O is fixed relative to support 0. Pulley 2 is identical to pulley 1 with its axle A free to move vertically by the use of a rope whose contact with both pulleys is sufficiently rough to prevent slipping. Pulley 2 supports a body 3 of mass M. The system $\{2,3\}$ is lifted by pulling on the cord with the vertical force F as shown. Assume that the cord is inextensible and of negligible mass.

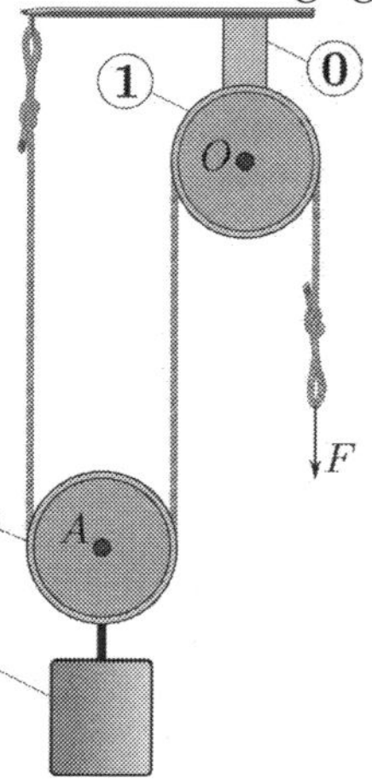

a. Find the acceleration of body 3 as a function of force F and in terms of the parameters m, M, R, J.

b. Find the minimal force F required to lift the system and discuss the advantage of the block and tackle.

Problem 4.31 Two identical uniform disks each of mass M and radius R are connected to a slender bar of mass m and length $2l$ by two identical massless rods of length r. All connections (pivots) are assumed frictionless. The assembled system is set on a horizontal support. The bar is displaced from its equilibrium and released without initial speed. The system is then observed to oscillate in such a way that the disks roll without slipping.

a. Find the accelerations of centers A and B in terms angle θ and its derivatives. Then find the equation which governs the evolution of θ.

b. Deduce the period of the small amplitude oscillations of the system (*linearize the ODE found in a)*). Examine the case $M \gg m$.

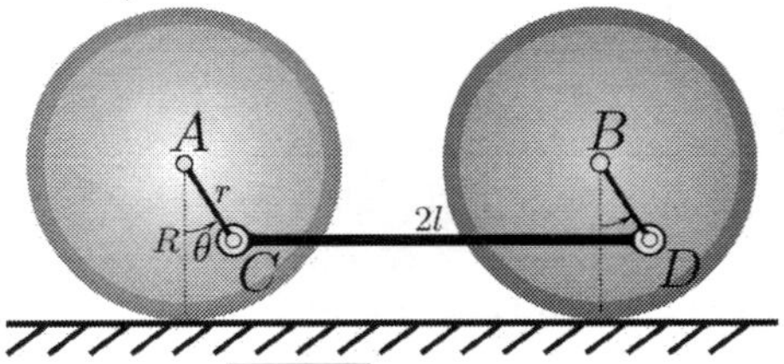

Answer: $T = 2\pi\sqrt{\frac{r}{g}\frac{3M}{m+3M}}$

Problem 4.32 A pile of ten identical cylinders (each of mass m and radius r) is assembled by interposing a thin massless plank between the lower and the upper pairs of cylinders. The cylinders axes are horizontal and parallel to one another. The pile is at rest on a fixed support when the lowest plank is pulled with an horizontal acceleration $\mathbf{a}$. Neglect the thickness of the plank.

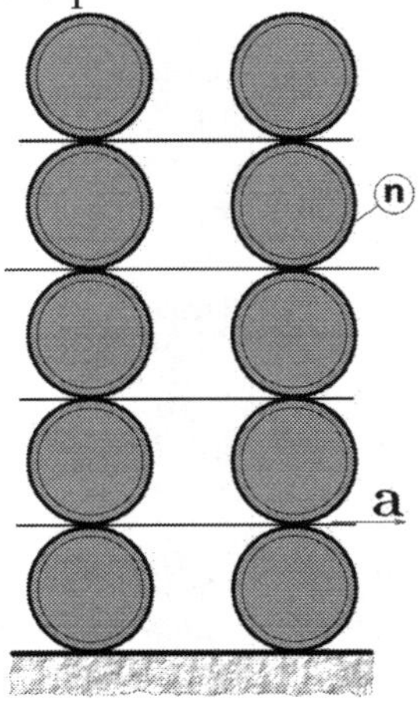

a. By considering the nth cylinder, write

(i) the no-slip equation at the contact point with the $(n-1)$th cylinder,

(ii) Euler's First Principle in the horizontal direction,

(iii) Euler's Second Principle about the mass center.

b. Show that the linear and angular accelerations of the nth cylinder can be found from those of the $(n-1)$th according to

$$\begin{bmatrix} a_n \\ r\alpha_n \end{bmatrix} = \begin{bmatrix} -3 & 2 \\ 4 & -3 \end{bmatrix} \begin{bmatrix} a_{n-1} \\ r\alpha_{n-1} \end{bmatrix}$$

c. Find the acceleration of the mass centers and angular accelerations of the ten cylinders.

Problem 4.33 A cylinder 1 of mass 0.35 kg, radius 5 cm, and height 8 cm is set in rotation about its vertical axis Oz. The pivots at A and B are assumed frictionless, so that the cylinder rotates at the constant angular velocity 60 rev/min. A ring 2 of mass 0.15 kg of radius 5 cm, positioned above the cylinder, is dropped so as to come in contact with it (the cylinder's upper base is slightly narrower than its lower base). The ring begins to slip and eventually becomes rigidly attached to the cylinder. The coefficient of kinetic friction between the two bodies is $\mu_k = 0.15$.

Find the final angular velocity of the system (neglect the initial speed of the ring when it first comes in contact with the cylinder).

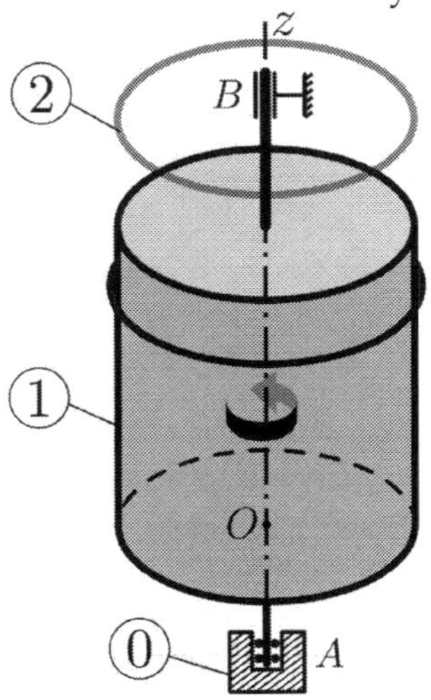

Problem 4.34 A parking barrier is comprised of 3 rigid bodies in planar motion in a vertical plane. Body 1 of mass m_1, mass center G_1 is in rotation about fixed point O. Its rotation is controlled by a motor of couple C (measured positively in the clockwise direc-

tion) and its orientation is defined by angle θ. Body 2 of mass m_2 and mass center G_2 is pinned to body 1 at A. Its orientation remains horizontal at all time due to its coupling with body 3 at pin C. Body 3 is pinned at fixed point B and is assumed of negligible mass. The 4 points O, A, B and C form a parallelogram at all time. All pin joints are assumed frictionless. The geometry is defined by the constants $|OA| = |BC| = 2l_1$, $|OG_1| = l_1$, $|AG_2| = l_2$, $|OB| = |AC| = r$. Denote by J_1 the moment of inertia of body 1 about axis Oz, and by J_2 the moment of inertia of body 2 about axis Az.

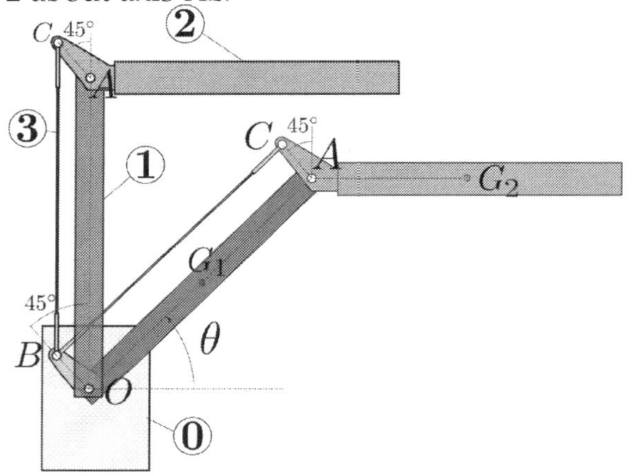

a. Find the force $\mathbf{F}_{3\to 2}$ which body 3 exerts on body 2 as a function of θ.

b. Find the motor's couple $\mathcal{C}$ necessary to reach a particular angular profile $\theta(t)$.

Answer: $\mathcal{C} = -(J_1 + 4m_2 l_2)\ddot{\theta} + (m_1 l_1 + 2m_2 l_2) g \cos\theta$

Problem 4.35 A dump truck is at rest on a horizontal road. The parking brakes have not been set by the operator. Hence the four wheels of the truck are free to rotate about their axles assumed frictionless. The operator then raises the dump box 1 which rotates about frictionless pivot A from angle $\theta_0 = 0$ to angle $\theta_1 = 55°$. The dump box (and its content) has mass center G and mass $m = 12.4$ tonnes. The truck's cab and frame 2 has mass center H and mass $M = 5.5$ tonnes. The mass of the four wheels is neglected. Use $h = 1.2$ m, $d = 3.2$ m, $r = 0.5$ m, $a = 1.1$ m.

a. Describe the motion of the mass center of the truck (including the dump box) during the rotation of the dump box.

b. Find the total displacement of mass center H during the rotation of the dump box.

Answer: 0.94 *cm (forward).*

Problem 4.36 A motorcycle and its rider (assumed rigidly connected) are running along a straight horizontal path. The mass of the system {motorcycle+rider} is M and its mass center is located at point G as shown in the Figure. The engine supplies a torque $\mathcal{C}$ to the backwheel. The geometry of the motorcycle is defined by the constants r, h, a and b. *Neglect the mass of the wheels.* Also denote by k_G the radius of gyration of the system about G.

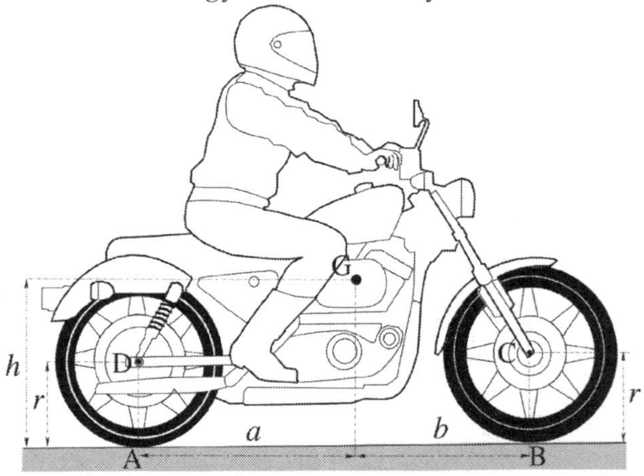

a. Find the value of $\mathcal{C}$ for the motorcycle to travel at constant velocity. Then find the corresponding reactions of the road on the wheels.

b. Now assume that the backwheel is driven at a given finite couple $\mathcal{C}$. Find the corresponding acceleration of mass center G and the acceleration of mass center G and the reaction forces on the front and back wheels (in terms of $\mathcal{C}$ and other parameters). What is the maximum torque $\mathcal{C}_{max}$ which must not be exceeded to prevent the motorcycle to rear its front wheel off the ground (the motorcycle is then said to do a wheelie)? What is the smallest coefficient of static friction μ_{min} between the wheels and the road in order for the rider to do a wheelie.

c. Assuming $C > C_{max}$, find the angular acceleration of the motorcycle immediately after the front wheel leaves the ground (in terms of C and other parameters).

Problem 4.37 A 15 kg circus dog 2 performs a trick which consists of standing on a large beach ball 1 of mass 6 kg and radius 0.55 cm in such a way that his paws are constantly located at a contact point I. More specifically, the dog walks by making small shuffling steps towards the highest point of the ball so as to maintain the angle of 35° between line CI and the vertical. The dog is modeled as a rigid body of mass center G such that line IG ($|IG| = 0.2$ m) remains vertical at all times. The ball 1 rolls without slipping on a fixed horizontal support.

a. Find the acceleration of center C.

b. Find the force exerted by the dog on the ball. Show that this force is not applied at I (*hint: show that the action of the dog on the ball also creates a non-zero moment at I*).

c. Assuming that the system is initially fixed, find the condition to be satisfied by the coefficient of (static) friction so that the slip of the ball does not occur at the start and at any subsequent time.

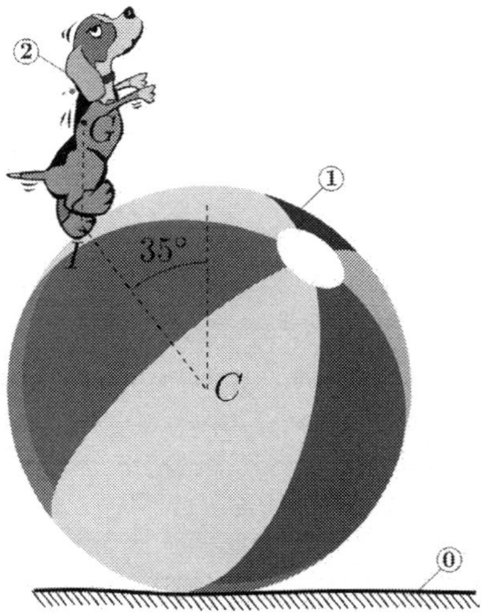

Answer: $a_C = 0.20\,g\ (\leftarrow)$

Problem 4.38 A rod of length $2l$ is set in motion without initial speed in such a way that its lower endpoint A can slide along a horizontal support while remaining in contact with corner point Q of a step of height h. The contacts at A and Q are frictionless. At $t = 0$, $\theta = \theta_0$ and $\dot{\theta} = 0$.

Assume that the rod immediately loses contact at A upon release. Find the acceleration of point A at $t = 0^+$. Then find the geometric condition governing parameters h, l, and θ_0 which guarantees that the rod slips while losing contact at A immediately after being released.

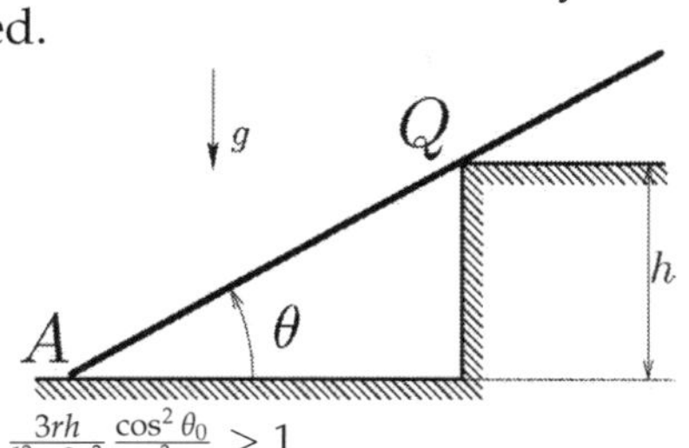

Answer: $\frac{3rh}{l^2+3r^2}\frac{\cos^2\theta_0}{\sin^3\theta_0} > 1$

Problem 4.39 A monkey of mass m climbs up a rope passing through a frictionless, massless pulley. The rope is assumed massless and inextensible. The other extremity of the rope is attached to a weight of mass M which rest on a horizontal support.

Under which condition will the monkey manage to lift the mass off the support?

Problem 4.40 Two gymnasts 1 and 2 of same mass m are hanging at the ends of a rope passing through a pulley 3 of frictionless shaft Oz. The rope is assumed massless and inextensible. It may move without slipping relative to the pulley. Initially both gymnasts are at the same level and the system is at rest.

a. Neglect the mass of the pulley. Gymnast 1 decides to climb up the rope, while gym-

nast 2 holds tightly to the rope. The rope stays vertical. Describe the vertical motions of the gymnasts relative to the ground. Will gymnast 2 stay at a fixed height from the ground? What if $m_2 > m_1$? Is the situation identical (assuming that the system is released from rest)?
b. Now we account for the pulley's inertia. Denote by I_O its moment of inertia relative to axis Oz. How is the situation of a) modified?

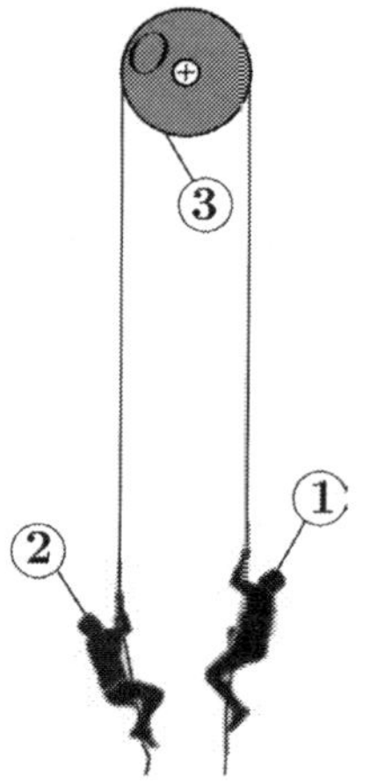

Problem 4.41 An aerial tramway Σ is supported by two fixed horizontal cables. The truck 1 of mass m supports a cabin 2 of mass M and mass center G. The cabin is free to rotate about A relative to the truck. The truck supports a set of wheels which can roll on the cables. It is pulled by a horizontal haulage cable which exerts a constant traction force $T\hat{\imath}$. The assumptions are as follows: the truck's wheels are massless, all sources of friction are neglected, and the angle θ made by the cabin 2 relative to the vertical is small. Denote by x the Cartesian coordinate of A along some horizontal axis Ox. The moment of inertia of the cabin about its axis of rotation Ay is $J_A = Mk_A^2$ and mass center G is located at distance L from A.
a. Apply Euler's First Principle to the tramway Σ to find a first equation of motion.
b. Apply Euler's Second Principle about A to the car 2 to find a second equation of motion.
c. Using the previous results show that angle θ is governed by the equation

$$(k_A^2 + L^2 \frac{M}{m+M})\ddot{\theta} + Lg\,\theta = \frac{LT}{m+M}$$

Show that the general solution is given by $\theta(t) = \theta_* + \theta_0 \cos \Omega t$ corresponding to small amplitude oscillations of period $\tau = 2\pi/\Omega$ about equilibrium position θ_*. Find θ_* and τ.
d. Modify the previous analysis by accounting for the p wheels supported by the truck. Assume that each wheel has mass m_W, radius r and moment of inertia $m_W k_W^2$ about its axle. Assume that the wheels roll without slipping on the supporting cables.

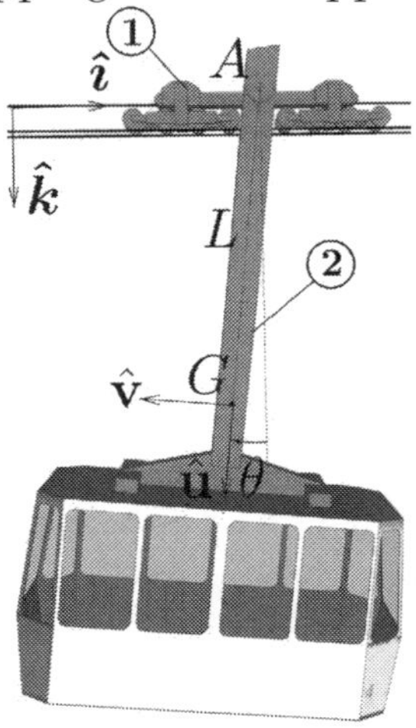

Problem 4.42 A rigid body is made of a hoop of radius r of center A and negligible mass and of a uniform slender rod AB of length r and mass m welded to the hoop. This body is set on a horizontal support, rod AB being horizontal. The body is released from rest. The contact is characterized by coefficient of static and kinetic friction $\mu_s = \mu_k = \mu$.

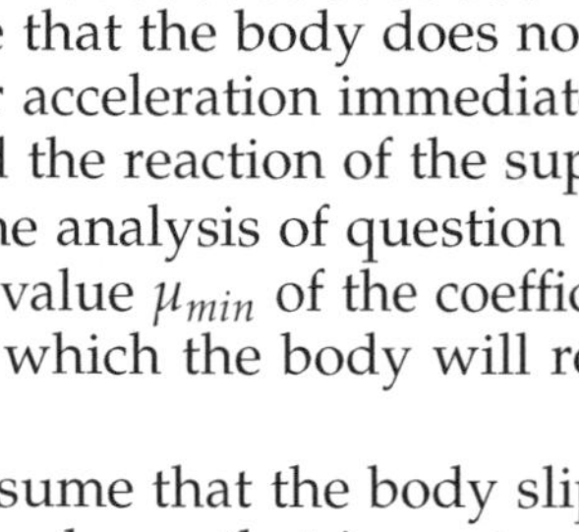

a. Assume that the body does not slip. Find its angular acceleration immediately after release. Find the reaction of the support.
b. From the analysis of question a), find the minimum value μ_{min} of the coefficient of friction μ for which the body will roll without slipping.
c. Now assume that the body slips immediately after release, that is, $\mu < \mu_{min}$. Find its angular acceleration and the reaction force immediately after release.
Answer: b) $\mu_{min} = 6/13$

Problem 4.43 It is well known that, in the absence of aerodynamic friction, two objects of differing mass, fall at same velocity under the effect of gravity and given the same initial conditions. Thus, it is surprising to find that, under some conditions, some parts of a rigid body may fall faster than in free fall. Consider the experiment shown below.

A rigid, homogeneous plank (of mass M) is pinned without friction at a fixed point O. Hence it is free to rotate about a horizontal axis Oy. At the other end of the plank, a small cup (of negligible mass) is attached to the plank. Finally a small ball (of mass m) is resting on top of the plank.

At time $t = 0$, the system is released from rest, inclined at angle θ_0 with the horizontal. It is then possible to observe, under the right conditions, that the cup falls faster than the ball and that the fall can be caught by the cup when the plank reaches the horizontal position. The geometric parameters are shown in the Figure. The mass moment of inertia of the plank about axis Oy is $\frac{1}{3}ML^2$.

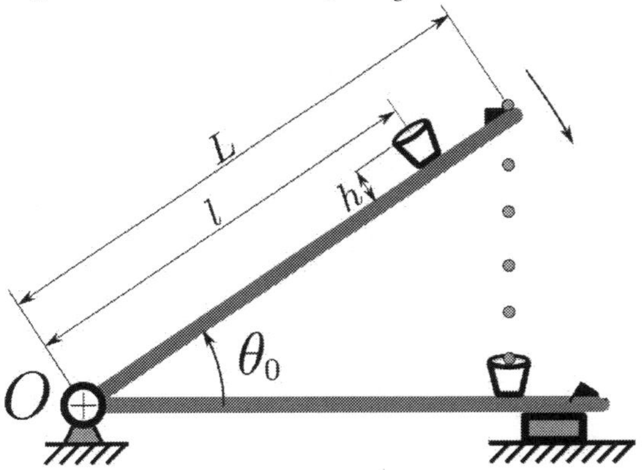

a. Explain how the ball may fall slower than the plank to eventually be caught up by the cup. *Hint: show that the ball can immediately lose contact with the plank if the initial angle* θ_0 *does not exceed a value* θ_*.

b. Find the time T taken by the plank to reach the horizontal position $\theta = 0$. Then find the time τ taken for the ball to reach the bottom of the cup. Verify that $T < \tau$ by finding the ratio τ/T. *Hint: since* $\theta_0 < \theta_*$*, use the small angle approximations* $\cos\theta \approx 1 - \frac{1}{2}\theta^2$ *and* $\sin\theta \approx \theta$.

Answer: a) $\theta_* \approx 35.3°$

Problem 4.44 A warehouse handler attempts to move a spool over a step by pulling on the cable with a horizontal force F. The spool is of cylindrical shape, of mass center G, mass 110 kg and radius of gyration $k_G = 0.25$ m.

a. Find the minimum force F to lift the spool off the ground. Under what condition is this result valid?

b. Assuming $F = 900$ N and no-slip at A, find the spool initial angular acceleration.

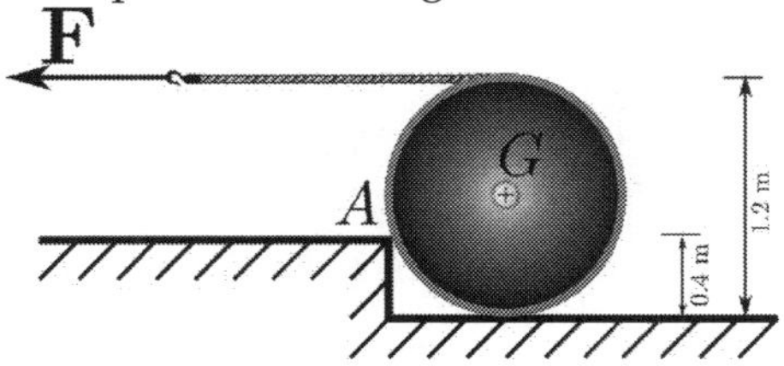

Problem 4.45 A plate 3 (of mass m_3) is placed on top of two homogeneous, full cylinders 1 (of mass m_1) and 2 (of mass m_2) which rest on a horizontal support. The mass centers A, B and C of the three bodies are coplanar. The initial configuration is shown in the figure, with $r_1 > r_2$. The friction is present at all contact points, characterized by the coefficients $\mu_s \approx \mu_k = \mu$. The system is released without initial speed.

a. Is there a configuration for which the system remains at rest?

b. Assume that the system cannot remain in equilibrium, and that friction is sufficiently large to prevent slipping at any of the contact points. Find the angular accelerations $\boldsymbol{\alpha}_1$ and $\boldsymbol{\alpha}_2$ of the cylinders and the acceleration $\mathbf{a}_C$ of mass center C, immediately after the system is released.

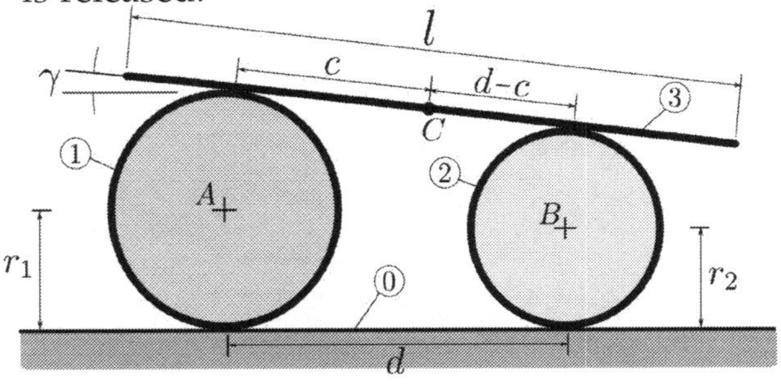

Problem 4.46 A plank of negligible mass is placed on top of a homogeneous, full cylinder 1 (of radius r, mass center A, and mass m) which rests on a horizontal support 0. A weight P (of mass M, treated as a point mass) is placed on the plank at a distance d of end point B. The plank may slip relative to the cylinder at K and relative to the support at B. The system is initially at rest. The friction is

present at all contact points, characterized by the same coefficients $\mu_s \approx \mu_k = \mu$.

a. Under what condition will the system remain in equilibrium?

b. The weight P exhibits a slight perturbation from its equilibrium position. Describe the ensuing motion of the system.

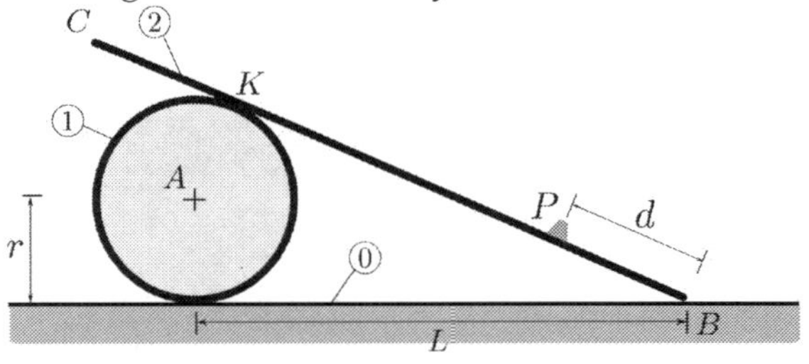

Problem 4.47 A slender, uniform rod 1 (of mass m, mass center G, length $2l$) can rotate freely in a vertical plane about point O. Its mass center G is connected to a counterweight 2 of mass M by a thin, massless cable which passes through massless pulley at A. The distance from O to A is l ($|OA| = |OG|$). The length of the cable is denoted L. Neglect all sources of friction.

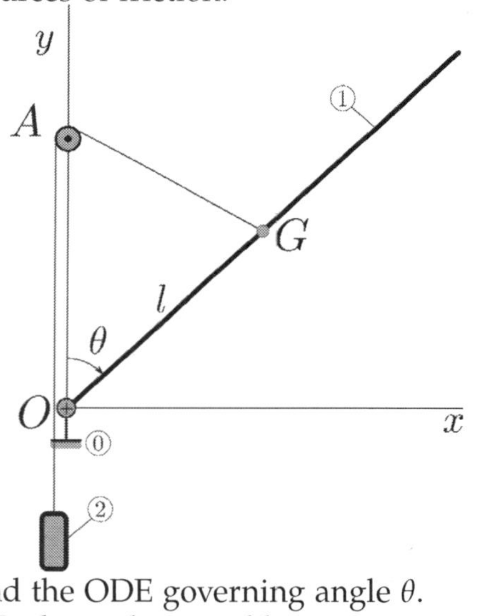

a. Find the ODE governing angle θ.

b. Deduce the equilibrium positions of the system and discuss their stability. Several cases will be considered, depending on whether the conditions $2l < L$ and $2M < m$ are satisfied or not.

Problem 4.48 A uniform slender rod of mass 0.35 kg and length 1.05 m is set in contact with the smooth surface of a cylinder maintained fixed.

The lower extremity A rests on a rough, horizontal support. Its inclination (measured from the vertical by angle θ) is slowly increased. When it reaches the value $\theta = 36°$, the rod is observed to slip. Assume that, at A, the friction coefficients satisfy $\mu_s \approx \mu_k = \mu$. Find the corresponding value of μ and the initial angular acceleration of the rod.

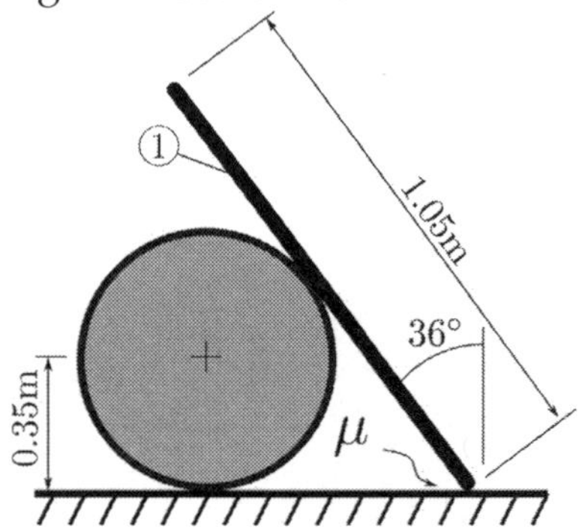

Problem 4.49 A uniform slender rod of mass m is set in contact with a smooth, vertical surface at its lower extremity A. It is then lowered so as to rest in contact with a nail at O, assumed perfectly polished. The nail is located at a distance x from the wall. Assume $|AG| = 4x$.

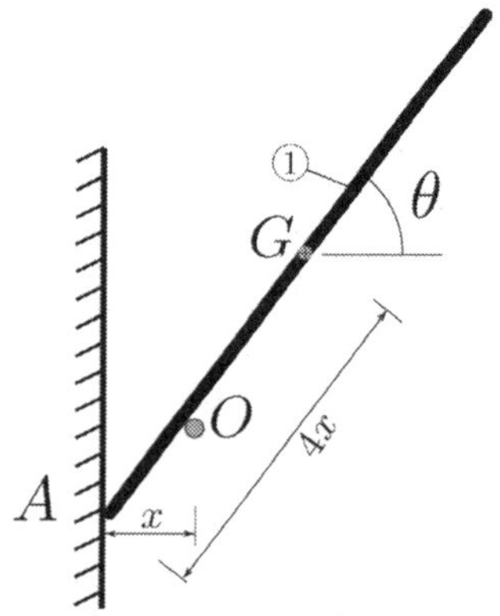

a. Find the inclination angle $\theta = \theta_*$ to set the rod in equilibrium.

b. Assume now that $\theta = \theta_* \pm 1\%$. Find the angular accelerations of the rod immediately after release from rest. Conclude as to the stability of the equilibrium position.

Answer: a) $\theta_* = 50.9°$.

Problem 4.50 A slender uniform bar of length $2l$ and mass m can rotate about a fixed horizontal axis Oz without friction. Its other extremity A is connected to a spring of constant k attached to a fixed point. The spring unstretched length is l_0. At equilibrium, the rod is horizontal and the spring is vertical.

a. Find the length of the spring at equilibrium.

b. Find the ODE which governs angle θ. Deduce the period of the small amplitude oscillations of the bar.

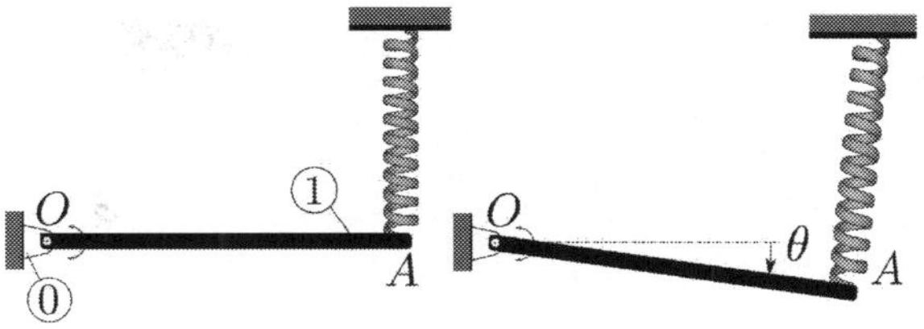

Problem 4.51 The goal of this problem is to find the relationship between the force applied by the rider on the pedals and the resulting acceleration, based on the following model:
1. The system bike/rider of total mass M has a mass center G whose motion is assumed rectilinear at velocity $v_G \hat{\mathbf{e}}_x$.
2. The frame/rider is supported by two wheels 1 and 2 which are connected by frictionless to the frame by frictionless pivots centered at O_1 and O_2. The wheels are assumed to roll without slipping relative to the road.
3. The front wheel 1 has mass center O_1, radius R_1, mass m_1, mass moment of inertia J_1 about its axis of rotation.
4. The rear wheel 2 of mass center O_2 is assumed identical to the front wheel (radius R_1, mass m_1, mass moment of inertia J_1 about its axis). The rear sprocket of radius R_2 is rigidly attached to the rear wheel.
5. The crankset 3 is connected to the frame by a frictionless pivot centered at O_3. Its mass center is O_3, its mass moment of inertia about its axis of rotation is J_3. The radius of the chainring is R_3. Each crank is of length R_4. The rider exerts a force of magnitude F alternately from from one pedal to the other, in the direction constantly perpendicular to the crank. Assume that only the top portion of the chain is taut.
Neglect air resistance and rolling friction.
a. Relate the velocity of mass center G to the angular velocity of the wheels. Then relate the angular velocity of the crankset to the angular velocity of the rear wheel.
b. Apply the principles to find four equations as follows:
1. Euler's first principle to system $\Sigma = \{\text{rider, bike}\}$,
2. Euler's second principle to the front wheel 1, to the rear wheel 2, and finally to the crankset 3.
c. Solve these equations to find the acceleration of mass center G as a function of F and the parameters. Use $M = 70$ kg, $J_1 = 0.7$ kg.m^2, $m_1 = 0.9$ kg, $R_1 = 0.35$ cm, $R_2 = 5$ cm, $R_3 = 10$ cm, $R_4 = 17.5$ cm, $J_3 = 0.02$ kg.m^2 and $F = 120$ N.

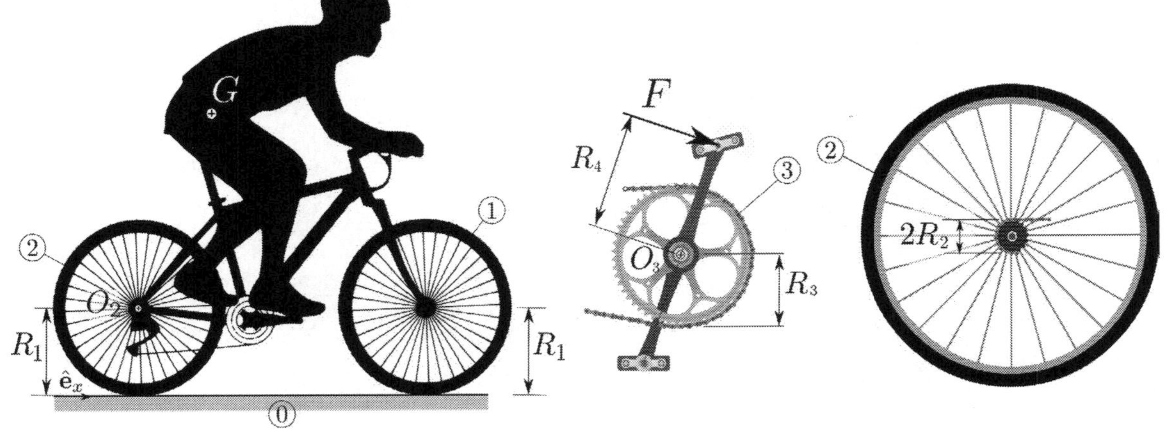

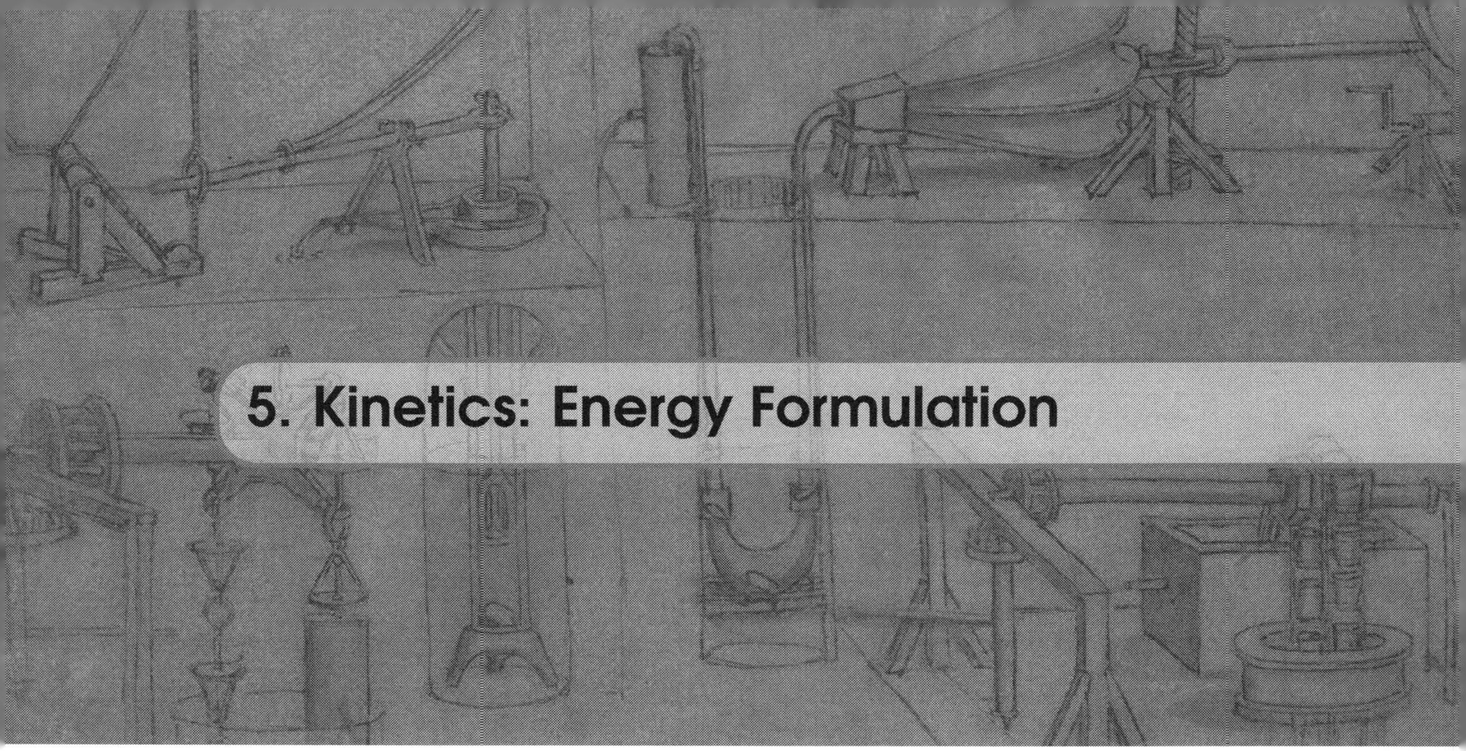

5. Kinetics: Energy Formulation

The notions of *work* and *kinetic energy* play a fundamental role in mechanics. In this chapter, we aim to derive energy formulations of the kinetics principles derived in Chapters 3 and 4, applicable to particles or rigid bodies. After reviewing the concepts of work for forces acting on particles (Section 5.1), we derive in a straightforward way the *Work-Energy Principle* (WEP) from Newton's Second Law (Section 5.2). When the work done by *non-conservative* forces vanishes, this principle takes the form of a first integral of motion known as the *Conservation of Mechanical Energy*. As an application of this conservation principle, we study the motion of particles in one-dimensional potentials (Section 5.3). We then consider the case of systems of particles and show that proper application of the WEP requires the contribution of *internal forces* between the particles (Section 5.4).

We then turn our attention to rigid bodies in planar motion. In Sections 5.5 and 5.6, we learn how to determine the kinetic energy of a rigid body and we extend the notion of work of a force acting on a particle to the work of a mechanical action exerted on a rigid body. We then derive the WEP for a rigid body (Section 5.7) or system of rigid bodies (Section 5.8) in motion in a Newtonian referential. In the latter case, *it is critical to account for the internal interactions between the bodies.*

In this chapter, we will learn

1. how to determine the work of a conservative or non-conservative force, or more generally of a mechanical action,
2. how to determine the kinetic energy of a body in planar motion,
3. how to apply the WEP to a particle, a rigid body or a system of rigid bodies,
4. the advantages and limitations of the WEP for the solution of kinetics problems.

5.1 Power, Work and Potential Energy

WITH the view of presenting an energy formulation of Newton's Second Law, we first define notions of power, work and potential in the context of particle motion. These notions will be extended to rigid body motion in Sections 5.5 and 5.6. Consider a particle P acted upon by a force $\mathbf{F}$ in a referential $\mathcal{E}$.

DEFINITION (POWER & WORK)

The (instantaneous) *power* generated by force $\mathbf{F}$ acting on a particle P in motion in a referential $\mathcal{E}$ is defined as the scalar product $\mathbf{F} \cdot \mathbf{v}_P$.

The *work* done by force $\mathbf{F}$ during the interval of time $t_1 \leq t \leq t_2$ is the integral defined by

$$W_{1\to 2} = \int_{t_1}^{t_2} \mathbf{F} \cdot \mathbf{v}_P \, dt = \int_{P_1}^{P_2} \mathbf{F} \cdot d\mathbf{r}_{OP} \tag{5.1}$$

where P_1 and P_2 are the positions taken by P at time t_1 and t_2.

Remark 1. The concepts of power and work depend on the choice of referential! The S.I. unit of power is the watt (W). The S.I. unit of work is the joule (J).

Remark 2. In the solution of problems, whenever multiple forces are at play, we will denote the work done by $\mathbf{F}$ as $W^F_{1\to 2}$.

Remark 3. Only the tangential component $F_T = \hat{\mathbf{e}}_T \cdot \mathbf{F}$ of force $\mathbf{F}$ contributes to its work: if $(\hat{\mathbf{e}}_T, \hat{\mathbf{e}}_N, \hat{\mathbf{e}}_B)$ is the Frenet basis at P, then $\mathbf{F} \cdot \mathbf{v}_P = F_T \dot{s}$ leading to the expression

$$W_{1\to 2} = \int_{s_1}^{s_2} F_T ds$$

Hence the sign of $W_{1\to 2}$ is given by the sign of F_T: if $F_T > 0$ ($F_T < 0$) along the path, the work done by $\mathbf{F}$ is positive (negative). The work done by $\mathbf{F}$ *vanishes* if $\mathbf{F}$ remains *normal* to the trajectory of P during the motion.

Consider now a *force field* in a referential $\mathcal{E}$, that is, at each location P of $\mathcal{E}$, a force $\mathbf{F}_P$ can be defined. For instance, this force field may be defined by specifying the components $X(x, y, z)$, $Y(x, y, z)$, and $Z(x, y, z)$ of $\mathbf{F}_P$ at a point P of Cartesian coordinates (x, y, z) on coordinate axes $Oxyz$ attached to $\mathcal{E}$. In general, the work $W_{1\to 2}$ done by such a force along the trajectory of a particle P between two instants t_1 and $t_2 > t_1$ is found as a curvilinear integral along the path described by P relative to $\mathcal{E}$

$$W_{1\to 2} = \int_{t_1}^{t_2} \mathbf{F}_P \cdot \mathbf{v}_P dt = \int_{P_1}^{P_2} \left(X(x,y,z)dx + Y(x,y,z)dy + Z(x,y,z)dz\right) \tag{5.2}$$

and is a function of the path taken by P starting at P_1 and ending at P_2. See Figure 5.1.

In some cases, it is possible that the work done by a time-independent force field is *only a function of the starting position P_1 and ending position P_2* of a particle P in motion in $\mathcal{E}$, and not a function of the actual path taken by P between P_1 and P_2. This property hinges on the concept of potential energy.

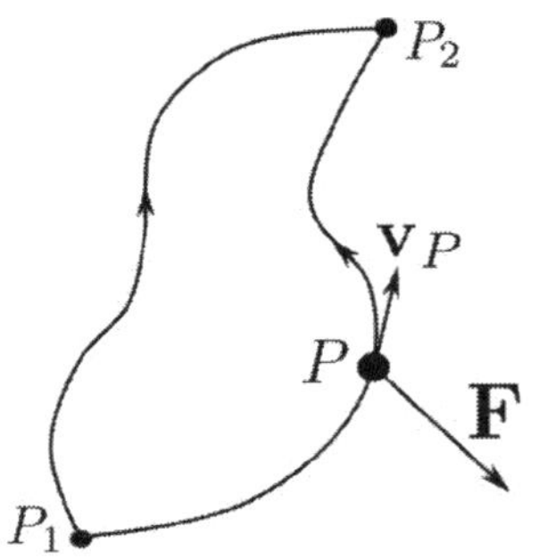

Figure 5.1 – *The work done by force* **F** *is a function of the path taken by P.*

Definition (Potential Energy)

If there exists a scalar function $U(P)$ such that the work done by force **F** takes the expression

$$W_{1\to 2} = \int_{P_1}^{P_2} \mathbf{F}_P \cdot d\mathbf{r}_{OP} = U(P_1) - U(P_2) \tag{5.3}$$

for all starting and ending positions of P, then **F** is said to be a *conservative* force, and the scalar field $U(P)$ is referred to as *potential energy*. The force field **F** is said to *derive from potential U*.

We can relate a conservative force to its potential energy as follows:

Conservative Force vs Potential Energy

If it exists, the potential energy U associated with conservative force **F** is defined by

$$\mathbf{F}_P = -\nabla U = -\frac{\partial U}{\partial x}\hat{\mathbf{e}}_x - \frac{\partial U}{\partial y}\hat{\mathbf{e}}_y - \frac{\partial U}{\partial z}\hat{\mathbf{e}}_z \tag{5.4}$$

Proof. Assuming (5.4), we can prove equation (5.3) in the following manner:

$$\begin{aligned} W_{1\to 2} &= -\int_{t_1}^{t_2} \left(\frac{\partial U}{\partial x}\dot{x} + \frac{\partial U}{\partial y}\dot{y} + \frac{\partial U}{\partial z}\dot{z}\right) dt = -\int_{t_1}^{t_2} \frac{d}{dt} U(x(t), y(t), z(t)) dt \\ &= U(P_1) - U(P_2) \end{aligned} \tag{5.5}$$

Conversely, it can be shown that, if the work done by **F** is independent of the path, then the force is of the type $\mathbf{F} = -\nabla U$. ∎

Remark 4. Equation (5.3) or (5.4) shows that potential U is only defined within an arbitrary additive constant. Since only differences of potential are evaluated, the value chosen for this constant is irrelevant.

Remark 5. The work done by a conservative force $\mathbf{F}_P$ along a closed trajectory is zero.

Here are two examples of conservative force fields encountered in practice:

1. Potential of a uniform force field. Assume that components (X, Y, Z) of **F** are independent of (x, y, z), then the potential solution of $\mathbf{F}_P = -\nabla U$ is given by $U = -Xx - Yy - Zz$. Consider for instance the constant gravitational force

$m\mathbf{g} = -mg\hat{\mathbf{e}}_z$ acting on a particle P of mass m, assuming that $\hat{\mathbf{e}}_z$ is *directed upward*, and that the gravitational acceleration g is constant. Then the corresponding potential is given by $U = mgz$.

2. Potential of an isotropic central force field. Assume a central force field, that is, a force $\mathbf{F}_P$ pointing toward a point O *fixed* in $\mathcal{E}$. See Figure 5.2. Also assume that its magnitude is function only of the radial distance $\rho = |\mathbf{r}_{OP}|$:

$$\mathbf{F}_P = -f(\rho)\hat{\mathbf{e}}_\rho$$

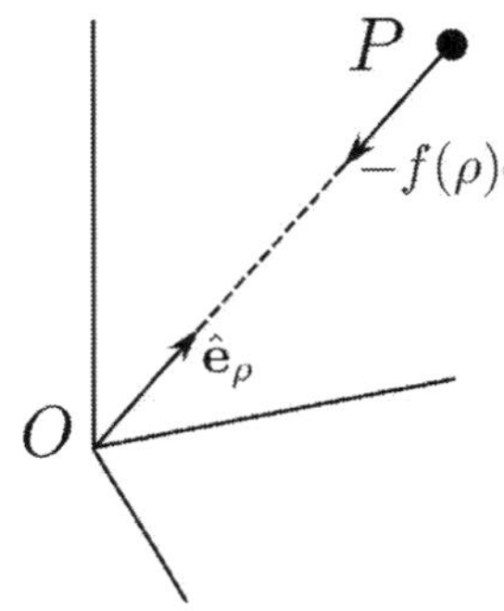

Figure 5.2 – *Central force.*

To find the work done by $\mathbf{F}_P$ we determine the following integral

$$W^{\mathbf{F}}_{1\to 2} = -\int_{t_1}^{t_2} f(\rho)\hat{\mathbf{e}}_\rho \cdot \mathbf{v}_P dt = -\int_{t_1}^{t_2} f(\rho)\hat{\mathbf{e}}_\rho \cdot (\dot{\rho}\hat{\mathbf{e}}_\rho + \rho\frac{d\hat{\mathbf{e}}_\rho}{dt})dt$$

Recall that $d\hat{\mathbf{e}}_\rho/dt$ is normal to $\hat{\mathbf{e}}_\rho$ at all time. We are left with the expression

$$W^{\mathbf{F}}_{1\to 2} = -\int_{\rho_1}^{\rho_2} f(\rho)d\rho = -\Delta U$$

where the function U defined as

$$U(\rho) = \int f(\rho)d\rho$$

is a potential function of the central force field. For instance, the gravitational force of a celestial body of spherical symmetry of mass M, center O and radius R acting on a particle of mass m is given by

$$\mathbf{F}_P = -\frac{GmM}{\rho^2}\hat{\mathbf{e}}_\rho$$

Then the corresponding potential is given by

$$U(P) = -\frac{GmM}{\rho}$$

as long as $\rho \geq R$.
As another example of a central isotropic force, consider the force exerted by a linear spring of constant k, unstretched length l_0 and attached at *fixed point* O of referential $\mathcal{E}$ on a particle P:

$$\mathbf{F}_P = -k(\rho - l_0)\hat{\mathbf{e}}_\rho$$

Then the corresponding potential is found to be

$$U = \frac{1}{2}k(\rho - l_0)^2$$

Remark 6. Whether the spring is compressed $\rho < l_0$ or stretched $\rho > l_0$, its potential energy takes the same expression.

PROPERTIES

①. For a *time-dependent force field*, it may still be possible to find a scalar field $U(t, x, y, z)$ such that $\mathbf{F} = -\nabla U$ relative to a particular referential $\mathcal{E}$. Is U a potential for $\mathbf{F}$ in referential $\mathcal{E}$? The answer is no, since in such a case equation (5.5) cannot be correct: in this case, the work done by $\mathbf{F}$ cannot be determined as the potential difference $U(P_1) - U(P_2)$.

②. The concept of potential energy depends on the choice of referential: a force field may derive from a potential in referential $\mathcal{E}$, but may *not* derive from a potential relative in referential $\mathcal{F}$.

5.2 Work-Energy Principle

CONSIDER a particle P of mass m moving in a Newtonian referential under the effect of resultant force $\mathbf{F}$. According to Newton's Second Law we have

$$m\mathbf{a}_P = \mathbf{F}$$

We then take the scalar product of this equation with velocity $\mathbf{v}_P$ and integrate from time $t = t_1$ to time $t = t_2$:

$$\int_{t_1}^{t_2} m\mathbf{a}_P \cdot \mathbf{v}_P dt = \int_{t_1}^{t_2} \mathbf{F} \cdot \mathbf{v}_P dt$$

We recognize in this last equation that
(i) $\mathbf{a}_P \cdot \mathbf{v}_P = \frac{d}{dt}(\frac{1}{2}m\mathbf{v}_P^2)$ and hence $\int_{t_1}^{t_2} m\mathbf{a}_P \cdot \mathbf{v}_P dt = \frac{1}{2}m\mathbf{v}_P^2(t_2) - \frac{1}{2}m\mathbf{v}_P^2(t_1)$,
(ii) $\int_{t_1}^{t_2} \mathbf{F} \cdot \mathbf{v}_P dt$ is the work done by resultant force $\mathbf{F}$ during the motion of P from $t = t_1$ and $t = t_2$ in $\mathcal{E}$.
This leads to the following definition:

Definition (Kinetic Energy)

The *kinetic energy* of P (in referential $\mathcal{E}$) is the scalar quantity defined by

$$K = \frac{1}{2} m \mathbf{v}_P^2 \tag{5.6}$$

We can now state the following result:

Work-Energy Principle (WEP)

Relative to a Newtonian referential, the work done by the resultant force $\mathbf{F}$ acting on P during the motion from $t = t_1$ and $t = t_2$ is equal to the change of the particle's kinetic energy:

$$\Delta K = \frac{1}{2} m \mathbf{v}_P^2(t_2) - \frac{1}{2} m \mathbf{v}_P^2(t_1) = W_{1\to 2} \tag{5.7}$$

This equation is known as the *Work-Energy Principle* (WEP).

Remark 7. The instantaneous version of this principle, known as the *kinetic energy theorem* can be stated as follows:

$$\frac{dK}{dt} = \mathbf{F} \cdot \mathbf{v}_P \tag{5.8}$$

Hence, *the time rate of change of kinetic energy is equal to the power generated by the resultant force acting on P.*

Alternative Form of the Work-Energy Principle. We can derive an alternative form of the Work-Energy Principle by writing the resultant force $\mathbf{F}$ as the sum $\mathbf{F}_c + \mathbf{F}_{nc}$ of conservative forces $\mathbf{F}_c$ deriving from a total potential U relative to $\mathcal{E}$ and of non-conservative forces $\mathbf{F}_{nc}$. Then we may write

$$\big(K(t_2) + U(P_2)\big) - \big(K(t_1) + U(P_1)\big) = W^{nc}_{1\to 2}$$

where $W^{nc}_{1\to 2}$ is the work done by non-conservative forces.
If we denote by $E = K + U$ the total *mechanical energy* of particle P, then the Work-Energy Principle can be stated as

$$E(t_2) - E(t_1) = W^{nc}_{1\to 2} \tag{5.9}$$

Finally if a particle P is in motion in a Newtonian referential under the sole effect of conservative forces or if the non-conservative forces acting on P do not work, then the mechanical energy of P is conserved during the motion

$$E(t) = \frac{1}{2} m(\dot{x}^2 + \dot{y}^2 + \dot{z}^2) + U(x, y, z) = \text{constant} \tag{5.10}$$

We call equation (5.10) a *first integral of motion* since it involves at most first-order time derivative of the coordinates.

ADVANTAGES AND LIMITATIONS OF THE WORK-ENERGY PRINCIPLE:

①. The Work-Energy Principle is an integrated form of the equation(s) of motion: it is very convenient for one-degree-of-freedom motions if the work of non-conservative forces is either neglected or can be evaluated. It may provide velocity versus position relationships without having to integrate the equation(s) of motion.

②. The Work-Energy Principle provides only one equation and is generally *not sufficient* to complete the solution of a problem. One must return to Newton's Second Law to obtain additional equations.

③. In practice, time $t = t_1$ in (5.7) is chosen as the time of the start of the motion. Time $t = t_2$ corresponds to a time at which a particular speed or position of the particle is sought. It can also be chosen as an arbitrary time $t > t_1$.

EXAMPLE PROBLEMS

GENERAL PROCEDURE: PARTICLE KINETICS BY ENERGY METHOD

Follow the guidelines given in Section 3.6.

① In Step 7, apply the Work-Energy Principle to the particle between an initial state at t_1 and a final state at t_2. For each force $\mathbf{F}$ acting on P determine the work $W^F_{1\to 2}$ by applying formula (5.1), unless the force is conservative and a potential energy is known. As an alternative, apply the Kinetic Energy Theorem to obtain an equation of motion (an ODE).

② In Step 8, the solution of a "Type II" problem should be immediate without the need to integrate an ODE unless the single equation given by the Work-Energy Principle involves too many unknowns and is insufficient to complete the solution. In that case, return to Newton's Second Law to extract additional equations.

Exercise 5.1 ▪ A crate P of mass 125 kg initially at rest on a horizontal support is pulled with a force of constant magnitude $F = 1.4$ kN and constant direction. The contact is characterized by the coefficient of kinetic friction $\mu = 0.55$.
Find the velocity of the crate when it has traveled a distance of 0.6 m. ▪

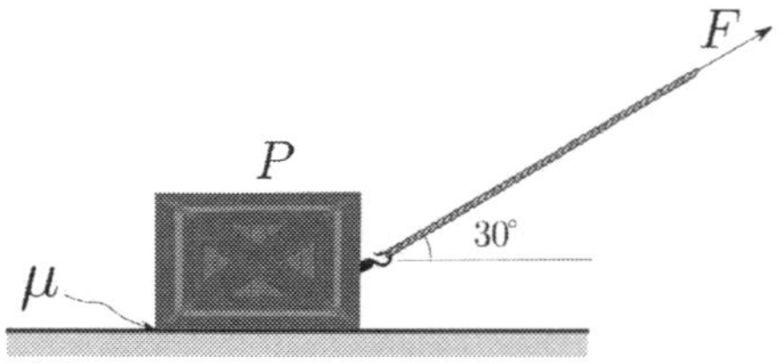

SOLUTION. STEP 1: The crate has a rectilinear motion along axis Ox. Its position is defined by the coordinate $x(t)$ with $x(0) = 0$. We seek the value $\dot{x} = \dot{x}_1$ when $x = x_1$ is known, given $\dot{x}(0) = 0$. See Figure 5.3.
STEP 2: We sketch a free-body diagram: The crate is subject to gravity $(-mg)\hat{\jmath}$, the normal reaction force $N\hat{\jmath}$ $(N > 0)$, the friction force $(-\mu N)\hat{\imath}$ (as long as the crate slips) and the constant applied force $F(\cos\beta\hat{\imath} + \sin\beta\hat{\jmath})$.

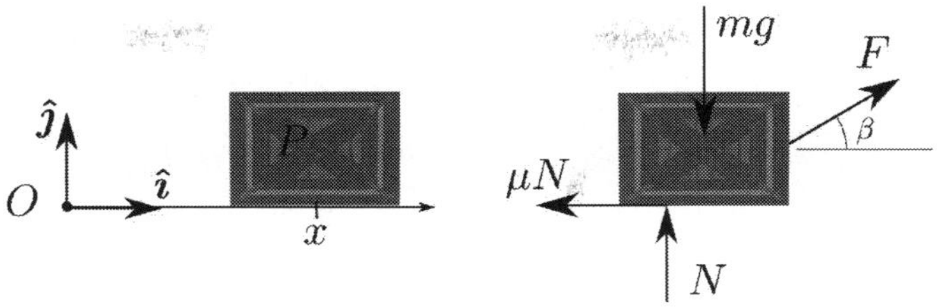

Figure 5.3 – *Problem 5.1.*

STEP 3: We apply the Work-Energy Principle from time $t = 0$ ($x = 0$, $\dot{x} = 0$) to time $t = t_1$ ($x = x_1$, $\dot{x} = \dot{x}_1$). The work done by the vertical component $(N + F\sin\beta - mg)\hat{\jmath}$ of the resultant force (normal to $\mathbf{v}_P = \dot{x}\hat{\imath}$) is zero. The remaining forces produce the work:

$$W^F_{0\to1} + W^{\mu N}_{0\to1} = \int_0^{x_1} (F\cos\beta)dx + \int_0^{x_1} (-\mu N)dx = (F\cos\beta)x_1 - \mu \int_0^{x_1} N dx$$

Application of the WEP then gives the equation

$$\tfrac{1}{2}m\dot{x}_1^2 = (F\cos\beta)x_1 - \mu \int_0^{x_1} N dx \qquad [1]$$

STEP 4: To complete the solution, we need to find the normal reaction force N. Application of Euler's first principle gives along the $\hat{\jmath}$-direction:

$$N + F\sin\beta - mg = 0$$

which shows that $N = mg - F\sin\beta$ remains constant along the motion. Now [1] gives

$$\boxed{\tfrac{1}{2}m\dot{x}_1^2 = \Big(F(\cos\beta + \mu\sin\beta) - \mu mg\Big)x_1}$$

We note that F must satisfy the bound $F > \mu mg/(\cos\beta + \mu\sin\beta)$ (to displace the crate) and $F < mg/\sin\beta$ (to keep the crate in contact with the support). Numerically, with $m = 125$ kg, $\mu = 0.55$, $F = 1.4$ kN, $x_1 = 0.6$ m, we find $\dot{x}_1 = \mathbf{2.98}$ m/s.

■

Exercise 5.2 ■■ Find the minimal launch speed v_0 of a small block P of mass 60 grams needed to reach a height of 1.25 m by sliding along the inclined flat support to position A. The contact is characterized by the coefficient of kinetic friction $\mu = 0.45$.
Repeat by treating the support as the arc of circle joining O and A and tangent at O with the horizontal axis. ■

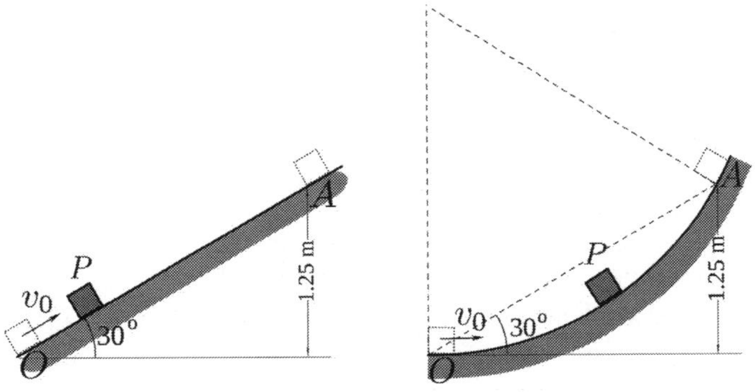

Figure 5.4 – *Problem 5.2.*

SOLUTION. CASE 1. We choose Cartesian coordinate axes Ox and Oy as shown in Figure 5.5. The motion is rectilinear and the velocity of P is given by $\mathbf{v}_P = \dot{x}\hat{\mathbf{e}}_x$. The initial and final conditions are: (i) at $t = 0$, $x = 0$ and $\dot{x} = v_0$, (ii) at $t = t_1$, $x = h/\sin\alpha$ and $\dot{x} = v_1 = 0$. The free-body diagram shows that P is subject to gravity (a conservative force) and the reaction force $N\hat{\mathbf{e}}_y - \mu N\hat{\mathbf{e}}_x$ due to the support. The component $N\hat{\mathbf{e}}_y$ does not work since it is normal to $\mathbf{v}_P$ along the motion. We can apply the WEP between $t = 0$ and $t = t_1$:

$$\tfrac{1}{2}mv_1^2 - \tfrac{1}{2}mv_0^2 = W_{0\to1}^{mg} + W_{0\to1}^{N} + W_{0\to1}^{\mu N} \qquad [1]$$

With $v_1 = 0$, $W_{0\to1}^{mg} = -mg\Delta z = -mgh$, and $W_{0\to1}^{N} = 0$ equation [1] becomes (since we are looking for the minimal value of v_0 to reach A, we set $v_1 = 0$ at A):

$$\tfrac{1}{2}mv_0^2 = mgh - W_{0\to1}^{\mu N} \qquad [2]$$

where the work done by the friction force takes the expression

$$W_{0\to1}^{\mu N} = -\int_0^{t_1} \mu N\hat{\mathbf{e}}_x \cdot \dot{x}\hat{\mathbf{e}}_x dt = -\mu \int_0^{h/\sin\alpha} N dx$$

At this point we cannot evaluate this integral since the normal reaction N is unknown. We must find N from Newton's Second Law. In view of the free-body diagram, we project the equation $m\mathbf{a}_P = m\mathbf{g} + N\hat{\mathbf{e}}_y - \mu N\hat{\mathbf{e}}_x$ on unit vector $\hat{\mathbf{e}}_y$ to find

$$N - mg\cos\alpha = 0 \qquad [3]$$

Equation [3] shows that N is a constant. We can now find the expression of the work done by the friction force:

$$W_{0\to1}^{\mu N} = -\mu N \frac{h}{\sin\alpha} = -\mu mgh\cot\alpha$$

Finally equation [2] gives the expression of v_0:

$$\boxed{v_0^2 = 2gh(1 + \mu\cot\alpha)} \qquad [4]$$

The result is, as expected, independent of mass m. With $h = 1.25$ m, $\mu = 0.45$ and $\alpha = 30°$, we find $v_0 = \mathbf{6.6}$ (m/s).

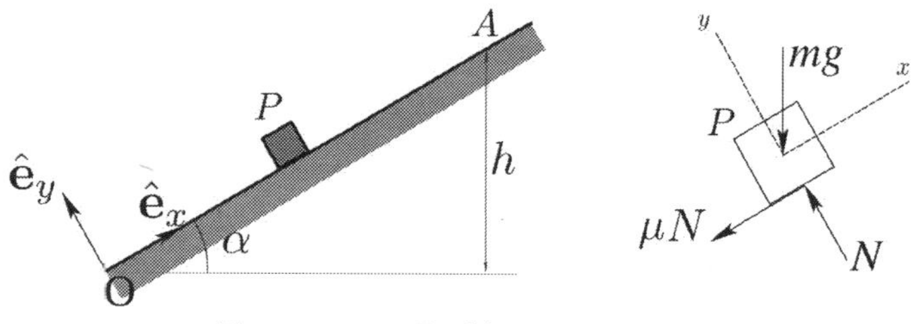

Figure 5.5 – *Problem 5.2. Case 1*

CASE 2. The path of P from O to A is now the arc circle of radius $R = h/(2\sin^2\alpha)$ and angle 2α. We choose the angle ϕ as shown in Figure 5.6 to parametrize the position of P. Angle ϕ also defines the orientation of the normal and tangential unit vectors at P. Its velocity takes the expression $\mathbf{v}_P = v\hat{\mathbf{e}}_T = R\dot{\phi}\hat{\mathbf{e}}_T$. The initial and final conditions are: (i) at $t = 0$, $\phi = 0$ and $R\dot{\phi} = v_0$, (ii) at $t = t_1$, $\phi = 2\alpha$ and $R\dot{\phi} = v = 0$. The free-body diagram shows that P is subject to gravity and the reaction force $N\hat{\mathbf{e}}_N - \mu N\hat{\mathbf{e}}_T$ due to the support. Again, the component $N\hat{\mathbf{e}}_N$ does not work since it is normal to $\mathbf{v}_P$ along the motion. The WEP between $t = 0$ and $t = t_1$ now gives

$$\tfrac{1}{2}mv_1^2 - \tfrac{1}{2}mv_0^2 = W_{0\to1}^{mg} + W_{0\to1}^{N} + W_{0\to1}^{\mu N} \qquad [1']$$

With $v_1 = 0$, $W_{0\to1}^{mg} = -mgh$, and $W_{0\to1}^{N} = 0$ equation [1′] becomes:

$$\tfrac{1}{2}mv_0^2 = mgh - W_{0\to1}^{\mu N} = mgh + \mu R \int_0^{2\alpha} N d\phi \qquad [2']$$

Again we need the normal reaction N which we find by projecting the second law

$$m(\dot{v}\hat{\mathbf{e}}_T + \frac{v^2}{R}) = m\mathbf{g} + N\hat{\mathbf{e}}_N - \mu N\hat{\mathbf{e}}_T$$

onto unit vector $\hat{\mathbf{e}}_N$: this gives

$$N = m\frac{v^2}{R} + mg\cos\phi \qquad [3']$$

We note here that N is no longer constant along the path and this seriously complicates the solution. Equation [2′] becomes

$$\tfrac{1}{2}v_0^2 = gh + \mu R \int_0^{2\alpha} (g\cos\phi + \frac{v^2}{R})d\phi$$

We can integrate the first term $g\cos\phi$ and replace v by $R\dot{\phi}$:

$$v_0^2 = 2gh + 2\mu gR\sin(2\alpha) + 2\mu R^2 \int_0^{2\alpha} \dot{\phi}^2 d\phi$$

With $R\sin(2\alpha) = h\cot\alpha$, we have the final expression

$$\boxed{v_0^2 = 2gh(1 + \mu\cot\alpha) + 2\mu R^2 \int_0^{2\alpha} \dot{\phi}^2 d\phi} \qquad [4']$$

Unfortunately the unknown integral $\int_0^{2\alpha} \dot{\phi}^2 d\phi$ does not allow for the determination of v_0 since the angular speed $\dot{\phi}$ is unknown along the path. This example demonstrates one of the limitations of the WEP: it does not always yield an answer even in the case of a one-degree-of-freedom motion.

Remark. In this problem we have verified that the work done by the friction force depends on the path joining O and A: this is a non-conservative force.

Remark. Compare [4′] with [4]: it takes a larger initial speed to reach A on the circular path, as expected.

Remark. To find $\dot{\phi}$ versus ϕ we need to return to Newton's Second Law, projected onto $\hat{\mathbf{e}}_T$:

$$m\dot{v} = -\mu N - mg\sin\phi$$

Using the expression [3′] of N and $\dot{v} = R\ddot{\phi}$ we obtain the equation of motion:

$$R\ddot{\phi} + \mu R\dot{\phi}^2 + g(\sin\phi + \mu\cos\phi) = 0 \qquad [5']$$

It is easy to see that [4′] can be obtained from [5′]: multiply [5′] by $\dot{\phi}$ and recognize

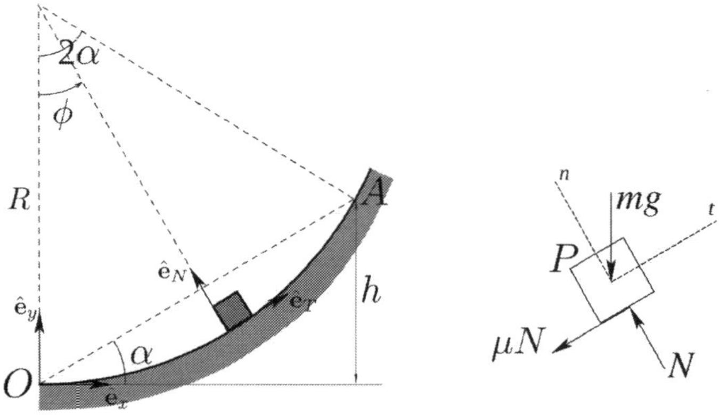

Figure 5.6 – *Problem 5.2: Case 2*

each term of the obtained equation as d/dt of a quantity: for example $\ddot{\phi}\dot{\phi} = \frac{d}{dt}(\dot{\phi}^2/2)$. After all, this is exactly how the Work-Energy Principle has been derived! To integrate [5′], let $u = \dot{\phi}^2$. Since we want u versus ϕ, we find $du/d\phi$:

$$\frac{du}{d\phi} = \frac{d}{dt}(\dot{\phi}^2)\frac{dt}{d\phi} = 2\dot{\phi}\ddot{\phi}\frac{1}{\dot{\phi}} = 2\ddot{\phi}$$

Now [5′] becomes:

$$\frac{du}{d\phi} + 2\mu u = -\frac{2g}{R}(\sin\phi + \mu\cos\phi)$$

This is a first-order linear ODE! We can find $u = \dot{\phi}^2$ vs ϕ in closed-form by imposing the condition $u = \dot{\phi} = 0$ at $\phi = 2\alpha$. Then it would be possible to evaluate the integral $\int_0^{2\alpha} \dot{\phi}^2 d\phi$ and complete the equation [4′] to find v_0. The courageous reader may complete the algebra to find v_0 numerically.

■

Exercise 5.3 ▪▪ A small ball P of mass m leaves a ramp without initial speed at a height h from horizontal axis Ax. It enters a circular track of radius R at point A. It may reach highest point B and exit the circular track at A to continue on horizontal track Ax. The contact between P and the tracks is unilateral and frictionless.
Study the motion of P as a function of h. In particular, show that, if $h/R > 5/2$, P travels around the entire loop.
■

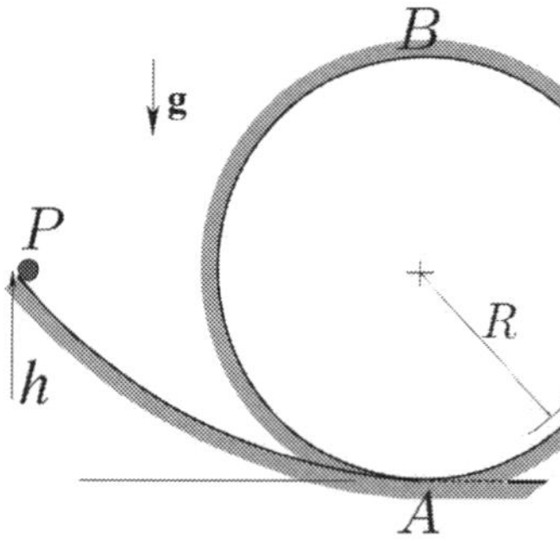

SOLUTION. We study the motion of P on its circular path by choosing the NT coordinate system as shown in Figure 5.7. Then the acceleration of P takes the expression

$$\mathbf{a}_P = \dot{v}\hat{\mathbf{e}}_T + \frac{v^2}{R}\hat{\mathbf{e}}_N$$

where the orientation of $(\hat{\mathbf{e}}_T, \hat{\mathbf{e}}_N)$ is defined by angle θ.
A free-body diagram shows that P is subject to gravity and the normal reaction force $N\hat{\mathbf{e}}_N$ (the contact is frictionless). Then application of Newton's Second Law

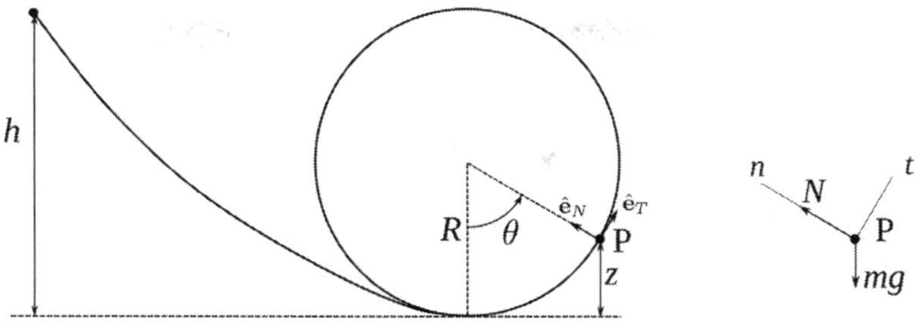

Figure 5.7 – *Problem 5.3 (Solution)*

$m\mathbf{a}_P = m\mathbf{g} + N\hat{\mathbf{e}}_N$ projected on $(\hat{\mathbf{e}}_T, \hat{\mathbf{e}}_N)$ yields two equations

$$m\dot{v} = -mg\sin\theta, \qquad m\frac{v^2}{R} = N - mg\cos\theta \qquad [1-2]$$

Recall that P stays in contact with its support as long as $N > 0$. To find out if this condition is satisfied we need to find N as a function of angle θ: thus we need to find speed v versus θ. Instead of integrating equation [1], we apply the WEP between $t = 0$ (when $v = 0$) and $t > 0$ (when P reaches arbitrary position defined by θ):

$$\frac{1}{2}mv^2 - \frac{1}{2}mv_0^2 = W^N_{0\to 1} + W^{mg}_{0\to 1} = mg(h - z) \qquad [3]$$

where we have used $W^N_{0\to 1} = 0$ (the contact is frictionless: $\mathbf{N}\cdot\mathbf{v}_P = 0$). Note that equation [3] is equivalent to writing conservation of energy $E = K + U = K_0 + U_0$ with $K = \frac{1}{2}mv^2$ and $U = mgz$. Then using $v_0 = 0$ and $z = R(1 - \cos\theta)$, we find the speed (squared) as a function of θ:

$$v^2 = 2g(h - R + R\cos\theta) \qquad [4]$$

Equation [4] shows that if P is released at height h, then P will possibly reach the maximum position $\theta_{max} = \cos^{-1}(1 - h/R)$ (as long as it does not lose contact). Using [4] we find the expression of N:

$$N = mg(2\frac{h}{R} - 2 + 3\cos\theta) \qquad [5]$$

N/mg decreases from the maximum value $2h/R + 1$ at $\theta = 0$ to the minimum value $2h/R - 5$ at $\theta = \pi$. We find that N vanishes at the position $\theta = \theta_*$ given by

$$\cos\theta_* = \frac{2}{3}(1 - \frac{h}{R}) \qquad [6]$$

as long as $h/R < 5/2$. At this position P reaches the speed v_* given by

$$v_*^2 = \frac{2}{3}gR(\frac{h}{R} - 1)$$

which requires the condition $h/R > 1$.

Remark. It is more challenging to integrate [1]: with $v = R\dot{\theta}$ the ODE becomes $R\ddot{\theta} = -g\sin\theta$. Using the technique of section C.2 we integrate to find $R(\dot{\theta}^2 - \dot{\theta}_0^2)/2 = g(\cos\theta - \cos\theta_0)$ with $\dot{\theta} = \dot{\theta}_0$ at $\theta = \theta_0$. The power of the WEP is that it performs this integration for us!

Now we can conclude that:

(i) if $h/R \leq 1$, P never loses contact and oscillates between the initial position and a maximum position given by $\theta_{max} = \cos^{-1}(1 - h/R)$.
(ii) if $1 < h/R \leq 5/2$, P reaches a maximum position on the circular path given by [6] and loses contact at the position given by $\cos\theta_* = \frac{2}{3}(1 - \frac{h}{R})$ (corresponding to height $z = \frac{2h}{3} + \frac{R}{3}$). For instance if $h/R = 2$ we find $\theta_* = 48.2°$.
(iii) if $h/R > 5/2$, P never loses contact and travels around the entire loop. ■

Exercise 5.4 ■ We are to design a skateboard ramp joining two points A and B of a vertical plane as shown in Figure 5.8: the skateboarder (modeled as a particle) is released from rest at A to reach position B *in the least amount of time.* We have the choice of three supports joining A and B: (1) the straight line AB, (2) the arc of circle with a vertical slope at A, and (3) an arc of inverted cycloid in the form

$$x(\theta) = R(\theta - \sin\theta), \qquad y(\theta) = R(1 - \cos\theta)$$

Find the general expression of time T_{AB} taken to travel on a path joining A and B. Then evaluate T_{AB} for each design path and conclude. ■

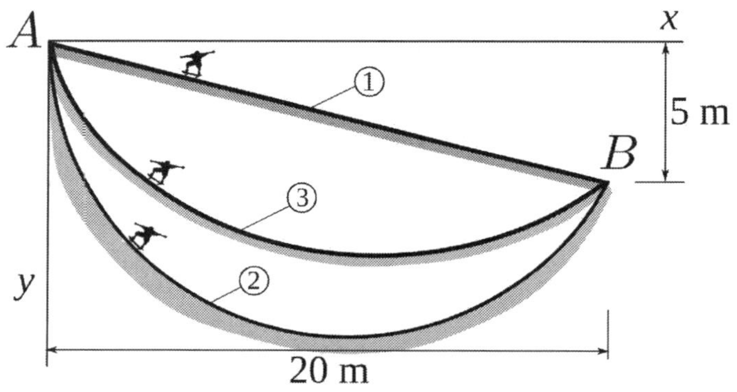

Figure 5.8 – *Problem 5.4*

Solution. On the coordinate axes Axy we denote as (a, b) the coordinates of B. See Figure 5.9. Since the contact is frictionless, the motion of P of coordinates (x, y) and speed v is characterized by conservation of energy: at any given time we have

$$\frac{1}{2}mv^2 = mgy$$

using the fact that $v = 0$ at $x = y = 0$ (point A). Denoting by s the arc-length measured from A to P, the speed of P is $v = ds/dt = \sqrt{2gy}$, assuming that $y \geq 0$ along the path. This gives us a relationship between the variables t, y and s: $dt = ds/\sqrt{2gy}$. Integration gives the total time taken to travel from A to B:

$$\boxed{T_{AB} = \int_0^{s_B} \frac{ds}{\sqrt{2gy}}} \qquad [1]$$

where s_B is the total distance traveled to B. In practice, equation [1] can be used to find time T_{AB} by finding the relationship between variables s and y for each chosen path described by P.

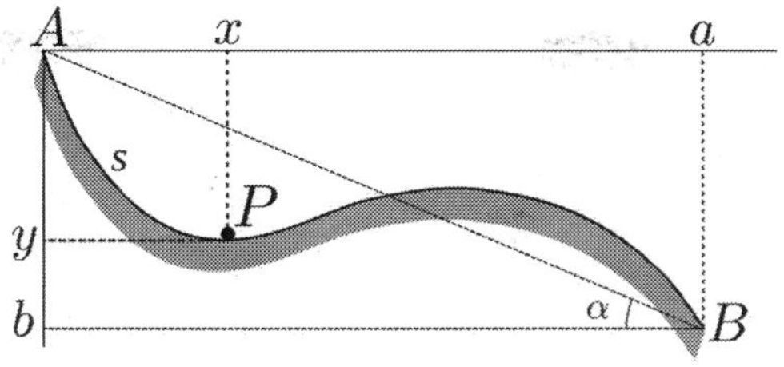

Figure 5.9 – *Problem 5.4 (Solution)*

PATH 1: the path is the straight line AB, making the angle α with the horizontal. In this case we have $dy = ds\sin\alpha$, with $\sin\alpha = b/\sqrt{a^2+b^2}$. Then equation [1] becomes

$$T_{AB} = \int_0^b \frac{dy}{\sin\alpha\sqrt{2gy}}$$

Using $\int dy/\sqrt{y} = 2\sqrt{y}$, we obtain, after simplifications

$$\boxed{T_{AB} = \sqrt{\frac{2}{g}}\sqrt{\frac{a^2+b^2}{b}}}$$

With $a = 20$ m and $b = 5$ m, we find $T_{AB} = \mathbf{4.16}$ seconds.
PATH 2: the arc of circle passing through A and with a vertical tangent at A is parametrized in the form (its center is necessarily on axis Ax)

$$x = r(1-\cos\phi), \qquad y = r\sin\phi$$

with $r > 0$ and $0 < \phi < 2\pi$. To guarantee that this curve passes through B we must solve the following equations:

$$r(1-\cos\Phi) = a, \qquad r\sin\Phi = b$$

This leads to $\tan(\Phi/2) = a/b$: for $a/b = 4$ we pick the root $\Phi = 2.65$ (radians) which leads to $r = 10.62$ m. Then to find T_{AB} from [1] we use the relationship $ds = rd\phi$ and find the expression

$$\boxed{T_{AB} = \sqrt{\frac{r}{2g}}\int_0^\Phi \frac{d\phi}{\sqrt{\sin\phi}}}$$

We rely on `Maple`™ for a numerical evaluation of T_{AB}: with $\Phi = 2.65$ (radians), $r = 10.62$ m, we find $T_{AB} = \mathbf{2.82}$ seconds.
PATH 3: the path is a cycloid described by

$$x(\theta) = R(\theta - \sin\theta), \qquad y(\theta) = R(1-\cos\theta)$$

with $R > 0$ and $0 < \theta < 2\pi$. To find the cycloid passing through B we must solve the nonlinear equations

$$R(\Theta - \sin\Theta) = a, \qquad R(1-\cos\Theta) = b$$

This can only be done numerically: with the help of a numerical solver, we find $R = 3.76$ m and $\Theta = 4.38$ (radians) for $a = 20$ m and $b = 5$ m. Next to find T_{AB} we must relate y, s and θ: we first find

$$ds^2 = dx^2 + dy^2 = R^2 d\theta^2(2 - 2\cos\theta) = 4R^2 d\theta^2 \sin^2\frac{\theta}{2}$$

leading to

$$\frac{ds}{\sqrt{2y}} = 2Rd\theta\frac{\sin\frac{\theta}{2}}{\sqrt{2y}} = R\dot{\theta}$$

Now we can find T_{AB}

$$\boxed{T_{AB} = \Theta\sqrt{\frac{R}{g}}}$$

With $R = 3.76$ m, $\Theta = 4.38$ (in radians) we find $T_{AB} = \mathbf{2.33}$ seconds.
CONCLUSION: Of the three choices, *the cycloid is the best design*. Its performance is quite impressive and unexpected: Figure 5.10 shows the positions of P on the straight and circular paths when P has reached B on the cycloid. It seems that a sharp (but not too sharp) drop below B with a portion climbing back to B is best.
This begs the question: of all possible designs, is the cycloid the optimal solution? This is the mathematical problem of determining the *brachistochrone curve* (from the Greek: *brakhisto* - the shortest, and *chronos* - time) or curve of fastest descent. This problem was first posed by Johann Bernoulli in 1696. The solution is indeed the cycloid and this can be shown by the method of *calculus of variations*.

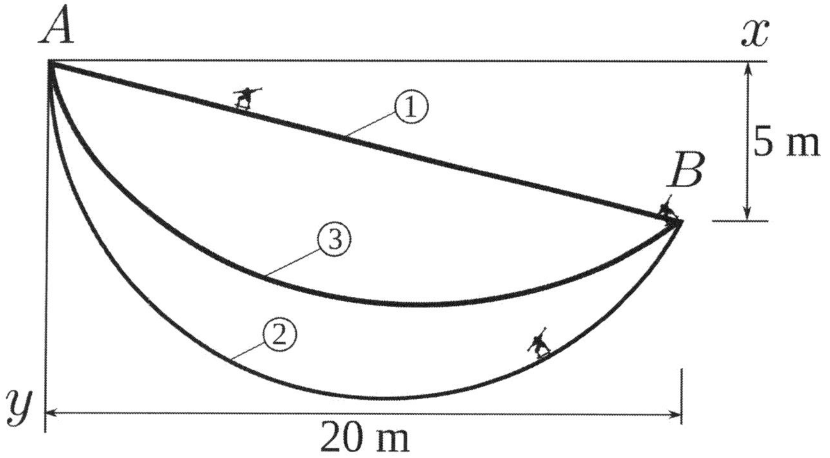

Figure 5.10 – *Problem 5.4 (Solution). The position of the skateboarders at t* = 2.33 *seconds.*

■

Exercise 5.5 ■■■ A pendulum is assembled by connecting a particle P of mass m to a fixed point O by a string of length l. After displacing the pendulum to the inclination angle θ_0, P is set in motion at initial speed v_0 normal to the vertical plane containing the pendulum.
Find the value which must be given to v_0 so that the pendulum reaches the maximum deviation $\theta_1 > \theta_0$.
Hint: find two first integrals based on mechanical energy and angular momentum. ■

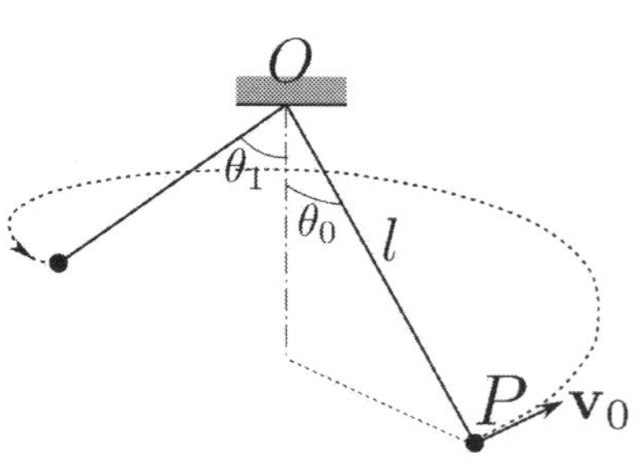

SOLUTION. We could solve this problem by applying Newton's Second Law in cylindrical coordinates and integrate this equation from $t = t_0$ to $t = t_1$. Instead, we will seek, as suggested, conservation laws in terms of mechanical energy and angular momentum. No integration will be needed.

Step 1. We set up the kinematics in the unit vectors $(\hat{\mathbf{e}}_r, \hat{\mathbf{e}}_\theta, \hat{\mathbf{e}}_\phi)$ with $\hat{\mathbf{e}}_r$ directed along line OP, $\hat{\mathbf{e}}_\theta$ normal to $\hat{\mathbf{e}}_r$ in the plane containing lines OP and the vertical. Then $\hat{\mathbf{e}}_\phi = \hat{\mathbf{e}}_\theta \times \hat{\mathbf{e}}_r$. Note that $(\hat{\mathbf{e}}_r, \hat{\mathbf{e}}_\theta, \hat{\mathbf{e}}_\phi)$ are not the standard cylindrical unit vectors. See Figure 5.11. Since $\mathbf{v}_P = d(l\hat{\mathbf{e}}_r)/dt$, we expect that the velocity is of the form $\mathbf{v}_P = v_\theta \hat{\mathbf{e}}_\theta + v_\phi \hat{\mathbf{e}}_\phi$. However at position $\theta = \theta_0$, the initial velocity is given by $\mathbf{v}_P = v_0 \hat{\mathbf{e}}_\phi$ since P is set in motion at initial speed v_0 normal to the vertical plane containing the pendulum. At the final position $\theta = \theta_1$, $\mathbf{v}_P = v_1 \hat{\mathbf{e}}_\phi$ since the pendulum reaches the maximum deviation.
Step 2. P is subject to the following forces: (i) tension $-T\hat{\mathbf{e}}_r$ (its moment about O vanishes), (ii) gravity $mg\hat{\mathbf{e}}_z$: its moment about O is $l\hat{\mathbf{e}}_r \times mg\hat{\mathbf{e}}_z = mgl\sin\theta \hat{\mathbf{e}}_\phi$.
Step 3. We apply the angular momentum form of the Second Law about O:

$$\frac{d}{dt}\mathbf{H}_O = mgl\sin\theta\hat{\mathbf{e}}_\phi$$

This shows that there is conservation of the angular momentum of P about O along $\hat{\mathbf{e}}_z$:

$$(\mathbf{H}_O \cdot \hat{\mathbf{e}}_z)(t_0) = (\mathbf{H}_O \cdot \hat{\mathbf{e}}_z)(t_1)$$

At $t = t_0$, we have $\mathbf{H}_O(t_0) = l\hat{\mathbf{e}}_r \times mv_0\hat{\mathbf{e}}_\phi = mlv_0\hat{\mathbf{e}}_\theta$ giving $(\mathbf{H}_O \cdot \hat{\mathbf{e}}_z)(t_0) = -mlv_0 \sin\theta_0$. Similarly at $t = t_1$: $(\mathbf{H}_O \cdot \hat{\mathbf{e}}_z)(t_1) = -mlv_1 \sin\theta_1$.

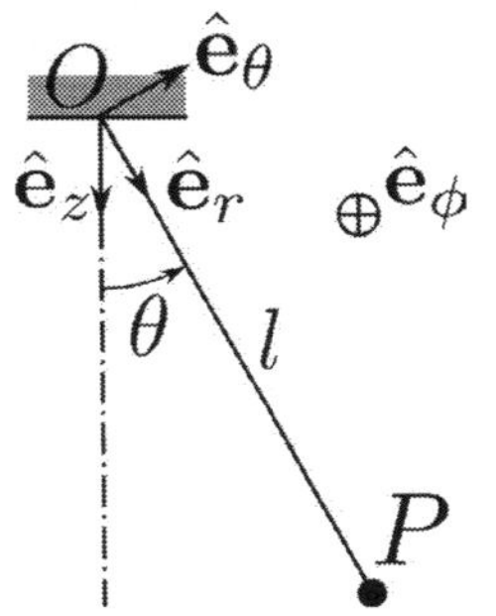

Figure 5.11 – *Problem 5.5 (Solution).*

We conclude that:

$$\boxed{v_0 \sin\theta_0 = v_1 \sin\theta_1} \qquad [1]$$

Step 4. We need to relate speed v_1 to initial speed v_0. We apply the WEP by noting that the tension force does not work:

$$\mathbf{T} \cdot \mathbf{v}_P = (-T\hat{\mathbf{e}}_r) \cdot \frac{d}{dt}(l\hat{\mathbf{e}}_r) = 0$$

Hence, we can state that the mechanical energy is conserved between $t = t_0$ to $t = t_1$:

$$\boxed{\tfrac{1}{2}mv_1^2 - \tfrac{1}{2}mv_0^2 = mgl(\cos\theta_1 - \cos\theta_0)} \qquad [2]$$

Step 5. We combine equations [1] and [2] to find the final result

$$\boxed{v_0^2 = 2gl\frac{\sin^2\theta_1}{\cos\theta_1 + \cos\theta_0}}$$

■

5.3 Application: Motion in a One-Dimensional Potential

As a special application of conservation of energy, we study in this section the motion of a particle in a one-dimensional potential. With simple arguments, we can show valuable qualitative and quantitative information of such motions.

ASSUMPTIONS. The motion of a particle P of mass m in a Newtonian referential $\mathcal{E}$ is described by a single independent coordinate $q(t)$: P has a single degree of freedom. Particle P does not necessarily move along a straight line, but rather can move along a curve, or can be constrained to move on a surface. Hence variable $q(t)$ may be a Cartesian coordinate $x(t)$, a radial distance $\rho(t)$ or even an angle $\theta(t)$.
Assume that the forces acting on P either do not work, or are conservative, that is, they derive from a potential $U(q)$ function of coordinate q only. Hence the motion of P is characterized by *conservation of the mechanical energy* $E = K + U$.

$$E = K + U(q) = \tfrac{1}{2}m\dot{q}^2 + U(q) = \text{constant} \tag{5.11}$$

The value of constant E in equation (5.11) is found from the initial conditions $q_0 = q(0)$ and $\dot{q}_0 = \dot{q}(0)$ prescribed to P:

$$E = \tfrac{1}{2}m\dot{q}_0^2 + U(q_0). \tag{5.12}$$

Since the kinetic energy K must remain positive or equal to zero, we see from the equation $U = E - K$ that the range of values described by variable $q(t)$ during the motion must satisfy the inequality

$$U(q) \leq E. \tag{5.13}$$

We will see that this condition has important consequences on the motion of P. Figure 5.12 shows some examples of potential functions.

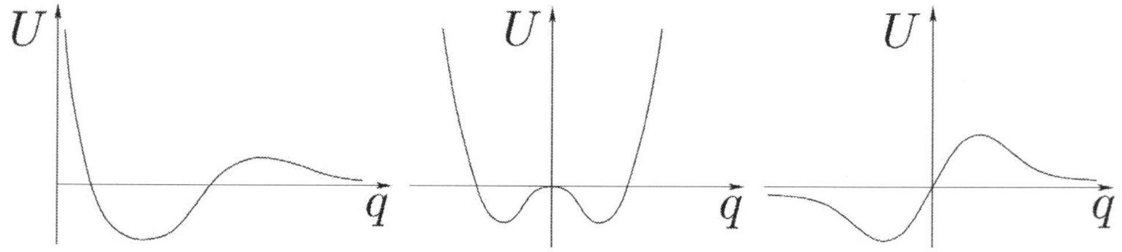

Figure 5.12 – *Example of potential functions $U(q)$.*

Remark 8. It is possible that, due to a constraint equation between q and some other coordinate describing the motion of P, both the kinetic and potential energy may include additional terms. The motion is still described by an "effective" potential energy function.

EQUILIBRIUM POSITIONS. Before examining the question of the possible motions described by P, we look for the conditions under which equilibrium can occur. The equation of motion is obtained by taking the time-derivative of equation (5.11):

$$m\ddot{q} + \frac{dU}{dq} = 0 \tag{5.14}$$

A solution $q(t) = q_e = $ constant is called an *equilibrium solution*: it satisfies $\dot{q} = \ddot{q} = 0$ and

and hence corresponds to a root of dU/dq

$$\frac{dU}{dq}(q_e) = 0. \tag{5.15}$$

We can then state that:

Equilibrium Positions

The equilibrium positions of a particle in motion in a one-dimensional potential correspond to the extrema of the potential energy.

For instance, for the potential shown in Figure 5.13, the equilibrium positions correspond to the values $q = q_1$ and q_2.

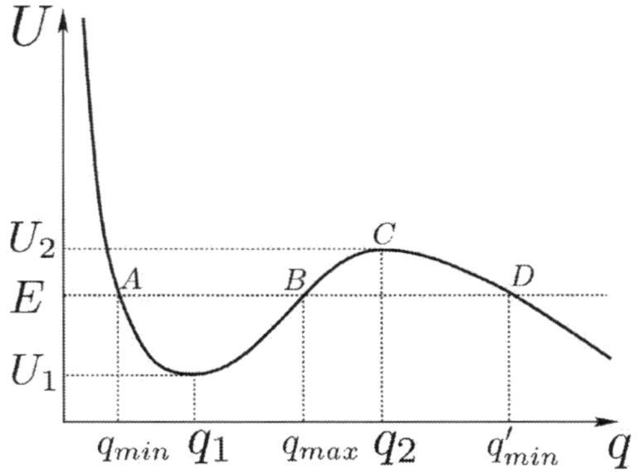

Figure 5.13 – *The equilibrium positions q_1 and q_2 correspond to the extrema of potential function $U(q)$. The minimum q_1 is stable. The maximum q_2 is unstable.*

Stability of an Equilibrium Point. Not all equilibrium positions are physically realizable: they may not be stable. We define a *stable* equilibrium position P_e at $q = q_e$ as a position such that motions of P starting near P_e remain near P_e at all times. More specifically, $q = q_e$ is stable if particle P is subjected to a "return" force F in the vicinity of $q = q_e$, that is, a force satisfying the conditions:

$$\begin{aligned} F(q) > 0 \qquad & \text{for} \quad q < q_e \\ F(q) < 0 \qquad & \text{for} \quad q > q_e \end{aligned} \tag{5.16}$$

for all $|q - q_e|$ sufficiently small. Particle P is subject to the force $F = -dU/dq$. To study the effect of force F on P near P_e, we expand F in a Taylor series about $q = q_e$:

$$F = -\frac{dU}{dq}(q_e) - (q - q_e)\frac{d^2U}{dq^2}(q_e) + \cdots = -(q - q_e)\frac{d^2U}{dq^2}(q_e) + \cdots \tag{5.17}$$

We can make the following conclusion:

Stability of Equilibrium Positions

The equilibrium position $q = q_e$ is *stable* if the potential U satisfies $\dfrac{d^2U}{dq^2}(q_e) > 0$, that is, if q_e corresponds to a *minimum* value of U. Conversely, the maxima of U correspond to unstable equilibrium positions.

Note that in the vicinity of a stable position $q = q_e$ the motion of P is governed by the linear differential equation

$$\ddot{q} + \Omega^2(q - q_e) = 0, \qquad \text{with} \quad \Omega^2 = \frac{1}{m}\frac{d^2U}{dq^2}(q_e)$$

the solution of which can be written as

$$q(t) = q_e + A\cos\Omega t + B\sin\Omega t.$$

The constants A and B are found by imposing the initial conditions of P. Hence, we see that the motion in the vicinity of the equilibrium position correspond to small amplitude harmonic oscillations about $q = q_e$ of period $T = 2\pi/\Omega$. Note that the period of these oscillations is independent of the energy level E. For the potential function shown in the Figure 5.13, $q = q_1$ is the only stable equilibrium.

Possible Motions. We can predict the qualitative nature of the possible motions of P by a graphical method. Consider the potential function U of Figure 5.13. Let us first assume that the initial conditions q_0 and $\dot{q}_0$ lead to a value of the energy E satisfying

$$U_1 = U(q_1) < E < U_2 = U(q_2).$$

We see graphically, that two possible motions guaranteeing $U(q) \leq E$ can arise. First, there exists a motion described by the segment AB corresponding to values of q satisfying the inequality

$$q_{\text{min}}(E) \leq q(t) \leq q_{\text{max}}(E).$$

as shown in the figure. The bounds $q_{\text{min}}(E)$ and $q_{\text{max}}(E)$ correspond to the minimum kinetic energy $T = 0$ and the maximum potential energy $U_{\text{max}} = E$. Hence, particle P oscillates between the extreme values q_{min} and q_{max} about the stable equilibrium position $q = q_1$. These oscillations are not linear though, and in general, are characterized by a period dependent on the energy level E, unless the motion is confined to a small neighborhood of q_1, that is, unless $E \approx U_1$. In this case, we say that *the motion is confined to a potential well*. It is easy to show from equation (5.11) that the period of oscillation is given by

$$T(E) = 2\sqrt{m}\int_{q_{\text{min}}(E)}^{q_{\text{max}}(E)} \frac{dq}{\sqrt{2(E - U(q))}}.$$

There exists a second possible motion described by the segment CD (with D located at $q = \infty$) corresponding to all values $q > q'_{\text{min}}(E)$: here the motion is *unbounded*. Even if P initially moves to the left on the q-axis (that is, $\dot{q}_0 < 0$), P can never get close to equilibrium position q_1: *it cannot overcome the potential barrier* $U_2 - E$. This second motion corresponds to the same value of energy E, but not to the same initial conditions!
Finally, assume that the initial conditions q_0 and $\dot{q}_0$ lead to a value of the energy E satisfying

$$E > U_2 = U(q_2).$$

Then, graphically, we see that the resulting motion satisfying $U(q) \leq E$ corresponds to all values of $q(t) \geq q''_{\min}(E)$, where the minimum bound $q''_{\min}$ corresponds to zero kinetic energy. For instance, if initially $q_0 > q_1$ and $\dot{q}_0 < 0$, then particle P passes through the stable equilibrium position q_1, reaches $q = q''_{\min}$, then bounces back never to return toward the stable equilibrium.

Application: Orbital Motion around a Celestial Body. As an application of energy methods, we examine the problem of the gravitational attraction of a large celestial body on a body P of much smaller mass m. The celestial body is assumed to have spherical material symmetry, to be of mass $M \gg m$, and of mass center O.

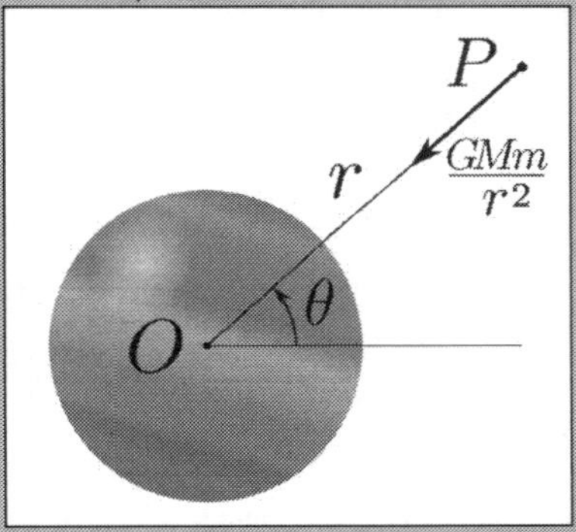

Figure 5.14 – *Particle P acted upon by a celestial body.*

Neglecting all interactions with other bodies, this is an example of a closed system, as studied in sections 3.8 and 4.2. Given the disparity in the two masses, the mass center of the system can be considered to be point O. With these assumptions, point O is necessarily in uniform rectilinear motion, and the motion of P is studied in the geocentric referential $\mathcal{G}$ of the celestial body (which is Newtonian).

Two fundamental laws govern the motion of P in $\mathcal{G}$: (i) conservation of its mechanical energy (the gravitational force of attraction is conservative), and (ii) conservation of its angular momentum about O (the force is a central force). From section 3.10, we know that the motion of P is planar, and that it can be studied in polar coordinates (r, θ) (of pole O). In this coordinate system, conservation of mechanical energy and of angular momentum can be expressed as $E = \frac{1}{2}m(\dot{r}^2 + r^2\dot{\theta}^2) - \frac{GMm}{r}$ and $r^2\dot{\theta} = C$, respectively. Both constants E and C can be expressed in terms of the initial conditions of P.

We can then write the kinetic energy term $\frac{1}{2}mr^2\dot{\theta}^2$ as $\frac{1}{2}mC^2/r^2$ and add it to the potential energy $U = -\frac{GMm}{r}$ to obtain an effective potential energy

$$U_{\text{eff}}(r) = mC^2/2r^2 - GMm/r \qquad (5.18)$$

which is sketched in Figure 5.15. The motion of P is then studied as the motion in a one-dimensional potential: recall that all possible motions must satisfy $U_{\text{eff}} \leq E$. We can then see that two possible motions are possible:

Case 1: $E = E_1 \geq 0$. The solution of $U_{\text{eff}}(r) \leq E$ leads to the interval $r_A \leq r \leq \infty$. The trajectory of P is unbounded, with a position A of shortest approach. The motion must correspond to the hyperbolic trajectory found in section 3.10. The special case of $E = 0$ corresponds to a parabolic motion.

Case 2: $E = E_2 < 0$. The solution of $U_{\text{eff}}(r) \leq E$ leads to the interval $r_B \leq r \leq r_C$. The two values r_B and r_C are solutions of the quadratic equation $E_2 r^2 + GMmr - mC^2/2 = 0$.

The special case of double root $r = r_D$ corresponds to a circular orbit. These values correspond to the minimum and maximum value of radial distance r ($\dot{r} = 0$) of an elliptic orbit.

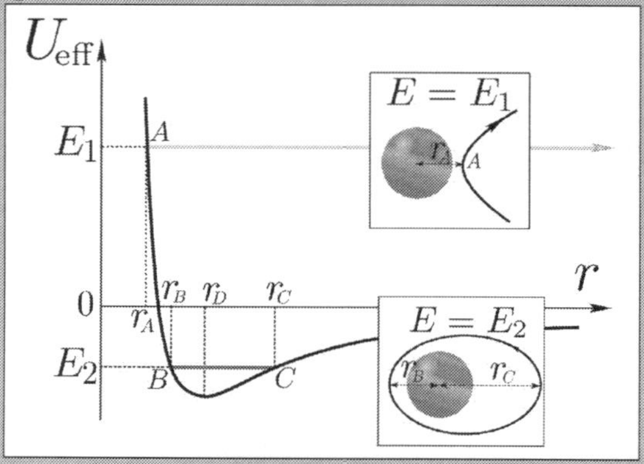

Figure 5.15 – *Effective potential energy.*

POLAR EQUATION OF TRAJECTORY. It is possible to obtain the polar equation $r(\theta)$ from the expression of the energy after using $dt = C^{-1}r^2 d\theta$ (defining the variable $u = 1/r$):

$$E = \tfrac{1}{2}mC^2((du/d\theta)^2 + u^2) - GMmu$$

Setting $dE/d\theta = 0$ (recall that the energy is constant), we obtain the equation:

$$d^2u/d\theta^2 + u = GM/C^2$$

whose solution[a] is given by the equation of a conic section

$$u = 1/r = (1 + e\cos(\theta - \theta_0))/p \qquad (5.19)$$

with $p = C^2/GM$ and (e, θ_0) are constants of integration. It is possible to relate the two constants e (the excentricity) and E (the energy). If the expression of u is substituted in the expression E, we find the relationship:

$$E = \tfrac{1}{2}m(\frac{GM}{C})^2(e^2 - 1) \qquad (5.20)$$

We can then relate the sign of E to the sign of $(e-1)$:
(i) if $0 \le e < 1$ ($E < 0$), the trajectory is elliptic (possibly circular).
(ii) if $e = 1$ ($E = 0$), the trajectory is parabolic.
(iii) if $e > 1$ ($E > 0$), the trajectory is hyperbolic.

ESCAPE VELOCITY. The escape velocity of body P is the minimum speed needed for P located at a distance r from center O to escape from the gravitational influence of the celestial body. Its magnitude v_e is found by guaranteeing that P will be found at $r = \infty$ (where the gravitational attraction vanishes) with zero (minimum) velocity: hence its energy $E = \frac{1}{2}mv^2 - GMm/r$ is zero, which leads to the expression:

$$v_e = \sqrt{\frac{2GM}{r}} \qquad (5.21)$$

On the surface of the Earth ($r = R_E = 6,370$ km), the escape velocity is about 11.2 km/s. At 9,000 km altitude, it decreases to 7.1 km/s. A mechanism by which planets or moons lose or retain their atmosphere can be obtained by comparing the escape velocity to the average thermal velocity of molecules $v_T = \sqrt{3k_BT/m}$, where k_B is Boltzmann's constant, T is temperature, and m is the mass of the molecule. On Earth, the escape velocity is sufficiently large to retain most molecules such as O_2, N_2 or CO_2. However, the thermal velocity of H_2 is much larger than the escape velocity, which explains the scarcity of H_2 molecules in our atmosphere.

[a]The solution of ODE $y'' + y = a$ is $y(x) = a + A\cos x + B\sin x$, where A and B are two constants.

EXAMPLE PROBLEMS

General Procedure: Motion in a 1D Potential

To study motions in 1D potentials, follow these guidelines:

① The motion of a particle in a one-dimensional potential hinges on conservation of energy: make sure that the forces acting on the particle are either conservative or do not work. The analysis can be done graphically, by finding the set of values of coordinate q which satisfy the inequality $U(q) \leq E$, for a given value E of the energy.

② Equilibrium points correspond to the minima (stable points) or maxima (unstable points) of the potential energy.

③ If the potential energy depends on more than one coordinate, it may still be possible to study the motion in terms of an effective one-dimensional potential U_{eff} by taking into account possible constraint equations between the coordinates.

④ One such equation can be provided by conservation of angular momentum, e.g. in central force motions.

Exercise 5.6 ▪ A collar P of mass $m = 0.2$ kg can slide along a smooth horizontal shaft. It is connected by a spring of constant $k = 0.4$ N/m and unstretched length $l_0 = 0.6$ m. Find the equilibrium positions and discuss their stability. Then find the period of small amplitude oscillations about each equilibrium positions.
Discuss the motion of P for the initial conditions $x_0 = 0$ and $\dot{x}_0 \neq 0$. ▪

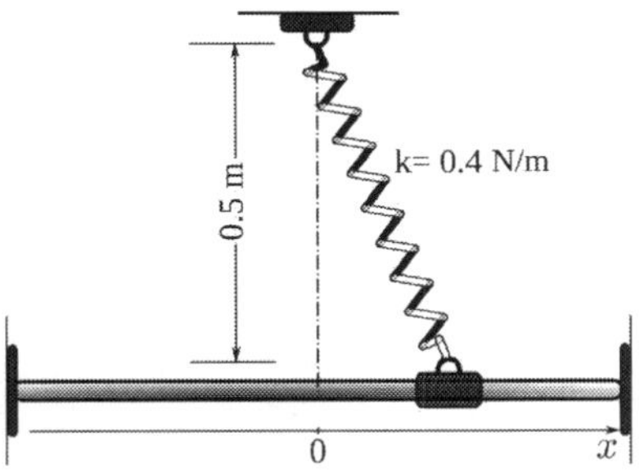

Solution. This problem can be solved as the motion of a particle in a one-dimensional potential: the only force which works is the elastic force whose potential is given by

$$U = \frac{k}{2}(l - l_0)^2$$

where l is the length of the spring when P is in position defined by coordinate x at time t (see Figure 5.16). To express U as a function of x, we find $l = \sqrt{h^2 + x^2}$. Conservation of energy can then be stated as

$$E = \tfrac{1}{2}m\dot{x}^2 + \frac{k}{2}\left(\sqrt{h^2 + x^2} - l_0\right)^2 = \text{Constant}$$

Note that the equation of motion is easily obtained by setting $dE/dt = 0$:

$$m\ddot{x} + k\left(1 - \frac{l_0}{\sqrt{h^2 + x^2}}\right)x = 0$$

This equation can also be found by projecting Newton's Second Law on Ox. The equilibrium positions are found as the zeros of dU/dx:

$$\frac{dU}{dx} = k\frac{x}{\sqrt{h^2+x^2}}\left(\sqrt{h^2+x^2} - l_0\right) = 0$$

This gives

$$\boxed{x_e = 0, \quad \pm\sqrt{l_0^2 - h^2} \quad \text{if } l_0 > h}$$

Two cases must be considered (see Figure 5.17):

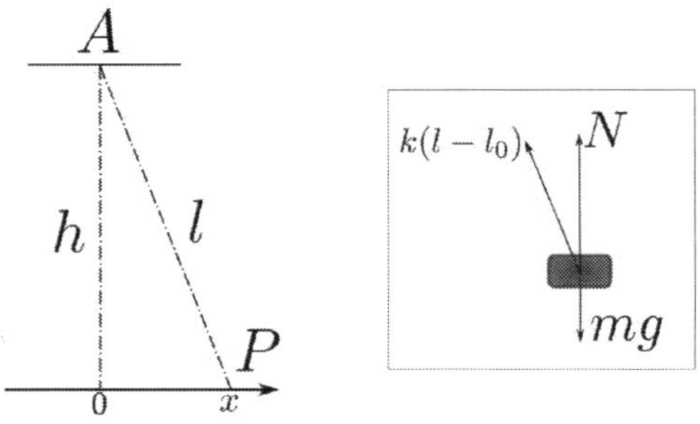

Figure 5.16 – *Problem 5.6*

Case 1. if $l_0 \leq h$, U has a single minimum at $x_e = 0$ corresponding to a stable equilibrium. See Figure 5.17 (right).
Case 2. if $l_0 > h$, U has two minima at $x_e = \pm(l_0^2 - h^2)^{1/2}$ corresponding to stable equilibriums, and a maximum $x_e = 0$ corresponding to a unstable equilibrium. See Figure 5.17 (left).
This can be verified by finding d^2U/d^2x at each equilibrium points:

$$\frac{d^2U}{d^2x} = k\frac{h-l_0}{h}, \quad \text{at } x = 0$$

$$\frac{d^2U}{d^2x} = k\frac{l_0^2 - h^2}{l_0^2}, \quad \text{at } x = \pm\sqrt{l_0^2 - h^2} \quad \text{if } l_0 > h$$

The period of small amplitude oscillations about $x_e = \pm(l_0^2 - h^2)^{1/2}$ is given by

$$\boxed{T = \frac{2\pi}{\Omega}, \qquad \Omega^2 = \frac{1}{m}\frac{d^2U}{dx^2}(x_e) = \frac{k}{m}\left(1 - \frac{h^2}{l_0^2}\right)}$$

For $h = 0.5$ m, $l_0 = 0.6$ m, $m = 0.2$ kg, $k = 0.4$ N/m, we find the stable positions $x_e = \pm\mathbf{0.33}$ m corresponding to the period of oscillations $T = \mathbf{8}$ s.

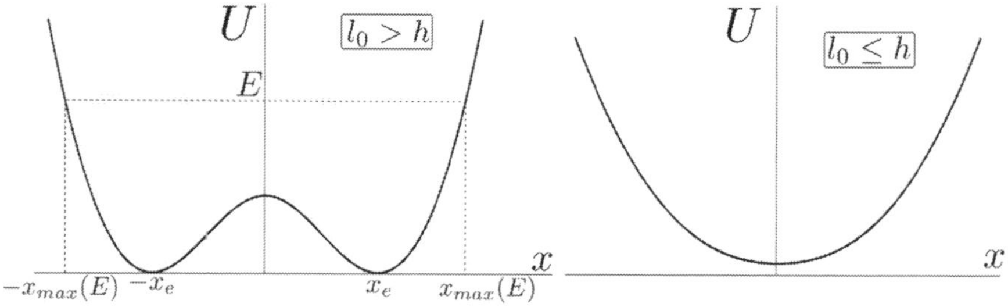

Figure 5.17 – *Problem 5.6*

With the initial conditions $\dot{x}_0 \neq 0$ and $x_0 = 0$ we find the corresponding energy level $E = \frac{1}{2}m\dot{x}_0^2 + \frac{k}{2}(h - l_0)^2$ which is larger than the potential level $U(0) = \frac{k}{2}(h - l_0)^2$. This implies that P will not be confined to one of the two potential wells. Instead P will exhibit large amplitude oscillations about $x = 0$:

$$-x_{max}(E) \leq x \leq x_{max}(E), \qquad \text{with } U(x_{max}) = E$$

See Figure 5.17 (left). ■

Exercise 5.7 ■■ A satellite of mass m is launched from point O of Earth (of mass M, radius R) at a velocity of magnitude $v_0 = \kappa v_e$, $\kappa < 1$ (where v_e is Earth's escape velocity) at an angle α with the radial line. The probe then describes an elliptic orbit under the effect of Earth's gravity to reach its apogee (the point of greatest distance from center C) at A.
a. Recall the expression of the escape velocity v_e. Find the distance r_A from center C to apogee A for $\kappa = 1/\sqrt{2}$ and $\alpha = 45°$. Find the speed reached at A.
b. The satellite is then sent on a new elliptic orbit by using a single engine impulse upon reaching A to increase the velocity from $\mathbf{v}_A$ to $\mathbf{v}_A + \Delta\mathbf{v}_A$. Find Δv_A for the new orbit be tangent to the first orbit at A and have a semi-major axis $2R$. ■

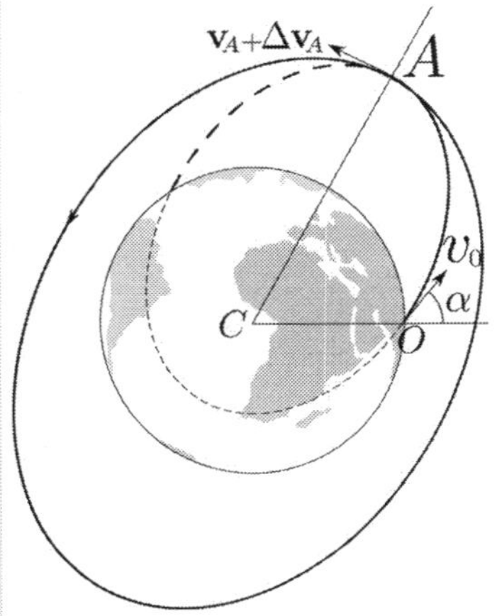

SOLUTION. **a.** The escape velocity v_e corresponds to the parabolic orbit with initial speed v_e at distance $r = R$ reaching infinity with zero velocity:

$$E_0 = \tfrac{1}{2}mv_e^2 - \frac{GMm}{R} = 0$$

leading to $\boxed{v_e = \sqrt{2GM/R}}$.
The motion of P is parametrized in polar coordinates (r, θ). See Figure 5.18. It is characterized by (i) conservation of mechanical energy $E = \frac{1}{2}m(\dot{r}^2 + r^2\dot{\theta}^2) - GMm/r$, and by (ii) conservation of angular momentum $mr^2\dot{\theta} = mRv_0 \sin\alpha$. These two expressions can be combined to lead to

$$E_1 = \tfrac{1}{2}m(\dot{r}^2 + \frac{R^2 v_0^2 \sin^2\alpha}{r^2}) - \frac{GMm}{r} = \tfrac{1}{2}mv_0^2 - \frac{GMm}{R}$$

At the apogee A, the radial distance is maximized ($\dot{r} = 0$):

$$\frac{R^2 v_0^2 \sin^2\alpha}{r_A^2} - \frac{2GM}{r_A} = v_0^2 - \frac{2GM}{R}$$

We then substitute $2GM/R = v_e^2$ and $v_0 = \kappa v_e$ to find the quadratic equation

$$\boxed{(1-\kappa^2)X^2 - X + \kappa^2 \sin^2\alpha = 0}$$

with $X = r_A/R$. In general, two roots are found: $X_1 < 1$ and $X_2 > 1$. We choose the root $X_2 > 1$ corresponding to position A. For $\kappa = 1/\sqrt{2}$ and $\alpha = 45°$, we find $r_A/R = \mathbf{1.707}$. The speed at A is found by conservation of angular momentum about

C: since $\mathbf{v}_A$ is normal to the radial line at A, we find $mr_A v_A = mRv_0 \sin\alpha$, giving

$$v_A = \kappa v_e \frac{R}{r_A} \sin\alpha$$

This gives $v_A = \mathbf{0.293}\, v_e$.

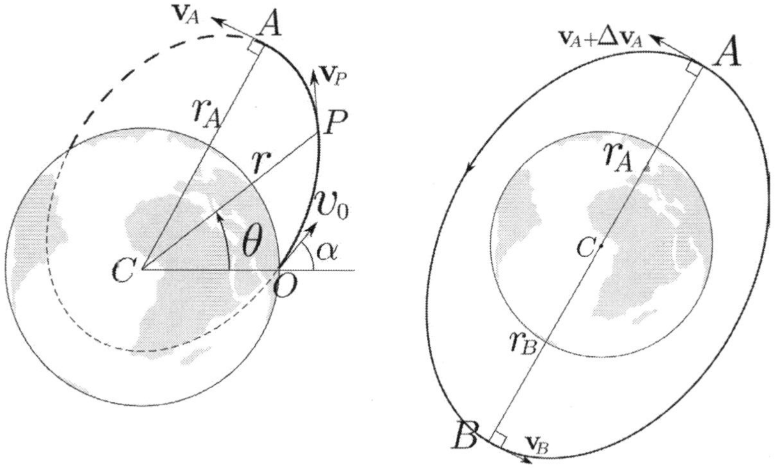

Figure 5.18 – *Problem 5.7 (Solution).*

b. The new elliptic trajectory of P is necessarily of major axis $AB = 2a = 4R$ since the velocity at A, $\mathbf{v}_A + \Delta\mathbf{v}_A$ is normal to radial line CA. The energy of P along this path takes the expression:

$$E_2 = \tfrac{1}{2}m(v_A + \Delta v_A)^2 - \frac{GMm}{r_A} = \tfrac{1}{2}mv_B^2 - \frac{GMm}{r_B} \qquad [1]$$

with $r_A + r_B = 4R$. We also impose conservation of angular momentum about C:

$$mr_A(v_A + \Delta v_A) = mr_B v_B = mC$$

If we substitute $v_A + \Delta v_A = C/r_A$ and $v_B = C/r_A$ in equation [1], we find

$$\frac{C^2}{r_A^2} - v_e^2\frac{R}{r_A} = \frac{C^2}{r_B^2} - v_e^2\frac{R}{r_B}$$

which gives constant C:

$$C = v_e\sqrt{R\frac{r_A r_B}{r_A + r_B}}$$

Now we can find $(v_A + \Delta v_A) = C/r_A$:

$$\boxed{\Delta v_A = v_e\sqrt{\frac{R}{r_A} - \frac{1}{4}} - v_A}$$

With $r_A/R = 1.707$ and $v_A = 0.292\, v_e$, we find $\Delta v_A = \mathbf{0.286}\, v_e$. ∎

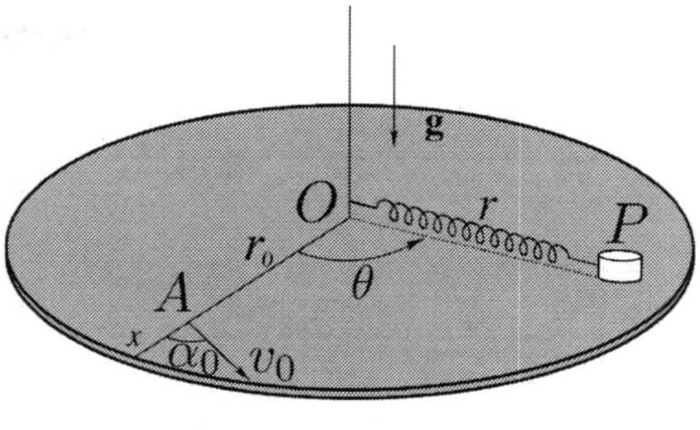

Exercise 5.8 ■■ A small object P of mass m can slide on a smooth horizontal support Oxy. It is connected to a linear spring of constant k and unstretched length l_0 attached at fixed point O. At time $t = 0$, particle P is at a distance r_0 from the origin and it is released with an initial velocity $\mathbf{v}_0$ making an angle α_0 with axis Ox.
Under which conditions governing the initial values r_0 and $\mathbf{v}_0$ does particle P have a circular trajectory of radius r_0 and center O? Is this motion stable? Find the period of the small oscillations about the circular path. ■

SOLUTION. Particle P is subject to a central force in the plane Oxy. The normal reaction of the support and gravitational force do not work. We can apply the results of section 5.3 with the elastic potential $U = \frac{1}{2}k(r - l_0)^2$. The energy conservation equation takes the form

$$E = \tfrac{1}{2}m(\dot{r}^2 + r^2\dot{\theta}^2) + \frac{1}{2}k(r - l_0)^2 = \tfrac{1}{2}m\dot{r}^2 + \frac{1}{2}k(r - l_0)^2 + \frac{mC^2}{2r^2} = \text{constant} \qquad [1]$$

where we have used the fact that $r^2\dot{\theta} = C = Cst$. Recall that, in a central force motion, the transverse acceleration $a_\theta = 0 = \frac{1}{r}\frac{d}{dt}(r^2\dot{\theta})$ leads to the condition $r^2\dot{\theta} = Cst$. The constant C is found from the initial conditions $r = r_0$, $\dot{r} = v_0 \cos\alpha_0$ and $r\dot{\theta} = v_0 \sin\alpha_0$ leading to $C = r^2\dot{\theta} = r_0 v_0 \sin\alpha_0$. From [1] and the expression of C, the effective potential $U_{eff}(r)$ is found to be

$$U_{eff}(r) = \frac{1}{2}k(r - l_0)^2 + \frac{mC^2}{2r^2}$$

It admits a single minimum at the value $r = r_e$ solution of

$$m\frac{C^2}{r_e^3} = k(r_e - l_0)$$

This solution corresponds to a stable circular path of center O and radius r_e. To impose $r_e = r_0$ gives the condition

$$\boxed{k(r_0 - l_0) = m\frac{v_0^2}{r_0}}$$

where we have used $C = r_0 v_0 \sin\alpha_0$ and $\alpha_0 = \pi/2$ since $\dot{r}$ must be imposed to zero at all time. This equation can be interpreted as a balance between the spring force and the inertial force.
If we denote by $\epsilon(t) = r(t) - r_0$ and assume that ϵ remains small, then the small-amplitude oscillations are governed by the equation

$$\ddot{\epsilon} + \Omega^2\epsilon = 0, \qquad \text{with} \quad \Omega^2 = \frac{1}{m}\frac{d^2U_{eff}}{dr^2}(r_0) = \frac{k}{m} + 3\frac{v_0^2}{r_0^2}$$

leading to the period

$$\boxed{T = 2\pi \sqrt{\frac{m}{k} \frac{1}{1 + \frac{3mv_0^2}{kr_0^2}}}}$$

■

5.4 Work-Energy Principle for a System of Particles

WE WISH to extend the Work-Energy Principle to the case of a system Σ of particles. It is sufficient to consider two particles A (of mass m_A) and B (of mass m_B). As we did in Section 3.7, we distinguish two types of forces acting on A and B:
- the *external forces* $\mathbf{F}_{ext\to A}$ (acting on A) and $\mathbf{F}_{ext\to B}$ (acting on B), and
- the *internal forces* $\mathbf{F}_{A\to B}$ and $\mathbf{F}_{B\to A} = -\mathbf{F}_{A\to B}$.
The question we ask here is whether the contribution of internal forces vanishes in the formulation of the Work-Energy Principle applied to system Σ as is the case in the derivation of Euler's first principle.
We apply the Work-Energy Principle to A and B between instants $t = t_1$ and $t = t_2$:

$$\Delta(\frac{1}{2}m_A v_A^2) = W_{1\to 2}^{F_{ext\to A}} + W_{1\to 2}^{F_{B\to A}}, \quad \Delta(\frac{1}{2}m_B v_B^2) = W_{1\to 2}^{F_{ext\to B}} + W_{1\to 2}^{F_{A\to B}} \tag{5.22}$$

Next we add these two equations to obtain:

$$\Delta(\frac{1}{2}m_A v_A^2 + \frac{1}{2}m_B v_B^2) = W_{1\to 2}^{F_{ext\to A}} + W_{1\to 2}^{F_{ext\to B}} + W_{1\to 2}^{F_{B\to A}} + W_{1\to 2}^{F_{A\to B}} \tag{5.23}$$

The terms in the left of equation (5.23) represent the variation of the system's kinetic energy ΔK_Σ. The first two terms in the right of (5.23) represent the work done by all external forces: we denote this term $W_{1\to 2}^{ext}$. It is then tempting to cancel out the last two terms in view of Newton's third law $\mathbf{F}_{A\to E} = -\mathbf{F}_{B\to A}$. But this is not the case, as is seen below:

$$W_{1\to 2}^{F_{B\to A}} + W_{1\to 2}^{F_{A\to B}} = \int_{t_1}^{t_2} \mathbf{F}_{A\to B} \cdot \mathbf{v}_B \, dt + \int_{t_1}^{t_2} \mathbf{F}_{B\to A} \cdot \mathbf{v}_A \, dt = \int_{t_1}^{t_2} \mathbf{F}_{A\to B} \cdot (\mathbf{v}_B - \mathbf{v}_A) \, dt$$

This last integral can be simplified by noting that

$$\mathbf{F}_{A\to B} = f_{AB}\hat{\mathbf{e}}_{AB}, \qquad \mathbf{v}_B - \mathbf{v}_A = \frac{d}{dt}(\mathbf{r}_{OB} - \mathbf{r}_{OA}) = \frac{d}{dt}\mathbf{r}_{AB} = \frac{d}{dt}(l_{AB}\hat{\mathbf{u}}_{AB})$$

where $\hat{\mathbf{e}}_{AB}$ is the unit vector pointing from A to B and l_{AB} denote the distance $|AB|$. Recall that $\hat{\mathbf{e}}_{AB}$ is normal to its time-derivative $\frac{d}{dt}\hat{\mathbf{e}}_{AB}$:

$$\hat{\mathbf{e}}_{AB} \cdot \frac{d}{dt}(l_{AB}\hat{\mathbf{e}}_{AB}) = \frac{d}{dt}l_{AB}$$

Then we obtain

$$\int_{t_1}^{t_2} \mathbf{F}_{A\to B} \cdot (\mathbf{v}_B - \mathbf{v}_A) \, dt = \int_{t_1}^{t_2} f_{AB}\hat{\mathbf{e}}_{AB} \cdot \frac{d}{dt}(l_{AB}\hat{\mathbf{e}}_{AB}) \, dt = \int_{l_1}^{l_2} f_{AB} dl_{AB}$$

where $l_2 = l_{AB}(t_2)$ and $l_1 = l_{AB}(t_1)$. So, unless particles A and B stay at a constant distance l_{AB} from each another, the last result shows that *the work done by the internal forces does not vanish.*

Definition (Internal Work)

The internal work $W^{int}_{1\to 2}$ defined as the sum of the works done by the action $\mathbf{F}_{A\to B}$ and the reaction $\mathbf{F}_{B\to A}$ takes the expression

$$W^{int}_{1\to 2} = \int_{l_1}^{l_2} f_{AB}\, dl_{AB} \tag{5.24}$$

It does not vanish, it is independent of the choice of referential, and it depends only on the relative motion of particles A and B.

We can now state the Work-Energy Principle applied to system Σ:

Work-Energy Principle for a System

The work done by the external forces applied to the system and by the internal forces within the system is equal to the variation of its kinetic energy:

$$\Delta K_\Sigma = W^{ext}_{1\to 2} + W^{int}_{1\to 2} \tag{5.25}$$

There lies a major difference between Euler's principles and the Work-Energy Principle when the latter is applied to a system of particles! In Euler's principles, the internal forces vanish. In the Work-Energy Principle, they must be taken into account.

Potential Energy of Interaction. If a scalar function $U_{A\leftrightarrow B}$ can be found such that $W^{int}_{1\to 2} = -\Delta U_{A\leftrightarrow B}$ then, the *internal forces are said to be conservative* and $U_{A\leftrightarrow B}$ is called *Potential Energy of Interaction*. This function can only depend on the distance l_{AB} separating the two interacting particles. Two relevant cases can be considered:

Elastic interaction: A and B are interconnected by a spring of constant k and unstretched length l_0. With $f_{AB} = -k(l_{AB} - l_0)$, we find

$$W^{int}_{1\to 2} = -\int_{l_1}^{l_2} k(l_{AB} - l_0)dl_{AB} = -\Delta(\frac{k}{2}(l_{AB} - l_0)^2).$$

This shows that the elastic potential energy of interaction is given by

$$U^e_{A\leftrightarrow B} = \frac{k}{2}(l_{AB} - l_0)^2$$

Gravitational interaction: A and B are in mutual gravitational attraction. With $f_{AB} = -Gm_A m_B/l^2_{AB}$, we find $W^{int}_{1\to 2} = \Delta(Gm_A m_B/l_{AB})$ giving the expression

$$U^g_{A\leftrightarrow B} = -\frac{Gm_A m_B}{l_{AB}}$$

of the gravitational potential energy of interaction.

Closed System of two particles. In Section 3.8, we found the main features which characterize a closed system Σ of two particles A and B in motion in a Newtonian referential $\mathcal{E}$:
(i) the mass center G has a uniform rectilinear motion,
(ii) within the centroidal referential $\mathcal{E}^*$, the motion of A and B can be predicted from the motion of the fictitious particle P of mass $m_A m_B/(m_A + m_B)$ defined by $\mathbf{r}_{GP} = \mathbf{r}_{AB}$ and acted upon by force $\mathbf{F}_{A\to B}$,
(iii) the motion of P in $\mathcal{E}^*$ satisfies the first integral $r^2\dot{\theta} = Cst = C$ where (r, θ) denote the polar coordinates of P in $\mathcal{E}^*$.

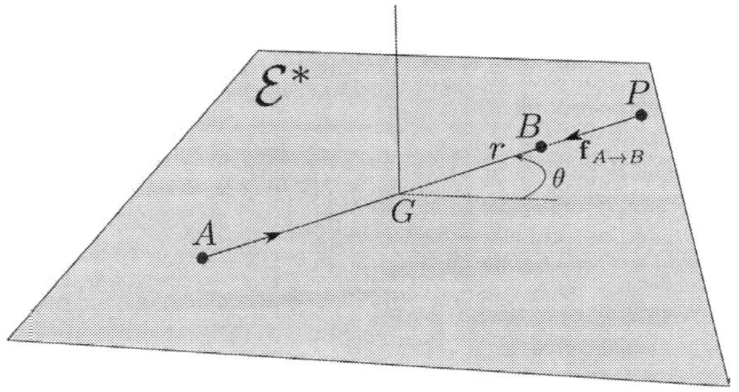

Figure 5.19 – *Motion of A, B and fictitious particle P in the centroidal referential $\mathcal{E}^*$.*

Let us assume that the force of interaction between A and B is conservative, with a corresponding potential energy of interaction $U_{A\leftrightarrow B}$. Since the centroidal referential $\mathcal{E}^*$ is Newtonian, we can apply the Work-Energy Principle (5.25) to system Σ relative to $\mathcal{E}^*$. Since $W_{1\to 2}^{ext} = 0$ and $W_{1\to 2}^{int} = -\Delta U_{A\leftrightarrow B}$, we obtain a third first integral of motion expressing the conservation of energy of system Σ relative to $\mathcal{E}^*$:

$$E = \tfrac{1}{2} m_A (v_A^*)^2 + \tfrac{1}{2} m_B (v_B^*)^2 + U_{A\leftrightarrow B} = \text{constant}$$

where v_A^* and v_B^* are the speeds of A and B relative to the centroidal referential. It is possible to express v_A^* and v_B^* in term of the speed v_P^* of the fictitious particle P by time-differentiating equation (3.39)

$$v_A^* = \frac{m_B}{m_A + m_B} v_P^*, \qquad v_B^* = \frac{m_A}{m_A + m_B} v_P^* \tag{5.26}$$

If $(r = |AB|, \theta)$ denote the polar coordinates of P, we can express E in terms of r only, with the use of $r^2\dot{\theta} = C$ since $(v_P^*)^2 = \dot{r}^2 + r^2\dot{\theta}^2 = \dot{r}^2 + C^2/r^2$:

$$E = \tfrac{1}{2} \frac{m_A m_B}{m_A + m_B} \left(\dot{r}^2 + \frac{C^2}{r^2} \right) + U_{A\leftrightarrow B}(r) = \text{constant} \tag{5.27}$$

This shows that the motion of the system is equivalent to a one-dimensional motion in the effective potential

$$U_{eff}(r) = \tfrac{1}{2} \frac{m_A m_B}{m_A + m_B} \frac{C^2}{r^2} + U_{A\leftrightarrow B}(r)$$

EXAMPLE PROBLEMS

General Procedure: Kinetics of System of Particles by Energy Method

To apply the Work-Energy Principle to a system of particles, it is advisable to proceed by applying the Work-Energy Principle to the individual particles separately (draw separate free-body diagrams), and by summing the resulting equations.

By following this method, the contribution of *both external and internal* forces is necessarily taken into account.

For each external force, find the corresponding work by applying the general formula (5.1) unless the force is conservative and a potential energy is known.

For each pair of internal action/reaction, find the internal work by summing the works of both action and reaction, unless a potential energy of interaction is known.

Exercise 5.9 ▪ Two crates A and B of mass m_A and m_B are connected by a rope-pulley system. The coefficient of kinetic friction between the inclined plane and crate A is μ.
The system is released from rest. Crate B is observed to move upward.
Find the speed of A when B rises a height H. Deduce a necessary condition for the motion to occur. ▪

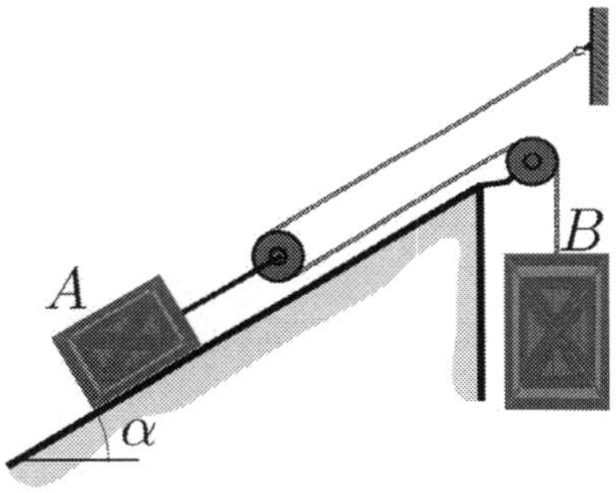

Solution. Step 1: The motions of both crates are rectilinear. We adopt a coordinate axis Ox directed downward along inclined plane to define the position of A as $\mathbf{r}_{OA} = x_A\hat{\boldsymbol{\imath}}$. We then adopt a coordinate axis Oz directed along the vertical downward to define the position of B is defined as $\mathbf{r}_{OB} = z_B\hat{\mathbf{k}}$. See Figure 5.20 (left). Their positions are not independent since the coordinates (x_A, z_B) satisfy the constraint equation: $2x_A + z_B = Cst$. This implies that:
1. the finite variations Δx_A and Δz_B from $t = 0$ to $t = t_1$ are dependent: $\Delta z_B = -2\Delta x_A = -H$,
2. the velocities $\mathbf{v}_A$ and $\mathbf{v}_B$ are also dependent: $\dot{z}_B = -2\dot{x}_A$.

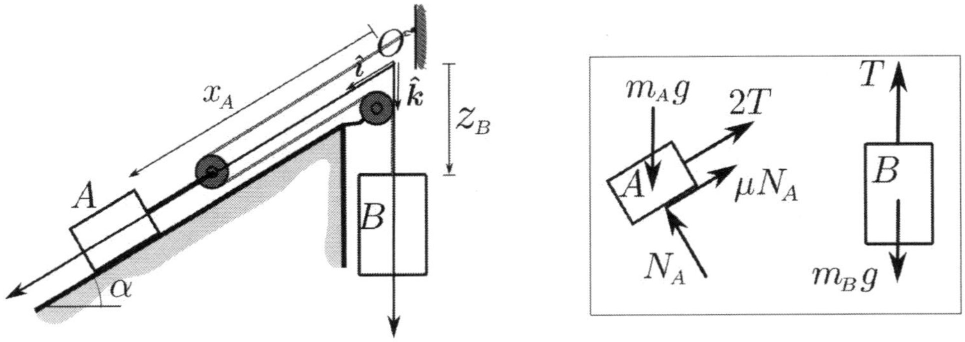

Figure 5.20 – *Problem 5.9 (Solution). Trajectory of P for $k = 0.1$ and $k = 2$ $(M/m = 1)$.*

Step 2: We sketch the free-body diagram for A: the forces acting on A are normal reaction N_A, the friction force μN_A, gravity $m_A g$ (conservative) and ropes' tensions $2T$. We then sketch the free-body diagram for B: the forces are gravity $m_B g$ (conservative) and rope's tension T. See Figure 5.20 (right).

STEP 3: We apply the WEP for A between $t = 0$ (when the system is released) to $t = t_1$ (when B rises a height H)

$$\tfrac{1}{2}m_A v_A^2 = m_A g \Delta x_A \sin\alpha - \int_0^{\Delta x_A} \mu N_A dx_A - \int_0^{\Delta x_A} 2T dx_A \quad [1]$$

We repeat for B:

$$\tfrac{1}{2}m_B v_B^2 = m_B g \Delta z_B - \int_0^{\Delta z_B} T dz_B \quad [2]$$

STEP 4: We now obtain the WEP for system $\{A, B\}$ by summing [1] and [2]

$$\tfrac{1}{2}m_A v_A^2 + \tfrac{1}{2}m_B v_B^2 = m_A g \Delta x_A \sin\alpha + m_B g \Delta z_B - \int_0^{\Delta x_A} \mu N_A dx_A - \int_0^{\Delta t} T(2\dot{x}_A + \dot{z}_B)dt$$

with $v_B^2 = 4v_A^2$, $\Delta z_B = -H$, $\Delta x_A = H/2$, $N_A = m_A g \cos\alpha$, and $2\dot{x}_A + \dot{z}_B = 0$.

$$\boxed{\tfrac{1}{2}(m_A + 4m_B)v_A^2 = (2m_A \sin\alpha - m_B)gH - \tfrac{1}{2}\mu m_A g H \cos\alpha}$$

where the work done by the internal forces (the tensions in the rope) vanishes. In order to obtain $v_A^2 > 0$, we need to impose condition:

$$\boxed{\mu < \frac{2(2m_A \sin\alpha - m_B)}{m_A \cos\alpha}}$$

■

Exercise 5.10 ■ Consider the system of problem 3.10. The system's position is defined by the coordinates (x, θ) with $x = 0$ and $\theta = \theta_0$ at $t = 0$ and with $\dot{x} = \dot{\theta} = 0$ at $t = 0$. Recall that the system satisfies the first integral

$$(M + m)\dot{x} + ml\dot{\theta}\cos\theta = 0 \quad [1]$$

a. Find another first integral of motion governing angle θ by showing that the energy of the system is conserved.
b. Find the period of the small amplitude oscillations of the system, that is, when the initial angle θ_0 is small. Examine the cases $M \gg m$ and $M \ll m$. ■

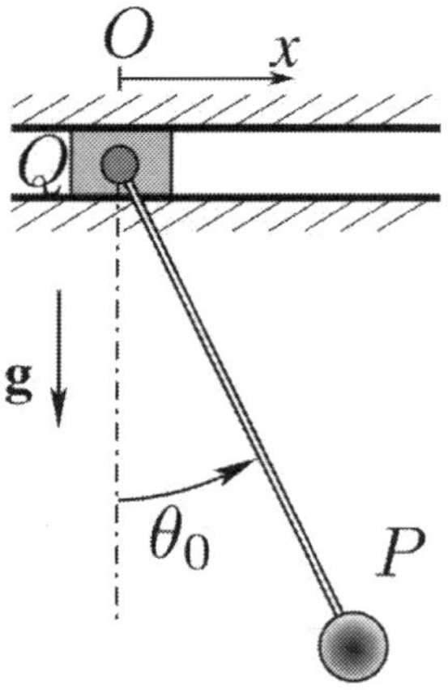

SOLUTION. **a.** To find another first integral of motion, we apply the WEP to particle Q and P separately. Particle Q is subject to the resultant force $\mathbf{F}_Q = (N + Mg)\hat{\mathbf{e}}_z + F\hat{\mathbf{u}}$, where F is the tension exerted by the rod. The kinetic energy of Q is $\frac{1}{2}M\dot{x}^2$. Since the only force producing work is $F\hat{\mathbf{u}}$, we find

$$\Delta(\tfrac{1}{2}M\dot{x}^2) = \int_0^t F\hat{\mathbf{u}} \cdot \mathbf{v}_Q dt = \int_0^x F\sin\theta\, dx \quad [2]$$

This shows that the energy of Q is not conserved due to the non-conservative tension force.

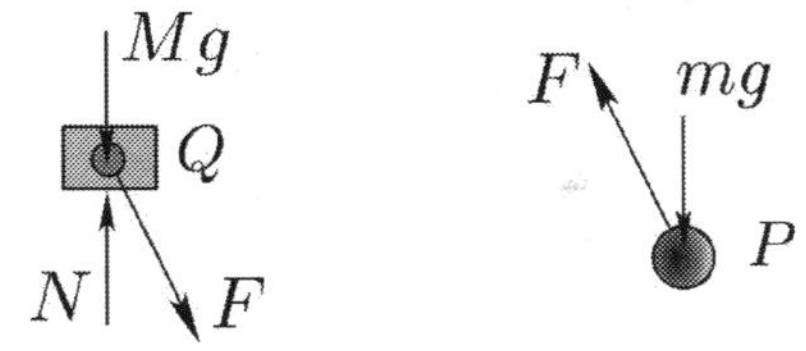

Figure 5.21 – *Problem 5.10 (Solution). Note that force is directed along line PQ because the rod connecting P and Q is assumed massless.*

Particle P is subject to the resultant force $\mathbf{F}_P = mg\hat{\mathbf{e}}_z - F\hat{\mathbf{u}}$, where $(-F\hat{\mathbf{u}})$ is the tension exerted by the rod on P. The kinetic energy of P is $\frac{1}{2}m(\dot{x}\hat{\mathbf{e}}_x + l\dot{\theta}\hat{\mathbf{v}})^2 = \frac{1}{2}m(\dot{x}^2 + l^2\dot{\theta}^2 + 2l\dot{x}\dot{\theta}\cos\theta)$ and its potential energy is $U = -mgl\cos\theta$. Then the WEP can be stated as

$$\Delta[\tfrac{1}{2}m(\dot{x}^2 + l^2\dot{\theta}^2 + 2l\dot{x}\dot{\theta}\cos\theta) - mgl\cos\theta] = \int_0^t (-F\hat{\mathbf{u}})\cdot\mathbf{v}_P dt = -\int_0^x F\sin\theta\, dx \qquad [3]$$

Again the energy of particle P is not conserved due to the non-conservative tension force.

However, comparing equations [2] and [3], we can eliminate the work done by the tension force by adding these two equations: now the work done by the internal forces vanishes because P and Q remains at constant distance to each other

$$\tfrac{1}{2}M\dot{x}^2 + \tfrac{1}{2}m(\dot{x}^2 + l^2\dot{\theta}^2 + 2l\dot{x}\dot{\theta}\cos\theta) - mgl\cos\theta = Cst = -mgl\cos\theta_0 \qquad [4]$$

which expresses the conservation of energy of system Σ. Note that adding [2] and [3] is equivalent to applying the WEP to the system.

Now we use equation [1] to eliminate $\dot{x}$ in equation [4] to obtain

$$\boxed{(1 + \frac{m}{M}\sin^2\theta)\dot{\theta}^2 = \frac{2g}{l}(1 + \frac{m}{M})(\cos\theta - \cos\theta_0)} \qquad [5]$$

We note that the left-hand-side of [5] remains positive leading to $\cos\theta \geq \cos\theta_0$ or $-\theta_0 \leq \theta \leq \theta_0$. This shows that pendulum QP oscillates about the vertical with maximum amplitude θ_0.

b. To find the period of small oscillations, we assume that the initial angle θ_0 is small: this implies that angle θ and its derivative $\dot{\theta}$ remain small at all time. First we derive the equation of motion (governing θ) by time-differentiating equation [5]

$$(1 + \frac{m}{M}\sin^2\theta)\ddot{\theta} + \frac{m}{M}\dot{\theta}^2\sin\theta\cos\theta + \frac{g}{l}(1 + \frac{m}{M})\sin\theta = 0$$

We approximate $\sin\theta$ by θ, and neglect all nonlinearities in θ and $\dot{\theta}$ to obtain a linear approximation of this equation

$$\ddot{\theta} + \frac{g}{l}(1 + \frac{m}{M})\theta = 0$$

The solution of this second-order linear ODE is $\theta = A\cos\Omega t + B\sin\Omega t$ with $\Omega^2 = \frac{g}{l}(1 + \frac{m}{M})$: the period of the oscillations of the system is then

$$\boxed{T = 2\pi\sqrt{\frac{l}{g(1 + \frac{m}{M})}}}$$

Two interesting limits can be examined from this result:

1. If $m \ll M$, then $T = 2\pi\sqrt{\frac{l}{g}}$, which is the equation of a pendulum suspended at a stationary point. Indeed, in this case, we expect that the velocity of the slider is negligible.

2. If $m \gg M$, then $T \to 0$: here the velocity of the cart $\dot{x}$ is much larger than the relative velocity $l\dot{\theta}$ of P. As P is released at angle θ_0, the slider Q slides "infinitely fast" in such a way as bringing the pendulum in its vertical equilibrium position. ■

Exercise 5.11 ■■■ Two particles P (mass m) and Q (mass M) are joined by an inextensible massless string of length l. P is constrained to move without friction on a horizontal support. The string can slide without friction through a small hole located at point O. Q is constrained to move along vertical axis Oz. At time $t = 0$, P is at a distance $r = r_0$ from O and is given a velocity $\mathbf{v}_0$ perpendicular to line OP.

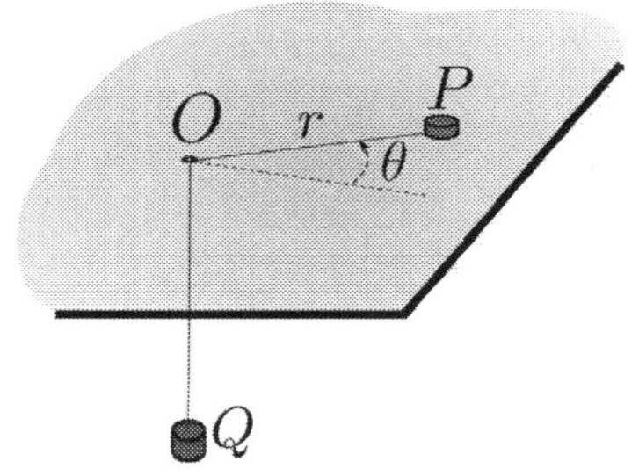

a. Method 1: Apply Newton's Second Law to show that the radial distance r is governed by

$$(m + M)\ddot{r} + m\frac{r_0^2 v_0^2}{r^3} = -Mg \qquad [1]$$

b. Method 2: Apply the Work-Energy Principle to the system of two particles to find a first integral of motion. Recover equation [1].

c. Using this first integral show that particle P oscillates between two concentric circles of center O in the horizontal support. ■

SOLUTION. **a**. The motion of particle P is constrained to move in horizontal plane $(O, \hat{\mathbf{e}}_x, \hat{\mathbf{e}}_y)$: we choose polar coordinates (r, θ) to parametrize its position. See Figure 5.22. On the corresponding unit vectors $(\hat{\mathbf{e}}_r, \hat{\mathbf{e}}_\theta)$ the position vector, the velocity and acceleration of P are given by

$$\mathbf{r}_{OP} = r\,\hat{\mathbf{e}}_r \qquad \mathbf{v}_P = \dot{r}\hat{\mathbf{e}}_r + r\dot{\theta}\hat{\mathbf{e}}_\theta$$
$$\mathbf{a}_P = (\ddot{r} - r\dot{\theta}^2)\hat{\mathbf{e}}_r + (r\ddot{\theta} + 2\dot{r}\dot{\theta})\hat{\mathbf{e}}_\theta.$$

The motion of particle Q is rectilinear, along the axis Oz directed downward: calling z the coordinate of Q on this axis of unit vector $\hat{\mathbf{e}}_z$, we have

$$\mathbf{r}_{OQ} = z\hat{\mathbf{e}}_z, \qquad \mathbf{v}_Q = \dot{z}\hat{\mathbf{e}}_z, \qquad \mathbf{a}_Q = \ddot{z}\hat{\mathbf{e}}_z$$

Particle P is subject to: (i) gravity $m\mathbf{g} = mg\hat{\mathbf{e}}_z$, (ii) the reaction force $\mathbf{R} = -N\hat{\mathbf{e}}_z$ $(N > 0)$ of the horizontal support: it is normal to the surface of contact since friction is neglected, (iii) the tension $\mathbf{T}_P = -T\hat{\mathbf{e}}_r$ $(T > 0)$, assuming the tension constant along the string.

Particle Q is subject to: (i) gravity $M\mathbf{g} = Mg\hat{\mathbf{e}}_z$, (ii) the tension $\mathbf{T}_Q = -T\hat{\mathbf{e}}_z$ $(T > 0)$, assuming the same tension applied on P, neglecting the friction of the sliding string through the hole at O.

Next we apply Newton's Second Law (i) for P: $m\mathbf{a}_P = m\mathbf{g} + \mathbf{R} + \mathbf{T}_P$ and (ii) for Q: $M\mathbf{a}_Q = M\mathbf{g} + \mathbf{T}_Q$ and project on basis $(\hat{\mathbf{e}}_r, \hat{\mathbf{e}}_\theta, \hat{\mathbf{e}}_z)$ to obtain four scalar equations for the five unknowns r, θ, z, N, T:

$$m(\ddot{r} - r\dot{\theta}^2) = -T, \quad r\ddot{\theta} + 2\dot{r}\dot{\theta} = 0, \quad N = mg, \quad M\ddot{z} = Mg - T \qquad [2-5]$$

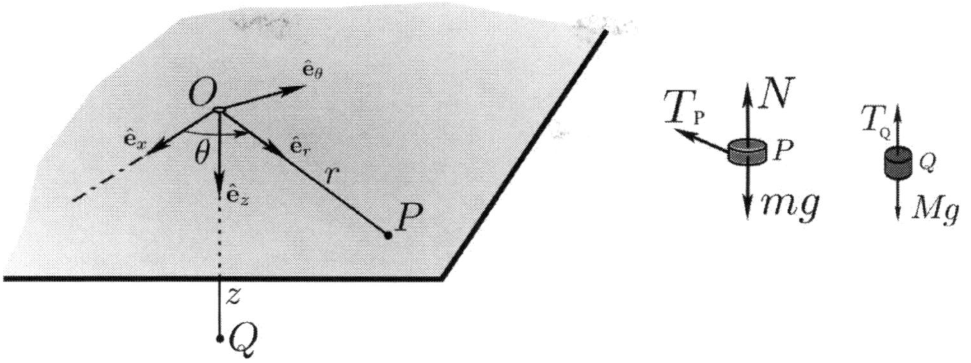

Figure 5.22 – *Problem 5.11 (Solution)*

with the initial conditions $r = r_0$, $\dot{r}_0 = 0$, $r_0\dot{\theta}_0 = v_0$ at time $t = 0$ (at time $t = 0$ we have $\mathbf{v}_P = r_0\dot{\theta}_0\hat{\mathbf{e}}_\theta = v_0\hat{\mathbf{e}}_\theta$).
We need an additional equation: P and Q do not move independently of each other, since they are connected by a string of constant length l

$$l = z + r = Cst \qquad [6]$$

Hence at $t = 0$, $z_0 = r_0 - l$, and $\dot{z}_0 = -\dot{r}_0 = 0$.
We can get an ODE governing $r(t)$ only by first realizing that equation [3] can be integrated to give a first integral equation:

$$\boxed{r^2\,\dot{\theta} = Cst = r_0 v_0} \qquad [7]$$

Remark. Equation [7] could have been found by realizing that the moment about O of the forces acting on P is zero: tension $\mathbf{T}_P$ always points toward O, whereas $\mathbf{R}$ and $m\mathbf{g}$ balance each other. This leads to $\mathbf{H}_O = \mathbf{r}_{OP} \times \mathbf{v}_P = -mr^2\dot{\theta}\hat{\mathbf{e}}_z = Cst$. This equation expresses the *conservation of angular momentum of P about O* as was learned in Section 4.1.
We can also eliminate T between equation [2] and [5]:

$$m(\ddot{r} - r\dot{\theta}^2) = M\ddot{z} - Mg$$

with $\ddot{z} = -\ddot{r}$ in view of equation [6]. Then upon using equation [7] to eliminate $\dot{\theta}$:

$$\boxed{(m + M)\ddot{r} - m\frac{r_0^2 v_0^2}{r^3} = -Mg} \qquad [1]$$

b. Let's apply the WEP for particle P, then particle Q. For particle P, the variation of the kinetic energy $\frac{1}{2}m\mathbf{v}_P^2$ is equal to the work done by the tension force ($\mathbf{T}_P \cdot \mathbf{v}_P = -T\dot{r}$):

$$\Delta(\tfrac{1}{2}m(\dot{r}^2 + r^2\dot{\theta}^2)) = -\int_{r_0}^{r} T dr \qquad [8]$$

For particle Q, the variation of the sum of the kinetic energy $\frac{1}{2}M\mathbf{v}_Q^2$ and the potential energy $(-Mgz)$ is equal to the work done by the tension force ($\mathbf{T}_Q \cdot \mathbf{v}_Q = -T\dot{z}$):

$$\Delta(\tfrac{1}{2}M\dot{z}^2 - Mgz) = -\int_{z_0}^{z} T dz \qquad [9]$$

Recall that $\dot{z} = -\dot{r}$. If we sum equations [8] and [9] we eliminate the work done by the tension force:

$$\Delta(\tfrac{1}{2}m(\dot{r}^2 + r^2\dot{\theta}^2) + \tfrac{1}{2}M\dot{z}^2 - Mgz) = 0$$

This shows that the energy $E = \frac{1}{2}m(\dot{r}^2 + r^2\dot{\theta}^2) + \frac{1}{2}M\dot{z}^2 - Mgz$ is a constant. After using equation [7] to replace $\dot{\theta}$ by r_0v_0/r^2 we obtain the first integral of motion

$$\boxed{E = \tfrac{1}{2}(m+M)\dot{r}^2 + \tfrac{1}{2}mv_0^2\frac{r_0^2}{r^2} + Mgr = \tfrac{1}{2}mv_0^2 + Mgr_0} \qquad [10]$$

This equation expresses the conservation of the energy of the system. It is easy to see that equation [1] is recovered by setting $dE/dt = 0$.

c. Equation [10] can be written as $E = K + U_{\text{eff}}(r)$ with

$$\boxed{K = \tfrac{1}{2}(m+M)\dot{r}^2, \qquad U_{\text{eff}}(r) = \tfrac{1}{2}mv_0^2\frac{r_0^2}{r^2} + Mgr, \qquad E = Mgr_0 + \tfrac{1}{2}mv_0^2.}$$

The motion of P can be understood as that of a particle in a one-dimensional potential. We can qualitatively predict the behavior of P by plotting the effective potential U_{eff}. This potential has the following properties: $U_{\text{eff}} \to \infty$ as $r \to 0$, U_{eff} reaches a minimum U_{min} at $r_e/r_0 = (mv_0^2/Mgr_0)^{1/3}$, and $U_{\text{eff}} \approx Mgr$ as $r \to \infty$.

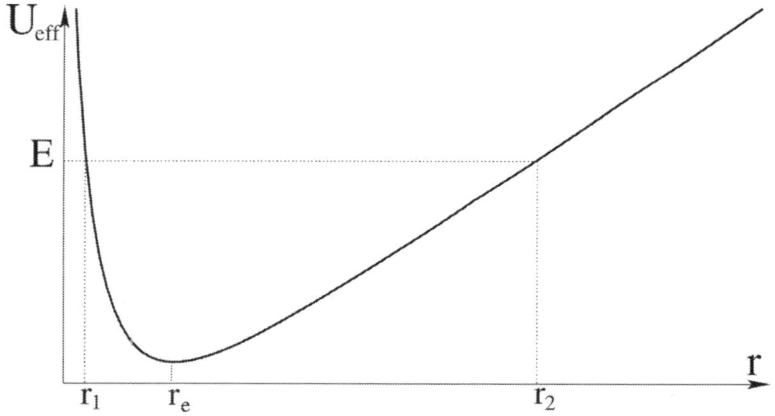

Figure 5.23 – *Problem 5.11 (Solution). Effective potential $U_{eff}(r)$.*

Recall that a motion of P must satisfy the inequality $U_{\text{eff}}(r) \leq E$. Hence if $E > U_{min}$, then the motion of P satisfies

$$r_1(E) \leq r \leq r_2(E)$$

where the values $r_1(E)$ and $r_2(E)$ correspond to zero kinetic energy and are solution of $U_{\text{eff}}(r) = E$. We find

$$r_1(E) = r_0, \qquad r_2(E) = \frac{r_0}{2}(k + \sqrt{k^2 + 4k})$$

where the dimensionless parameter k is defined as

$$\boxed{k = \frac{\frac{1}{2}mv_0^2}{Mgr_0}}$$

This parameter is seen to be the ratio of the initial kinetic energy of P to the initial potential energy of Q. The equality $r_1 = r_2$ gives the condition satisfied by the initial conditions for the system to reach the equilibrium position $r = r_e$

$$k = 1/2, \qquad r_e = r_0$$

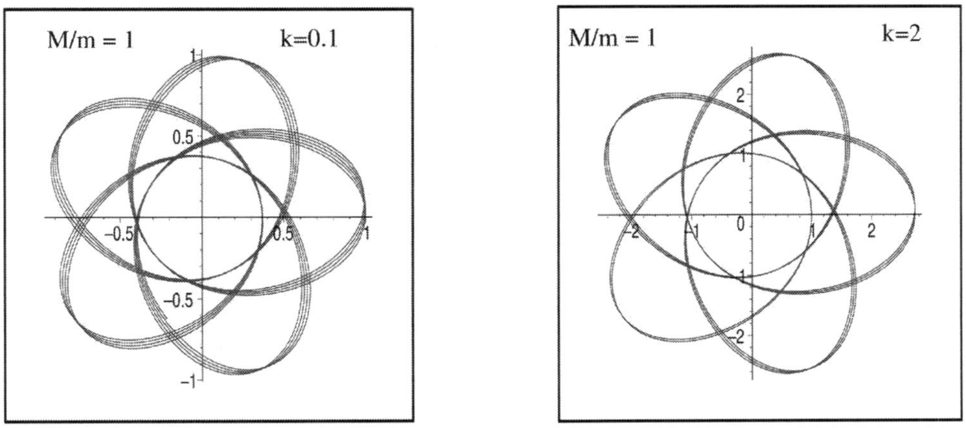

Figure 5.24 – *Problem 5.11 (Solution). Trajectory of P for $k = 0.1$ and $k = 2$ $(M/m = 1)$.*

In Conclusion:
1. If $k = 1/2$, the solution of equation [9] is the *stable* equilibrium $r = r_0$: in this case, particle P moves uniformly along the circle of center O and radius $r = r_0$, while particle Q is stationary.
2. If $0 < k < 1/2$, then $r_2 < r_1 = r_0$, particle P oscillates between the two concentric circles of center O and radii $r = r_2$ and r_0. Note that $r(t)$ is periodic, but that the motion of P in the horizontal plane is not necessarily periodic. See an example in Figure 5.24. In the limit $k \to 0$ (for instance if $m \ll M$ or if $v_0 \to 0$), the radius r_2 of the inner circle tends to zero.
3. If $1/2 < k < k_{max} = mv_0^2/2Mgl$, then $r_2 > r_1 = r_0$, particle P oscillates between the inner circle of center O of radius r_0 and the outer circle of center O of radius r_2. The radius r_2 of the outer circle increases as k increases. ■

5.5 Kinetic Energy of a Rigid Body

In the next three sections, we aim to formulate an energy method applicable to rigid bodies in planar motion. The result, still known as the *Work-Energy Principle*, is particularly useful to find first integrals of motion in an efficient manner. Our first task is to derive a general expression of the kinetic energy $K_\mathcal{B} = \frac{1}{2}\int_\mathcal{B} \mathbf{v}_P^2\, dm$ of a rigid body $\mathcal{B}$ in planar motion. Referring to Figure 5.25, consider the rectangular axes $Axyz$ attached to $\mathcal{B}$ with unit vectors $(\hat{\mathbf{x}}, \hat{\mathbf{y}})$ in the plane of motion: given a point A, recall that the velocity of a generic point P of coordinates (x, y, z) can be related to the velocity of point A according to

$$\mathbf{v}_P = \mathbf{v}_{A\in\mathcal{B}} + \omega_\mathcal{B}\hat{\mathbf{z}} \times (x\hat{\mathbf{x}} + y\hat{\mathbf{y}} + z\hat{\mathbf{z}}) = \mathbf{v}_{A\in\mathcal{B}} + \omega_\mathcal{B}(x\hat{\mathbf{y}} - y\hat{\mathbf{x}})$$

giving

$$\mathbf{v}_P^2 = \mathbf{v}_{A\in\mathcal{B}}^2 + 2\omega_\mathcal{B}\mathbf{v}_{A\in\mathcal{B}} \cdot (x\hat{\mathbf{y}} - y\hat{\mathbf{x}}) + \omega_\mathcal{B}^2(x^2 + y^2)$$

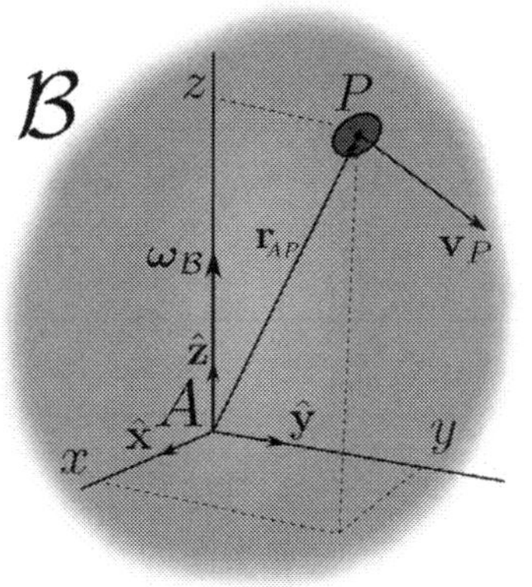

Figure 5.25 – *Rigid body $\mathcal{B}$ in planar motion.*

This leads to the following expression of $K_{\mathcal{B}}$:

$$K_{\mathcal{B}} = \frac{1}{2} m \mathbf{v}^2_{A\in\mathcal{B}} + \omega_{\mathcal{B}} \mathbf{v}_{A\in\mathcal{B}} \cdot \int_{\mathcal{B}} (x\hat{\mathbf{y}} - y\hat{\mathbf{x}}) dm + \frac{1}{2}\omega^2_{\mathcal{B}} \int_{\mathcal{B}} (x^2 + y^2) dm$$

We recognize the integral $\int_{\mathcal{B}}(x^2 + y^2)dm$ as the body's mass moment of inertia I_{Azz} about axis Az. We may also write the second term as $\mathbf{v}_{A\in\mathcal{B}} \cdot (\boldsymbol{\omega}_{\mathcal{B}} \times m\mathbf{r}_{AG})$ which is seen to vanish in two cases:
(i) when $A = G$ is the body's mass center,
(ii) when $\mathbf{v}_{A\in\mathcal{B}} = \mathbf{0}$: this occurs whenever body $\mathcal{B}$ rotates about axis Az.

Kinetic Energy of a Rigid Body

The kinetic energy of a rigid body $\mathcal{B}$ in planar motion takes the expression

$$K_{\mathcal{B}} = \frac{1}{2} m \mathbf{v}^2_{A\in\mathcal{B}} + \mathbf{v}_{A\in\mathcal{B}} \cdot (\boldsymbol{\omega}_{\mathcal{B}} \times m\mathbf{r}_{AG}) + \frac{1}{2} I_{Azz}\omega^2_{\mathcal{B}} \qquad (5.28)$$

where A is an arbitrary point of $\mathcal{B}$. Two special cases are useful in practice:

$$K_{\mathcal{B}} = \frac{1}{2} m \mathbf{v}^2_G + \frac{1}{2} I_{Gzz}\omega^2_{\mathcal{B}} \qquad (5.29)$$

valid for all planar motion, and

$$K_{\mathcal{B}} = \frac{1}{2} I_{Azz}\,\omega^2_{\mathcal{B}} \qquad (5.30)$$

when body $\mathcal{B}$ rotates about axis Az (at all time or possibly instantaneously).

Properties

①. The kinetic energy of body $\mathcal{B}$ depends on the choice of (Newtonian) referential $\mathcal{E}$ and should be denoted $K_{\mathcal{B}/\mathcal{E}}$ whenever needed.

②. The kinetic energy is an extensive quantity: the kinetic energy K_Σ of a system Σ of two (or more) rigid bodies 1, 2, ... is the sum $K_1 + K_2 + \cdots$.

③. In practice, formula (5.29) is used unless the body is in rotational motion, in which case (5.30) is used.

④. If a body 2 rolls without slipping relative a body 1 itself in motion relative to a referential 0, then formula (5.30) cannot be used to find $K_{2/0}$ by choosing A as the contact point between 1 and 2 since $\mathbf{v}_{A\in 2/0} \neq \mathbf{0}$. We can, however, use (5.30) to find kinetic energy $K_{2/1}$ of body 2 relative to body 1 since $\mathbf{v}_{A\in 2/1} \neq \mathbf{0}$.

5.6 Work done by Mechanical Actions Exerted on a Rigid Body

Our second task is to derive a general expression of the work done by mechanical actions exerted on a rigid body $\mathcal{B}$. Consider first the action of body forces, that is, forces acting throughout the volume of $\mathcal{B}$: assume that an infinitesimal volume dV enclosing arbitrary point P is acted upon by a force $\mathbf{f}_P dV$. Then the infinitesimal work done by these forces during time dt is found to be

$$dW = \int_{\mathcal{B}} \mathbf{f}_P \cdot \mathbf{v}_P dV dt = \int_{\mathcal{B}} \mathbf{f}_P \cdot (\mathbf{v}_{A\in\mathcal{B}} + \boldsymbol{\omega}_{\mathcal{B}} \times \mathbf{r}_{AP}) dV dt$$

leading to two terms:

$$\int_{\mathcal{B}} \mathbf{f}_P \cdot \mathbf{v}_{A\in\mathcal{B}} dV dt = \mathbf{F} \cdot \mathbf{v}_{A\in\mathcal{B}}\, dt$$

denoting by $\mathbf{F} = \int_{\mathcal{B}} \mathbf{f}_P dV$ the resultant of the body forces, and [1]

$$\int_{\mathcal{B}} \mathbf{f}_P \cdot (\boldsymbol{\omega} \times \mathbf{r}_{AP}) dV dt = \boldsymbol{\omega}_{\mathcal{B}} \cdot \int_{\mathcal{B}} (\mathbf{r}_{AP} \times \mathbf{f}_P) dV dt = \boldsymbol{\omega}_{\mathcal{B}} \cdot \mathbf{M}_A\, dt$$

denoting $\mathbf{M}_A = \int_{\mathcal{B}} \mathbf{r}_{AP} \times \mathbf{f}_P dV$ the resultant moment about A of the body forces. We also note that the same result would be obtained for contact forces $\mathbf{f}^s_Q dA$ acting on the boundary of $\mathcal{B}$ by replacing the volume integrals by surface (or line) integrals:

$$\begin{aligned} dW &= \int_{\partial\mathcal{B}} \mathbf{f}^s_Q \cdot \mathbf{v}_Q dA dt = \int_{\partial\mathcal{B}} \mathbf{f}^s_P \cdot (\mathbf{v}_{A\in\mathcal{B}} + \boldsymbol{\omega}_{\mathcal{B}} \times \mathbf{r}_{AP}) dA dt \\ &= (\mathbf{F}^s \cdot \mathbf{v}_{A\in\mathcal{B}} + \boldsymbol{\omega}_{\mathcal{B}} \cdot \mathbf{M}^s_A)\, dt \end{aligned}$$

denoting $\mathbf{F}^s$ and $\mathbf{M}^s_A$ the resultant force and resultant moment about A of the contact forces.

Work done by Body or Contact Forces

The total work done by body (contact) forces acting on a rigid body $\mathcal{B}$ from time t_1 to time t_2 can be calculated from the formula

$$W_{1\to 2} = \int_{t_1}^{t_2} (\mathbf{v}_{A\in\mathcal{B}} \cdot \mathbf{F} + \boldsymbol{\omega}_{\mathcal{B}} \cdot \mathbf{M}_A)\, dt \tag{5.31}$$

[1] Here we use the property of the triple scalar product (see Appendix B) $\mathbf{A} \cdot (\mathbf{B} \times \mathbf{C}) = \mathbf{C} \cdot (\mathbf{A} \times \mathbf{B})$ for 3 vectors $\mathbf{A}$, $\mathbf{B}$ and $\mathbf{C}$.

where $\mathbf{F}$ is the resultant of the body (contact) forces and $\mathbf{M}_A$ is the resultant moment about arbitrary point A of the body (contact) forces. The velocities $\mathbf{v}_{A\in\mathcal{B}}$ and $\boldsymbol{\omega}_{\mathcal{B}}$ are evaluated relative to the referential in which the work $W_{1\to 2}$ is calculated.

Remark 9. Note that, to apply (5.31) properly, the same point must be used to evaluate velocity $\mathbf{v}_{A\in\mathcal{B}}$ and moment $\mathbf{M}_A$. We can show that the quantity $\mathbf{v}_{A\in\mathcal{B}} \cdot \mathbf{F} + \boldsymbol{\omega}_{\mathcal{B}} \cdot \mathbf{M}_A$ is an invariant, that is, it is independent of the choice of A.

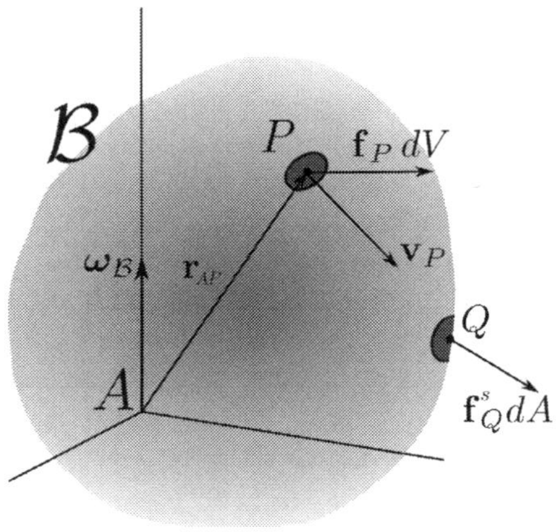

Figure 5.26 – *Rigid body $\mathcal{B}$: an elementary volume dV at P and elementary surface dA at Q is subject to body force $\mathbf{f}_P dV$ and contact force $\mathbf{f}^s_Q dA$, respectively.*

We can now use this general formula for special cases of importance in practice:

1. INTERIOR FORCES: recall that the interior forces acting within a rigid body satisfy

$$\mathbf{F}_{int} = \mathbf{0}, \qquad \mathbf{M}_{A,int} = \mathbf{0} \tag{5.32}$$

as a consequence of the law of action and reaction. Then formula (5.31) shows that $W^{int}_{1\to 2} = 0$: *the work done by internal forces acting within a rigid body vanishes.*

2. GRAVITATIONAL FORCES: for a body $\mathcal{B}$ of mass m and mass center G acted upon by the body forces $\mathbf{f}_P = \rho\mathbf{g}$ whose resultant force and resultant moment are given by

$$\mathbf{F} = \int_{\mathcal{B}} \rho \mathbf{g} dV = m\mathbf{g}, \qquad \mathbf{M}_G = \int_{\mathcal{B}} \mathbf{r}_{GP} \times \mathbf{g}\rho dV = \Big(\int_{\mathcal{B}} \mathbf{r}_{GP}\, dm\Big) \times \mathbf{g} = \mathbf{0} \tag{5.33}$$

the corresponding work is obtained by choosing $A = G$ in (5.31)

$$W^g_{1\to 2} = \int_{t_1}^{t_2} m\mathbf{g} \cdot \mathbf{v}_G\, dt = -mg\Delta Z_G \tag{5.34}$$

with coordinate Z_G of G measured on a fixed vertical axis OZ directed upward. Hence the work done by gravitational forces can be found from the potential energy $U^g = mgZ_G$.

3. WORK DONE BY A COUPLE $\mathcal{C}$: assume that body $\mathcal{B}$ is exerted upon by forces satisfying

$$\mathbf{F} = \mathbf{0}, \qquad \mathbf{M}_A = \mathcal{C}\hat{\mathbf{z}} \tag{5.35}$$

Then, the work done by these forces is found to be, according to (5.31)

$$W^{\mathcal{C}}_{1\to 2} = \int_{t_1}^{t_2} \mathcal{C}\hat{\mathbf{z}} \cdot \boldsymbol{\omega}_{\mathcal{B}}\, dt = \int_{t_1}^{t_2} \mathcal{C} d\theta \tag{5.36}$$

with $\boldsymbol{\omega}_{\mathcal{B}} = \dot{\theta}\hat{\mathbf{z}}$.

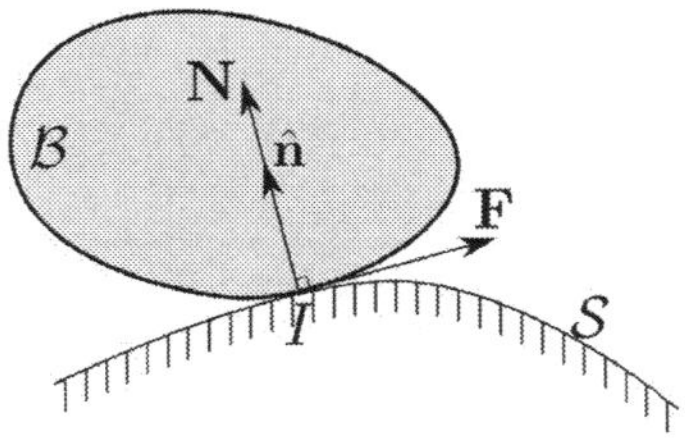

Figure 5.27 – *Rigid body $\mathcal{B}$ in contact at I with rigid surface $\mathcal{S}$.*

4. WORK DONE BY CONTACT FORCES: assume that body $\mathcal{B}$ is in contact with a surface $\mathcal{S}$ at point I. Since the contact force satisfies

$$\mathbf{R} = N\hat{\mathbf{n}} + \mathbf{F}, \qquad \mathbf{M}_I = \mathbf{0} \tag{5.37}$$

then, the corresponding work (relative to referential $\mathcal{E}$)

$$W^{c}_{1\to 2} = \int_{t_1}^{t_2} (N\hat{\mathbf{n}} + \mathbf{F}) \cdot \mathbf{v}_{I\in\mathcal{B}/\mathcal{E}}\, dt \tag{5.38}$$

does not vanish in general. In fact it may not be negative. *If in addition the body does not slip relative to a surface $\mathcal{S}$* **fixed** *relative to $\mathcal{E}$, then $\mathbf{v}_{I\in\mathcal{B}/\mathcal{E}} = \mathbf{0}$ and the work done by the reaction force $\mathbf{R}$ vanishes.*

PROPERTIES

①. In case of a frictionless contact with surface $\mathcal{S}$, that is, when $\mathbf{F} = \mathbf{0}$, the work done by the normal force $N\hat{\mathbf{n}}$ may not vanish since in general $N\hat{\mathbf{n}} \cdot \mathbf{v}_{I\in\mathcal{B}/\mathcal{E}} \neq 0$ unless the surface $\mathcal{S}$ is *fixed* relative to $\mathcal{E}$.

②. If the surface $\mathcal{S}$ of contact is fixed, then the slip velocity $\mathbf{v}_{I\in\mathcal{B}/\mathcal{E}}$ is colinear and opposed to the friction force $\mathbf{F}$: in that case $W^{c}_{1\to 2} = \int_{t_1}^{t_2} \mathbf{F} \cdot \mathbf{v}_{I\in\mathcal{B}/\mathcal{E}}\, dt$ is negative. The force $\mathbf{F}$ is *dissipative*.

③. A mechanical action is said to be *conservative* if its work done during an arbitrary time interval $t_1 \leq t \leq t_2$ can be found from a potential energy $\mathcal{U}$ as

$$W^{cons}_{1\to 2} = \mathcal{U}(t_1) - \mathcal{U}(t_2) \tag{5.39}$$

where function $\mathcal{U}$ depends only on the position of body $\mathcal{B}$.

④. The work done by a force $\mathbf{F}_A$ acting through a point A of $\mathcal{B}$ is given by $W_{1\to 2} = \int_{t_1}^{t_2} \mathbf{v}_{A\in\mathcal{B}/\mathcal{E}} \cdot \mathbf{F}_A \, dt$. In particular the expression found in Section 5.1 for the work done by an elastic force still holds: it may be determined by an elastic potential energy.

5.7 Work-Energy Principle for a Rigid Body

CONSIDER a rigid body $\mathcal{B}$ in motion in a *Newtonian* referential $\mathcal{E}$ exerted upon by external mechanical actions of resultant $\mathbf{F}_{ext\to\mathcal{B}}$ and resultant moment $\mathbf{M}_{G,ext\to\mathcal{B}}$ about the body's mass center G. According to formula (5.31) the work done by these actions during time interval $t_1 \leq t \leq t_2$ can be found as

$$W_{1\to 2}^{ext} = \int_{t_1}^{t_2} \left(\mathbf{v}_{G/\mathcal{E}} \cdot \mathbf{F}_{ext\to\mathcal{B}} + \boldsymbol{\omega}_{\mathcal{B}/\mathcal{E}} \cdot \mathbf{M}_{G,ext\to\mathcal{B}}\right) dt \tag{5.40}$$

Furthermore, according to Euler's First and Second Principles, we have

$$\mathbf{F}_{ext\to\mathcal{B}} = m\mathbf{a}_{G/\mathcal{E}}, \qquad \mathbf{M}_{G,ext\to\mathcal{B}} = \frac{d}{dt}\mathbf{H}_G \tag{5.41}$$

Using these expressions we find that equation (5.40) becomes

$$W_{1\to 2}^{ext} = \int_{t_1}^{t_2} \left(\mathbf{v}_{G/\mathcal{E}} \cdot m\mathbf{a}_{G/\mathcal{E}} + \boldsymbol{\omega}_{\mathcal{B}/\mathcal{E}} \cdot \frac{d}{dt}\mathbf{H}_G\right) dt \tag{5.42}$$

We then use two simple results (dropping the dependence upon $\mathcal{E}$):

$$\mathbf{v}_G \cdot \mathbf{a}_G = \frac{d}{dt}\left(\frac{\mathbf{v}_G^2}{2}\right), \quad \omega_{\mathcal{B}} \cdot \frac{d}{dt}\mathbf{H}_G = \omega_{\mathcal{B}}\hat{\mathbf{z}} \cdot \frac{d}{dt}\left(\omega_{\mathcal{B}}(I_{Gxz}\hat{\mathbf{x}} + I_{Gyz}\hat{\mathbf{y}} + I_{Gzz}\hat{\mathbf{z}})\right) = \frac{d}{dt}\left(I_{Gzz}\frac{\omega_{\mathcal{B}}^2}{2}\right)$$

to arrive at

$$W_{1\to 2}^{ext} = \int_{t_1}^{t_2} \frac{d}{dt}\left(m\frac{\mathbf{v}_G^2}{2} + I_{Gzz}\frac{\omega_{\mathcal{B}}^2}{2}\right) dt = \Delta\left(\tfrac{1}{2}m\mathbf{v}_G^2 + \tfrac{1}{2}I_{Gzz}\omega_{\mathcal{B}}^2\right) \tag{5.43}$$

We recognize the quantity $\frac{1}{2}m\mathbf{v}_G^2 + \frac{1}{2}I_{Gzz}\omega_{\mathcal{B}}^2$ as the kinetic energy of rigid body $\mathcal{B}$ according to (5.29). We conclude with the following result:

WORK-ENERGY PRINCIPLE (SINGLE RIGID BODY)

Relative to a Newtonian referential, the work done by all external mechanical actions exerted on a rigid body $\mathcal{B}$ from time t_1 to time t_2 is equal to the change of the body's kinetic energy over the same time interval

$$K_{\mathcal{B}}(t_2) - K_{\mathcal{B}}(t_1) = W_{1\to 2}^{ext} \tag{5.44}$$

Properties

①. The Work-Energy Principle can only be applied relative to a Newtonian referential: both the work $W^{ext}_{1\to 2}$ and the kinetic energy $K_{\mathcal{B}}$ must be determined relative to this referential.

②. The instantaneous version of the Work-Energy Principle is known as the *Kinetic Energy Theorem*:

$$\frac{d}{dt}K_{\mathcal{B}} = \mathcal{P}^{ext} \tag{5.45}$$

where $\mathcal{P}^{ext} = \mathbf{v}_G \cdot \mathbf{F}_{ext\to\mathcal{B}} + \boldsymbol{\omega}_{\mathcal{B}} \cdot \mathbf{M}_{G,ext\to\mathcal{B}}$ is the (instantaneous) *power* of the external actions.

③. The external actions can be divided into two categories: (i) non-conservative actions, (ii) conservative actions which derive from potential energies. Then, the Work-Energy Principle can be stated as follows:

$$W^{nc}_{1\to 2} = \Delta(K_{\mathcal{B}} + U) \tag{5.46}$$

where $W^{nc}_{1\to 2}$ is the work done by all non-conservative actions and U is the sum of all potential energies associated with the conservative actions.

④. If the work done by all non-conservative actions vanishes, then the mechanical energy of the body $E_{\mathcal{B}} = K_{\mathcal{B}} + U$ is conserved during the motion: this leads to a *first integral of motion*.

⑤. The Work-Energy Principle is an integrated form of the equations of motion. It is quite useful for Type II problems of kinetics: given initial conditions, it provides velocity versus position relationships without having to integrate ODEs.

⑥. The Work-Energy Principle provides only one equation and is generally not sufficient to complete the solution of a kinetics problem. One must return to Euler's principles to obtain additional equations.

EXAMPLE PROBLEMS

General Procedure: Rigid Body Kinetics by Energy Method

Follow the guidelines given in Section 4.8.

- In Step 7, apply the Work-Energy Principle to the body between an initial state at t_1 and a final state at t_2. For each "action" determine the work $W^{action}_{1\to 2}$ by carefully applying formula (5.31) unless the action is conservative and a potential energy is known.
- In Step 8, the solution of a Type II problem may be accomplished without the need to integrate an ODE unless the single equation provided by the Work-Energy Principle involves too many unknowns: it is then insufficient to complete the solution. In that case, return to Euler's First or Second Principle to extract one or more equations.

Exercise 5.12 ▪ A can modeled as a uniform, full cylinder of mass 1.2 kg and radius 6 cm is placed in contact with a flat support inclined at the angle 20° with the horizontal. Released from rest, the can is then observed to roll without slipping along the steepest line of the support. The contact is characterized by the static and kinetic friction coefficients $\mu_s = 0.6$ and $\mu_k = 0.45$.

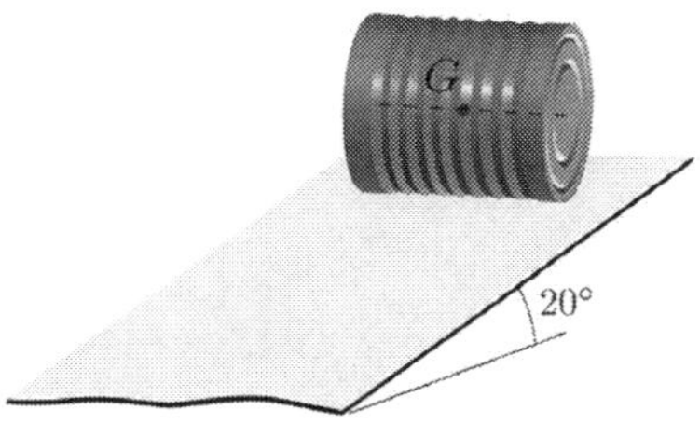

a. Find the speed of the can's mass center G when it has traveled a distance of 2 m. Find the corresponding travel time.
b. Find the reaction force exerted by the support and verify that the condition of rolling without slipping is satisfied. ▪

SOLUTION. **a.** We first define a basis $(\hat{\imath}, \hat{\jmath}, \hat{\mathbf{k}})$ as shown in Figure 5.28 (left) attached to the support. We define as R the can's radius and β the angle of inclination. The axis Ox is directed along the steepest line of incline with origin O defined as the initial position of mass center G. The motion of G is rectilinear: its position is defined by coordinate x_G. This leads to $\mathbf{v}_G = \dot{x}_G \hat{\imath}$. Defining the body's angular velocity as $\boldsymbol{\omega} = \omega \hat{\mathbf{k}}$, the no-slip condition at contact point I gives

$$\mathbf{v}_G = \omega \hat{\mathbf{k}} \times \mathbf{r}_{IG} = \omega \hat{\mathbf{k}} \times (-R\hat{\jmath}) = R\omega \hat{\imath}$$

and the condition

$$\dot{x}_G = R\omega \qquad [1]$$

Next we sketch a free-body diagram in Figure 5.28 (right): the body is subject to
- the contact force $-N\hat{\jmath} + F\hat{\imath}$ at I: this force does not work since $\mathbf{v}_{I\in\mathcal{B}} = \mathbf{0}$,
- the gravitational force $m\mathbf{g}$: this force is conservative and its work can be found from potential energy $U = -mgx_G \sin\beta$ since G drops a height $x_G \sin\beta$ when it travels a distance x_G.

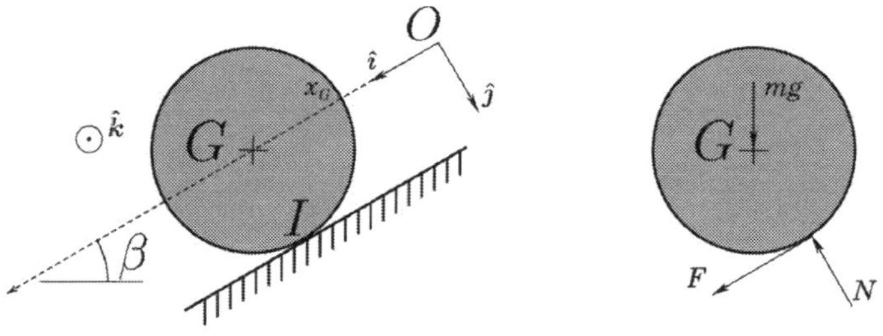

Figure 5.28 – *Problem 5.12*

Next we apply the WEP between $t = 0$ and $t > 0$:

$$\Delta(K + U) = W_{0\to t}^{N+F} = 0$$

showing that the energy $E = K + U$ is conserved during the motion

$$K + U = Cst \qquad [2]$$

where the constant Cst is found from the initial conditions. The can's kinetic energy K is found by applying formula (5.30) with $A = I$ since I is the instantaneous center of rotation of the body

$$K = \frac{1}{2} I_{Izz} \omega^2 = \frac{1}{2}\left(\frac{1}{2} mR^2 + mR^2\right)\omega^2 = \frac{3}{4} m\dot{x}_G^2$$

taking into account relationship [1]. Now [2] becomes

$$\frac{3}{4}m\dot{x}_G^2 - mgx_G\sin\beta = 0 \qquad [3]$$

where constant Cst is found by setting $x_G = 0$ and $\dot{x}_G = 0$ at time $t = 0$. This gives a relationship between $\dot{x}_G$ and x_G:

$$\boxed{\dot{x}_G = \left(\frac{4gx_G}{3}\sin\beta\right)^{1/2}}$$

When $x_G = 2$ m and $\beta = 20°$, we find $\dot{x}_G = \mathbf{3}$ **m/s**. Equation [3] can be time-differentiated to show that the acceleration $\ddot{x}_G$ is constant:

$$\ddot{x}_G = \frac{2}{3}g\sin\beta \qquad [4]$$

Two integrations give x_G versus time:

$$\boxed{x_G = \frac{1}{3}gt^2\sin\beta}$$

With $x_G = 2$ m, we find $t = \mathbf{1.34}$ **s**.
b. To find the reaction of the support we apply Euler's First Principle

$$m\mathbf{a}_G = m\ddot{x}_G\hat{\imath} = \mathbf{R}_{ext} = (mg - N)\hat{\jmath} + F\hat{\imath}$$

giving 2 equations

$$\frac{2}{3}mg\sin\beta = F, \qquad N = mg$$

using equation [4] for $\ddot{x}_G$. The can rolls without slipping as long as the condition $|F| < \mu_s N$, that is

$$\boxed{\beta < \beta_c = \sin^{-1}(\frac{3}{2}\mu_s)}$$

With $\mu_s = 0.6$ we find $\beta_c = \mathbf{64}°$: the condition is satisfied. ■

Exercise 5.13 ▪ A yoyo of mass m, mass center G and radius of gyration k about its axis is released without initial speed from an origin O. It then drops along a vertical line by unwinding the massless, inextensible string wrapped around the yoyo's inner hub of radius r. Assume that the string remains vertical during the motion and neglect its thickness. The position of G is defined by its coordinate $0 \le x \le L$ along the vertical axis Ox, where L denotes the string's length.
a. Find $\dot{x}$ versus x and parameters r, k, L, etc.
b. Find the tension T in the string as a function of x.
c. Find the time τ taken for the string to completely unwind for $r = 5$ mm, $k = 20$ mm, $m = 120$ g and $L = 60$ cm. ■

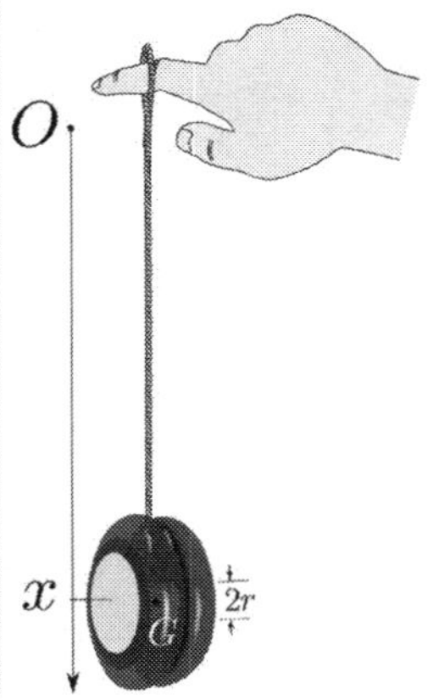

SOLUTION. **a.** To find $\dot{x}$ vs x we will apply the WEP to show that the energy of the yoyo is conserved. First we choose a basis $(\hat{\imath}, \hat{\jmath}, \hat{\mathbf{k}})$ of fixed unit vectors as shown in Figure 5.29 (left). Mass center G is in rectilinear motion along axis Ox with velocity $\mathbf{v}_G = \dot{x}\hat{\imath}$. The yoyo's angular velocity is written $\boldsymbol{\omega} = \omega\hat{\mathbf{k}}$. The velocities $\dot{x}$ and ω are not independent since the point of contact J of the string with the hub must satisfy

$\mathbf{v}_{J\in\text{string}} = \mathbf{v}_{J\in\mathcal{B}}$ (the string does not stretch!):

$$\mathbf{v}_{J\in\mathcal{B}} = \mathbf{v}_G + \boldsymbol{\omega} \times \mathbf{r}_{GJ} = \dot{x}\hat{\boldsymbol{\imath}} + \omega\hat{\mathbf{k}} \times r\hat{\boldsymbol{\jmath}} = (\dot{x} - r\omega)\hat{\boldsymbol{\imath}} = \mathbf{0}$$

giving

$$\dot{x} = r\omega \qquad [1]$$

Next we sketch a free-body diagram in Figure 5.29 (bottom): the body is subject to
- the tension $-T\hat{\boldsymbol{\imath}}$ at J: this force does not work since $\mathbf{v}_{J\in\mathcal{B}} = \mathbf{0}$,
- the gravitational force $mg\hat{\boldsymbol{\imath}}$: this force is conservative and its work can be found from potential energy $U = -mgx$

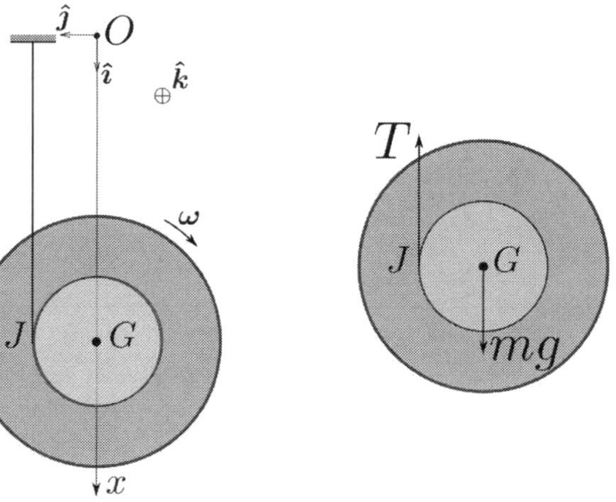

Figure 5.29 – *Problem 5.13 (Solution)*

This shows that the yoyo's energy is conserved: $\Delta(K + U) = W^T_{0\to 1} = 0$

$$K + U = Cst \qquad [2]$$

where the constant Cst is found from the initial conditions. The yoyo's kinetic energy K is found by applying formula (5.30) with $A = J$ since J is the instantaneous center of rotation of the body

$$K = \frac{1}{2}I_{Jzz}\omega^2 = \frac{1}{2}(mk^2 + mr^2)\omega^2 = \frac{1}{2}m(\frac{k^2}{r^2} + 1)\dot{x}^2$$

using the parallel axis theorem and equation [1]. Now equation [2] becomes

$$\frac{1}{2}m(\frac{k^2}{r^2} + 1)\dot{x}^2 - mgx = Cst = 0$$

where we have used $x = \dot{x} = 0$ at $t = 0$. Finally we find the relationship

$$\boxed{\dot{x} = \sqrt{2gx}\Big(1 + \frac{k^2}{r^2}\Big)^{-1/2}} \qquad [3]$$

b. To find the tension T, we apply Euler's First Principle

$$m\mathbf{a}_G = \mathbf{F}_{ext\to\mathcal{B}} = (mg - T)\hat{\boldsymbol{\imath}}$$

giving

$$T = mg - m\ddot{x} = mg - m\sqrt{2g}\frac{\dot{x}}{2\sqrt{x}}(1 + \frac{k^2}{r^2})^{-1/2} = mg - mg(1 + \frac{k^2}{r^2})^{-1}$$

where we have obtained acceleration $\ddot{x}$ by time-differentiating equation [3]. This gives

$$\boxed{T = mg \frac{k^2}{k^2 + r^2}}$$

c. To find the time τ taken for the yoyo to completely unwind, we need to impose $x(\tau) = L$. We have to integrate ODE [3] which is separable:

$$\int_0^L \frac{dx}{\sqrt{x}} = \sqrt{2g}\left(1 + \frac{k^2}{r^2}\right)^{-1/2} \tau$$

giving

$$\boxed{\tau = \sqrt{\frac{2L}{g} \frac{r^2 + k^2}{r^2}}}$$

It is also possible to use

$$\ddot{x} = g - T/m = g \frac{r^2}{r^2 + k^2} = a_G = Cst$$

Then integration gives $L = a_G \tau^2/2$. With $r = 5$ mm, $k = 20$ mm, and $L = 60$ cm , we find $\tau = \mathbf{1.44}$ s.

■

Exercise 5.14 ■ An axisymmetric body of mass m, mass center G and radius of gyration k about its axis Gz rolls without slipping on a cylindrical fixed support of center O and radius $R + r$. It is released at position $\theta = \theta_0$ without initial speed. The body then oscillates in a vertical plane.
a. Find the equation governing angle θ.
b. Deduce the period of small amplitude oscillations. ■

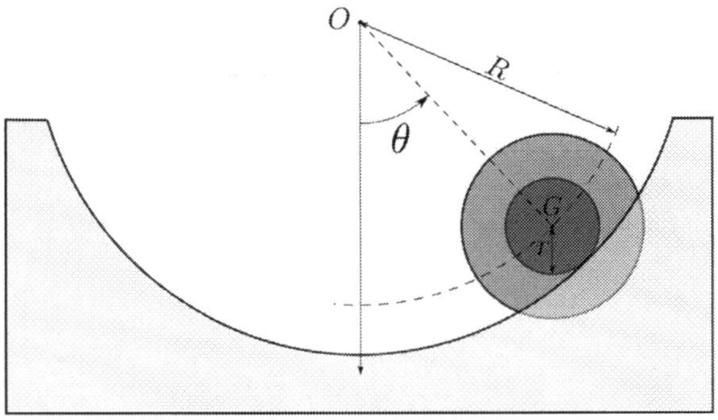

Figure 5.30 – *Problem 5.14*

SOLUTION. **a.** We proceed as in problem 5.13: we can show that the energy of the body is conserved. Taking the time-derivative of $E = Cst$ will yield the equation of motion. First we choose a set of unit vectors $(\hat{\mathbf{u}}, \hat{\mathbf{v}})$ best suited for the circular motion of mass center G. Also define unit vector $\hat{\mathbf{k}} = \hat{\mathbf{u}} \times \hat{\mathbf{v}}$. Figure 5.31 (right). From $\mathbf{r}_{OG} = R\hat{\mathbf{u}}$ we obtain $\mathbf{v}_G = R\dot{\theta}\hat{\mathbf{v}}$. We then apply the no-slip condition at contact point I

$$\mathbf{v}_{I\in\mathcal{B}} = \mathbf{0} = \mathbf{v}_G + \omega\hat{\mathbf{k}} \times \mathbf{r}_{GI} = R\dot{\theta}\hat{\mathbf{v}} + \omega\hat{\mathbf{k}} \times r\hat{\mathbf{u}} = (R\dot{\theta} + r\omega)\hat{\mathbf{v}}$$

denoting the body's angular velocity $\boldsymbol{\omega} = \omega\hat{\mathbf{k}}$. This gives the kinematic relationship

$$\omega = -\frac{R}{r}\dot{\theta} \qquad [1]$$

Next we sketch a free-body diagram in Figure 5.31 (left): the body is subject to
- the reaction force $-N\hat{\mathbf{u}} + F\hat{\mathbf{v}}$ at I: this force does not work thanks to condition $\mathbf{v}_{I\in\mathcal{B}} = \mathbf{0}$,
- the gravitational force $m\mathbf{g}$: this force is conservative and its work can be found from potential energy $U = -mgR\cos\theta$

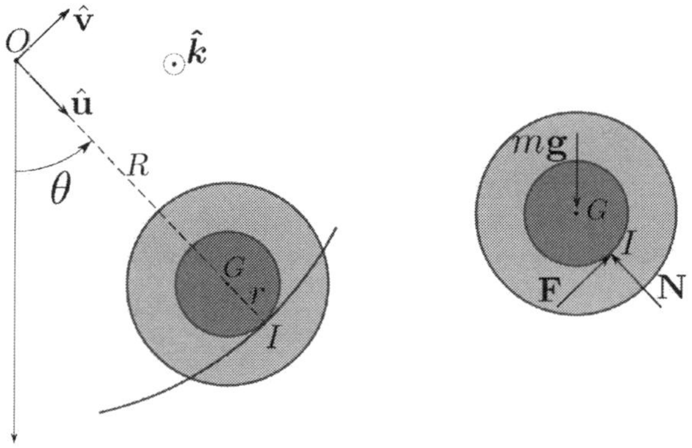

Figure 5.31 – *Problem 5.14 (Solution)*

This shows that the body's energy is conserved: $\Delta(K + U) = W_{0\to1}^{N+F} = 0$

$$K + U = Cst \qquad [2]$$

where the constant Cst is found from the initial conditions. The body's kinetic energy K is found by applying formula (5.29) or (5.30)

$$K = \frac{1}{2}mk^2\omega^2 + \frac{1}{2}mv_G^2 = \frac{1}{2}m(k^2 + r^2)\frac{R^2}{r^2}\dot{\theta}^2$$

after using [1]. Then [2] becomes

$$\frac{1}{2}m(k^2 + r^2)\frac{R^2}{r^2}\dot{\theta}^2 - mgR\cos\theta = -mgR\cos\theta_0$$

using $\theta = \theta_0$ and $\dot{\theta} = 0$ at time $t = 0$. This equation is valid at all time: we can time-differentiate it to obtain the equation of motion

$$\boxed{(k^2 + r^2)\frac{R^2}{r^2}\ddot{\theta} + gR\sin\theta = 0} \qquad [3]$$

b. Equation [3] is a pendulum equation: it can be linearized by approximating $\sin\theta$ by

θ for $\theta \ll 1$. Then [3] becomes

$$\ddot{\theta} + \frac{g}{R}\frac{r^2}{k^2+r^2}\theta = 0$$

This equation is of the type $\ddot{\theta} + \Omega^2\theta = 0$ with $\Omega^2 = \frac{g}{R}\frac{r^2}{k^2+r^2}$. The period of small amplitude oscillations is given by

$$\boxed{T = \frac{2\pi}{\Omega} = 2\pi\sqrt{\frac{g}{R}\frac{r^2}{k^2+r^2}}}$$

■

Exercise 5.15 ■■ A solid homogeneous sphere of mass center G, mass m and radius r is in motion in a vertical plane by pivoting about the corner point O of a fixed rigid boundary. The sphere is released from the position $\theta = 0$ from rest (the equilibrium is unstable). This motion occurs as long as the friction forces can overcome the effect of gravity. Eventually the sphere slips or loses contact. The contact is characterized by static friction coefficient μ.

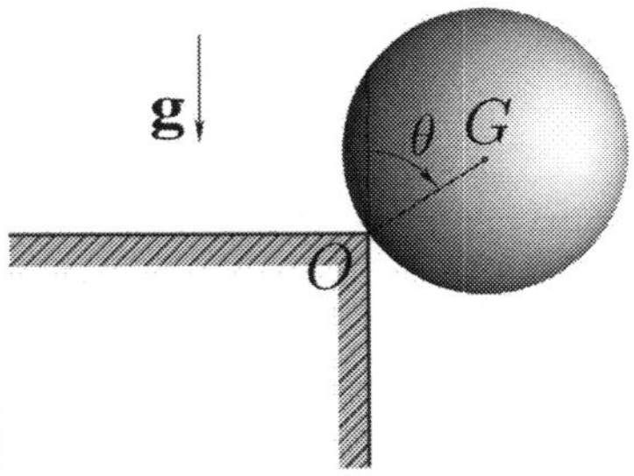

a. Find the reaction force $\mathbf{R}_O$ of the support on the sphere as a function of angle θ.
b. Use the expression of $\mathbf{R}_O$ to find out whether the sphere first slips or loses contact. At which value of angle θ does this occur for $\mu = 0.5$? Describe the subsequent motion of the sphere. ■

SOLUTION **a.** To find the reaction force $\mathbf{R}_O$ of the support on the sphere as a function of angle θ, we will apply Euler's first principle: this will require the expression of acceleration $\mathbf{a}_G$. However this quantity involves both $\dot{\theta}$ and $\ddot{\theta}$. To find $\dot{\theta}$ and $\ddot{\theta}$ in terms of θ we will apply the WEP. (The alternative is to apply Euler's Second Principle: this would give us an ODE which we would need to integrate.)
First we choose a basis $(\hat{\mathbf{u}}, \hat{\mathbf{v}}, \hat{\mathbf{k}})$ as shown in Figure 5.32 (left) suitable for the circular motion of G. Then from $\mathbf{r}_{OG} = r\hat{\mathbf{u}}$ we obtain $\mathbf{v}_G = r\dot{\theta}\hat{\mathbf{v}}$ and $\mathbf{a}_G = r\ddot{\theta}\hat{\mathbf{v}} - r\dot{\theta}^2\hat{\mathbf{u}}$. Next we sketch a free-body diagram in Figure 5.32 (center): the body is subject to
- the reaction force $\mathbf{R}_O = N\hat{\mathbf{u}} + F\hat{\mathbf{v}}$ at O: this force does not work since O is a fixed point of the body $\mathbf{v}_{O\in\mathcal{B}} = \mathbf{0}$,
- the gravitational force $m\mathbf{g}$: this force is conservative and its work can be found from potential energy $U = mgr\cos\theta$
We can now apply the WEP: $\Delta(K+U) = W^{R_O}_{0\to 1} = 0$. The sphere's energy is conserved:

$$K + U = Cst \quad [1]$$

where the constant Cst is found from the initial conditions. The body's kinetic energy K is found by applying formula (5.30) with $A = O$ since the body rotates about O

$$K = \frac{1}{2}I_{Ozz}\omega^2 = \frac{1}{2}(\frac{2}{5}mr^2 + mr^2)\dot{\theta}^2 = \frac{7}{10}mr^2\dot{\theta}^2$$

we have used the parallel axis theorem to find I_{Ozz} and the expression $\boldsymbol{\omega} = \dot{\theta}\hat{\mathbf{k}}$ of the body's angular velocity.

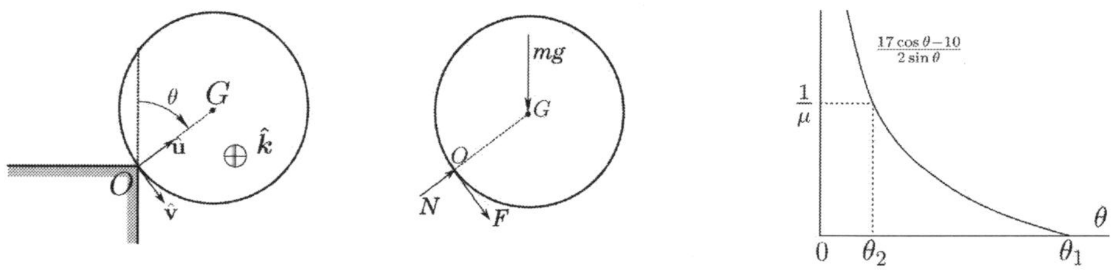

Figure 5.32 – *Problem 5.15 (Solution)*

Now equation [1] becomes

$$\frac{7}{10}mr^2\dot{\theta}^2 + mgr\cos\theta = Cst = mgr$$

where we have used $\dot{\theta} = 0$ when $\theta = 0$. We then find the relationship

$$\boxed{\dot{\theta}^2 = \frac{10}{7}\frac{g}{r}(1 - \cos\theta)} \qquad [2]$$

We can take the time-derivative of [2] to obtain $\ddot{\theta}$: we find

$$\boxed{\ddot{\theta} = \frac{5}{7}\frac{g}{r}\sin\theta} \qquad [3]$$

Now we apply Euler's First Principle to find the reaction force $\mathbf{R}_O$: we project $m\mathbf{a}_G = \mathbf{R}_O + m\mathbf{g}$ on $(\hat{\mathbf{u}}, \hat{\mathbf{v}})$ to find

$$-mr\dot{\theta}^2 = N - mg\cos\theta, \qquad mr\ddot{\theta} = F + mg\sin\theta \qquad [4-5]$$

We then use [2-3] to substitute the expressions of $\dot{\theta}^2$ and $\ddot{\theta}$ in [4-5] to find

$$\boxed{N = \frac{1}{7}mg(17\cos\theta - 10), \qquad F = -\frac{2}{7}mg\sin\theta} \qquad [6-7]$$

b. The sphere stays in contact as long as $N > 0$. With the expression of [6] this imposes

$$\theta < \theta_1 = \cos^{-1}\frac{10}{17} = 54^\circ$$

The sphere rolls without slipping as long as $|F| < \mu N$ which gives the condition using [6-7]

$$\frac{1}{\mu} < \frac{17\cos\theta - 10}{2\sin\theta} \qquad [8]$$

The simple sketch of Figure 5.32 (right) shows that condition [8] is equivalent to $\theta < \theta_2$ with $\theta_2 < \theta_1$. This proves that *the sphere will first slip before losing contact*. For $\mu = 0.5$ we find numerically $\theta_2 = \mathbf{41.8^\circ}$. For $\theta > \theta_2$, the sphere will roll and slip while still maintaining contact with O. Eventually contact will be lost. ∎

5.8 Work-Energy Principle for a System of Rigid Bodies

IN THIS SECTION, we generalize the Work-Energy Principle to the case of a system of rigid bodies. It is sufficient to consider a system $\Sigma = \{1, 2\}$ of two bodies 1 and 2 in motion relative to a Newtonian referential 0. The two bodies interact (mostly due to contact) and the forces of interaction satisfy:

$$\mathbf{F}_{1\to 2} = -\mathbf{F}_{2\to 1}, \qquad \mathbf{M}_{A,1\to 2} = -\mathbf{M}_{A,2\to 1} \tag{5.47}$$

due to the law of action and reaction (A is an arbitrary point). By taking into account equations (5.47) we have seen that the actions between the two bodies cancel each other out in the formulation of Euler's First and Second Principles to system Σ. This may lead us to believe that this will also occur in the formulation of the Work-Energy Principle. But this is not the case as we are about to show. The first step consists in applying the Work-Energy Principle to body 1 and body 2 from initial time t_i to final time t_f as follows:

$$\Delta K_{1/0} = \int_{t_i}^{t_f} (\mathbf{v}_{A\in 1/0} \cdot \mathbf{F}_{ext\to 1} + \boldsymbol{\omega}_{1/0} \cdot \mathbf{M}_{A,ext\to 1})dt + \int_{t_i}^{t_f} (\mathbf{v}_{A\in 1/0} \cdot \mathbf{F}_{2\to 1} + \boldsymbol{\omega}_{1/0} \cdot \mathbf{M}_{A,2\to 1})dt \tag{5.48}$$

$$\Delta K_{2/0} = \int_{t_i}^{t_f} (\mathbf{v}_{B\in 2/0} \cdot \mathbf{F}_{ext\to 2} + \boldsymbol{\omega}_{2/0} \cdot \mathbf{M}_{B,ext\to 2})dt + \int_{t_i}^{t_f} (\mathbf{v}_{B\in 2/0} \cdot \mathbf{F}_{1\to 2} + \boldsymbol{\omega}_{2/0} \cdot \mathbf{M}_{B,1\to 2})dt \tag{5.49}$$

where for each body we have separated the work done by external actions from the work done by internal actions. The second step consists of summing equations (5.48) and (5.49):

$$\begin{aligned}\Delta(K_{1/0} + K_{2/0}) = &\int_{t_i}^{t_f} (\mathbf{v}_{A\in 1/0} \cdot \mathbf{F}_{ext\to 1} + \boldsymbol{\omega}_{1/0} \cdot \mathbf{M}_{A,ext\to 1})dt \\ &+ \int_{t_i}^{t_f} (\mathbf{v}_{B\in 2/0} \cdot \mathbf{F}_{ext\to 2} + \boldsymbol{\omega}_{2/0} \cdot \mathbf{M}_{B,ext\to 2})dt \\ &+ \int_{t_i}^{t_f} \Big((\mathbf{v}_{A\in 2/0} - \mathbf{v}_{A\in 1/0}) \cdot \mathbf{F}_{1\to 2} + (\boldsymbol{\omega}_{2/0} - \boldsymbol{\omega}_{1/0}) \cdot \mathbf{M}_{A,1\to 2}\Big)dt\end{aligned} \tag{5.50}$$

where:

1. The sum $K_{1/0} + K_{2/0}$ represents the kinetic energy $K_{\Sigma/0}$ of the system,
2. The work done by the external actions on bodies 1 and 2 can be added to obtain the *external work*:

$$W_{i\to f}^{ext} = \int_{t_i}^{t_f} (\mathbf{v}_{A\in 1/0} \cdot \mathbf{F}_{ext\to 1} + \boldsymbol{\omega}_{1/0} \cdot \mathbf{M}_{A,ext\to 1})dt + \int_{t_i}^{t_f} (\mathbf{v}_{B\in 2/0} \cdot \mathbf{F}_{ext\to 2} + \boldsymbol{\omega}_{2/0} \cdot \mathbf{M}_{B,ext\to 2})dt$$

3. The two internal terms can be grouped together by taking into account (5.47). Furthermore

we can use the kinematic identities learned in Section 2.10:

$$\mathbf{v}_{A\in 2/0} - \mathbf{v}_{A\in 1/0} = \mathbf{v}_{A\in 2/1}, \qquad \omega_{2/0} - \omega_{1/0} = \omega_{2/1}$$

to obtain the general expression of *internal work*:

Internal Work of an Action/Reaction

The work done by a pair of internal action/reaction only depends upon the relative motion of the two interacting bodies and not on the choice of referential. It can be found according to the formula

$$W_{i\to f}^{int} = \int_{t_i}^{t_f} \left(\mathbf{v}_{A\in 2/1} \cdot \mathbf{F}_{1\to 2} + \omega_{2/1} \cdot \mathbf{M}_{A,1\to 2}\right) dt \qquad (5.51)$$

Remark 10. Recall that, for a given "action" on body i the quantity (power)

$$\mathbf{v}_{A\in i/j} \cdot \mathbf{F}_{action} + \omega_{i/j} \cdot \mathbf{M}_{A,action}$$

is an invariant, that is, it is independent of the choice of A.

We can now state the Work-Energy Principle for Σ.

Work-Energy Principle (System of Rigid Bodies)

Relative to a Newtonian referential, the sum of the work done by all external mechanical actions exerted on a system of rigid bodies and the work done by all internal actions from time t_i to time t_f is equal to the variation of the system's kinetic energy

$$\Delta K_\Sigma = W_{i\to f}^{ext} + W_{i\to f}^{int} \qquad (5.52)$$

where $\Delta K_\Sigma = K_\Sigma(t_f) - K_\Sigma(t_i)$.

Remark 11. The instantaneous version of the Work-Energy Principle is known as the *Kinetic Energy Theorem*:

$$\frac{d}{dt} K_\Sigma = \mathcal{P}^{ext} + \mathcal{P}^{int} \qquad (5.53)$$

where $\mathcal{P}^{ext}$ and $\mathcal{P}^{int}$ are the (instantaneous) *powers* of the external and internal actions.

The external work $W_{i\to f}^{ext}$ is determined as we learned in Section 5.6 for any external action exerted on a specific body of system Σ in motion relative to referential 0. It is found by integrating the instantaneous power $\mathbf{v}_{A\in body/0} \cdot \mathbf{F}_{ext\to body} + \omega_{body/0} \cdot \mathbf{M}_{A,ext\to body}$ over time, choosing point A conveniently.

To determine the internal work involving a particular interaction between body 1 and body 2, formula (5.51) shows that one must account for their relative kinematics. This is illustrated on the following cases encountered in practice:

1. Point Contact between two Bodies: Assume that two bodies of system Σ are in

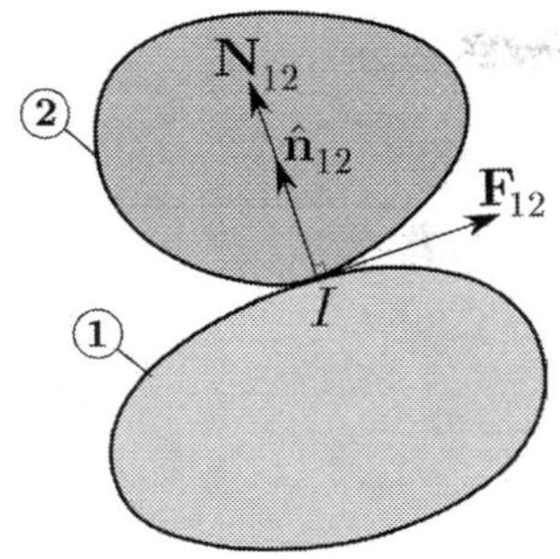

Figure 5.33 – *Two rigid bodies in contact at I.*

point-contact at a common point I. Then the action of body 1 on body 2 satisfies

$$\mathbf{R}_{1\to 2} = N_{12}\hat{\mathbf{n}}_{12} + \mathbf{F}_{12}, \qquad \mathbf{M}_{I,1\to 2} = \mathbf{0} \tag{5.54}$$

if rolling friction is neglected. Vector $\hat{\mathbf{n}}_{12}$ denotes the unit normal at I pointing from body 1 to body 2 (see Section 4.6). Then the corresponding internal work takes the expression

$$W^{int}_{i\to f} = \int_{t_i}^{t_f} (N_{12}\hat{\mathbf{n}}_{12} + \mathbf{F}_{12}) \cdot \mathbf{v}_{I\in 2/1}\, dt = \int_{t_i}^{t_f} \mathbf{F}_{12} \cdot \mathbf{v}_{I\in 2/1}\, dt \tag{5.55}$$

where we have used the fact that $N_{12}\hat{\mathbf{n}}_{12} \cdot \mathbf{v}_{I\in 2/1} = 0$. This work vanishes if $\mathbf{v}_{I\in 2/1} = \mathbf{0}$ during the time interval $t_i \leq t \leq t_f$, that is, if body 2 rolls without slipping relative to body 1.

2. Frictionless Joint between two Bodies: Assume that two bodies 1 and 2 of system Σ are interconnected by one of the joints described in Section 4.7. An energy-based criterion to characterize a *frictionless joint* can be stated by imposing the corresponding internal power of the contact mechanical actions vanishes

$$\mathbf{v}_{A\in 2/1} \cdot \mathbf{F}_{1\to 2} + \boldsymbol{\omega}_{2/1} \cdot \mathbf{M}_{A,1\to 2} = 0 \tag{5.56}$$

at any time, and for all allowable motions of body 2 relative to body 1. It is easy to show that this condition leads to the those stated in Section 4.7 to characterize frictionless joints. For instance, consider the *circular slider* of Figure 4.22 (case d). The motion of body 2 relative to body 1 satisfies

$$\mathbf{v}_{O\in 2/1} = \mathbf{0}, \qquad \boldsymbol{\omega}_{2/1} = \omega_{2/1}\hat{\mathbf{k}}.$$

Then equation (5.56) guaranteeing a frictionless joint leads to

$$\omega_{2/1}\hat{\mathbf{k}} \cdot \mathbf{M}_{O,1\to 2} = 0$$

We recover the condition found in Section 4.7.

3. Elastic coupling between two Bodies: Assume that bodies 1 and 2 are connected by an elastic spring of constant k and unstretched length l_0. Then the corresponding internal work done by the elastic force $\mathbf{F}_{1\to 2} = -k(l - l_0)\hat{\mathbf{u}}$ which body 1 exerts on body 2 is found to be

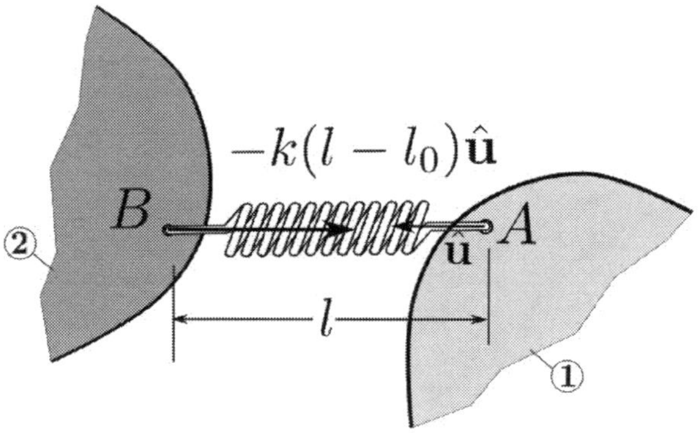

Figure 5.34 – *Two rigid bodies elastically coupled.*

$$W_{i\to f}^{int} = \int_{t_i}^{t_f} \mathbf{v}_{B\in 2/1} \cdot \mathbf{F}_{1\to 2}\, dt = -\int_{t_i}^{t_f} \frac{d}{dt}(l\hat{\mathbf{u}}) \cdot k(l-l_0)\hat{\mathbf{u}}\, dt$$
$$= -\int_{t_i}^{t_f} \frac{d}{dt}\left(\frac{k}{2}(l-l_0)^2\right)dt = -\Delta U_{1\leftrightarrow 2} \quad (5.57)$$

Not surprisingly, the internal work can be found from the potential energy

$$U_{1\leftrightarrow 2} = \frac{k}{2}(l-l_0)^2$$

called the elastic *potential of interaction* between the two bodies.

4. Internal Torque: Assume that body 1 exerts on body 2 a mechanical action (torque) which satisfies

$$\mathbf{F}_{1\to 2} = \mathbf{0}, \qquad \mathbf{M}_{A,1\to 2} = C_{12}\hat{\mathbf{z}}$$

then the corresponding internal work is found to be

$$W_{i\to f}^{int} = \int_{t_i}^{t_f} C_{12}\hat{\mathbf{z}} \cdot \boldsymbol{\omega}_{2/1}\, dt = \int_{t_i}^{t_f} C_{12} d\theta_{21}$$

with angular $\boldsymbol{\omega}_{2/1} = \dot{\theta}_{21}\hat{\mathbf{z}}$ expressed in terms of an angle θ_{21} defining the orientation of body 2 relative to body 1. If couple C_{12} is a known function of angle θ_{21}, then the internal work can be expressed in terms of a potential $W_{i\to f}^{int} = -\Delta U_{1\leftrightarrow 2}$. For instance a torsional spring mounted between the two bodies provides the couple $C_{12} = -c\theta_{21}$: then $U_{1\leftrightarrow 2} = \frac{c}{2}\theta_{21}^2$.

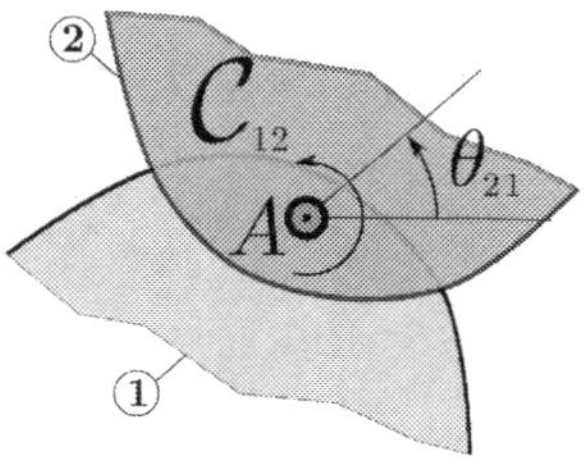

Figure 5.35 – *Internal torque.*

EXAMPLE PROBLEMS

General Procedure: Kinetics of Rigid Bodies by Energy Method

Follow the guidelines outlined in Section 4.8 for each rigid body.

- In Step 6, sketch a free-body diagram for each body of the system so as to clearly show both internal and external actions. Alternatively, sketch a diagram showing all exterior actions into the system and internal interactions within the system.
- In Step 7, apply the Work-Energy Principle to the system between an initial state at t_i and a final state at t_f. For each external "action", determine the corresponding work $W^{ext}_{i\to f}$ by applying formula (5.31). For each internal action/reaction, find the corresponding work $W^{int}_{i\to f}$ by applying formula (5.51). If the external or internal action is conservative, find the work by using the corresponding potential energy. Recall that, when two bodies of a system are connected by a frictionless joint, the corresponding internal work done by the contact mechanical actions vanishes.
- If an equation of motion (as opposed to a first integral) is sought, then apply the energy method in the form of the Kinetic Energy Theorem.

Exercise 5.16 ▪ The mechanism composed of two identical slender bars 1 and 2 and a roller 3 is dropped from rest in a vertical plane at the position $\theta = \theta_0$. The bars each have mass m and length $2l$. They are connected by a linear spring of constant k which is unstretched in the initial position. The roller is of radius r, mass M and radius of gyration R about its axis. It is assumed to roll without slipping on the fixed support 0. Assume that all joints are frictionless.
a. Determine the angular velocity of the bars at the instant they become horizontal ($\theta = 0$).
b. Find the condition which must be satisfied for the system to reach its final position. ▪

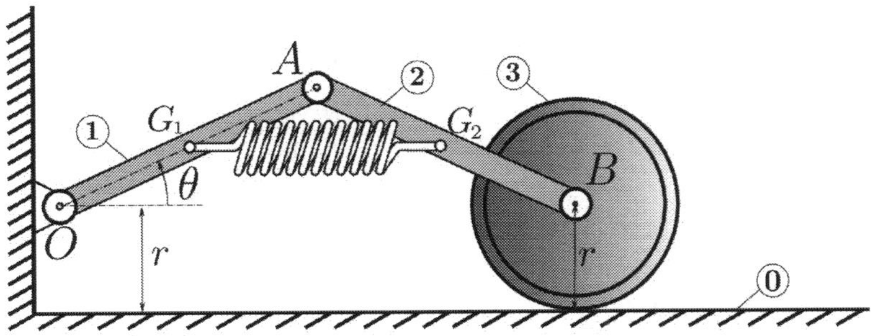

Figure 5.36 – *Problem 5.16*

Solution. **a.** First, we note that the system has one degree-of-freedom represented by angle θ: this is a consequence of the equal lengths of the two bars $|OA| = |AB|$ and the no-slip rolling motion of the roller. It is easy to show that the position of the system is entirely specified by angle θ. After defining the unit vectors $(\hat{\imath}, \hat{\jmath}, \hat{\mathbf{k}})$ as shown in Figure 5.37, we first identify the kinematic properties of the system when it reaches the position $\theta = 0$. First, we find that (at all time)

$$\omega_1 = -\dot{\theta}\hat{\mathbf{k}} = -\omega_2 = \dot{\theta}\hat{\mathbf{k}}$$

When $\theta = 0$ given the direction of velocities $\mathbf{v}_A$ and $\mathbf{v}_B$ we find that the instantaneous center of rotation of bar 2 is located at the intersection of line $(A, \hat{\imath})$ and line $(B, \hat{\jmath})$: since the 2 lines intersects at B, we deduce that $\mathbf{v}_B = 0$. Since B is the center of roller 3, the no-slip condition leads to at $t = t_f$

$$\omega_3 = 0.$$

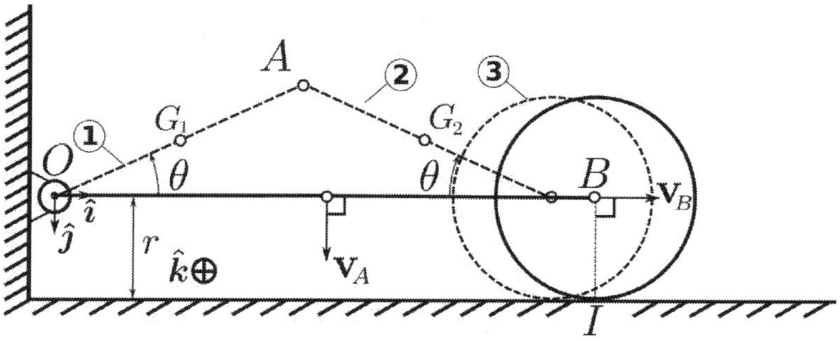

Figure 5.37 – *Problem 5.16 (Solution)*

The variation of the kinetic energy of the system between $t = t_i$ $(\theta = \theta_0)$ and $t = t_f$ $(\theta = 0)$ is then given by

$$\Delta K_\Sigma = K_\Sigma(t_f) - K_\Sigma(t_i) = K_\Sigma(t_f) = K_1(t_f) + K_2(t_f) + K_3(t_f) = K_1(t_f) + K_2(t_f)$$

where we have used $K_3(t_f) = \frac{1}{2}MR^2\omega_3^2(t_f) + \frac{1}{2}M\mathbf{v}_B^2(t_f) = 0$. We also find the expressions of the kinetic energies of bodies 1 and 2 at $t = t_f$:

$$K_1(t_f) = \tfrac{1}{2}I_{Ozz}\omega_1^2 = \frac{2}{3}ml^2\dot{\theta}_f^2, \qquad K_2(t_f) = \tfrac{1}{2}I_{Bzz}\omega_2^2 = \frac{2}{3}ml^2\dot{\theta}_f^2 \qquad [1]$$

Next we sketch the free-body diagram of each body in Figure 5.38. The system is subject to the following external actions:
-the reaction at O which does not work since the pivot is frictionless: $\boldsymbol{\omega}_{1/0} \cdot \mathbf{M}_{O,0\to 1} = 0$,
-the gravitational forces acting on the bars whose work is given by $W_{i\to f}^{2mg} = -2mg\Delta y_G = 2mgl\sin\theta_0$,
-the gravitational forces acting on the roller whose work is $W_{i\to f}^{Mg} = -Mg\Delta y_B = 0$,
-the reaction of the support at I whose work is $W_{i\to f}^{N+F} = \int_{t_i}^{t_f}(-N\hat{\jmath} + F\hat{\imath}) \cdot \mathbf{v}_{I\in 3/0}dt = 0$ since the roller does not slip.
The system is also subject to internal actions:
- the reaction of the pivot at A which does not work since the pivot is frictionless: $\boldsymbol{\omega}_{2/1} \cdot \mathbf{M}_{A,1\to 2} = 0$,
- the reaction of the pivot at B which does not work since the pivot is frictionless: $\boldsymbol{\omega}_{3/2} \cdot \mathbf{M}_{B,2\to 3} = 0$,
- the spring force between rods 1 and 2 whose work is found from the elastic potential $U_{1\leftrightarrow 2} = \frac{k}{2}(L - L_0)^2$ with $L = |G_1G_2| = 2l\cos\theta$ and $L_0 = 2l\cos\theta_0$: $W_{i\to f}^{int} = -\Delta U_{1\leftrightarrow 2}$.

We can now apply the Work Energy Principle to system Σ between $t = t_i$ ($\theta = \theta_0$) and $t = t_f$ ($\theta = 0$) using [1] and the expressions of the external and internal works found above:

$$\underbrace{\frac{4}{3}ml^2\dot{\theta}_f^2}_{\Delta K_\Sigma} = \underbrace{2mgl\sin\theta_0}_{W_{i\to f}^{ext}} \underbrace{- \frac{k}{2}(2l - 2l\cos\theta_0)^2}_{W_{i\to f}^{int}}$$

This gives the final angular velocity of the rods:

$$\boxed{\dot{\theta}_f^2 = \frac{3g}{2l}\sin\theta_0 - \frac{3k}{2m}(1-\cos\theta_0)^2} \qquad [2]$$

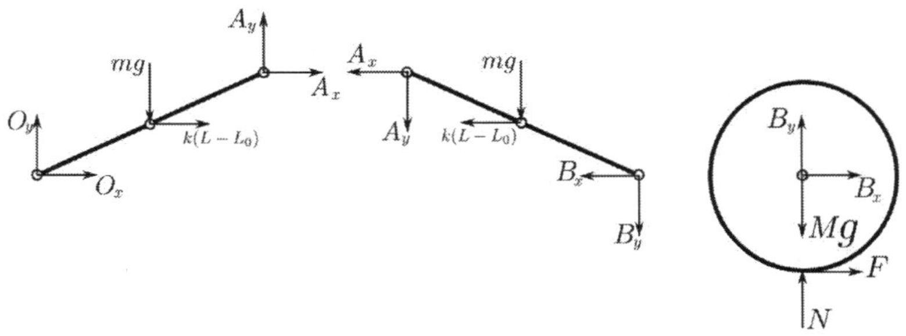

Figure 5.38 – *Problem 5.16 (Solution): in dashed lines $t = t_i$, in solid lines $t = t_f$.*

b. Equation [2] shows that the system reaches the final position if the effects of gravitational forces exceed the effect of the elastic force:

$$\boxed{\frac{mg}{kl} \geq \frac{(1-\cos\theta_0)^2}{\sin\theta_0}}$$

otherwise, the expression [2] giving $\dot{\theta}_f^2$ has a negative r.h.s. ■

Exercise 5.17 ■■ Figure 5.39 shows a system Σ comprised of two rigid bodies 1 and 2 in motion relative to a referential 0 assumed Newtonian. The axis $\Delta = (O, \hat{\mathbf{z}}_0)$ is directed along the vertical upward. Rigid body 1 is connected to a fixed support by frictionless pivots of axis Δ. Rigid body 2 is free to rotate relative to 1 about axis Δ. The two bodies are given the initial angular velocities $\boldsymbol{\omega}_{1/0} = \omega_{01}\hat{\mathbf{z}}_0$ and $\boldsymbol{\omega}_{2/0} = \omega_{02}\hat{\mathbf{z}}_0$ (ω_{01} and ω_{02} are two positive constants). After a finite time τ, it is observed that the two bodies rotate at the same angular velocity $\boldsymbol{\omega}_{1/0} = \boldsymbol{\omega}_{2/0} = \Omega\hat{\mathbf{z}}_0$ due to the frictional effects between the two bodies. Denote by I_1 and I_2 the moment of inertia of bodies 1 and 2 about axis Δ.
a. Find the angular velocity Ω in terms of ω_{01}, ω_{02} and the constants I_1 and I_2.
b. Find the variation of the kinetic energy K_Σ of the system during the time interval $0 \leq t \leq \tau$. Justify the sign of this quantity. ■

Solution. **a.** Since the pivot connections between 0 and 1 are frictionless, the contact forces exerted on body 1 satisfy $\mathbf{M}_{O,0\to 1} \cdot \hat{\mathbf{z}}_0 = 0$. Furthermore the only other external forces acting on the system are the gravitational forces on the two bodies directed along $\hat{\mathbf{z}}_0$. This necessarily implies that the resultant external moment about O satisfies

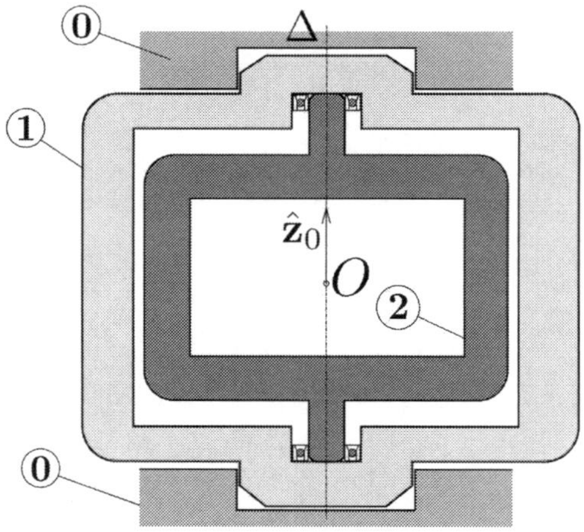

Figure 5.39 – *Problem 5.17*

$\mathbf{M}_{O,ext\to\Sigma} \cdot \hat{\mathbf{z}}_0 = 0$. According to Euler's Second Principle applied to system Σ, this leads to

$$\frac{d}{dt}\mathbf{H}_{O,\Sigma} \cdot \hat{\mathbf{z}}_0 = 0$$

Hence the $\hat{\mathbf{z}}_0$-component of the system's angular momentum $\mathbf{H}_{O,\Sigma/0}$ is conserved:

$$\mathbf{H}_{O,\Sigma} \cdot \hat{\mathbf{z}}_0 = \mathbf{H}_{O,1} \cdot \hat{\mathbf{z}}_0 + \mathbf{H}_{O,2} \cdot \hat{\mathbf{z}}_0 = Cst$$

with $\mathbf{H}_{O,1} \cdot \hat{\mathbf{z}}_0 + \mathbf{H}_{O,2} \cdot \hat{\mathbf{z}}_0 = (I_1 + I_2)\Omega$ at time $t = \tau$ and $\mathbf{H}_{O,1} \cdot \hat{\mathbf{z}}_0 = I_1\omega_{01}$, $\mathbf{H}_{O,2} \cdot \hat{\mathbf{z}}_0 = I_2\omega_{02}$ at time $t = 0$. This gives angular velocity Ω in terms of ω_{01}, ω_{02}:

$$\boxed{\Omega = \frac{I_1\omega_{01} + I_2\omega_{02}}{I_1 + I_2}} \qquad [1]$$

b. For bodies in rotations about axis Δ the kinetic energies K_1, K_2 and K_Σ are given by

$$K_1 = \tfrac{1}{2}I_1\omega_1^2, \qquad K_2 = \tfrac{1}{2}I_2\omega_2^2, \qquad K_\Sigma = \tfrac{1}{2}I_1\omega_1^2 + \tfrac{1}{2}I_2\omega_2^2$$

The variations of these quantities during during the time interval τ is then:

$$\Delta K_1 = \tfrac{1}{2}I_1(\Omega^2 - \omega_{01}^2) = \tfrac{1}{2}\frac{I_1 I_2}{(I_1 + I_2)^2}(\omega_{02} - \omega_{01})[(2I_1 + I_2)\omega_{01} + I_2\omega_{02}]$$

$$\Delta K_2 = \tfrac{1}{2}I_2(\Omega^2 - \omega_{02}^2) = \tfrac{1}{2}\frac{I_1 I_2}{(I_1 + I_2)^2}(\omega_{01} - \omega_{02})[(2I_2 + I_1)\omega_{02} + I_1\omega_{01}]$$

leading to

$$\boxed{\Delta K_\Sigma = \Delta K_1 + \Delta K_2 = -\tfrac{1}{2}\frac{I_1 I_2}{(I_1 + I_2)}(\omega_{02} - \omega_{01})^2}$$

First we see that $\Delta K_\Sigma < 0$: this can be explained by the Work-Energy Principle applied to system Σ during time interval $0 \leq t \leq \tau$:

$$\Delta K_\Sigma = W_{0\to\tau}^{ext} + W_{0\to\tau}^{int} = W_{0\to\tau}^{int}$$

where the work $W_{0\to\tau}^{ext}$ done by the external actions (the weights and the contact forces

0→1) vanishes. The work done by the internal forces takes the expression

$$W^{int}_{0\to\tau} = \int_0^\tau \boldsymbol{\omega}_{2/1} \cdot \mathbf{M}_{O,1\to 2} dt = \int_0^\tau (\omega_2 - \omega_1)\hat{\mathbf{z}}_0 \cdot \mathbf{M}_{O,1\to 2} dt$$

where $(\omega_2 - \omega_1)\hat{\mathbf{z}}_0 \cdot \mathbf{M}_{O,1\to 2} < 0$ at any time since the friction forces exerted by 1 on 2 must oppose the rotation of 2 relative to 1. This shows that $\Delta K_\Sigma < 0$.
Note that the variations ΔK_1 and ΔK_2 can be positive or negative depending on the sign of $\omega_{01} - \omega_{02}$: application of the Work-Energy Principle to body 1, then body 2 gives

$$\Delta K_1 = -\int_0^\tau \omega_1 \hat{\mathbf{z}}_0 \cdot \mathbf{M}_{O,1\to 2}\, dt$$

$$\Delta K_2 = \int_0^\tau \omega_2 \hat{\mathbf{z}}_0 \cdot \mathbf{M}_{O,1\to 2}\, dt$$

Assume that $\omega_{02} > \omega_{01} > 0$ then $\hat{\mathbf{z}}_0 \cdot \mathbf{M}_{O,1\to 2} < 0$ from $t = 0$ until $t = \tau$: this implies that $\int_0^\tau \hat{\mathbf{z}}_0 \cdot \mathbf{M}_{O,1\to 2}\, dt$ is negative

$$\Delta K_1 > 0, \qquad \Delta K_2 < 0$$

Conversely if $\omega_{02} > \omega_{01} > 0$ at $t = 0$, then $\Delta K_1 < 0$ and $\Delta K_2 > 0$. ■

Exercise 5.18 ■■■ In this speed reducer, input shaft 1 terminates in a gear wheel of radius r_1. It is coupled to the output shaft 3 by a satellite gear 2 whose axis is mounted on the carrier rigidly connected to shaft 3. Gear 2 is also meshed to a stationary ring gear 4. The dimension of the gear train is defined by the parameters r_1 and r_2. A motor exerts a couple $C_{in}\hat{\mathbf{x}}_0$ on shaft 1. The output shaft exerts a couple $C_{out}\hat{\mathbf{x}}_0$ to an output device. Shaft 1 has moment of inertia I_1 about axis $(O, \hat{\mathbf{x}}_0)$. (Only half of gear 4 is shown in the sketch).

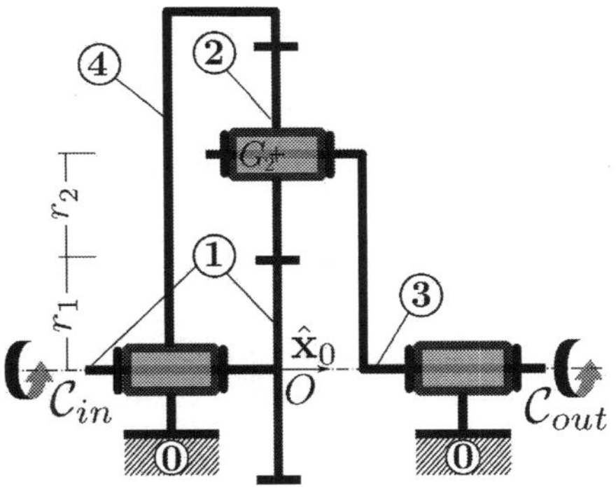

Shaft 3 has moment of inertia I_3 about axis $(O, \hat{\mathbf{x}}_0)$. Gear 2 has mass m_2, mass center G_2 and moment of inertia I_2 about axis $(G_2, \hat{\mathbf{x}}_0)$. Assume that all joints are frictionless and neglect the effect of gravity.
a. Find the ratio of angular velocities ω_3/ω_1.
b. Find the ODE which governs the angular velocity $\omega_1(t)$ as a function of couple C_{in} and C_{out} and the system's parameters. ■

SOLUTION **a.** Using the notations of Figure 5.40 (left), the relationship between ω_3 and ω_1 is found by writing the no-slip conditions at I (between 1 and 2) and at J (between 4 and 2):

$$\underbrace{\omega_1 \hat{\mathbf{x}}_0 \times r_1 \hat{\mathbf{y}}_3}_{\mathbf{v}_{I\in 1/0}} = r_1\omega_1 \hat{\mathbf{z}}_3 = \underbrace{\mathbf{v}_{G_2} + \omega_2 \hat{\mathbf{x}}_0 \times (-r_2 \hat{\mathbf{y}}_3)}_{\mathbf{v}_{I\in 2/0}} = \mathbf{v}_{G_2} - r_2\omega_2 \hat{\mathbf{z}}_3 \qquad [1]$$

$$\underbrace{\mathbf{0}}_{\mathbf{v}_{J\in 4/0}} = \underbrace{\mathbf{v}_{G_2} + \omega_2 \hat{\mathbf{x}}_0 \times (r_2 \hat{\mathbf{y}}_3)}_{\mathbf{v}_{J\in 2/0}} = \mathbf{v}_{G_2} + r_2\omega_2 \hat{\mathbf{z}}_3 \qquad [2]$$

with $\mathbf{v}_{G_2} = (r_1 + r_2)\omega_3 \hat{\mathbf{z}}_3$. Comparing [1] and [2] gives

$$\boxed{\frac{\omega_3}{\omega_1} = \frac{r_1}{2(r_1 + r_2)}} \qquad [3]$$

Also note that $\omega_2 = -\frac{r_1+r_2}{r_2}\omega_3 = -\frac{r_1}{2r_2}\omega_1$.

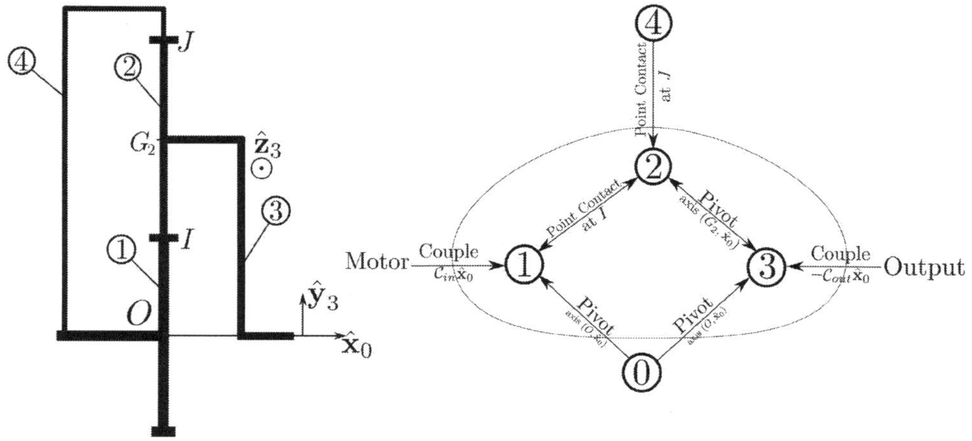

Figure 5.40 – *Problem 5.18 (Solution)*

b. Denote by Σ the system $\{1, 2, 3\}$. Since we are looking for the ODE which governs the angular velocity $\omega_1(t)$, we apply the Kinetic Energy Theorem, that is:

$$\frac{d}{dt}K_\Sigma = \mathcal{P}^{ext} + \mathcal{P}^{int} \qquad [4]$$

To find the kinetic energy of the system, we find the kinetic energy of each body:
- body 1 in rotation about axis $(O, \hat{\mathbf{x}}_0)$:

$$K_1 = \tfrac{1}{2}I_1\omega_1^2$$

- body 2 in instantaneous rotation about axis $(J, \hat{\mathbf{x}}_0)$:

$$K_2 = \tfrac{1}{2}(I_2 + m_2 r_2^2)\omega_2^2 = \frac{1}{8}(I_2 + m_2 r_2^2)\frac{r_1^2}{r_2^2}\omega_1^2$$

- body 3 in instantaneous rotation about axis $(O, \hat{\mathbf{x}}_0)$:

$$K_3 = \tfrac{1}{2}I_3\omega_2^2 = \frac{1}{8}I_3\frac{r_1^2}{(r_1 + r_2)^2}\omega_1^2$$

To find the powers $\mathcal{P}^{ext}$ and $\mathcal{P}^{int}$, we sketch a diagram in Figure 5.40 (right) showing all exterior actions into the system and internal interactions within the system. Single arrows correspond to external actions and contribute to the power

$$\begin{aligned}\mathcal{P}^{ext} &= \underbrace{\mathcal{C}_{in}\omega_1}_{\text{input motor}} \; \underbrace{-\mathcal{C}_{out}\omega_3}_{\text{output device}} + \underbrace{\omega_1\hat{\mathbf{x}}_0 \cdot \mathbf{M}_{O,0\to 1}}_{\text{pivot connection}} + \underbrace{\omega_3\hat{\mathbf{x}}_0 \cdot \mathbf{M}_{O,0\to 3}}_{\text{pivot connection}} + \underbrace{\mathbf{F}_{4\to 2} \cdot \mathbf{v}_{J\in 2/0}}_{\text{point contact at } J} \\ &= (\mathcal{C}_{in} - \frac{r_1}{2(r_1 + r_2)}\mathcal{C}_{out})\omega_1\end{aligned}$$

where we have used [3] and the fact that (i) the pivot connections are frictionless: $\hat{\mathbf{x}}_0 \cdot \mathbf{M}_{O,0\to 1} = \hat{\mathbf{x}}_0 \cdot \mathbf{M}_{O,0\to 2} = 0$ and (ii) there is no-slip at J. Double arrows correspond to interactions within Σ and contribute to the internal power

$$\mathcal{P}^{in} = \underbrace{\omega_{2/3}\hat{\mathbf{x}}_0 \cdot \mathbf{M}_{G_2,3\to 2}}_{\text{pivot connection}} + \underbrace{\mathbf{F}_{1\to 2} \cdot \mathbf{v}_{I\in 2/1}}_{\text{point contact at } I} = 0$$

where we have used the fact that (i) the pivot connections 2/3 is frictionless: $\hat{\mathbf{x}}_0 \cdot \mathbf{M}_{G_2,3\to 2} = 0$ and (ii) there is no-slip at I: $\mathbf{v}_{I\in 2/1} = \mathbf{0}$. Now equation [4] becomes

$$\boxed{\left(I_1 + \frac{r_1^2}{4r_2^2}I_2 + \frac{1}{4}m_2 r_1^2 + \frac{r_1^2}{4(r_1+r_2)^2}I_3\right)\frac{d\omega_1}{dt} = \mathcal{C}_{in} - \frac{r_1}{2(r_1+r_2)}\mathcal{C}_{out}}$$

This final result shows the benefit of the energy approach over the use of Euler's principles for a single-degree-of-freedom system. ■

Exercise 5.19 ■■■ The figure shows a push-cam mechanism: a camshaft 1 of mass m_1, radius R and mass center G is mounted on a shaft in rotation at constant angular velocity Ω about axis $(O, \hat{\mathbf{z}}_0)$ perpendicular to the plane of motion. A piston 2 of mass m_2 is in contact at a point J of the cam and can translate along the vertical direction $(O, \hat{\mathbf{x}}_0)$. Its connection to the fixed housing 0 is frictionless. A spring of constant k is mounted in compression between the piston and the housing. The camshaft is subject to a couple $\mathcal{C}\hat{\mathbf{z}}_0$. Its rotation is defined by angle θ. The spring is unstretched when the cam reaches the position $\theta = 3\pi/2$. Denote by e the distance $|OG|$ and by I_1 the moment of inertia of the camshaft about axis $(O, \hat{\mathbf{z}}_0)$.

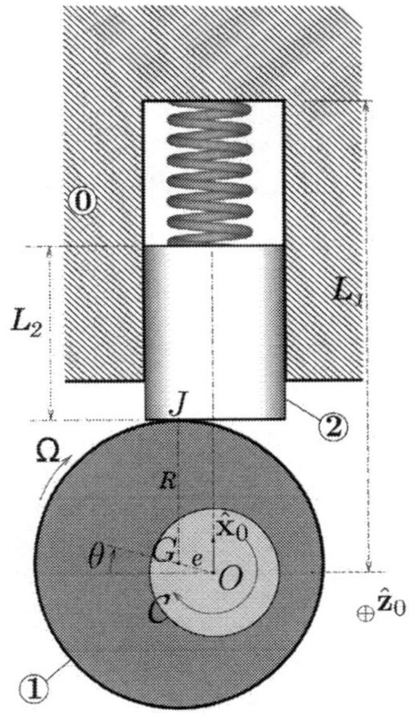

a. Find the slip velocity $\mathbf{v}_{J\in 2/1}$ as a function of θ.

b. The contact force at J between the cam and the piston is modeled as the sum of a normal and tangential components as follows

$$\mathbf{F}_{1\to 2} = N_{12}\hat{\mathbf{x}}_0 - \mu \mathbf{v}_{J\in 2/1}$$

where μ is a positive constant. Apply the Work-Energy Principle to find the couple $\mathcal{C}$ in terms of angle θ and the constants Ω, e, R, k, μ, etc. *Neglect the effect of gravity.* ■

SOLUTION **a.** Define unit vectors $(\hat{\mathbf{u}}, \hat{\mathbf{v}})$ such that $\mathbf{r}_{OG} = e\hat{\mathbf{u}}$ and $\mathbf{v}_{G/0} = e\Omega\hat{\mathbf{v}}$. Then we can find $\mathbf{v}_{J\in 1/0}$ from $\mathbf{v}_G$:

$$\mathbf{v}_{J\in 1/0} = \mathbf{v}_{G/0} + \Omega\hat{\mathbf{z}}_0 \times \mathbf{r}_{GJ} = e\Omega\hat{\mathbf{v}} + R\Omega\hat{\mathbf{y}}_0$$

The velocity $\mathbf{v}_{J\in 2/0}$ is that of the translating piston:

$$\mathbf{v}_{J\in 2/0} = \frac{d}{dt}(e\sin\theta + R)\hat{\mathbf{x}}_0 = e\Omega\cos\theta\hat{\mathbf{x}}_0$$

The slip velocity $\mathbf{v}_{J\in 2/1}$ is then found to be

$$\mathbf{v}_{J\in 2/1} = \mathbf{v}_{J\in 2/0} - \mathbf{v}_{J\in 1/0} = e\Omega\cos\theta\hat{\mathbf{x}}_0 - e\Omega\hat{\mathbf{v}} - R\Omega\hat{\mathbf{y}}_0$$

which simplifies to

$$\boxed{\mathbf{v}_{J\in 2/1} = -\Omega(R - e\sin\theta)\hat{\mathbf{y}}_0} \qquad [1]$$

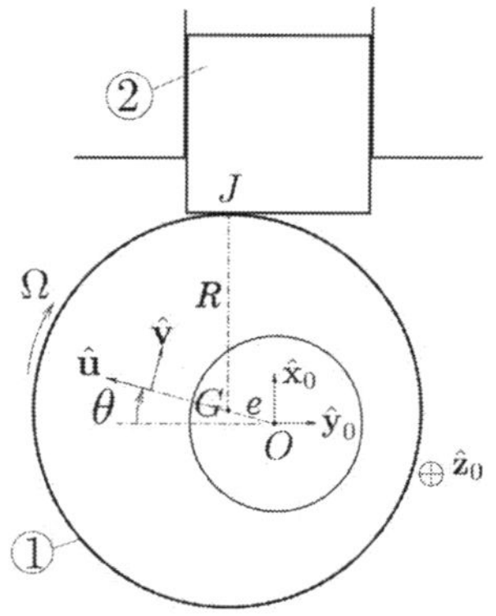

Figure 5.41 – *Problem 5.19 (Solution)*

b. We apply the Work-Energy Principle to the system Σ between $t = 0$ and $t > 0$:

$$\Delta(K_1 + K_2) = W^{ext}_{0\to t} + W^{int}_{0\to t} \qquad [2]$$

with
- the kinetic energy of the system given by

$$K_1 + K_2 = \frac{1}{2}I_1\Omega^2 + \frac{1}{2}m_2\mathbf{v}^2_{2/0} = \frac{1}{2}I_1\Omega^2 + \frac{1}{2}m_2 e^2\Omega^2\cos^2\theta$$

- the work done by the external actions on the system:

$$W^{ext}_{0\to t} = W^{C}_{0\to t} + W^{spring}_{0\to t} = \int_0^t C\Omega dt - \Delta(\frac{k}{2}(L - L_0)^2)$$

with $L = L_1 - L_2 - R - e\sin\theta$ and the unstretched length $L_0 = L_1 - L_2 - R + e$ (the work done by all contact actions is zero since the joints are frictionless)
- the work done by the internal contact force

$$W^{int}_{0\to t} = \int_0^t \mathbf{F}_{1\to 2}\cdot\mathbf{v}_{J\in 2/1}dt = -\int_0^t \mu\mathbf{v}^2_{J\in 2/1}dt = -\int_0^t \mu\Omega^2(R - e\sin\theta)^2 dt$$

using equation [1]. Equation [2] becomes

$$\Delta(\frac{1}{2}I_1\Omega^2 + \frac{1}{2}m_2e^2\Omega^2\cos^2\theta + \frac{k}{2}e^2(1+\sin\theta)^2)$$
$$= \int_0^t \Omega C dt - \int_0^t \mu\Omega^2(R - e\sin\theta)^2 dt \quad ([3])$$

We can then obtain couple C by taking the time-derivative of [3]: this is equivalent to applying the Kinetic Energy Theorem to obtain a power equation

$$\Omega C = \mu\Omega^2(R - e\sin\theta)^2 + \frac{d}{dt}\Big(\frac{1}{2}I_1\Omega^2 + \frac{1}{2}m_2e^2\Omega^2\cos^2\theta + \frac{k}{2}e^2(1+\sin\theta)^2\Big)$$

With $\dot{\theta} = \Omega = Cst$, we find after simplifications

$$\boxed{C = \mu\Omega(R - e\sin\theta)^2 - m_2e^2\Omega^2\cos\theta\sin\theta + ke^2(1+\sin\theta)\cos\theta}$$

Remark: It is far more labor intensive to find this result by using Euler's principles. ∎

PROJECT 10: SWING AMPLIFICATION

Consider a child sitting on a swing and denote by G the mass center of the system child/swing. The motion of the system is assumed identical to that of its mass center G located at a distance l from the point O at which he rotates. The swing ropes are assumed to be always taut. The child is released in the position given by angle $\theta_0 > 0$ (assuming positive angles are measured in the clockwise direction) without initial velocity. All friction effects are neglected. In order to amplify his motion, the child raises the mass center G by a distance h any time the ropes reach the vertical position $\theta = 0$, and brings back G at the distance l from point O once the maximum oscillation angle $\theta = \theta_n$, $n = 1, 2, ...$ is reached. Assume that this up and down motion relative to the swing is instantaneous. The motion of G is shown during the first half-cycle from $\theta_0 > 0$ to $\theta_1 < 0$. More generally, the motion of G is characterized by

$$|OG| = l, \qquad \text{if} \quad \theta\dot{\theta} < 0$$
$$|OG| = l - h, \qquad \text{if} \quad \theta\dot{\theta} > 0$$

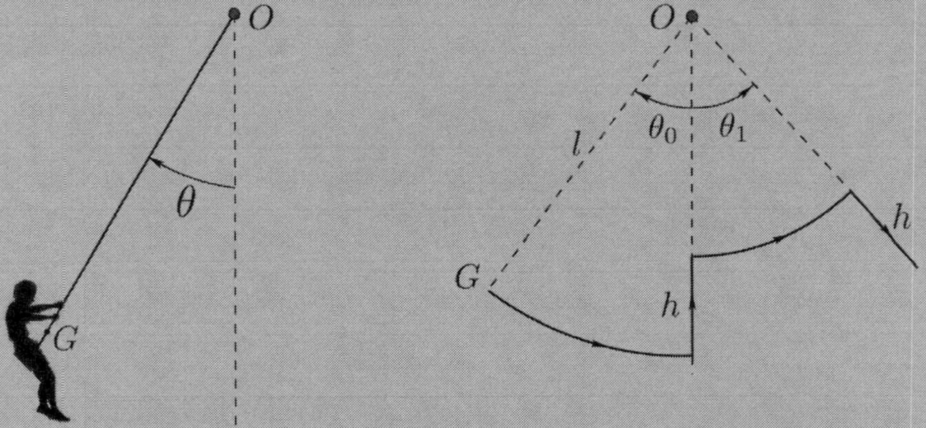

Part I: Analysis

I.1. Show that the total energy $E = \frac{1}{2}mv_G^2 - mgr\cos\theta$, $r = |OG|$, cannot possibly be conserved.

I.2. By consideration of the angular momentum of the system, show that any two consecutive amplitude angles θ_n and θ_{n+1} satisfy

$$\int_{\theta_n}^{\theta_{n+1}} r^3 \sin\theta \, d\theta = 0 \qquad [1]$$

Using [1], show that $|\theta_{n+1}| > |\theta_n|$ and derive the amplitude θ_n in terms of θ_0 and $k = h/l$.

I.3. Assume the initial angle θ_0 small so that $\sin\theta_0 \approx \theta_0$. How many passages across the vertical position does it take for the swing to reach the horizontal position? Use $\theta_0 = 20°$ and $k = 0.05$.

Part II: Experimentation

An example of swing amplification is provided by the *Botafumeiro*, a swinging metal incense container suspended to the ceiling of the Santiago de Compostela cathedral. It is a pendulum whose length is varied by a system of cords and pulleys. This system allows eight operators (*tiraboleiros*) to amplify its amplitude by periodically decreasing and increasing its length at the lowest and highest points of the oscillations, respectively. With mass of 54 kg and a height of 21 m, it reaches a maximum amplitude of 82 degrees reached after 17 cycles in about 80 seconds.

Build a simple pendulum with variable length. Shorten the pendulum length by a fixed amount Δl whenever the pendulum passes the lowest point. Release it by the same amount when the pendulum reaches its maximum amplitude. Observe the oscillations. Measure the maximum amplitude and the number of cycles taken to reach this value. Comment.

PROJECT 11: Transport of Pipes on Conveyor Belt

A cylindrical pipe 1 of radius R, mass m and mass center G is set without initial speed on a conveyor belt 2 inclined at angle θ and moving at constant speed v_C. The contact is characterized by static and kinetic friction coefficients $\mu_s = \mu_k = \mu$. Neglect the thickness of the pipe.

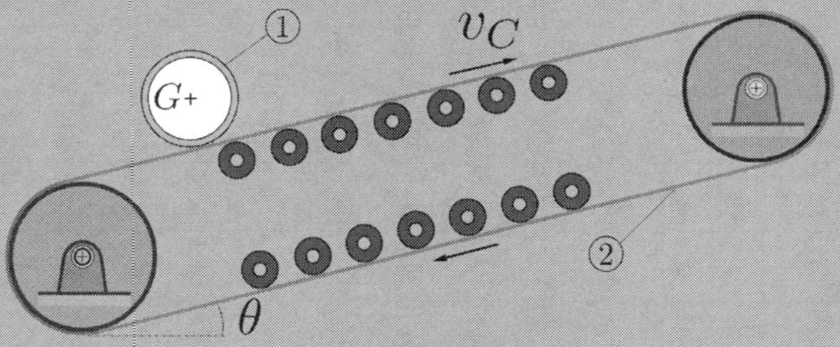

1. Show that the pipe will initially slip relative to the conveyor belt.
2. Apply Euler's principles to the pipe in its initial phase of motion. Under what condition will it roll uphill? Assuming this condition satisfied, show that the pipe will cease to slip at some time t_s. Find time t_s and the corresponding distance L_s traveled by the pipe.
3. Apply the Work-Energy Principle to find the work done by the friction force from $t = 0$ to $t = t_s$. Is this work negative? comment!
4. Study the motion of the pipe for $t > t_s$: find the speed of mass center G versus time. Describe the motion of the pipe for $t > t_s$.
5. A conveyor belt must be used to transport pipes to a height $h = 5$ m at the inclination angle at the maximum speed $v_C = 4$ m/s and with a coefficient of friction of $\mu = 0.7$. Can a conveyor be run so that the pipes are successfully carried to their destination?

5.9 Problems

Problem 5.1 A ship of mass m travels at constant speed $v = v_0$ along a rectilinear trajectory. At $t = 0$, the engine is stopped and the ship eventually comes to a stop after traveling a distance D due to the frictional effects of water, modeled by the force $-kv$.

a. By applying Newton's second law, find the distance D and the work done by the friction force from time $t = 0$ to the time the ship comes to rest.

b. Show that this result is in accordance with the Work-Energy Principle.

Problem 5.2 A small block P of mass 150 g is suspended by a string of length 1.4 m to a nail attached to a vertical wall. What velocity should P be given so that the string first becomes slack at the angle $\theta_1 = 135°$?

Find the string's tension as a function of θ, $0 \leq \theta \leq \theta_1$. Assume that the contact of P with the wall is frictionless.

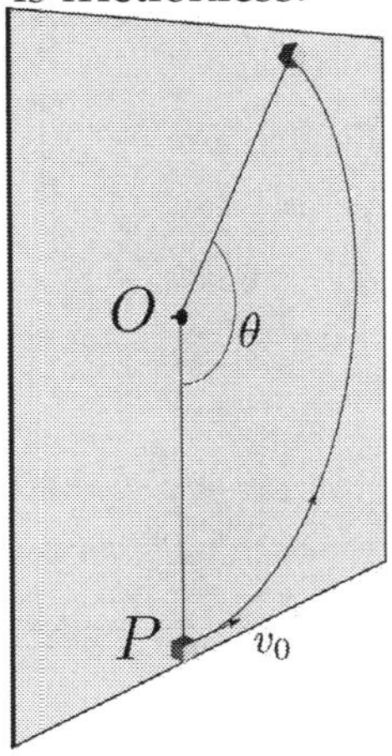

Answer: $v_0 = 7.5$ *m/s*

Problem 5.3 A bungee jumper of mass 85 kg jumps off a bridge at the level of point O of a vertical axis Oz without initial speed. Before reaching the level $z = z_A = 25$ m of point A his path is that of a particle P in free fall. Upon reaching A, the elastic cord of constant $k = 150$ N.m^{-1} starts to stretch. Neglect the mass of the cord.

a. Find the maximum drop z_{max} reached by the jumper.

b. Find the maximum speed reached by P and the maximum height P will fly upwards as the cord recoils.

c. Find the time taken to reach $z = z_{max}$.

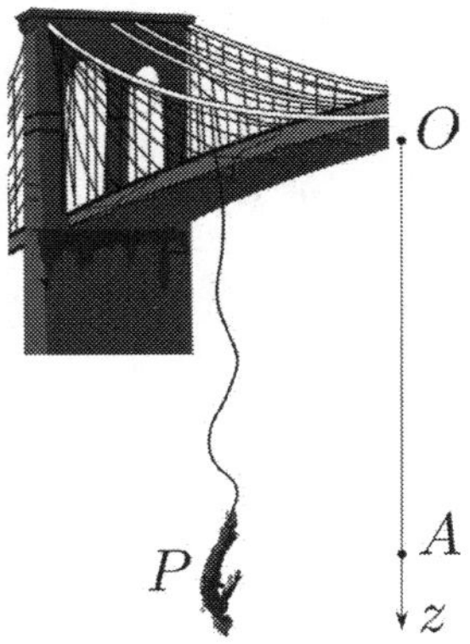

Answer: c). $T = 3.68$ *s*

Problem 5.4 A particle P is set on the apex O of a parabola of equation $2py = x^2$. This position being unstable, P slides downward under the effect of gravity. The contact is assumed frictionless. The axis of the parabola Oy is vertical.

Find the reaction of the support on P as a function of position. Conclude that P cannot lose contact with the parabola.

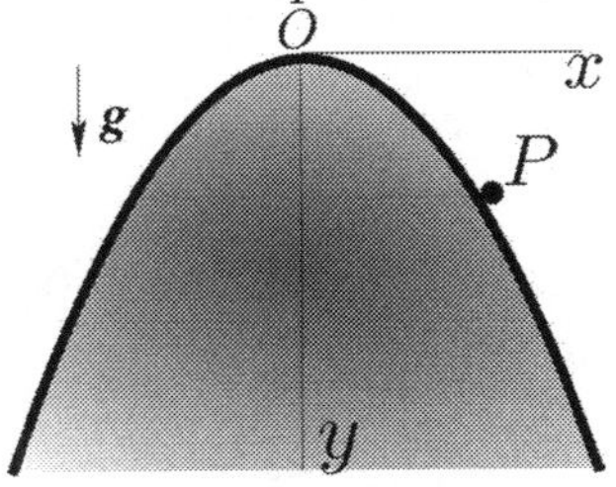

Answer: $N = mg\cos^3\theta$ *with* $\tan\theta = x/p$

Problem 5.5 A block P of mass M is tied to a rope of length L and mass m. Initially P is at rest at the level of the attachment point of the cord. P is then dropped without initial speed. Neglect the cord's elasticity.

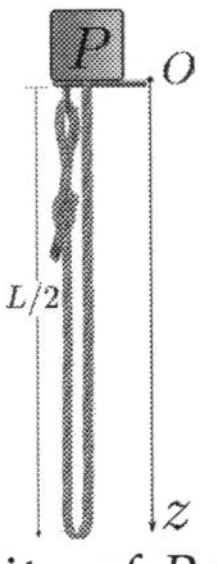

a. Find the velocity of P when P has fallen a distance $z = L$.

b. Find the acceleration of P when P has fallen a distance $z = L$.

Hint: show that the energy E of the system

rope/block is conserved. Then write $dE/dt = 0$ to find the equation of motion of P.

Answer: b). $a_P = g(1 + \frac{m(m+4M)}{8M^2})$

Problem 5.6 A child of mass M climbs a hill from A to B, pulling a sled of mass m. The location of point B relative to point A is given by distance D, and height H. Find the work done by the child to bring his sled from A to B, assuming that
(i) the path is piecewise linear,
(ii) the kinetic friction coefficient between the sled and the snow is μ,
(iii) the child leaves A and arrives at B at zero speed,
(iv) the force applied by the child to the sled is directed parallel to the slope.

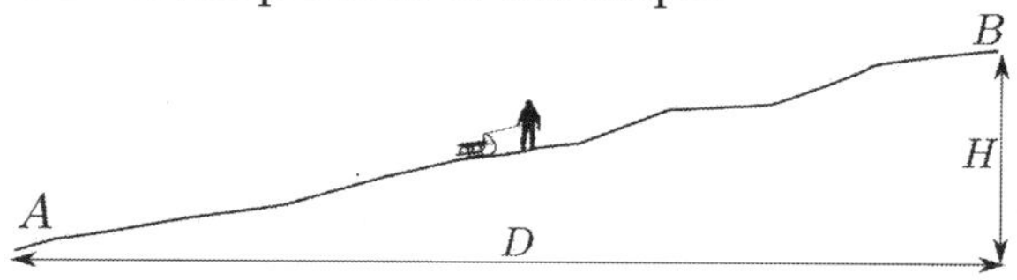

Problem 5.7 A rigid rod of length $2L$ supports two identical small spherical bodies A and B of mass m at its endpoints. It is placed on top of a cylindrical support of radius R and exhibits oscillations about its horizontal position without slipping at contact point I. The position of the rod is defined by angle θ between the vertical and line OI. When $\theta = 0$, the contact point I coincides with the midpoint of AB.

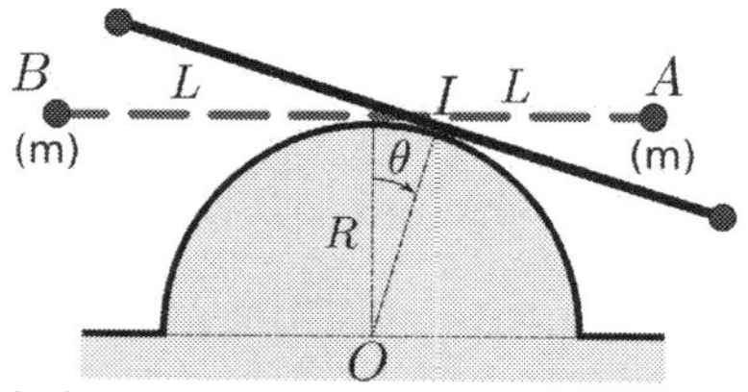

a. Find the kinetic energies of A and B as a function of θ, $\dot{\theta}$, R and L.
b. Show that the energy E of the system $\{A, B, \text{rod}\}$ is conserved.
c. Find the period of small amplitude oscillations of the system about its equilibrium position. Use $R = 1.1$ m, $L = 0.75$ m.
Hint: write $dE/dt = 0$ and use the approximation $\cos\theta \approx 1$ and $\sin\theta \approx \theta$.

Answer: c). $T = 1.4$ s.

Problem 5.8 The goal of this problem is to find a curved support in a vertical plane for the frictionless sliding motion of a particle P with the following property: leaving from origin O without initial speed, the time taken to reach *any position* on the curve must be equal to the time that P would take to slide along the straight line connecting O to this position. Adopt polar coordinates (r, θ) as shown and assume that $r = r_0$ at the angle $\theta = \pi/4$.
a. Find the time t taken to reach position (r, θ) on the curve $r(\theta)$. Then find the time taken on the straight line connecting O to position (r, θ). Equate these two expressions to find the equation

$$r \cot 2\theta = \frac{dr}{d\theta}$$

b. Using $\int \cot\theta d\theta = \ln|\sin\theta|$, find the curve $r(\theta)$ which passes through point $(r_0, \pi/4)$. Sketch this curve in the interval $0 \le \theta \le \pi/2$.
c. Find the time taken to reach $r_0 = 12$ m and the time taken to return to O.

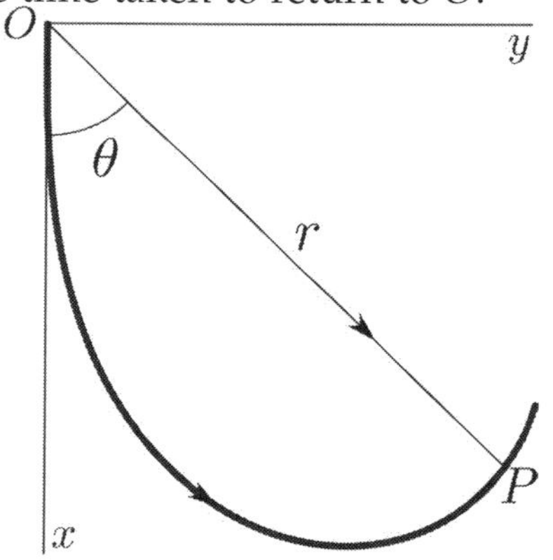

Answer: Answer: c). $T = 3.46$ s.

Problem 5.9 Two identical spherical balls A and B are connected to a vertical shaft by slender rods of length l and negligible mass. The system is initially at rest, with the balls in contact with the shaft. The shaft is then slowly set in rotation to reach a constant angular velocity ω. Neglect the size of the balls and the diameter of the shaft.
a. Apply Newton's Second Law to find the equilibrium positions $\theta = \theta_e$ taken by the balls. Under what condition will the balls lose contact with the shaft?
b. Plot θ_e versus $\kappa = \frac{g}{l\omega^2}$. Find θ_e for $\kappa = 1.2$.
c. Now assume that angle θ is varying and find the equation governing θ. Deduce that the system is conservative. Find the potential energy and discuss the stability of the

equilibrium positions.

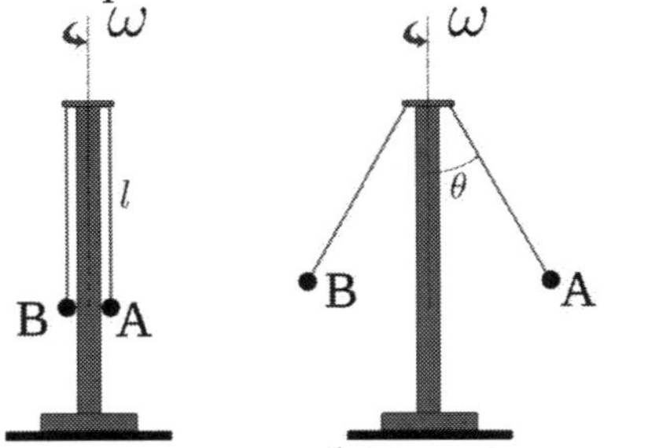

Answer: $\theta_e = 0$ *or* $\theta_e = \cos^{-1}\kappa$ $(\kappa \leq 1)$

Problem 5.10 A small ring P of mass m can slide along a circular wire of radius r and center O contained in a vertical plane Oxz. P is also connected to a spring of constant k and unstretched length l_0 attached at point A a distance r above O. Assume that $l_0 < r\sqrt{2}$. The position of P is defined by angle θ with $-\pi/2 < \theta < \pi/2$.

a. Show that P can take three possible equilibrium positions.

b. Choose $l_0/r = \sqrt{3}/2$ and $mg/kl_0 = 1/\sqrt{3}$. Find the equilibrium positions and discuss their stability. Then find the period T of small amplitude oscillations about each stable equilibrium positions in terms of the period $T_0 = 2\pi\sqrt{\frac{r}{g}}$. Discuss the motion of P for the initial conditions $\theta = 0$ and $\dot{\theta}_0 \neq 0$.

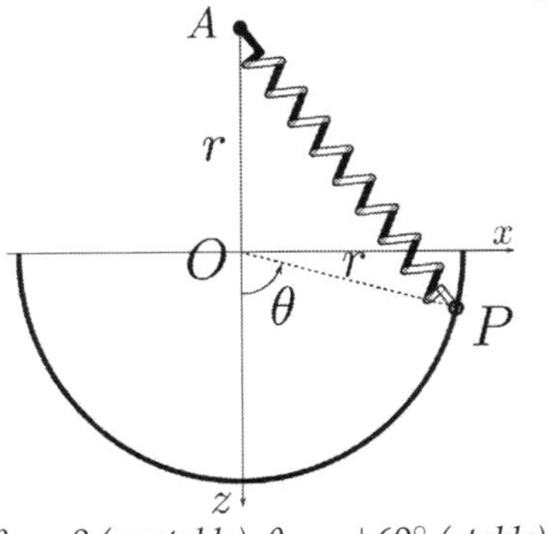

Answer: $\theta_e = 0$ *(unstable),* $\theta_e = \pm 60^\circ$ *(stable).* $T/T_0 = 2$

Problem 5.11 A heavy object P of mass $m = 750$ g is attached to the rim of a fixed cylinder of radius $R = 1$ m by an inextensible string of length $l = 2\pi$ m. At $t = 0$, the string is taut and P is released without initial speed. P is then set into swinging motion in a vertical plane, while the string winds around the cylinder.

Find the velocity of P when $\theta = 180^\circ$. Show that the string becomes slack at some angle $\theta = \theta_*$.

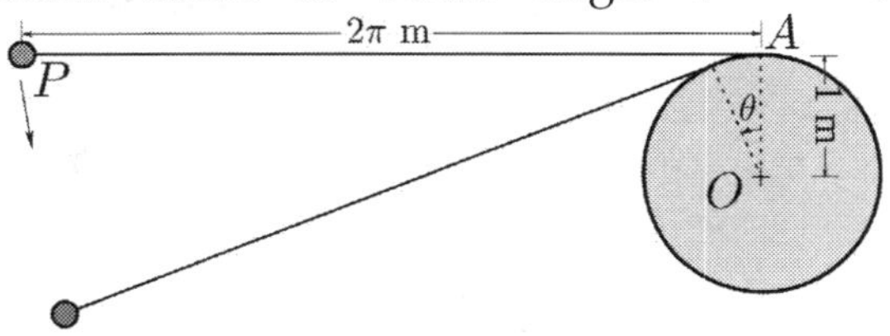

Problem 5.12 Two bodies A (of mass m) and B (of mass $2m$) are connected by a massless rod of length $3l$. The system is placed in a vertical position with A in contact with a horizontal, smooth support. The system is then released from rest.

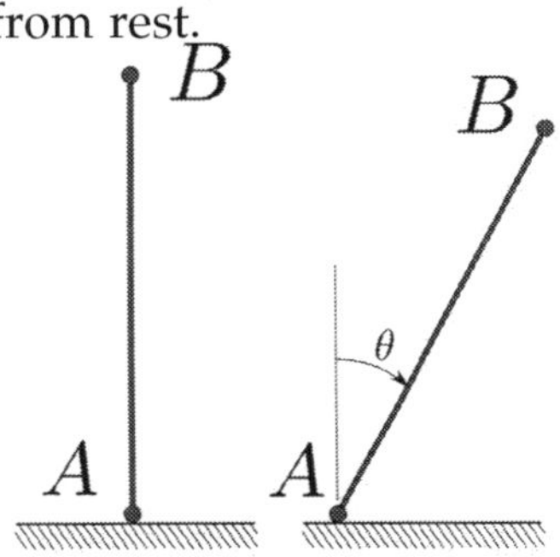

a. Describe the motion of the mass center G of the system.

b. Find the angular speed $\dot{\theta}$ as a function of angle θ.

c. Find the speed of A and B just before the rod hits the support.

Answer: c). $v_A = 0$, $v_B = \sqrt{6gl}$.

Problem 5.13 A chain of length L and mass m is held at its extremity P on a horizontal support, while its other extremity Q hangs at a distance l below the support. The chain is then released without initial speed and slips off the support. Denote by $z(t)$ the position of Q on a vertical axis Oz at time t, with $z(0) = l$. Neglect all friction and denote $\omega^2 = g/L$.

a. Find the kinetic and potential energies of the chain at time t.

b. Derive the equation of motion of the chain. Solve it to obtain $z(t)$.

c. Find the time T taken by the chain to leave the table for $m = 0.5$ kg, $L = 1.1$ m and $l_0 = 0.2$ m.

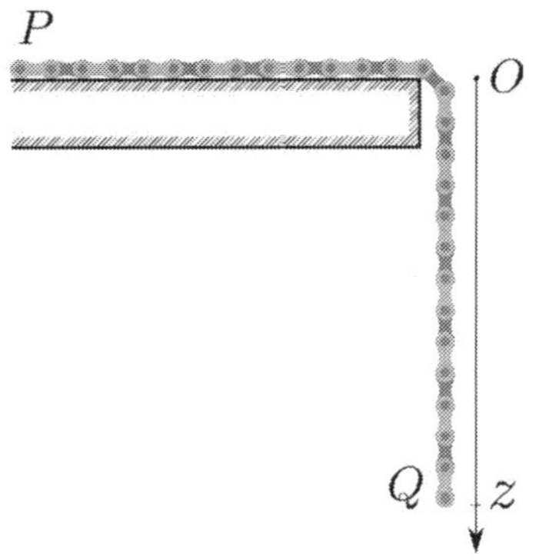

Answer: c). $T = 0.80$ s.

Problem 5.14 A hollow hoop of mass center G, radius r and mass M is in motion in a vertical plane. A small spherical object A of mass m is stuck inside the hoop. Its position is defined by angle θ relative to the vertical. The system is then given an initial position $\theta = \theta_0$ and is released from rest. It then oscillates by rolling without slipping relative to its support.

a. Find the equation which governs angle θ. Find the period of small amplitude oscillations.

b. Find the condition which the coefficient of static friction μ_s must satisfy for this motion to be guaranteed (assuming $\theta_0 \ll 1$).

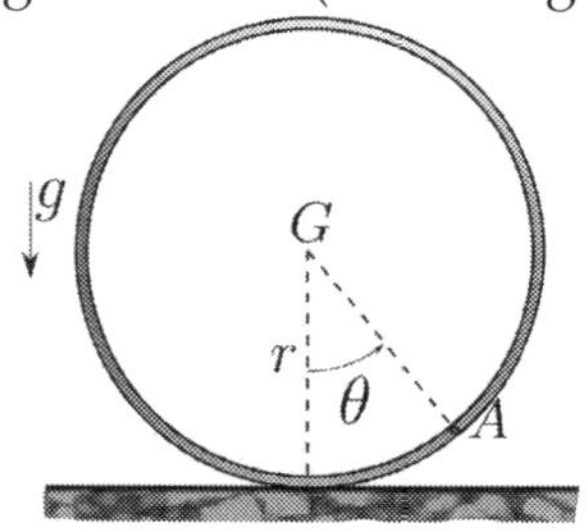

Answer: $T = 2\pi\sqrt{\frac{2Mr}{mg}}$

Problem 5.15 A (uniform, full) cylinder of mass 40 kg and radius 0.5 m is resting on the flatbed of a truck when the truck starts up at constant acceleration a. The cylinder eventually falls off the truck. The contact is characterized by the coefficients of static and kinetic friction $\mu_s = \mu_k = 0.1$.

a. For $a = 2.0$ m/s^2 the cylinder is observed to roll without slipping. Find the work done by the friction forces (relative to the referential attached to the ground) during the entire time of travel of the cylinder on the flatbed.

b. Repeat question a) for acceleration $a =$ 3.2 m/s^2 which causes the cylinder to slip.

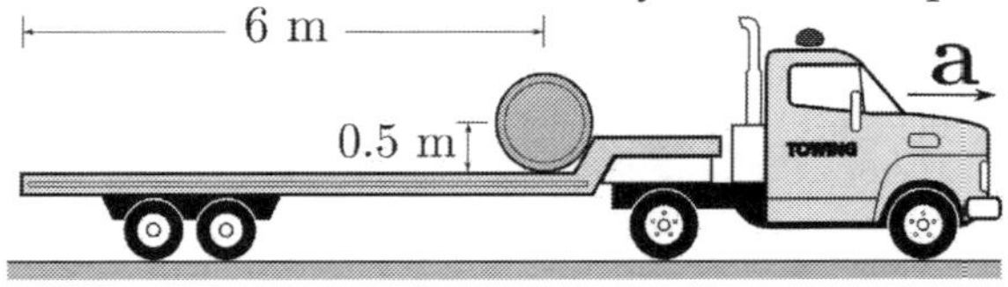

Answer: a) $W = 240$ J, b) $W = 312.26$ J

Problem 5.16 A uniform cylinder of mass m, radius r and mass center G is given an initial angular velocity ω_0 in the direction shown. It is then gently put in contact with a horizontal fixed support. The cylinder rolls forward (and slips) until it stops slipping.

Find the work done by the friction forces by two methods:

a. by applying the work-energy principle,

b. by the direct application of the integral formula giving the work done by a force.

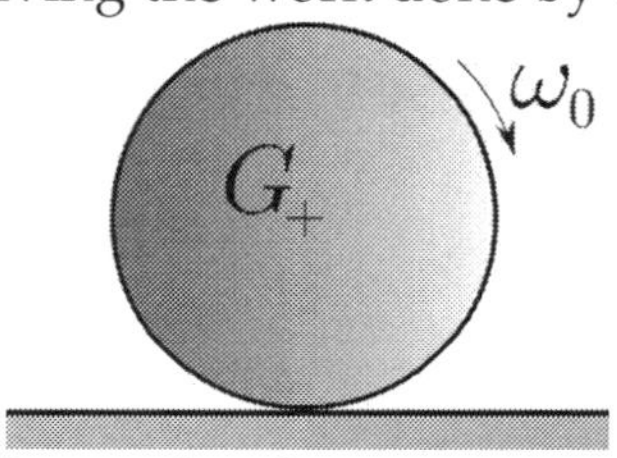

Answer: $W^F_{0\to 1} = -\frac{1}{6}m\omega_0^2$

Problem 5.17 A cylinder of mass center G and radius r is set on top of a cylindrical, stationary support of center O and radius $R > r$. It is released from rest. The equilibrium is unstable, and the cylinder begins to roll down the support without slipping under the effect of gravity. The position of point G is defined by angle θ. Denote by μ the coefficient of static friction.

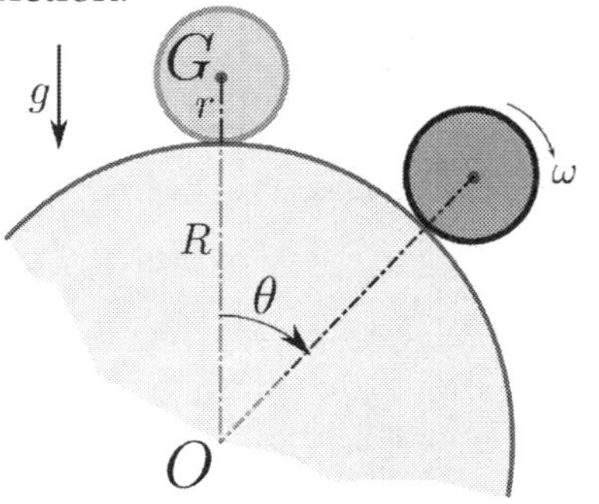

a. Find the angular speed $\dot{\theta}$ versus θ. Deduce the expression of the reaction force acting on the cylinder.

b. Find the angle θ_1 at which the cylinder loses contact with its support.

c. Find the angle θ_2 at which the cylinder starts to slip.

d. Conclude as to whether the cylinder slips or loses contact first.

Problem 5.18 A uniform slender rod of mass 1.5 kg is pinned at O to a frictionless pivot. The other endpoint is connected to a linear spring which is initially unstretched. The rod is released from rest from the horizontal position ($\theta = 0$).

Find the spring constant so that the rod reaches the position $\theta_1 = 30°$ with the angular velocity $\dot{\theta}_1 = 1.6$ rad/s. Then find the maximum amplitude reached by the rod.

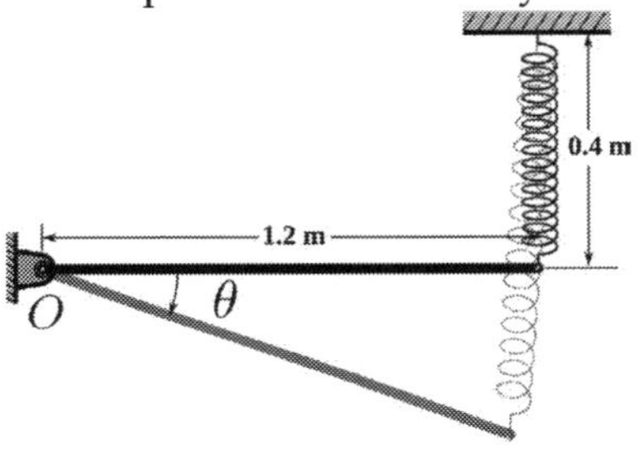

Problem 5.19 A billiard ball of mass m and radius r is set in motion on a horizontal support with the initial conditions (v_0, ω_0) as shown in the Figure. Denote by μ the coefficient of kinetic friction at the contact.

Find a condition satisfied by the initial conditions for the ball to return to its initial position. Then find the work done by the friction forces during the motion.

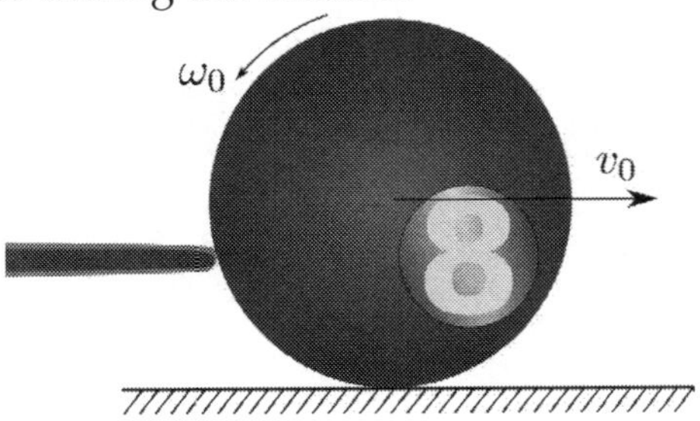

Problem 5.20 Two rigid bodies 1 and 2 are connected by an inextensible string passing through a massless pulley 3 of negligible size. Body 1 is a block of mass m_1. Body 2 is a disk of mass m_2, radius r, mass center A. It is assumed to roll without slipping on its support. Body 1 can slide relative to its smooth support. The system is released from rest.

Study the motion of the system by finding:

a. the accelerations of body 1 and center A,

b. the speed of body 1 when center A travels a distance L.

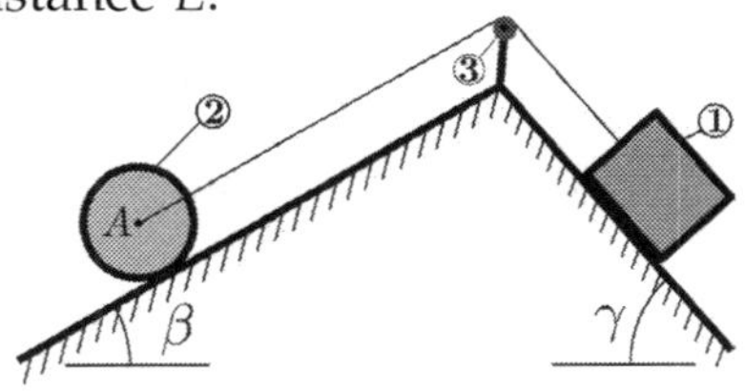

Problem 5.21 A spherical ball of mass m and radius r leaves a ramp without initial speed at a height h from horizontal axis Ax. It enters a circular loop of radius R at point A. It may reach highest point B, and exit the circular support at A to continue on horizontal support Ax. Assume that the ball rolls without slipping.

Study the motion of the ball as a function of h. In particular find the minimum value of h for the sphere to travel completely around the loop without losing contact.

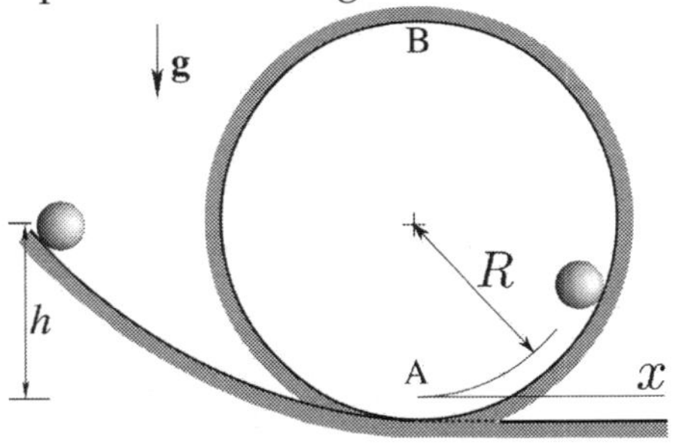

Problem 5.22 Both pulleys 1 and 2 have mass m, radius R and moment of inertia $J = kmR^2$ about their axis. The rope, assumed inextensible and of negligible mass, does not slip relative to the pulleys. Pulley 2 supports a body 3 of mass $m_3 = 2m$. Body 4 of mass $m_4 = 2m$ is attached to the free extremity of the cord. The pulleys are assumed to rotate about their shaft without friction. The system is released from rest in the configuration shown in the Figure. Assuming that the cord does not swing, find at any time $t > 0$

a. the kinetic energy of the system,

b. the potential energy of the system,

c. the acceleration of body 4,

d. the velocity of body 3 when body 4 has traveled a distance H.

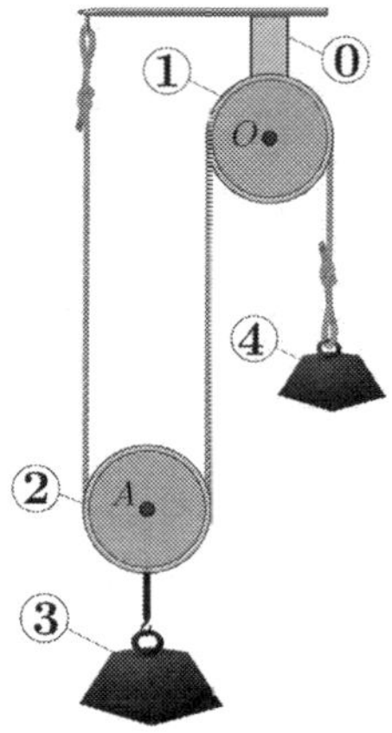

Problem 5.23 An aerial tramway Σ is supported by two fixed horizontal cables. The truck 1 of mass m supports a cabin 2 of mass M and mass center G. The cabin is free to rotate about A relative to the truck. The truck supports a set of wheels assumed massless which can roll on the cables. The oscillations of the tramway are modeled by two coordinates: Cartesian coordinate x for the rectilinear motion of A along the $\hat{\imath}$ direction, angle θ for the rotation of the cabin. Assume that angle θ is small, that is, make the approximation $\cos\theta \approx 1$ and $\sin\theta \approx \theta$. The moment of inertia of the cabin about its axis of rotation Ay is $J_A = Mk_A^2$ and mass center G is located at distance L from A. All sources of friction are neglected.

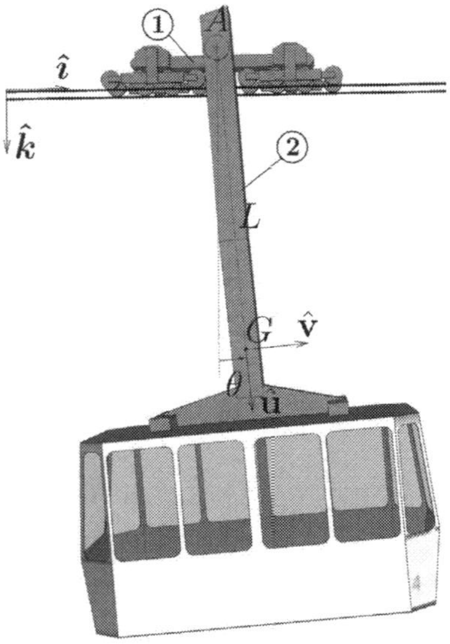

a. Apply the work-energy to the tramway to find a first-integral of motion.

b. Apply Euler's second principle about A to the car to find a second equation.

c. Using the previous results find the period of small amplitude oscillations.

Problem 5.24 A snowcat moves in the horizontal direction at speed $v(t)$. It is comprised of the body 1 of mass m_1, of two wheel assemblies each of mass m_2 and radius of gyration k_2 about their axis, and of two tracks 3 each of mass m_3. The tracks do not slip relative to the wheels or the ground. The back wheels are subject to a constant drive torque C.

Find the equation which governs speed v.

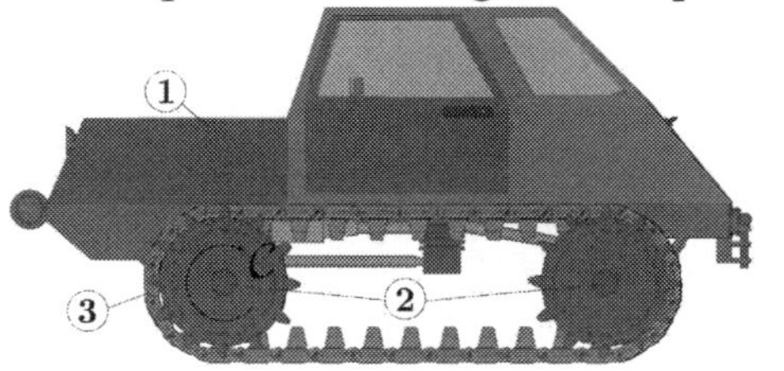

Problem 5.25 A hinge made of two identical plates, each of mass m, side length l, connected by a frictionless pivot at A is placed on top of a smooth horizontal support. It is released from rest when A is at a distance h from the support.

a. Apply the work-energy principle to the hinge to find the angular velocity $\dot{\theta}$ as a function of θ and the geometric constants.

b. Find the velocity of point A when it comes in contact with the support. Then find the corresponding work done by the reaction at A.

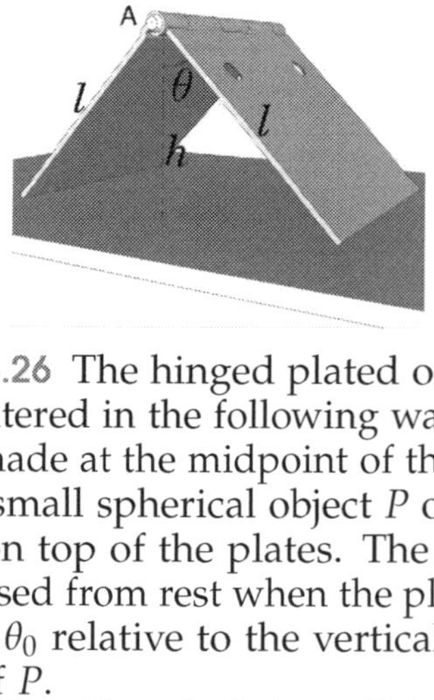

Problem 5.26 The hinged plated of problem 5.25 are altered in the following way: a small notch is made at the midpoint of the hinge A so that a small spherical object P of mass M can rest on top of the plates. The system is then released from rest when the plates make the angle θ_0 relative to the vertical. Neglect the size of P.

a. Show that if angle θ_0 is sufficiently large ($\theta_0 \geq \theta_{min}$) particle P will be in free fall as soon as the plates slide on their support. Find angle θ_{min}.

b. Assume $\theta_0 = \theta_{min}$: find the height of P measured above the plane when the hinge hits the plane ($\theta = \pi/2$).

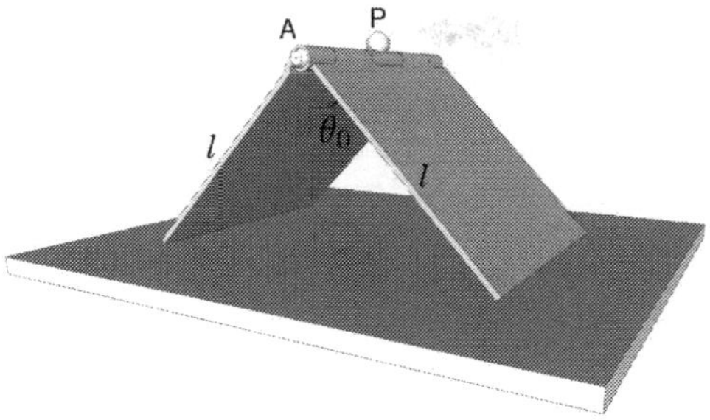

Problem 5.27 In this speed reducer, the carrier 1 supports the double satellite gear 3 which is meshed with two co-axial ring gears 2 and 4. Ring gear 4 is stationary.

A motor exerts a couple $C_1\hat{\mathbf{z}}_0$ on the shaft of carrier 1. The output shaft of ring gear 2 exerts a couple $C_2\hat{\mathbf{z}}_0$ to an output device. Carrier 1 has moment of inertia I_1 about axis $(O, \hat{\mathbf{z}}_0)$. Satellite gear 3 has moment of inertia I_3 about axis $(A, \hat{\mathbf{z}}_0)$, mass m_3, and mass center A. Gear 2 has moment of inertia I_2 about axis $(O, \hat{\mathbf{z}}_0)$. All joints are assumed frictionless and the effect of gravity is neglected.

a. Find the ratio of angular velocities ω_2/ω_1 in terms of the parameters r_1, r_3' and r_3''.

b. Find the o.d.e. which governs the angular velocity $\omega_1(t)$ as a function of couple C_1 and C_2 and the system's parameters. Deduce the ratio C_2/C_1 in steady-state regime.

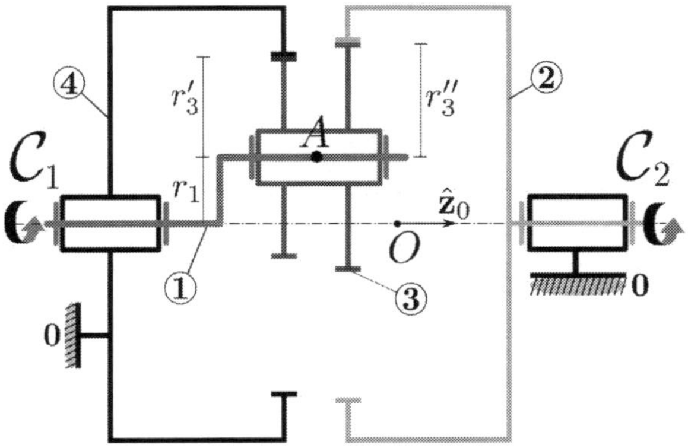

Problem 5.28 The figure displayed below shows a tall slender industrial chimney (flue gas stack). It is initially at rest. Due to a disturbance it collapses by rotation about ground point O. Its position is defined by angle θ and the unit vectors $(\hat{\mathbf{u}}, \hat{\mathbf{v}})$. For simplicity the chimney is modeled as a slender rod of mass m, length L and mass center G. Assume that G is located at distance $L/2$ from O.

a. Apply the Work Energy Principle to find the equation governing the angle θ and $\dot{\theta}$. Find the speed of G when $\theta = \pi/2$.

b. Find the reaction force at O in the form $\mathbf{R} = R_r\hat{\mathbf{u}} + R_\theta\hat{\mathbf{v}}$.

c. The chimney is unlikely to complete its fall without breaking apart. To find out how this occurs, we look at the effect of a section of the chimney located above some point P on the lower section below P. We model this as a force $\mathbf{R}_P = R_{Pr}\hat{\mathbf{u}} + R_{P\theta}\hat{\mathbf{v}}$ and a couple C (directed in the clockwise direction). Assume that P is located at a distance r from O. Draw a free-body diagram for the section of chimney below P and find the components R_{Pr}, $R_{P\theta}$ and couple C versus $u = r/L$. Admitting that the failure mechanism is based on torsion (as opposed to shear or traction), find the likely location $r = r_1$ where the chimney will break apart.

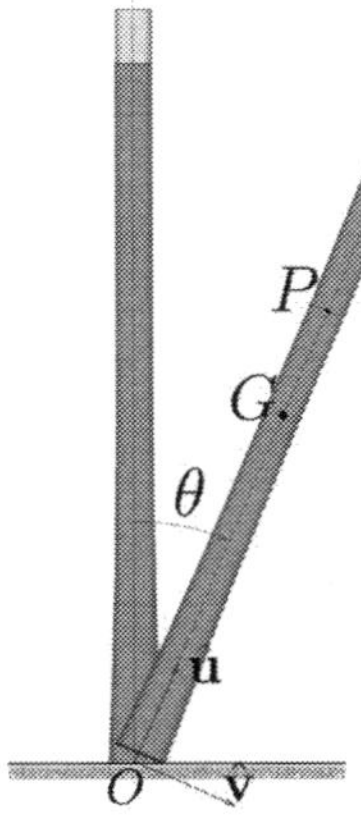

Answer: $r_1 = L/3$

Problem 5.29 A rigid body is made of a hoop of radius r of center A and negligible mass and of a uniform slender rod AB of length r and mass m welded to the hoop. This body is set on a horizontal support, rod AB being horizontal. The body is released from rest. Friction is assumed large enough to prevent the body from slipping during its motion.

Find the angular velocity of the body when the rod AB reaches the vertical position.

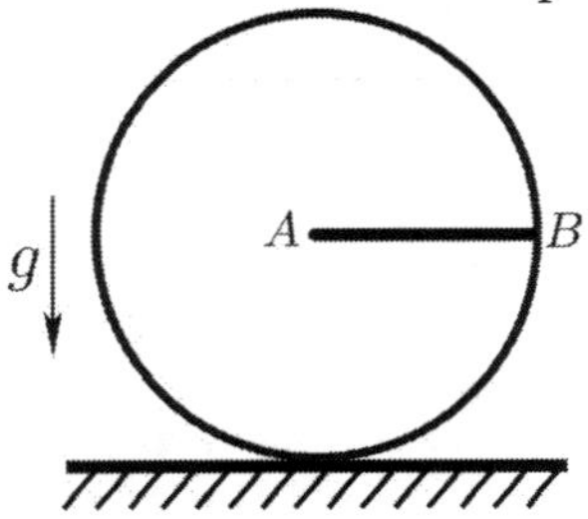

Problem 5.30 A small spherical ball of mass m is in motion under the effect of gravity in the interior of a cone of vertical axis Oz and half-angle α.

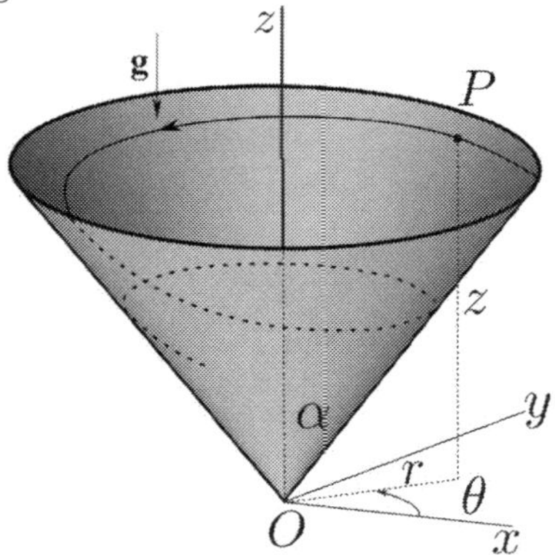

The contact is frictionless. Its initial height is $z = h > 0$, and its initial velocity is of magnitude v_0, tangent to the cone, and perpendicular to axis Oz. Hence $\mathbf{v}_0$ lies in the horizontal plane $z = h$. Assume $\theta = 0$ at $t = 0$.

a. Apply the Work-Energy Principle to show that P moves in a one-dimensional potential given by

$$U_{eff} = \frac{1}{2}mv_0^2\frac{h^2}{z^2} + mgz$$

b. Sketch the function U_{eff} vs z. Show that U_{eff} admits a minimum at $z = z_e$. Show that this value corresponds to a circular motion of P. Find the period of this motion. Is the motion stable?

c. Using function U_{eff}, give a qualitative description of the motion away from $z = z_e$. Are these motions periodic? Sketch the paths of P on the surface of the cone.

Problem 5.31 A space probe P rotates about the Sun (centered at O) along an elliptic trajectory of semi-major axis a. Show that the mechanical energy $E = \frac{1}{2}mv_P^2 - GmM/r$ of the probe takes the expression

$$E = -\frac{GmM}{2a}$$

where G is the universal gravitational constant and M is the mass of the Sun. *Hint: Find the energy at the locations A and B of the major axis and apply conservation of angular momentum about O.*

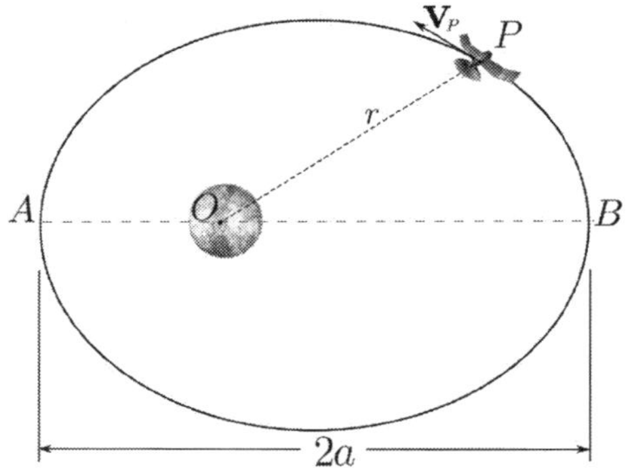

Problem 5.32 The Hohmann transfer orbit is an elliptical orbit used to transfer a spacecraft from one circular orbit to another with the use of two engine impulses, one to move a spacecraft onto the transfer orbit and a second to move off it. This transfer orbit is tangent to both departure and arrival circular orbits.

Consider a spacecraft P orbiting Earth (of center O, mass M) along a circular trajectory of radius r_1. We wish to transfer P to the concentric circular orbit of radius $r_2 > r_1$. Show that the necessary increase in speed Δv_1 to give the spacecraft by an instantaneous impulse at A is given by

$$\Delta v_1 = \sqrt{\frac{GM}{r_1}}\Big(\sqrt{\frac{2r_2}{r_1 + r_2}} - 1\Big)$$

Then find the increase in speed Δv_2 to leave the transfer orbit at B to the circular orbit of radius r_2. *Hint: Use the result of problem 5.31.*

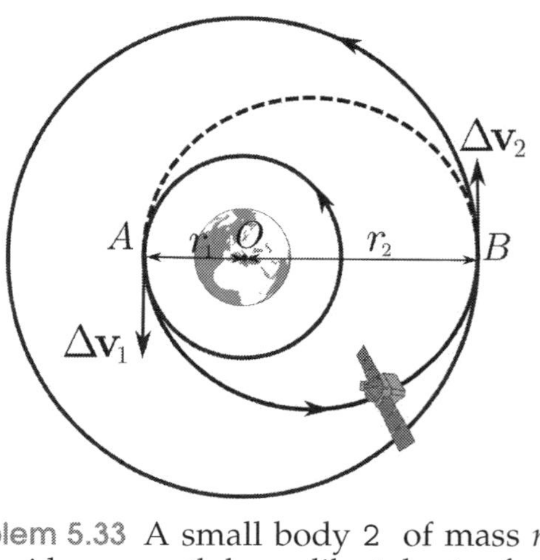

Problem 5.33 A small body 2 of mass m is set inside a smooth hoop-like tube 1 of mass M, mass center C and of radius R. Initially ($t = 0$), body 1 is at rest on a horizontal support 0. The system is then released without

initial speed when body 2 is at the level of C. The contact between the two bodies is assumed frictionless.

At a later time t_1, body 2 is at its lowest possible position. Friction between body 1 and the support is large enough to prevent slipping during the motion. We are interested in finding the velocity $\mathbf{v}_2$ of body 2, velocity $\mathbf{v}_C$ of body 1, and the force $\mathbf{N}_{12}$ exerted by body 1 on body 2 at this time. Assume that the motions take place in a vertical plane. Treat body 2 like a hollow cylinder.

a. Justify whether the mechanical energy of body 1, body 2, or system $\Sigma = \{1,2\}$ is conserved during the motion. If so, write the corresponding equation(s).

b. Apply Euler's principles to body 1 and Newton's second law to body 2 to find a relationship between $\mathbf{a}_C$ and the horizontal component of $\mathbf{a}_2$ valid at all time. Deduce a relationship between $\mathbf{v}_C$ and $\mathbf{v}_2$ by integration.

c. Use the results of a) and b) to find $\mathbf{v}_2$ and $\mathbf{v}_C$ at t_1.

d. Find N_{12} by applying Newton's second law to body 2.

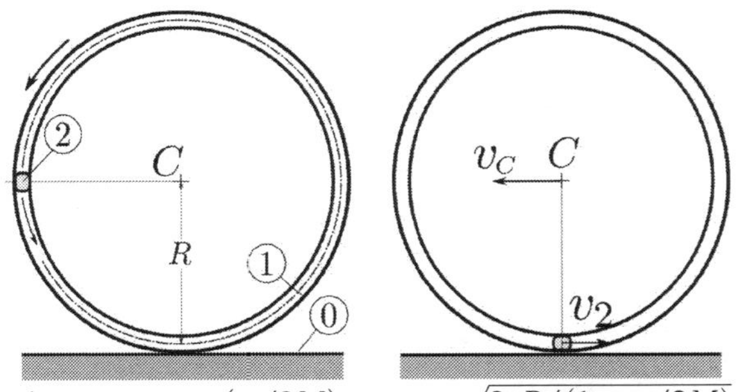

Answer: $v_C = (m/2M)v_2$, $v_2 = \sqrt{2gR/(1+m/2M)}$, $N_{12} = mg(3+m/M)$.

Problem 5.34 A heavy door 1 of mass $m_1 = 275$ kg is hinged about fixed point O. It is free to rotate in a vertical plane. A point B of its upper edge is connected to a counterweight 2 by a cable which goes around a frictionless pulley at A. Neglect the size and mass of the pulley and treat body 1 as a uniform plate of mass center G. The geometry is defined by $h = 1.25$ m and $l = |OB| = 1.8$ m.

a. Find the mass m_2 of the counterweight so that the system is in equilibrium at position $\theta = 0$.

b. Find the ODE which governs angle $\theta(t)$.

c. From the result of b) determine whether the equilibrium position $\theta = 0$ is stable.

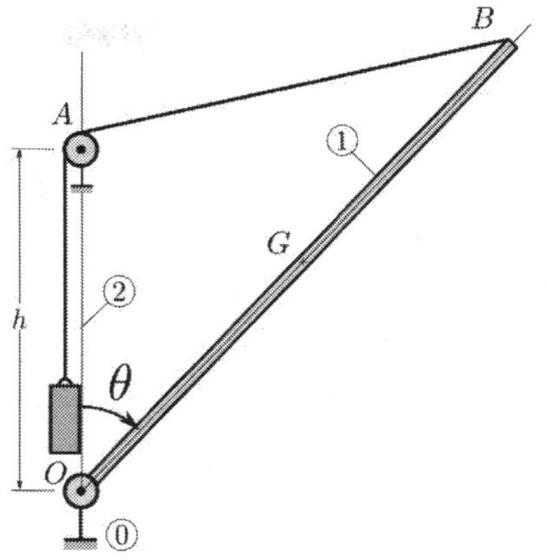

Problem 5.35 Near-earth objects such as asteroids pause a significant danger to Earth. It is therefore of interest to find out under which condition an asteroid will strike Earth with advance knowledge of its motion. An asteroid P (of mass m, and of negligible size) is approaching Earth as shown in the sketch. The axes Oxy are attached to the geocentric referential $\mathcal{G}$ of Earth (of center O, of mass M_E, and radius R_E). Assume that P is close enough to Earth to be only acted upon by Earth's gravitation. At some point very far from Earth, the velocity of P is known to be $v_\infty \hat{\mathbf{e}}_x$ and its position is defined by distance b (known as the "impact parameter"). The position at the point A of closest approach to Earth is defined by the distance $r_A = |OA|$.

a. Recall that the angular momentum of P about O is conserved. Deduce a relationship between the speed v_A at A and v_∞.

b. Using conservation of mechanical energy, find the distance r_A of closest approach to Earth as a function of v_∞, b, and the product GM_E.

c. Deduce a condition for the asteroid to avoid striking the planet.

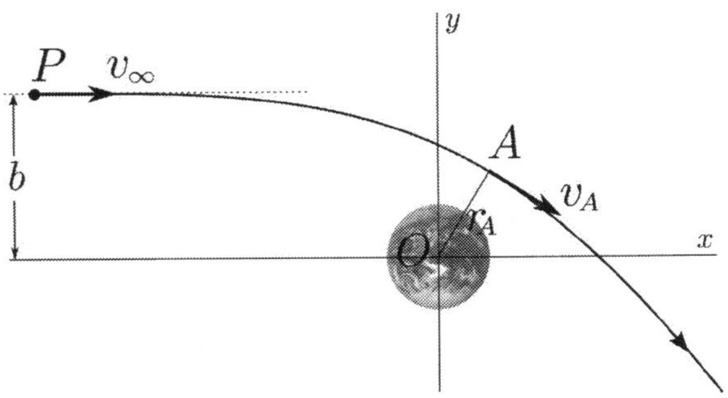

Problem 5.36 A space probe P of mass m is in a circular orbit of radius r_0 and constant speed v_0 around a planet of center O, mass

M and radius R. The probe wishes to escape from the gravitational field of the planet by the use of engine impulses. Two options of identical cost are possible.

a. In the first option, a single engine impulse is used to bring the probe from speed v_0 to speed $3v_0$. Show that the probe escapes from the planet's gravity and find its speed v_∞ at infinity (neglecting the presence of other gravitational fields).

b. In the second option, the probe uses two engine impulses. A first impulse reduces the speed from v_0 to $v_0/2$ (without directional change) to send the probe along an elliptical orbit. Once the probe passes at its perigee A at the speed v_A, a second impulse brings the speed from v_A to $v_A + 3v_0/2$ to send the probe along a hyperbolic trajectory. Find the new speed v'_∞ at infinity. Which of the two options is preferable? *Hint: Use the result of problem 5.31 to find velocity v_A.*

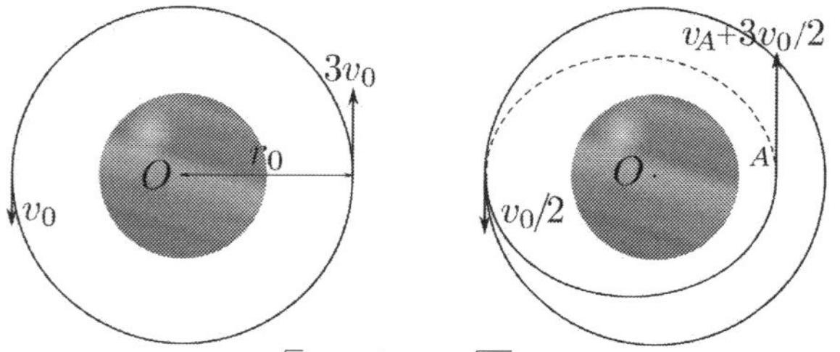

Answer: $v_\infty = \sqrt{7}v_0$, $v'_\infty = \sqrt{11}v_0$.

Problem 5.37 This problem proposes a simple physical explanation of the crack made by a whip, caused by the section of the whip moving faster than sound, thereby creating a small sonic boom. The sketch shown below shows a whip modeled as 1-dimensional body of mass per unit length μ and length $l = 2\pi R$. It is wrapped around a massless cylinder of center C and radius R. At any given time, it is made of a section OI as rest on a flat support Ox and a moving section IA in contact with the cylinder without slipping. The angle θ measures the position of endpoint A. At time $t = 0$, $\theta = 2\pi$ and the center C is given a velocity $v_0\hat{\mathbf{e}}_x$. We wish to find the velocity of C in the limit $\theta \to 0$. Neglect the effect of gravity.

a. Justify that the mechanical energy of the whip is conserved.

b. Show that the kinetic energy of the whip is given by $K = \mu R^3\dot{\theta}^2(\theta - \sin\theta)$ as a function of θ. Deduce $\dot{\theta}$ and v_C as a function of angle θ.

c. Examine the limit $\theta \to 0$. Conclude.

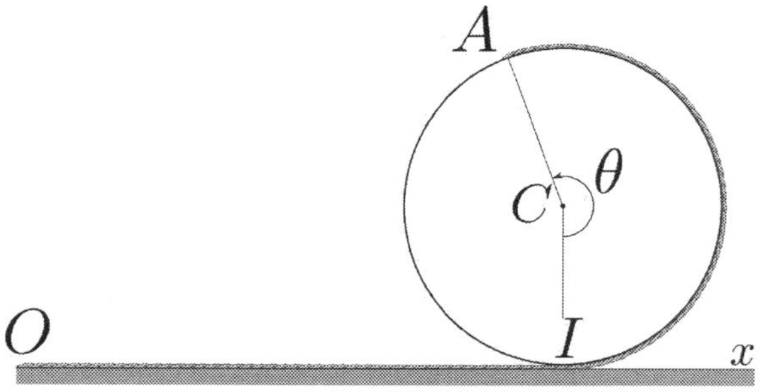

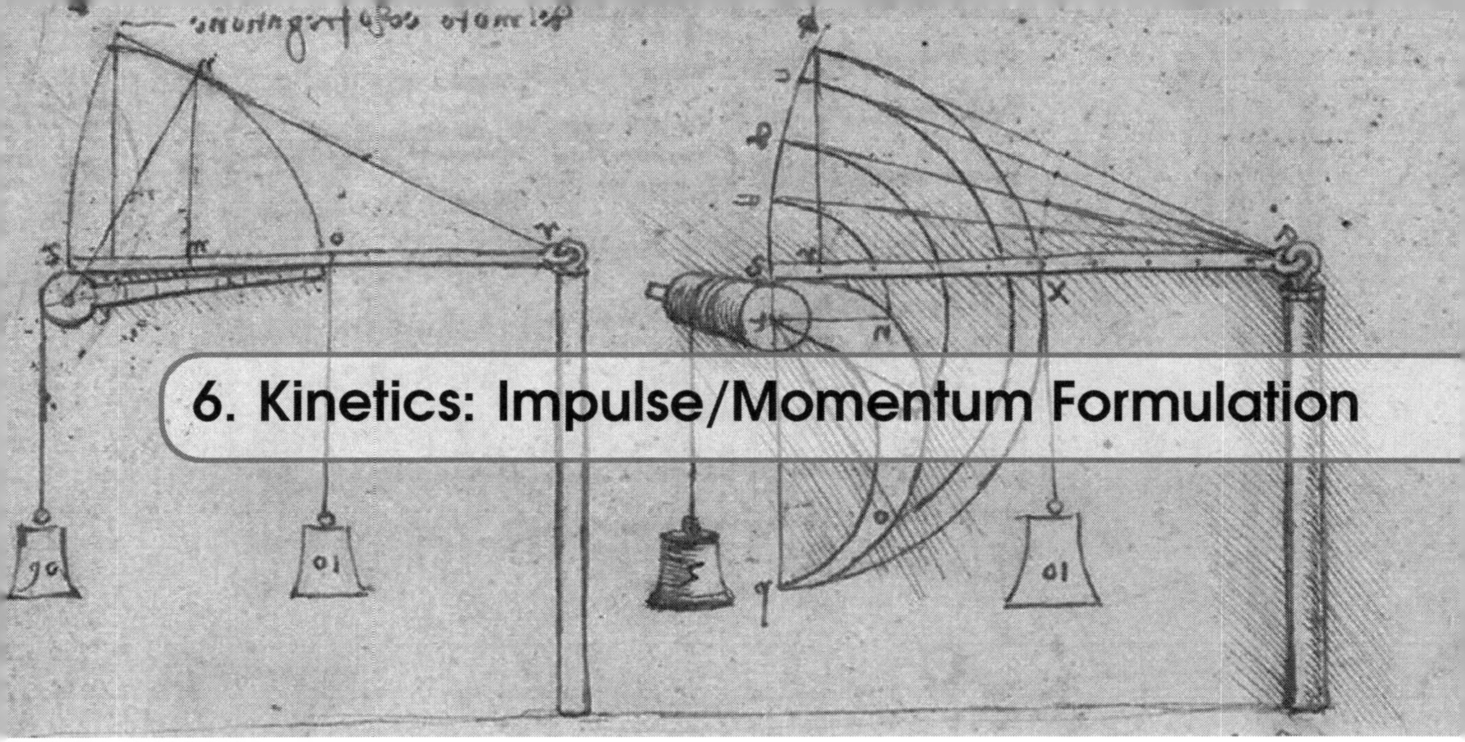

6. Kinetics: Impulse/Momentum Formulation

In this final chapter, we derive impulse/momentum forms of Euler's principles. They are easily obtained by integration over a finite time interval. On one side of the equation, we obtain the variation of the linear or angular momentum of the system. On the other side of the equation, we obtain the integral of the total external force or moment applied to the system (over time). We call these integrals (linear or angular) *impulses*. Hence, it is readily seen that the resulting formulations (Sections 6.1 and 6.3) are not fundamentally different from the original forms of Euler's principles. They can nevertheless be useful when rigid bodies are exhibiting instantaneous velocity changes due to the effect of *impulsive forces* (or moments), that is, forces (or moments) of very large magnitude, which act over a very short interval of time. Such forces (moments) cause significant changes in the linear (angular) momentum of the system. Such is the case during the *impact* of rigid bodies. In Section 6.2, we develop a first *theory of impact*, whereby the bodies are modeled geometrically as smooth, small, spherical bodies. To predict the velocities of the particles immediately after impact from their values immediately before impact, we are faced with an indeterminate problem. Hence, we will need to introduce a constitutive law, known as Newton's Law of Impact, which characterizes the ability of the bodies to recover their initial state after impact, in terms of an empirical parameter called *coefficient of restitution*. This will enable us to achieve the solution of simple impact problems. In Section 6.4, we then present a second, more advanced, theory of impact for rigid bodies.

In this chapter, we will learn

1. to distinguish impulsive forces from non-impulsive forces,
2. to solve impact problems modeled as particle or rigid bodies,
3. to use the Impulse/Momentum Principles for systems exhibiting instantaneous velocity changes under the effect of impulsive forces.

6.1 Impulse/Momentum Formulation for System of Particles

In some applications, alternative forms of Euler's principles can be useful: by consideration of a finite time interval, impulse/momentum formulations can be easily obtained. Consider a system Σ (of mass m, and mass center G) of N particles in motion relative to a Newtonian referential. To find the impulse/momentum forms of Euler's principles, we simply integrate these vector equations over a finite time interval $t_1 \leq t \leq t_2$ to obtain:

Euler's Principles (Impulse/Momentum forms)

The (linear) impulse of the external forces imparted to system Σ is equal to the change of the system's linear momentum over some interval of time $t_1 \leq t \leq t_2$:

$$\underbrace{m\mathbf{v}_G(t_2) - m\mathbf{v}_G(t_1)}_{\text{variation of linear of momentum}} = \underbrace{\int_{t_1}^{t_2} \mathbf{F}_{ext\to\Sigma}\, dt}_{\text{impulse of external forces}} \tag{6.1}$$

The angular impulse imparted to system Σ by all external forces is equal to the change of the system's angular momentum over some interval of time $t_1 \leq t \leq t_2$:

$$\underbrace{\mathbf{H}_G(t_2) - \mathbf{H}_G(t_1)}_{\text{variation of angular of momentum}} = \underbrace{\int_{t_1}^{t_2} \mathbf{M}_{G,ext\to\Sigma}\, dt}_{\text{angular impulse of external forces}} \tag{6.2}$$

where both angular impulse and angular momentum are taken about the system's mass center G.

Properties

The impulse/momentum formulations (6.1-6.2) are just integrated versions of Euler's principles, and hence do not offer anything fundamentally new. These equations are generally not very useful since the integrands of the impulses are generally unknown, unless one or all components of the external forces or moments are zero. In that case, we recover the conservation laws of linear or angular momentum learned in previous chapters.

Remark 1. A equation similar to (6.2) can also be written about a fixed point O.

Remark 2. It is possible to write a impulse/momentum equation for each particle P_i (of mass m_i, linear momentum $m_i\mathbf{v}_i$) of the system:

$$m_i\mathbf{v}_i(t_2) - m_i\mathbf{v}_i(t_1) = \int_{t_1}^{t_2} \mathbf{F}_{ext\to\Sigma}\, dt + \int_{t_1}^{t_2} \sum_{j=0}^{N} \mathbf{F}_{j\to i}\, dt \tag{6.3}$$

where we have written the impulse in terms of external and internal contributions.

6.2 A Simple Theory of Impact

CONSIDER two small bodies P_1 of mass m_1 and P_2 of mass m_2 undergoing an impact. During the impact, the bodies are subject to interactions occurring over a very short timespan and in very confined space. These interactions are responsible for the rapid change of their velocities from the values $\mathbf{v}_1^-$ and $\mathbf{v}_2^-$ just before impact ($t = t^-$) to the values $\mathbf{v}_1^+$ and $\mathbf{v}_2^+$ just after impact ($t = t^+$). See Figure 6.1. The force of interaction between the two bodies is characterized by a large magnitude so that its impulse over vanishingly small interval $\Delta t = t^+ - t^-$ remains finite:

$$\lim_{\Delta t \to 0} \int_{t^-}^{t^+} \mathbf{F}_{1\to 2}\, dt \neq 0$$

We call such a force, *an impulsive force*. Non-impulsive forces are characterized by negligible impulses over the duration of the impact.

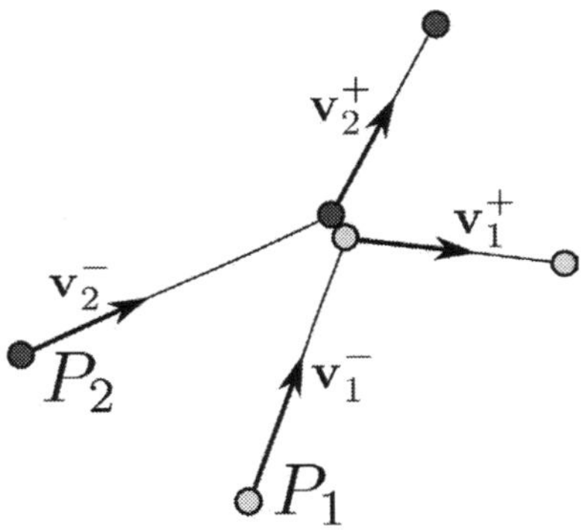

Figure 6.1 – *Impact between two particles.*

The goal of any theory of impact is to predict the velocities of the particles immediately after impact from their values immediately before impact. Unfortunately, the force of interaction $\mathbf{F}_{1\to 2} = -\mathbf{F}_{2\to 1}$ is generally unknown. The equations which we are about to derive are insufficient to fully solve this problem without additional empirical information.

Let us write Newton's Second Law in impulse/momentum form for each particle during the time of impact:

$$m_1\mathbf{v}_1^+ - m_1\mathbf{v}_1^- = \int_{t^-}^{t^+} (\mathbf{F}_{ext\to 1} + \mathbf{F}_{2\to 1})dt \tag{6.4}$$

$$m_2\mathbf{v}_2^+ - m_2\mathbf{v}_2^- = \int_{t^-}^{t^+} (\mathbf{F}_{ext\to 2} + \mathbf{F}_{1\to 2})dt \tag{6.5}$$

Let us make the assumption that all external forces are non-impulsive: $\int_{t^-}^{t^+} \mathbf{F}_{ext\to 1} dt = \int_{t^-}^{t^+} \mathbf{F}_{ext\to 2} dt \approx \mathbf{0}$. Then in view of $\mathbf{F}_{1\to 2} = -\mathbf{F}_{2\to 1}$, we can conclude that:

IMPACT (CONSERVATION OF LINEAR MOMENTUM)

If the impulse of all external forces acting on the system of two particles is negligible during the duration of impact, then the system behaves like a closed system and its linear momentum is conserved:

$$\underbrace{m_1\mathbf{v}_1^- + m_2\mathbf{v}_2^-}_{\text{just before impact}} = \underbrace{m_1\mathbf{v}_1^+ + m_2\mathbf{v}_2^+}_{\text{just after impact}} \tag{6.6}$$

Assuming that the four vectors $(\mathbf{v}_1^-, \mathbf{v}_2^-)$, $(\mathbf{v}_1^+, \mathbf{v}_2^+)$ are coplanar, (6.6) yields two scalar equations to solve for four unknowns. The problem requires additional hypotheses to be fully solvable. This is addressed in the following model.

MODEL OF SMOOTH SPHERICAL BODIES. The two particles are modeled as two spherical bodies centered at P_1 and P_2. The *line of impact* is defined as the line joining their centers: it is also the line normal to the surfaces in contact and directed along the unit vector $\hat{\mathbf{e}}_N$. Let $\hat{\mathbf{e}}_T$ be the unit tangent vector to the surfaces in contact. See Figure 6.2. Assuming that the surfaces in contact are smooth, the force of interaction $\mathbf{F}_{1\to 2}$ is directed along $\hat{\mathbf{e}}_N$. Equations (6.4-6.5) show that the linear momentum of each particle is conserved along the tangential direction:

$$m_1\mathbf{v}_1^- \cdot \hat{\mathbf{e}}_T = m_1\mathbf{v}_1^+ \cdot \hat{\mathbf{e}}_T, \quad m_2\mathbf{v}_2^- \cdot \hat{\mathbf{e}}_T = m_2\mathbf{v}_2^+ \cdot \hat{\mathbf{e}}_T \tag{6.7}$$

This adds two equations. However the problem is still indeterminate: an equation is missing. Note that equations (6.6) and (6.7) are not independent!

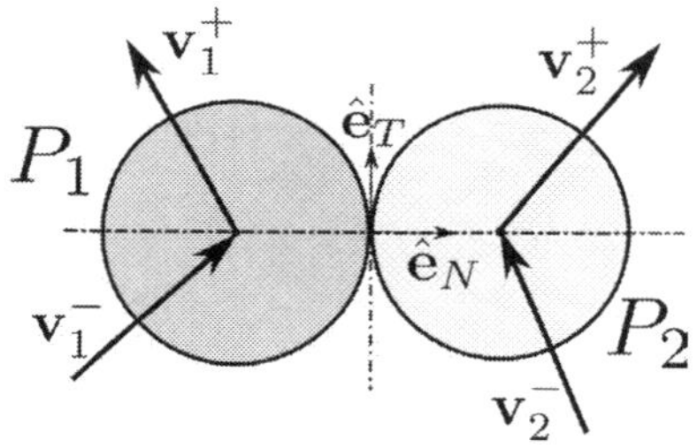

Figure 6.2 – *Impact between smooth spheres. If the incoming velocities are colinear to the line of impact, the impact is said to be direct. Otherwise, it is indirect.*

To address this indetermination, we must introduce an empirical measure of the capacity of the two bodies to rebound from each other. In the most idealized situation, we can assume that the work done by the internal forces during the phase of deformation of the bodies is exactly canceled out by the work of the same forces during the restoration phase: this implies that the kinetic energy of the system is conserved (there is no change of potential of energy since the positions of the bodies is unchanged during impact):

$$\tfrac{1}{2}m_1 v_1^{-2} + \tfrac{1}{2}m_2 v_2^{-2} = \tfrac{1}{2}m_1 v_1^{+2} + \tfrac{1}{2}m_2 v_2^{+2} \tag{6.8}$$

A more realistic measure is given by the *coefficient of restitution*.

DEFINITION (COEFFICIENT OF RESTITUTION)

Two bodies in impact are characterized by a coefficient of restitution defined as the ratio of the impulse of the force of interaction during restoration ($t^* \leq t \leq t^+$) to that during deformation ($t^- \leq t \leq t^*$):

$$e = \frac{\int_{t^*}^{t^+} F_{1\to 2}\, dt}{\int_{t^-}^{t^*} F_{1\to 2}\, dt} = \frac{\hat{\mathbf{e}}_N \cdot (\mathbf{v}_2^+ - \mathbf{v}_1^+)}{\hat{\mathbf{e}}_N \cdot (\mathbf{v}_1^- - \mathbf{v}_2^-)} \tag{6.9}$$

Proof. The second equality of (6.9) can be shown by applying the impulse/momentum equations for each particle for the phases of deformation ($t^- \leq t \leq t^*$) and of restoration ($t^* \leq t \leq t^+$). At $t = t^*$, it is assumed that the two particles reach the same normal velocity $v_* \hat{\mathbf{e}}_N$. To simplify the analysis, we further assume that the impact is direct, that is, the incoming and departing velocities are all colinear with the unit normal vector $\hat{\mathbf{e}}_N$. Then we write $\mathbf{v}_1^- = v_1^- \hat{\mathbf{e}}_N$, $\mathbf{v}_2^- = v_2^- \hat{\mathbf{e}}_N$, $\mathbf{v}_1^+ = v_1^+ \hat{\mathbf{e}}_N$, and $\mathbf{v}_2^+ = v_2^+ \hat{\mathbf{e}}_N$. During the deformation phase ($t^- \leq t \leq t^*$) we find

$$m_2(v_* - v_2^-) = \int_{t^-}^{t^*} F_{1\to 2} dt = -m_1(v_* - v_1^-)$$

Likewise, during the restoration phase ($t^* \leq t \leq t^+$),

$$m_2(v_2^+ - v_*) = \int_{t^*}^{t^+} F_{1\to 2} dt = -m_1(v_1^+ - v_*)$$

Eliminating v_* in these expressions leads to

$$\int_{t^-}^{t^*} F_{1\to 2} dt = \frac{m_1 m_2}{m_1 + m_2}(v_1^- - v_2^-), \qquad \int_{t^*}^{t^+} F_{1\to 2} dt = \frac{m_1 m_2}{m_1 + m_2}(v_2^+ - v_1^+)$$

This leads to the expression of $e = (v_2^+ - v_1^+)/(v_1^- - v_2^-)$. For oblique impact, the analysis is identical, replacing the velocities by their normal components. ■

PROPERTIES

①. It is possible to show that $0 \leq e \leq 1$. The case $e = 1$ corresponds to an *elastic impact*: using the expression of e and the momentum conservation equation (6.7) along $\hat{\mathbf{e}}_T$, it is straightforward to show that $e = 1$ implies conservation of kinetic energy (6.8).

②. The case $e = 0$ corresponds to *perfectly inelastic impact*. In a direct impact, it implies that the particles stick together since $e = 0$ is equivalent to $\mathbf{v}_2^+ = \mathbf{v}_1^+$.

③. The coefficient of restitution depends on the two materials which come into contact. It is also a function of many other factors, such as the relative initial velocities of the bodies.

EXAMPLE PROBLEMS

General Procedure: Impact Problems

① First identify the line of impact to define unit vectors $(\hat{\mathbf{e}}_T, \hat{\mathbf{e}}_N)$. Write the velocities before and after impact on basis $(\hat{\mathbf{e}}_T, \hat{\mathbf{e}}_N)$ unless one of the particles has a prescribed direction.

② As long as all external forces are non-impulsive, the solution is typically found by applying (1) conservation of linear momentum of the system, (2) conservation of linear momentum of each particle in the tangential direction, (3) the knowledge of the coefficient of restitution.

③ If one of the particles is constrained to move in a particular direction due to a string or rigid boundary, then the reaction force or tension must be impulsive: conservation of linear momentum of the system must be restricted to that direction (in the absence of friction).

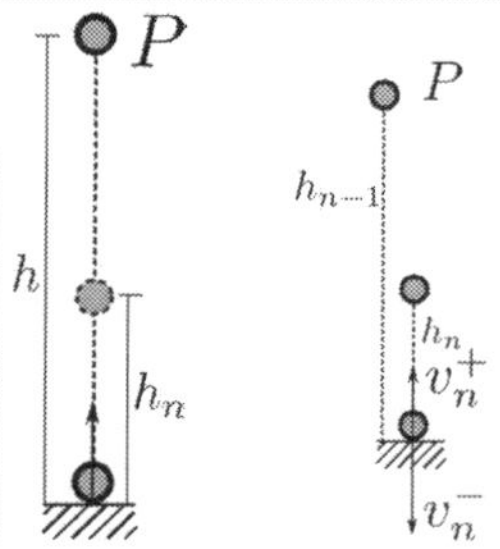

Exercise 6.1 ▪ A bouncy ball is dropped from the height h without initial speed above a smooth horizontal surface. The coefficient of restitution between the support and the ball is e. After the nth bounce, the ball reaches the height h_n in the time τ_n. Find h_n and τ_n versus n. Examine the limit $n \to \infty$. ▪

Solution. Consider the nth impact of the ball: the velocity v_n^+ after impact is related to the velocity v_n^- before impact by the relationship:

$$v_n^+ = e v_n^-$$

The height h_n reached by the ball starting with the speed v_n^+ immediately after impact is given by $gh_n = \frac{1}{2} v_n^{+2} = \frac{1}{2} e^2 v_n^{-2}$ using conservation of energy. Height h_{n-1} is related to the speed v_n^- immediately before impact according to $gh_{n-1} = \frac{1}{2} v_n^{-2}$. This gives a relationship between h_n and h_{n-1}:

$$h_n = e^2 h_{n-1} = e^4 h_{n-2} = \ldots = e^{2n} h$$

The time taken to drop distance h_{n-1} and climb distance h_n is given by

$$\sqrt{\frac{2h_{n-1}}{g}} + \sqrt{\frac{2h_n}{g}} = (1+e)\sqrt{\frac{2h_{n-1}}{g}} = (1+e)e^{n-1}\sqrt{\frac{2h}{g}}$$

So the total time taken to reach h_n is

$$\boxed{\tau_n = (1+e)(e^{n-1} + e^{n-2} + \cdots + e + 1)\sqrt{\frac{2h}{g}} = (1+e)\frac{1-e^n}{1-e}\sqrt{\frac{2h}{g}}}$$

where we have used the identity $1 + e + e^2 + \cdots + e^{n-1} = (1-e^n)/(1-e)$. In the limit

$n \to \infty$, we find that $h_n \to 0$ since $e < 1$ and

$$\boxed{\tau_n \to \tau_\infty = \frac{1+e}{1-e}\sqrt{\frac{2h}{g}}}$$

Remark. The ball will travel a total distance $h_\infty = h_0 \frac{1+e^2}{1-e^2}$.

■

Exercise 6.2 ■ Two identical ($m_1 = m_2$) pucks sliding on an air table collide at known incident speeds v_1^- and v_2^-. The direction of their velocities relative to the line of impact is given by the angles θ_1 and θ_2 as shown. The surfaces are assumed smooth and the bodies are characterized by the coefficient of restitution $e \neq 0$. Find the velocities of the pucks immediately after impact. ■

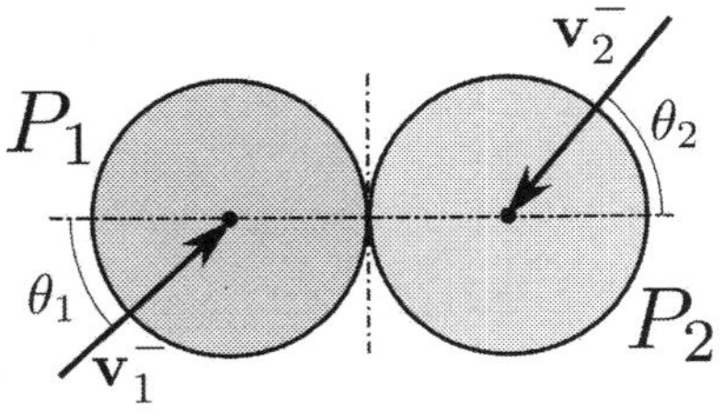

Solution. We adopt unit vectors $(\hat{\imath}, \hat{\jmath})$ with $\hat{\imath}$ directed along the line of impact. We write the velocities $(\mathbf{v}_1^-, \mathbf{v}_2^-)$ on $(\hat{\imath}, \hat{\jmath})$:

$$\mathbf{v}_1^- = v_1^-(\cos\theta_1 \hat{\imath} + \sin\theta_1 \hat{\jmath}), \qquad \mathbf{v}_2^- = -v_2^-(\cos\theta_2 \hat{\imath} + \sin\theta_2 \hat{\jmath})$$

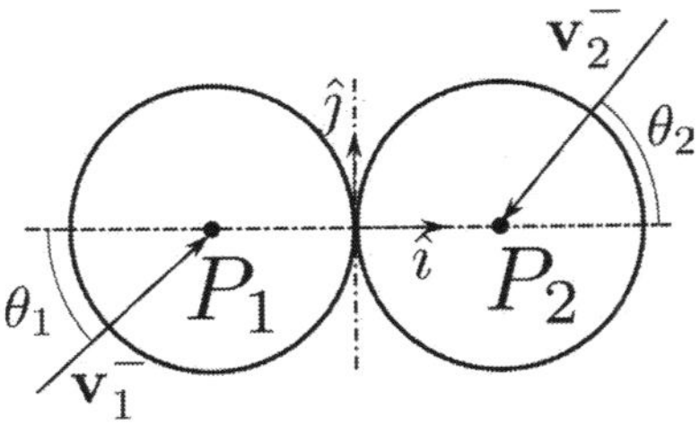

Figure 6.3 – *Problem 6.2*

Next, we write conservation of linear momentum of the system:

$$\mathbf{v}_1^- + \mathbf{v}_2^- = \mathbf{v}_1^+ + \mathbf{v}_2^+$$

and project on $(\hat{\imath}, \hat{\jmath})$:

$$v_1^- \cos\theta_1 - v_2^- \cos\theta_2 = v_{1x}^+ + v_{2x}^+, \qquad v_1^- \sin\theta_1 - v_2^- \sin\theta_2 = v_{1y}^+ + v_{2y}^+ \qquad [1-2]$$

Next we write conservation of linear momentum of each particle in the $\hat{\jmath}$-direction:

$$v_1^- \sin\theta_1 = v_{1y}^+, \qquad -v_2^- \sin\theta_2 = v_{2y}^+ \qquad [3-4]$$

since the interacting force is directed on the $\hat{\imath}$-direction. Note that equations [3-4] imply equation [2]. We still need one more equation: it is found by using the expression of e

$$e = \frac{v_{2x}^+ - v_{1x}^+}{v_1^- \cos\theta_1 + v_2^- \cos\theta_2} \qquad [5]$$

By solving [1]-[3-5] we find the expressions of (v_{1x}^+, v_{1y}^+) and (v_{2x}^+, v_{2y}^+):

$$\boxed{v_{1x}^+ = \frac{1-e}{2} v_1^- \cos\theta_1 - \frac{1+e}{2} v_2^- \cos\theta_2, \qquad v_{1y}^+ = v_1^- \sin\theta_1}$$

$$\boxed{v_{2x}^+ = \frac{1+e}{2} v_1^- \cos\theta_1 - \frac{1-e}{2} v_2^- \cos\theta_2, \qquad v_{2y}^+ = -v_2^- \sin\theta_2}$$

■

Exercise 6.3 ■■ A spherical body P_1 of mass m_1 is dropped from a height h above the smooth surface of a wedge-shaped cart P_2 of mass m_2 which may move freely along the horizontal support thanks to rollers. Both bodies are initially at rest. $m_1 = m_2 = 0.5$ kg, $h = 0.8$ m, $e = 0.8$ and $\beta = 45°$.
Given the coefficient of restitution e between the two bodies, find the velocities of the bodies immediately after impact. Find the condition for P_1 to bounce upward. ■

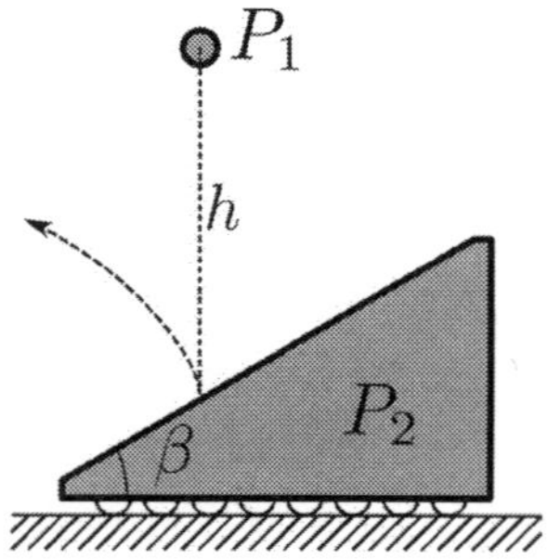

SOLUTION. We adopt unit vectors $(\hat{\imath}, \hat{\jmath})$ with $\hat{\imath}$ directed along the horizontal and unit vectors $(\hat{\mathbf{e}}_T, \hat{\mathbf{e}}_N)$ with $\hat{\mathbf{e}}_N$ along the line of impact. See Figure 6.4 (left). First we write the velocities $(\mathbf{v}_1^-, \mathbf{v}_2^-)$ before impact on $(\hat{\imath}, \hat{\jmath})$:

$$\mathbf{v}_1^- = -\sqrt{2gh}\hat{\jmath}, \qquad \mathbf{v}_2^- = \mathbf{0}$$

and the velocities $(\mathbf{v}_1^+, \mathbf{v}_2^+)$ after impact

$$\mathbf{v}_1^+ = v_{1T}^+ \hat{\mathbf{e}}_T + v_{1N}^+ \hat{\mathbf{e}}_N, \qquad \mathbf{v}_2^+ = v_2^+ \hat{\imath}$$

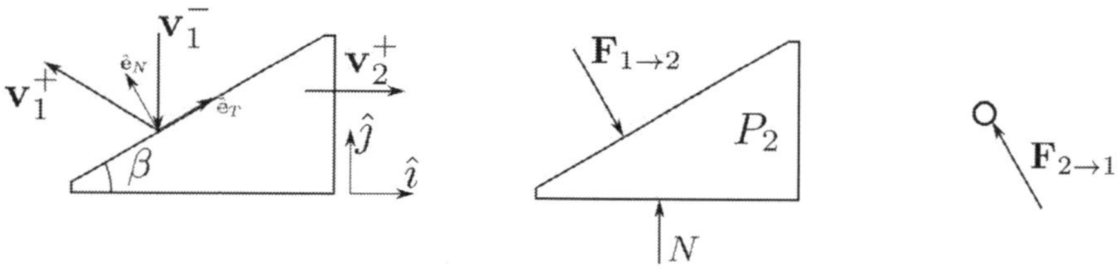

Figure 6.4 – *Problem 6.3*

Next we sketch free-body diagrams for P_1 and P_2 on Figure 6.4 (right). We cannot neglect the impulse of the reaction force exerted by the support on the cart: the normal force N is an impulsive force. Based on these diagrams, we can only write the conservation of linear momentum of the system along the horizontal direction. This gives

$$m_1(v_{1T}^+ \hat{\mathbf{e}}_T + v_{1N}^+ \hat{\mathbf{e}}_N) \cdot \hat{\imath} + m_2 v_2^+ = 0$$

which gives

$$m_1(v_{1T}^+ \cos\beta - v_{1N}^+ \sin\beta) + m_2 v_2^+ = 0 \qquad [1]$$

Next we write the conservation of linear momentum of P_1 along the $\hat{\mathbf{e}}_T$ direction:

$$-\sqrt{2gh}\sin\beta = v_{1T}^+ \qquad [2]$$

Remark. We cannot write the conservation of linear momentum of P_2 along the $\hat{\mathbf{e}}_T$ direction due to reaction force N.

Finally we use the knowledge of the coefficient of restitution e

$$e = \frac{\hat{\mathbf{e}}_N \cdot (\mathbf{v}_2^+ - \mathbf{v}_1^+)}{\hat{\mathbf{e}}_N \cdot (\mathbf{v}_1^- - \mathbf{v}_2^-)} = \frac{-v_2^+ \sin\beta - v_{1N}^+}{-\cos\beta\sqrt{2gh} - 0} \qquad [3]$$

We have found three equations to solve for three unknowns $(v_{1T}^+, v_{1N}^+, v_2^+)$:

$$\boxed{v_2^+ = (1+e)\frac{\sin\beta\cos\beta}{\frac{m_2}{m_1} + \sin^2\beta}\sqrt{2gh}}$$

$$\boxed{v_{1T}^+ = -\sqrt{2gh}\sin\beta, \qquad v_{1N}^+ = \cos\beta\frac{e\frac{m_2}{m_1} - \sin^2\beta}{\frac{m_2}{m_1} + \sin^2\beta}\sqrt{2gh}}$$

To find out if P_1 bounces in the upward direction we need to find the $\hat{\jmath}$ component of $\mathbf{v}_1^+$:

$$v_{1y}^+ = v_{1T}^+ \sin\beta - v_{1N}^+ \cos\beta = \sqrt{2gh}\frac{e\frac{m_2}{m_1}\cos^2\beta - (1 + \frac{m_2}{m_1})\sin^2\beta}{\frac{m_2}{m_1} + \sin^2\beta}$$

We need to impose $v_{1y}^+ > 0$ and find the condition

$$\boxed{\tan^2\beta < e\frac{m_2}{m_1 + m_2}} \qquad [4]$$

For $m_1 = m_2 = 0.5$ kg, $h = 0.8$ m, $e = 0.8$ and $\beta = 45°$, we find $v_2^+ = \mathbf{2.37}$ m/s, $v_{1T}^+ = \mathbf{2.80}$ m/s, $v_{1N}^+ = \mathbf{0.56}$ m/s. We find that condition [4] is not satisfied and that $\mathbf{v}_1^+$ points downward at the angle 33.7° relative to the horizontal. ■

6.3 Impulse/Momentum Formulation for Rigid Bodies

We have already obtained impulse/momentum forms of Euler's principles for systems of particles. Consider now a system Σ (of mass m, and mass center G) of one or more rigid bodies in motion relative to a Newtonian referential. To find the impulse/momentum forms of Euler's principles, we again integrate these vector equations over a finite time interval $t_1 \le t \le t_2$ to obtain:

EULER'S PRINCIPLES (IMPULSE/MOMENTUM FORMS)

The (linear) impulse of the external forces imparted to system Σ is equal to the change of the system's linear momentum over some interval of time $t_1 \le t \le t_2$:

$$\underbrace{m\mathbf{v}_G(t_2) - m\mathbf{v}_G(t_1)}_{\text{variation of linear of momentum}} = \underbrace{\int_{t_1}^{t_2} \mathbf{F}_{ext\to\Sigma}\, dt}_{\text{impulse of external forces}} \qquad (6.10)$$

The angular impulse imparted to system Σ by all external actions is equal to the change of the system's angular momentum over some interval of time $t_1 \leq t \leq t_2$:

$$\underbrace{\mathbf{H}_G(t_2) - \mathbf{H}_G(t_1)}_{\text{variation of angular of momentum}} = \underbrace{\int_{t_1}^{t_2} \mathbf{M}_{G,ext\to\Sigma}\, dt}_{\text{angular impulse of external forces/couples}} \tag{6.11}$$

where both angular impulse and angular momentum are taken about the system's mass center G.

PROPERTIES

①. The impulse/momentum formulations (6.10-6.11) are identical to those found in Sections 6.1. Note that the internal contributions to the impulses do not come into play. These equations are generally not very useful since the integrand of the impulses is generally unknown, unless one or all components of the resultant force or moment are zero. In that case, we recover the conservation laws of Sections 3.7 and 4.9.

②. It is not possible to give a version of (6.11) about an arbitrary point A due to the extra term $\mathbf{v}_A \times m\mathbf{v}_G$ present in equation (4.43). This term cannot be integrated in closed form. The exception is the case of a body in rotation about some axis Oz: then it is easy to show that

$$I_{Ozz}(\omega_{\mathcal{B}}(t_2) - \omega_{\mathcal{B}}(t_1)) = \int_{t_1}^{t_2} \hat{\mathbf{k}} \cdot \mathbf{M}_{O,ext\to\mathcal{B}}\, dt$$

with $\omega_{\mathcal{B}} = \boldsymbol{\omega}_{\mathcal{B}} \cdot \hat{\mathbf{k}}$.

③. The equations (6.10-6.11) can be useful in situations whereby a rigid body $\mathcal{B}$ exhibits a sudden change of linear or angular momentum at some time $t = t_0$. The equations are then applied just before t_0 ($t = t_0^-$) and just after t_0 ($t = t_0^+$) to give

$$m\mathbf{v}_G^+ - m\mathbf{v}_G^- = \int_{t_0^-}^{t_0^+} \mathbf{F}_{ext\to\mathcal{B}}\, dt, \qquad \mathbf{H}_G^+ - \mathbf{H}_G^- = \int_{t_0^-}^{t_0^+} \mathbf{M}_{G,ext\to\mathcal{B}}\, dt$$

where $\mathbf{v}_G^+ = \mathbf{v}_G(t_0^+)$, $\mathbf{v}_G^- = \mathbf{v}_G(t_0^-)$, etc. If the impulses can be neglected in view of the fact that the time of integration is vanishingly small ($t_0^+ - t_0^- \to 0$), important relationships between the states of the system at t_0^+ and t_0^- will be obtained.

④. The impulse/momentum formulations of Euler's principles will be useful for the study of systems undergoing impacts, as will be learned in Section 6.4.

The following problems involve application of Euler's principles in one form or another. However, they all have in common a sudden change of linear or angular momentum and require careful application of the impulse/momentum formulation.

Exercise 6.4 ▪▪ A box of side length b, height h, and mass m is dropped without initial speed from a negligible height onto a horizontal conveyor belt moving at constant speed

v_0 so that its bottom face becomes suddenly in contact with the belt. The friction between the box and the belt is so large that the box cannot slip.

a. Show that the box cannot translate along with the conveyor belt but must begin to tip about an edge immediately after contact. Find the angular velocity of the box immediately after contact with the belt. *Hint: apply the first principles in impulse/momentum form during a small interval starting immediately before contact and ending immediately after contact.*

b. How should speed v_0 be chosen to prevent the box to completely tip over? ■

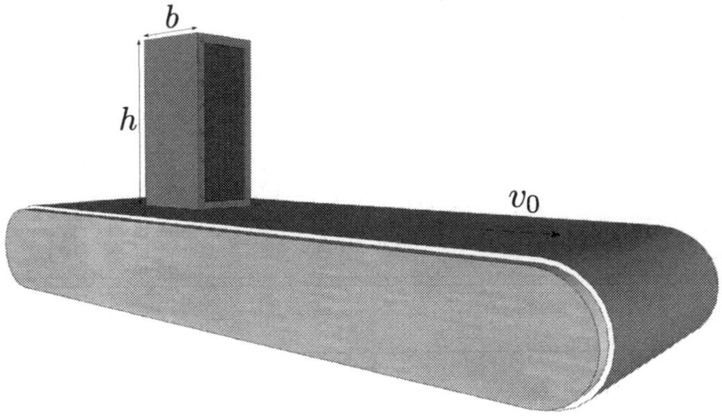

Figure 6.5 – *Problem 6.4*

SOLUTION **a.** First we note that referential moving with the belt, denoted 0, is Newtonian, since its motion is a uniform rectilinear translation. We attach a basis $(\hat{\imath}, \hat{\jmath}, \hat{\mathbf{k}})$ to referential 0 as shown in Figure 6.6. Just before contact (at $t = 0^-$) the box kinematics satisfies $\mathbf{v}_G^- = -v_0\hat{\imath}$ and $\boldsymbol{\omega}^- = \mathbf{0}$. Its linear and angular momenta before contact take the values:

$$\mathbf{L}^- = -mv_0\hat{\imath}, \qquad \mathbf{H}_G^- = \mathbf{0}$$

Assuming that the box neither slip nor tip, the linear and angular momenta of the box immediately after contact (at $t = 0^+$) now take the values

$$\mathbf{L}^+ = \mathbf{0}, \qquad \mathbf{H}_G^+ = \mathbf{0}$$

since $\mathbf{v}_G^+ = \mathbf{0}$ and $\boldsymbol{\omega}^+ = \mathbf{0}$. Now we apply Euler's principles in impulse/momentum form between $t = 0$ and $t = \epsilon$ ($\epsilon \ll 1$):

$$\mathbf{L}^+ - \mathbf{L}^- = \int_0^\epsilon \mathbf{F}_{ext\to 1} dt = mv_0\hat{\imath}, \qquad \mathbf{H}_G^+ - \mathbf{H}_G^- = \int_0^\epsilon \mathbf{M}_{G,ext\to 1} dt = \mathbf{0} \qquad [1-2]$$

Upon substituting $\mathbf{F}_{ext\to 1} = F\hat{\imath} + (N - mg)\hat{\jmath}$ and $\mathbf{M}_{G,ext\to 1} = (\frac{h}{2}F - xN)\hat{\mathbf{k}}$ in [1-2] we find the equation

$$\frac{h}{2}mv_0 = \epsilon xmg$$

Since ϵ is arbitrarily small, this result is *not acceptable. This shows that the box must immediately tip about I with angular velocity* $\boldsymbol{\omega}^+ = \omega^+\hat{\mathbf{k}} \neq \mathbf{0}$: to find ω^+, we apply Euler's Second Principle about I to eliminate the reaction at I

$$\mathbf{H}_I^+ - \mathbf{H}_I^- = \int_0^\epsilon \mathbf{r}_{IG} \times m\mathbf{g}\, dt \to 0 \quad \text{as } \epsilon \to 0$$

giving

$$\mathbf{H}_I^+ - \mathbf{H}_I^- = \mathbf{0} = I_{Izz}\omega^+\hat{\mathbf{k}} - \frac{h}{2}mv_0\hat{\mathbf{k}}$$

with $I_{Izz} = I_{Gzz} + \frac{m}{4}(h^2 + b^2) = \frac{m}{3}(h^2 + b^2)$. This gives ω^+:

$$\boxed{\omega^+ = \frac{3h}{2(h^2 + b^2)} v_0}$$

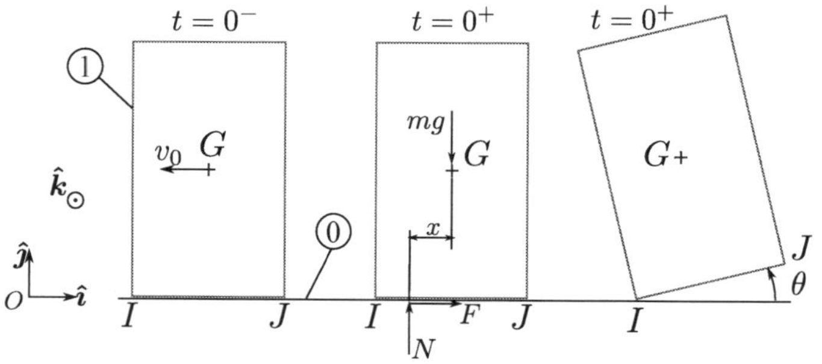

Figure 6.6 – *Problem 6.4 (Solution)*

b. The box will not completely tip over if the mass center G does not pass directly over I. This requires v_0 to not exceed a particular value. To find this value we apply the Work Energy Principle between the state $(\theta = 0, \dot{\theta} = \omega^+)$ and the state $(\theta = \theta_1, \dot{\theta}_1)$ with angle $\tan\theta_1 = \frac{b}{h}$ corresponding to the position of G directly above I. Energy is conserved since the reaction force at I does not work: $K + U = Cst$ gives

$$\frac{m}{6}(h^2 + b^2)\omega^{+^2} + mg\frac{h}{2} = \frac{m}{6}(h^2 + b^2)\dot{\theta}_1^2 + \frac{1}{2}mg\sqrt{h^2 + b^2}$$

By imposing $\dot{\theta}_1^2 < 0$ we guarantee that the box does not reach the critical position $\theta = \theta_1$:

$$\omega^{+^2} < 3g\frac{\sqrt{h^2 + b^2} - h}{h^2 + b^2}$$

or in terms of v_0

$$\boxed{v_0^2 < \frac{4}{3}\frac{h^2 + b^2}{h^2}(\sqrt{h^2 + b^2} - h)g}$$

■

Exercise 6.5 ■■■ A squirrel S of mass m is at rest at the bottom of a wheel 1 of mass M, mass center O and radius of gyration k (about its axis). The wheel is free to rotate about a frictionless pivot of horizontal axis Oz. Suddenly, the squirrel runs up the wheel at constant speed v_0 (relative to the wheel). At any given time t, the position of S (treated as a particle) is defined by its polar coordinates (r, θ), where distance r is a given constant.

a. Find the relationship between $\dot{\theta}$, the angular velocity ω of the wheel, and speed v_0.

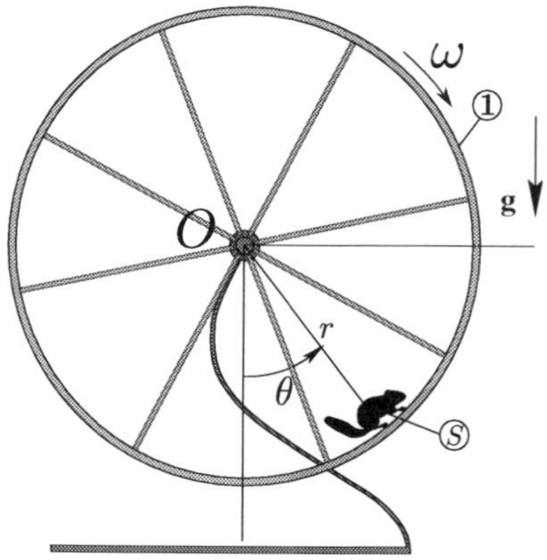

b. Show that the angular acceleration of the wheel takes the expression

$$\boldsymbol{\alpha} = -\frac{mrg\sin\theta}{mr^2 + Mk^2}\hat{\mathbf{k}}$$

c. With $m = 125$ g, $M = 500$ g, $v_0 = 3$ m/s and $r = k = 40$ cm, find out whether the squirrel can reach the top of the wheel ($\theta = \pi$) assuming no loss of contact. ■

SOLUTION **a.** First we define unit vectors $(\hat{\mathbf{u}}, \hat{\mathbf{v}}, \hat{\mathbf{k}})$ corresponding to the motion of S. See Figure 6.7 (left). We can relate the absolute velocity of particle S to its relative velocity according to

$$\mathbf{v}_{S/0} = \mathbf{v}_{S/1} + \mathbf{v}_{S\in 1/0} \qquad [1]$$

Since wheel 1 is in rotation about O with angular velocity $\boldsymbol{\omega}_1 = -\omega\hat{\mathbf{k}}$, the transport velocity of S is found to be $\mathbf{v}_{S\in 1/0} = \boldsymbol{\omega}_1 \times \mathbf{r}_{OS} = -r\omega\hat{\mathbf{v}}$. With $\mathbf{v}_{S/0} = r\dot{\theta}\hat{\mathbf{v}}$ and $\mathbf{v}_{S/1} = v_0\hat{\mathbf{v}}$, equation [1] leads to the relationship between $\dot{\theta}$, ω and v_0:

$$\boxed{r\dot{\theta} = v_0 - r\omega} \qquad [2]$$

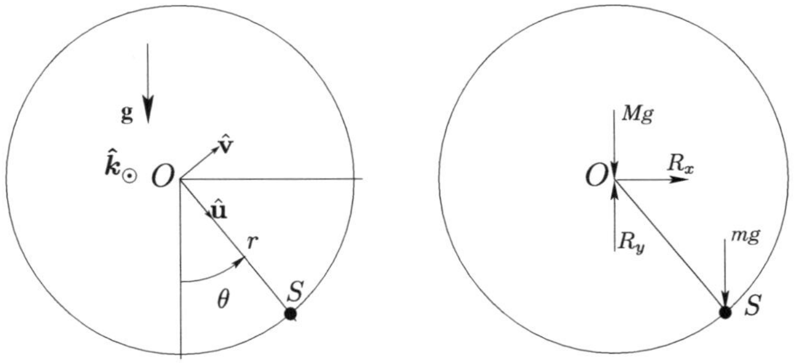

Figure 6.7 – *Problem 6.5 (Solution)*

b. By considering of the system $\{1, S\}$, we eliminate the force of interaction between the wheel and the squirrel. The system's free-body diagram of Figure 6.7 (bottom) shows that the forces acting on the system are the weights of 1 and S and the reaction of the pivot on the wheel. The latter is applied at O since the pivot is frictionless. This leads us to apply Euler's second principle about O to the system (which eliminates the contribution of the reaction $\mathbf{R}$ at O):

$$\frac{d}{dt}(\mathbf{H}_{O,1} + \mathbf{H}_{O,S}) = \mathbf{r}_{OS} \times m\mathbf{g} \qquad [3]$$

We can express the angular momentum of wheel 1 and particle S as follows

$$\mathbf{H}_{O,1} = Mk^2\boldsymbol{\omega}_1 = -Mk^2\omega\hat{\mathbf{k}}, \qquad \mathbf{H}_{O,S} = \mathbf{r}_{OS} \times m\mathbf{v}_S = mr\hat{\mathbf{u}} \times r\dot{\theta}\hat{\mathbf{v}} = mr^2\dot{\theta}\hat{\mathbf{k}}$$

Then equation [3] gives (denoting $\alpha = \dot{\omega}$)

$$-Mk^2\alpha + mr^2\ddot{\theta} = -mrg\sin\theta \qquad [4]$$

Equation [2] can be time-differentiated to give $\ddot{\theta} = -\alpha$. Then equation [4] becomes

$$\boxed{\boldsymbol{\alpha}_1 = -\alpha\hat{\mathbf{k}} = -\frac{mrg\sin\theta}{mr^2 + Mk^2}\hat{\mathbf{k}}} \qquad [5]$$

c. To find out if S can reach the position $\theta = \pi$, we need to find the ODE governing angle θ and integrate it. This equation is given by [5]:

$$\ddot{\theta} = -\frac{mrg}{mr^2 + Mk^2}\sin\theta$$

It is a pendulum equation and a first integral is easily obtained (See Appendix C, Section 3):

$$\frac{\dot{\theta}^2}{2} = \frac{mrg}{mr^2 + Mk^2}\cos\theta + Cst \qquad [6]$$

where Cst is a constant of integration.

Remark. Equation [6] cannot be obtained by conservation of energy! why?

To find the unknown constant in [6], we need the value of $\dot{\theta} = \dot{\theta}_0$ when $\theta = 0$. Note that the quantity $\dot{\theta}$ is discontinuous at $t = 0$: it takes the value $\dot{\theta} = 0$ when $t \to 0^-$ and the value $\dot{\theta} = \dot{\theta}_0$ when $t \to 0^+$. To find $\dot{\theta}_0$, we apply Euler's Second Principle in impulse/momentum form to the system between $t = 0^-$ (just before the start of the motion) and $t = \epsilon$ (just after the start of the motion) and let $\epsilon \to 0$:

$$\Delta(-Mk^2\omega + mr^2\dot{\theta}) = -\int_0^{\epsilon} mrg\sin\theta\, dt$$

In the limit $\epsilon \to 0$ we find

$$-Mk^2\omega_0 + mr^2\dot{\theta}_0 = 0 \qquad [7]$$

since $\omega = \dot{\theta} = 0$ when $t \to 0^-$. Also according to [2] we have $r\dot{\theta}_0 = v_0 - r\omega_0$: [7] now gives

$$\dot{\theta}_0 = \frac{1}{1 + \frac{mr^2}{Mk^2}}\frac{v_0}{r} \qquad [8]$$

Now we can find the constant in [6] by letting $\dot{\theta} = \dot{\theta}_0$ and $\theta = 0$:

$$Cst = \frac{\dot{\theta}_0^2}{2} - \frac{mrg}{Mk^2 + mr^2}$$

To find the angular velocity $\dot{\theta} = \dot{\theta}_1$ of S at position $\theta = \pi$ we apply [6] and find

$$\dot{\theta}_1^2 = \dot{\theta}_0^2 - \frac{4mrg}{Mk^2 + mr^2}$$

We see that S can reach the position $\theta = \pi$ if the condition $\dot{\theta}_0^2 \geq \frac{4mrg}{Mk^2+mr^2}$ is satisfied (otherwise $\dot{\theta}_1^2 < 0$). Using the expression of $\dot{\theta}_0$ from [8] we find the lower bound which speed v_0 must exceed to reach the top position:

$$\boxed{v_0 \geq v_{0,min} = \sqrt{4gr\frac{mr^2}{Mk^2}\left(1 + \frac{mr^2}{Mk^2}\right)}}$$

With $m = 125$ g, $M = 500$ g and $r = k = 40$ cm we find $v_{0,min} =$ **2.21** m/s. With $v_0 = 3$ m/s, *we conclude that the squirrel can reach the top position* $\theta = \pi$. ■

Exercise 6.6 ▪▪▪ A long bar of uniform mass m and square cross-section of side length b is placed in contact with a plane inclined with an angle β with the horizontal. It is displaced from equilibrium and starts to "roll" in a irregular manner down the plane. It is assumed that its faces are slightly concave so that it is only in contact along its horizontal edges. Friction is sufficiently large that the edges of the bar never slip during the motion. Denote by ω^- and ω^+ the bar's angular velocities just before and just after an edge comes in contact with the plane.

a. Find the relationship between ω^+ and ω^-.

b. What is the minimum value K^-_{min} of the kinetic energy before impact, necessary for the bar to come again in contact with the support?

c. If it is observed that the bar "rolls" indefinitely, find the terminal kinetic energy K^∞ of the bar.

d. Using the results of a) and b) find the minimum angle β_{min} above which the rolling motion of the bar will continue indefinitely. ▪

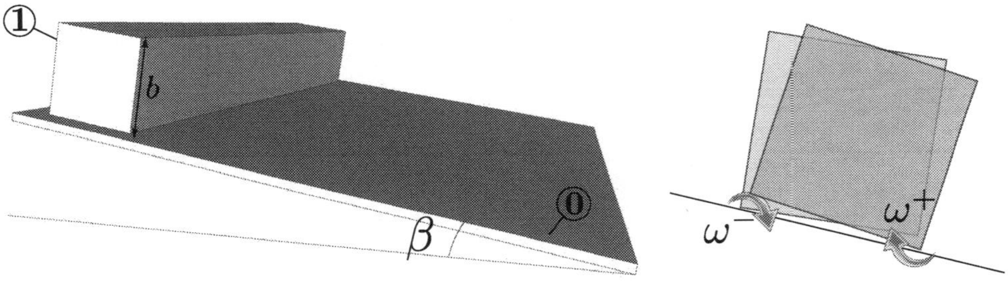

Figure 6.8 – *Problem 6.6*

SOLUTION **a.** Consider the time interval $0 \leq t \leq \epsilon$ ($\epsilon \to 0$) during which the bar rotates about a new edge Iz at angular velocity $\boldsymbol{\omega}^+ = \omega^+\hat{\mathbf{k}}$ and its mass center G has the velocity $\mathbf{v}_G^+ = \omega^+\hat{\mathbf{k}} \times \mathbf{r}_{IG}$. Just before impact ($t \to 0^-$) the bar rotates about edge Jz at angular velocity $\boldsymbol{\omega}^- = \omega^-\hat{\mathbf{k}}$ and its mass center G has the velocity $\mathbf{v}_G^- = \omega^-\hat{\mathbf{k}} \times \mathbf{r}_{JG}$. See Figure 6.9. During the motion, the bar is subject to the reaction $\mathbf{R}_I$ at corner point I and to the effect of gravity. To eliminate the contribution of $\mathbf{R}_I$, we apply Euler's Second Principle about I in momentum/impulse form

$$\mathbf{H}_I^+ - \mathbf{H}_I^- = \int_0^\epsilon \mathbf{r}_{IG} \times m\mathbf{g}\, dt \to 0, \qquad \text{as } \epsilon \to 0$$

In the limit $\epsilon \to 0$ we have equality of angular momentum about I before and after impact $\mathbf{H}_I^+ = \mathbf{H}_I^-$. We know that $\mathbf{H}_G = \frac{m}{6}b^2\omega\hat{\mathbf{k}}$ and we can find the angular momentum about I by using the relationship $\mathbf{H}_I = \mathbf{H}_G + \mathbf{r}_{IG} \times m\mathbf{v}_G$: we find just after impact

$$\mathbf{H}_I^+ = \frac{m}{6}b^2\omega^+\hat{\mathbf{k}} + \frac{m}{2}b^2\omega^+\hat{\mathbf{k}} = \frac{2m}{3}b^2\omega^+\hat{\mathbf{k}}$$

and just before impact

$$\mathbf{H}_I^- = \mathbf{H}_G^- + \mathbf{r}_{IG} \times m\mathbf{v}_G^- = \frac{m}{6}b^2\omega^-\hat{\mathbf{k}}$$

since the vectors $\mathbf{r}_{IG}$ and $\mathbf{v}_G^-$ are colinear. The equality of angular momentum about I

gives

$$\boxed{\omega^+ = \frac{1}{4}\omega^-} \qquad [1]$$

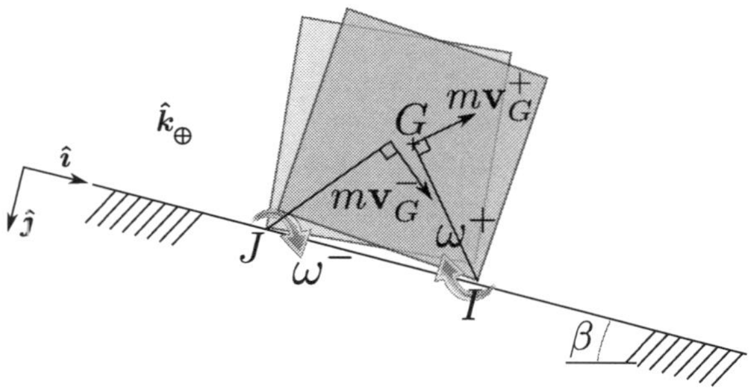

Figure 6.9 – *Problem 6.6 (Solution)*

b. For the bar to continue to "roll" and impact the support along a new edge, the mass center G must be able to reach the position directly above the corner point I. See Figure 6.10 (left). The bar must then rotate by angle $\frac{\pi}{4} - \beta$ about I. During this rotation the mass center G goes up a vertical distance

$$\Delta Z_G = \frac{\sqrt{2}}{2}b(1 - \cos(\frac{\pi}{4} - \beta))$$

To find the minimum kinetic energy, K^+_{min} we apply the Work-Energy Principle during this finite rotation, setting the final kinetic energy to 0 (the reaction $\mathbf{R}_I$ at I does not work, the mechanical energy is conserved)

$$K^+_{min} = mg\frac{\sqrt{2}}{2}b(1 - \cos(\frac{\pi}{4} - \beta)) \qquad [2]$$

Next we need to relate the kinetic energy K^- immediately before the impact at I to the kinetic energy K^+ immediately after the impact at I (the kinetic energies are purely rotational):

$$K^- = \tfrac{1}{2}I_{Jzz}\omega^{-2}, \qquad K^+ = \tfrac{1}{2}I_{Izz}\omega^{+2}$$

Using [1] we find $K^+ = \frac{1}{16}K^-$ since $I_{Jzz} = I_{Izz}$. Then using [2] we find minimum kinetic energy K^-_{min} before impact necessary for the bar to roll

$$\boxed{K^-_{min} = 8\sqrt{2}mgb(1 - \cos(\frac{\pi}{4} - \beta))} \qquad [3]$$

c. Let's apply the Work-Energy Principle during a quarter rotation of the bar, between the state just after the nth impact and the state just before the $(n+1)$th impact: during this rotation the mass center has dropped a vertical distance of $b\sin\beta$ and the bar has lost the potential energy $mgb\sin\beta$

$$K^-_{n+1} - K^+_n = mgb\sin\beta$$

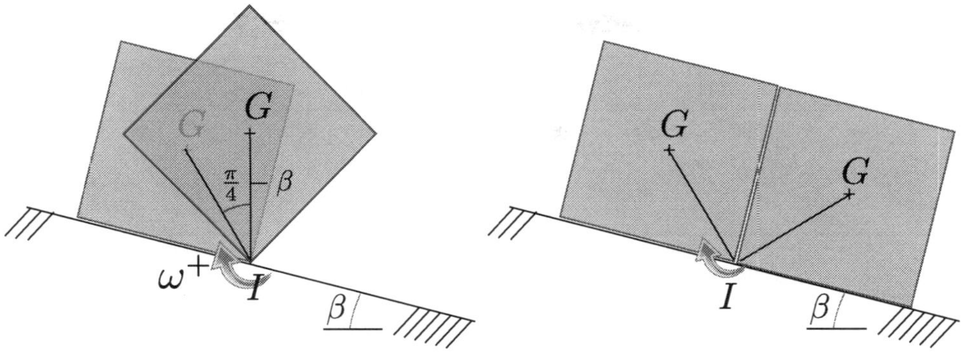

Figure 6.10 – *Problem 6.6 (Solution)*

with $K_n^+ = \frac{1}{16}K_n^-$ according to [1]. This gives

$$K_{n+1}^- = \frac{1}{16}K_n^- + mgb\sin\beta$$

As time increases the kinetic energies K_n^- and K_{n+1}^- approach the common value K^∞:

$$K^\infty = \frac{1}{16}K^\infty + mgb\sin\beta$$

giving

$$\boxed{K^\infty = \frac{16}{15}mgb\sin\beta}$$

d. For the rolling motion to continue indefinitely we need to impose $K^\infty \geq K_{min}^-$:

$$\frac{16}{15}mgb\sin\beta \geq 8\sqrt{2}mgb(1-\cos(\frac{\pi}{4}-\beta))$$

After simplifications, this give the inequality

$$\cos\beta + \frac{17}{15}\sin\beta \geq \sqrt{2}$$

This equality gives the value of β_{min}: it is found numerically to be $\beta_{min} = \mathbf{27.9°}$. If $\theta > 27.9°$ and if the initial kinetic energy is sufficiently large, *the bar will "roll" indefinitely.* ∎

6.4 A Second Theory of Impact

We learned in Section 6.2 that impact between two material systems involves forces of interaction of very large magnitude and acting over very short intervals of time. We identified three phases during an impact:

(i) the initial phase immediately before impact characterized by impact velocities,

(ii) the impact phase characterized by a phase of deformation and a phase of restoration, and

(iii) the final phase immediately after impact characterized by rebound velocities.

In this section, we aim to give a more advance theory of impact. We are not interested in the description of the phase of impact ($t_0^- < t < t_0^+$) during which complex processes take place, but rather in the relationships between the states of the system immediately before ($t = t_0^-$) and immediately after impact ($t = t_0^+$). At these states, the body is assumed to be rigid. A reasonable assumption to make is that the position of a body $\mathcal{B}$ undergoing an impact is unchanged during the very short time interval $t_0^- < t < t_0^+$. However, we expect that its velocity characteristics exhibit discontinuities:

$$\mathbf{v}_G^- \neq \mathbf{v}_G^+, \qquad \boldsymbol{\omega}_{\mathcal{B}}^- \neq \boldsymbol{\omega}_{\mathcal{B}}^+$$

where the $-$ and $+$ superscripts refer to the states of the system immediately before and after impact. We must also distinguish between impulsive and non-impulsive forces during impact.

Definition (Impulsive Force)

A force $\mathbf{F}$ is said to be *impulsive* during an impact of duration $t_0^+ - t_0^-$ if its impulse

$$\mathbf{I}(t_0) = \lim_{t_0^- \to t_0^+} \int_{t_0^-}^{t_0^+} \mathbf{F}\, dt \qquad (6.12)$$

is finite.

Impulsive forces are forces of large magnitude and acting over very short duration. The contribution of non-impulsive forces during impacts is negligible. Impulsive forces acting on a rigid body $\mathcal{B}$ are generally of two types: (i) applied impulsive forces due to obstacles or projectiles, (ii) impulsive forces due to constraints, such as the impulse of a reaction force occurring at a joint constraining $\mathcal{B}$, or due to the sudden tightening of a string attached to $\mathcal{B}$. The latter usually come about in response to impulsive forces applied to constrained rigid bodies.

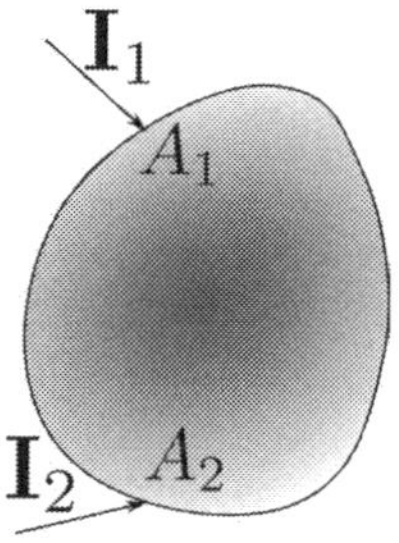

Figure 6.11 – *Body $\mathcal{B}$ acted upon by impulses $\mathbf{I}_1$ at A_1, $\mathbf{I}_2$ at A_2, ... at time t_0.*

The equations governing the impact of a rigid body at time t_0 are found by application of the impulse/momentum forms of Euler's principles during the duration ($t_0^- < t < t_0^+$) of the impact. Assume that body $\mathcal{B}$ is subject to the impulsive forces $\mathbf{F}_1$ at A_1 of impulse $\mathbf{I}_1$, $\mathbf{F}_2$ at A_2 of impulse $\mathbf{I}_2$, ... as shown in Figure 6.11. Then, the impulse/momentum form of

Euler's First Principle gives

$$\underbrace{\int_{t_0^-}^{t_0^+} m\mathbf{a}_G \, dt}_{m\mathbf{v}_G^+ - m\mathbf{v}_G^-} = \int_{t_0^-}^{t_0^+} \mathbf{F}_{ext\to\mathcal{B}} \, dt = \mathbf{I}_1(t_0) + \mathbf{I}_2(t_0) + \cdots$$

where only the impulsive forces are accounted for. Similarly the impulse/momentum form of Euler's Second Principle taken about an *arbitrary* point A gives:

$$\mathbf{H}_A^+ - \mathbf{H}_A^- + \int_{t_0^-}^{t_0^+} \mathbf{v}_A \times m\mathbf{v}_G \, dt = \int_{t_0^-}^{t_0^+} \mathbf{r}_{AA_1} \times \mathbf{F}_1 dt + \int_{t_0^-}^{t_0^+} \mathbf{r}_{AA_2} \times \mathbf{F}_2 dt + \cdots$$

Taking into account the fact that the positions of the body and of point A are unchanged during the impact, two simplifications can be made in the limit $t_0^- \to t_0^+$:

- the integral $\int_{t_0^-}^{t_0^+} \mathbf{v}_A \times m\mathbf{v}_G \, dt$ must vanish,
- the impulsive moment $\int_{t_0^-}^{t_0^+} \mathbf{r}_{AA_1} \times \mathbf{F}_1 \, dt = \mathbf{r}_{AA_1} \times \int_{t_0^-}^{t_0^+} \mathbf{F}_1 \, dt$ is equal to the moment of the impulse: $\mathbf{r}_{AA_1} \times \mathbf{I}_1$.

The leads to the following result.

Impact Equations for a Rigid Body

During an impact at time t_0, a rigid body $\mathcal{B}$ of mass m and mass center G and acted upon by external impulsive forces of impulses $\mathbf{I}_k$ at point A_k, $k = 1, 2, \ldots$ exhibits a change of linear momentum and angular momentum given by

$$m\mathbf{v}_G^+ - m\mathbf{v}_G^- = \mathbf{I}_1(t_0) + \mathbf{I}_2(t_0) + \cdots \tag{6.13}$$

and

$$\mathbf{H}_A^+ - \mathbf{H}_A^- = \mathbf{r}_{AA_1} \times \mathbf{I}_1(t_0) + \mathbf{r}_{AA_2} \times \mathbf{I}_2(t_0) + \cdots \tag{6.14}$$

where A is an arbitrary point.

These equations are generally not sufficient to solve for the kinematic unknowns $(\mathbf{v}_G^+, \omega^+)$ after impact from the knowledge of the kinematic data $(\mathbf{v}_G^-, \omega^-)$ before impact. Consider the planar motion of a body $\mathcal{B}$ impacting a surface $\mathcal{S}$ at a point A, denoting $(\hat{\boldsymbol{\tau}}, \hat{\mathbf{n}})$ the tangential/normal unit vectors at A as shown in Figure 6.12.

Given $(\mathbf{v}_G^-, \omega^-)$, the problem involves five unknowns consisting of the components of $\mathbf{v}_G^+$, ω^+ and of the impulse $\mathbf{I}_A$ at A. It cannot be fully solved by the 3 equations found by applying (6.13-6.14). One additional equation can be written by applying Coulomb laws:

Case 1: *if friction at A is neglected*, the tangential component of the impulse at A vanishes:

$$\mathbf{I}_A \cdot \hat{\boldsymbol{\tau}} = 0$$

Case 2: *if friction characterized by the coefficients of static and kinetic friction (μ_s, μ_k) is taken into account*, then either the body slips relative to surface $\mathcal{S}$ and

$$|\mathbf{I}_A \cdot \hat{\boldsymbol{\tau}}| = \mu_k |\mathbf{I}_A \cdot \hat{\mathbf{n}}| \qquad (\mathbf{v}_{A\in\mathcal{B}/\mathcal{S}}^+ \cdot \hat{\boldsymbol{\tau}} \neq 0)$$

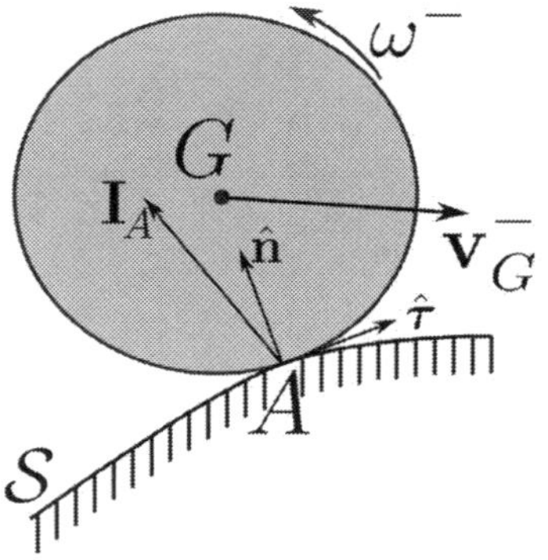

Figure 6.12 – *Body impacting surface $\mathcal{S}$: $(\hat{\boldsymbol{\tau}}, \hat{\mathbf{n}})$ are the unit tangential and normal vectors at the contact point A.*

or the body does not slip immediately after impact

$$\mathbf{v}^+_{A\in\mathcal{B}/\mathcal{S}} \cdot \hat{\boldsymbol{\tau}} = 0 \quad \text{and} \quad |\mathbf{I}_A \cdot \hat{\boldsymbol{\tau}}| < \mu_s |\mathbf{I}_A \cdot \hat{\mathbf{n}}|$$

One equation is still missing: it can be provided by Newton's Law of Impact.

DEFINITION (NEWTON'S LAW OF IMPACT)

The ratio of the normal relative velocity between the two points of $\mathcal{B}$ and $\mathcal{S}$ coming in contact at A immediately after impact to that immediately before impact is a constant known as the *coefficient of restitution*:

$$\mathbf{v}^+_{A\in\mathcal{B}/\mathcal{S}} \cdot \hat{\mathbf{n}} = -e\,\mathbf{v}^-_{A\in\mathcal{B}/\mathcal{S}} \cdot \hat{\mathbf{n}} \tag{6.15}$$

where $0 \le e \le 1$ is only a function of the nature of the materials in contact.

APPLICATION: CENTER OF PERCUSSION OF A ROTATING RIGID BODY. Consider a rigid body $\mathcal{B}$ of mass m and mass center G supported by a pivot connection of axis Oz. See Figure 6.13. The body is at rest when it is suddenly subject to an impulse $\mathbf{I}_A$ applied to a point A in the plane of motion of G. Since point O must remain fixed after impact, the impulse $\mathbf{I}_O$ of the reaction force at O cannot be neglected: hence, the reaction at O is an impulsive force. Large values of $\mathbf{I}_O$ can be destructive to the pivot. Hence we ask if it is possible to make impulse $\mathbf{I}_O$ vanish by a proper location of the line of action of the applied impulse $\mathbf{I}_A$.

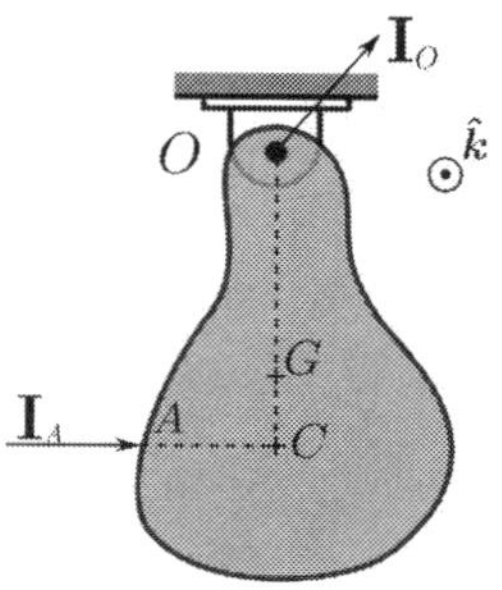

Figure 6.13 – *Center of percussion of body $\mathcal{B}$.*

Definition (Center of Percussion)

We call *center of percussion* the point C of line OG through which the line of action of impulse $\mathbf{I}_A$ must pass in order to cancel out the impulse of the reaction at O.

To determine the location of C we apply equations (6.13-6.14) choosing O as the point about which moments are taken:

$$m\mathbf{v}_G^+ - m\mathbf{v}_G^- = m\mathbf{v}_G^+ = \mathbf{I}_A + \mathbf{I}_O$$

and

$$\mathbf{H}_O^+ - \mathbf{H}_O^- = \mathbf{H}_O^+ = \mathbf{r}_{OA} \times \mathbf{I}_A$$

taking into account the conditions $\mathbf{v}_G^- = \mathbf{0}$ and $\boldsymbol{\omega}^- = \mathbf{0}$ before impact. After impact we can write $\mathbf{v}_G^+ = \boldsymbol{\omega}^+ \times \mathbf{r}_{OG}$ and $\mathbf{H}_O^+ \cdot \hat{\mathbf{k}} = J_O \omega^+$ denoting by J_O the moment of inertia of the body about axis Oz. This leads to the equation

$$m(\mathbf{r}_{OA} \cdot \mathbf{r}_{OG} - J_O)\boldsymbol{\omega}^+ = \mathbf{r}_{OA} \times \mathbf{I}_O$$

We can then see that, if the location of A satisfies the condition $\mathbf{r}_{OA} \cdot \mathbf{r}_{OG} - J_O = 0$, then the impulse at O vanishes. We conclude with the following result:

Center of Percussion

The center of percussion C of a body of mass m, mass center G and radius of gyration k_O about axis Oz is given by

$$|OC||OG| = \frac{J_O}{m} = k_O^2 \tag{6.16}$$

Properties

①. Using symmetry arguments, point O can be considered to be the center of percussion of the body in rotation about axis Cz passing through the center of percussion.

②. The center of percussion of "point" mass pendulum is located at the point itself since $k_O = |OG|$.

③. If an object such as a hammer, a baseball bat or a tennis racket is used to strike another object, it should be held at a point O a distance $|OC|$ (given by (6.16)) away from the point of impact C so as to minimize the reaction force on the operator's wrist.

EXAMPLE PROBLEMS

General Procedure: Impact Problems (rigid bodies)

Problems involving rigid body impact typically amount to finding the kinematic unknowns $(\mathbf{v}_G^+, \boldsymbol{\omega}^+)$ immediately after impact, given $(\mathbf{v}_G^-, \boldsymbol{\omega}^-)$ immediately before impact. After carefully identifying the kinematic quantities $(\mathbf{v}_G^-, \boldsymbol{\omega}^-)$ and the impulsive forces at impact, the problem is solved by application of (i) the impact equations (Euler's principles in impulse/momentum form), (ii) Newton's Impact Law (given a coefficient of restitution), and (iii) Coulomb friction laws (given coefficients of friction).

Exercise 6.7 ▪▪ A billiard ball of mass m, mass center G, and radius r is at rest on a horizontal support when it is subject to a impulse $\mathbf{I}_A$ directed horizontally at a height h from the support and contained in the plane of motion of G.

Analyze the motion of the ball immediately after impact. Describe the motion of the ball after impact. ▪

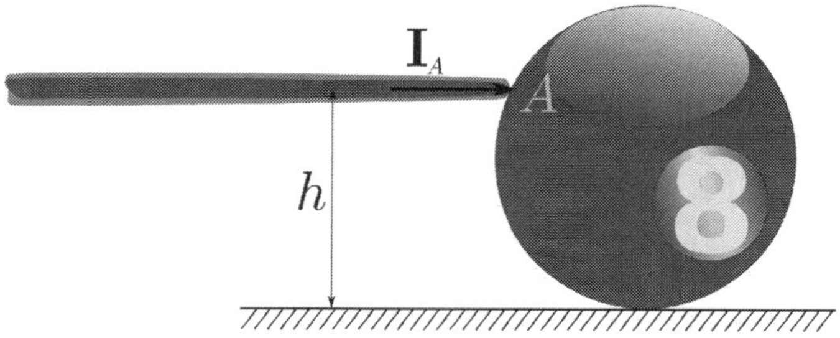

Figure 6.14 – *Problem 6.7*

Solution. During the impact, the ball is subject to two impulses: the applied impulse $\mathbf{I}_A$ at A and the impulse $\mathbf{I}_J$ of the reaction force at contact point J (see Figure 6.15). Application of equations (6.13-6.14) gives (choosing G as the point about which moments are taken) on $(\hat{\imath}, \hat{\jmath}, \hat{\mathbf{k}})$:

$$mv_G^+ = I_A + \mathbf{I}_J \cdot \hat{\imath}, \qquad 0 = \mathbf{I}_J \cdot \hat{\jmath}, \qquad \frac{2}{5} mr^2 \omega^+ = (h - r) I_A \qquad [1-3]$$

taking into account $\mathbf{v}_G^- = \mathbf{0}$, $\boldsymbol{\omega}^- = \mathbf{0}$ and writing $\mathbf{v}_G^+ = v_G^+ \hat{\imath}$, $\boldsymbol{\omega}^+ = \omega^+ \hat{\mathbf{k}}$. Equation [2] necessarily implies $\mathbf{I}_J \cdot \hat{\imath} = 0$ according to Coulomb laws (the ball necessarily slips), leading to $\mathbf{I}_J = \mathbf{0}$: *the reaction force at J is non-impulsive*. Then substitution of $I_A = mv_G^+$ in [3] gives

$$\boxed{r\omega^+ = \frac{5}{2} \frac{h - r}{r} v_G^+} \qquad [4]$$

and the corresponding slip velocity immediately after impact:

$$\boxed{\mathbf{v}_{J \in 1/0}^+ = \mathbf{v}_G^+ + \boldsymbol{\omega}^+ \times \mathbf{r}_{GJ} = \frac{7r - 5h}{2r} v_G^+ \hat{\imath}} \qquad [5]$$

The results given by equations [4-5] lead to the following conclusions summarized in Figure 6.16:

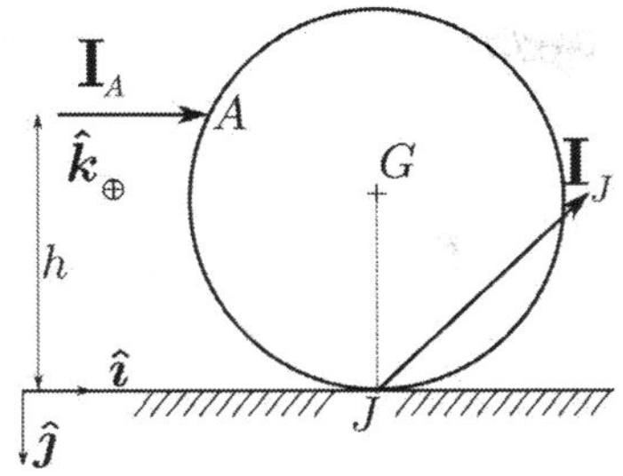

Figure 6.15 – *Problem 6.7*

- if $h > \frac{7}{5}r$, then the slip velocity is directed in the $(-\hat{\imath})$ direction, and the friction force will increase the speed of G while decreasing the clockwise rotational speed. Eventually the ball will stop slipping.
- if $h = \frac{7}{5}r$, the ball rolls without slipping: the ball will continue to roll without slipping at the constant speed v_G^+ and constant angular speed ω^+ in the clockwise direction.
- if $h < \frac{7}{5}r$, the slip velocity is directed in the $(+\hat{\imath})$ direction, the friction force will decrease the speed of G. Two additional cases are found: (1) if $h > r$, the friction force increases the clockwise angular speed, (2) if $h < r$, the ball rolls forward with backspin: the friction force decreases the counterclockwise angular speed. Eventually the ball will stop slipping. ■

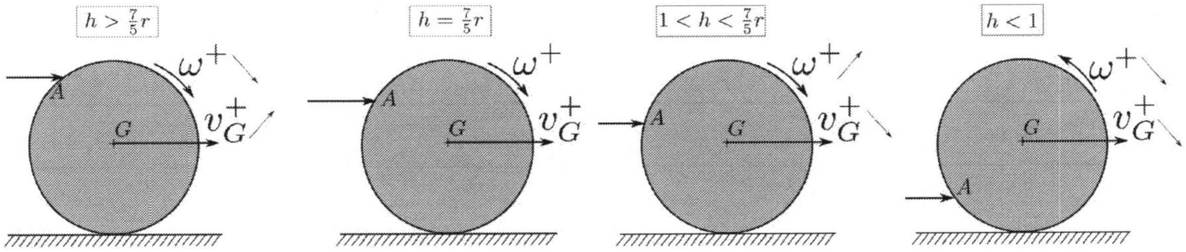

Figure 6.16 – *Problem 6.7 (Solution)*

Exercise 6.8 ▪ A ballistic pendulum 1 used to measure the speed of a bullet 2 is supported by a frictionless pivot of horizontal axis Oz. Its mass is M, mass center G ($|OG| = L$), and radius of gyration k_O about O. The bullet of mass $m \ll M$ and velocity v_0 strikes the pendulum at a distance l from O, and becomes embedded in the pendulum's soft material.

Find the maximum amplitude θ_1 of oscillations of the pendulum as a function of v_0. Assume that angle θ_1 is small. ■

Solution. First, we determine the angular velocity of the pendulum immediately after impact: to eliminate the impulse imparted by the bullet to the pendulum, we consider the system $\Sigma = \{1,2\}$ at the time t_0 of impact. The external forces acting on Σ are the gravitational forces and the contact forces acting on the pivot connection. If we apply Euler's Second Principle (in impulse/momentum form) about O, we eliminate the moment of the impulse of the reaction at O. The impulse of force $m\mathbf{g}$ is negligible and

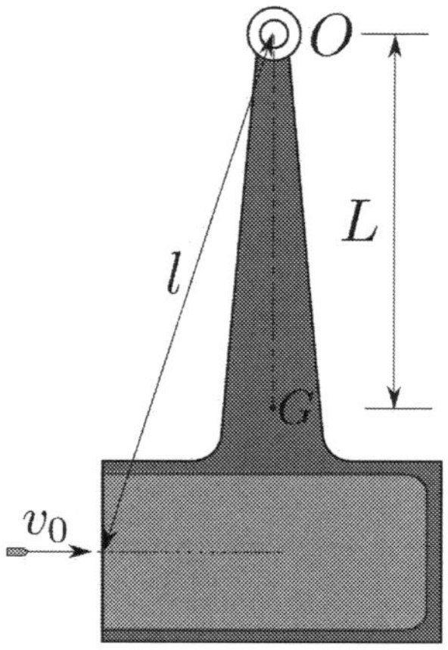

Figure 6.17 – *Problem 6.8*

that of $M\mathbf{g}$ has a zero moment at time t_0. We are left with

$$(\mathbf{H}^+_{O,\Sigma} - \mathbf{H}^-_{O,\Sigma}) \cdot \hat{\mathbf{k}} = \int_{t_0^-}^{t_0^+} \hat{\mathbf{k}} \cdot \mathbf{M}_{O,ext\to\Sigma} dt = 0$$

The system's angular momentum about axis Oz is conserved: $\mathbf{H}^+_{O,1} \cdot \hat{\mathbf{k}} = \mathbf{H}^-_{O,2} \cdot \hat{\mathbf{k}}$. This gives the initial angular velocity $\dot{\theta}^+$ at time t_0:

$$\boxed{Mk_0^2\dot{\theta}^+ = mlv_0} \qquad [1]$$

Then, to find the maximum amplitude θ_1 at time t_1 (corresponding to the value

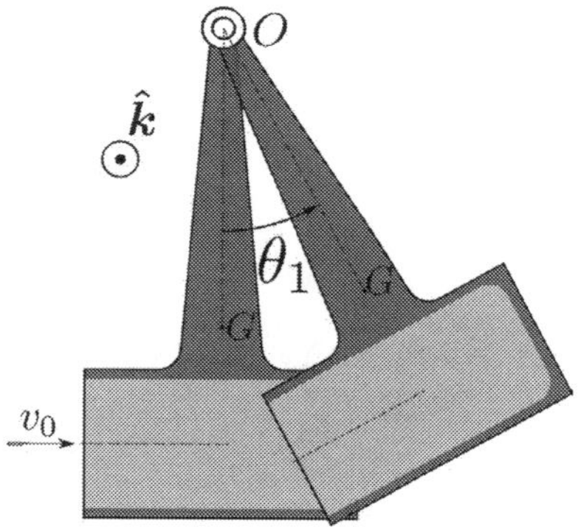

Figure 6.18 – *Problem 6.8*

$\dot{\theta}_1 = 0$), we apply the Work-Energy Principle to system Σ between t_0^+ and t_1:

$$K_\Sigma(t_1) - K_\Sigma(t_0) = W^{ext}_{0\to 1} = -\Delta U = -MgL(1 - \cos\theta_1)$$

where we have neglected the contribution of the bullet's potential energy. The kinetic energy of Σ is entirely rotational:

$$K_\Sigma(t_0) = \tfrac{1}{2}Mk_0^2\dot{\theta}^{+^2}, \qquad K_\Sigma(t_1) = 0$$

neglecting the bullet's mass. This gives

$$\tfrac{1}{2}Mk_0^2\Big(\frac{mlv_0}{Mk_0^2}\Big)^2 = MgL(1-\cos\theta_1) \approx \tfrac{1}{2}MgL\theta_1^2$$

after using equation [1] for $\dot{\theta}^+$, and the small angle approximation $1-\cos\theta \approx \frac{1}{2}\theta^2$. This gives the following approximation of the maximum amplitude

$$\boxed{\theta_1 = \frac{m}{M}\frac{l}{k_0}\sqrt{\frac{v_0^2}{gL}}}$$

■

Exercise 6.9 A ball of mass m, radius r, mass center G, and radius of gyration k_G about one of its diameters, is thrown with a horizontal speed v_0 and counterclockwise angular speed ω_0. Its initial position is located at a height H from a horizontal rigid obstacle.
Assuming the contact is characterized by a coefficient of restitution e and that the ball does not slip during the impact, find the condition for the ball to bounce backward.

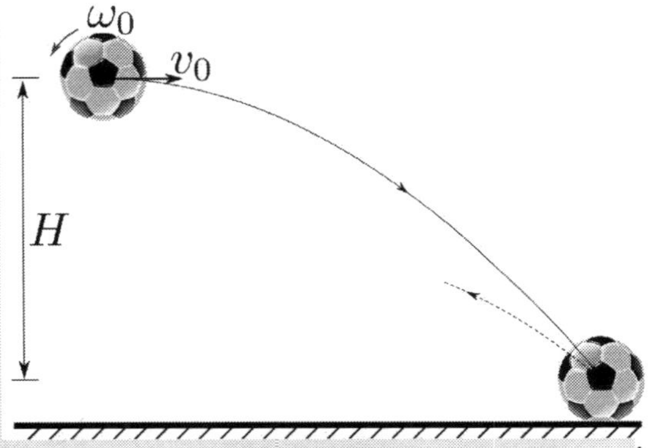

Determine the condition satisfied by the coefficient of static friction μ which guarantees that the ball bounces without slipping. ■

SOLUTION. Step 1: First we need to find the kinematic unknowns $\mathbf{v}_G^-$ and ω^- reached by the ball just before impact given the initial conditions (v_0, ω_0). We adopt a set of unit vectors $(\hat{\imath}, \hat{\jmath}, \hat{\mathbf{k}})$ as shown in Figure 6.19. The velocity of the mass center and the angular velocity are written in the following way:

$$\mathbf{v}_G = \dot{x}\hat{\imath} + \dot{y}\hat{\jmath}, \qquad \boldsymbol{\omega} = \omega\hat{\mathbf{k}}$$

with $\dot{x} = v_0 > 0$, $\dot{y} = 0$ and $\omega = \omega_0 > 0$ at time $t = 0$. During the phase of free-fall, the ball is only subject to gravity and its motion is governed by the equations

$$\ddot{x} = 0, \qquad \ddot{y} = -g, \qquad mk_G^2\dot{\omega} = 0$$

Integration gives

$$\dot{x} = v_0, \qquad \dot{y} = -gt, \qquad \omega = \omega_0 \qquad [1]$$

Mass center G describes the parabola given by

$$x = v_0 t, \qquad y = H - \tfrac{1}{2}gt^2$$

assuming that $x = 0$ and $y = H$ at time $t = 0$. At the time $t = t_0$ of impact we have $y = 0$ giving $t_0 = \sqrt{\frac{2H}{g}}$. From [1] we obtain the conditions reached by the ball just before impact:

$$\dot{x}^- = v_0, \qquad \dot{y}^- = \sqrt{2gH}, \qquad \omega^- = \omega_0 \qquad [2]$$

Step 2: given $(\dot{x}^-, \dot{y}^-, \omega^-)$ we need to find the kinematic quantities $(\dot{x}^+, \dot{y}^+, \omega^+)$

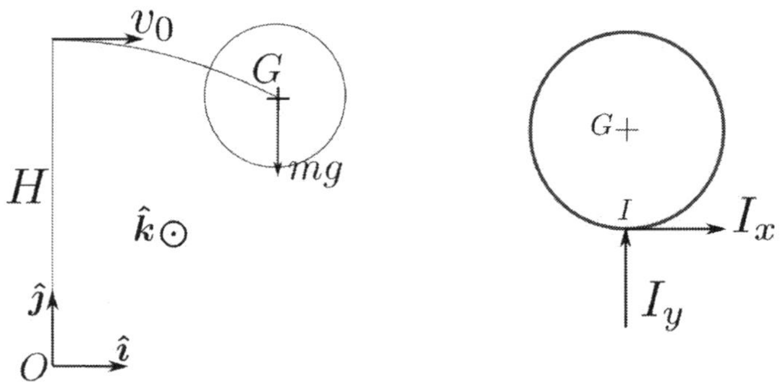

Figure 6.19 – *Problem 6.9 (Solution)*

immediately after impact. During impact, the ball is subject to the impulse $\mathbf{I} = I_x\hat{\imath} + I_y\hat{\jmath}$ from the obstacle acting at the point of contact. See Figure 6.19. We then apply the impact equations (6.13-6.14) on $(\hat{\imath}, \hat{\jmath}, \hat{\mathbf{k}})$ to find (choosing G as the point about which moments are taken):

$$m(\dot{x}^+ - \dot{x}^-) = I_x, \qquad m(\dot{y}^+ - \dot{y}^-) = I_y, \qquad mk_G^2(\omega^+ - \omega^-) = rI_x \qquad [3-5]$$

This gives 3 equations for 5 unknowns $(\dot{x}^+, \dot{y}^+, \omega^+, I_x, I_y)$. Two additional equations are needed. First we use the knowledge of the coefficient of restitution e: according to Newton's Law of Impact (6.15) the normal components of $\mathbf{v}_{I\in Ball}$ before and after impact satisfy

$$\dot{y}^+ = -e\dot{y}^- \qquad [6]$$

A second equation is provided by imposing the no-slip condition immediately after impact:

$$\dot{x}^+ + r\omega^+ = 0 \qquad [7]$$

We can now solve equations [3-7] to find

$$\dot{x}^+ = \frac{r}{r^2 + k_G^2}(rv_0 - k_G^2\omega_0), \qquad \dot{y}^+ = -e\dot{y}^-, \qquad \omega^+ = \frac{k_G^2\omega_0 - rv_0}{r^2 + k_G^2} \qquad [8-11]$$

$$I_x = -m\frac{k_G^2}{r^2 + k_G^2}(r\omega_0 + v_0), \qquad I_y = -m(e+1)\dot{y}^- \qquad [12-13]$$

Step 3: we need to impose $\dot{x}^+ < 0$ to guarantee that the ball bounces backward. Using equation [8] we find the condition

$$\boxed{rv_0 < k_G^2\omega_0}$$

Equation [7] of no-slip at t_0^+ is only valid if the impulse satisfies the condition $|I_x| < \mu I_y$. Using the expressions of [12-13] we find

$$\boxed{\frac{k_G^2}{r^2 + k_G^2}(r\omega_0 + v_0) < \mu(e+1)\sqrt{2gH}}$$

after using the expression of $\dot{y}^-$ given by [2]. ∎

6.5 Problems

Problem 6.1 Consider two pendulums of identical length l. Sphere P_1 is of mass m_1. Sphere P_2 is of mass m_2. The two spheres are initially at rest in the vertical position and in contact at a point O. Sphere P_1 is then displaced, the pendulum making a small angle θ_0 with the vertical, and released without initial speed. The impact is elastic. *(use* $\sin\theta_0 \approx \theta_0$ *and* $(1-\cos\theta_0) \approx \frac{1}{2}\theta_0^2$*).*

a. Find the velocities of P_1 and P_2 immediately after impact.

b. Find the maximum angles reached by the pendulums after the first impact.

c. Does the second impact take place at O?

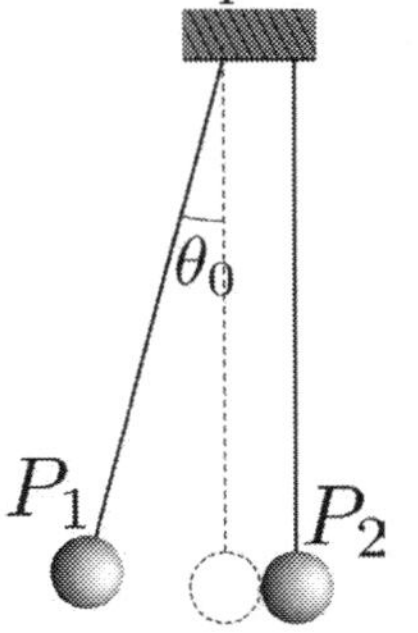

Problem 6.2 A small body P of mass m is tied to an inelastic string of length l, attached to point O at a distance h from O. The body is dropped without initial speed. Find the velocity of P just before and after the string becomes taut (neglect the impulse of the gravitational force). At which distance from the support will P climb, assuming that the string does not break?

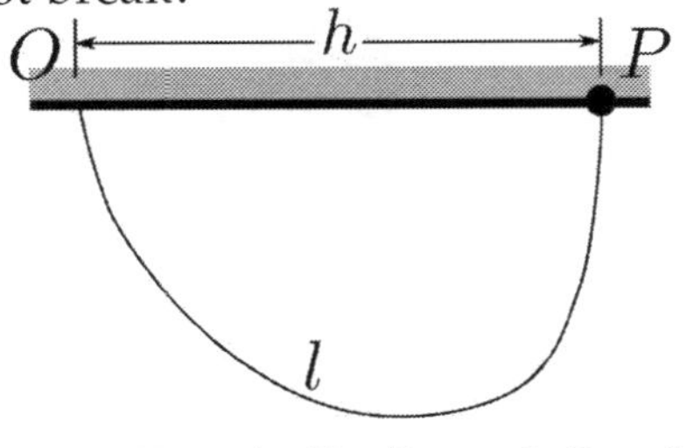

Problem 6.3 Two balls P_1 and P_2 of identical size and mass m_1 and $m_2 \ll m_1$, respectively, are dropped from the same height h, P_1 slightly directly below P_2. Ball P_1 bounces off a horizontal support a fraction of an instant before P_2 and rebounds on ball P_2.

Find the height H reached by P_2 assuming all impacts elastic.

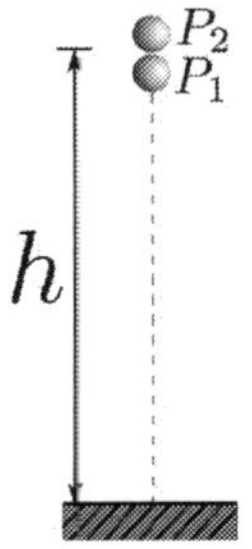

Answer: $H = 9h$

Problem 6.4 An elastic rope of negligible mass, unstretched length l_0 is attached at a point O. A small object P_1 of mass m_1 is attached at the other end. The length of the rope becomes $2l_0$ at rest. A ring-like object P_2 of mass m_2 is released from O without initial speed. It slides freely down the rope and eventually hits P_1. The impacts are assumed elastic.

a. Find the mass ratio m_2/m_1 needed to allow the second impact between P_1 and P_2 to take place at the same location as that of the first impact.

b. With this value of m_2/m_1, find the maximum length reached by the rope and the shortest distance between O and P_2.

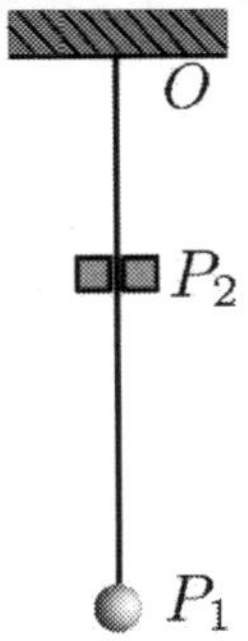

Problem 6.5 A billiard ball P_1 of mass m and initial velocity $v_1^- = 1.2$ m/s collides with another ball P_2 initially at rest, and of mass $2m$. The directions of the velocities of the balls after impact are known.

a. Find the speeds v_1^+ and v_2^+ of P_1 and P_2 after impact.

b. Was the impact elastic?

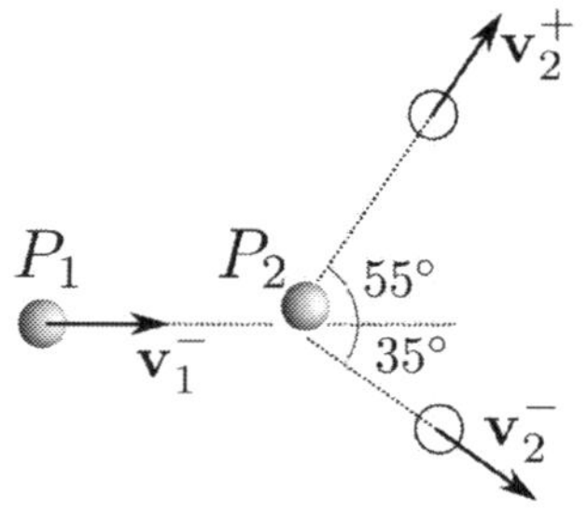

Problem 6.6 A uniform rod of mass M, and length l can rotate in a vertical plane about a frictionless pivot of axis Oz. It is initially at rest in the position $\theta = 0$ when it is hit by a bullet of mass m and traveling horizontally at speed v_0. The bullet becomes instantly embedded into the rod at end point Q.

a. Find the initial angular velocity $\dot{\theta}_0$ of the rod. Show that there is a loss of kinetic energy of the system {rod, bullet} between $t = 0^-$ (just before impact) and $t = t^+$ (just after impact).

b. Find the condition satisfied by v_0 for the rod to reach a maximum amplitude $\theta = \theta_1$. Then find angle θ_1. Describe the motion of the rod if this condition is not satisfied.

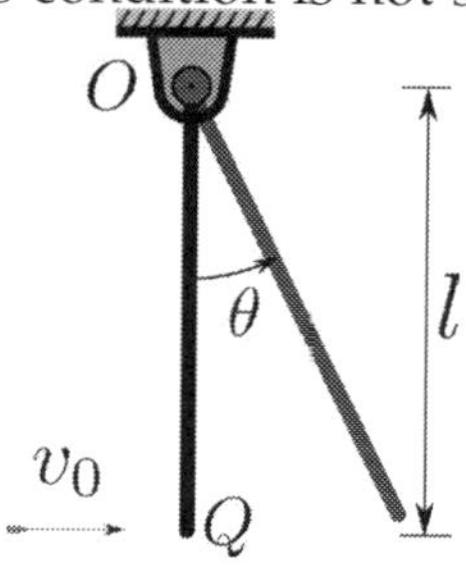

Problem 6.7 A crate of mass m, height h, width b, and mass center G slides down a plane inclined at angle β with the horizontal. It reaches the speed v_0 when it hits a step-like obstacle at O.

Find the maximum value which speed v_0 must not exceed for the crate to not completely tip over.

Treat the crate as a uniform cuboid and neglect the height of the obstacle.

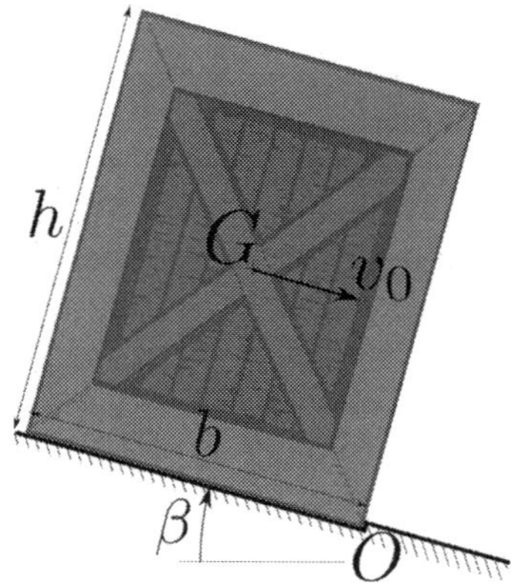

Problem 6.8 A uniform slender rod of length l and mass m is at rest on a smooth horizontal plane. It is then subject to an impulsive force of impulse $\mathbf{I}$ at a point A located a distance $\frac{l}{4}$ away from the closest end point.

Find the instantaneous center of rotation of the rod immediately after impact.

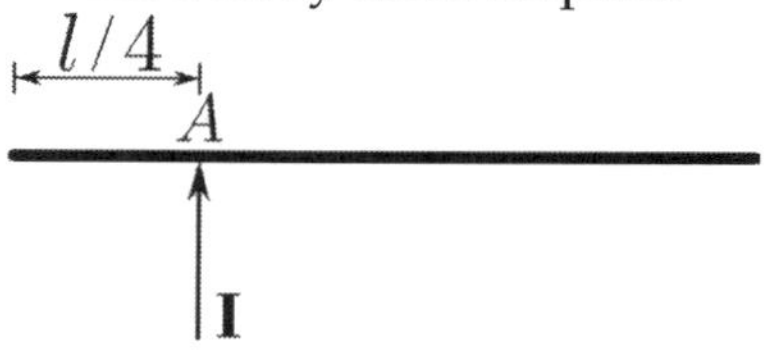

Problem 6.9 A uniform disk of mass m and radius r rolls without slipping on a horizontal plane at speed v_0. It then hits a step-like obstacle of height h at the corner point Q. The friction is large enough to prevent the disk to slip at Q.

Find the minimum speed v_{0min} for the disk to "roll" over the step.

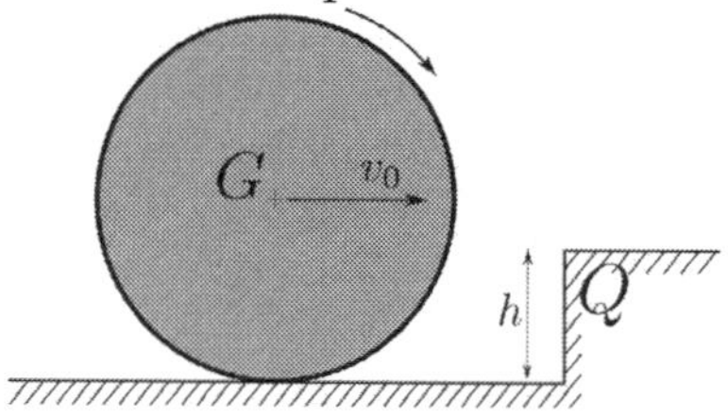

Answer: $v_{0min}^2 = 3gh/(3/2 - h/r)$

Problem 6.10 A slender rod of mass m, mass center G, and radius of gyration k about an axis Gz is in motion in a vertical plane. At time t_0 the rod's angular velocity ω^- and the position of its instantaneous center of rotation I are known. At $t = t_0$, the rod hits an obstacle at a point A located at a distance x from G. The contact is assumed frictionless and perfectly inelastic ($e = 0$). Point I is assumed to be located along the direction of the rod at a distance l from G.

a. Find the impulse $\mathbf{I}$ at A exerted by the obstacle, the velocity $\mathbf{v}_G^+$ and the angular velocity ω^+ of the rod immediately after impact.

b. Show that there exists a value x^* of x for which $\mathbf{v}_G^+ = \mathbf{0}$ and $\omega^+ = \mathbf{0}$. Denote by A^* the corresponding location of A (center of percussion).

c. Show that there exist two values x_1 and x_2 (and corresponding points A_1 and A_2) for which the impulse $\mathbf{I}$ reaches maximal magnitude.

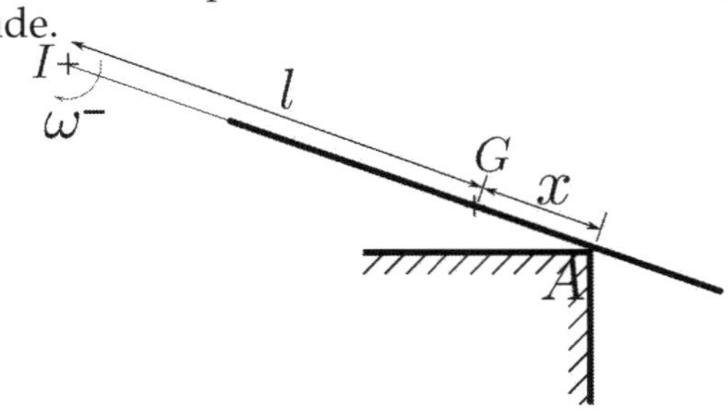

Problem 6.11 A slender uniform rod of mass m, length L, and mass center G is suspended to a horizontal support at O by a massless, inextensible string of length l. Initially the rod is placed in contact with the support, with the two attachment points O and A coinciding and with the string slack. The rod is then released without initial speed. The rod drops until the string tautens. The drop is so short that the orientation of the rod has not sensibly changed. Hence, the rod is subject to an impact-like process. The impact is assumed elastic.

a. Find the velocity of mass center G and the rod's angular velocity immediately after impact.

b. Find the corresponding impulse $\mathbf{I}$ imparted by the string on the rod.

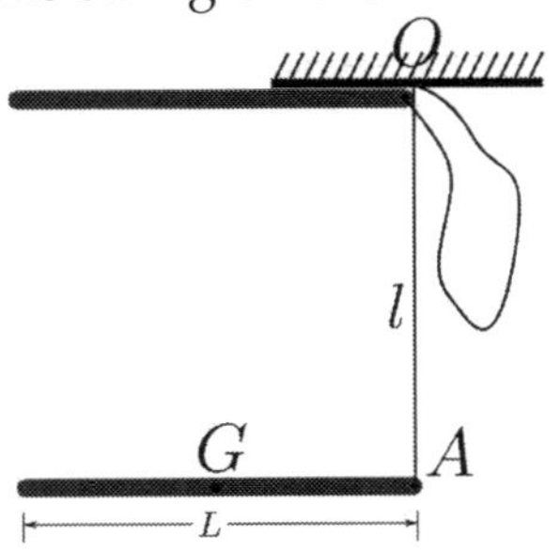

Problem 6.12 A bat 1 of uniform mass m_1 is in motion over a smooth surface at constant angular velocity ω_0 and constant centroidal velocity $\mathbf{v}_G = \mathbf{v}_0$. At the instant its axis is perpendicular to velocity $\mathbf{v}_0$, the bat hits a small spherical body 2 of mass m_2 at rest on the surface. The impact is assumed elastic. Denote k_G the radius of gyration of the bat about G.

Find the mass ratio m_2/m_1 which allows the bat to become stationary immediately after impact, while the ball takes a rectilinear path.

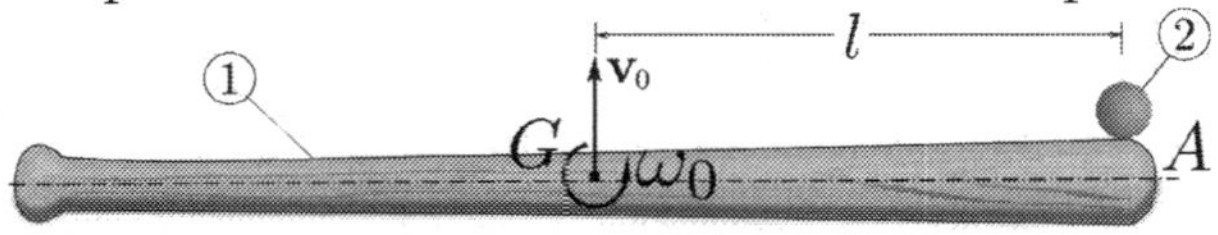

Problem 6.13 A slender rod 1 of uniform mass m, and length $2l$ slides on a smooth surface at constant angular velocity ω_0 and constant centroidal velocity $\mathbf{v}_G = \mathbf{v}_0$. At the instant its axis is perpendicular to velocity $\mathbf{v}_0$, the extremity A of the rod hits a fixed cylindrical obstacle 0 whose axis is perpendicular to the plane of motion. The impact is characterized by the coefficient of restitution e. Assume that friction at contact point A is negligible.

Find the angular velocity and the centroidal velocity of the rod immediately after impact.

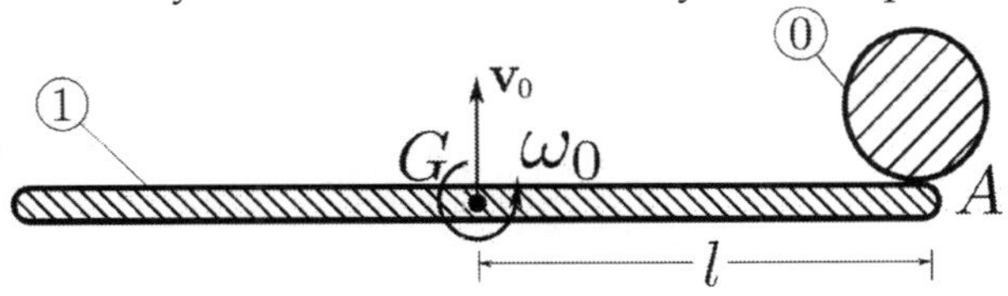

Problem 6.14 A soccer ball of mass m, radius r, mass center G, and radius of gyration k_G about one of its diameters, is thrown with a horizontal speed v_0 and zero angular velocity. Its initial position is located at a height H from a horizontal rigid obstacle. The impact is elastic. The friction at the contact point is characterized by $\mu_s \approx \mu_k = \mu$.

a. Assuming that the ball bounces without slipping, find the velocity of mass center G and angular velocity of the ball immediately after impact. Under what conditions are these results valid?

b. Assuming that the ball bounces with slip, find the velocity of mass center G and angular velocity of the ball immediately after impact.

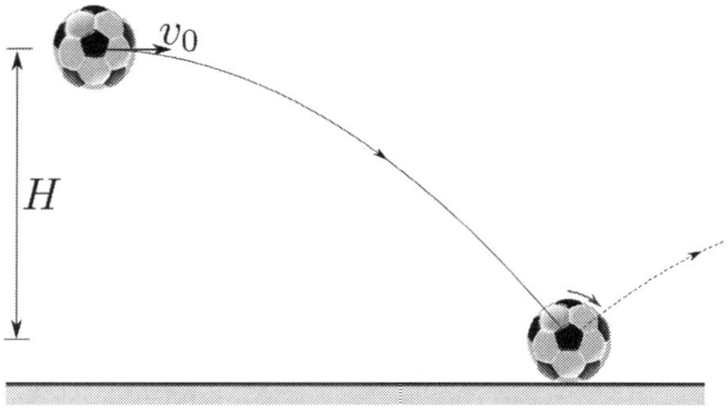

Problem 6.15 A uniform rod of mass 5.4 kg, and length 0.45 m can rotate in a vertical plane about a frictionless pivot of axis Oz. It is initially at rest in the vertical position when it is hit by a 30-gram bullet traveling horizontally at speed $v_0 = 570$ m/s. The bullet becomes instantly embedded into the rod at end point Q.

a. Assuming $h = 0.15$ m, find the angular velocity of the rod immediately after impact and the impulse reaction at O.

b. Find the value of h so that the impulse reaction at O vanishes.

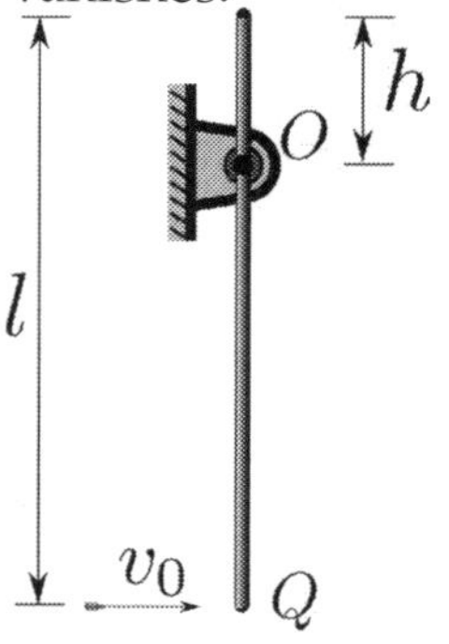

Appendix

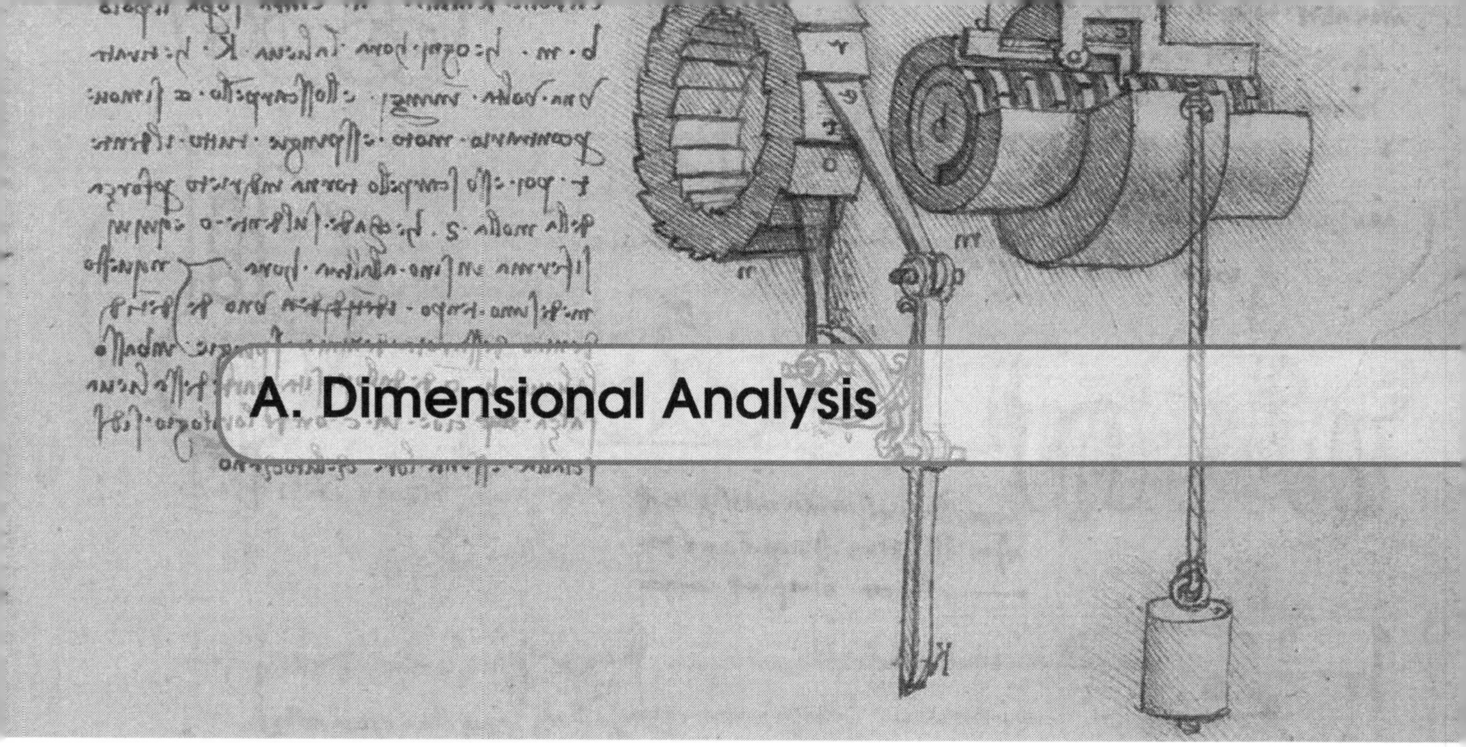

A. Dimensional Analysis

The quantities which describe physical phenomena are characterized by their *dimension*. In mechanics, only three fundamental dimensions play a role: length, time and mass. They are denoted as L, T and M, respectively. The dimension of a physical quantity Q, denoted $[Q]$, can always be expressed in terms of L, T and M:

$$[Q] = L^{\alpha} T^{\beta} M^{\gamma}$$

where the exponents α, β and γ are in general simple rational numbers. Do not confuse the notion of dimension with that of *unit*! Also note that a physical quantity may be dimensionless and still possess a unit: for instance, angle θ have S.I. unit of *radians*, yet $[\theta] = 1$.

The table displayed below gives a list of typical physical quantities encountered in this course and their corresponding dimension:

quantity (S.I. unit)	dimension
velocity (m/s)	LT^{-1}
acceleration (m/s^2)	LT^{-2}
angular velocity (rad/s)	T^{-1}
angular acceleration (rad/s^2)	T^{-2}
force (N)	MLT^{-2}
linear momentum (kg.m/s)	MLT^{-1}
angular momentum (kg.m^2/s)	ML^2T^{-1}
power (W)	ML^2T^{-3}
work (J)	ML^2T^{-2}
moment of inertia (kg.m^2)	ML^2

The task of performing the dimensional analysis of a relationship consists of replacing all physical quantities taking place in this relationship by their dimension. This allows to check for the *homogeneity* of the relationship. Since both sides of a relationship $A = B$ must have the same dimension, checking that $[A] = [B]$ provides a way to verify that a result is valid. Of course, if a relationship has been found to be homogeneous, one should not expect that it is necessarily correct. However, if it has not been found homogeneous, it is incorrect: this gives us an indication that there is a mistake somewhere in our derivation (perhaps a mere algebraic mistake). Note that checking for homogeneity of expressions can be done at all steps of a derivation.

Here are key rules to abide by, when checking the homogeneity of a relationship:

1. If two physical quantities are added, they must have the same dimension: $C = A + B$ implies $[C] = [A] = [B]$.
2. In expressions involving usual mathematical functions f such as exp, ln, sin, cos, tan, $\cdots$, the arguments of these functions must be dimensionless: if $B = f(A)$, then $[A] = [B] = 1$. Consider for instance the function $\sin x$ which can be written as the expansion $x - x^3/3! + x^5/5! + \cdots$. This implies that x, x^3, x^5, ... have the same dimension: this is only possible if x is dimensionless. Then $[\sin x] = [x] = 1$.
3. The dimension of a derivative $C = dB/dA$ is given by $[C] = [B]/[A]$. Similarly the dimension of the integral $C = \int B\, dA$ is given by $[C] = [B] \cdot [A]$.

> EXAMPLE 1. Check that the formula $\tau = 2\pi\sqrt{\frac{l}{g}}$ is homogeneous, if τ represents the period of small oscillations of a pendulum of length l, and if g is the gravitational acceleration.

Since l and g have dimension L and LT^{-2}, respectively, we find that

$$[2\pi(l/g)^{1/2}] = [l]^{1/2}/[g]^{1/2} = L^{1/2}/(L/T^{-2})^{1/2} = T$$

This is indeed the dimension of a time.

> EXAMPLE 2. The Cartesian coordinate x of a point satisfies the o.d.e. $\ddot{x} + \lambda x = 0$ where $\ddot{x} = d^2x/dt^2$ and λ is a parameter. Find the dimension of λ.

Since the quantities $\ddot{x}$ and λx must have the same dimension, we find $[\lambda x] = [\ddot{x}] = LT^{-2}$. This implies that $[\lambda] \cdot [x] = LT^{-2}$ and that $[\lambda] = T^{-2}$.

> EXAMPLE 3. The gravitational force of attraction between two bodies of mass m and M separated by the distance d takes the expression $F = GmM/d^2$. Find the dimension of the gravitational constant G.

We write $[G] = [F].[d]^2/[mM] = MLT^{-2}.L^2/M^2 = M^{-1}L^3T^{-2}$

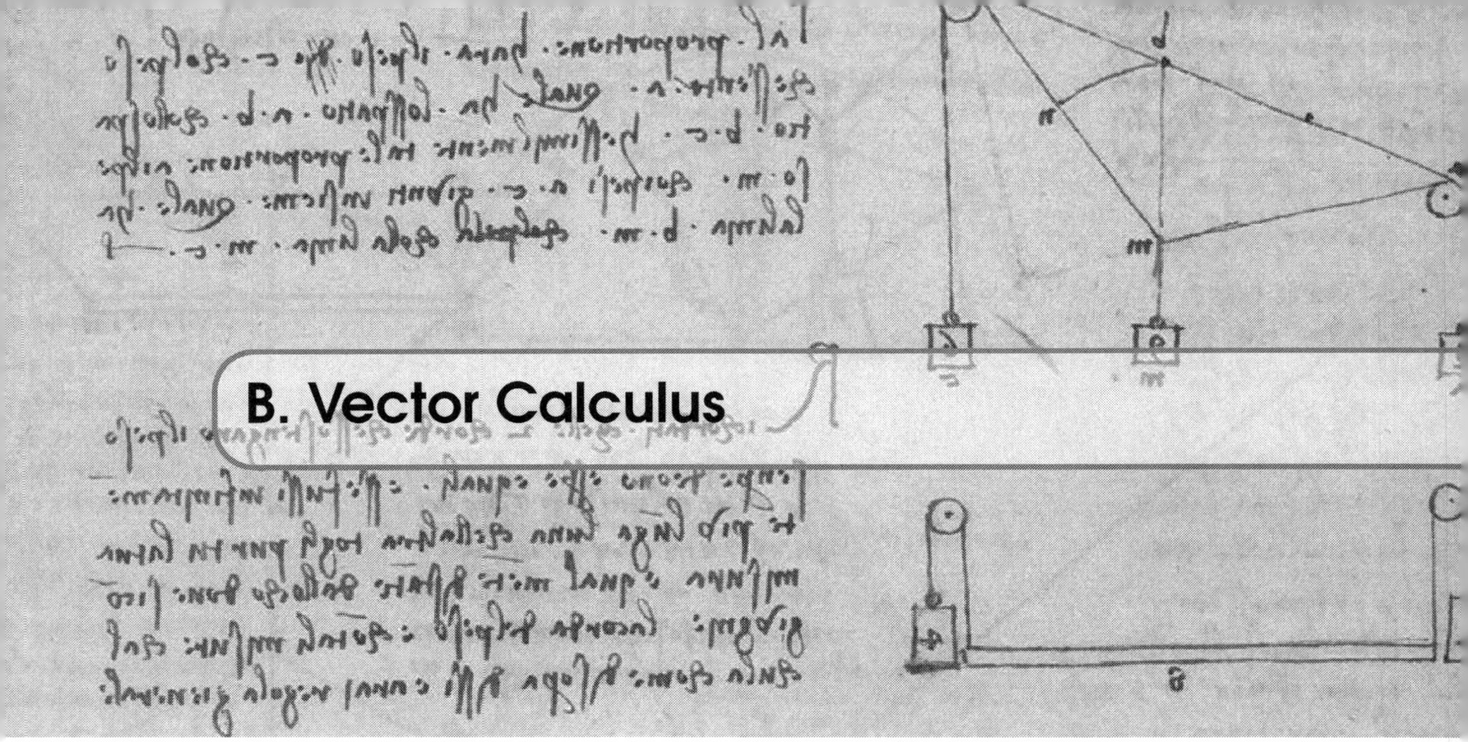

B. Vector Calculus

In mechanics, physical quantities can be represented as scalars or as vectors. Vectors differ fundamentally from scalars in that, to fully define them, one must indicate their *magnitude, their direction and sense, as well as their point of application.* We represent vectors graphically by arrows and use boldface notation to distinguish vector $\mathbf{u}$ from the scalar u. The magnitude (or *norm*) of vector $\mathbf{u}$ is denoted $|\mathbf{u}|$. Unit vectors, that is vectors of magnitude one, are denoted as $\hat{\mathbf{u}}$. While vectorial quantities are often defined at specific points in space, we also encounter *vector fields*, that is, vector functions of the space variables.

B1. Properties

Definition (Equality)

Two vectors $\mathbf{u}$ and $\mathbf{v}$ are equal if and only if they are parallel, have the same sense and the same magnitude: we then write $\mathbf{u} = \mathbf{v}$.

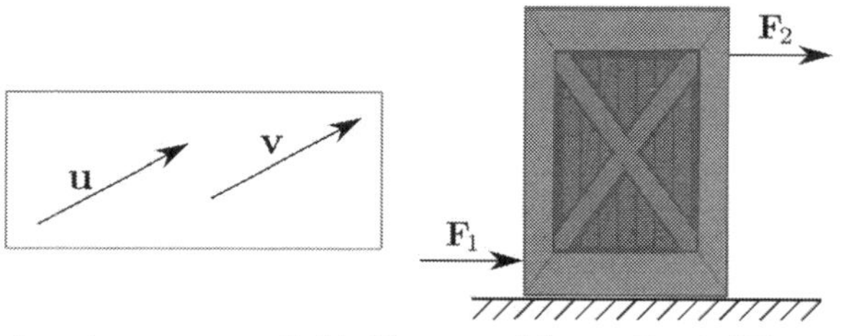

Figure B.1 – *Two equal vectors:* $\mathbf{u} = \mathbf{v}$ *(left). Two equal forces* $\mathbf{F}_1$ *and* $\mathbf{F}_2$ *exerted on a body (right).*

Hence the starting point of a vector is immaterial. However, two vectors can be equal and have a different point of application. A typical example in mechanics consists of the representation of forces as vectorial quantities. It is well known that two forces $\mathbf{F}_1$ and $\mathbf{F}_2$, of differing point of application, can be equal as vectors, and yet have a very different effect on the body upon which they are acting. See Figure B.1.

Two basic operations are defined on vectors.

DEFINITION (MULTIPLICATION BY A SCALAR)

The product of scalar λ by vector $\mathbf{u}$ is the vector parallel to $\mathbf{u}$, of magnitude $|\lambda||\mathbf{u}|$, in the same sense of $\mathbf{u}$ if $\lambda > 0$, opposed to $\mathbf{u}$ if $\lambda < 0$.

DEFINITION (ADDITION OF VECTORS)

The sum of two vectors $\mathbf{u}$ and $\mathbf{v}$ is obtained by constructing a triangle with $\mathbf{u}$ and $\mathbf{v}$ forming two sides, $\mathbf{v}$ adjoined to $\mathbf{u}$: then the sum $\mathbf{u}+\mathbf{v}$ is the vector starting at the origin of $\mathbf{u}$ and ending at the arrow of $\mathbf{v}$. See Figure B.2 (left).

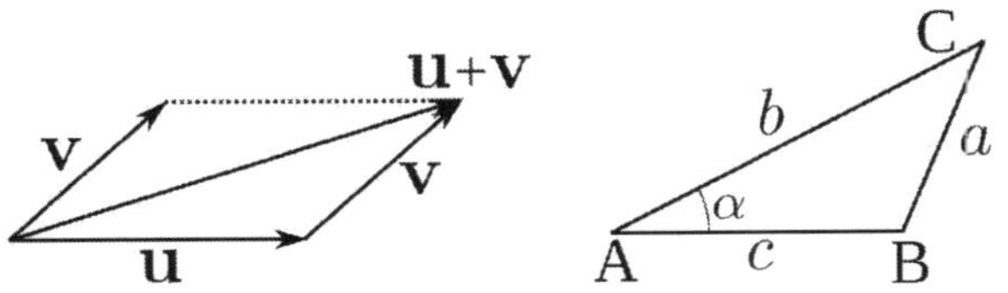

Figure B.2 – *Addition of vectors (left). Law of Cosines (right).*

PROPERTIES

(i) The vector $(-\mathbf{u})$ opposed to $\mathbf{u}$ is the vector satisfying $\mathbf{u}+(-\mathbf{u}) = \mathbf{0}$.
(ii) Commutativity: $\mathbf{u}+\mathbf{v} = \mathbf{v}+\mathbf{u}$.
(iii) Distributivity: $(\mathbf{u}+\mathbf{v})+\mathbf{w} = \mathbf{u}+(\mathbf{v}+\mathbf{w})$.
(iv) Multiplication by a scalar: $\lambda(\mathbf{u}+\mathbf{v}) = \lambda\mathbf{u}+\lambda\mathbf{v}$.
(v) Two vectors $\mathbf{u}$ and $\mathbf{v}$ are parallel, or *colinear*, if there exists a non-zero scalar λ such that $\mathbf{v} = \lambda\mathbf{u}$.

B2. Scalar Product

DEFINITION (SCALAR PRODUCT)

The *scalar product*, or *dot product*, $\mathbf{u}\cdot\mathbf{v}$ is the scalar defined by $\mathbf{u}\cdot\mathbf{v} = |\mathbf{u}||\mathbf{v}|\cos\alpha$, where $\alpha = (\mathbf{u},\mathbf{v}) \in [0,\pi]$ is the angle measured between $\mathbf{u}$ and $\mathbf{v}$. The scalar product between two vectors can be positive, negative or zero.

PROPERTIES

(i) Commutativity: $\mathbf{u}\cdot\mathbf{v} = \mathbf{v}\cdot\mathbf{u}$
(ii) Distributivity: $\mathbf{u}\cdot(\mathbf{v}+\mathbf{w}) = \mathbf{u}\cdot\mathbf{v}+\mathbf{v}\cdot\mathbf{w}$
(iii) Multiplication by a scalar λ: $\mathbf{u}\cdot(\lambda\mathbf{v}) = \lambda\mathbf{u}\cdot\mathbf{v}$
(iv) If $\mathbf{u}\cdot\mathbf{v} = 0$, the non-zero vectors $\mathbf{u}$ and $\mathbf{v}$ are said to be *orthogonal*.
(v) The scalar $\mathbf{u}^2 = \mathbf{u}\cdot\mathbf{u} = |\mathbf{u}|^2$ is the square of the magnitude of $\mathbf{u}$.
(vi) The square of the magnitude of $(\mathbf{u}+\mathbf{v})$ can be obtained from the following identity

$$(\mathbf{u}+\mathbf{v})^2 = \mathbf{u}^2 + 2\mathbf{u}\cdot\mathbf{v}+\mathbf{v}^2 \qquad \text{(B.1)}$$

(vii) A consequence of (B.1) is the *Law of Cosines* in triangle ABC (see Figure B.2, right):

$$a^2 = b^2 + c^2 - 2bc\cos\alpha \qquad \text{(B.2)}$$

(viii) The projection of vector $\mathbf{v}$ on a line Δ directed along *unit* vector $\hat{\mathbf{u}}$ is the vector defined as (see Figure B.3, left)

$$\mathbf{v}_\perp = (\mathbf{v} \cdot \hat{\mathbf{u}})\hat{\mathbf{u}} \tag{B.3}$$

An example in mechanics of the use of scalar product is given by the notion of power, defined as $\mathbf{v}_P \cdot \mathbf{F}$, generated by a force $\mathbf{F}$ applied at a point P of velocity $\mathbf{v}_P$.

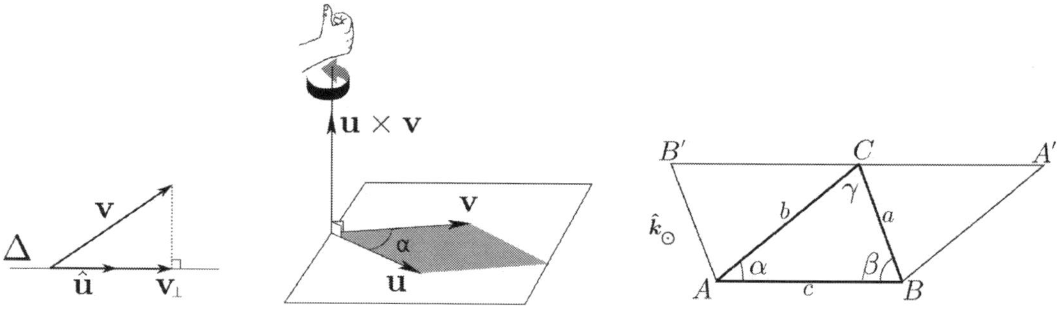

Figure B.3 – *Projection of vector* $\mathbf{v}$ *on line* Δ *(left). Vector product* $\mathbf{u} \times \mathbf{v}$ *(center). Law of Sines (right).*

B3. Vector Product

Definition (Vector Product)

The *vector product*, or *cross product*, between two vectors $\mathbf{u}$ and $\mathbf{v}$ is the vector denoted $\mathbf{u} \times \mathbf{v}$ such that:

(i) $\mathbf{u} \times \mathbf{v}$ is perpendicular to the plane containing both $\mathbf{u}$ and $\mathbf{v}$,
(ii) the set $(\mathbf{u}, \mathbf{v}, \mathbf{u} \times \mathbf{v})$ is oriented in the right-handed direction,
(iii) its magnitude is given by $|\mathbf{u} \times \mathbf{v}| = |\mathbf{u}||\mathbf{v}| \sin\alpha$, where $\alpha \in [0, \pi]$ is the angle between $\mathbf{u}$ and $\mathbf{v}$.

See Figure B.3 (center).

Note that $|\mathbf{u} \times \mathbf{v}|$ represents the area of the parallelogram formed by the two vectors. We can use this property of cross-products to show an important geometric property of triangles: consider the triangle ABC of Figure B.3 (right) and the cross-product $\mathbf{r}_{AB} \times \mathbf{r}_{BC}$:

$$\mathbf{r}_{AB} \times \mathbf{r}_{AC} = bc \sin\alpha\, \hat{\mathbf{k}} = \mathcal{A}_{ABA'C}\hat{\mathbf{k}}$$

where $\mathcal{A}_{ABA'C}$ is the area of parallelogram $ABA'C$. Similarly we have

$$\mathbf{r}_{BC} \times \mathbf{r}_{BA} = ac \sin\beta\, \hat{\mathbf{k}} = \mathcal{A}_{ABCB'}\hat{\mathbf{k}}$$

where $\mathcal{A}_{ABCB'}$ is the area of parallelogram $ABCB'$. Since $\mathcal{A}_{ABA'C} = \mathcal{A}_{ABCB'}$, we arrive at the identity known as the *Law of Sines*:

$$\frac{a}{\sin\alpha} = \frac{b}{\sin\beta} = \frac{c}{\sin\gamma} \tag{B.4}$$

where the second equality is obtained by simple permutations.

PROPERTIES

(i) Skew-symmetry: $\mathbf{u} \times \mathbf{v} = -\mathbf{v} \times \mathbf{u}$, and hence $\mathbf{u} \times \mathbf{u} = \mathbf{0}$
(ii) Bilinearity: $\mathbf{u} \times (\mathbf{v} + \mathbf{w}) = \mathbf{u} \times \mathbf{v} + \mathbf{u} \times \mathbf{w}$, $(\mathbf{u} + \mathbf{v}) \times \mathbf{w} = \mathbf{u} \times \mathbf{w} + \mathbf{v} \times \mathbf{w}$
(iii) Multiplication by a scalar λ: $\mathbf{u} \times (\lambda \mathbf{v}) = \lambda \mathbf{u} \times \mathbf{v}$
(iv) If 3 vectors satisfy the equality $\mathbf{u} \times \mathbf{v} = \mathbf{u} \times \mathbf{w}$, then $\mathbf{v} = \mathbf{w} + \lambda \mathbf{u}$ where λ is an indeterminate scalar.
(v) Given two vectors $\mathbf{u}$ and $\mathbf{v}$ we have

$$(\mathbf{u} \cdot \mathbf{v})^2 + (\mathbf{u} \times \mathbf{v})^2 = \mathbf{u}^2 \mathbf{v}^2$$

(vi) Given 3 vectors $\mathbf{u}$, $\mathbf{v}$ and $\mathbf{w}$, the triple vector product is the vector defined as $\mathbf{u} \times (\mathbf{v} \times \mathbf{w})$. The following formula can be shown

$$\mathbf{u} \times (\mathbf{v} \times \mathbf{w}) = (\mathbf{u} \cdot \mathbf{w})\mathbf{v} - (\mathbf{u} \cdot \mathbf{v})\mathbf{w}$$

An example of the use of vector products in mechanics is given by the notion of moments: given a vector $\mathbf{u}$ bound at point B (that is, of point of application B), the *moment* of $\mathbf{u}$ about point A is the vector

$$\mathbf{M}_A = \mathbf{r}_{AB} \times \mathbf{u}$$

B4. Coordinate System Representation

In practice, a coordinate system is used to resolve vectors on basis of unit vectors. There is no unique way of doing this as shown in Figure B.4: the unit vectors $(\hat{\mathbf{e}}_x, \hat{\mathbf{e}}_y, \hat{\mathbf{e}}_z = \hat{\mathbf{e}}_x \times \hat{\mathbf{e}}_y)$ define a Cartesian basis, while unit vectors $(\hat{\mathbf{e}}_r, \hat{\mathbf{e}}_\theta = \hat{\mathbf{e}}_z \times \hat{\mathbf{e}}_r, \hat{\mathbf{e}}_z)$ define a cylindrical basis. Given a vector $\mathbf{u}$, let $\hat{\mathbf{e}}_\rho = \mathbf{u}/|\mathbf{u}|$ be the unit vector defined by the direction of $\mathbf{u}$. Then, $\mathbf{u}$ can be resolved as

$$\mathbf{u} = u_x \hat{\mathbf{e}}_x + u_y \hat{\mathbf{e}}_y + u_z \hat{\mathbf{e}}_z$$

or as

$$\mathbf{u} = u_r \hat{\mathbf{e}}_r + u_z \hat{\mathbf{e}}_z$$

or, finally as

$$\mathbf{u} = u_\rho \hat{\mathbf{e}}_\rho$$

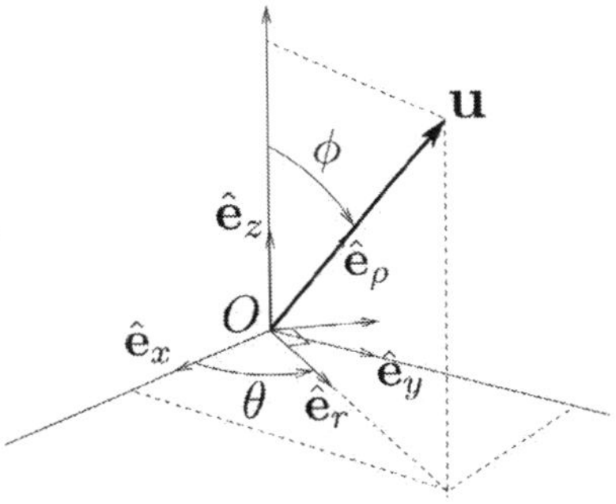

Figure B.4 – *Representation of vector* **u** *on coordinate systems.*

It is straightforward to relate these expressions by projection from one basis to the other, using the notion of scalar product. For instance, we can relate component u_ρ to cylindrical components (u_r, u_θ) or to Cartesian components (u_x, u_y, u_z) as follows

$$u_\rho = \hat{\mathbf{e}}_\rho \cdot \mathbf{u} = \hat{\mathbf{e}}_\rho \cdot (u_r \hat{\mathbf{e}}_r + u_z \hat{\mathbf{e}}_z) = u_r \sin\phi + u_z \cos\phi$$

or

$$u_\rho = \hat{\mathbf{e}}_\rho \cdot (u_x \hat{\mathbf{e}}_x + u_y \hat{\mathbf{e}}_y + u_z \hat{\mathbf{e}}_z) = u_x \sin\phi \cos\theta + u_y \sin\phi \sin\theta + u_z \cos\phi$$

where the angles θ and ϕ are defined in Figure B.4.

Finally, given two arbitrary vectors $\mathbf{u} = u_x \hat{\mathbf{e}}_x + u_y \hat{\mathbf{e}}_y + u_z \hat{\mathbf{e}}_z$ and $\mathbf{v} = v_x \hat{\mathbf{e}}_x + v_y \hat{\mathbf{e}}_y + v_z \hat{\mathbf{e}}_z$ resolved on Cartesian basis $(\hat{\mathbf{e}}_x, \hat{\mathbf{e}}_y, \hat{\mathbf{e}}_z)$, we can determine their scalar product and vector product according to

$$\mathbf{u} \cdot \mathbf{v} = u_x v_x + u_y v_y + u_z v_x, \quad \mathbf{u} \times \mathbf{v} = (u_y v_z - u_z v_y)\hat{\mathbf{e}}_x + (u_z v_x - u_x v_z)\hat{\mathbf{e}}_y + (u_x v_y - u_y v_x)\hat{\mathbf{e}}_z$$

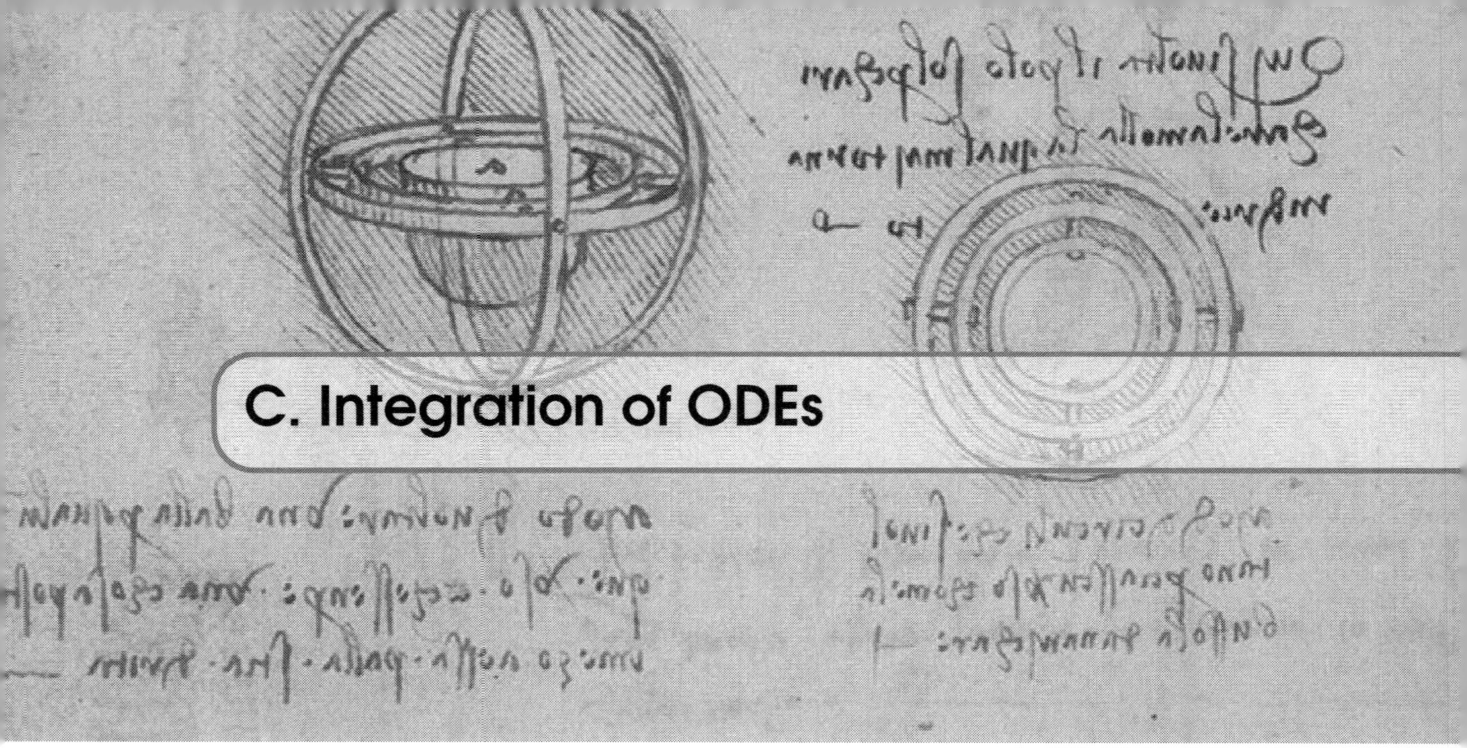

C. Integration of ODEs

THE GOAL of this section is to review a few basic techniques for the integration of simple ordinary differential equations (ODE) encountered in this course.

C1. Separable First-Order Equations

One often encounters first-order ODEs, that is, equations of the type

$$F(x, y, y') = 0$$

with $y' = dy/dx$. The equation is *separable* if it can be written in the form

$$g(y)\, y' = f(x) \tag{C.1}$$

The equation is solved by separating the variable x from the variable y, that is, by putting (C.1) in the form $g(y)dy = f(x)dx$. Then the solution of (C.1) is found to be

$$\int_{y_0}^{y} g(Y)dY = \int_{x_0}^{x} f(X)dX \tag{C.2}$$

or, denoting F and G primitives of f and g,

$$G(y) - G(y_0) = F(x) - F(x_0) \tag{C.3}$$

where we have imposed $y(x_0) = y_0$. Note that, if the equation possesses more that two variables, it cannot be separable.

EXAMPLE 1. The equation $yy' + x = 0$ is separable: $ydy = -xdx$ leads to $y^2 = -x^2 + C$ where C is a constant of integration.

EXAMPLE 2. Denoting $\dot{\theta} = d\theta/dt$, $\ddot{\theta} = d^2\theta/dt^2$, consider the second-order equation

$$\ddot{\theta} \sin\theta \cos\theta = \dot{\theta}^2 (\sin^2\theta + 2\cos^2\theta)$$

Find $\dot{\theta}$ vs θ with the initial conditions $\dot{\theta}(0) = 2\sqrt{3}$ and $\theta(0) = \pi/6$. We try to separate in the following way

$$\frac{\ddot{\theta}}{\dot{\theta}} = \frac{\sin^2\theta + 2\cos^2\theta}{\sin\theta\cos\theta}\dot{\theta}, \text{ or } \qquad \frac{d\dot{\theta}}{\dot{\theta}} = \frac{\sin^2\theta + 2\cos^2\theta}{\sin\theta\cos\theta}d\theta$$

We can now integrate:

$$\int_{2\sqrt{3}}^{\dot{\theta}} \frac{du}{u} = \int_{\pi/6}^{\theta} \frac{\sin^2 v + 2\cos^2 v}{\sin v\cos v}dv$$

With $\int \frac{du}{u} = \ln(u)$ and $\int \frac{\sin^2 v + 2\cos^2 v}{\sin v \cos v}dv = \ln\frac{\cos v}{\sin^2 v}$ we obtain

$$\boxed{\dot{\theta} = \frac{\cos\theta}{\sin^2\theta}}$$

C2. Linear ODEs

A linear, non-homogeneous ODE is solved in three steps:

1. First, one searches for the general solution y_h of the *homogeneous* equation.
2. Second, a *particular solution* y_p of the *non-homogeneous* equation is found.
3. Third, the general solution is found as the sum of the general solution of the homogeneous equation and the particular solution: $y = y_h + y_p$.

The expression of y_h will involve one or more constants of integration depending on the order of the equation. These constants are found by imposing the initial conditions to the general solution y.

Example 1. Solve $\dot{y} + ky = h$ for $y(t)$ satisfying $y(0) = y_0$, and where k and h are two non-zero constants. The corresponding homogeneous equation is $\dot{y} + ky = 0$: its solution $y_h = C\exp(-kt)$ is found by separation of the variables, where C is constant of integration. A particular solution is given by the constant $y_p = \frac{h}{k}$. Then the general solution is $y = y_h + y_p = C\exp(-kt) + \frac{h}{k}$. The constant C is found by imposing $y(0) = y_0$: $C = -\frac{h}{k} + y_0$. The final solution is then given by $\boxed{y(t) = (-\frac{h}{k} + y_0)e^{-kt} + \frac{h}{k}}$

Example 2. Solve $\ddot{x} - a^2x = b\cos(\omega t)$ for $x(t)$ satisfying $x(0) = \dot{x}(0) = 0$. First, we solve the homogeneous equation $\ddot{x} - a^2x = 0$: we find two solutions of the type e^{-at} and e^{at}. The general homogeneous solution [1] is then $x_h = Ae^{-at} + Be^{at}$. A particular solution can be found of the type $x_p = C\cos(\omega t)$: substitution in the equation $\ddot{x} - a^2x = b\cos(\omega t)$ gives $-\omega^2C - a^2C = b$, leading to $C = -b/(a^2 + \omega^2)$. Then, the general solution is $x = x_h + x_p = Ae^{-at} + Be^{at} + C\cos(\omega t)$. To find the constants A and B we impose $x(0) = 0$ and $\dot{x}(0) = 0$: this gives $A + B + C = 0$ and $A - B = 0$. Finally, the solution is given by $\boxed{x(t) = \frac{b}{a^2+\omega^2}\cosh(at) - \frac{b}{a^2+\omega^2}\cos(\omega t)}$.

Example 3. Solve $\ddot{x} = x + 2e^{-t}$ with $x(0) = 0$ and $\dot{x}(0) = -1$. As in example 2, the general homogeneous solution is given by $x_h = Ae^{-t} + Be^{t}$. A particular solution can be found of the type $x_p = kte^{-t}$: substitution in the equation $\ddot{x} - x = 2e^{-t}$ gives $k = -1$. The general solution is then $x = x_h + x_p = Ae^{-t} + Be^{t} - te^{-t}$. We find A and

[1] The general solution of the homogeneous equation $\ddot{x} + p\dot{x} + qx = 0$ (where p and q are two constants) is sought by seeking $x(t)$ in the form e^{rt}: in general, two independent solutions $r_{1,2}$ are found, leading to $x(t) = c_1e^{r_1t} + c_2e^{r_2t}$.

B by imposing the initial conditions $x(0) = 0$ and $\dot{x}(0) = 1$, leading to $A + B = 0$ and $-A + B - 1 = -1$: this gives $B = A = 0$. The final solution is given by $\boxed{x(t) = -te^{-t}}$.

C3. Nonlinear Second-order ODEs

Many ODEs in dynamics are of the type $F(\ddot{x}, \dot{x}, x, t) = 0$. They are representative of mechanical systems with a single degree of freedom $x(t)$. If the time variable t is explicit in the equation, the behavior of the solution can be extremely complex, and an exact solution is hopeless. If the time variable t is not explicit in the equation, then an exact solution may be possible. An important class of problems involves equations of the type:

$$\ddot{x} + \frac{dU}{dx} = 0 \tag{C.4}$$

given a function $U(x)$. To solve this equation, we define $y = \dot{x}$. The equation becomes $\dot{y} + \frac{dU}{dx} = 0$ and can be written in the form

$$y\frac{dy}{dt}\frac{dt}{dx} + \frac{dU}{dx} = 0, \quad \text{or} \quad y\frac{dy}{dx} + \frac{dU}{dx} = 0$$

This last equation is separable and immediately gives

$$\tfrac{1}{2}y^2 + U(x) = C \tag{C.5}$$

where C is a constant of integration. Equation (C.5) is called a *first integral* of (C.4). The constant C can be interpreted as the energy of the system, defined as the sum of the kinetic energy $\frac{1}{2}\dot{x}^2$ (setting the mass to one) and the potential energy $U(x)$. It is then possible to express x as a function of time by a simple quadrature.

Another way to obtain solution (C.5) is to pre-multiply both sides of (C.4) by $\dot{x}$ and recognize $\dot{x}\ddot{x} = \frac{d}{dt}(\frac{1}{2}\dot{x}^2)$ and $\dot{x}\frac{dU}{dx} = \frac{d}{dt}(U(x))$. Then (C.4) is equivalent to $\frac{d}{dt}(\frac{1}{2}\dot{x}^2 + U(x)) = 0$ and equation (C.5) is recovered.

EXAMPLE 1. Find a first integral of the pendulum equation $\ddot{\theta} + \omega^2 \sin\theta = 0$. The potential is easily found to be $U = -\omega^2\cos\theta$. The first integral is then

$$\boxed{\tfrac{1}{2}\dot{\theta}^2 - \omega^2\cos\theta = C}$$

where constant C is found from initial conditions.

EXAMPLE 2. Find a first integral for the equation

$$m(\theta)\ddot{\theta} + \tfrac{1}{2}m'(\theta)\dot{\theta}^2 + U'(\theta) = 0$$

denoting $m' = dm/d\theta$, $U' = dU/d\theta$. First we pre-multiply both sides of the equation by $\dot{\theta}$. Then we write

$$m(\theta)\dot{\theta}\ddot{\theta} = \frac{d}{dt}(\tfrac{1}{2}m(\theta)\dot{\theta}^2) - \tfrac{1}{2}m'(\theta)\dot{\theta}^3$$

So the original equation becomes $\frac{d}{dt}(\frac{1}{2}m(\theta)\dot{\theta}^2 + U(\theta)) = 0$ which gives the first integral $\boxed{\frac{1}{2}m(\theta)\dot{\theta}^2 + U(\theta) = Cst}$.

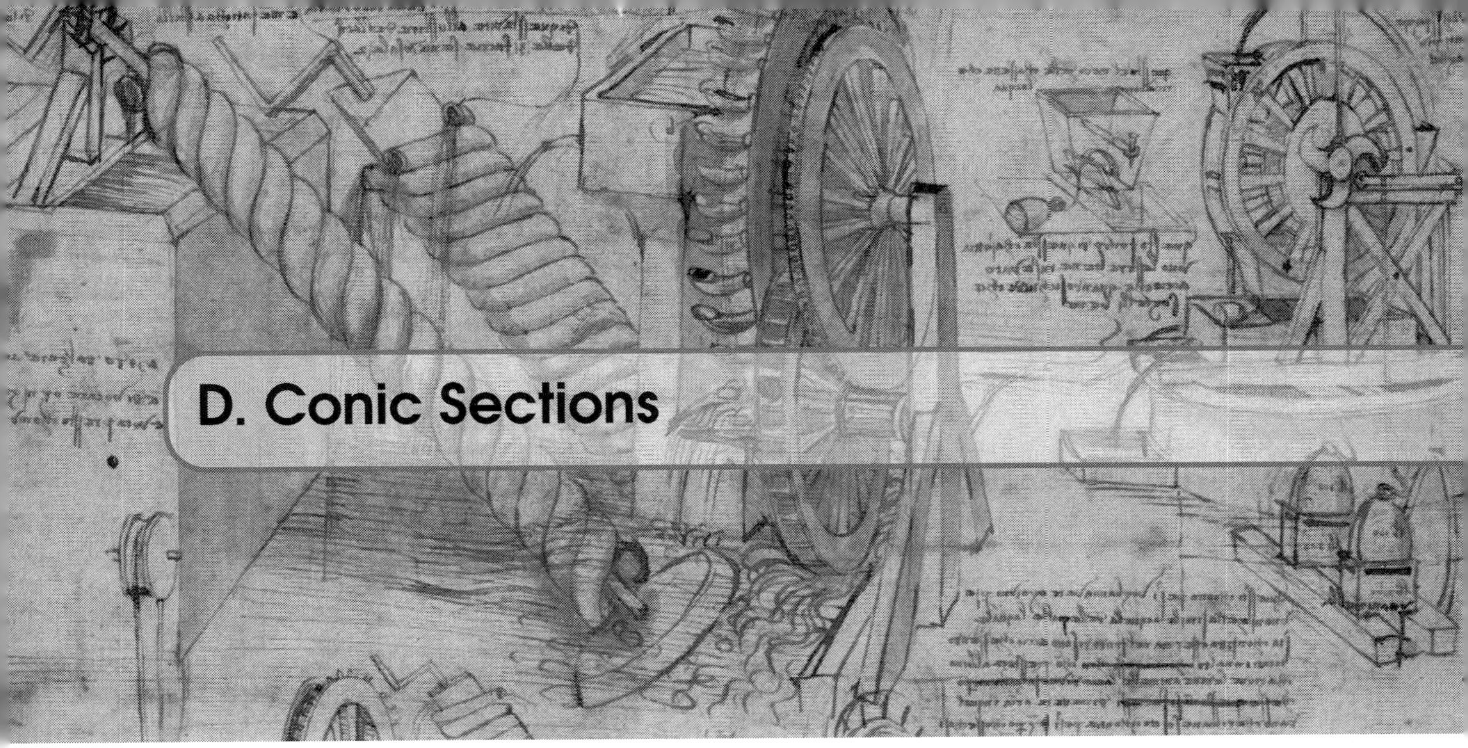

D. Conic Sections

CONIC sections are planar curves which play an important role in mechanics, most notably in celestial mechanics.

DEFINITION (CONIC SECTION)

Given a point F and a line Δ, a *conic section* can be defined as the locus of points P such that the ratio of distance $|PF|$ to point F to distance $|PH|$ to line Δ is constant

$$\frac{|PF|}{|PH|} = e \tag{D.1}$$

where H is the projection of P on Δ. Point F, line Δ and constant e are referred to as the *focus*, the *directrix*, and the *eccentricity* of the conic section. See Figure D.1.

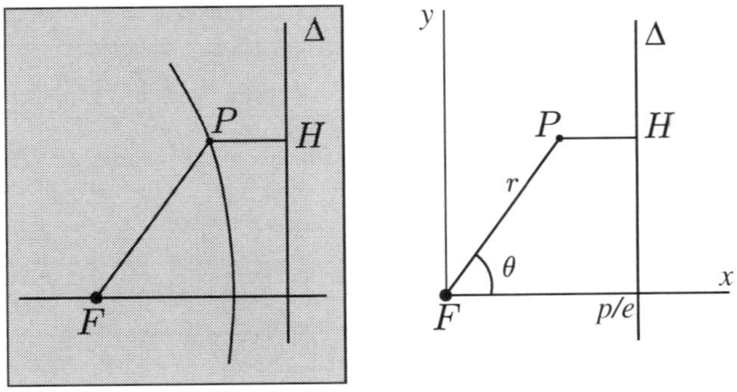

Figure D.1 – *Conic section of focus F and directrix Δ.*

The case $e = 1$ corresponds to *parabolas*. For $e \neq 1$, a simple symmetry argument can be used to show that there exists a second set (F', Δ'): in this case we obtain *ellipses* for $e < 1$ and *hyperbola* for $e > 1$ as *bifocal* conic sections whose center is the midpoint of segment FF'.

POLAR/CARTESIAN EQUATION OF CONICS. We define the Cartesian axes Fxy with axis Fy is parallel to directrix Δ. Let (r, θ) be the polar coordinates (r, θ) of P. Define p/e the distance between F and Δ. See Figure D.1. It can then be shown that equation (D.1) is equivalent to

$$\boxed{r = \frac{p}{1 + e\cos\theta}} \tag{D.2}$$

With $x = r\cos\theta$ and $y = r\sin\theta$, the Cartesian equation of conics is given by

$$\boxed{(1 - e^2)x^2 + 2epx + y^2 = p^2} \tag{D.3}$$

or, for $e \neq 1$,

$$\frac{(x + \frac{ep}{1-e^2})^2}{\frac{p^2}{(1-e^2)^2}} + \frac{y^2}{\frac{p^2}{1-e^2}} = 1 \tag{D.4}$$

CONCLUSION:

- Case 1: $\boxed{e = 1}$, the equation $2epx + y^2 = p^2$ is that of a *parabola* of axis Fx.
- Case 2: $\boxed{e < 1}$, equation (D.4) is that of an *ellipse* of the center C located at $(-c = -ep/(1 - e^2), 0)$, with semi-axis major a and semi-axis minor b given by

$$\boxed{a = \frac{p}{1 - e^2}, \qquad b = \frac{p}{\sqrt{1 - e^2}}}$$

Note that $a^2 = b^2 + c^2$. One can then prove that the points of the ellipse satisfy

$$|FP| + |F'P| = 2a$$

For $e = 0$, the ellipse becomes a *circle* of center F and radius p.

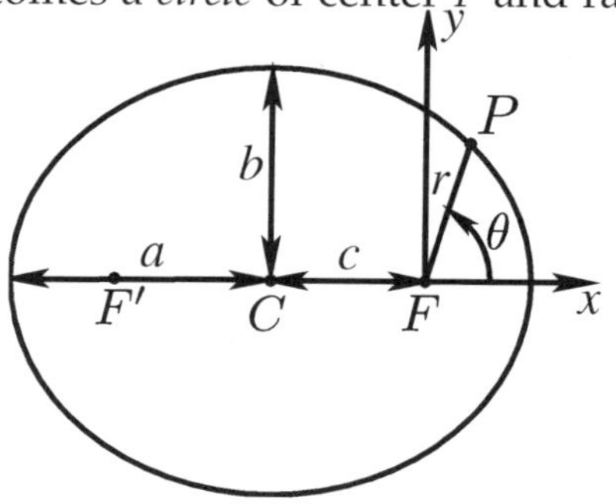

Figure D.2 – *Ellipse of semi-axis major a and semi-axis minor b.*

- Case 3: $\boxed{e > 1}$, the conic is a *hyperbola* of center $C(-c, 0)$ and asymptotes $(x + c)/a = \pm y/b$ with

$$\boxed{a = \frac{p}{1 - e^2}, \qquad b = \frac{p}{\sqrt{e^2 - 1}}}$$

In this case we have $a^2 + b^2 = c^2$ and $|FP| - |F'P| = 2a$

Index

Made in United States
Orlando, FL
04 November 2026

90202229R00243